Engineering Materials

Wilfried G.J.H.M. van Sark · Lars Korte
Francesco Roca (Eds.)

Physics and Technology of Amorphous-Crystalline Heterostructure Silicon Solar Cells

Springer

Dr. Wilfried G.J.H.M. van Sark
Utrecht University
Copernicus Institute
Science Technology and Society
Budapestlaan 6
3584 CD Utrecht
The Netherlands
E-mail: w.g.j.h.m.vansark@uu.nl

Dr. Lars Korte
Helmholtz-Zentrum Berlin für
Materialien und Energie
Inst. Silizium-Photovoltaik
Kekuléstraße 5
12489 Berlin
Germany
E-mail: korte@helmholtz-berlin.de

Dr. Francesco Roca
ENEA - Agenzia Nazionale per
le Nuove Tecnologie, l'Energia e
lo Sviluppo Economico Sostenibile
Unità Tecnologie Portici,
Localitá Granatello
P. le E. Fermi
80055 Portici
Napoli
Italy
E-mail: franco.roca@enea.it

ISBN 978-3-642-27007-9 ISBN 978-3-642-22275-7 (eBook)

DOI 10.1007/978-3-642-22275-7

Engineering Materials ISSN 1612-1317

Softcover reprint of the hardcover 1st edition 2011

Typeset & Cover Design: Scientific Publishing Services Pvt. Ltd., Chennai, India.

Printed on acid-free paper

9 8 7 6 5 4 3 2 1

springer.com

Preface

The development of hydrogenated amorphous (a-Si:H) / crystalline silicon (c-Si) heterojunction (SHJ) solar cells has recently accelerated tremendously. This is not just triggered by the recent expiration of core patents of Sanyo Electric Company, but most of all due to the high efficiency that has been proven to be achievable in practice (being close to the theoretical limit for c-Si) and the very advanced architectures that can be realized with this technology, such as fully back contacted solar cells with very thin wafers. The low temperature processing and reduction of materials resources is bringing grid parity rapidly within reach, even in countries with little solar irradiation, and this way of processing is highly cost competitive with the 'classic' c-Si solar cells with diffusion processed junctions. SHJ photovoltaic technology merges the best of the worlds of both high efficiency crystalline silicon technology and thin film technology. Institutes and companies entering this field have found that high conversion efficiencies can quickly be accomplished based on the nearly complete elimination of surface defect states.

A consortium of 12 partners has been working together in the HETSI project (in full: heterojunction solar sells based on a-Si/c-Si), funded by the European Commission in the framework of the 7th Research Framework Programme from 2008 to 2011. In the scope of this project, a workshop was held at Utrecht University in 2010, to present and discuss the status as well as the issues in amorphous-crystalline heterojunction silicon solar cells.

At this workshop the idea was born to collect all the present understanding as well as the ongoing innovations in a book, as one of the broad dissemination activities of HETSI. The result is a comprehensive collection of the knowledge available at the most prestigious laboratories in Europe involved in SHJ solar cell research. It is an authoritative review of present-day research topics and future opportunities in this field. It is an invaluable asset to anyone who is involved in this field, but also to the increasing numbers of researchers and industrialists who are entering this rapidly evolving solar photovoltaic technology.

Ruud E.I. Schropp

Debye Institute for Nanomaterials Science
Section Nanophotonics
Faculty of Science
Utrecht University

Acknowledgements

The editors would like to thank all the many authors and co-authors that have contributed to this book. It is their knowledge, which gives the book the value it has. We also would like to thank all institutions and individuals, who granted permission to publish figures, supplied data for this book or provided valuable feedback.

This book originated from a workshop organized at Utrecht University in February 2010 within the framework of the project HETSI (heterojunction solar cells based on a-Si/c-Si), which ran from February 2008 until February 2011, and was funded by the European Commission in the framework of the 7th Research Framework Programme. Partners in this project were: Institut National de l'Energie Solaire (INES, FR), Centre National de la Recherche Scientifique (CNRS, FR), Energieonderzoek Centrum Nederland (ECN, NL), Utrecht University (UU, NL), Agenzia Nazionale per le Nuove Tecnologie, l'Energia e lo Sviluppo Economico Sostenibile (ENEA, IT), Interuniversity MicroElectronics Centrum (IMEC, BE), Institut de Microtechnologie - Ecole Polytechnique Fédérale de Lausanne (EPFL, CH), Helmholtz-Zentrum Berlin für Materialien und Energie (HZB, DE), SOLON SE (DE), Photowatt SAS (FR), Q-Cells SE (DE), and ALMA Consulting Group SAS (FR). In the workshop many experts presented an overview of the state-of-the-art in physics and technology of amorphous-crystalline heterostructure silicon solar cells, including a hands-on training session on computer modelling of cells. In this book, the presentations have been converted in comprehensive chapters. To our opinion, thanks to the many contributors that are world-renowned experts in their respective fields, the book as a whole contains a thorough overview of amorphous-crystalline heterostructure silicon solar cells, from the fundamental physical principles to the experimental and modelling details. We hope that it will serve as a reference base for the ever-growing scientific and industrial community in the photovoltaics field.

Statements of views, facts and opinions as described in this book are the responsibility of the author(s).

Wilfried van Sark
Lars Korte
Francesco Roca

There is one forecast of which you can already be sure: someday renewable energy will be the only way for people to satisfy their energy needs. Because of the physical, ecological and (therefore) social limits to nuclear and fossil energy use, ultimately nobody will be able to circumvent renewable energy as the solution, even if it turns out to be everybody's last remaining choice. The question keeping everyone in suspense, however, is whether we shall succeed in making this radical change of energy platforms happen early enough to spare the world irreversible ecological mutilation and political and economic catastrophe.

Hermann Scheer (1944 – 2010), Energy Autonomy: The Economic, Social and Technological Case for Renewable Energy, Earthscan, London, UK, 2007, page 29.

Table of Contents

List of Contributors

Heike Angermann
Helmholtz-Zentrum Berlin für
Materialien und Energie GmbH,
Institut für Silizium-Photovoltaik,
Kekuléstraße 5,
D-12489 Berlin,
Germany
angermann@helmholtz-berlin.de

Rudolf Brüggemann
Institut für Physik, Carl von Ossietzky
Universität Oldenburg,
26111 Oldenburg,
Germany
rudi.brueggemann@uni-oldenburg.de

Domenico Caputo
Department of Electronic Engineering
Rome University "Sapienza",
Via Eudossiana 18,
00139 Rome,
Italy
caputo@die.uniroma1.it

Giampiero de Cesare
Department of Electronic Engineering
Rome University "Sapienza",
Via Eudossiana 18,
00139 Rome,
Italy
decesare@ die.uniroma1.it

Thibaut Desrues
CEA-INES, Savoie Technolac,
50 avenue du lac Léman - BP258,
F-73375 Le Bourget du Lac – Cedex,
France
thibaut.desrues@cea.fr

Djicknoum Diouf
Laboratoire de Génie Electrique de
Paris, CNRS UMR8507,
SUPELEC; Univ Paris-Sud,
UPMC Univ Paris 06,
11 rue Joliot-Curie, Plateau de Moulon,
F-91192 Gif-sur-Yvette Cedex,
France
djicknoum.diouf@lgep.supelec.fr

Ivan Gordon
imec,
Photovoltaics/Solar Cell Technology,
Kapeldreef 75,
B-3001 Leuven,
Belgium
Ivan.Gordon@imec.be

Simona De Iuliis
ENEA - Research Center Casaccia,
Via Anguillarese 301,
00123 Rome,
Italy
simona.de.iuliis@enea.it

Massimo Izzi
ENEA - Research Center Casaccia,
Via Anguillarese 301,
00123 Rome,
Italy
massimo.izzi@enea.it

Jean-Paul Kleider
Laboratoire de Génie Electrique de
Paris, CNRS UMR8507,
SUPELEC; Univ. Paris-Sud,
UPMC Univ. Paris 06,
11 Rue Joliot-Curie, Plateau de Moulon,
F-91192 Gif-sur-Yvette Cedex,
France
jean-paul.kleider@lgep.supelec.fr

Lars Korte
Helmholtz-Zentrum Berlin für Materialien und Energie GmbH, Institut für Silizium-Photovoltaik, Kekuléstraße 5, D-12489 Berlin, Germany
korte@helmholtz-berlin.de

Caspar Leendertz
Helmholtz-Zentrum Berlin für Materialien und Energie GmbH, Institut für Silizium-Photovoltaik, Kekuléstraße 5, D-12489 Berlin, Germany
caspar.leendertz@helmholtz-berlin.de

Christophe Longeaud
Laboratoire de Génie Electrique de Paris, CNRS UMR8507, SUPELEC; Univ Paris-Sud, UPMC Univ Paris 06, 11 rue Joliot-Curie, Plateau de Moulon, F-91192 Gif-sur-Yvette Cedex, France
longeaud@lgep.supelec.fr

Delfina Muñoz
CEA-INES, Savoie Technolac, 50 avenue du lac Léman - BP258, F-73375 Le Bourget du Lac – Cedex, France
delphina.munoz@cea.fr

Barry O'Sullivan
imec, Photovoltaics/Solar Cell Technology, Kapeldreef 75, B-3001 Leuven, Belgium
Barry.Osullivan@imec.be

Niels Posthuma
imec, Photovoltaics/Solar Cell Technology, Kapeldreef 75, B-3001 Leuven, Belgium
Niels.Posthuma@imec.be

Jörg Rappich
Helmholtz-Zentrum Berlin für Materialien und Energie GmbH, Institut für Silizium-Photovoltaik, Kekuléstraße 5, D-12489 Berlin, Germany
rappich@helmholtz-berlin.de

Jatin K. Rath
Utrecht University, Debye Institute for Nanomaterials Science, Section Nanophotonics, P.O. Box 80000, 3508 TA Utrecht, The Netherlands
j.k.rath@uu.nl

Pierre-Jean Ribeyron
CEA-INES, Savoie Technolac, 50 avenue du lac Léman - BP258, F-73375 Le Bourget du Lac – Cedex, France
pierre-jean.ribeyron@cea.fr

Francesco Roca
ENEA - Agenzia Nazionale per le Nuove Tecnologie, l'Energia e lo Sviluppo Economico Sostenibile Unità Tecnologie Portici, Localitá Granatello P. le E. Fermi 80055 Portici Napoli Italy
franco.roca@enea.it

Pere Roca i Cabarrocas
Laboratoire de Physique des Interfaces et des Couches Minces, CNRS Ecole Polytechnique, 91128 Palaiseau, France
pere.roca@polytechnique.edu

Florian Ruske
Helmholtz-Zentrum Berlin für Materialien und Energie GmbH, Institut für Silizium-Photovoltaik, Kekuléstraße 5, D-12489 Berlin, Germany
florian.ruske@helmholtz-berlin.de

Wilfried G.J.H.M. van Sark
Utrecht University, Copernicus Institute,
Science, Technology and Society,
Budapestlaan 6,
3584 CS Utrecht,
The Netherlands
w.g.j.h.m.vansark@uu.nl

Ruud E.I. Schropp
Utrecht University, Debye Institute for
Nanomaterials Science,
Section Nanophotonics,
P.O. Box 80000,
3508 TA Utrecht,
The Netherlands
r.e.i.schropp@uu.nl

Luca Serenelli
ENEA - Research Center Casaccia,
Via Anguillarese 301,
00123 Rome,
Italy
luca.serenelli@enea.it

Rolf Stangl
Helmholtz-Zentrum Berlin für
Materialien und Energie GmbH,
Institut für Silizium-Photovoltaik,
Kekuléstraße 5,
D-12489 Berlin,
Germany
stangl@helmholtz-berlin.de

Mario Tucci
ENEA - Research Center Casaccia,
Via Anguillarese 301,
00123 Rome,
Italy
mario.tucci@enea.it

Stefaan De Wolf
Ecole Polytechnique Fédérale de
Lausanne (EPFL), Institute of
Microengineering (IMT), Photovoltaics
and thin-film electronics laboratory
(PVlab),
Breguet 2,
2000 Neuchâtel,
Switzerland
stefaan.dewolf@epfl.ch

Miro Zeman
Delft University of Technology,
Photovoltaic Materials and Devices group,
Mekelweg 4,
2628 CD, Delft,
The Netherlands
M.Zeman@tudelft.nl

Dong Zhang
Delft University of Technology,
Photovoltaic Materials and Devices group,
Mekelweg 4,
2628 CD, Delft,
The Netherlands
D.Zhang@tudelft.nl

List of Abbreviations, Units, and Signs

1D	:	one dimensional
2D	:	two dimensional
3D	:	three dimensional
4-BrB	:	4-bromobenzene
4-NB	:	4-nitrobenzene
4-NBDT	:	4-nitrobenzene diazonium tetrafluoroborate
AC (ac)	:	alternate current
ACJ-HIT	:	artificially constructed junction-heterojunction with intrinsic thin film
AD	:	analog to digital
AFORS-HET	:	automat for simulation of heterostructures
AFM	:	atomic force microscopy
AIST	:	National Institute of Advanced Industrial Science and Technology (Japan)
ALD	:	atomic layer deposition
AM1.5	:	air mass 1.5
AM1.5G	:	air mass 1.5, global
APCVD	:	chemical vapour deposition at atmospheric pressure
APM	:	ammonia/hydrogen peroxide mixture
AR	:	anti-reflection
ARC	:	anti-reflection coating
AS	:	admittance spectroscopy
a-Si:H	:	hydrogenated amorphous silicon
a-SiC:H	:	hydrogenated amorphous silicon carbide
a-SiO:H	:	hydrogenated amorphous silicon oxide
ATR	:	attenuated total reflection
ATR-FTIR	:	attenuated total reflection Fourier transform infrared (spectroscopy)
BACG	:	back amorphous-crystalline silicon heterojunction
BEHIND	:	back enhanced heterostructure with interdigitated contact
BHD	:	Brooks-Harring-Dingle
BP	:	band pass
BSF	:	back surface field
CB	:	conduction band
CBM	:	conduction band maximum
CDMR	:	capacitance detected magnetic resonance
CFSYS	:	constant final state yield spectroscopy

CIGS	:	copper indium gallium selenide
CNRS	:	Centre National de la Recherche Scientifique
CPM	:	hydrochloric acid/hydrogen peroxide mixture
CPM	:	constant photocurrent mode
CS	:	capacitance spectroscopy
c-Si	:	crystalline silicon
CV / C-V	:	capacitance-voltage
CVD	:	chemical vapour deposition
cw	:	continuous wave
CZ	:	czochralski
DB	:	dangling bond
DBR	:	dielectric Bragg reflector
DC (dc)	:	direct current
DH	:	dihydride
DIN	:	Deutsches Institut für Normung
DIW	:	dionised water
DOS	:	density of electronic states
ECN	:	Energieonderzoek Centrum Nederland
EDMR	:	electrically detected magnetic resonance
EFG	:	edge-defined film-fed-growth
EL	:	electroluminescence
EMA	:	effective medium approximation
ENEA	:	Agenzia Nazionale per le Nuove Tecnologie, l'Energia e lo Sviluppo Economico Sostenibile
EWT	:	emitter wrap through
EPFL	:	Ecole Polytechnique Fédérale de Lausanne
epi-Si	:	epitaxially grown crystalline silicon
EPR	:	electronic paramagnetic resonance
EQ	:	equilibrium
EQE	:	external quantum efficiency
ESR	:	electron spin resonance
FE	:	front emitter
FF	:	fill factor
FPD	:	flat panel displays
FSF	:	front surface field
FSRV	:	front surface recombination velocity
FTIR	:	fourier-transform infrared
FTIR-SE	:	fourier-transform infrared ellipsometry
FZ	:	float zone
GB	:	grain boundary
HETSI	:	heterojunction solar cells based on a-Si c-Si
HF	:	hydrofluoric acid
HIT	:	heterojunction with intrinsic thin-layer solar cell
HJ	:	heterojunction
HP	:	hot plate
HR-TEM	:	high resolution transmission electron microscopy

HSM	:	high stretching mode
HWCVD	:	hot wire chemical vapour deposition
HZB	:	Helmholtz-Zentrum Berlin für Materialien und Energie
IBBC	:	interdigitated backside buried contact
IBC	:	interdigitated back-contact
IBC-SiHJ	:	interdigitated back contact silicon heterojunction
IBC-HJ	:	interdigitated back contact heterojunction
I/E	:	iodine/ethanol
imec	:	Interuniversity MicroElectronics Centre
INES	:	Institut National de l'Energie Solaire
IP	:	internal photoemission
IPA	:	isopropyl alcohol
IPE	:	internal photo emission
ISE	:	Institut für Solare Energiesysteme
IQE	:	internal quantum efficiency
ISFH	:	Institut für Solarenergieforschung Hameln
ITO	:	tin-doped indium oxide
IV / I-V	:	current-voltage
IZO	:	indium zinc oxide
KOH	:	potassium hydroxide
LBL	:	layer by layer
LBSF	:	local back surface field
LCD	:	liquid crystal display
LID	:	light induced degradation
LPCVD	:	low pressure chemical vapour deposition
LSM	:	low stretching mode
MH	:	monohydride
MIGS	:	metal-induced gap states
MIS	:	metal insulator semiconductor
MOCVD	:	metal organic chemical vapour deposition
MOS	:	metal oxide semiconductor
MPL	:	modulated photoluminescence
MS	:	magnetron sputtering
MTCE	:	multitunneling with successive recombination through carrier capture or reemission into the band
MW	:	microwave
μc-Si:H	:	hydrogenated microcrystalline silicon
μPCD	:	microwave photo conductive decay
μW-PCD	:	microwave detected photoconductance decay
MWT	:	metal wrap through
NB	:	nitrobenzene
NDMR	:	noise detected magnetic resonance
NUV-PES	:	near ultraviolet photoelectron spectroscopy
ODMR	:	optically detected magnetic resonance
OECE	:	oblique evaporation of contact
PC	:	personal computer

PC	:	planar conductance
PCD	:	photoconductance decay
PDS	:	photothermal deflection spectroscopy
PECVD	:	plasma enhanced chemical vapour deposition
pEDMR	:	pulsed electrically detected magnetic resonance
PERL	:	passivated emitter and rear locally diffused
PERT	:	passivated emitter rear totally diffused
PES	:	photoelectron spectroscopy
PESC	:	passivated emitter solar cell
PL	:	photoluminescence
PLD	:	pulsed laser deposition
PMMA	:	poly methyl methacrylate
pm-Si :H	:	polymorphous silicon
por-Si	:	porous silicon
PRECASH	:	point rear emitter crystalline/amorphous silicon heterojunction
PS	:	photoyield spectroscopy
PV	:	photovoltaics
PVD	:	physical vapour deposition
QSSPC	:	quasi-steady-state photoconductance
RCA	:	radio corporation of america
RCPCD	:	resonance-coupled photoconductive decay
RE	:	rear emitter
RECASH	:	rear emitter crystalline/amorphous silicon heterojunction
RF	:	radio frequency
RT	:	room temperature
SAF	:	Salpetersäure – Ammoniumfluorid – Flusssäure (etch mixture of nitric acid, 70% HNO_3, ammonia fluoride, 40% NH_4F, and hydrofluoric acid, 50% HF)
SC	:	semiconductor
SCR	:	space charge region
SDPC	:	spin dependent photoconductivity
SDT	:	spin dependent transport
SE	:	spectroscopic ellipsometry
SE	:	selective emitter
SEM	:	scanning electron microscopy
SHJ	:	crystalline silicon heterojunction
SlSF	:	Schwefelsäure – Salpetersäure - Flusssäure (etch mixture of sulphuric acid, 96% H_2SO_4, nitric acid, 70% HNO_3, and hydrofluoric acid, 50% HF)
SOD	:	spin-on dopant
SPM	:	sulphuric peroxide mixture
SPV	:	surface photovoltage
SR	:	spectral response
SRH	:	Shockley-Read-Hall

TBAF	:	tetrabutylamonium hexafluorophosphate
TCO	:	transparent conductive oxide
TDS	:	thermal desorption spectroscopy
TE	:	texture etch
TFT	:	thin film transistor
TFT-LCD	:	thin film transistor-liquid crystal display
TH	:	trihydride
TLM	:	transfer length method
TR	:	transient
TRMC	:	transient microwave conduction
UNSW	:	University of New-South Wales
UPS	:	ultraviolet photoelectron spectroscopy
UU	:	Utrecht University
UV-NIR	:	ultraviolet-near infrared
UV-VIS	:	ultraviolet-visible
VB	:	valence band
VBM	:	valence band maximum
VHF	:	very high frequency
VFP	:	voltage filling pulse method
VIGS	:	virtual induced gap states
XPS	:	x-ray photoelectron spectroscopy

Chapter 1
Introduction – Physics and Technology of Amorphous-Crystalline Heterostructure Silicon Solar Cells

Wilfried van Sark[1], Lars Korte[2], and Francesco Roca[3]

[1] Utrecht University, Copernicus Institute, Science, Technology and Society, Budapestlaan 6, 3584 CD Utrecht, The Netherlands
[2] Helmholtz-Zentrum Berlin GmbH, Department Silicon Photovoltaics, Kekuléstraße 5, D-12489 Berlin, Germany
[3] ENEA - Agenzia Nazionale per le Nuove Tecnologie, l'Energia e lo Sviluppo Economico Sostenibile - Unità Tecnologie Portici, Localitá Granatello, P. le E. Fermi, 80055 Portici, Napoli, Italy

1.1 General Introduction

Although photovoltaic solar energy technology (PV) is not the sole answer to the challenges posed by the ever-growing energy consumption worldwide, this renewable energy option can make an important contribution to the economy of each country. According to the *New Policies Scenario* of the "World Energy Outlook 2010" published in November 2010 by the International Energy Agency (IEA) [1], it is to be expected that the share of renewable energies in global energy production increases threefold over the period 2008-2035, and that almost one third of global electricity production will come from renewables by 2035, thus catching up with coal. The "Solar Generation 6" report of the European Photovoltaic Industry association published in October 2010 [2] predicts in its *Solar Generation Paradigm Shift* Scenario that by 2050, PV could generate enough solar electricity to satisfy 21% of the world electricity needs, i.e. a total of up to 6750 TWh of solar PV electricity in 2050, coming from an installed capacity of 4670 GW in 2050. This is to be compared with 40 GW installed in the world at the end of 2010 [3].

After the first solar cell was demonstrated in silicon 55 years ago [4] the cost has declined by a factor of nearly 200, and high-throughput mass-production compatible processes are omnipresent all over the globe. More than 90% of the current production uses first generation PV wafer based crystalline Silicon (c-Si), a technology with the ability to continue to reduce its cost at its historic rate [5,6]. The direct production costs for crystalline silicon modules are expected to be around 1 €/Wp in 2013, below 0.75 €/Wp in 2020 and lower in the long term, as stated in the Strategic Research Agenda of the European Photovoltaic Technology Platform [7].

W.G.J.H.M. van Sark et al. (Eds.): Physics & Tech. of Amorphous-Crystalline, EM, pp. 1–12.
springerlink.com

However the challenge of developing photovoltaic technology to a cost-competitive alternative for established fossil-fuel based energy sources remains enormous and new cell concepts based on thin films of various types of organic and inorganic materials are entering the market. Thin film silicon (TFS), cadmium telluride (CdTe), copper indium selenide (CIS) generally are denoted as the second generation of PV technologies and are currently considered a very interesting market alternative to crystalline silicon. Advanced thin film approaches such as dye-sensitized titanium oxide (TiO_2) and blends of polythiophene and C_{60} (P3HT:PCBM) [8] are showing fast progress. World-record solar cell efficiencies are regularly updated, see e.g. [9], and some interesting initiatives related to their industrialization and commercialization have recently been undertaken.

For large scale PV deployment in large power plants or in building integrated applications it is a prerequisite that the performance of solar energy systems is enhanced by assuring low cost in production and long term reliability (>25 years). This requires the following issues to be addressed: 1) increase of the efficiency of solar irradiation conversion; 2) decrease of the amount of materials that are used, while these materials should be durable, stable, and abundant on earth; and 3) reduction of the manufacturing and installation cost.

The fantastic boom of thin film technology in recent years can suggest further development on the medium to long term due to the application of innovative concepts to conventional materials and developments of new classes of thin film materials stemming from nanotechnologies, photonics, optical metamaterials, plasmonics and new semiconducting organic and inorganic sciences, most of them recognized as next (third) generation approaches.

On the other hand the growth of the PV industry is also requesting well proven technology in order to sustain the emerging market; here, crystalline silicon has a long history of 'pulling rabbits out of the hat' [5].

Today, the industry has reached a new level of scale that is mobilizing vast new resources, enthusiasm, skills, and energy in order to reduce wafer thickness, enhance efficiency and improve processes related to substrate cleaning, junction realization, surface passivation, contact realization. We see that PV's historic price reduction is a result from the combined effects of step-by-step evolutionary improvements in a wide variety of areas rather than one or two huge breakthroughs [5,6]. For example, processes such as dry texturing, spray-on phosphorus doping sources or impurity gettering have become standard, while last but not least actions related to increase the factory size and automation further lead to cost reductions ("economies of scale").

In contrast, larger values of the conversion efficiency of PV technology have been reached with the realization of sophisticated crystalline silicon (c-Si) cell structures, involving numerous and very complicated steps. This approach inevitably implies an increase of costs, which is not compatible with industrial production requirements that demand simple, high-throughput and reproducible processes.

In order to realize reliable devices characterized by high efficiency and low cost, an approach has been developed on the basis of amorphous/crystalline silicon heterojunction solar cells (SHJ), which combines wafer and thin film technologies. In this area impressive results were achieved by Sanyo Electric with the so called a-Si/c-Si Heterojunction with Intrinsic Thin layer (HIT) solar cell [10,11]. This technology showed excellent surface passivation (open circuit voltage (V_{oc}) values of around 730 mV) and the highest power conversion efficiency to date for a cell size of 100.4 cm^2: 23.0% was obtained [11].

1.2 Amorphous Crystalline Heterojunction Solar Cells

The design of the silicon hetero-junction solar cell is based on an emitter and back surface field (BSF) that are produced by low temperature growth of ultra-thin layers of amorphous silicon (*a*-Si:H) on both sides of a thin crystalline silicon wafer-base, less than 200 μm in thickness, where electrons and holes are photogenerated. The low temperature *a*-Si:H deposition lowers the thermal budget in the production of the cell (see Fig. 1.1), and at the same time will allow for high-throughput production machinery. Taken together, this can lead to a considerable lowering of manufacturing costs thus opening opportunities for the production of GWp/year manufacturing plants to sustain the booming PV market.

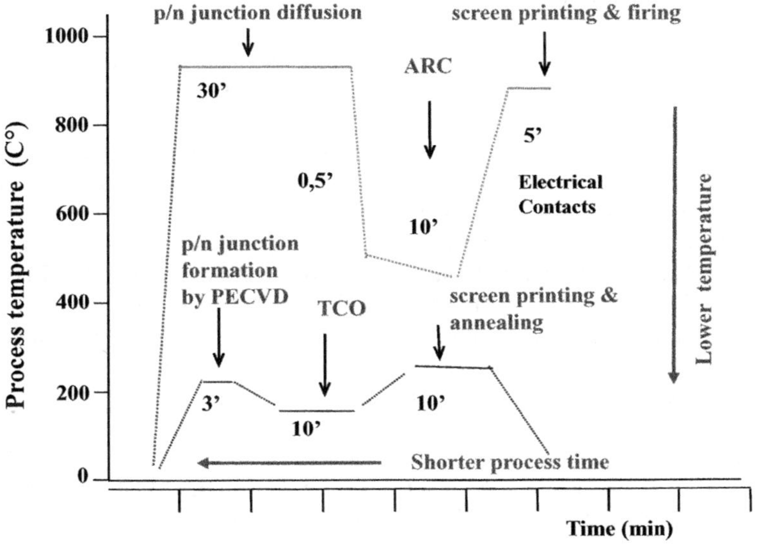

Fig. 1.1 Authors' estimated thermal budget and process time for the conventional c-Si technology (top curve) and SHJ technology (bottom curve).

The idea of making solar cells from silicon heterojunctions is a rather old one: It was first published in 1974 by Walther Fuhs and coworkers from the University of Marburg (Germany) [12]. However, it turned out that to realize the V_{oc} potential > 700 mV inherent to the heterojunction concept, it is mandatory to include additional, very thin (of the order of 10 nm) undoped – so called intrinsic – *a*-Si:H buffer layers between the wafer and the doped (emitter or BSF) *a*-Si:H layers. Briefly, the reason is that the defect density in *a*-Si:H increases strongly with doping, and this leads to an increase in interface defect density at the *a*-Si:H/*c*-Si junction, thus to enhanced recombination and a lower V_{oc}. This finding is the essence of a patent filed by Sanyo in 1991, which can be seen as the "core patent"

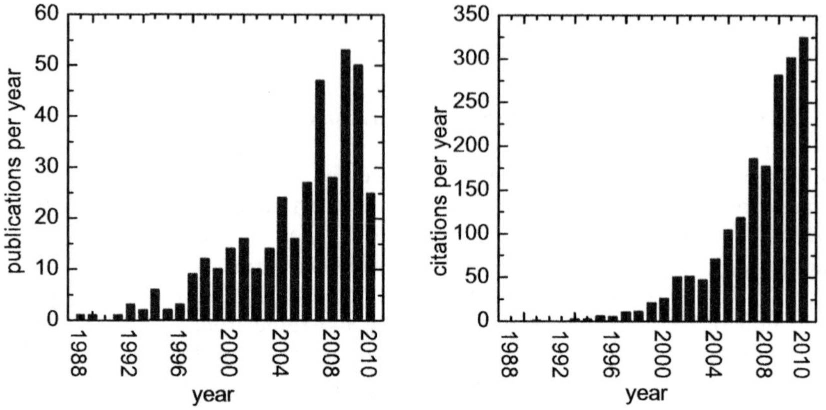

Fig. 1.2 Development of the number of both publications and citations related to silicon heterojunction solar cells over time [14].

for the subsequent successful commercialization of their so-called "HIT" concept. This patent has expired in 2010. A more in-depth discussion of the intellectual property aspect can be found in [13].

As a consequence, over the last decade, there have been many encouraging results on developing alternative concepts making use of *a*-Si:H/*c*-Si heterojunctions for high efficiency cells, such as omitting the undoped buffer and lowering the doping levels in the emitter and BSF, working on p-type *c*-Si substrates (the HIT cell is produced on n-type material), or on modifications to the *a*-Si:H layers like using *a*-Si:H/*μc*-Si stacks, *a*-SiC:H etc. This is reflected in the steadily increasing number of publications and citations related to *a*-Si:H/*c*-Si heterojunction solar cells, cf. Fig. 1.2[1]. Still, it appears that among other factors, the expiry of the mentioned "core patent(s)" has contributed significantly to the strongly increased interest in HIT-type cells seen in the last few years.

Today, many research groups and industries are pursuing intense R&D to further develop the a-Si:H/c-Si heterojunction technology. One such consortium has received funding from the European Commission in the framework of the 7th Research Framework Programme to develop a knowledge base and optimized device structure based on new insights in the physics and technology of wafer-based silicon heterojunction devices, within the project *"Heterojunction Solar Cells based on a-Si c-Si"* (HETSI) [15] [2].

[1] The database used for this analysis does not contain the proceedings of the European photovoltaic conferences prior to ~ 2008.

[2] The partners (acronym, country) are Institut National de l'Energie Solaire (INES, FR), Centre National de la Recherche Scientifique (CNRS, FR), Energieonderzoek Centrum Nederland (ECN, NL), Utrecht University (UU, NL), Agenzia Nazionale per le Nuove Tecnologie, l'Energia e lo Sviluppo Economicamente Sostenibile (ENEA, IT), Interuniversity MicroElectronics Centrum (IMEC, BE), Institut de Microtechnologie - Ecole Polytechnique Fédérale de Lausanne (EPFL, CH), Helmholtz-Zentrum Berlin für Materialien und Energie (HZB, DE), SOLON SE (DE), Photowatt SAS (FR), Q-Cells SE (DE), and ALMA Consulting Group SAS (FR).

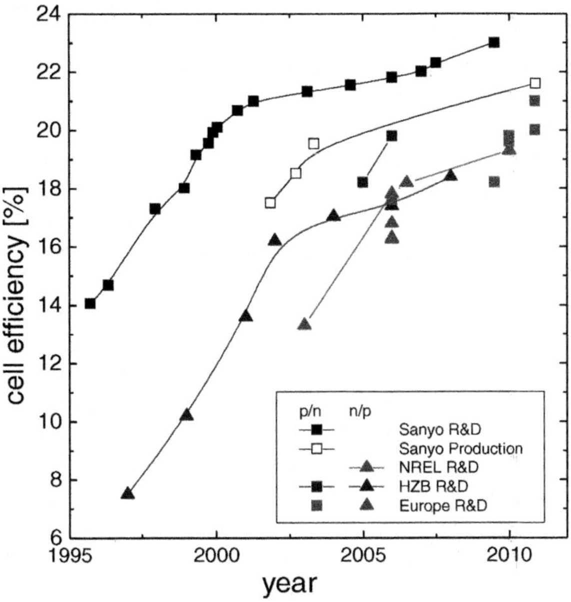

Fig. 1.3 Development of *a*-Si:H/*c*-Si heterojunction cell efficiency vs. time. Both (n)*a*-Si:H/(p)*c*-Si and (p)*a*-Si:H/(n)*c*-Si cell structures are shown.

The reported cell efficiencies have developed accordingly: Fig. 1.3 gives a (non-exhaustive) overview on the progress over time, where the distinction is made between (n)*a*-Si:H/(p)*c*-Si type cells and the "canonical" (p)*a*-Si:H/(n)*c*-Si structure as used by Sanyo. There is evidence for the gap in cell efficiencies between the two doping sequences being due to differences in fundamental device physics (carrier mobilities, band offsets), cf. Chapter 6 in this book. Furthermore, it is apparent that the Sanyo HIT cell has a significant lead on the reported cell efficiencies, by ~2% absolute at the time of writing. Nevertheless, others are covering lost ground at a fast pace: The latest reported cell efficiencies from NREL (US) are 18.2% on n-type and, interestingly, 19.3% (V_{oc} of 678 mV) on p-type wafers [16]. In Europe, the highest efficiencies reported so far are 21.0% obtained at Roth & Rau Switzerland in cooperation with EPFL Neuchâtel [17] and up to 19.6 % (20 % on 100 cm²) with a V_{oc} up to 718 mV on industrially relevant surfaces, i.e. large area 148 cm² pseudo-square n-type *c*-Si industrial wafers [18]. Recently Sanyo reported on opportunities to reach impressive efficiencies over 23% based on the utilization of very thin wafers (<100 μm) [19].

The realization of high quality a-Si:H/c-Si heterojunctions is not a trivial process requiring a very deep knowledge of several chemical and physical aspects on which the interface formation and the doped layers growth is based. Surface cleaning and/or preparation are critical, and chemistry and physics of the gas phase interaction during plasma deposition or treatment is another key issue [20]. Different process schemes affect structural quality of deposited films, surface morphology, roughness, surface reactivity and surface composition. The kinetics of impinging plasma particles and the formation of chains and islands of radicals on the surface dramatically change electrical and optical properties of the deposited films including the optical gap, activation energy, band offset, band bending, gap state and interface state density.

After formation of the a-Si:H/c-Si heterojunction, the cell is contacted using a ~80 nm thin transparent conductive oxide (TCO) layer and a metal grid on the front. The TCO is typically InO doped with Sn (ITO) or ZnO doped with Al. Often, a TCO is also used to form a dielectric mirror on the back side of the cell. Thus, to understand and optimize the whole a-Si:H/c-Si solar cell, also the influence of the TCOs on the optoelectronic properties of the cell has to be considered: Due to its high doping, the TCO behaves electronically like a metal with rather poor charge carrier mobility, and the electronic behavior of the TCO/a-Si:H junction is usually assumed as similar to a metal-semiconductor junction. The TCO work function plays an important role for the band alignment in the TCO/a-Si:H/c-Si structure and for charge carrier transport across the heterojunctions. Furthermore, TCO deposition on the about 10 nm thin a-Si:H is usually done using sputter processes; here, the possibility of damaging the delicate a-Si:H/c-Si interface during this sputter process should be taken into consideration and has to be accounted for during process optimization.

1.3 HETSI Workshop

A workshop has been organized at Utrecht University in February 2010 by the HETSI Consortium, at which many experts in the field presented an overview of the state-of-the-art in physics and technology of amorphous-crystalline heterostructure silicon solar cells, including a hands-on training session on computer modeling of cells. Over 80 attendees coming from different organizations and countries around the globe experienced an informal atmosphere with ample interaction possibilities.

In this book, the contributors to this workshop have written on their expertise, and we believe that as a whole, the book contains a broad overview of amorphous-crystalline heterostructure silicon solar cells, from the fundamental physical principles to the experimental and modeling details. It is intended to serve the strongly growing scientific and industrial PV community, not limited to silicon heterojunctions.

1.4 Guide to the Reader

The content of this book is organized as follows: Chapter 2 (Miro Zeman and Dong Zhang) introduces the heterojunction concept: The best wafer-based homojunction and heterojunction crystalline silicon solar cells are compared, and the advantages of heterojunction silicon solar cells related to the processing of the junction and solar cell operation are explained. The current status of SHJ R&D is outlined, summarizing the different approaches by institutes world-wide and comparing to Sanyo's HIT cell concept. This sets the stage for the subsequent Chapters 3-10 that follow loosely the processing steps of an actual silicon heterojunction cell. Chapters 11-14 then deal with characterization and modelling of SHJ cells, followed by two chapters on modelling and realization of interdigitated back contact silicon heterojunction (IBC-SHJ) cells. The final chapter 17 closes this book by arguing that silicon heterojunction cells are a smart choice for the high efficiency cell of the future.

Chapter 3 (Heike Angermann and Jörg Rappich) discusses the wet-chemical pre-treatment of c-Si wafers. This is a mandatory processing step to achieve a low density D_{it} of surface states on the wafer, which influences strongly the passivation quality at the a-Si:H/c-Si interface. The influence of these treatments on surface morphology and electronic interface properties is discussed for a wide scope of materials comprising not only a-Si:H, but also Si oxides (SiO_x), Si nitride (a-SiN_x:H) and Si carbide (a-SiC:H), which are frequently applied in Si heterostructure solar cells. An important aspect is the stability of wet-chemical surface passivation during storage in ambient air, which is found to be strongly influenced by the preparation-induced surface morphology. As shown for various heterojunction structures, the effect of optimized wet-chemical pre-treatments can be preserved during the subsequent soft PECVD growth of a-Si:H, a-SiN_x:H or a-SiC:H.

Chapter 4 (Jörg Rappich) is also devoted to c-Si surface preparation, but focuses on advanced concepts of using electrochemistry approaches for c-Si surface passivation, such as electropolishing in the current oscillating regime in diluted HF solutions. In addition, the use of *in-situ* photoluminescence and surface photovoltage is put forth as non-destructive technique to monitor the electronic surface properties during electrochemical oxidation, hydrogenation, and grafting of organic molecules and ultra-thin polymeric layers.

Chapter 5 (Pere Roca i Cabarrocas) provides an overview of the many deposition processes presently in use for the deposition or growth of amorphous and microcrystalline silicon. It is pointed out that the choice of the deposition technique may help to favour a particular type of film precursor, in particular SiH_3 which is often considered as the most suitable to obtain device grade material. The growth process and film properties are mainly controlled by the surface and subsurface reactions: a growth zone exists close to the film surface, where cross-linking reactions leading to bulk-like formation take place. It is suggested that film properties are governed neither by the film precursor, nor by the deposition technique. The chapter closes with the issue of substrate dependence of the growth process, which is of special importance in the case of heterojunction solar cells.

Chapter 6 (Lars Korte) discusses the electronic properties of the ultrathin a-Si:H layers used in SHJ cells and their interface to the c-Si wafer. The well-known properties of thick (several 10–100 nm) a-Si:H layers such as those used in a-Si:H *pin* cells are briefly summarized. Subsequently, it is shown how for ultra-thin a-Si:H on c-Si substrates the density of occupied valence band and defect states $N_{occ}(E)$ and the position of the Fermi level in the band gap can be measured. The measured *a*-Si:H properties are correlated to the band bending in the *c*-Si absorber, to charge carrier recombination at the *a*-Si:H/*c*-Si interface and to solar cell open circuit voltage V_{oc}. The current state-of-the-art of *c*-Si surface passivation by (i)*a*-Si:H is reviewed. Furthermore, the use of temperature-dependent current-voltage measurements on complete *a*-Si:H/*c*-Si solar cells to extract information on recombination and transport is discussed. The chapter also shows how an important parameter of the *a*-Si:H/*c*-Si junction, the band offset in the valence and conduction band edges, can be determined using a special variant of photoelectron spectroscopy.

Chapter 7 (Stefaan De Wolf) takes a closer look at the a-Si:H/c-Si interface passivation and its correlation to the a-Si:H properties: The relevant literature on *c*-Si surfaces is briefly reviewed, including the effect of hydrogenation of surface states. The physical passivation mechanism of intrinsic *a*-Si:H is elucidated, and it is concluded that it stems from chemical surface state passivation, i.e. saturation of Si dangling bonds by hydrogen, similar to defect passivation in the *a*-Si:H bulk. For these films, it is also argued how epitaxial growth may detrimentally influence the passivation quality. The effect of doping on the amorphous films is discussed, and an explanation is proposed for the experimental fact that *a*-Si:H/*c*-Si interface passivation decreases when (p)*a*-Si:H or stacks of (*p*/*i*)*a*-Si:H are deposited, as compared to passivation by (i)*a*-Si:H alone. The HIT cell concept is thus understood as providing a compromise between doping and surface-passivation by employing an intrinsic buffer layer, between the doped film and the wafer.

Still within the context of interface recombination, Chapter 8 (Rudolph Brüggemann) discusses how photoluminescence (PL) and electroluminescence (EL) from amorphous/crystalline silicon heterostructures can be used for the characterisation of precursor structures for solar cell optimisation and for the study of related physical aspects. It is shown that the luminescence yield, or more precisely the deduced quasi-Fermi level splitting, is directly related to the open-circuit voltage of the device which itself is limited by factors like the interface recombination rate. The usefulness of contactless PL and EL techniques for investigations of the SHJ physics as well as for process control are thus highlighted.

Chapter 9 (Florian Ruske) deals with the next step of fabricating a typical SHJ cell, namely the deposition of transparent conductive oxides (TCOs) – typically ITO or ZnO:Al – on top of the *a*-Si:H in order to provide light trapping and a sufficient lateral conductivity towards the metal of the grid fingers. The optical properties of these films strongly depend on the electrical transport properties, especially the carrier concentration. The details of this mutual dependency are discussed using models for optical absorption, and it is shown that it is advantageous to use materials with moderate carrier concentrations. Non-vacuum and vacuum deposition techniques for TCOs are discussed, with a focus on magnetron sputtering, a process belonging to the latter class. It is shown how the additional challenges posed by the use of sputtered TCOs in SHJ, i.e. the low thickness of the films and the low deposition temperature, can be handled.

The final step of cell fabrication, the deposition of metal contacts, is discussed in Chapter 10 (Mario Tucci, Luca Serenelli, Simona De Iuliis, Massimo Izzi). Here, the doping of amorphous films is discussed together with the possibility to enhance the amorphous film conductivity by using chromium silicide formation on top of doped films. A finite difference numerical model is used to describe the a-Si:H/c-Si heterojunction solar cell in which both contacts are made by amorphous films, and a detailed investigation is presented comparing experimental current voltage characteristics of heterojunction contacts with the numerical models. TCOs and the formation of contacts by screen printing are discussed, and three examples of heterojunction solar cells are proposed using different approaches to form the contacts.

The following five chapters deal with characterization and modelling of SHJ cells: Chapter 11 (Jatin Rath) describes the standard electrical characterization techniques of SHJ solar cells which should elucidate the link between improvements in cell parameters obtained via process development and the microscopic nature of the functioning of the SHJ device. Although the SHJ cell is a bulk device, the parts of the SHJ cell that control the charge transport are limited to very thin regions. Characterization of such thin layers, in particular defect densities, conductivity, carrier recombination, is a complex issue. The chapter discusses the origin of the so-called S-type character in the I-V characteristics. Also, experimental methods to determine the band offset and the tunneling behavior at the spikes in the bands are described. Determining interface states is difficult to perform, however, electrically detected magnetic resonance (EDMR) or spin dependent photoconductivity (SDPC) is described as a potentially powerful technique to measure these states.

In Chapter 12 (Jean-Paul Kleider), a technique to determine the band offsets in *a*-Si:H/*c*-Si heterojunctions from electrical measurements is discussed. The chapter starts by recalling the principal models for band lineup at interfaces, with particular emphasis on Anderson's electron affinity rule and Tersoff's branching point alignment theory. The principal electrical characterization tools based on capacitance and admittance measurements are presented, and the main potential problems and sources of uncertainty when applying the C-V technique to the *a*-Si:H/*c*-Si system are addressed. Finally, a simple technique based on the measurement of the planar conductance of *a*-Si:H/*c*-Si structures is presented, and the determination of band offsets from such measurements and related modeling on both (p)*a*-Si:H/(n)*c*-Si and (n)*a*-Si:H/(p)*c*-Si structures is discussed. Note that the results obtained here compare favourably with those in Chapter 6, obtained with a completely different technique.

Chapter 13 (Rolf Stangl and Caspar Leendertz) discuss the approaches for numerical modelling of SHJ cells, and Chapter 14 (Caspar Leendertz and Rolf Stangl) gives a "hands-on" introduction to using a concrete simulation software, AFORS-HET (Automat for Simulation of Heterostructures), for this purpose: Chapter 13 outlines the basic equations for the optical and electrical calculations used in AFORS-HET, then focuses on the detailed description of the equations needed to calculate the recombination via defects in the semiconductor layers. Then, Chapter 14 describes the physical models and material parameters needed to simulate an *a*-Si:H/*c*-Si solar cell with AFORS-HET, and a simulation study showing the dependence of solar cell characteristics on emitter doping, i-layer thickness and interface quality is presented. The AFORS-HET user interface is introduced and a step-by-step explanation of how to define a structure and how to simulate a solar cell under different external conditions is given, so that the interested reader can repeat the simulation study.

While AFORS-HET is limited to simulations in one spatial dimension (1D), the simulation issue is taken one step further by discussing 2D simulations in Chapter 15 (Djicknoum Diouf, Jean Paul Kleider, Christophe Longeaud). These are carried out to investigate interdigitated back contact silicon heterojunction (IBC-SHJ) solar cells. A comparative study between the IBC-SHJ structure based on *n*-type and *p*-type *c*-Si is discussed. Similar to the 1D case (front and rear contacted cells), the results on IBC indicate that the key parameters to achieve high efficiency are a high *c*-Si substrate quality, low surface recombination velocity especially at the front surface, and *a*-Si:H/*c*-Si interfaces with low recombination velocity.

The properties of actual IBC-SHJ cells realized in different labs world-wide are discussed in Chapter 16 (Niels Posthuma, Barry O'Sullivan, Ivan Gordon): The advantages of such cells are outlined, e.g. the high current density since no metal contacts are present at the front of the cell and easier series interconnection between various cells at module level. After a discussion of conventional homojunction IBC concepts, which have shown 21 to 23% energy conversion efficiency on large area industrially produced cells, SHJ-IBC cells are introduced, and the research on implementing the heterojunction emitter at the rear of the wafer, which has a rather short history of only about five years, is presented. It is concluded that the current SHJ-IBC cells, with cell structures and processing that are not optimized and are typically fabricated on small area, are just the start of a new development.

The book closes with Chapter 17 (Delfina Muñoz, Thibaut Desrues, Pierre-Jean Ribeyron) that takes a look at the big picture: First, it outlines the current state of the photovoltaics market and discusses how the market share of high efficiency cell concepts such as the SHJ can be expected to develop in the future. Then, all process steps discussed in detail in the previous chapters are briefly reviewed and put into context. Finally, the question is answered whether the SHJ is a good choice with respect to other, competing high efficiency concepts. It is concluded that the advantages of the SHJ, i.e. high efficiency with comparably simple, low temperature processing steps, easy module integration, cost reduction potential in conjunction with thin wafers etc., make silicon heterojunction cells indeed a smart choice for the high efficiency cell of the future.

We trust that with the contents laid out in this book, it will find widespread use in the photovoltaics community, both industrial and academic because it covers a broad range of scientific and technical aspects related to silicon heterojunction technology. Particularly, parts of the book are well-suited for use in (under)graduate courses, thus we hope that this publication can serve as a means to train students in those skills that are in high demand to sustain the growth in the photovoltaic industry. In fact the SHJ is among the more effective concepts used to tackle reduction of material consumption (g/Wp), for which new manufacturing technologies are in development that carefully consider costs, high throughput and yield, and integrated industrial processing. It is expected that within a few years these developments will lead to costs that are competitive with traditional technologies that generate electricity. All of these concepts are needed for the challenge to reach a carbon neutral society by the middle of this century.

References

[1] International Energy Agency (IEA): World Energy Outlook 2010 (November 2010), ISBN 978-92-64-08624-1, `http://www.worldenergyoutlook.org/` (accessed July 20, 2011)

[2] European Photovoltaic Energy Association (EPIA), Greenpeace: Solar Generation 6 – Executive Summary (October 2010), `http://www.eupvplatform.org/publications/strategic-research-agenda-implementation-plan.html` (accessed July 20, 2011)

[3] European Photovoltaic Energy Association (EPIA): Global market outlook until 2015 (2011)

[4] Chapin, D.M., Fuller, C.S., Pearson, G.L.: A new silicon *p-n* junction photocell for converting solar radiation into electrical power. J. Appl. Phys. 25, 676–677 (1954)

[5] Swanson, R.M.: A Vision for Crystalline Silicon Photovoltaics. Prog. Photovolt: Res. Appl. 14, 443–453 (2006)

[6] Van Sark, W.G.J.H.M., Alsema, E.A., Junginger, H.M., De Moor, H.H.C., Schaeffer, G.J.: Accuracy of progress ratios determined from experience curves: the case of photovoltaic technology development. Prog. Photovolt: Res. Appl. 16, 441–453 (2008)

[7] EU PV technology platform. A strategic research agenda for photovoltaic solar energy technology (2007), `http://www.eupvplatform.org/index.php?eID=tx_nawsecuredl&u=0&file=fileadmin%2FDocuments%2FPVPT_SRA_Complete_070604.pdf&t=1305037849&hash=ba3874f927ade8486701939c98d06cdf` (accessed May 9, 2011)

[8] Goetzberger, A., Hebling, C., Schock, H.-W.: Photovoltaic materials, history, status and outlook. Mater. Sci. Eng. R 40, 1–46 (2003)

[9] Green, M.A., Emery, K., Hishikawa, Y., Warta, W.: Solar efficiency tables (version 36). Prog. Photovolt: Res. Appl. 18, 346–352 (2010)

[10] Tanaka, M., Taguchi, M., Matsuyama, T., Sawada, T., Tsuda, S., Nakano, S., Hanafusa, H., Kuwano, Y.: Jpn. J. Appl. Phys. 31, 3518–3522 (1992)

[11] Mishima, T., Taguchi, M., Sakata, H., Maruyama, E.: Development status of high-efficiency HIT solar cells. Solar Energy Materials and Solar Cells 95, 18–21 (2011)

[12] Fuhs, W., Niemann, K., Stuke, J.: Heterojunctions of amorphous silicon and silicon single crystals. In: Tetrahedrally Bonded Amorphous Semiconductors, AIP Conference Proceedings, vol. 20, pp. 345–350 (1974)

[13] Chunduri, S.K.: A HIT for all? Photon International, 130–140 (December 2010)

[14] Based on citation report data extracted in December 2010 from ISI Web of Knowledge, databases SCI-EXPANDED and CPCI-S (December 2010)

[15] `http://www.hetsi.eu` (2010) (accessed July 20, 2011)

[16] Wang, Q., Page, M.R., Iwaniczko, E., Xu, Y., Roybal, L., Bauer, R., To, B., Yuan, H.-C., Duda, A., Hasoon, F., Yan, Y.F., Levi, D., Meier, D., Branz, H.M., Wang, T.H.: Efficient heterojunction solar cells on p-type crystal silicon wafers. Appl. Phys. Lett. 96, 013507 (2010)

[17] Lachenal, D., Andrault, Y., Bätzner, D., Guerin, C., Kobas, M., Mendes, B., Strahm, B., Tesfai, M., Wahli, G., Buechel, A., Descoeudres, A., Choong, G., Bartlomé, R., Barraud, L., Zicarelli, F., Bôle, P., Fesquet, L., Damon-Lacoste, J., de Wolf, S., Ballif, C.: High Efficiency Silicon Heterojunction Solar-Cell Activities in Neuchâtel, Switzerland. In: Proc. 25th European Photovoltaic Solar Energy Conference and Exhibition / 5th World Conference on Photovoltaic Energy Conversion, pp. 1272–1275 (2010)

[18] Muñoz, D., Ozanne, A.S., Harrison, S., Danel, A., Souche, F., Denis, C., Favier, A., Desrues, T., Martin de Nicolás, S., Nguyen, N., Hickel, P.E., Mur, P., Salvetat, T., Moriceau, H., Le-Tiec, Y., Kang, M.S., Kim, K.M., Janin, R., Pesenti, C., Blin, D., Nolan, T., Kashkoush, I., Ribeyron, P.J.: Towards high efficiency on full wafer a-Si:H/c-Si heterojunction solar cells: 19.6% on 148cm^2. In: Proceedings of the 35th PVSC, Hawaii, pp. 39–43 (2010)

[19] Kinoshita, T., Fujishima, D., Yano, A., Ogane, A., Tohoda, S., Matsuyama, K., Nakamura, Y., Tokuoka, N., Kanno, H., Sakata, H., Taguchi, M., Maruyama, E.: The approaches for High Efficiency HIT solar cell with very thin (<100 μm) Silicon wafer over 23%. In: Proc. 26th European Photovoltaic Solar Energy Conference and Exhibition (2011)

[20] Roca, F., Bobeico, E., Della Noce, M., Delli Veneri, P., Lancellotti, L., Formisano, F., Mercaldo, L., Morvillo, P., Giangregorio, M.M., Bianco, G.V., Sacchetti, A., Losurdo, M., Bruno, G.: Key issues for the improvement of the interface and emitter quality in a-Si:H/c-Si heterojunction solar cells. In: Proc. 21st European Photovoltaic Solar Energy Conference and Exhibition, pp. 1556–1560 (2006)

Chapter 2
Heterojunction Silicon Based Solar Cells

Miro Zeman and Dong Zhang

Delft University of Technology, EEMCS Building
Photovoltaic Materials and Devices group
Mekelweg 4
2628 CD, Delft
The Netherlands

Abstract. Heterojunction (HJ) silicon solar cells use crystalline silicon wafers for both carrier transport and absorption, and amorphous and/or microcrystalline thin silicon layers for passivation and junction formation. The top electrode is comprised of a transparent conductive oxide (TCO) layer in combination with a metal grid. Heterojunction silicon solar cells have attracted a lot of attention because they can achieve high conversion efficiencies, up to 25%, while using low temperature processing, typically below 200 °C for the complete process. Low processing temperature allows handling of silicon wafers of less than 100 μm thick while maintaining a high yield.

In this chapter the best wafer-based homojunction and heterojunction crystalline silicon solar cells are compared, and the advantages of heterojunction silicon solar cells related to the processing of the junction and solar cell operation are explained.

The development and recent status of HIT (Heterojunction with Intrinsic Thin-layer) silicon solar cells at the company Sanyo are presented. In order to reduce cost of the HIT solar cells, Sanyo is focusing on reducing the thickness of the silicon wafer. In 2009 the company demonstrated 22.8% conversion efficiency and record high open circuit voltage of 0.743 V on a solar cell based on a 98 μm thick wafer with a total area of 100.3 cm^2.

Achievements from other research groups such as Tokyo Institute of Technology (Tokyo Tech) and the National Institute of Advanced Industrial Science and Technology (AIST) in Japan, the National Renewable Energy Laboratory (NREL) in the U.S.A., Helmholtz Zentrum Berlin (HZB) and Frauhofer institute for Solar Energy Systems (Frauhofer ISE) in Germany, L'Institut National de l'Energie Solaire (INES) in France, Neuchatel PV-lab of Ecole Polytechnique Federale de Lausanne (EPFL) in Switzerland, National Agency for New Technologies, Energy and the Environmentand (ENEA) in Italy and Mingdao University in China are presented. The research activities and results achieved with heterojunction silicon solar cells in the Netherlands are also reported.

Challenges to further improve the performance of heterojunction silicon solar cells by minimizing the optical, recombination, and resistance losses in

W.G.J.H.M. van Sark et al. (Eds.): Physics & Tech. of Amorphous-Crystalline, EM, pp. 13–43.
springerlink.com

heterojunction silicon solar cells are discussed. These challenges deal with wafer cleaning, suppression of epitaxial growth, controlling thin silicon layer thickness, reduction of absorption losses in thin silicon layers and transparent conductive oxide, surface texturing and the improvement of grid electrodes.

2.1 Introduction to Silicon Photovoltaic Technologies

The term photovoltaics (PV) refers to the direct energy conversion of solar radiation into electricity. PV is attracting a large academic and industrial interest, and is considered by many to be the most promising energy generation technology for the future. PV systems are able to supply electrical power ranging from a few hundred watts to tens of megawatts. The energy conversion takes place in solar cells, which are usually made of a semiconductor material such as silicon. Silicon-based solar cells have become the dominant PV technology because silicon exhibits good stability, a well balanced set of physical, chemical and electronic properties, and also because of the economic benefits arising from the microelectronics industry [1].

Silicon-based solar cells can be divided into two main groups: homojunction wafer-based crystalline silicon (c-Si) solar cells and thin-film silicon solar cells. Wafer-based c-Si solar cells dominated the PV market in 2008 with an overall share of 87%, and feature a high module efficiency of 12 to 20% and a long-time warranty of 10 to 25 years [2]. However, cost reduction is the main challenge for wafer-based c-Si solar cells due to the use of expensive wafers and the requirement of high temperature processing during junction formation.

Thin-film silicon solar cells based on hydrogenated amorphous silicon (a-Si:H) and hydrogenated microcrystalline silicon (μc-Si:H) are promising candidates for low-cost PV technology due to their low material consumption and low temperature processing in comparison to wafer-based c-Si solar cells. Moreover, thin-film silicon solar cells can be fabricated on a range of substrates, including flexible metal foils. However, a low module efficiency of around 6 to 9% is the principle limitation for this PV technology, which is a result of the poor electronic properties of the absorbers, such as a low carrier lifetime. Improvement of the efficiency of thin-film silicon solar cells is an essential requirement if the technology is to remain competitive with other PV technologies.

The heterojunction silicon solar cell approach benefits from combining both wafer-based c-Si and thin-film silicon solar cells. Heterojunction silicon solar cells can achieve high conversion efficiencies while using thin-film silicon processes to lower the cost in comparison with c-Si solar cells.

2.2 Motivation for Developing Heterojunction Silicon Solar Cells

In the thermodynamic approach the process of conversion of solar energy into electrical energy can be described in two steps. In the first step the energy of the solar radiation is converted into chemical energy in a suitable semiconductor material.

The chemical energy is related to the concentration of photo-generated electrons and holes in the semiconductor absorber. In the next step the chemical energy of the photo-generated charge carriers is converted into electrical energy by separating negatively charged electrons and positively charged holes from each other, and collecting them at the electrodes.

In general a solar cell contains an absorber layer in which photons of the incident radiation are absorbed, thereby generating electron-hole pairs. In order to separate the electrons and holes from each other, so-called "semi-permeable membranes" can be attached to the both sides of the absorber [3]. The important requirement for the semi-permeable membranes is that they selectively allow only one type of charge carrier to pass through. An important issue for designing an efficient solar cell is that the electrons and holes generated in the absorber layer must reach the membranes. This requires that the thickness of the absorber layer is smaller than the diffusion lengths of the charge carriers.

A membrane that lets electrons pass and blocks holes is a material that has a large conductivity for electrons and a small conductivity for holes. An example of such material is an n-type semiconductor, in which a large electron conductivity with respect to the hole conductivity is caused by a large difference in electron and hole concentrations. Electrons can easily move through the n-type semiconductor while the transport of holes, which are the minority carriers in such material, is very limited due to the recombination process. The opposite holds for electrons in a p-type semiconductor, which is an example of a hole membrane.

In order to minimize the injection of holes from the absorber into the n-type semiconductor an energy barrier should be introduced in the valence band, ΔE_V, at the interface between the n-type semiconductor and the absorber (see Fig. 2.1). Ideally, this can be achieved by choosing an n-type semiconductor that has a larger band gap than that of the absorber, where the energy difference between the band gaps is fully accommodated in the valence band of the two materials. Similarly, the injection of electrons from the absorber into the p-type semiconductor can be suppressed by use of a p-type semiconductor with a larger band gap than that of the absorber, with the band offset contained fully within the conduction band, ΔE_C. The requirement of having the band offset in the conduction band means that the electron affinity, X_e, of the p-type semiconductor is smaller that the electron affinity of the absorber. The additional advantage of applying membrane materials with large band gaps is to allow a larger fraction of photons in the solar spectrum to be transmitted through the membranes to the absorber.

The asymmetry between the electronic structure of n-type and p-type semiconductors is the basic requirement for photovoltaic energy conversion. Figure 2.1 shows a schematic band diagram of an illuminated ideal solar cell structure with an absorber and semi-permeable membranes. The terminals, i.e. the electrodes, of the solar cell are attached to the membranes. We refer to the structure between the terminals as a junction, and this solar cell structure is denoted as a single junction solar cell. When the absorber and membrane materials have different semiconductor properties, such as different energy band gaps, we describe the junction as a heterojunction. When the absorber and doped layers are based on the same material, for example c-Si, we denote the junction as a homojunction.

The quasi-Fermi level for electrons, E_{FC}, and the quasi-Fermi level for holes, E_{FV}, are used to describe the illuminated state of the solar cell. The energy difference between the quasi-Fermi levels is a measure of the efficient conversion of radiation energy into chemical energy. In Fig. 2.1 the illuminated solar cell is shown at the open-circuit condition, which is when the terminals of the solar cell are not connected to each other and therefore no electric current can flow through an external circuit. Under this condition, a voltage difference can be measured between the terminals of the solar cell. This voltage is denoted the open-circuit voltage, V_{oc}, and it is an important parameter to consider when characterizing the performance of solar cells.

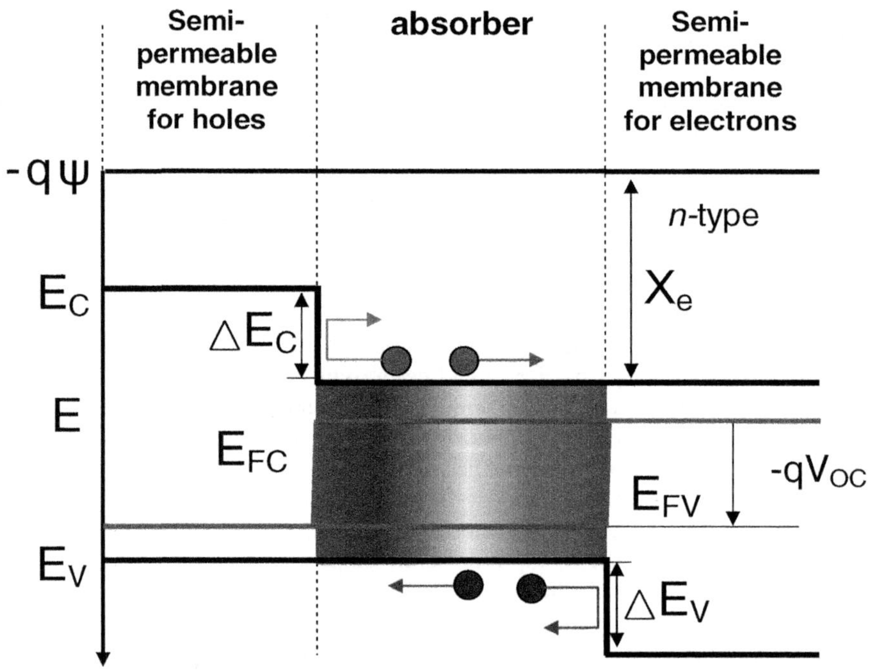

Fig. 2.1 Schematic band diagram of an idealized heterojunction solar cell structure at the open- circuit condition.

In summary, in a heterojunction solar cell the injection of one type of charge carriers from the absorber into membrane materials, in which they become minority carriers and recombine, can be suppressed. This can result in a more efficient use of photo-generated carriers and consequently a higher photocurrent from the cell.

In practical heterojunction solar cells, the band offsets between different materials are accommodated in both the conduction band and the valence band. This can result in the formation of transport barriers between the absorber and the membrane for the majority carriers. This is illustrated in Fig. 2.2, which shows a heterojunction formed between n-type c-Si with a band gap of 1.1 eV and p-type a-Si:H with a band gap of 1.7 eV. A transport barrier is formed for the holes at the interface between the two materials. The holes can drift through narrow 'spike' barriers by tunneling, trap-assisted tunneling and/or thermionic emission.

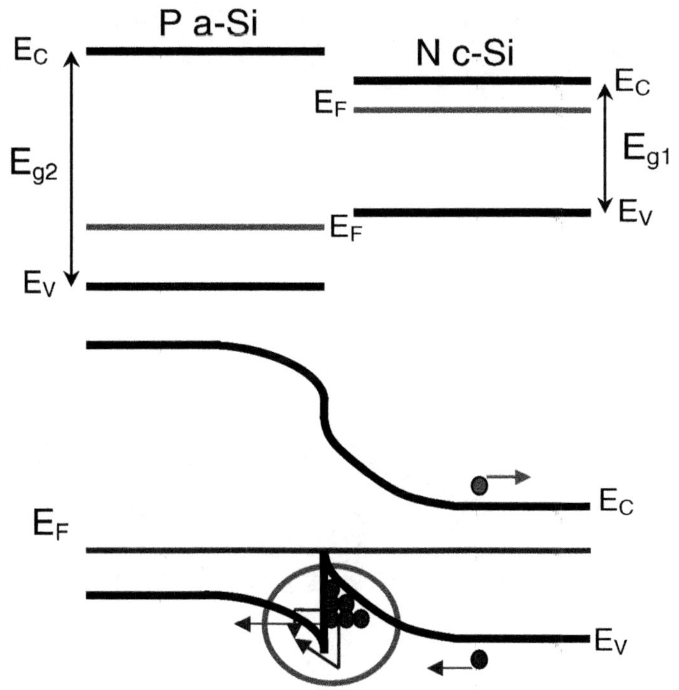

Fig. 2.2 Schematic band diagram of a practical a-Si:H/c-Si heterojunction. Carrier transport through the energy barrier at the interface is highlighted by the red circle.

2.3 Comparison of Homojunction and Heterojunction c-Si Based Solar Cells

In Fig. 2.3a the schematic structure of the best homojunction c-Si solar cell is presented. This is the PERL (passivated emitter with rear locally diffused) c-Si solar cell, which achieved an efficiency of 25% [4]. In the PERL solar cell, the high-quality wafer surface is textured to form an "inverted pyramids" structure in order to reduce surface reflection and to increase internal reflection on the rear side [1]. Silicon oxide is thermally grown on both sides of the wafer to passivate surface defects. Small openings in the silicon oxide are made to provide access to connect the metallic contacts to silicon regions that have been heavily doped. The small heavily doped areas can reduce the recombination caused by metallic contacts, and make it possible to decrease the distance between openings to reduce the lateral resistance. Moreover, the surface of the PERL cell is coated with MgF_2/ZnS double antireflective layers to further reduce reflection.

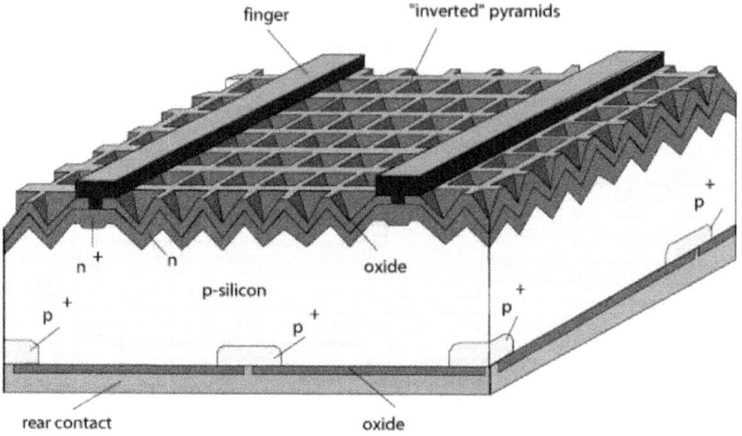

(a)

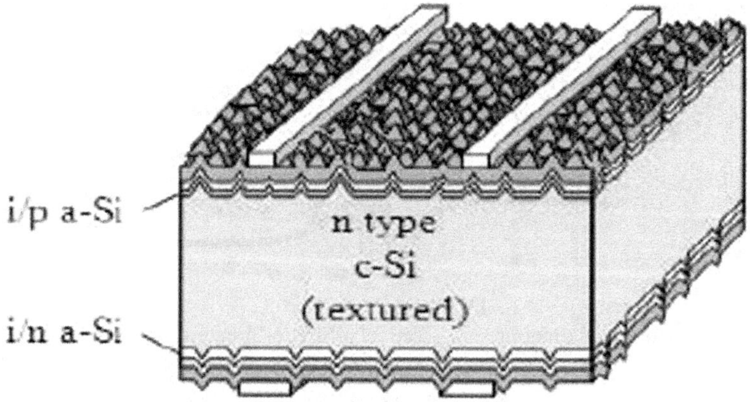

(b)

Fig. 2.3 Solar cell structures of a) PERL c-Si solar cell made by UNSW [4] and b) the HIT solar cell made by SANYO [5].

In Fig. 2.3b the schematic structure of the best heterojunction c-Si solar cell, known as the HIT solar cell, is presented [5]. The n-type c-Si wafer is randomly textured to provide effective light trapping. Intrinsic a-Si:H is deposited on both sides of the wafer for passivation. P-type a-Si:H is deposited on one side as an emitter, while n-type a-Si:H is deposited on the other side to form the back surface field (BSF). TCO layers are required to enhance carrier transport to the contacts because the a-Si:H layers are thin and highly resistive. Fabrication of the PERL c-Si solar cell involves complicated and demanding processing steps. Optical lithography is required for patterning surfaces, local oxidation and local dopant diffusion. High-temperature processing such as thermal oxidation at 1000 °C is required. By comparison, the fabrication of HIT solar cells is relatively simple.

The wafer surface is randomly textured, and the emitter and BSF are formed by depositing a-Si:H and/or μc-Si:H layers at a temperatures below 250 °C in a PECVD (plasma enhanced chemical vapor deposition) process.

Table 2.1 Comparison of external parameters between the PERL c-Si solar cell made by UNSW [4] and the HIT solar cells made by SANYO [5].

	PERL c-Si solar cell	HIT solar cell
Efficiency record	25%	23%
Jsc (mA/cm^2)	42.7	39.5
Voc (mV)	705	729
FF	0.828	0.8
Area (cm^2)	4	100

In Table 2.1 the external parameters of the PERL and HIT c-Si solar cells are compared. These parameters are the short-circuit current density, J_{sc}, open-circuit voltage, V_{oc}, and fill factor, *FF*. The HIT solar cell features excellent surface passivation due to the a-Si layer, and so has a higher V_{oc} than the PERL cell. However, the incorporation of TCO and a-Si:H layers in HIT solar cells causes light absorption losses, which result in a lower J_{sc}. It should be noted that although the record efficiency of the HIT solar cell is lower than the PERL cell, it was obtained on a much larger solar cell area.

Although the heterojunction silicon solar cell comprises both a-Si:H and c-Si materials, it does not exhibit a strong performance degradation under light exposure, as is the case for thin-film a-Si:H solar cells, or a strong temperature dependence of the performance, as is the case for wafer-based c-Si solar cells. In Fig. 2.4a one can observe that there is no light-induced degradation after 5 hours of high-intensity illumination. This is because the a-Si:H layers in heterojunction silicon solar cells are very thin (only several nanometers) and so provide a negligible contribution to the overall power generation [6]. The heterojunction silicon solar cell exhibits a smaller drop in performance with increasing the temperature in comparison with conventional c-Si solar cells, as shown in Fig. 2.4b. It has been observed that a solar cell with improved surface passivation, and a correspondingly higher V_{oc}, exhibits an improved temperature dependence [7]. However, the reasons for this are not yet clear, and so further work is required to explain this phenomenon.

The process requirements of heterojunction silicon solar cells have several advantages in comparison to those of the homojunction c-Si solar cell. The thermal budget during the heterojunction formation is considerably reduced compared to homojunction formation. The deposition temperature of a-Si:H layers and TCO front contacts is usually less than 250 °C. The time required to form the junction and deposit contact layers is also shorter for heterojunctions than for conventional c-Si solar cells (Fig. 2.5). Wafer bowing is suppressed due to the low processing temperature of the heterojunction silicon solar cell and its symmetric structure.

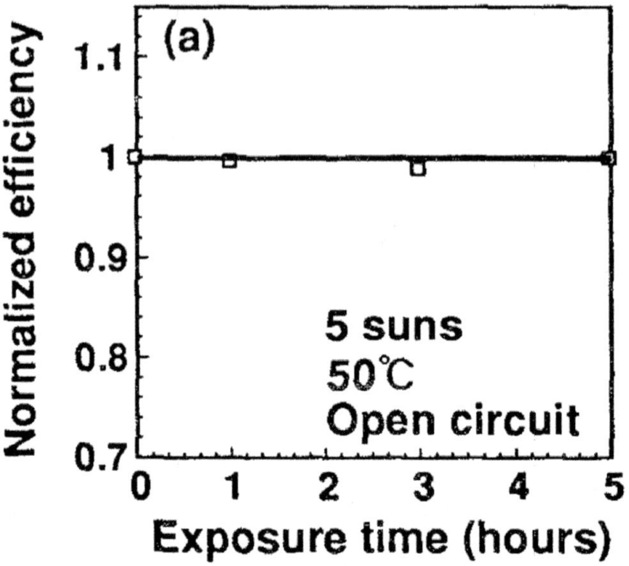

(a)

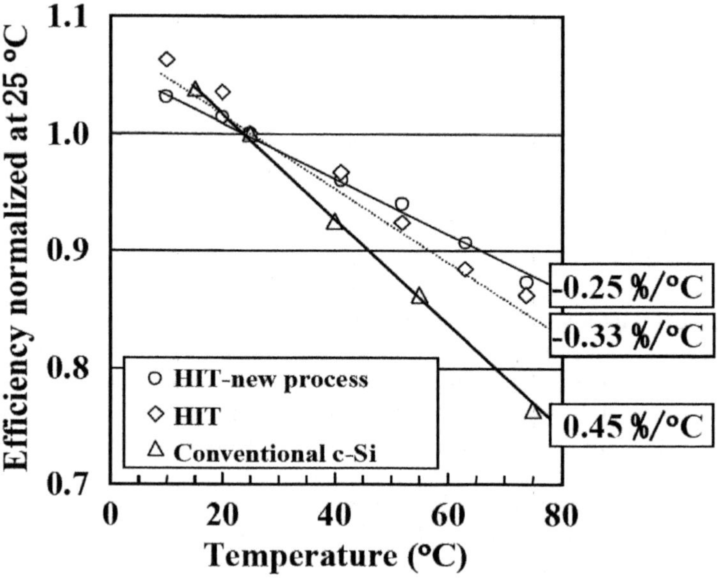

(b)

Fig. 2.4 Stability of heterojunction silicon solar cells regarding a) light-induced degradation[6] and b) temperature dependence [7].

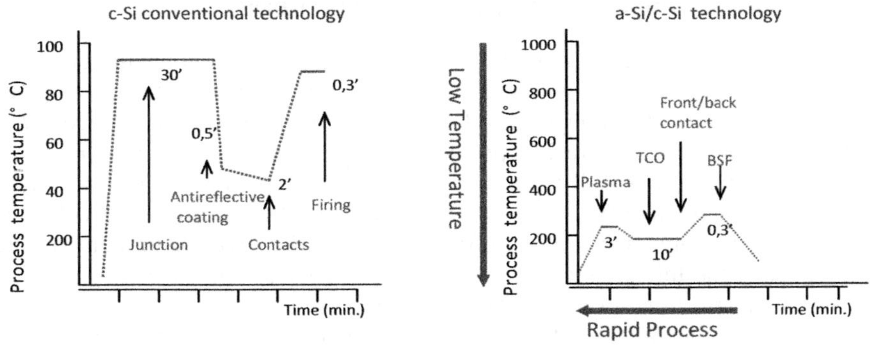

(a) (b)

Fig. 2.5 Industrial processing temperature and time for a) conventional c-Si solar cells and b) heterojunction solar cells [8].

This enables use of thinner wafers, which results in a reduction of the material cost. Low temperature processing and excellent surface passivation of wafers result in a higher effective lifetime of minority carriers. This makes the use of lower quality wafers feasible, which contributes to further reduction of the material cost.

2.4 Performance of Heterojunction Silicon Solar Cells

Research on heterojunction silicon solar cells has been carried out in many research institutes around the world, motivated by the potential of achieving a high efficiency at a low cost. The parameters of leading heterojunction silicon solar cells fabricated at several companies and institutes are reported in Table 2.2.

Table 2.2 Achievements on heterojunction silicon solar cells in some institutes (FZ: float zone; CZ: Czochralski).

	FZ/CZ	Area (cm^2)	Jsc (mA/cm^2)	Voc (mV)	FF (%)	Efficiency (%)
Sanyo [5]	nCZ	100	39.5	729	80	23
AIST [9]	nCZ	0.2	35.6	656	75	17.5
HZB [10]	nFZ	1	39.3	639	79	19.8
	pFZ	1	36.8	634	79	18.5
Roth&Rau and EPFL [11]	nFZ	4	37	729	77.9	21
NREL [12]	pFZ	0.9	35.9	678	78.6	19.1
	nFZ	0.9	35.3	664	74.5	17.2
INES [13]	nFZ	148	35.6	718	76.7	19.6
ENEA [14]	pCZ	2.25	37.1	600	76.3	17
Tokyo tech [15]	pFZ	0.908	36.7	668	73.1	17.9
MingDao University [16]	p	1	34.8	645	73	16.4
Fraunhofer ISE [17]	nFZ	4	37.8	667	78.6	19.8

2.4.1 Development of HIT Solar Cells at Sanyo

Sanyo is the forerunner in this research field, and presently holds the efficiency record for a heterojunction device. We shall give an overview of the development of the HIT solar cell at Sanyo in order to follow their route towards record high efficiency solar cells. Figure 2.6a presents the schematic structure of the first heterojunction silicon solar cell reported by Sanyo. A p-type a-Si:H layer was deposited directly onto an n-type c-Si wafer to form the heterojunction solar cell. With this solar cell Sanyo achieved an efficiency of 12.3%. Figure 2.6b demonstrates the effect of the thickness of the p-type a-Si:H layer on the external parameters of the heterojunction solar cell. The cell exhibited a relatively low V_{oc} and FF because of its high defect-state density at the a-Si:H/c-Si interface. An increase of the p-type a-Si:H layer thickness results in a decrease of J_{sc}, due to absorption losses in the a-Si:H.

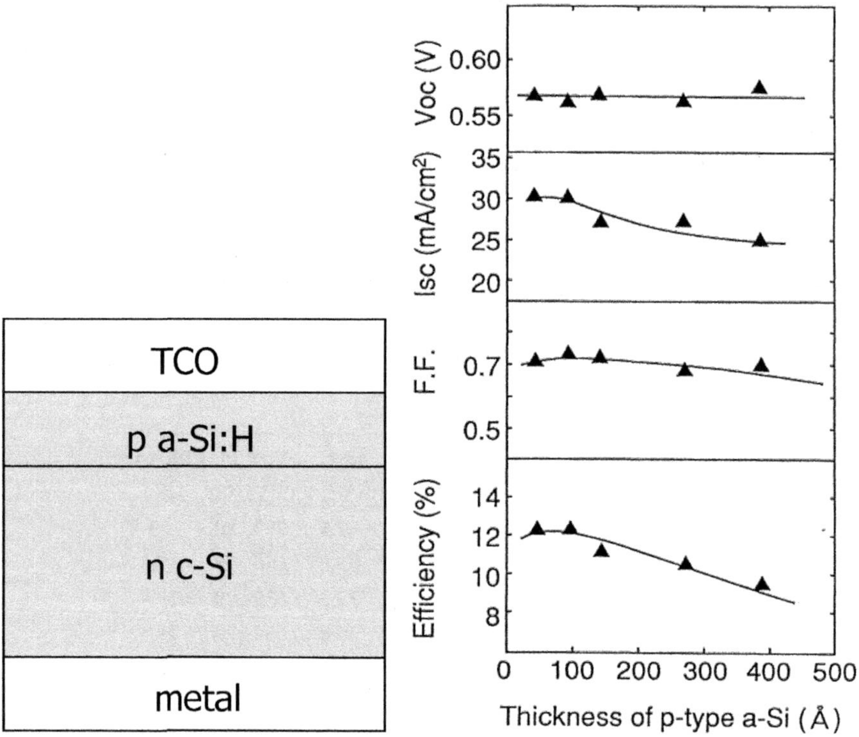

(a) (b)

Fig. 2.6 a) Schematic structure of heterojunction a-Si:H/c-Si solar cells made by Sanyo and b) performance of the solar cell as a function of p-type a-Si:H thickness [18].

Figure 2.7a shows a solar cell structure in which Sanyo incorporated a thin intrinsic a-Si:H layer between the n-type c-Si wafer and the p-type a-Si:H. Sanyo named this structure the ACJ-HIT (artificially constructed junction-heterojunction with intrinsic thin film) solar cell. The aim of including an intrinsic a-Si:H layer was to passivate the dangling bonds on the c-Si surface. As a result, the a-Si:H/c-Si interface defect-state density was significantly reduced. The inclusion of the intrinsic layer resulted in an enhancement of the V_{oc} and FF, as shown in Fig. 2.7b. From Fig. 2.7b it is also clear that increasing the thickness of the intrinsic a-Si:H results in a lower J_{sc} because of the absorption losses in the a-Si:H layers. The FF also decreases because of the high resistivity of the intrinsic a-Si:H, which acts as a transport barrier. Figure 2.7b demonstrates that there is an optimal thickness for both the p-type and the intrinsic a-Si:H layers. The highest efficiency of 14.8% was obtained for a HIT solar cell with a 4 nm thick intrinsic a-Si:H layer.

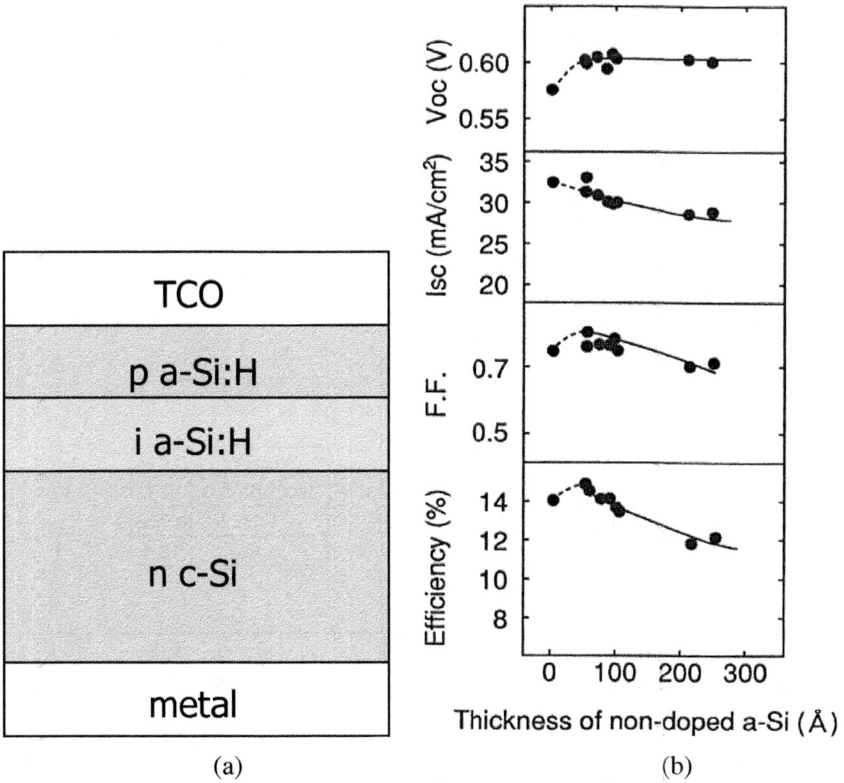

Fig. 2.7 a) Schematic structure of ACJ-HIT solar cells made by Sanyo and b) performance of the solar cell as a function of intrinsic a-Si:H thickness [18].

The performance of the HIT solar cell was further improved by the introduction of textured wafer surfaces for light-trapping, and the inclusion of a BSF on the back side of the solar cell. This solar cell structure is shown in Fig. 2.8a. Randomly textured surfaces help to reduce surface reflection and increase the average optical path length inside the wafer, which increases J_{sc}. However, surface texturing also increases the overall surface area, which can result in an increase of the surface defect-state density. A comparison of the V_{oc} of solar cells without (Fig. 2.7b) and with (Fig. 2.8b) surface texturing shows that the V_{oc} has not been strongly affected. This is because the increased surface defect-state density is compensated by the BSF, which reduces the carrier recombination at the backside of the wafer. The inclusion of surface texturing and the BSF enabled HIT solar cell efficiencies as high as 18.1%.

In the next step Sanyo applied double-sided passivation to the wafer. This resulted in an increase of the V_{oc} to 0.717 V and of efficiency to 21.3%, as shown in Fig. 2.9. Furthermore, Sanyo designed a highly symmetrical HIT solar cell structure. The symmetry in the solar cell structure helps to suppress thermal and mechanical stresses in the wafers during the fabrication process. The symmetrical structure also enables illumination of the device from both sides. Sanyo demonstrated that a correctly orientated HIT module can generate more power when illuminated on both sides than a module illuminated from one side. The different possibilities to position the module in order to take advantage of double-sided illumination are illustrated in Fig. 2.10.

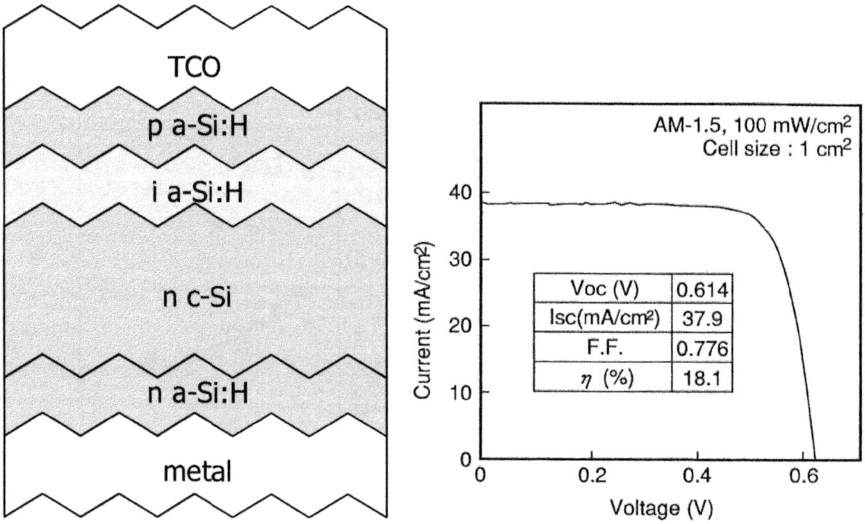

(a) (b)

Fig. 2.8 a) Schematic structure and b) I-V characteristic of the HIT solar cell with textured surfaces and BSF [18].

By 2008, Sanyo had achieved HIT solar cell efficiencies as high as 22.3%, which was independently confirmed by AIST (Fig. 2.11). Sanyo outlined three key steps that were instrumental for this achievement [21]. The first step was the further improvement in the quality of the a-Si:H/c-Si heterojunction. This was achieved by improved cleaning of the c-Si wafer before deposition of a-Si:H layers, use of high quality a-Si:H layers, and reducing the plasma and thermal damage caused to the c-Si surface during processing. The second step was the improvement of the grid electrodes; the aspect ratio of the grid fingers, which is the ratio of the height of the finger to its width, should be as high as possible in order to reduce the area shaded by the grid electrodes. The third step was the reduction of absorption in the a-Si:H and TCO layers by applying wide-gap a-Si:H alloys, and enhancing the carrier mobility of the TCO instead of the carrier concentration.

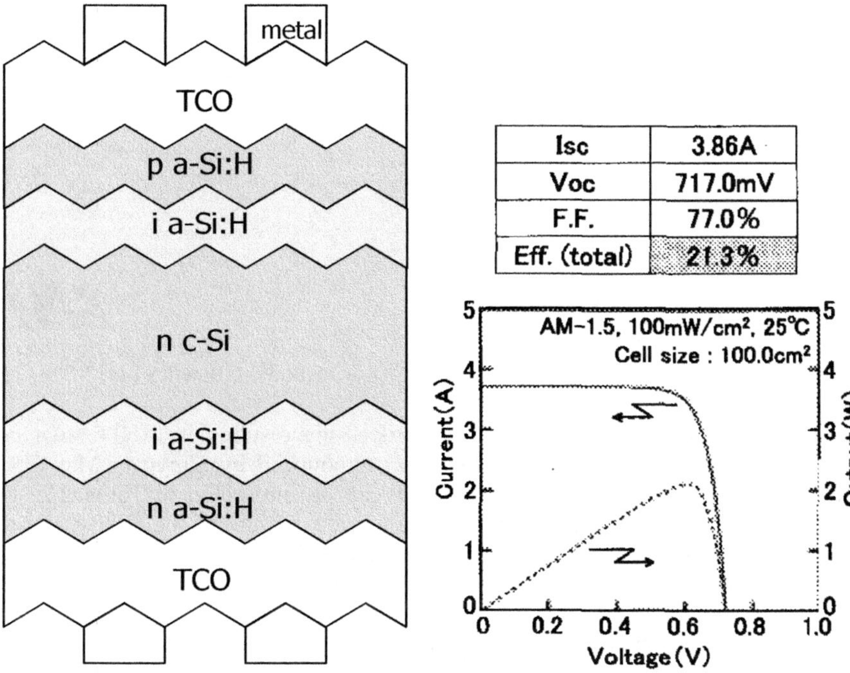

(a) (b)

Fig. 2.9 a) Schematic symmetric structure of the HIT solar cell with double-sided passivation and b) its I-V characteristic [19].

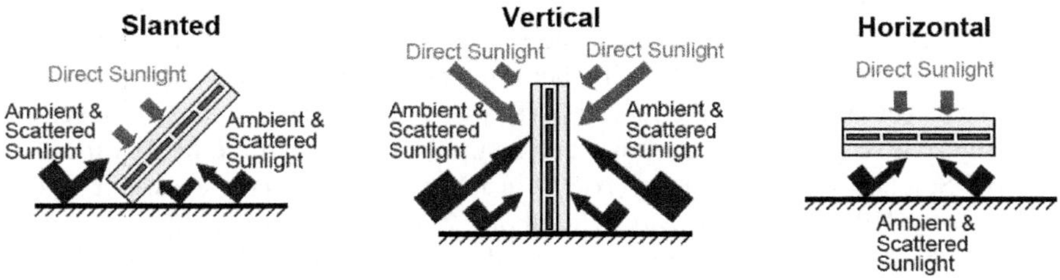

Fig. 2.10 Light absorption from both sides of a HIT module designed by Sanyo [20].

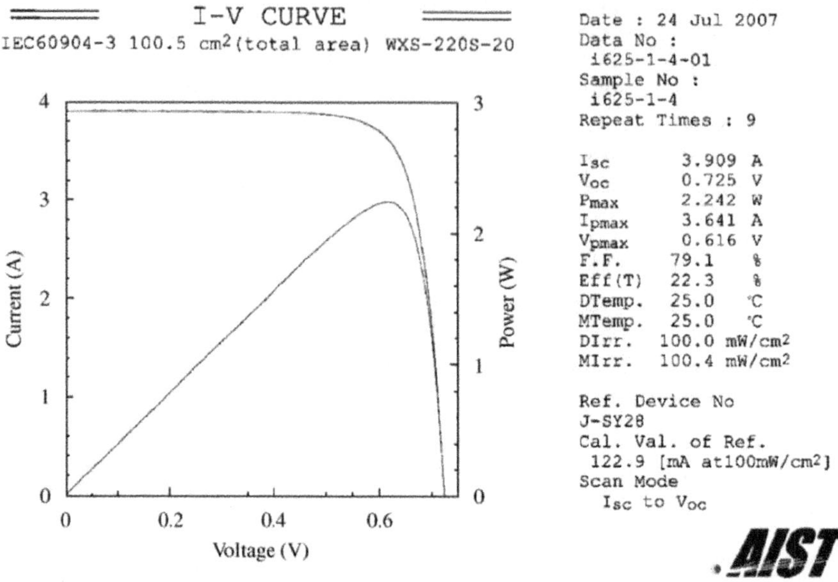

Fig. 2.11 The I-V characteristic of the HIT solar cell with 22.3% efficiency [21].

Further optimization of materials and processing resulted in a HIT solar cell with a record efficiency of 23%, which was announced by Sanyo in May 2009. The external parameters of this record HIT cell are presented in Table 2.2. The low temperature processing and highly symmetric structure of the HIT solar cells enables the use of ultra thin wafers. Sanyo presented a HIT solar cell with 22.8% efficiency using wafers only 98 μm thick [22]. The V_{oc} of this solar cell is 743 mV which is the highest V_{oc} demonstrated by any c-Si solar cell.

Sanyo began commercial production of HIT solar cells in 1997. Its production capacity continues to increase [23], which suggests a huge market potential for HIT solar cells.

2.4.2 Research Results of Heterojunction Silicon Solar Cells

The high performance and low-temperature processing of Sanyo HIT solar cells have attracted interest of companies and researchers all over the world. Several institutes achieved good results with heterojunction silicon solar cells; however, none have yet approached the record performance of the HIT solar cell developed by Sanyo. Several research groups, such as the National Institute of Advanced Industrial Science and Technology (AIST) in Japan, the National Renewable Energy Laboratory (NREL) in the U.S.A., Helmholtz Zentrum Berlin (HZB) and Neuchatel PV-lab of Ecole Polytechnique Federale de Lausanne (EPFL), have investigated innovative approaches for obtaining high performance heterojunction silicon solar cells. Their important work has contributed to a better understanding of heterojunction solar cell processing and operation.

AIST conducted a detailed investigation of the deposition conditions required for epitaxial growth during PECVD deposition of a-Si:H [24]. Instead of a sharp interface between c-Si and the a-Si:H passivation layer, epitaxial growth results in an interface region of mixed phases with an increased density of interface defect states. Epitaxial growth during the deposition of a-Si:H resulted in a deteriorated performance of heterojunction solar cells, and particularly affected the V_{oc}. It was demonstrated that a high deposition temperature (>140 °C) during the deposition of a-Si:H led to epitaxial growth [25]. Other deposition conditions, such as power and the nature of the substrate surface, also have impact on epitaxial growth [24, 26], although the exact mechanism is still not clear. AIST found that epitaxial growth can be effectively suppressed by using an a-SiO:H alloy instead of a-Si:H [9]. AIST achieved an efficiency of 17.5% with an a-SiO:H passivation layer in their heterojunction silicon solar cell. A schematic structure of the AIST heterojunction silicon solar cell is shown in Fig. 2.12a. The performance of the heterojunction solar cell decreased with the reduction of wafer thickness (Fig. 2.12b) partially because of the polishing damage during wafer thinning [9].

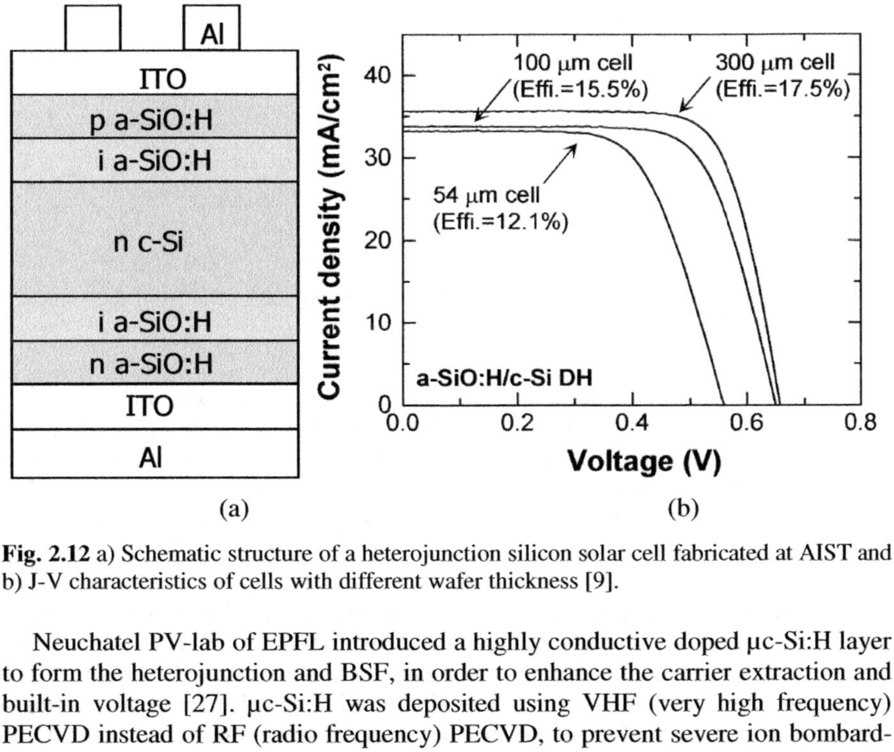

Fig. 2.12 a) Schematic structure of a heterojunction silicon solar cell fabricated at AIST and b) J-V characteristics of cells with different wafer thickness [9].

Neuchatel PV-lab of EPFL introduced a highly conductive doped μc-Si:H layer to form the heterojunction and BSF, in order to enhance the carrier extraction and built-in voltage [27]. μc-Si:H was deposited using VHF (very high frequency) PECVD instead of RF (radio frequency) PECVD, to prevent severe ion bombardment of the growing surface. Using this solar cell structure with flat interfaces the Neuchatel group achieved an efficiency of 19.1% [27]. Moreover, S. Olibet *et al.* [28] developed a model for a-Si:H/c-Si interface recombination based on amphoteric dangling-bond defects. The injection dependence of a-Si:H/c-Si interface recombination was well reproduced using this model. In 2010, Roth&Rau and

EPFL together presented heterojunction solar cells (Fig. 2.13a) with a high efficiency of 21% (Fig. 2.13b) in the so-called S-Cube reactor designed by Roth&Rau [11]. High quality passivation of 9.6 ms measured carrier lifetime at the carrier density of 10^{15} cm^{-3} is obtained in the passivated wafer with only 7 nm intrinsic a-Si:H.

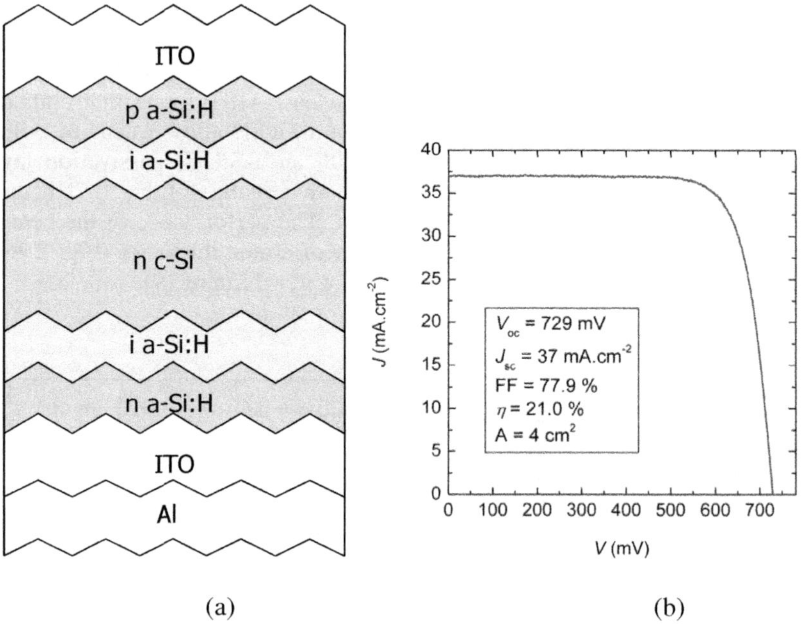

Fig. 2.13 a) Schematic structure of heterojunction silicon solar cells from the Neuchatel PV-lab and Roth&Rau b) J-V characteristics [11].

In most heterojunction silicon solar cells an intrinsic a-Si:H is used as the surface passivation layer, and this layer has been considered essential to achieve high performance. However, Helmholtz Zentrum Berlin (HZB) prepared high quality heterojunction solar cells without an intrinsic a-Si:H passivation layer [10]. The structure of the cell is presented in Fig. 2.14a [10]. They cleaned the wafer surfaces using RCA (Radio Corporation of America) cleaning procedure and textured them by application of a KOH/IPA solution. The HZB group demonstrated that smoothing procedures after texturing and cleaning resulted in improved surface uniformity with less micro-roughness, and an improved overall device performance. Additionally, a hydrogen post-treatment was discovered to be beneficial for improving the quality of the a-Si:H thin film and the surface passivation. Without an intrinsic a-Si:H passivation layer in the solar cell, HZB achieved efficiencies as high as 19.8% on surface-textured wafers. However, the doped a-Si:H cannot passivate the surfaces of c-Si wafers as well as intrinsic a-Si:H [30]. As such, the interface defect-state density was expected to be high, resulting in a relatively low V_{oc}.

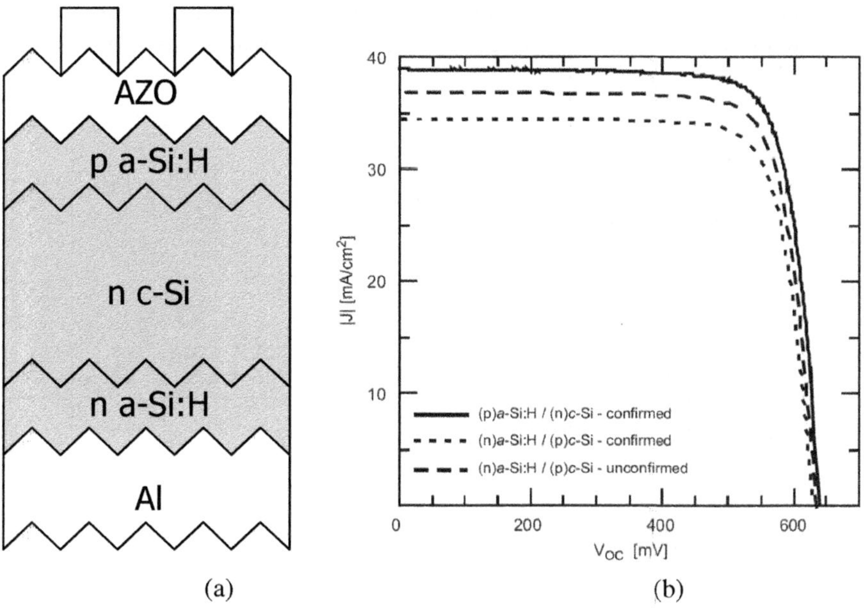

Fig. 2.14 a) Schematic structure of a heterojunction silicon solar cell fabricated at HZB and b) J-V characteristics of cells with standard HF dip as the last cleaning step prior to a-Si:H deposition confirmed by ISE Freiburg (full and dotted lines), cell with optimized oxidation and etching pre-treatment (dashed line) [10].

PECVD is the most common method to deposit a-Si:H for heterojunction silicon solar cells. Wang *et al.* investigated the use of the HWCVD (hot-wire chemical vapor deposition) method for depositing a-Si:H based layers. HWCVD offers several advantages to PECVD [31]. For example, the thermal pyrolysis of silane avoids ion bombardment of the surface, and the generated atomic hydrogen readily passivates the surface. The group achieved an efficiency of 19.1% for a heterojunction silicon solar cell fabricated by HWCVD-deposited a-Si:H on textured wafer surfaces (Fig. 2.15) [12].

In 2010, INES [13] in France presented their achievement of heterojunction solar cells with over 19.5% efficiency. The structure is illustrated in Fig. 2.16a. Since aluminum is deposited all over the back side of the solar cell, high lateral conductance of TCO is not as critical as that in the front of the device. Therefore, boron doped zinc oxide (ZnO:B) is utilized to replace ITO due to its low absorption in the infrared region and compatible work function to n-type a-Si:H.

Conductivity of the doped layers is very important to obtain high FF in heterojunction silicon solar cells. Tucci *et al.* [14] in ENEA presented that chromium silicide can be formed on n-type a-Si:H by depositing and removing the chromium layer. This treatment can increase the conductivity of n-type a-Si:H by one order of magnitude without any significant change of its optical properties. The structure of the solar cell is presented in Fig. 2.17a. Aluminium at the back of the solar cell serves both as back contact and as the source for BSF formation by annealing. As illustrated in Fig. 2.17b, the solar cell with chromium-treated n-type a-Si;H has a higher FF.

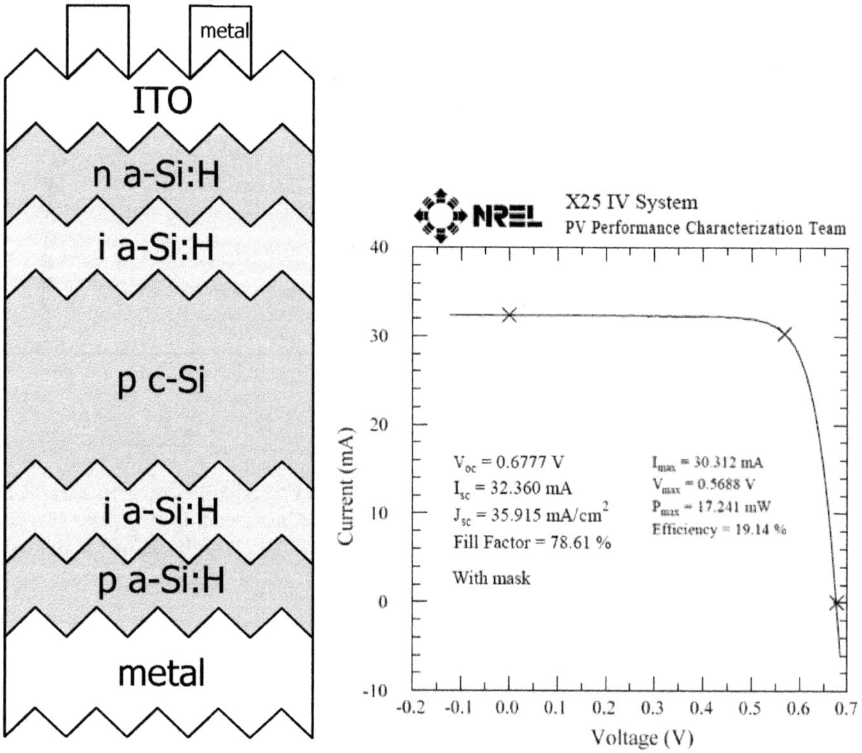

(a) (b)

Fig. 2.15 a) Schematic structure of heterojunction silicon solar cells made at NREL and b) I-V characteristic [12].

The absorption at short wavelengths in the a-Si:H emitter is one of the major limits for heterojunction silicon solar cells to achieve high J_{sc} similarly to the PERL c-Si solar cells. Hydrogenated nanocrystalline cubic silicon carbide (nc-3C-SiC:H), which has significantly lower absorption coefficients than a-Si:H alloys, has been developed as the emitter for heterojunction solar cells [15]. In the heterojunction silicon solar cell of Fig. 2.18a, high J_{sc} has been achieved on the flat wafer (Fig. 2.18b) by reduction of absorption losses in the short-wavelength region.

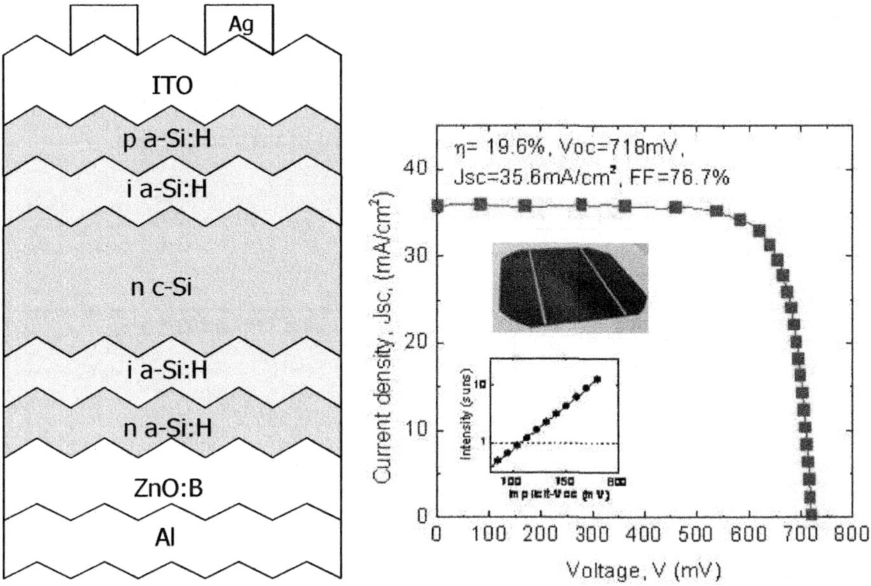

(a) (b)

Fig. 2.16 a) Schematic structure of heterojunction silicon solar cells made at INES and b) I-V characteristics [13].

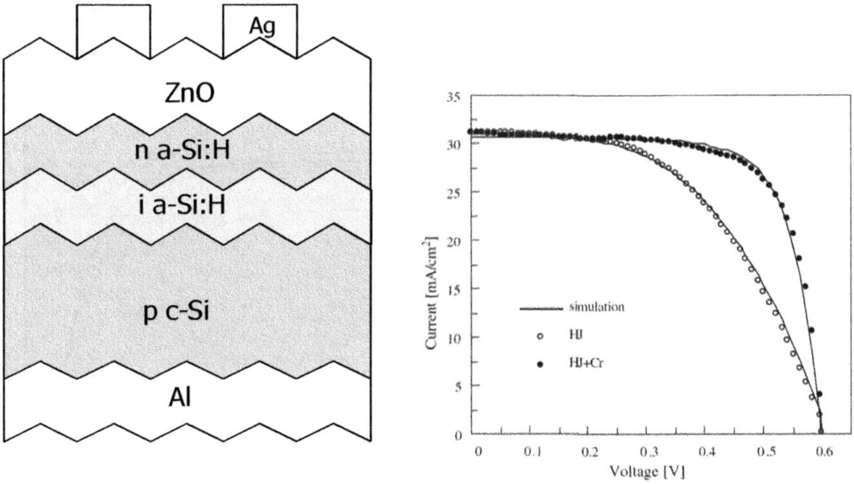

(a) (b)

Fig. 2.17 a) Schematic structure of heterojunction silicon solar cells made at ENEA and b) I-V characteristics and influence of chromium silicide on the performance of the device [14].

Lien *et al.* [32, 16] used HWCVD to deposit doped μc-Si:H as emitters and BSF of heterojunction solar cells (Fig. 2.19a). Hydrogen treatment of the silicon wafer surface has been studied in order to increase the performance of the device. Instead of using sputtered ITO, ITO deposited by electron-beam evaporation has been optimized for heterojunction solar-cell application. In Fig. 2.19b, the emitter thickness of 15 nm gives the optimum performance of the final device.

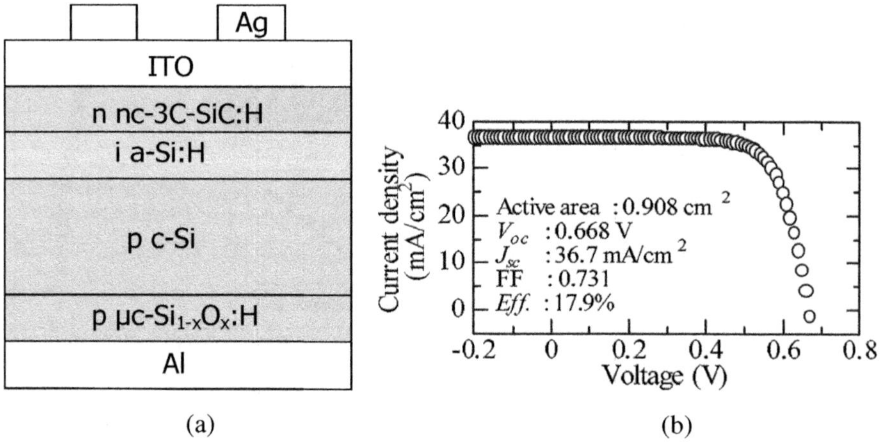

Fig. 2.18 a) Schematic structure of heterojunction silicon solar cells made at Tokyo Institute of Technology and b) I-V characteristic [15].

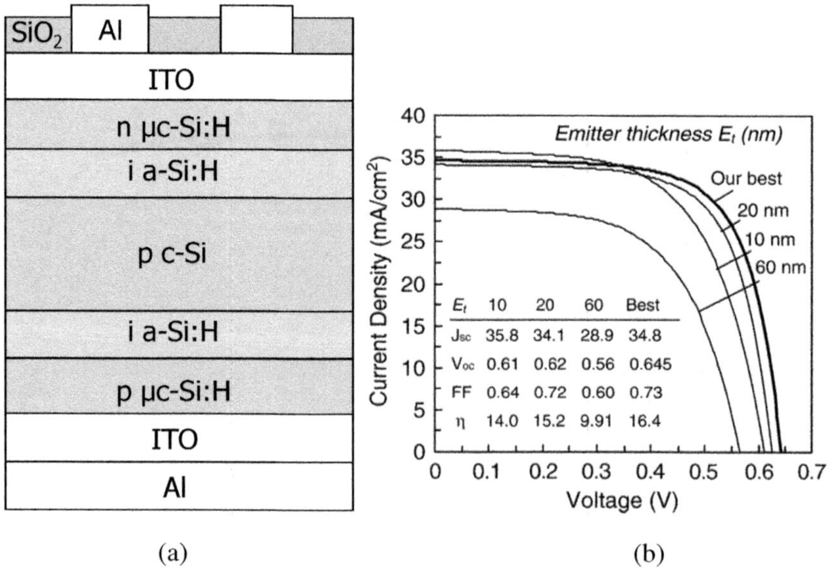

Fig. 2.19 a) Schematic structure of heterojunction silicon solar cells made at Mingdao University and b) I-V characteristic regarding different thickness of emitters [32].

In order to avoid the absorption losses in the ITO and a-Si:H at the front, the heterojunction solar-cell structure with the emitter on the rear side has been designed [17]. The solar cell structure is shown in Fig. 2.20. Phosphorus diffusion was used to form the front surface field (FSF). Thermally grown SiO_2 and PECVD-deposited SiN_x are used as passivating layers and antireflection coatings. A high efficiency of 19.8% has been obtained on this kind of solar cells.

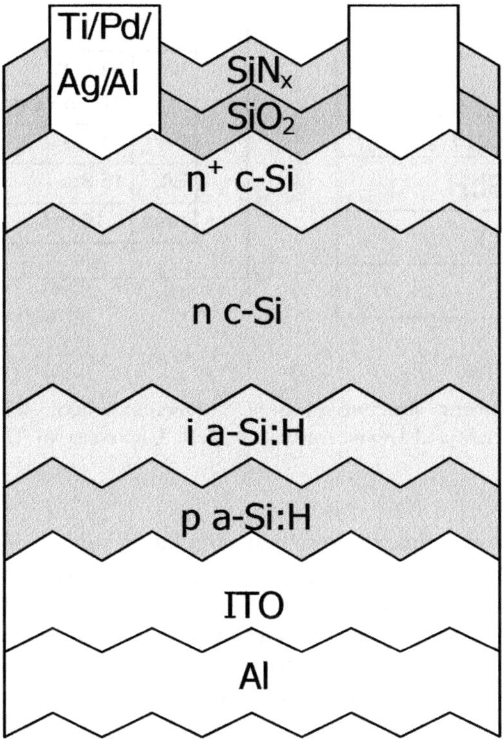

Fig. 2.20 Schematic structure of heterojunction silicon solar cells designed at Fraunhofer ISE.

Several research groups in the Netherlands have been developing heterojunction silicon solar cell technology. Delft University of Technology, in cooperation with ECN (Energy research Centre of the Netherlands), began research on heterojunction silicon solar cells in 2009. The heterojunction solar cell was fabricated by using the structure illustrated in Fig. 2.21, and achieved an efficiency of 15.8%. Utrecht University and ECN also cooperate on fabricating heterojunction silicon solar cells. Their best heterojunction silicon solar cells, with an efficiency of 16.4%, have the following structure: Ag/ITO/p+ a-Si:H/ i a-Si:H/n-type c-Si wafer/ BSF/Ag. The FZ (100) oriented wafer received an isotexturing treatment and the BSF was formed by diffusion.

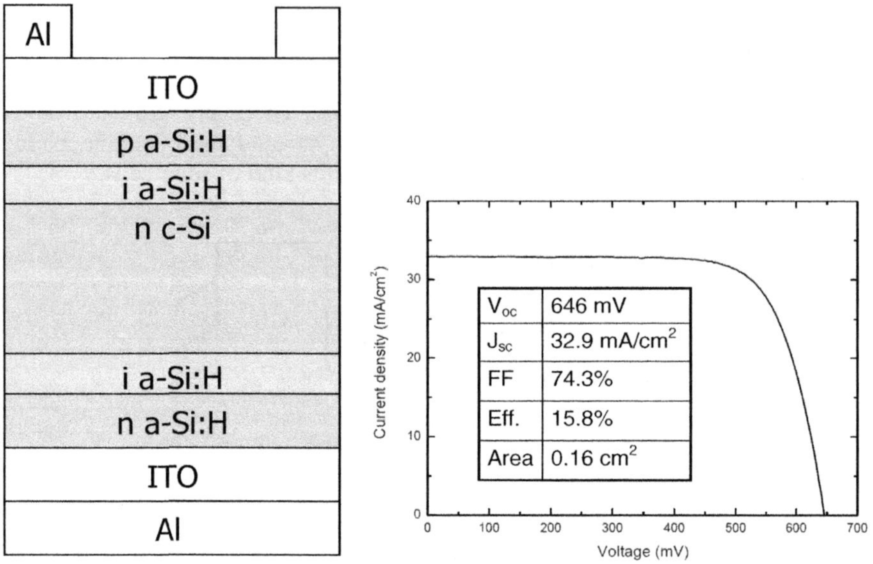

(a) (b)

Fig. 2.21 a) Schematic structure of a heterojunction silicon solar cell made in the Photovoltaic Materials and Devices group at Delft University of Technology and b) J-V characteristics.

Besides all the experimental research, simulation studies of the performance of heterojunction silicon solar cells are very important in order to understand the operation of the solar cells and optimize the design of solar-cell structures. The AFORS-HET software developed by HZB in Germany has been widely utilized for the simulation study of heterojunction silicon solar cell (see also Chapters 13 and 14 in this book). Dao *et al.* [33] simulated the influence of BSF, passivation using intrinsic a-Si:H, densities of interface defects, resistivity of p-type c-Si substrate and work function of TCO on solar-cell performance. Depending on density of oxygen defects in c-Si bulk and interface defects, Zhao *et al.* [34] proposed an optimal substrate resistance for obtaining good performance of heterojunction silicon solar cells. Kanevce and Metzger [39] incorporated ITO as an active semiconductor layer into the simulations and investigated an impact of tunneling through the ITO/p interface.

2.5 Challenges to Improve Heterojunction Silicon Solar Cells

It took Sanyo over 20 years to develop a HIT solar cell with 23% efficiency, which is currently the highest efficiency heterojunction silicon solar cell in the world. There is no doubt that the performance of silicon solar cells can be further improved. Such improvements will mainly focus on: i) the optical losses that limit the J_{sc}, ii) the recombination losses that mainly influence the V_{oc} and iii) the resistance losses affecting the *FF*. These losses are schematically illustrated in Fig. 2.22.

Several solutions have been considered to reduce these losses. To reduce the optical losses the approaches have been: the surface-texturing of wafers to provide efficient light trapping, optimization of TCO and a-Si:H layers to reduce their absorption, and the aspcct ratio of the grid electrodes has been increased to reduce the shaded area. To reduce recombination losses the cleaning of wafer surfaces prior to a-Si:H deposition is very important, as this removes recombination centers from the surface, such as metallic contamination and particles. The interface defect-state density can be reduced by saturating the dangling bonds on the wafer surface with hydrogen termination, and using high-quality a-Si:H deposition. The resistance losses can be suppressed by decreasing the series resistance of the device. In this respect highly conductive TCO and good ohmic contacts at the contact interfaces can contribute to the minimization of resistance. In the following sections some of the critical issues concerning the fabrication of heterojunction silicon solar cells are discussed in more detail.

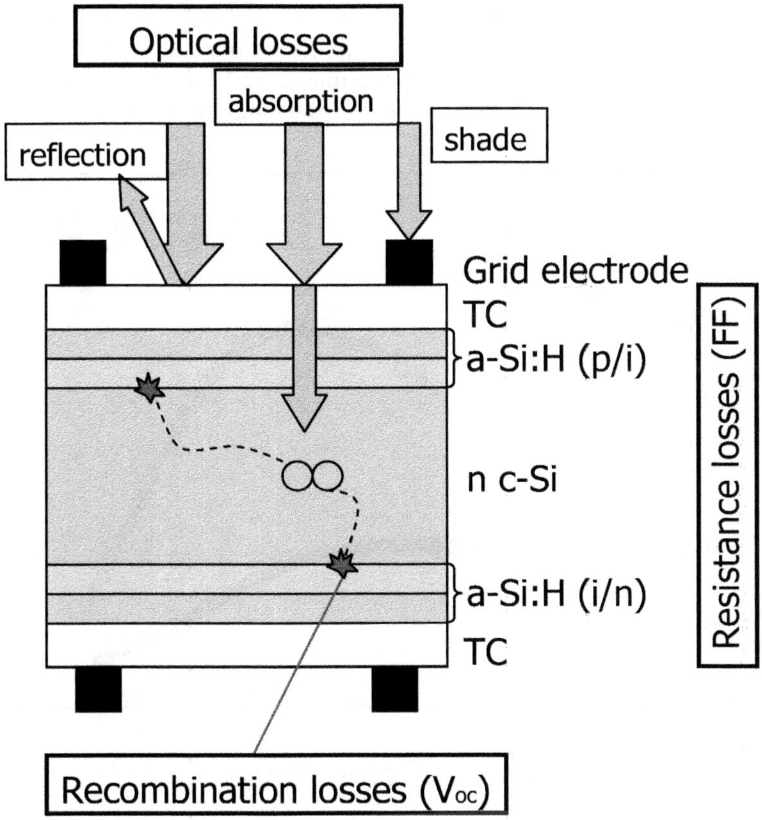

Fig. 2.22 Illustration of losses that occur in heterojunction silicon solar cells.

2.5.1 Wafer Cleaning

Wafer cleaning before the deposition of a-Si:H has two effects. One is to remove the particles and metallic contamination from the wafer surfaces. The other is to partially passivate the dangling bonds on the surface with hydrogen. Cleaning is a crucial first step to reducing the a-Si:H/c-Si interface state density. The effect of different cleaning procedures can be investigated by measuring the carrier lifetime of cleaned wafers which have been passivated with identical a-Si:H films. The carrier recombination in the bulk region of high-quality silicon wafers can be considered as negligible, and so the measurement of carrier lifetime indicates the surface recombination, and hence the quality of the cleaning process. Figure 2.23 shows the comparison of three different cleaning methods carried out at Delft University of Technology. The (100) oriented FZ c-Si wafers were treated in three different ways before the deposition of intrinsic a-Si:H layers on both sides of the wafer. The first wafer was cleaned using a standard RCA clean, and the second wafer was cleaned using the standard DIMES cleaning procedures involving concentrated nitric acid. All three wafers were dipped in HF to remove the native oxide layer, and this was the only treatment performed on the third wafer. After the pretreatments, a 120 nm thick intrinsic a-Si:H layer was deposited on both sides of the wafers, using identical deposition conditions for each run. The carrier lifetime was measured using a Sinton lifetime tester [35], to evaluate the passivation quality. The wafer cleaned using the standard RCA process exhibited the

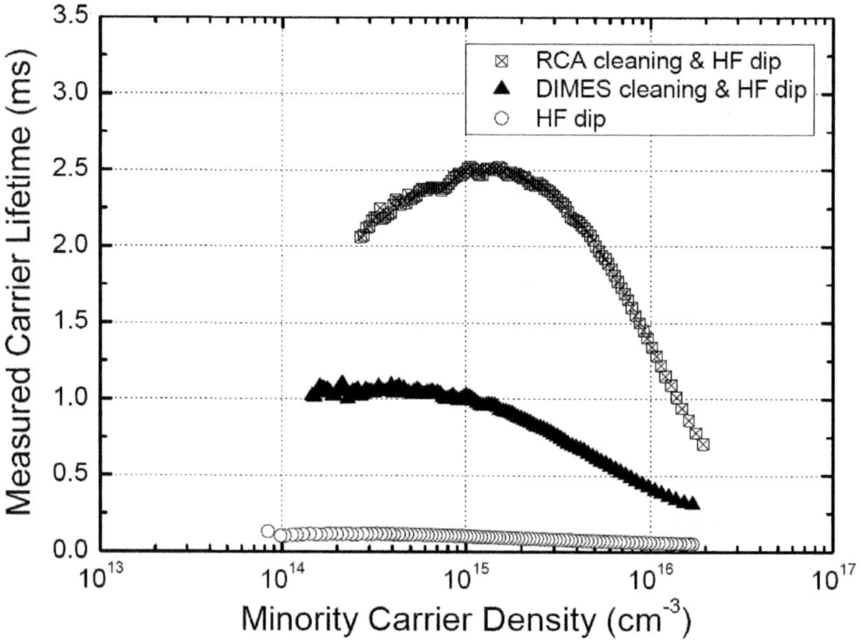

Fig. 2.23 The carrier lifetimes of a-Si:H passivated (100) Si wafers cleaned by three different methods.

highest carrier lifetimes, and hence the best passivation, as shown in Fig. 2.23. The lowest carrier lifetimes were observed for the wafer which only received a HF dip treatment.

Typically dry cleaning processes are preferred for industrial applications. However, the plasma used in dry cleaning can damage the wafer surface [36]. Therefore, soft wet-chemical cleaning is the dominant cleaning method for heterojunction solar cells, because it results in a smoother wafer surface and a correspondingly lower interface state density [37].

Fig. 2.24 SEM image of the pyramidal structure for light trapping used in heterojunction silicon solar cells [36].

2.5.2 Wafer Surface Texturing

The wafers used in heterojunction silicon solar cells should be as thin as possible in order to reduce material costs. However, absorption of light is strongly reduced in thin wafers. The application of light trapping techniques has become an important issue in heterojunction silicon solar cells. The role of light trapping is to keep the physical thickness of the absorber layer as thin as possible and to maximize its effective optical thickness. Texturing of the wafer surface is a common light-trapping method used in c-Si based solar cells. For single crystalline silicon wafers, a "random pyramid" texture is commonly used. This texture is typically formed by treating the c-Si wafer with a base solution. The <111> and <100> crystalline orientations etch at different rates, which results in a random pyramidal surface structure (Fig. 2.24) in which the <111> crystalline plane is exposed. However, this treatment is not suitable for the multicrystalline wafer due to its distributed crystalline orientation on the surface. Additionally, surface texturing creates additional challenges such as achieving a uniform deposition of an ultrathin a-Si:H film, and ensuring that the surface is correctly cleaned. Therefore additional investigation of surface cleaning and a-Si:H deposition parameters are required in order to optimize heterojunction solar cell fabrication on textured substrates.

2.5.3 Epitaxial Growth at the Heterojunction Interfaces

The deposition of a-Si:H layers should be performed at an elevated temperature to minimize defects in the material. However, epitaxial growth may occur when a-Si:H is deposited onto a c-Si wafer at a high temperature (Fig. 2.25a) [25]. This deteriorates the performance of the heterojunction silicon solar cell, predominantly due to a reduction of V_{oc}. Ion impinging induced by RF power can help to prevent the epitaxial growth, but it can also result in surface damage if the power is too high. Therefore, there are tradeoffs for both temperature and RF power (Fig. 2.25b). Epitaxial growth can be avoided by optimizing the RF power and growth temperature [24].

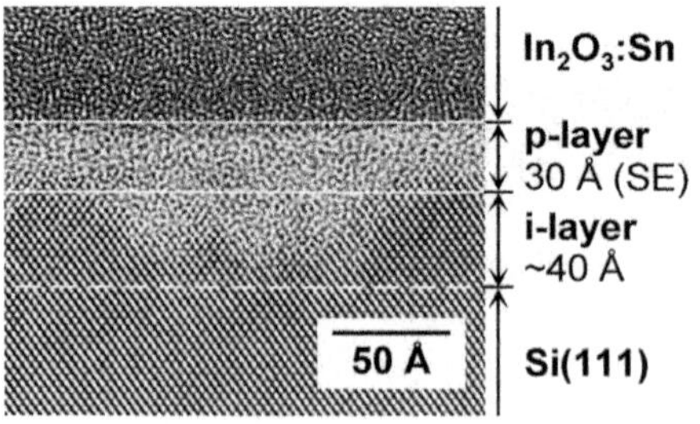

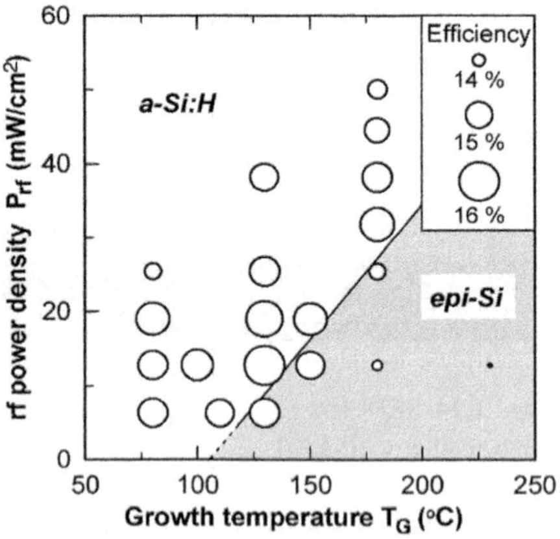

Fig. 2.25 a) Cross-sectional TEM image showing the epitaxial growth of a-Si:H [24], and b) the efficiency of heterojunction silicon solar cells corresponding to deposition temperature and RF power density.

An alternative method to avoid epitaxial growth, which has been used by AIST and Tokyo Institute of Technology, is to use a-SiO:H to form the heterojunction with c-Si [9, 38]. Silicon oxide has good passivation properties, and epitaxial growth can be effectively suppressed because the coordination number of oxygen is only two.

2.5.4 Controlling Thickness of a-Si:H Based Layers

As already illustrated in Figs. 2.6b and 2.7b, Sanyo investigated the effect of a-Si:H thickness on the external parameters of heterojunction silicon solar cells. Because of the absorption losses in a-Si:H films and the high resistance of intrinsic a-Si:H, both the doped and intrinsic a-Si:H layers should be as thin as possible to maximise J_{sc} and FF of the solar cells. However, the doped a-Si:H layer should be thick enough to avoid complete depletion in the layer, which would lead to a reduction in the built-in voltage. Furthermore, the intrinsic a-Si:H must exceed a certain minimum thickness in order to provide high quality passivation [24]. Simulation results suggest that tunneling plays a very important role for carrier transport through both the p/n heterointerface and the ITO/p-type a-Si:H interface, particularly in case of heterojunction solar cells based on n-type c-Si wafers [39]. Therefore, optimization of the thicknesses of both intrinsic and doped a-Si:H layers is required. In-situ ellipsometry can be used to control the thickness of the layers during the deposition [40].

2.5.5 Reducing Absorption Losses in a-Si:H and TCO Layers

The a-Si:H layers predominantly absorb short wavelength photons (< 500 nm), and longer wavelength photons can be absorbed by free carriers in the TCO. The light absorbed in a-Si:H and TCO will not contribute to the current of the solar cell. These losses are illustrated in Fig. 2.26 by comparing the internal quantum efficiency (IQE) of a HIT solar cell to that of a PERL c-Si solar cell. The use of a material with a wider band gap, such as a-SiC:H, in place of a-Si:H may help to reduce absorption losses. Current TCO materials require a high free-carrier density to achieve an acceptable conductivity. This trade-off between high conductivity and low absorption can only be solved by developing new TCO materials with increased free carrier mobility.

2.5.6 Improvement of Grid Electrodes

The industrial method of applying grid electrodes is to use a screen-printing process. Applied resin-bonded Ag paste typically exhibits the shape of a grid line with a spreading area, as shown in Fig. 2.27a. The shaded area should be minimized in order to maximize transmission of light into the solar cell. If the cross-sectional area of the grid electrode is kept constant (in order to maintain a high conductance) the shaded area can be reduced by minimizing the spreading area and increasing the aspect ratio of the grid electrode, i.e. the width (w) should be reduced and correspondingly the height (h) should be increased to keep the area of the cross-section constant. In order to accomplish these two approaches, the screen-printing process parameters and the properties of the paste, such as its viscosity and rheology, need to be optimized. Figure 2.27b shows the optimum shape of the grid electrode [21].

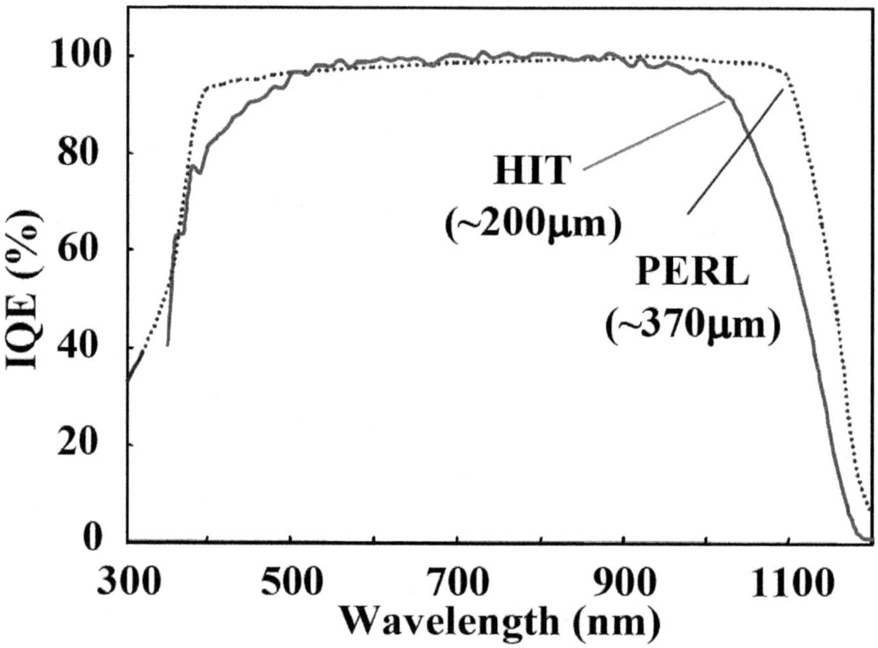

Fig. 2.26 IQE spectrums of the PERL c-Si solar cell and the HIT solar cell [7].

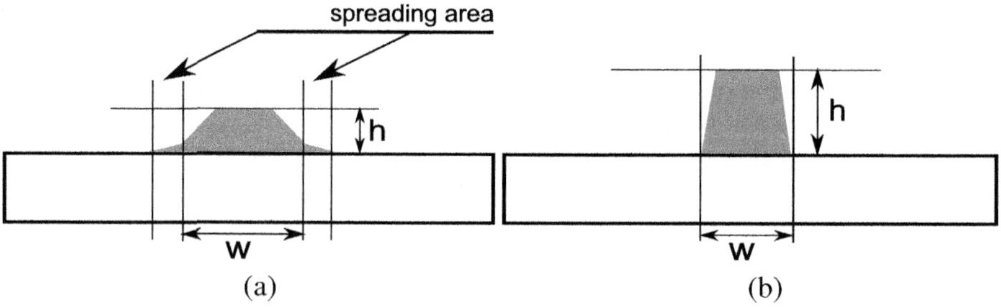

Fig. 2.27 Schematic diagrams of a) a conventional grid electrode with a spreading area and b) an ideal grid electrode without a spreading area and a high aspect ratio.

2.6 Conclusion

Heterojunction silicon solar cells have comparably high performance to conventional c-Si solar cells, but offer numerous advantages including a less complex and lower-temperature fabrication process. Sanyo is the forerunner in this field, and holds the record efficiency of 23% for HIT solar cells. Many other institutes have demonstrated excellent achievements in heterojunction silicon solar cells. In order to further enhance the efficiency of heterojunction silicon solar cells, some challenging problems need to be investigated and solved, as indicated in the above sections.

References

[1] Green, M.: Crystalline Silicon Solar cells. In: Archer, M., Hill, R. (eds.) Crystalline Silicon Solar Cells, p. 149. Imperial College Press, London (2001)

[2] Watt, G., Fechner, H.: Photovoltaic market and industry trends – latest results from the IEA PVPS programme. E&I Elektrotechnik und Informationstechnik 126, 328 (2009)

[3] Würfel, P.: Physics of Solar Cells: From Principles to New Concepts. Wiley-WCH, Weinheim (2005)

[4] Honsberg, C., Bowden, S.: High efficiency solar cells. PVCDROM, `http://pvcdrom.pveducation.org/manufact/labcells.htm`

[5] Mishima, T., Taguchi, M., Sakata, H., Maruyama, E.: Development status of high-efficiency HIT solar cells. Sol. Energ. Mat. Sol. 95, 18 (2010)

[6] Sawada, T., Terada, N., Tsuge, S., Baba, T., Takahama, T., Wakisaka, K., Tsuda, S., Nakano, S.: High-efficiency a-Si/c-Si heterojunction solar cell. In: Proceedings of the 24th IEEE Photovoltaic Specialists Conference (1994)

[7] Maruyama, E., Terakawa, A., Taguchi, M., Yoshimine, Y., Ide, D., Baba, T., Shima, M., Sakata, H., Tanaka, M.: Sanyo's challenges to the development of high-efficiency HIT solar cells and the expansion of HIT business. In: Proceedings of the 4th WCPEC (2006)

[8] Roca, F., Cárabe, J., Jäger-Waldau, A.: Silicon heterojunction cells R&D in Europe. In: Proceedings of the 19th EU-PVSEC (2004)

[9] Fujiwara, H., Sai, H., Kondo, M.: Crystalline Si Heterojunction Solar Cells with the Double Heterostructure of Hydrogenated Amorphous Silicon Oxide. Jpn. J. Appl. Phys. 48, 064506 (2009)

[10] Korte, L., Conrad, E., Angermann, H., Stangl, R., Schmidt, M.: Advances in a-Si:H/c-Si heterojunction solar cell fabrication and characterization. Sol. Energ. Mat. Sol. 93, 905 (2009)

[11] Lachenal, D., Andrault, Y., Bätzner, D., Guerin, C., Kobas, M., Mendes, B., Strahm, B., Tesfai, M., Wahli, G., Buechel, A., Descoeudres, A., Choong, G., Bartlome, R., Barraud, L., Zicarelli, F., Bôle, P., Fesquet, L., Damon-Lacoste, J., Wolf, S.D., Ballif, C.: High efficiency silicon heterojunction solar cell activities in Neuchatel, Switzerland. In: Proceedings of the 25th EU-PVSEC (2010)

[12] Wang, Q., Page, M., Iwaniczko, E., Xu, Y., Roybal, L., Bauer, R., To, B., Yuan, H., Duda, A., Yan, Y.: Crystal silicon heterojunction solar cells by hot-wire CVD. In: Proceedings of the 33rd IEEE Photovoltaic Specialists Conference (2008)

[13] Munoz, D., Ozanne, A., Harrison, S., Danel, A., Souche, F., Denis, C., Favier, A., Desrues, T., Nicolás, S.M., Nguyen, N., Hickel, P., Mur, P., Salvetat, T., Moriceau, H., Le-Tiec, Y., Kang, M., Kim, K., Janin, R., Pesenti, C., Blin, D., Nolan, T., Kashkoush, I., Ribeyron, P.: Towards high efficiency on full wafer a-Si:H/c-Si heterojunction solar cells: 19.6% on 148cm2. In: Proceedings of the 35th IEEE Photovoltaic Specialists Conference (2010)

[14] Tucci, M., Cesare, G.: 17% efficiency heterostructure solar cell based on p-type crystalline silicon. J. Non-Cryst. Solids 338, 663 (2004)
[15] Miyajima, S., Irikawa, J., Yamada, A., Konagai, M.: High-quality nanocrystalline cubic silicon carbide emitter for crystalline silicon heterojunction solar cells. Appl. Phys. Lett. 97, 023504 (2010)
[16] Lien, S.: Characterization and optimization of ITO thin films for application in heterojunction silicon solar cells. Thin Solid Films 518, S10 (2010)
[17] Bivour, M., Meinhardt, C., Pysch, D., Reichel, C., Ritzau, K., Hermle, M., Glunz, S.: n-type silicon solar cells with amorphous/crystalline silicon heterojunction rear emitter. In: Proceedings of the 35th IEEE Photovoltaic Specialists Conference (2010)
[18] Tanaka, M., Taguchi, M., Matsuyama, T., Sawada, T., Tsuda, S., Nakano, S., Hanafusa, H., Kuwano, Y.: Development of new a-Si/c-Si heterojunction solar cells: ACJ-HIT (artificially constructed junction-heterojunction with Intrinsic thin-layer). Jpn. J. Appl. Phys. 31, 3518 (1992)
[19] Tanaka, M., Okamoto, S., Tsuge, S., Kiyama, S.: Development of hit solar cells with more than 21% conversion efficiency and commercialization of highest performance hit modules. In: Proceedings of the 3rd WCPEC (2003)
[20] Sanyo: HIT double technology (assessed 2009), http://us.sanyo.com/dynamic/LinkLinstingItems/Files/ITDoublePresentation-1.pdf
[21] Tsunomura, Y., Yoshimine, Y., Taguchi, M., Baba, T., Kinoshita, T., Kanno, H., Sakata, H., Maruyama, E., Tanaka, M.: Twenty-two percent efficiency HIT solar cell. Sol. Energ. Mat. Sol. 93, 670 (2009)
[22] Kawai, M., Microdevices, N.: Sanyo claims 98 micron-thick HIT solar cell with 22.8% efficiency. Tech-on (assessed 2009), http://techon.nikkeibp.co.jp/english/NEWS_EN/20090923/175532/
[23] Osborne, M.: Sanyo targets 600MW HIT solar cell production with new plant (assessed 2009), http://www.pv-tech.org/news/_a/sanyo_targets_600mw_hit_solar_cell_production_with_new_plant/
[24] Fujiwara, H., Kondo, M.: Impact of epitaxial growth at the heterointerface of a-Si:H/c-Si solar cells. Appl. Phys. Lett. 90, 013503 (2007)
[25] Gielis, J., Oever, P., Hoex, B., Sanden, M., Kessels, W.: Real-time study of a-Si:H/c-Si heterointerface formation and epitaxial Si growth by spectroscopic ellipsometry, infrared spectroscopy, and second-harmonic generation. Phys. Rev. B 77, 205329 (2008)
[26] Levi, D., Teplin, C., Iwaniczko, E., Yan, Y., Wang, T., Branz, H.: Real-time spectroscopic ellipsometry studies of the growth of amorphous and epitaxial silicon for photovoltaic applications. J. Vac. Sci. Technol. A 24, 1676 (2006)
[27] Olibet, S.: Properties of interfaces in amorphous/crystalline silicon heterojunctions. PhD thesis. IMT, Neuchatel University (2008)
[28] Olibet, S., Vallat-Sauvain, E., Ballif, C.: Model for a-Si:H/c-Si interface recombination based on the amphoteric nature of silicon dangling bonds. Phys. Rev. B 76, 035326 (2007)
[29] Fesquet, L., Olibet, S., Vallat-Sauvain, E., Shah, A., Ballif, C.: High quality surface passivation and heterojunction fabrication by VHF-PECVD deposition of amorphous silicon on crystalline Si: Theory and experiments. In: Proceedings of the 22th EU-PVSEC (2007)
[30] De Wolf, S., Kondo, S.: Nature of doped a-Si:H/c-Si interface recombination. J. of Appl. Phys. 105, 103707 (2009)

[31] Wang, T., Iwaniczko, E., Page, M., Levi, D., Yan, Y., Branz, H., Wang, Q.: Effect of emitter deposition temperature on surface passivation in hot-wire chemical vapor deposited silicon heterojunction solar cells. Thin Solid Films 501, 284 (2006)
[32] Lien, S., Wu, B., Liu, J., Wuu, D.: Fabrication and characteristics of n-Si/c-Si/p-Si heterojunction solar cells using hot-wire CVD. Thin Solid Films 516, 747 (2008)
[33] Dao, V.A., Heo, J., Choi, H., Kim, Y., Park, S., Jung, S., Lakshminarayan, N., Yi, J.: Simulation and study of the influence of the buffer intrinsic layer, back-surface field, densities of interface defects, resistivity of p-type silicon substrate and transparent conductive oxide on heterojunction with intrinsic thin-layer (HIT) solar cell. Solar Energy 84, 777 (2010)
[34] Zhao, L., Li, H., Zhou, C., Diao, H., Wang, W.: Optimized resistivity of p-type Si substrate for HIT solar cell with Al back surface field by computer simulation. Sol. Energy 83, 812 (2009)
[35] Sinton, R., Cuevas, A.: Contactless determination of current–voltage characteristics and minority-carrier lifetimes in semiconductors from quasi-steady-state photoconductance data. Appl. Phys. Lett. 69, 2510 (1996)
[36] Tucci, M., Rosa, R., Roca, F.: CF4/O2 dry etching of textured crystalline silicon surface in a-Si:H/c-Si heterojunction for photovoltaic applications. Sol. Energ. Mat. Sol. 69, 175 (2001)
[37] Angermann, H., Rappich, J.: Surface States and Recombination Loss on Wet-Chemically Passivated Si Studied by Surface Photovoltage (SPV) and Photoluminescence (PL). Sol. St. Phen. 134, 41 (2007)
[38] Sritharathikhun, J., Jiang, F., Miyajima, S., Yamada, A., Konagai, M.: Optimization of p-type hydrogenated microcrystalline silicon oxide window layer for high-efficiency crystalline silicon heterojunction solar cells. Jpn. J. Appl. Phys. 48, 101603 (2009)
[39] Kanevce, A., Metzger, W.: The role of amorphous silicon and tunneling in heterojunction with intrinsic thin layer (HIT) solar cells. J. Appl. Phys. 105, 094507 (2009)
[40] Fujiwara, H., Kondo, M.: Real-time monitoring and process control in amorphous/crystalline silicon heterojunction solar cells by spectroscopic ellipsometry and infrared spectroscopy. Appl. Phys. Lett. 86, 032112 (2005)

Chapter 3
Wet-Chemical Conditioning of Silicon Substrates for *a*-Si:H/*c*-Si Heterojunctions

Heike Angermann and Jörg Rappich

Helmholtz-Zentrum Berlin für Materialien und Energie GmbH,
Institut für Silizium-Photovoltaik, Kekuléstraße 5, D-12489 Berlin, Germany

Abstract. The influence of wet-chemical silicon (Si) substrate pre-treatments on surface morphology and electronic interface properties is discussed for various hetero interfaces of crystalline Si (c-Si) and Si oxides (SiO_x), or amorphous materials such as Si (*a*-Si:H), Si nitride (*a*-SiN_x:H) and Si carbide (*a*-SiC:H), which are typically applied in Si heterostructure solar cells. Combined application of surface sensitive techniques, the field-modulated surface photovoltage (SPV), ex-situ and in-situ photoluminescence (PL) measurements, atomic force microscopy (AFM), scanning electron microscopy (SEM), spectroscopic ellipsometry in the ultra-violet and visible region (UV-VIS-SE) and Fourier-Transform infrared ellipsometry (FTIR-SE), total hemispherical UV-NIR-reflectance measurements, microwave detected photo-conductance decay (μW-PCD) and quasi-steady-state photo conductance (QSSPC) provides detailed information about the influence of wet-chemical treatments on preparation induced micro-roughness, surface charge, energetic distribution of interface states $D_{it}(E)$ and the resulting interface recombination behaviour of wet-chemically passivated Si substrates with special surface morphology. The stability of wet-chemical surface passivation during storage in ambient air is found to be strongly influenced by the preparation-induced surface morphology. As shown for various heterojunction structures, the effect of optimized wet-chemical pre-treatments can be preserved during the subsequent soft plasma enhanced chemical vapour deposition of *a*-Si:H, *a*-SiN_x:H or *a*-SiC:H. As demonstrated for selected examples, the results of these investigations could be successfully used to enhance the energy conversion efficiency of heterojunction solar cells prepared on flat, saw damage etched and textured Si substrates. Implementation of optimised wet-chemical surface pre-treatments prior to *a*-Si:H deposition in (ZnO/*a*-Si:H(n)/*c*-Si(p)/Al) heterojunction solar cells with pyramidal texturisation increased significantly the solar cell parameters I_{sc}, V_{oc}, fill factor and enhanced the solar cell efficiency from 17.4% (confirmed) to 18.4%.

3.1 Introduction

In heterojunction solar cells the *a*-Si:H layer is used to form the p/n junction. Therefore, the textured substrate surface directly becomes part of the electronic

W.G.J.H.M. van Sark et al. (Eds.): Physics & Tech. of Amorphous-Crystalline, EM, pp. 45–94.
springerlink.com

interface. A crucial pre-condition to avoid recombination losses of charge carriers on these hetero-interfaces is to optimise the wet-chemical substrate pre-treatment. Methods to passivate interfaces, which were developed for microelectronic device technologies, have been extended to solar cell manufacturing in the past. These methods, however, have been optimised for polished substrates and are not applicable to textured surfaces. Textured mono- and polycrystalline substrates are commonly used to optimise the light trapping properties in Si solar cells. Additionally, the interface area is increased and consequently interface defects become more and more critical to the quality of subsequent processing like thin oxide film preparation and deposition of epitaxial or passivation layers.

The recombination losses of charge carriers on Si interfaces are mainly controlled by surface charge, and the density and character of rechargeable interface states D_{it} [1]. These electronic interface states result from stretched and dangling bond defects localised in a very small interlayer extended over only a few Å. Therefore, the density of these states is strongly related to the surface morphology and micro-roughness. To reduce interface recombination losses, different approaches are utilised to minimise the density of electronically active defects at the Si interface: (i) the removal of damaged regions from the Si surface by wet-chemical smoothing [2], (ii) the saturation of dangling bonds at the surface and near surface region by hydrogen (H-termination) [3] or other substituents [4, 5] and (iii) the engineering of the surface band bending to separate the electronic junction from the crystallographic interface [6].

This chapter reports on combined optimisation of wet-chemical surface pre-treatment and application of thin passivating layers: oxides, *a*-Si:H, a-SiN_x:H, or *a*-SiC:H which are utilised to achieve these goals. The final aim of chemical pre-treatments is the removal of damaged regions and the saturation of dangling bonds at the interface/surface by single bonded species (i.e. hydrogen or carbon [7]) or well defined and bond angle fitting layers (i.e. wet-chemical oxides).

The preparation of intrinsic and doped a-Si:H layers by plasma enhanced chemical vapour deposition (PECVD), however, pre-supposes well-ordered, oxygen-free substrate surfaces. Therefore this chapter is predominantly focused on wet-chemical smoothing and H-termination of various crystalline Si solar cell substrates and on the characterisation of the resulting electronic interface properties and their stability against re-oxidation in ambient air.

3.2 Wet-Chemical Substrate Preparation in Solar Cell Manufacturing

Wet-chemical substrate surface preparation is used in solar cell manufacturing mainly for three purposes: (i) removal of saw damage, (ii) texturisation and (iii) surface conditioning for the subsequent passivation and/or p/n junction and contact formation.

In order to evaluate in detail the effect of these processes on the solar cell performance, we applied standard wet-chemical processes, typically used in solar cell or microelectronic device manufacturing as well as newly developed wet-chemical methods. The substrates were treated first to remove the saw damage and to produce the macroscopic crystallographic configuration, followed by special sequences of wet-chemical oxidation and etching treatments to reduce the microscopic surface roughness. To establish the final surface configuration, the passivation by hydrogen (H-termination) and wet-chemical oxides were investigated in detail.

Fundamental investigations of the wet-chemical etching behaviour and the resulting electronic properties were carried out using polished monocrystalline p- and n-type Si(100) and Si(111) float zone (FZ) substrates. The influence of wet-chemical treatments on the electronic interface properties and the solar cell parameters were also tested for saw damage etched and textured Czochralski (CZ) solar cell substrates, as well as for multicrystalline and polycrystalline EFG (Edge-defined Film-fed-Growth) substrates.

3.2.1 Saw Damage Etch and Wafer Texturisation Methods

In order to avoid expensive and time consuming lapping and polishing processes, wet-chemical etching procedures e.g. in alkaline potassium hydroxide (KOH) solution are preferred in solar cell manufacturing, to remove saw damage from the as-cut wafers. Wet-chemical etching is additionally applied in high efficiency Si solar cells to enhance anti-reflection properties. To reduce manufacturing expense it should be advantageous to combine saw damage removal and the formation of light trapping structures in a single step process.

A well-established process in Si solar cell manufacturing is the anisotropic etching of Si(100) in KOH / isopropyl alcohol (IPA) solution at 80 °C [8, 9]. Randomly distributed pyramids, prepared by anisotropic etching, are utilized to optimize the light trapping properties [10].

In order to reduce chemical consumption, acidic solutions are employed that are based on a mixture of hydrofluoric acid and nitric acid (HF/HNO_3) for simultaneous saw-damage removal and surface texturisation. Applying acidic etchants, the removal of ~ 30-50 μm of Si reveals the characteristic surface morphology of isotropic etched monocrystalline substrates. Moreover, acidic solutions can be advantageously used for multi-crystalline substrates, due to their isotropic etching behaviour [11].

We investigated the effect of different standard single and double step wet-chemical processes on the resulting electronic interface properties. The as-cut wafers were cleaned with Puratron and deionised water (DIW) (T = 80°C) in an ultrasonic system. Afterwards, the saw damage was etched back using diluted KOH for different treatment times, resulting in an etch removal of 12 to 30 μm on each side. Randomly distributed pyramids (2 ... 10 μm), were formed on as cut, saw damage etched and polished substrates by anisotropic etching in KOH/IPA solution at 80 °C for different treatment times. A second process for wet-chemical texturisation of Si surfaces, discussed here, has been transferred from the wafer thinning technology, where the final treatment of the wafer backside is the application of a texture etch to roughen the Si surface and to enhance the adhesion of the backside metallization. In semiconductor manufacturing the texture etch (TE), typical composition SISF 80-10-10 (mixture of sulphuric acid, 96% H_2SO_4, nitric acid, 70% HNO_3, and hydrofluoric acid, 50% HF) is almost exclusively applied in single wafer spin-processes [12]. An alternative wet-etching process was tested, developed at the Honeywell TSD lab in Seelze, in order to improve the economics in manufacturing of monocrystalline Si solar cells [12]. This special etchant, named SAF-etch (mixture of nitric acid, 70% HNO_3, ammonia fluoride, 40% NH_4F, and hydrofluoric acid, 50% HF) although being of acidic nature shows a clearly anisotropic etch behaviour. The SAF-etch leads to a much faster etch rate of 100 μm/min compared to 2-3 μm/min for the TE-etch (determined according to DIN 50453-1).

3.2.2 *Wet-Chemical Surface Conditioning*

The subsequent preparation of thin-film structures requires defect- and contamination-free Si substrate surfaces as starting point. In order to decrease the surface contaminations as well as the preparation-induced micro-roughness on the textured substrate, the final surface conditioning has to be carefully optimised in consideration of substrate configuration and subsequent interface passivation. Wet-chemical cleaning processes for Si wafers can be classified according to the final surface condition into two groups: the surface passivation (i) by H-termination or (ii) by thin wet-chemical oxide layers, resulting in hydrophobic and hydrophilic surfaces, respectively. Which substrate surface conditioning can successfully be applied to Si solar cells largely depends on the details of the device structure and the kind of subsequent layer deposition.

In order to investigate the effect of wet-chemical surface conditioning on electronic interface properties, at first all substrates were cleaned using the standard process of the Radio Corporation of America (RCA) [13]. Afterwards, wet-chemical oxide layers were prepared in the following solutions: (1) ammonia / hydrogen peroxide mixture, APM (RCA I); (2) hydrochloric acid / hydrogen peroxide mixture, CPM (RCA II); (3) deionised water (DIW) at 80°C (hot water oxidation) [2, 14]; (4) boiling solution of sulphuric acid / hydrogen peroxide (1:1 H_2SO_4/H_2O_2) for 10 minutes [3], and (5) nitric acid (HNO_3 68%) at 120°C [15]. The H-termination was achieved by removing the prepared wet-chemical oxides in concentrated ammonia fluoride (NH_4F 40 %, pH=7.8) solution for 4 or 6.5 minutes or alternatively in diluted hydrofluoric acid (HF 1 %) solution for different treatment times (30 – 180 s) followed by rinsing in DIW at room temperature (RT) for different treatment times and / or drying in a N_2 stream.

Different sequences of wet-chemical oxidation and H-termination procedures were tested in order to reduce the surface roughness of structured surfaces in the atomic range. A systematic increase in the surface micro-roughness was achieved by etching in concentrated HF (48%) solution. In order to improve the stability of surface passivation by H-termination as well as wet-chemical oxides, the samples have been stored in clean-room air (25 °C, humidity approximately 50%).

3.3 Surface and Interface Characterisation Methods

Very sensitive electrical methods are necessary to measure the densities of electronic states D_{it} ranging from 10^{10} $cm^{-2}eV^{-1}$ to 10^{14} $cm^{-2}eV^{-1}$, which significantly define the electronic properties of the Si surfaces and interfaces. We utilized in-situ / ex-situ photoluminescence (PL) [16, 17], large signal surface photo voltage technique (SPV) [18-20] and microwave detected photoconductance decay (μW-PCD) [21, 22] and quasi-steady-state photo conductance (QSSPC) [23] to establish correlations between the wet-chemical substrate preparation, the resulting charge density, energetic distribution of rechargeable interface states and recombination behaviour of the resulting Si interfaces.

Interface recombination losses were investigated by transient microwave conduction (TRMC) measurements after excitation by a light pulse at a wavelength of $\lambda = 1064$ nm [24]. A commercial setup (Sinton Consulting WCT-100) was employed, where a photoflash is used to generate excess charge carriers in the device under test. Their density was detected in real time via the detuning of an oscillating circuit inductively coupled to the charge carriers while they decay via recombination. The microwave detected photoconductance decay (μW-PCD) was used to determine the spatially resolved minority charge carrier lifetime [21].

Surface micro-roughness and oxide thickness in the atomic scale were determined by ex-situ UV-VIS-spectroscopic ellipsometry (SE) using a J.A.Woollam Co. VASE spectrometer [25]. The hemispherical UV-NIR-reflectance measurements were performed using a double-beam spectrophotometer (Perkin-Elmer) with an integrating sphere [26].

Atomic Force Microscopy (AFM) images were taken in contact mode (F = 80nN) or dynamic force mode with the XE-100 from PSIA using cantilevers from Budget-sensors. Either BS-Multi-75-AL (Si) or BS-ElectriMulti75 (CrPt covered) with a force constant of 3 N/m and a resonance frequency of 75 kHz were used. Scanning Electron Microscopy (SEM) investigations were carried out using a HITACHI S-4100 scanning electron microscope with a cold field emission gun.

For all Si solar cells, I-V curves were measured under AM 1.5 illumination in a solar simulator equipped with a cold stage, which was used to keep the cell temperature at 25 °C.

3.3.1 Ex-situ SPV

The sensitivity of most of the defect-specific spectroscopic methods is not sufficient to detect the small number of interface defects that influence the electronic interface properties. Therefore, highly sensitive electrical methods such as capacitance-voltage (CV) or SPV measurements are necessary to determine electronic states and charges. In this paper large signal surface photo voltage measurements of H-terminated, HF-etched, and differently oxidized Si surfaces are utilised, to investigate the influence of important wet-chemical cleaning and passivation procedures on surface charge and energetic distribution of states. This SPV technique has the general advantage that the measurements can be carried out repeatedly during the wet-chemical treatment without any contact preparation. The measurements were carried out immediately after wet-chemical treatment, using a mica foil dielectric spacer in the same experimental set-up as recently specified in ref [20]. A transparent electrode and the sample together with the thin mica layer form a capacitor. The insulator capacity C_i is measured externally by a capacitance bridge. A pulsed laser diode (pulse length: 150 ns, wavelength $\lambda = 904$ nm and photon flux $\Phi \cong 10^{21}$ $s^{-1}cm^{-2}$) was used to excite electron-hole pairs. The illumination intensity was high enough to obtain high injection conditions, to achieve flat band conditions. The surface potential Φ_s was obtained from the large-signal photovoltage pulse recorded with a transient recorder (time resolution: 5 ns) as described in [19].

The energetic distributions of the interface states $D_{it}(E)$ were determined from a series of photovoltage pulses at different field voltages U_F between a transparent electrode and the Si wafer, as firstly described by Heilig [18] in 1968. For this purpose, a varying electric field perpendicular to the surface was applied, which changes the surface potential Φ_s continuously as a function of the field voltage. The duration of a field voltage pulse was 100 ms, followed by an inverted but otherwise identical pulse, to minimise effects due to charge accumulation [27]. Due to screening effects, the influence of the field voltage U_F on the surface potential Φ_s depends on the charge Q_{it} trapped in interface states. A change in field voltage dU_F in this Metal Insulator Semiconductor (MIS) system leads to a change in the voltage drop at the insulator dU_i and of the surface potential $d\Phi_S$, eq. (3.1).

$$dU_F = dU_i + d\Phi_s \tag{3.1}$$

Taking charge neutrality into account one gets

$$Q_f + Q_g + Q_{it} + Q_{sc} = 0 \tag{3.2}$$

and

$$dQ_g + dQ_{it} + dQ_{sc} = 0 \tag{3.3}$$

for the charging of the system due to a voltage change, where Q_f is the fixed charge, Q_g the influenced charge on the field electrode, Q_{it} the charge in rechargeable interface states and Q_{sc} the interface charge, i.e. the projection of the complete space charge onto the interface, determined by the space charge function F:

$$Q_{sc} = \left(2n_i \cdot \varepsilon_{si} \cdot kT\right)^{\frac{1}{2}} F \tag{3.4}$$

where ε_{si} is the static dielectric constant of Si. Q_{sc} is evaluated as a function of Φ_s as described in [20]. From the definitions of the interface state density

$$D_{it}\left(\Phi_s\right) = -\frac{1}{q}\frac{dQ_{it}}{d\Phi_s} \tag{3.5}$$

and the insulator capacity

$$C_i = dQ_d / dU_i \tag{3.6}$$

and eqs. (3.1-3.4) the following dependence for the density of states as a function of the change in surface potential is derived [28]:

$$D_{it}(\Phi_s) = \frac{1}{q} C_i \left(\frac{dU_F}{d\Phi_s} - 1 \right) + \frac{dQ_{sc}(\Phi_s)}{q\, d\Phi_s} \tag{3.7}$$

Finally, Q_{sc} can be evaluated from eq. (3.4) as function of the surface potential Φ_s.

3.3.1 In-situ / ex-situ PL

The change of recombination behaviour during wet-chemical treatments was monitored by in-situ PL measurements. I_{PL} is related to the band recombination of the light induced charge carriers generated by the dye laser pulse via the bandgap. The change in I_{PL} is due to defect formation (quenching by non-radiative recombination) or defect passivation (PL enhancement). For in-situ PL measurements a tuneable dye laser (typical wavelength 500 nm, pulse length: 5 ns, pulse energy 60 µJ/pulse) excited the PL light which passed an interference filter and was measured by an integrating InGaAs charge coupled detector. The chemical cell consisted of Teflon. Fast exchange of the solution can be performed by inlet and outlet tubes. For more details see [29, 30]. For more information see Chapters 4 and 8 in this book.

3.3.2 Ex situ UV-VIS Spectroscopic Ellipsometry (UV-VIS-SE)

Spectroscopic ellipsometry (SE) in the ultraviolet and visible (UV-VIS) spectral range is a well-known surface-sensitive and non-destructive method for detecting roughness and sub-monolayer coverage during processing of silicon surfaces for microelectronic devices.

Because there is no clear definition of 'micro-roughness', the term will be used here to denote atomic scale irregularities on the Si surface or at the Si/SiO_x interface. The optical effect of a microscopically rough surface can be described accurately [31] by a Bruggeman effective medium approximation [32] (EMA) layer. For microscopically rough wafers, a two-layer model was used that consists of bulk c-Si [33], and a Bruggeman EMA layer consisting of 50 % bulk c-Si and 50 % voids. Wafers with a thin oxide layer were modelled by a layer of SiO_2 on top of c-Si. Here, the data from the initially H-terminated sample before oxidation were used to include the small fraction of surface roughness remaining after the different H-termination procedures [25, 26]. It is impossible however to distinguish the effects of surface roughness and of a thin oxide layer by UV-VIS ellipsometric measurements alone. Therefore, the hydrogen coverage of Si (100) and Si (111) surfaces was directly measured by sensing the Si–H and Si–H_2 vibrational resonance by Fourier-transform infrared ellipsometry [25] which has the general advantage that vibrational modes can be directly observed through characteristic, energetically sharp structures in the measured spectra [26].

The ex situ UV-VIS SE measurements, were carried out using a J. A. Woollam Co. VASE rotating analyzer ellipsometer. For a short duration of the measurement, the photon energy range was limited to 3.2 – 4.5 eV as the imaginary part of the pseudo-dielectric function at the E_2 silicon critical point near 4.25 eV is of highest sensitivity to surface modifications and overlayers [34]. The angle of incidence was 77.00° ± 0.02°. The measured complex ratio of Fresnel reflection coefficients was transformed into the complex pseudo-dielectric function $<\varepsilon>$. Assuming models of plane-parallel homogeneous layers, values for the relative surface roughness $<d_r>$, and the relative thickness of the native oxide $<d_{ox}>$, respectively, were calculated.

3.4 Preparation-Induced Electronic Interface Properties Investigated on Polished Si-Substrates

In order to prepare H-terminated Si substrate surfaces with excellent electronic properties and to preserve the surface passivation during the technological process, we investigated special combinations of wet-chemical etching and cleaning procedures for various solar cell applications. The next section, exemplary reports on investigations of electronic and morphological interface properties after different conventional and newly developed wet-chemical cleaning and passivation methods, carried out on polished Si(111) and Si(100) substrates. On these surfaces correlation between structural imperfections, interface state densities, interface recombination losses and stability of surface passivation has been established. The results of these investigations were successfully applied to optimise surface treatments for saw damage etched substrates, textured with by isotropic or anisotropic etching, as well as multicrystalline EFG wafers. These results are discussed in the following sections.

3.4.1 *Effect of Wet-Chemical Pre-treatment on $U_{Ph,max}$, $D_{it}(E)$ and Recombination Losses Demonstrated for Polished a-Si:H/c-Si(n) and a-SiN$_x$:H /c-Si(n) Interfaces*

The influence of wet-chemical treatments on the surface electronic properties is first demonstrated using the example of polished n-type float zone (FZ) Si(111) substrates (3 Ωcm) after various hydrophilic and hydrophobic processes, which were tested for an inverted *a*-Si:H/*c*-Si heterojunction structure solar cell, with front side passivation by *a*-SiN$_x$:H and a p-type *a*-Si:H emitter on the rear side [35]. The change of surface charges, interface state densities and interface recombination losses were repeatedly determined (i) after different substrate pre-treatments, (ii) after the following preparation of an *a*-SiN$_x$:H passivation layer in a PECVD reactor (AK 1000 from Roth & Rau AG) operating with a microwave (MW) excited plasma by means of a linear antenna array at a frequency of 2.45 GHz and (iii) after subsequent deposition of *a*-Si:H emitter layers - consisting of a stack of ~3 nm (i)*a*-Si:H and ~8 nm (p$^+$)*a*-Si:H.

Fig. 3.1 shows SPV pulses $U_{Ph}(t)$ obtained on these substrates after wet-chemical oxidation in RCA I (curve 1), in hot DIW (curve 2), in H_2SO_4/H_2O_2 (curve 3), in HNO_3 (curve 6) and after H-termination by HF (1%) (curve 4) and subsequent DIW rinsing at RT (curve 5). The maximal values $U_{Ph,\ max}$ of $U_{Ph}(t)$ were found to be strongly influenced by the preparation induced surface charge [20, 36]. The time decays of U_{Ph}, which yield information about interface recombination behaviour [37], also show significant differences between oxidised (Fig. 3.1, curves 1, 2, 3, 6) and H-terminated (Fig. 3.1, curves 4, 5) surfaces. The decreasing slope of the U_{Ph} decay can be taken as measure for the decrease in interface recombination achieved by wet-chemical H-termination.

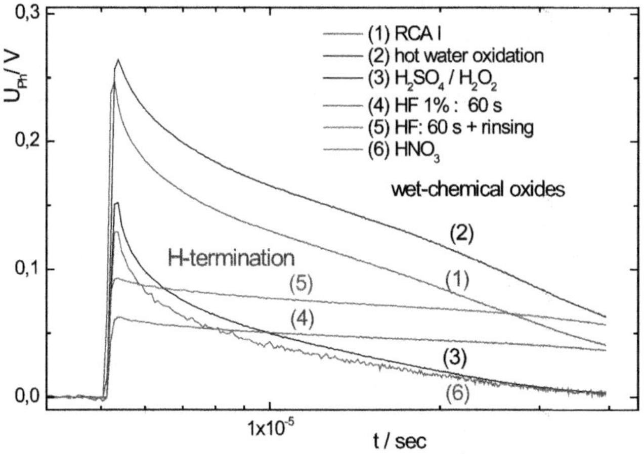

Fig. 3.1 Surface photovoltage pulses obtained on n-type FZ Si after wet-chemical oxidation in RCA I (1), in hot DIW (2), in H_2SO_4/H_2O_2 (3), in HNO_3 (6) and after H-termination by HF (1%) (4) and subsequent DIW rinsing at RT (5).

Fig. 3.2a presents photovoltage vs. fieldvoltage, $U_{ph}(U_F)$ plots measured by field modulated SPV on the same samples. The energetic distributions of rechargeable interface states $D_{it}(E)$ on these polished substrates, calculated by using eq. 3.6 – 3.7, are given in Fig. 3.2b. The minimum value $D_{it,min}$ of the energetic distribution of rechargeable interface states $D_{it}(E)$ (Fig. 3.2b) and its energetic position is commonly used as a technological parameter of the electronic surface quality.

All conventional wet-chemical oxidation treatments result in high values of $D_{it,min}$ attributed to the rapid and irregular wet-chemical oxide growth [38]. The standard RCA cleaning process (see ref. [13]) shown in Fig. 3.2b consisting of RCA I (curve 1) and RCA II treatment (not shown here), as well as the wet-chemical oxidation using H_2SO_4:H_2O_2 (curve 3) and HNO_3 (curve 6) cause high densities of interface states $D_{it,min} > 5*10^{12}\ cm^{-2}eV^{-1}$.

Solely thin wet-chemical oxide layers prepared in deionised water at 80 °C (Fig. 3.2*b*, curve 2) exhibit significantly lower interface state densities of about $D_{it,min} \cong 1*10^{12}$ cm^{-2}eV^{-1} [39]. Applying this hot water oxidation process on previously H-terminated Si(100) and atomically flat Si(111) surfaces, thin oxide layers with interface state densities of $D_{it,min} < 5*10^{11}$eV^{-1}cm^{-2} were found, which are significantly lower than conventionally prepared chemical and thermal oxides in the thickness range of 1 to 3 nm. We assume that a very slow layer-by-layer oxide growth leads to a well-ordered interface during the formation of the first monolayers [38].

Wet-chemical H-termination in HF 1% solution leads to lower values of $D_{it,min} \leq 3 \times 10^{11}$ cm^{-2}eV^{-1} on the same polished substrates (Fig. 3.2b, curve 4). A subsequent rinse (5 s) in deionised water at room temperature (RT) does not increase the density of interface states (Fig. 3.2b, curve 5) but changes the surface charge by the appearance of polarized ≡Si–OH bonds as possible side reaction of the H-termination. This leads to different surface Fermi level positions and thus changes the band bending, as shown in the difference of maximum values of surface photovoltage $U_{Ph,max}$ (Fig. 3.1, curves 4 and 5).

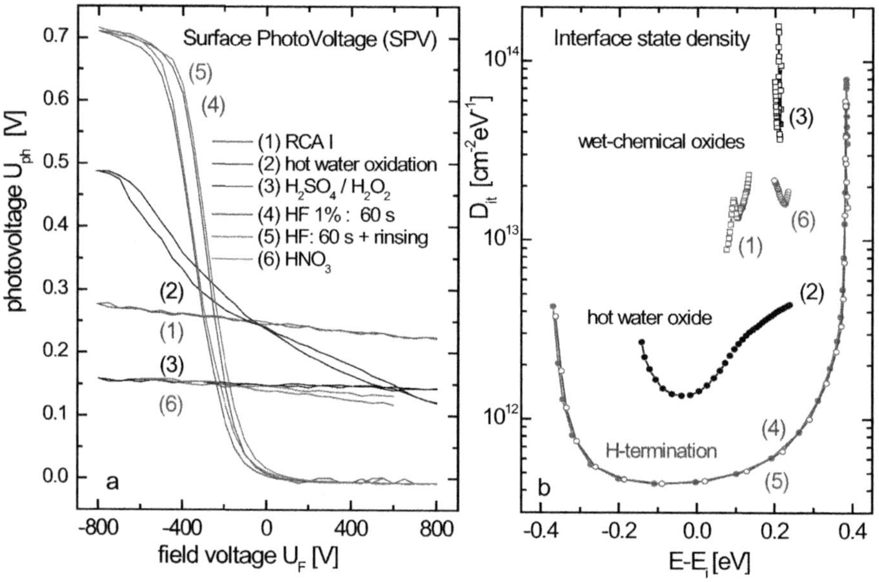

Fig. 3.2 (a) Phovoltage vs. field voltage $U_{ph}(U_F)$ plots and (b) the calculated interface state distributions $D_{it}(E)$ on n-type FZ Si after wet-chemical oxidation in RCA I (1), in hot DIW (2), in H_2SO_4/H_2O_2 (3), in HNO_3 (6) and after H-termination by HF (1%) (4) and subsequent DIW rinsing at RT (5).

In wafer-based high efficiency Si solar cells hydrogenated amorphous silicon (*a*-Si:H) as well as silicon nitride (*a*-SiN_x:H) layers are utilised to passivate the Si surface and/or to form the rear surface field. As is well known from the microelectronics and sensor technology, the electronic properties of Si interfaces are strongly influenced by the chemical integrity and morphological structure of the substrate surface. Therefore, increasing attention should be also focused on the monitoring and control of interface properties after pre-cleaning treatments and subsequent *a*-SiN_x:H deposition. The effect of optimised wet-chemical treatment was verified for front side passivation by a-SiN_x:H prepared on polished n-type FZ wafers [36].

Fig. 3.3 shows $U_{Ph,\,max}$ determined on the same polished substrates (i) after surface pre-treatments as described before: oxidation by RCA I (Fig. 3.3, curve 1), hot water (Fig. 3.3, curve 2), and H_2SO_4/H_2O_2 (Fig. 3.3, curve 3), and after H-termination by HF 1 % (Fig. 3.3, curve 4) and HF 1% + rinsing (Fig. 3.3, curve 5), and (ii) after subsequent front side passivation by *a*-SiN_x:H, and (iii) after removing the wet-chemical oxides on the rear side and subsequent deposition of the rear side *a*-Si:H hetero emitter. On differently treated substrates (Fig. 3.1) the preparation-induced surface charge result in different values of $U_{Ph,max}$ as shown in Fig. 3.3 (i). After *a*-SiN_x:H deposition on the front side no variation in $U_{Ph,max}$ between the different treatments was found as shown in Fig. 3.3 (ii), because the field-effect passivation is based on the band bending caused by the positive fixed charge in the *a*-SiN_x:H layer which determines the Fermi level position. After removal of wet-chemical oxides and *a*-Si:H deposition on the rear side, however, a strong influence of the surface pre-treatment on $U_{Ph,max}$ was found on the rear side *a*-Si:H/*c*-Si, as well as on the front side *a*-SiN_x:H/*c*-Si interface, as can be seen in Fig. 3.3 (iii) [36].

In Fig. 3.4 microwave detected photo conductance decays are plotted, obtained on the same inverted a-Si:H/c-Si heterojunction structure, after front side passivation by *a*-SiN_x:H (ii) and subsequently removing wet-chemical oxides in HF 1% and deposition of an *a*-Si:H emitter on the rear side (iii) [36]. After the front side *a*-SiN_x:H deposition, no relation was obtained between $D_{it}(E)$ (Fig. 3.2b) and interface recombination (Fig. 3.4a), due to the field effect passivation. So far, the lowest interface recombination losses were achieved on the *a*-SiN_x:H/*c*-Si structure by wet-chemical oxidation in H_2SO_4/H_2O_2 (Fig. 3.3 (ii), curve 3).

In Fig. 3.4b microwave detected photo conductance decays are given as obtained after removing all wet-chemical and native oxides by HF 1% dip for 30 s, and subsequent deposition of the rear side *a*-Si:H emitter layer by plasma enhanced chemical vapour deposition (PECVD). The *a*-Si:H layer consisted of a stack of ~3nm (i)*a*-Si:H and ~8 nm (p^+)*a*-Si:H.

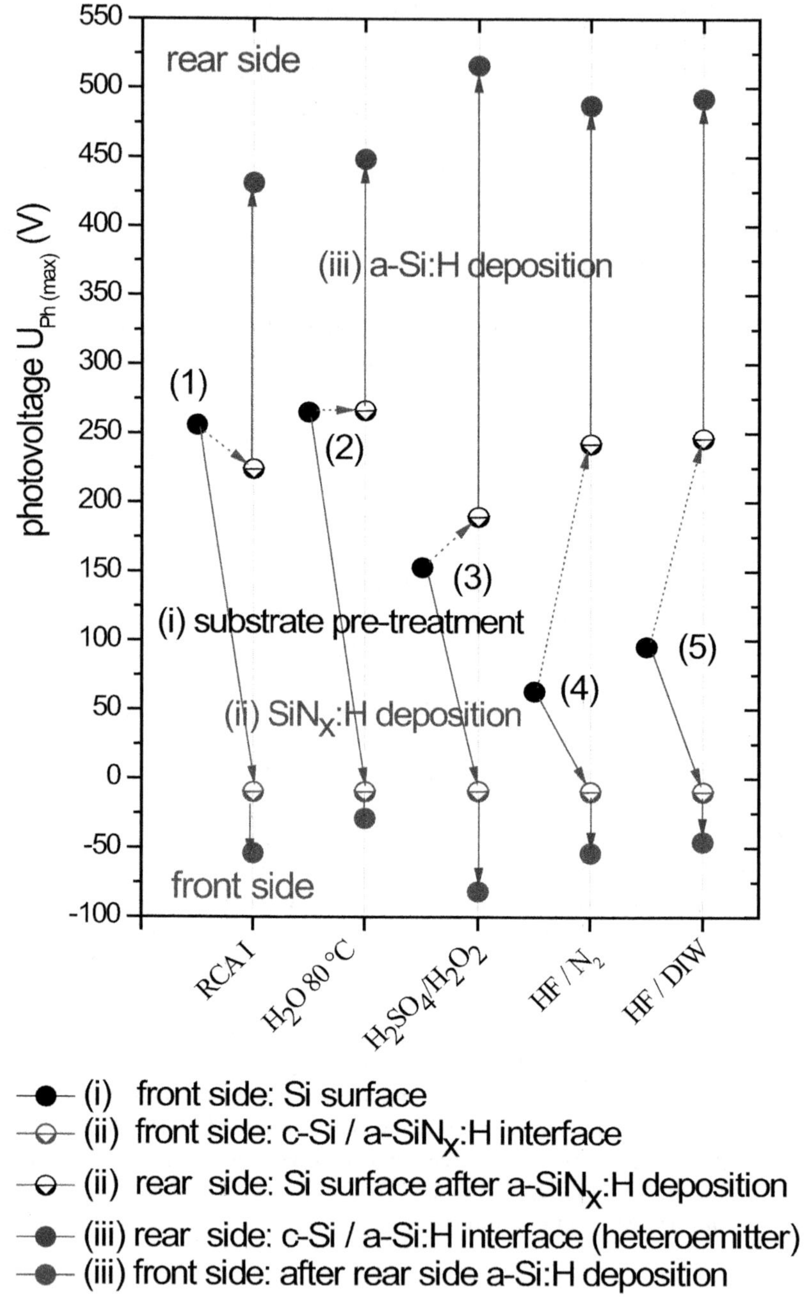

Fig. 3.3 Maximal values of $U_{Ph,max}$ of n-type FZ Si interfaces (i) after surface pre-treatments: oxidation by RCA I (1), hot water (2), and H_2SO_4/H_2O_2 (3), H-termination by HF 1 % (4) and HF 1% + DIW rinsing (5), (ii) after subsequent front side passivation by *a*-SiN_x:H, and (iii) after subsequent HF 1% and *a*-Si:H deposition on rear side.

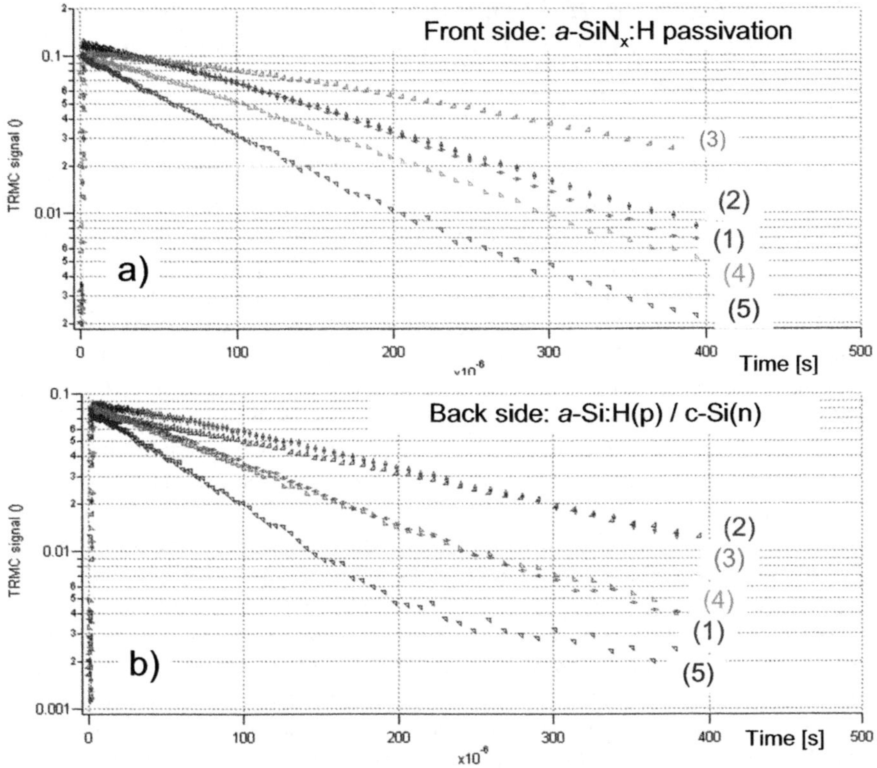

Fig. 3.4 Microwave detected photo conductance decays measured on inverted *a*-Si:H/*c*-Si heterojunction structures [36]: (a) front side passivation by *a*-SiN_x:H (ii) deposited after oxidation by RCA I (1), hot water (2) or H_2SO_4/H_2O_2 (3), or after H-termination by HF 1% (4) or HF 1% with subsequent DIW rinsing (5) and (b) *a*-Si:H(p) emitter deposited on the back side (iii) after removing wet-chemical oxides in HF 1%.

Lowest recombination losses on the so prepared *a*-SiN_x:H/*c*-Si interface were observed after oxidation in H_2SO_4/H_2O_2 or hot water and subsequent dip in HF 1% (Fig. 3.4b, curves 2, 3). Subsequent rinsing in DIW at RT significantly increases the recombination velocities on both interfaces (Fig. 3.4a and b, curves 5).

Both preparation-induced surface charges and rechargeable interface states, as well as various further effects of wet-chemical substrate surface conditioning were found to influence the recombination losses on *a*-SiN_x:H/*c*-Si and *a*-Si:H/*c*-Si interfaces in different ways [36]. Highest $U_{Ph,max}$ and best energy conversion efficiency η =13.96% [35] on test solar cells with inverted *a*-Si:H/*c*-Si heterojunction structures were obtained by H_2SO_4/H_2O_2 pre-treatment (Fig. 3.3 and Fig. 3.4, curves 3).

3.4.2 Wet-Chemical H-Termination Processes Optimised for Polished p-Type Si(111), Si(100) and Multi-crystalline EFG Substrates

In the last section we have demonstrated the influence of surface pre-treatment on the interface recombination losses, in this section we focus on details of the H-termination process. As shown in Fig. 3.4b, recombination losses on *a*-Si:H/*c*-Si interfaces prepared on H-terminated substrate surfaces are strongly influenced by the different sequences of wet-chemical oxidation, etching and rinsing steps.

U-shaped energetic distributions of interface state density D_{it} (E) and low minimum values $D_{it,min}$ can be observed (Fig. 3.2b, curves 4 and 5) on H-terminated Si surfaces. The typically U-shaped distribution of D_{it} and the high D_{it} value near the band edges, which were experimentally observed on Si/SiO_2 interfaces, have been attributed to specific defect centres in the oxide layer [40]. They can be understood as superposition of two groups of rechargeable states: i) the states near the band edges result from strained (Si_3≡Si—Si≡Si_3) bonds and decay roughly exponentially into the band gap, and ii) the groups of states symmetrically distributed around a minimum near midgap are correlated to dangling bond defects (Si_3≡Si–), which are back-bonded to Si atoms only [41, 42].

These defect groups were identified by comparison to electronic model and also by Electronic Paramagnetic Resonance (EPR) measurements (P_b–centres) [41, 43], and have been identified mainly as unpaired sp^3 orbitals on trivalently bonded interface Si atoms or Si(111)-oriented Si dangling bonds pointing out of the interface into the oxide [44]. The two related defect groups on the Si(100)/SiO_2 and Si(111)/SiO_2 interfaces are the so called P_{B0} and P_{B1} centres, respectively.

The influence of surface orientations, micro-roughness and grain boundaries on the efficiency of various H-termination procedures was previously studied in detail on polished Si(111) and Si(100) substrates as well as on polycrystalline EFG Si, commonly used to reduce substrate costs for photovoltaic applications [14, 26, 45, 46].

Fig. 3.5 summarises the results of these investigations obtained on p-type (a) Si(111), (b) Si(100) and on (c) EFG substrates [14]. All wet-chemical H-termination procedures include two essential steps: the formation of a Si/SiO_2 interface by oxidising solutions and the subsequent removal of this oxide layer by HF- or NH_4F-containing solutions. The interface state distributions $D_{it}(E)$ shown in Fig. 3.5 were determined from SPV measurements carried out under clean room conditions in dry nitrogen (N_2) atmosphere immediately after wet-chemical oxidation in RCA solutions, H_2SO_4/H_2O_2 or DIW at 80°C and subsequent HF (1%) or NH_4F (40 %, *p*H=7.8) treatments. The frequently used RCA-treatment [13], followed by the HF (1%) dip (60 s), leads to $D_{it,min} \cong 1{\cdot}10^{11}cm^{-2}eV^{-1}$ on monocrystalline Si(111) and Si(100) surfaces (Fig. 3.2, Fig. 3.5a and b, curves 1), but to significantly higher values on EFG substrates (Fig. 3.5c, curve 1). For comparison, the standard H-termination procedure described by Chabal et al. [3], which uses H_2SO_4:H_2O_2 solution for oxidation followed by etching in NH_4F results in lower values of $D_{it,min} \cong 5{\cdot}10^{10}cm^{-2}eV^{-1}$ on Si(111) (Fig. 3.5a, curve 2) and higher values of $D_{it,min} > 2{\cdot}10^{11}cm^{-2}eV^{-1}$ on Si(100) (Fig. 3.5b, curve 2), respectively. This standard H-termination procedure causes very high values of $D_{it,min} > 1 \cdot 10^{12} cm^{-2}eV^{-1}$ on EFG substrates (Fig. 3.5c, curve 2), obviously due to the anisotropic etching of areas without Si(111) orientation and grain boundaries by NH_4F solution.

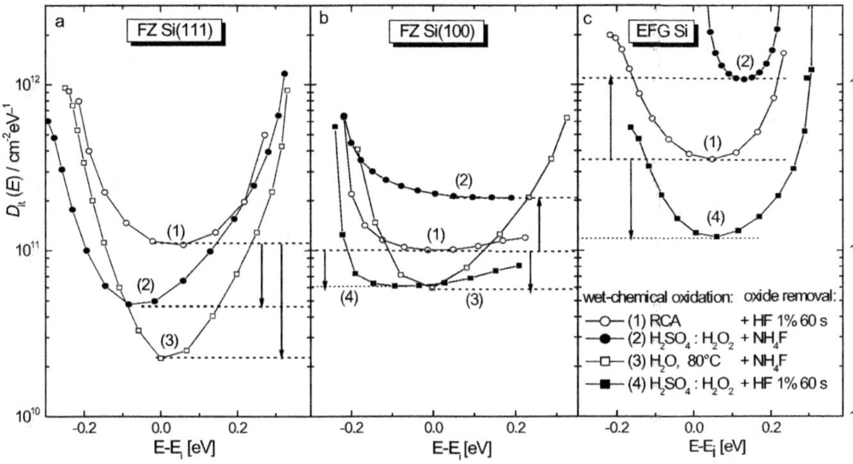

Fig. 3.5 $D_{it}(E)$ obtained on hydrogen (H) -terminated p-type substrates (a) Si(111), (b) Si(100) and (c) EFG-Si prepared by wet-chemical oxidation by RCA II (1), H_2SO_4/H_2O_2 (2 and 4) or hot-water (*3*), and subsequent oxide removal by HF (1%) (1 and 4) or H_4F (40 %, pH=7.8) (2 and 3) treatment, respectively [2]. [with permission of Elsevier].

Up to now, the lowest values of $D_{it,min} \cong 1\text{-}2 \cdot 10^{10} cm^{-2}eV^{-1}$ on Si(111) (Fig. 3.5a, curve 3) and $D_{it,min} \cong 5 \cdot 10^{10} cm^{-2}eV^{-1}$ on Si(100) (Fig. 3.5b, curve 3) were achieved by oxidation in DIW at 80 °C [14] combined with an oxide removal in NH_4F solution.

The NH_4F treatment, however, leads to anisotropic etching on the Si(100) surface [47] resulting in Si(111) facets as high as 7 nm [48]. Therefore, in order to avoid surface unevenness on Si(100) and EFG substrates, the final etching step in NH_4F solution was replaced by an HF (1%) dip for 60 s. Thereby $D_{it,min}$ of about $5 \cdot 10^{10} cm^{-2}eV^{-1}$ on Si(100) surfaces (Fig. 3.5b, curve 4) and of $1 \cdot 10^{11} cm^{-2}eV^{-1}$ on EFG substrates could be achieved. This is a significant reduction with respect to the NH_4F treatment (Fig. 3.5c, curve 4).

3.4.3 Influence of Preparation-Induced Surface Micro-roughness on $D_{it,min}$ Investigated for HF- and NH_4F Treated Surfaces

As demonstrated in Fig. 3.5, the preparation-induced D_{it} of polished H-terminated Si surfaces results from the course of two different chemical processes: the wet-chemical oxidation of the surface as well as the etching process of silicon oxide and the Si substrate in HF-containing solutions, respectively. The wet-chemical oxidation shifts the Si/SiO_2 interface into the Si bulk. Afterwards a few Ångstrom thick Si surface layer can be removed by the etch-back of wet-chemical oxide layers in NH_4F or HF (1%). This principle can be utilised in order to remove contaminations and to reduce surface micro-roughness on polished monocrystalline, on polycrystalline and nano-structured substrates. However, the pre-condition is that both processes – the wet-chemical oxidation as well as the oxide removal – have to be carefully optimised with respect to the oxidation and etching behaviours of the substrates with different surface orientations.

Our previously reported results have shown, that the surface charge and D_{it} of polished Si substrates are mainly determined by the preparation-induced surface micro-roughness and surface coverage [14, 25, 38, 39, 46]. Fig. 3.6 merges the results of these investigations which have been carried out by combined UV-VIS-SE and SPV measurements. Thereby, correlations between the $D_{it,min}$ and the preparation-induced surface effective micro-roughness $<d_r>$ on differently pre-treated polished Si(111) surfaces after H-termination in NH_4F (4,5 min) as well as after HF (48%) -etching by different treatment times were established.

At first, the effect of various oxidising solutions was investigated in detail. To verify the influence of the Si/SiO_2 interface roughness, from differently wet-chemically oxidised Si surfaces the oxide layers were carefully eliminated by NH_4F-treatment which is known to prevent additional roughness during the etching process. As shown in Fig. 3.6, curve 1, on H-terminated p-type FZ Si(111) surfaces prepared in this manner, both the remaining surface micro-roughness $<d_r>$ and $D_{it,min}$ were found to be significantly influenced by the previously applied wet-chemical oxidation solution.

Immediately after (i) conventional RCA II process [13], the NH_4F treatment yields a comparatively high $<d_r>$ of approximately 3 Å and a $D_{it,min}$ of about $1\text{-}3\cdot10^{11}cm^{-2}eV^{-1}$ (Fig. 3.5a, curve 1). Applying (ii) boiling H_2SO_4:H_2O_2 as oxidising solution [3], the standard H-termination process (ii) in NH_4F, usually leads to a $<d_r>$ of about 2 - 3 Å and a $D_{it,min}$ of $5\cdot10^{10}cm^{-2}eV^{-1}$ (Fig. 3.5a, curve 2). A special H-termination process (iii) carried out by oxidation in DIW at 80 °C [14] and subsequent careful oxide removal in NH_4F solution at N_2 atmosphere provides the lowest $D_{it,min}$ of $1\text{-}2\cdot10^{10}cm^{-2}eV^{-1}$ (Fig. 3.5a, curve 3) on atomically flat Si(111) surfaces with a $<d_r>$ of about 1 Å [38]. On H-terminated Si(100) surfaces, generally higher values of $<d_r>$ of about 4 Å [38] and $D_{it,min}$ were observed (Fig. 3.5b, curves 1-3).

The influence of surface micro-roughness on an atomic scale was further evaluated utilising a set of p- and n-type Si(111) samples with a systematic and well defined increase in surface roughness treated in HF (48 %) solution (5 to 180 s). As demonstrated in Fig. 3.6 (curves 2 and 3) the micro-roughness as well as the $D_{it,min}$ of n- and p-type surfaces strongly increase during the HF (48%) etching time. On n-type Si, a stepwise increase in $D_{it,min}$ was observed if $<d_r>$ affects one, two, and three monolayers (Fig. 3.6, curve 2). After exposure time of 5 s, $<d_r>$ already reached one monolayer and $D_{it,min} \approx 3\cdot10^{11}cm^{-2}eV^{-1}$. Finally, the $D_{it,min}$ increased up to about $2\cdot10^{12}cm^{-2}eV^{-1}$ when the effective surface micro-roughness reached three monolayers. On HF (48%) etched p-type silicon samples, similar relationships between surface micro-roughness and density of interface states were found (Fig. 3.6, curve 3) [25].

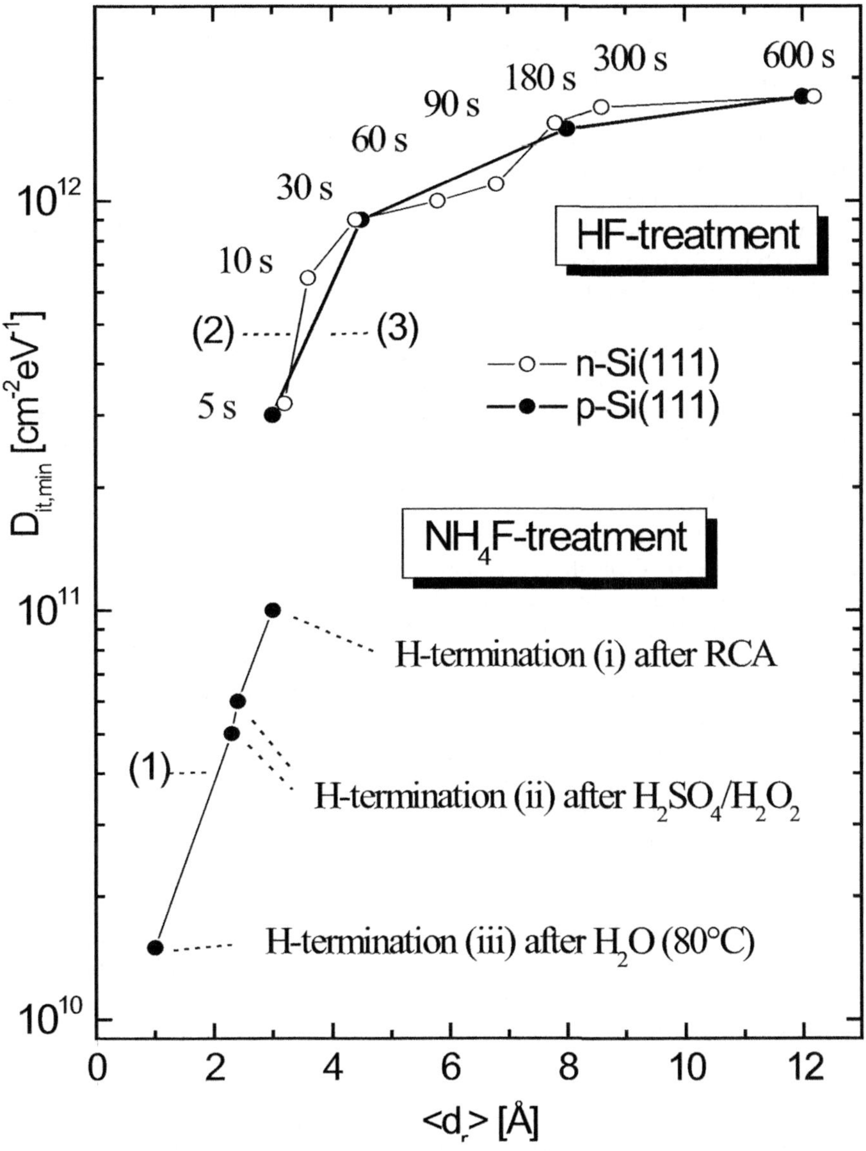

Fig. 3.6 $D_{it,min}$ as a function of the preparation-induced surface micro-roughness $<d_r>$ obtained by spectroscopic ellipsometry (UV-VIS-SE) on differently pre-treated Si(111) surfaces immediately after H-termination in NH_4F solution (1) and on n- and p-type Si(111) after HF (48%) etching (5 s to 10 min)(2 and 3) [46] [with permission of Elsevier].

3.4.4 Optimisation of Oxide Removal by In-situ Photoluminescence Measurements during HF Treatment on Polished Si Substrates

Summarising these results on the preparation-induced micro-roughness and energetic distribution of surface states it is clear that they depend on the course of three different chemical processes: (i) the formation Si/SiO_2 interface in oxidising solutions, (ii) the etch-back of the silicon oxide layer as well as (iii) the etching of the underlying Si substrate surface in the HF containing solutions. To optimise the duration of etching steps, the changes in the recombination behaviour during these wet-chemical treatments can be monitored by pulsed in-situ PL measurements using a dye laser ($\lambda = 500$ nm, $\tau_{pulse} = 6$ ns, and 60 μJ pulse energy) [49].

Fig. 3.7 shows I_{PL} of differently prepared wet-chemical SiO_2 layers as a function of the etching time in HF 1% for different n-type substrates: Si(111) (Fig. 3.7, curves 1 and 2), Si(100) (Fig. 3.7, curves 3 and 4), randomly distributed pyramids with Si(111) facets on textured surfaces (Fig. 3.7, curves 5 and 6).

The oxide layer was either prepared by (i) the RCA process [13] (Fig. 3.7, curves 1, 3, and 5) or by (ii) an oxidation process in H_2SO_4:H_2O_2 [3] (Fig. 3.7, curves 2, 4, and 6). Typically, three phases can be distinguished which are correlated to the thinning of the oxide layer (I), oxide removal and H-termination (II), and the etching attack of the H-terminated Si surface (III). No change in I_{PL} was observed during the initial phase of oxide etching (I). Further etching leads to an increase in I_{PL} (II) due to the removal of oxide induced defects at the interface coupled with H-termination, which leads to the highest value of I_{PL} (lowest density of recombination centres) followed by a decrease in I_{PL} (III) caused by the etching induced defects on the H-terminated surface.

On polished Si(111) substrates, the maximum value of I_{PL} in 1% HF was reached after an etching time of 180 s on both RCA treated (Fig. 3.7, curve 1) and wet-chemically oxidised substrates (Fig. 3.7, curve 2). On polished Si(100) substrates faster reaction rates of oxide thinning (I) and oxide removal (II) were observed, so that the maximum value of I_{PL} in 1% HF was already reached after an etching time of 30 s after RCA (Fig. 3.7, curve 3) and of 20 s after H_2SO_4:H_2O_2 (Fig. 3.7, curve 4) treatments, respectively. The kinetics of oxide thinning (I) and oxide removal (II) on Si(100) substrates textured by anisotropic etching of pyramids with Si(111) facets (Fig. 3.7, curves 1 and 2) were found to be comparable to that of polished Si(111). The maximum value of I_{PL} in 1% HF was reached after an etching time of 90 s on RCA treated (Fig. 3.7, curve 5) and of 120 s on substrates oxidised in H_2SO_4:H_2O_2 solution (Fig. 3.7, curve 6). On all n-type substrates extended HF (1%) etching times (III) lead to a decrease in I_{PL} caused by etching induced dangling bond defects on the H-terminated surface e.g. on step facets or kink sites. Particularly on textured surfaces a strong quenching of the I_{PL}, due to a fast defect formation, was observed after complete removal of RCA oxides (Fig. 3.7, curve 5). Similar effects were observed on p-type Si(100) solar cell substrates textured with Si(111) pyramids [50].

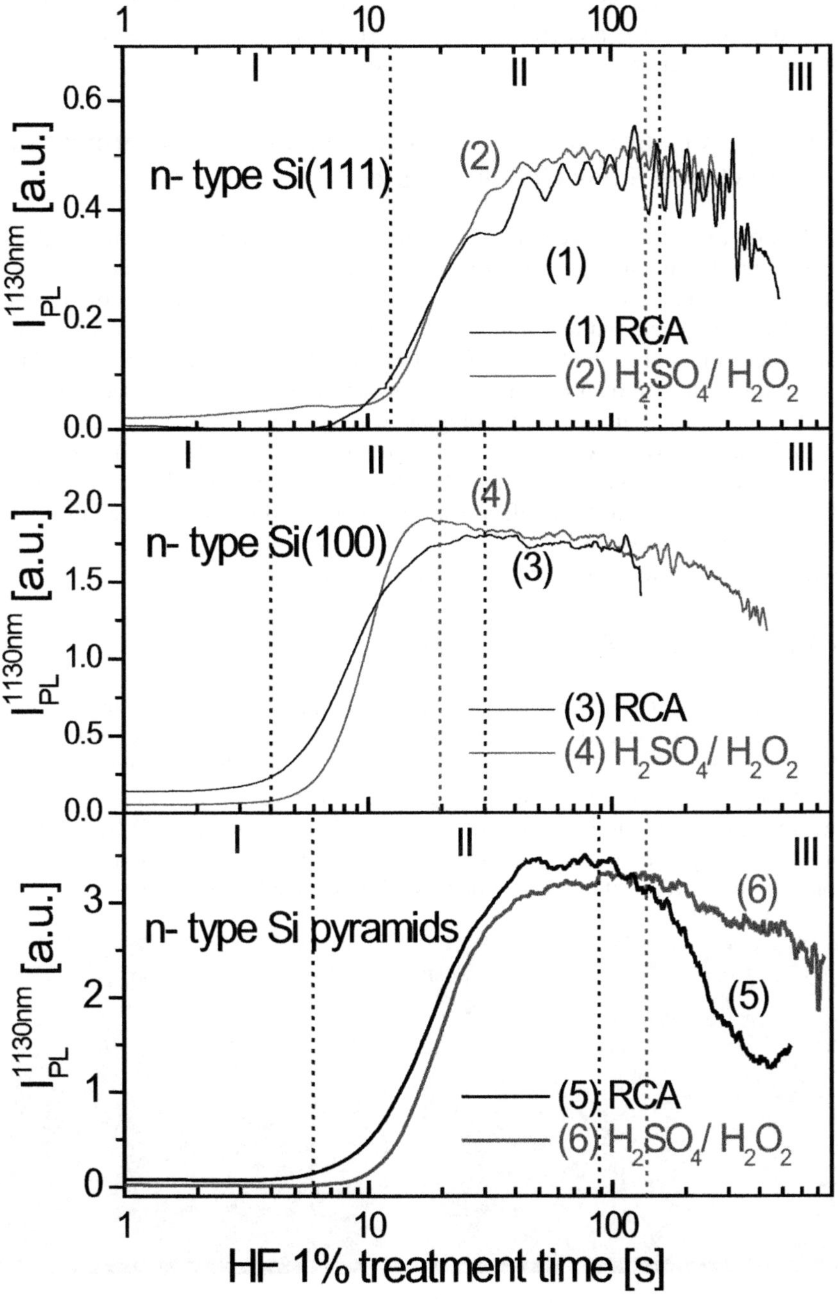

Fig. 3.7 Time dependence of PL intensities (I_{PL}) measured in-situ on n-type substrates [49]: polished Si(111) (1, 2), polished Si(100) (3, 4), Si(111) pyramids (5, 6), during the oxide removal in HF (1%): subsequent (i) to RCA II [13] (1, 3, 5) or (ii) to wet-chemical oxidation process in H_2SO_4:H_2O_2 (1:1) at 120°C [3] (2, 4, 6). [with permission of Springer].

As demonstrated here, in-situ PL measurements can be applied as a very sensitive tool, to monitor preparation-induced Si surface electronic properties during the wet-chemical processing and to optimise the treatment times.

3.5 Optimisation of Wet-Chemical Surface Pre-treatment for Textured Solar Cell Substrate

In the previous section, we studied in detail the correlations between wet-chemical surface pre-treatments and electronic interface properties on polished Si(111) and Si(100) surfaces. In this section we will concentrate on the surface conditioning of saw damage etched and textured wafers commonly used as substrates in solar cell manufacturing. The final aim of wet-chemical conditioning of Si solar cell substrates is the removal of contaminations and damaged surface regions, the fabrication of special surface morphologies to reduce light reflectance and interface recombination losses and the passivation of the preparation induced interface states. Therefore it has to be carefully optimised concerning the subsequent layer deposition. In solar cell manufacturing, wet-chemical etching procedures are preferred to remove saw-damage (Fig. 3.8) from the as-cut wafers in order to avoid expensive and time consuming lapping and polishing processes.

3.5.1 Removal of Saw Damage

The complete removal of saw damage is a crucial precondition to reduce recombination losses on solar cell substrates. SEM micrographs (cross section) show saw damaged regions (Fig. 3.8a) and cracks (Fig. 3.8b) typically observed on as-cut c-Si substrates. The thickness of the damaged surface layer is dependent on the wafer saw technology. It can be removed (i) by saw damage etching, (ii) by combination of saw damage etching and subsequent anisotropic etching of pyramids, or (iii) solely by anisotropic etching of pyramids with extended etching times.

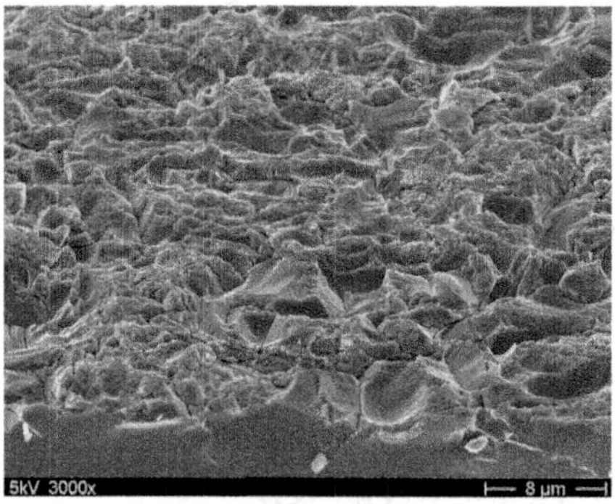

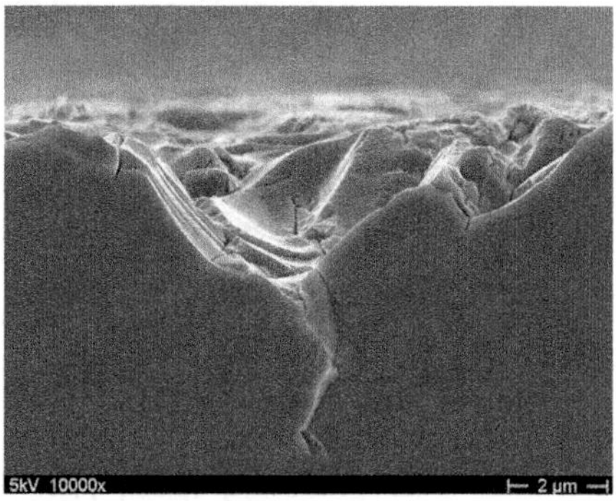

a) x 3000, cross section tildet by 30°   b) x 10000, cross section °

Fig. 3.8 SEM micrographs of as-cut Si wafer substrate surfaces.

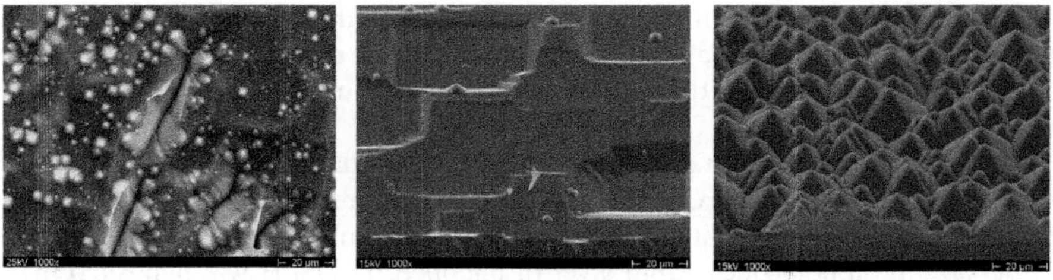

a) x 1000, top view

b) x 1000, cross section tildet by 30°

c) x 1000, cross section tildet by 30°

Fig. 3.9 SEM micrographs of Si solar cell substrate surfaces after inchoate saw damage etch (a), after completed saw damage etch (b) in KOH, and after formation of random distributed pyramids in KOH/IPA solution (c).

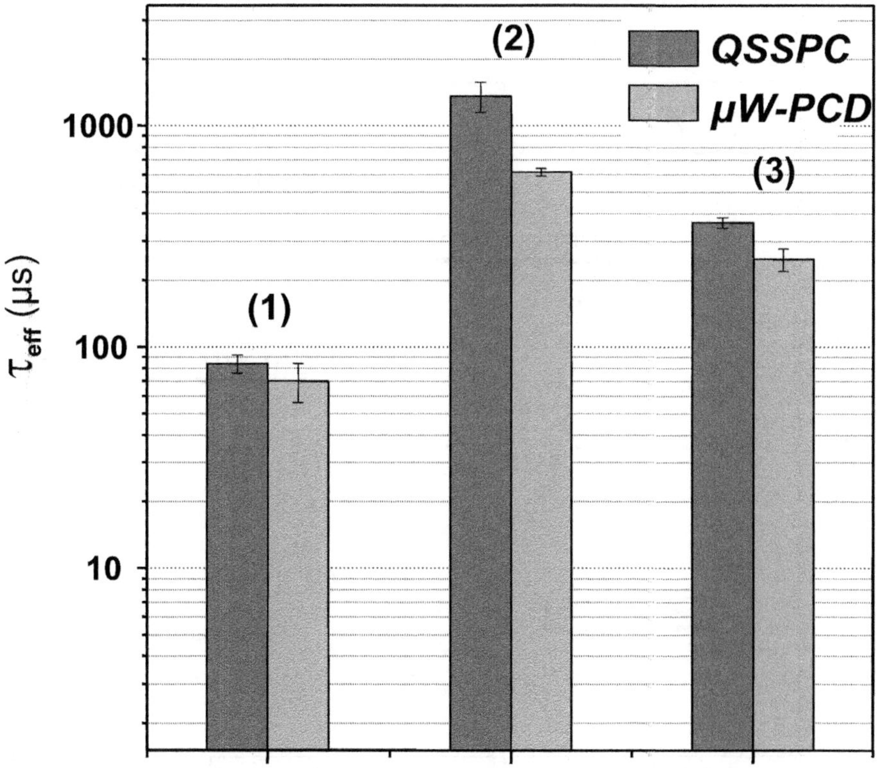

Fig. 3.10 Minority charge carrier life time measurements on n-Si (3 Ωcm) substrates which were passivated by an intrinsic a-Si:H layer after inchoate (1) and complete (2) saw damage etch, and after texturisation by random pyramids (3). Surface pre-treatment: RCA and HF 1% dip [51].

Fig. 3.9 presents SEM micrographs of Si solar cell substrates, after inchoate saw damage etch (Fig. 3.9a) and after completed saw damage etch (Fig. 3.9b) in KOH and after formation of random distributed pyramids in KOH/IPA solution (Fig. 3.9c).

In Fig. 3.10 the effect of saw damage removal on charge carrier life times (τ_{eff}) obtained by μW-PCD and QSSPC measurements is presented. The samples were saw damage etched and textured n-type wafers (3 Ωcm). After RCA cleaning and HF (1%) dip on both sides a thin layer of undoped (i) *a*-Si:H was deposited [51]. After inchoate saw damage etch the remaining cracks during the initial formation of pyramids (Fig. 3.9a) cause low values of τ_{eff} (Fig. 3.10, bars 1). The highest value of τ_{eff} (Fig. 3.10, bars 2) was obtained after complete removal of saw damage and formation of a macroscopically smooth surface as shown in Fig. 3.9b.

Generally, lower charge carrier life times (Fig. 3.10, bars 3) were observed on randomly distributed pyramids (see Fig. 3.9c) even though the saw damage was previously completely removed.

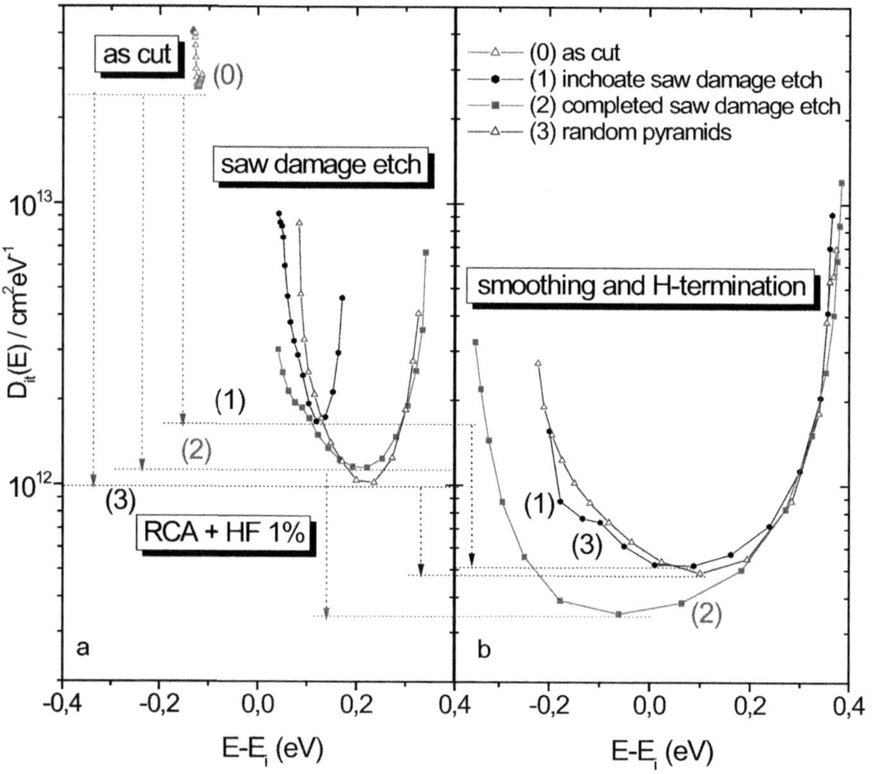

Fig. 3.11 $D_{it}(E)$ measured on n-type Si substrates after inchoate (1) and completed (2) saw damage etch and after texturisation by random pyramids (3), and after subsequent a) HF 1% and b) wet-chemical smoothing or H-termination, compared to the as-cut surface (0).

Fig. 3.11a summarises $D_{it}(E)$ curves measured after H-termination in 1% HF of differently etched n-type Si substrates after inchoate (curve 1) and completed (curve 2) saw damage etch, and after texturisation by random pyramids (curve 3). Fig. 3.11b shows the $D_{it}(E)$ curves of the respective surfaces after wet-chemical smoothing and H-termination. For comparison, the very high surface state density of $D_{it,min} > 10^{13}cm^{-2}eV^{-1}$ typically obtained on as-cut wafers is shown in Fig. 3.11a as curve (0). The complete etch-back of saw damage in KOH or KOH / IPA solutions and subsequent RCA- HF (1%) treatment lead to a strong decrease in $D_{it,min}$ of one order of magnitude (Fig. 3.11a, curves 2,3), due to removal of crystallographic defects in the near surface range of about 30 μm. Inchoate saw damage etch results in higher $D_{it,min}$, due to a remaining defect rich layer. In comparison with polished surfaces, on all types of un-polished surfaces the standard RCA+ HF (1%) treatment [13] results in $D_{it,min} > 10^{12}cm^{-2}eV^{-1}$.

However, by subsequent wet-chemical smoothing and H-termination (Fig. 3.11b) on all textured substrate surfaces a significant further decrease in $D_{it,min}$ down to $\leq 5 \cdot 10^{11}cm^{-2}eV^{-1}$ can be achieved, due to the removal of surface micro-roughness in the atomic scale as described in detail in section 3.5.4.

3.5.2 Lowering of Interface Recombination Losses on Saw Damage Etched Substrates

Saw damage etched substrate surfaces are commonly used for solar cell interfaces without high optical surface-finish requirements, for reasons of economy and / or in order to avoid an increase in defect densities by further surface texturisation. Recently Laades et. al. [22] reported that the wet-chemical surface conditioning has to be carefully optimised with respect to the different substrate configurations (doping type, surface texture, passivation film) by applying combined spatially resolved microwave detected photoconductance decay (μW-PCD) and SPV measurements.

Fig. 3.12 summarises the results of these investigations on saw damage etched p- and n-type CZ-Si solar cell substrates (see Fig. 3.9b) after different wet-chemical treatments and subsequent passivation by iodine/ethanol (I/E) or by *a*-SiN_x:H [22]. Although the passivation of *a*-SiN_x:H/*c*-Si interfaces is based on the field effect caused by a fixed positive charge in the *a*-SiN_x:H film, a strong influence of wet-chemical treatment, differing for p- and n-type substrates, was observed (Fig. 3.12, *a*-SiN_x:H).

The passivation in I/E is based on saturation of dangling bond defects, hence similar results are obtained on n-and p-type substrates (Fig. 3.12, I/E). The effect of various sequences of wet-chemical processes, RCA I and RCA II, HF (1%) and NH_4F etching, as well as DIW rinsing at 80°C and at RT on the charge carrier life times (τ_{eff}) was investigated in detail.

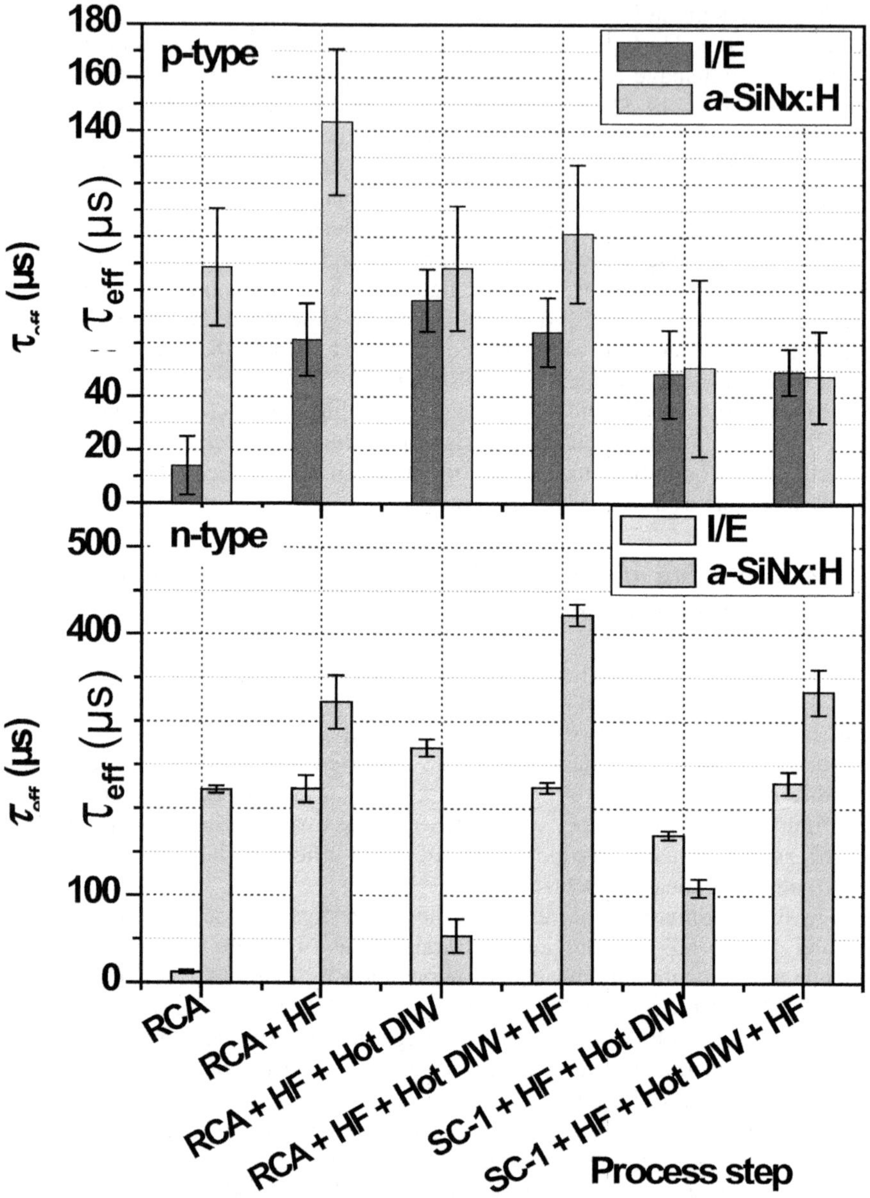

Fig. 3.12 Lifetime distribution over the area on damage etched p- and n-type CZ-Si solar cell substrates after different wet-chemical treatments and passivation with iodine/ethanol (I/E) or by a-SiN_x:H [22].

The removal of oxides remaining from the RCA process in 1% HF solution was shown to be an essential step to achieve high values of τ_{eff} on all kinds of substrates and passivation methods. Subsequent rinsing in hot DIW (80°C, 5 min) leads to a hydrophilic surface by moderate oxidation. This process significantly increases τ_{eff} on- n-type substrates as measured on the hydrophilic surface after I/E passivation and on the resulting hydrophobic surface after removing the oxide in HF 1% and passivation by a-SiN_x:H. As shown in Fig. 3.12, the omission of SC 2 process leads to comparably low values of τ_{eff} for all substrates due to the recombination losses caused by remaining contaminations, especially resulting from metal impurities existing in the less pure feedstock or by introduction during ingot and solar cell fabrication.

3.5.3 Surface Morphology, Optical Reflectance and $D_{it}(E)$ after Wet-Chemical Surface Texturisation

Various texturisation schemes applying acidic [52-54] as well as alkaline [10, 55] solutions were recently reported, which are utilised in high efficiency solar cells to enhance anti-reflection properties through multiple bounce incidence of light at the front surface, and the path length for absorption and internal reflection at the back surface. The influence of texturisation on optical reflection, surface microroughness and surface state densities will be exemplified here by correlated SEM, total hemispherical UV-NIR-reflectance, SPV and PL measurements after isotropic and anisotropic etching of polished and as-cut monocrystalline Si(100) and Si(111) surfaces.

Fig. 3.13 shows SEM micrographs of Si(100) substrates (a) as-cut and after various texturisation that were obtained by applying (b) isotropic texture etch (TE) [12], (c) a special acidic etchant (SAF) [11] and (d) anisotropic standard alkaline etchant KOH/IPA [9]. Classical wet-chemical processes applying strong acidic and alkaline etchants are still state-of-the-art to texturize Si solar cell substrates. The number and size of the random distributed pyramids, obtained by anisotropic etching in KOH/IPA (Fig. 3.13d), can be influenced by specific process modifications like changing the concentration of the alkaline etchant, temperature, and the presence of chemical additives. An isotropic TE process transferred from the wafer thinning technology leads to a strongly roughened surface of characteristic morphology (Fig. 3.13b) [12]. However, both kinds of processes have certain disadvantages. KOH solution requires a substantial amount of energy to keep the etch baths at the required temperature. Another issue is the limited solubility of the reaction product, potassium fluorosilicate (K_2SiF_6), which can be a limiting factor in bath lifetime and throughput [56].

The TE process is critical in preparation as well as in use, due to the corrosive components and heat development. The alternative SAF-etch process [56], although being of acidic nature shows a clearly anisotropic etching behaviour (Fig. 3.13c).

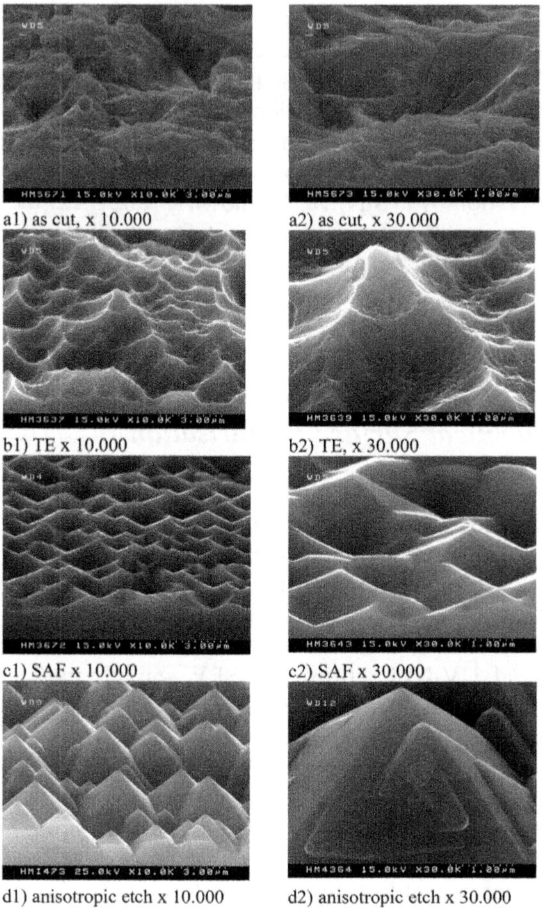

Fig. 3.13 SEM micrographs, (cross section tilted by 30°) of Si(100) substrates: (a) as-cut, and after various texturisation technologies, (b) texture etch (TE) technology [12], (c) a special acidic etchant (SAF) [56] and (d) anisotropic standard alkaline etchant KOH/IPA [9].

Reflectance spectra between 200 and 1200 nm wavelength are given in Fig. 3.14 as typically recorded by total hemispherical UV-NIR-reflectance measurements on the polished Si(100) surface (Fig. 3.14, curve 0) and on the as-cut surface (Fig. 3.14, curve 1), after saw damage etch in KOH solution (Fig. 3.14, curve 2), after isotropic TE process (Fig. 3.14, curve 3), after special acidic etch SAF (Fig. 3.14, curve 4) and after anisotropic etching in KOH/IPA (Fig. 3.14, curve 5), respectively. The reflectance of the rough as-cut wafer surface (Fig. 3.14, curve 1) increases during the etching processes, which were applied for removal of saw damage (see also Fig. 3.13a). For instance, processing in KOH

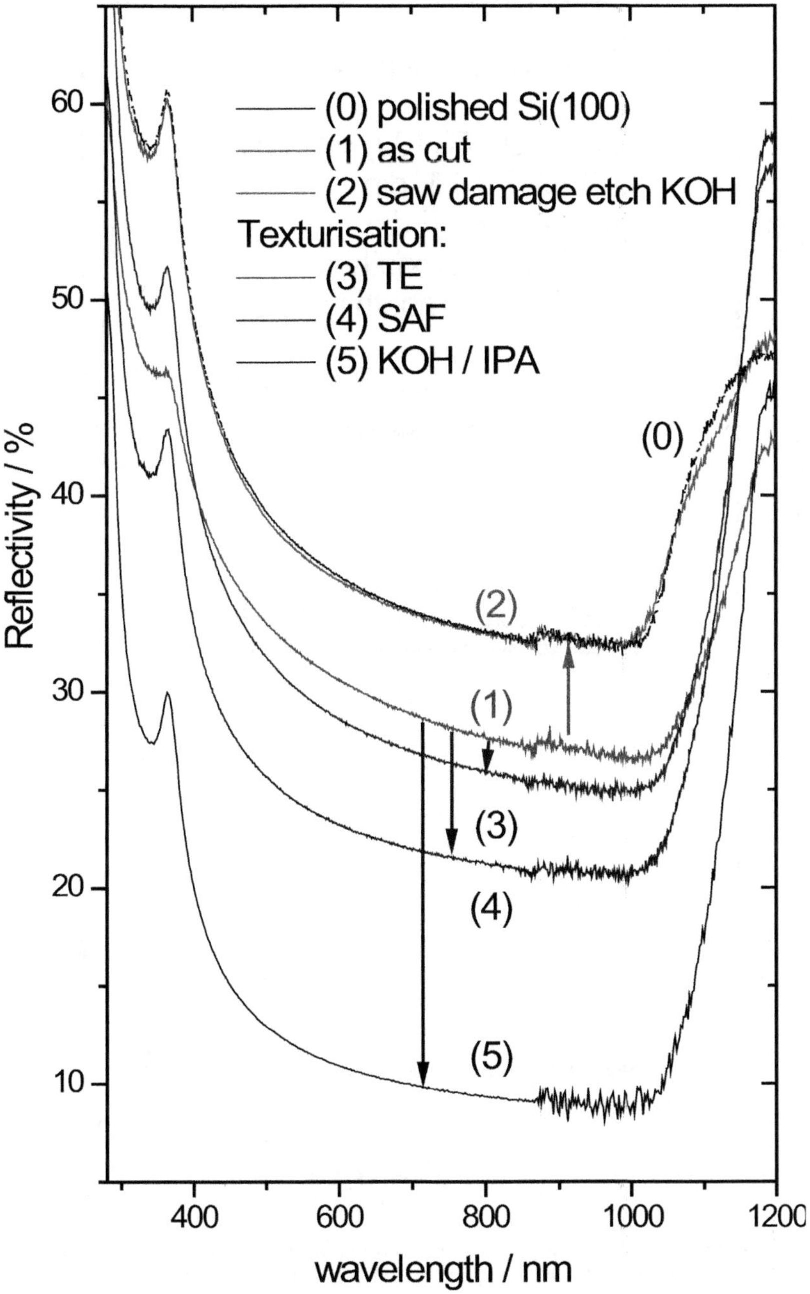

Fig. 3.14 Reflectance spectra recorded by total hemispherical UV-NIR-reflectance measurements on the polished Si(100) surface (0), and on the as-cut surface (1), after saw damage etch in KOH solution (2), after isotropic texture etching (TE) [12] (3), after special acidic etchant SAF [56] (4) and after anisotropic etch in KOH/IPA [9] (5).

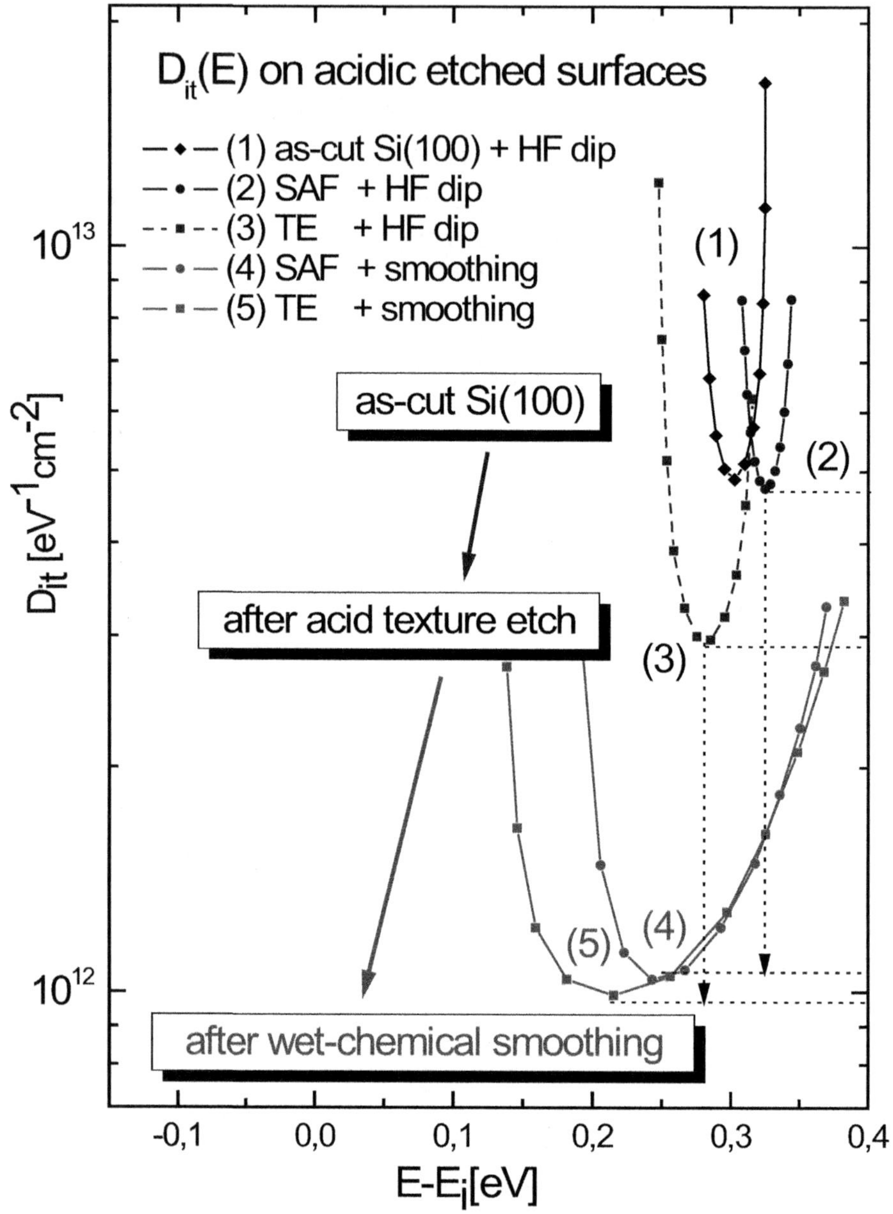

Fig. 3.15 $D_{it}(E)$ obtained on n-type monocrystalline Si(100) wafer surfaces as-cut (1), after texturisation by SAF (2, 4) and TE (3, 5). Two types of final surface treatment were applied: (i) HF 1% subsequent to RCA treatment (HF dip) (1, 2, 3,) and (ii) wet-chemical smoothing and H-termination (4, 5).

solution results in surfaces with characteristic flat areas (Fig. 3.13b) and reflectance spectra (Fig. 3.14, curve 2) similar to those of polished Si surfaces (Fig. 3.14, curve 0). Therefore for all applications, which need light trapping substrate structures, a simultaneous or subsequent texturisation is inevitable.

Applying acidic etchants, the removal of ~ 30-50 µm of Si reveals the characteristic roughened surface morphology of TE (Fig. 3.13b) and SAF etching (Fig. 3.13c), respectively. The comparison of the spectra of the SAF etched surfaces (Fig. 3.14, curve 3) and the TE etched surfaces (Fig. 3.14, curves 4) show a higher enhancement of absorption for the TE etched sample [56]. Alkaline etchants, e.g. KOH-IPA, provide highly textured surfaces on Si(100) substrates by anisotropic etching (Fig. 3.13d) and yield a significantly improved advancement of light trapping properties (Fig. 3.14, curve 5).

Fig. 3.15 shows $D_{it}(E)$ curves obtained by SPV measurements on n-type substrates of the non-etched as-cut Si(100) surface (Fig. 3.15, curve 1) and after texturisation in SAF (Fig. 3.15, curves 2,4) and TE (Fig. 3.15, curves 3,5). As final steps, two types of final surface treatment were applied: (i) HF 1% subsequent to standard RCA treatment (Fig 3.15, curves 1, 2, 3,) and (ii) wet-chemical smoothing and H-termination (Fig 3.15, curves 4, 5) [56].

The high micro-roughness on the as-cut and textured wafer (Fig. 3.13a2 and 3.13b2) leads to a high level of surface irregularities and high densities of surface states near the band edges and in the lower part of the gap, i.e. closer to the valence band edge, and results in significantly narrowed $D_{it}(E)$ distributions (Fig. 3.15, curves 1, 2 ,3). As demonstrated here on both SAF and TE etched surfaces, a significant decrease in $D_{it,min} \approx 1 \cdot 10^{12} cm^{-2} eV^{-1}$ was achieved (Fig. 3.15, curves 4, 5) by applying a wet-chemical smoothing and H-termination procedure using oxidation in H_2SO_4/H_2O_2 solution as specified in section 3.4.2.

Summarising these results, it was shown that the strongest decrease in surface reflectance can be achieved by anisotropic etching of pyramids in KOH/IPA solution (Fig. 3.14, curve 5). Moreover, surface texturisation by acidic solutions, shown here for TE and SAF, generally lead to higher light reflectance (Fig. 3.14, curves 3, 4) and interface state densities (Fig. 3.15, curves 4, 5), compared to the KOH/IPA etched substrates (Fig. 3.11b, curve 3). Therefore, for high efficient Si hetero structure solar cells, normally anisotropic etched substrates with random pyramids should be preferred [9].

3.5.4 Optimisation of Wet-Chemical Surface Pre-treatment for Substrates with Random Pyramids

So far we have studied the influence of wet-chemical pre-treatments on micro-roughness and electronic interface properties of polished and saw damage etched Si substrates. However, in many cases light trapping structures on the substrate surface are necessary for good cell performance. In this section, we focus on the investigation of the influence of the increased surface area and micro-roughness on the resulting density of interface states and recombination behaviour on textured surfaces.

Table 3.1 Optimization of wet-chemical smoothing processes for p-type Si substrates with randomly distributed pyramids.

Fig. 3.16b, curve	Cleaning	Oxide removal	Wet-chemical oxidation	Oxide removal
(2)	RCA	HF 1%		
(3)	RCA	NH_4F (48%)		
(4)	RCA	HF 1%	H_2SO_4/H_2O_2	HF 1%
(5)	RCA	HF 1%	H_2SO_4/H_2O_2	NH_4F (48%)
(6)	RCA	NH_4F (48%)	H_2SO_4/H_2O_2	HF 1%

The anisotropic etching in KOH/IPA changes the initial surface orientation of the Si(100) substrate to Si(111) orientated well-ordered pyramid facets (Fig. 3.13d). The complete removal of contaminations and the passivation of rechargeable states on these substrate interfaces are very important issues for the further improvement of the energy conversion efficiency of Si-based solar cells. After etching in alkaline solutions typically lower values of $D_{it}(E)$ are obtained after subsequent standard RCA and HF (1%, 60 s) treatment [13] compared to surface textures prepared by acidic solutions (Fig. 3.13b and c).

Also the anisotropic etching of pyramids leads to a strong increase in crystallographic surface irregularities as shown in Fig. 3.13d2. Therefore, according our recently reported results, on substrates textured with pyramids, the preparation-induced surface charge, surface state density and interface recombination loss of charge carriers can be additionally reduced by optimised wet-chemical pre-treatments. Special wet-chemical smoothing and H-termination processes were applied [57-59] to remove the damaged surface and to decrease the micro-roughness on the Si(111) facets.

In order to eliminate bulk effects the optimisation of surface pre-treatment will be exemplified here for random pyramids prepared on polished p-type FZ-Si samples with high bulk carrier lifetime [57]. Fig. 3.16 shows the changes in $D_{it}(E)$ on these Si surfaces after RCA cleaning and different sequences of wet-chemical oxidation and oxide removal: (a) during the formation of pyramids, (b) during wet-chemical smoothing (Table 3.1), and (c) during completed removal of wet-chemical oxides (Table 3.2).

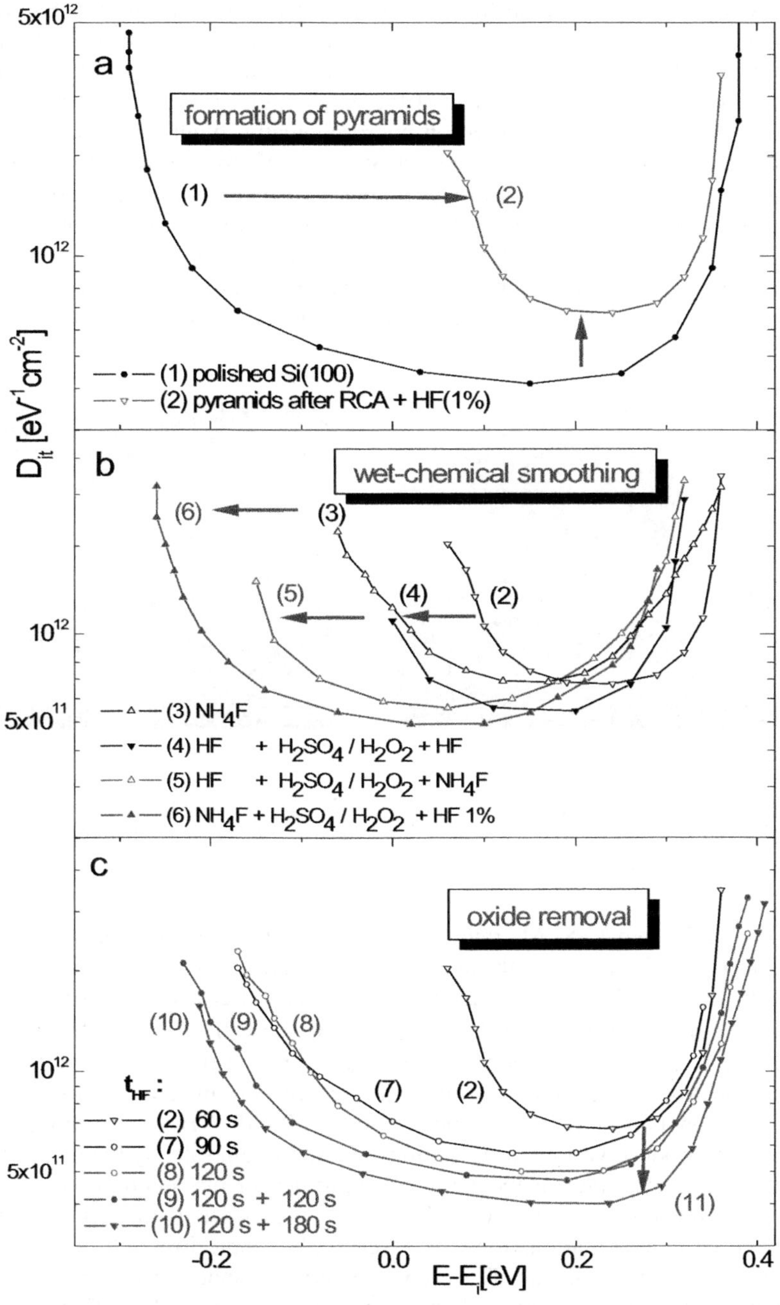

Fig. 3.16 Change in $D_{it}(E)$ on p-type Si surfaces after RCA cleaning and different sequences of wet-chemical oxidation and oxide removal: a) during the formation of pyramids, b) during wet-chemical smoothing (Table 3.1), and c) during completed removal of wet-chemical oxides (Table 3.2).

As shown in Fig. 3.16a the formation of pyramids on initially H-terminated polished Si(100) surfaces (curve 1) increases D_{it} and results in significantly narrowed energetic distributions (curve 2) by additional appearance of states near the band edges and in the lower half of the gap, i.e. closer to the valence band edge. These states, typically observed on microscopically rough surfaces, are related to strained bond defects and defects on Si atoms of lower stage of oxidation, Si^{1+} ($Si_2O{\equiv}Si-$) e.g. formed by hydroxyl groups [20]. As demonstrated in Fig. 3.16b the defect density on these surface states typically obtained after standard RCA + HF dip pre-treatments (Fig. 3.16b, curve 2), can be stepwise reduced by wet-chemical smoothing applying sequences of oxidation in H_2SO_4/H_2O_2 and oxide removal in HF (1%) (Fig 3.16, curves 4 to 10) or in NH_4F (48 %, 4.5 min) (Fig. 3.16b, curves 3,5,6) as given in table 3.1 [58].

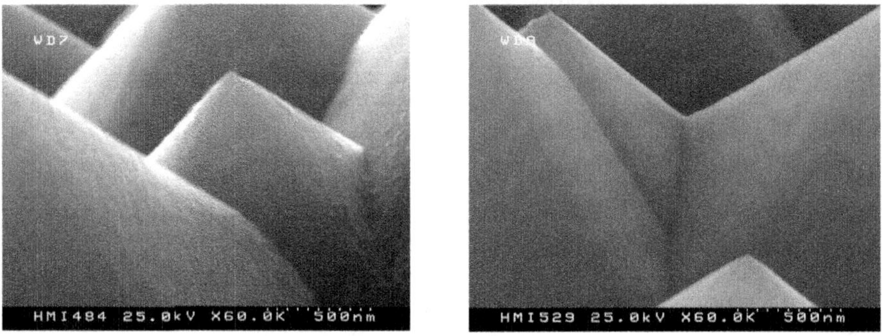

a b

Fig. 3.17 SEM micrographs (60000x, tilted by 30°) of Si substrates with randomly distributed pyramids. [58] (a) after H-termination by NH_4F (48 %), and (b) after subsequent wet-chemical oxidation in H_2SO_4/H_2O_2 + HF dip (120 s).

Fig. 3.17 shows SEM micrographs (magnification 60000, tilted by 30°) obtained on Si wafer with randomly distributed pyramids after standard H-termination by NH_4F solution [3] (Fig. 3.17a) and after additional subsequent wet-chemical oxidation in H_2SO_4/H_2O_2 and HF dip for 120 s (Fig. 3.17b) [58]. Various wet-chemical treatments, using NH_4F containing solutions, have been developed in microelectronics technology in order to prepare H-terminated, atomically smooth surfaces on polished Si(111) substrates [3, 60-63]. NH_4F containing solutions, however, produce various compositions of ammonium salts by the reaction with Si and SiO_2. Depending on the level of impurities in the solutions, these reaction products form deposits on the surface. This process seems to start particularly on surface particle contaminations, or on crystallographic irregularities of the structured surfaces (Fig. 3.17a). These contaminations cannot be removed even by intensive water rinse, because of the poor solubility of $(NH_4)_2SiF_6$ in water [64]. This behaviour leads to strong increase in D_{it} on the initially rough RCA treated surface (Fig. 3.16b, curve 3). The application of NH_4F solution after previous smoothing of the Si/SiO_2 interface by oxidation in H_2SO_4/H_2O_2 leads to a reduction in D_{it}, mainly in the lower half of the gap (Fig. 3.16b, curve 5).

Best results were obtained by smoothing in NH_4F solution with a subsequent wet-chemical oxidation step in H_2SO_4/H_2O_2, followed by an HF dip (Fig. 3.16b, curve 6), which obviously dissolves these contaminations in H_2SO_4/H_2O_2 as shown in Fig. 3.17b.

The density of states in the upper half of the gap, i.e. closer to the conductivity band edge, results from dangling bonds of the next higher stage of oxidation, Si^{2+} (SiO_2≡Si–) and were found to be strongly related to the stage of Si surface oxidation [20]. In order to further elucidate the influence of wet-chemical oxide removal, the HF treatment times were systematically stepped up after RCA as well as after H_2SO_4/H_2O_2 (Table 3.2). The extension of the HF treatment time after the RCA process from 60 s (Fig. 3.16c, curve 2), to 90 s (Fig. 3.16c, curve 7) and to 120 s (Fig. 3.16c, curve 8) leads to a stepwise reduction in D_{it} and broadening of $D_{it}(E)$. Additional H_2SO_4/H_2O_2 oxidation and HF etching steps (120 or 180 s) (Fig. 3.16c, curves 9 and 10) lead to a further reduction in D_{it} and broadening of $D_{it}(E)$ curves at the upper half of the gap [2].

Table 3.2 Optimization of HF 1% treatment times for p-type Si substrates with randomly distributed pyramids

Fig. 3.16c curve	Oxide removal after RCA process in HF (1%)	Oxide removal after H_2SO_4/H_2O_2 in HF (1%)
(2)	60 s	
(7)	90.s	
(8)	120 s	
(9)	120 s	60 s
(10)	120 s	120 s
(11)	120 s	180 s

The influence of both wet-chemical smoothing and oxide removal on interface recombination losses was investigated by ex-situ PL measurements. Fig. 3.18 shows PL intensities (I_{PL}) on solar cell substrates with Si(111) pyramids. As described in section 3.2.2 the quenching or enhancement of I_{PL} is due to defect formation (non-radiating recombination) or defect passivation.

Very low I_{PL}, due to high recombination losses, was obtained after applying NH_4F solution subsequent to the RCA process (Fig. 3.18, curve 1) compared to the HF dip 90 s (Fig. 3.18, curve 2). A further increase in I_{PL} after wet-chemical smoothing (Fig. 3.18, curve 3) and HF dip 120 s (Fig. 3.18, curve 4) indicates that the wet-chemical smoothing and complete oxide removal effectively enhance the surface passivation.

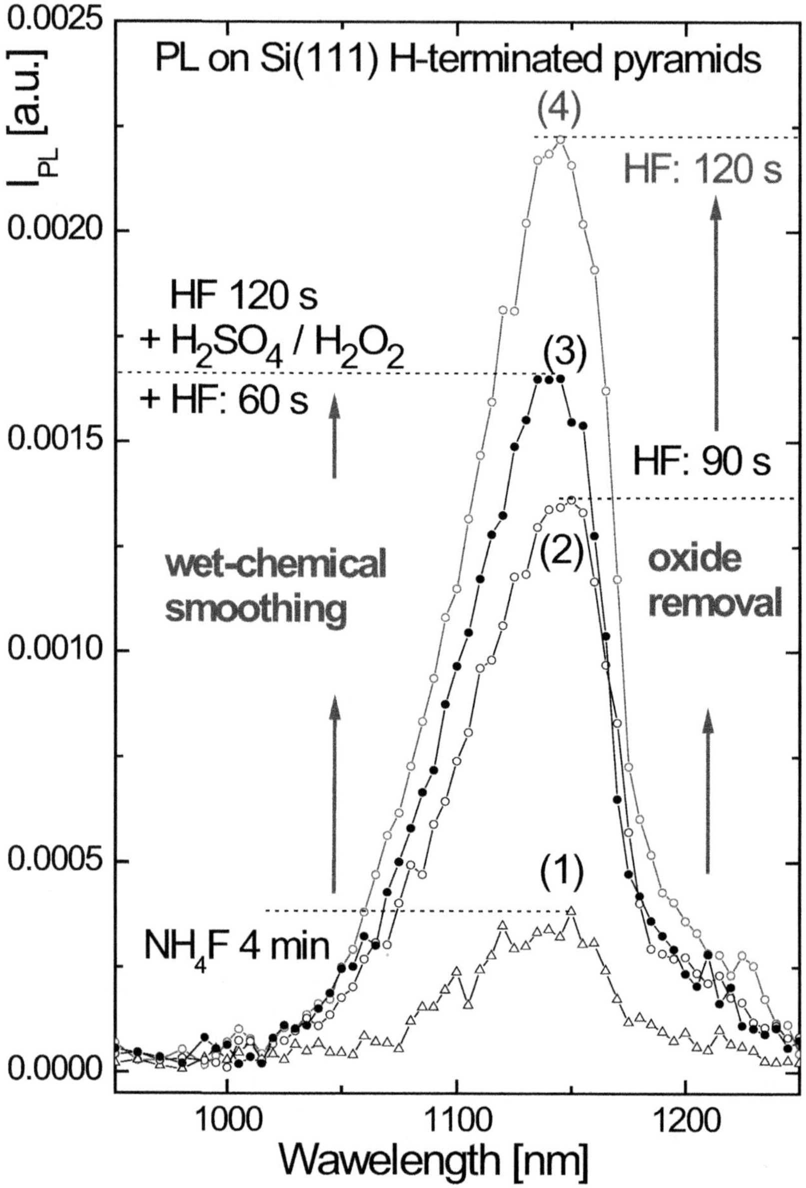

Fig. 3.18 PL intensities measured ex-situ on p-type Si substrates with randomly distributed upside pyramids: after RCA process and subsequent H-termination by NH_4F (1), after RCA process and subsequent H-termination by HF dip 90 s (2), after wet-chemical smoothing and final HF dip 60 s (3) after wet-chemical smoothing and final HF dip 120 s (4).

Additionally, the changes in the recombination behaviour during the etch-back of wet-chemical oxides in the HF (1%) were monitored by in-situ PL measurements as shown in Fig. 3.7 [49] in order to find the optimal HF treatment time. In this way, energetic distributions $D_{it}(E)$ on the smoothed pyramids (Fig. 3.16, curve 10) were achieved which are comparable to that on the initial flat Si(100) surface (Fig. 3.16, curve 1).

3.6 Stability of Si Surface Passivation during Storage in Ambient Atmosphere

The stability of surface passivation on wet-chemically treated wafers over time up to the subsequent layer preparation step is an important parameter in the technological cell fabrication process. The H-terminated Si surface strongly limits oxidation in clean room air but does not completely inhibit native oxidation. The stability of hydrogen passivation was shown to be influenced mainly by substrate morphology and the wet-chemical preparation method used [65-67].

3.6.1 Native Oxidation of H-Terminated Substrates

A strong influence of various etching and smoothing procedures on the stability of the H-terminated vicinal Si(111) surfaces under ambient atmosphere was recently shown by M. Kolibal et al. by X-ray photoelectron spectroscopy and AFM measurements [67]. Best results (i.e. stability over 3 h) were obtained after wet-chemical smoothing by buffered HF etching and NH_4F (72°C).

Relations between surface micro-roughness, native oxide growth and the electronic properties during the initial phase of oxidation on H-terminated polished mono-crystalline and μc-Si:H surfaces were established by SPV, spectroscopic ellipsometry measurements in the ultraviolet and visible (UV-VIS) region and Fourier-transform infrared ellipsometry (FTIR-SE) [25]. A continuous reduction in the number of $Si_3 \equiv Si{-}H$ and $Si_2{=}Si\,({-}H)_2$ bonds was observed during the formation of a first oxide monolayer on Si(111) and Si(100), respectively, accompanied by the native oxide growth [26]. These Si-H vibrations completely disappear when the effective thickness $<d_{ox}>$ of the native oxide film achieves one monolayer. D_{it} of the H-terminated Si surfaces as well as the long-time stability of the H-termination were found to primarily depend on the preparation-induced surface morphology.

Fig. 3.19 shows the increase in D_{it} on initially H-terminated Si(100) and poly-crystalline EFG Si substrates observed during storage in clean-room air [57]. On Si(100) immediately after preparation typical values of $D_{it,min}$ of about $1{\cdot}10^{11}cm^{-2}eV^{-1}$ were obtained (Fig. 3.19, curve 1). A strong increase in defect density was observed (Fig. 3.19, curve 2) during storage in clean-room air. After 180 min. storage time in air when, according to the UV-VIS SE data, the effective oxide coverage exceeds one monolayer the native oxide growth in air causes a high $D_{it,min}$ of about $2{\cdot}10^{12}\,cm^{-2}eV^{-1}$ (Fig. 3.19, curve 3). This value is too high to be successfully used for further preparation of thin hetero-structures cells.

On EFG substrates, with higher value of $D_{it,min}$ of approx. $3{\cdot}10^{11}cm^{-2}eV^{-1}$ on the initially H-terminated surface (Fig. 3.19, curve 4) also higher values of $D_{it,min}$ (Fig. 3.19, curve 5, 6) of about $5{\cdot}10^{12}cm^{-2}eV^{-1}$ were obtained after storage on air. On all initially H-terminated Si surfaces, a considerable increase in $D_{it,min}$ was observed during the initial phase of oxidation that leads to the growth of the first oxide mono-layer.

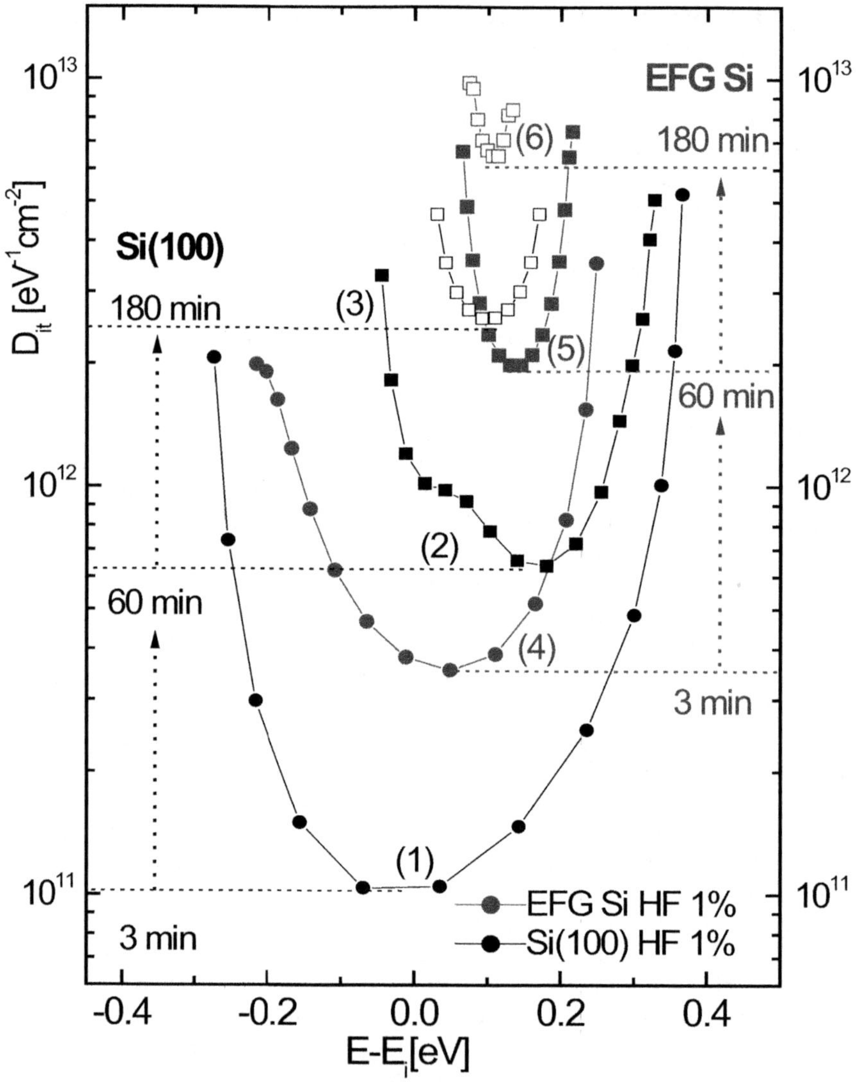

Fig. 3.19 Change in $D_{it}(E)$ on H-terminated polished Si(100) (1, 2, 3) and EFG-Si substrates (4, 5, 6) during storage in clean-room air (exposure times: 3, 60 and 180 minutes) [57]. [with permission of Elsevier].

Fig. 3.20 shows the relation between the initial phase of oxidation (t_{ini}) in clean-room air, the initial $D_{it,min}$, and the surface micro-roughness on H-terminated Si(111) prepared by hot water oxidation + NH_4F (<d_r> 1 Å), RCA + NH_4F (<d_r> 3 Å), and HF 48 % treatments (<d_r> 4.5, 8, 12 Å). The duration of the initial phase of oxidation (t_{ini}) was determined by UV-VIS SE measurements as the exposure time in clean-room air until the oxide-thickness on the initially H-terminated surface reached one monolayer [14]. The time t_{ini} was found to be inversely proportional to the initial $D_{it,min}$ in the semi logarithmic plot. Highest stability was found on the atomically smooth monocrystalline Si(111) surface ($t_{ini} \cong 48$ h). On microcrystalline surfaces, t_{ini} was found to 3.5 h, i.e. significantly shorter [25].

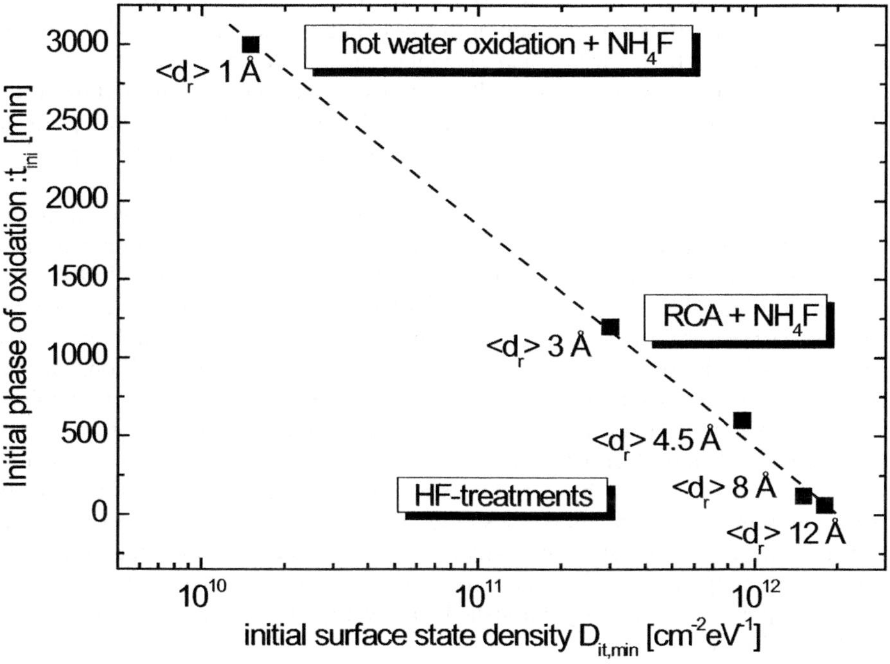

Fig. 3.20 Duration of the initial phase of oxidation, t_{ini}, in clean-room air as a function of the initial $D_{it,min}$ on H-terminated Si(111) prepared by (i) by hot water oxidation + NH_4F (<d_r> 1 Å), (ii) by RCA + NH_4F (<d_r> 3 Å), (iii) by HF 48 % treatments (<d_r> 4,5 … 12 Å). The time t_{ini} was obtained by UV-VIS SE measurements of the oxide-thickness reaching one monolayer [14] [with permission of Elsevier].

Similar findings on Si(100) substrates polished and textured with random pyramids were recently reported by Zhao et al. [68] obtained by monitoring the time decay of the effective minority carrier lifetime via the microwave photoconductive decay (μPCD) method during etch-back of RCA oxides in HF 1% solution and subsequent storage in air. Thereby a completed removal of RCA oxides was achieved after 3 min treatment time on polished p-type Si (100) substrates, compared to 6 min on substrates textured with random pyramids for application in the heterojunction with intrinsic thin-layer solar cell (HIT). A rapid decrease in the effective minority carrier life time was observed during the initial phase of oxidation in air, corresponding to an increase in surface states. A significantly shorter duration of the initial phase of oxidation in air was found on the polished Si(100) surfaces in contrast to that observed on textured surfaces [68].

These experimental results indicate a strong influence of surface morphology on native oxidation in air. The initial oxidation reactions start on atomic defects and steps because of the higher reactivity of the polarised Si back-bonds in $Si(-H)_2$ and Si–OH surface groups, where oxidants, such as oxygen and water in air, are able to break these Si back-bonds on the rough surface regions rather than the Si≡Si–H bonds on the atomically flat Si(111) terraces. The insertion of oxygen, results in a strong polarization of Si≡Si–O back-bonds, which leads to further nucleophilic reactions with their neighbouring Si atoms forming the first monolayer of native oxide. When all Si atoms of the top layer are oxidized, the $Si_{bulk}{\equiv}Si_{bulk}{-}O$ back-bonds of the second monolayer are broken by subsequent insertion of oxygen and a layer-by-layer growth of native oxide films takes place on Si surfaces.

To inhibit native oxide growth, H-terminated substrates are often stored in dry nitrogen atmosphere. However, a very slow initial native oxidation was also observed in N_2 atmosphere, probably due to adsorbed water molecules [69].

Summarising these results, the duration of initial phase of oxidation, which can be used as measure for the stability of H-terminated Si surfaces in ambient atmosphere, was found to be not only influenced by the clean room conditions and the handling procedures, but also by the substrate bulk properties, surface morphology and the preparation-induced surface micro-roughness und surface coverage.

The re-oxidation of H-terminated Si surfaces can be slowed down for a short time (30 min … 2 days) by excellent smoothing and complete saturation of the dangling bonds by hydrogen or by storage in dry N_2 atmosphere until subsequent layer deposition. Other surface passivation methods, however, are required to inhibit completely the native oxidation of substrate surface for longer time duration. An example is the recently published passivation by organically modified Si surfaces [70, 71].

3.6.2 Surface Passivation by Wet-Chemically Prepared Oxide Layers

As is well known from semiconductor device manufacturing, thermally prepared SiO_2 layers are characterised by high chemical and electrical stability and provide excellent surface passivation. According our previously reported results, also special wet-chemically formed SiO_2 layers do not degrade the surface morphology and electronic interface properties during storage in clean room air for a long time [57]. The preparation of these passivation layers requires the formation well ordered wet-chemical Si/SiO_2 interface on an atomic scale starting from the undamaged, H-terminated Si surface. The influence of growth and removal of thin wet-chemical oxide layers on the Si surface morphology is exemplified by AFM images (2μm x 2μm) of an initially H-terminated Si(111) surface [57].

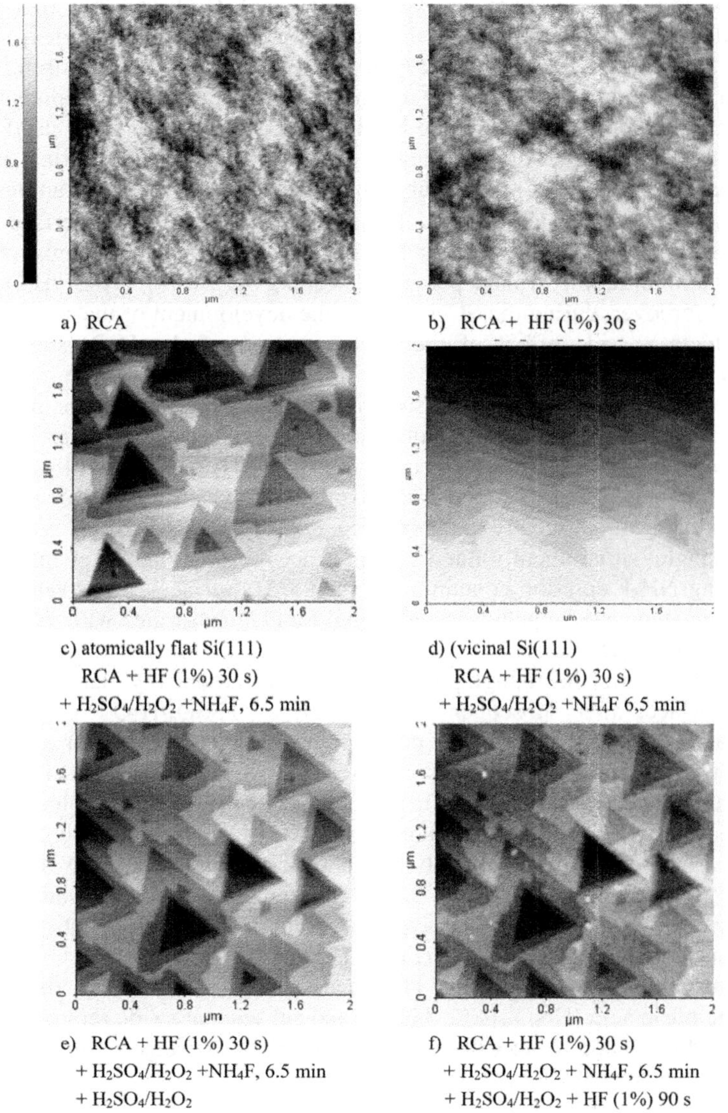

a) RCA

b) RCA + HF (1%) 30 s

c) atomically flat Si(111)
RCA + HF (1%) 30 s)
+ H_2SO_4/H_2O_2 +NH_4F, 6.5 min

d) (vicinal Si(111)
RCA + HF (1%) 30 s)
+ H_2SO_4/H_2O_2 +NH_4F 6,5 min

e) RCA + HF (1%) 30 s)
+ H_2SO_4/H_2O_2 +NH_4F, 6.5 min
+ H_2SO_4/H_2O_2

f) RCA + HF (1%) 30 s)
+ H_2SO_4/H_2O_2 + NH_4F, 6.5 min
+ H_2SO_4/H_2O_2 + HF (1%) 90 s

Fig. 3.21 AFM images (2μm x 2μm) of an initially H-terminated Si(111) surface (a) after RCA treatment, b) after subsequent HF dip (1%, 30 s), c) after subsequent wet-chemical smoothing (H_2SO_4/H_2O_2 + NH_4F (6.5 min) of atomically flat and d) vicinal Si(111), e) after subsequent wet-chemical oxidation in H_2SO_4/H_2O_2 and long time storage in air, and f) after final removal of the wet-chemical passivation oxide.

Fig. 3.21 shows AFM images of the topography (2μm x 2μm) of an initially H-terminated polished Si(111) surface (a) after RCA treatment, (b) after subsequent HF dip (1%, 30 s) and (c) after subsequent wet-chemical smoothing applying H_2SO_4/H_2O_2 + NH_4F (48%, 6.5 min) of atomically flat or (d) vicinal Si(111), (e) after subsequent wet-chemical oxidation in H_2SO_4/H_2O_2 and long time storage in air and (f) after final removing of the wet-chemical passivation oxide. As shown in Fig. 3.21a the RCA process, using hydrogen peroxide (H_2O_2) as a strong oxidising agent, results in microscopically rough interfaces covered with unintentionally grown and often contaminated oxide layers ($\langle d_{ox} \rangle$ 10 .. 15 Å). H_2O_2 itself is very stable, however, the stability was found to be very sensitive to certain metallic contaminations (Cu and Fe) in the sub-ppb range and to non-metallic anionic components (Cl^-) of the solution itself [72]. The increase in the interface micro-roughness during the RCA process has been demonstrated to mainly originate from the ammonium-hydrogen peroxide cleaning step SC-1 ($NH_4F/H_2O_2/H_2O_2$) [73, 74]. Moreover, during the SC-1 process the development of the so-called light point defects as side effect of the decomposition of the H_2O_2 solution was observed [72].

As shown in Fig. 3.21b, after subsequent HF dip, the Si surfaces is microscopically rough due to the initially non-uniformly oxidised surface (see Fig. 3.21a). By subsequent wet-chemical smoothing applying H_2SO_4/H_2O_2 + NH_4F (48%, 6.5 min) an H-terminated Si(111) surface can be prepared which is characterised by a strong geometry of atomically flat areas (Fig. 3.21c).

The triangles of atomically flat Si(111) terraces result from the etching process by applying NH_4F solution containing oxygen [75]. Surfaces with evenly distributed atomic steps can be achieved on vicinal Si(111) substrates with various miscuts as shown in Fig. 3.21d.

Fig. 3.21e shows the topography of an initially atomically flat H-terminated Si(111) surface, after subsequent re-oxidation by H_2SO_4/H_2O_2 solution under clean-room conditions and storage under ambient air conditions for a couple of months. The morphology of the wet-chemically oxidised interface was fond to be stable during a few months. The unchanged surface morphology of the H-terminated surface can be re-established by removal of the native oxide layer and the particle contamination, applying a short HF dip (1 % HF, 90 s) instead of NH_4F treatment as presented in Fig. 3.21f. This smoothing and passivation method can also be successfully utilised to avoid ammonium salts contaminations particularly on structured Si(111) surfaces (Fig. 3.17a and b).

Recently, atomically flat Si(111) surfaces were achieved by oxidation in azeotropic nitric acid (HNO_3 68%, 121°C) and subsequent oxide removal in NH_4F solution, which leads to an extreme reduction of the leakage current density of ultrathin oxide layers [15].

3.7 Influence of Optimised Wet-Chemical Treatment on Energy Conversion Efficiency of Test Solar Cells

In amorphous/crystalline heterojunction solar cells, the interface between *c*-Si and the *a*-Si:H layer is used to form the p/n junction, so that the substrate surface directly becomes part of the electronic interface, whose recombination activity determines the cell performance. As recently shown, the solar cell performance degrades significantly if the area density of D_{it} exceeds 10^{11} cm^{-2} [76]. The final aim of the final wet-chemical pre-treatment after texturisation is the removal of damaged regions of the Si surface as well as the saturation of dangling bonds at the surface and in the near-surface region by H-termination.

3.7.1 Amorphous Silicon Carbide (a-SiC:H)/c-Si Heterojunction Solar Cells

The influence of preparation induced surface micro-roughness on the performance of amorphous silicon carbide (*a*-SiC:H)/*c*-Si heterojunction solar cells for different wet-chemical pre-treatments was recently shown by Becker et al. [77]. Wet-chemical oxidation in hot-water and HNO_3 (69%, 110° C) as described in section 3.4.1 and subsequent oxide removal in NH_4F solution were successfully applied to enhance the open-circuit voltage (V_{oc}: 636.3 mV) as well as the pseudo fill factor (FF: 75.4 %) and efficiency (η: 18.5 %) of *a*-SiC:H/*c*-Si heterojunction solar cells prepared on p-type substrates with random pyramids.

3.7.2 Amorphous Silicon a-Si:H/c-Si Heterojunction Solar Cells

The influence of optimised surface pre-treatments, as described in section 3.5.4, was investigated for amorphous/crystalline heterojunction solar cells which were prepared on substrates with pyramidal light trapping structures. To verify the effect of wet-chemical pre-treatment on solar cells performance, standard RCA and HF treatments as well as smoothing and H-termination procedures, optimised for p-and n-type textured substrates were applied prior to the *a*-Si:H emitter deposition. The results of these investigation obtained on (n)*a*-Si:H/(p)*c*-Si [50, 59] as well as on (p)*a*-Si:H/(n)*c*-Si type [78] heterojunction solar cells were recently reported.

On p-type substrates with random pyramids (ZnO/*a*-Si:H/*c*-Si/BSF/Al) heterojunction solar cells were prepared after differed substrate pre-treatments [59]. For each of two groups of samples, (i) pre-treated by RCA + HF dip and (ii) subsequent wet-chemical smoothing by H_2SO_4/H_2O_2 + HF (1%) treatment for optimised etching time, 24 cells of 1 cm^2 were processed.

The cells were characterised by current-voltage-measurements under standard AM 1.5 illumination. From these curves, cell parameters and the efficiency for the conversion of light into electrical power were extracted.

Fig. 3.22 shows a histogram depicting the distribution of the solar cell efficiencies for both the RCA cleaned and the smoothened textured c-Si substrate surface. In addition to the histogram, the result of fitting normal (Gaussian) distributions to the data is shown. These curves should be regarded mainly as a guide to the eye, due to the limited number of samples. A pronounced increase in efficiency was achieved by the optimised wet-chemical smoothing and H-termination of the p-type substrate. The mean value increases from 16.5 to 17.8 %, this is a relative increase by 8 %. The best efficiency of this series (η =18.4 %) was also obtained for a cell with a smoothened *a*-Si:H/*c*-Si interface. The increase in efficiency is mainly due to a higher I_{sc} (mean improve 5.0 %) and to a less pronounced extent, an increased fill factor (2.2%), while the change in the open circuit voltage V_{oc} is rather small (0.5%). Previous simulation studies [79], however, have shown that changes in the density of interface states should influence primarily the open circuit voltage. Additional investigations are under way to clarify the reason for this behaviour.

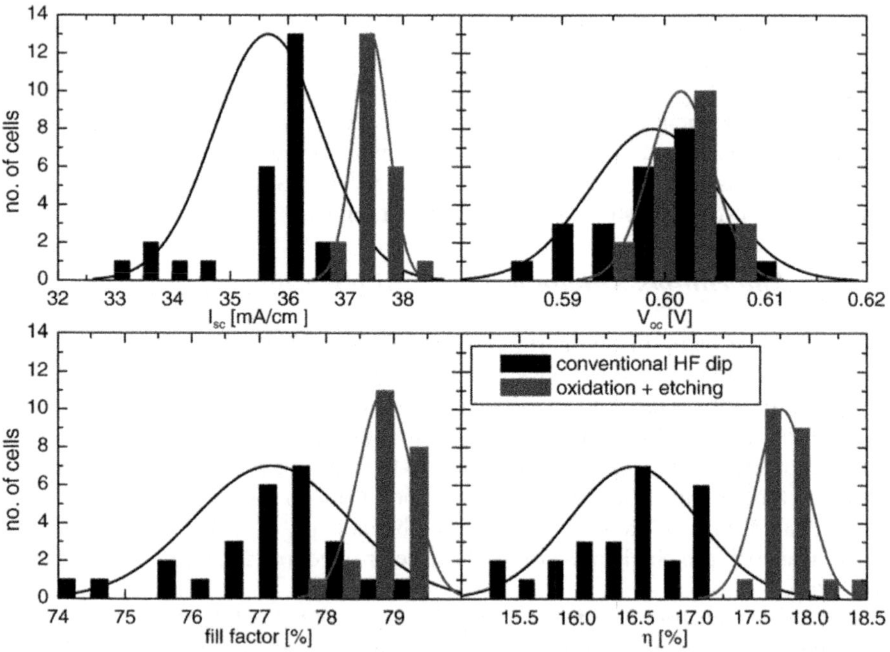

Fig. 3.22 Histogram of the basic solar cell parameters short cut current I_{sc}, open circuit voltage V_{oc}, fill factor and efficiency η, for two groups of 1 cm^2 (n)*a*-Si:H/(p)*c*-Si solar cells [59] [with permission of Elsevier]. black bars: conventional HF treatment after cleaning of the c-Si substrate; red bars: additional smoothening of the textured substrate surface; lines: normal distributions fitted to the data.

3.7.3 a-Si:H/c-Si Heterojunction Solar Cells with (i)a-Si:H Buffer Layers

Excellent passivation of the interface forming the heterojunction can also be reached by the inclusion of (i)*a*-Si:H buffer layers [80, 81]. Fig. 3.23 demonstrates the effects of optimised wet-chemical surface pre-treatment (Fig. 3.23 a) and additionally insertion of a 10 nm intrinsic *a*-Si:H buffer layer (Fig. 3.23 b) incorporated at the emitter and the back surface field (BSF) on the fill factor (FF) and open circuit voltage (V_{oc}) exemplarily for (n)a-Si:H/(p)c-Si type heterojunction solar cells [50]. The additional application of an intrinsic buffer layer results an improvement of the V_{oc} by ~50 mV, leading to efficiencies enhanced by ~3% (Fig. 3.23 c) for these ZnO/(n,i)*a*-Si:H/(p)c-Si/(i,p+)*a*-Si:H/Al test solar cells, as compared to cells with non-optimised substrate treatment and without (i)*a*-Si:H buffer layer. Similar results were also obtained on (ZnO/*a*-Si:H(p[,i])/c-Si(n)/*a*-Si:H ([i,]n])/Al) test solar cells prepared on textured n-type substrates [78]. The histograms over cells demonstrate that for all investigated substrates, the samples with the smoothened *a*-Si:H/*c*-Si interface show much narrower distributions of the respective parameters around the mean value.

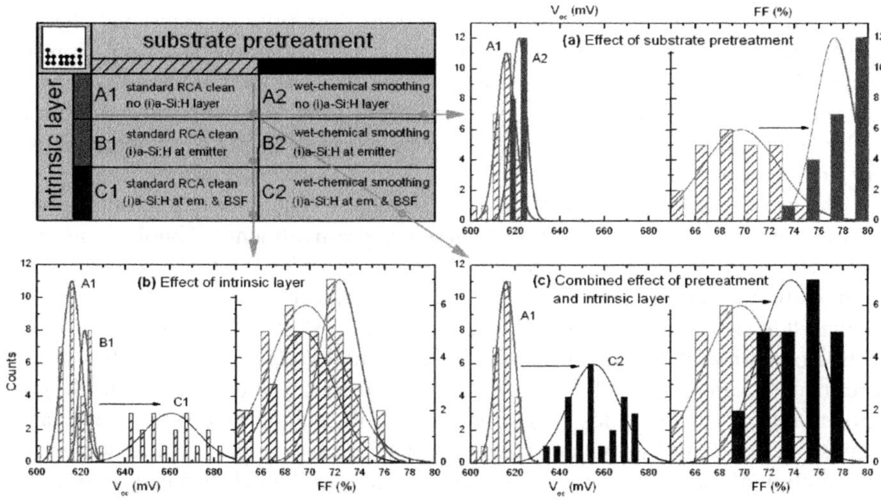

Fig. 3.23 Top left: Matrix depicting the investigated combinations of substrate pretreatments (left to right) and insertion of (i)*a*-Si:H buffer layers at the *a*-Si:H/*c*-Si interface (top to bottom) [50] [with permission of Elsevier]. Each group consists of 10 to 20 cells processed under identical conditions. a), b), and c) are histograms of V_{oc} and the FF obtained from the *I-V* curves under AM 1.5 illumination. a) a comparison of cells without intrinsic layers where the pre-treatment was changed; b) the effect of an (i)*a*-Si:H layer at the front and back *a*-Si:H/*c*-Si interface, respectively, c) the combination of both optimised pre-treatment and (i)*a*-Si:H buffer layers on both cell interfaces.

This behaviour indicates a better reproducibility of the solar cell processing when textured substrates are processed by an additional smoothing step and optimised wet-chemical oxide removal.

3.8 Summary and Outlook

In this chapter, it was shown that the optimised Si substrate surface state after wet-chemical conditioning can be preserved and transferred into amorphous silicon (*a*-Si:H)/*c*-Si, silicon nitride (*a*-SiN_x:H)/*c*-Si and silicon carbide (*a*-SiC:H)/*c*-Si interfaces, typically applied in silicon heterostructure solar cells. The resulting interface recombination loss on the hetero-junctions prepared by soft *a*-Si:H deposition were found to be significantly reduced applying wet-chemical smoothing procedures followed by thorough oxide removal with optimised HF (1%) treatment times.

It is important to apply non-aggressive oxidising solutions under controlled conditions to avoid an increase in the interface roughness during the oxidation process. To this end, smoothing procedures for polished and textured Si(111) substrates were developed applying sequences of NH_4F treatment, wet-chemical oxidation in H_2SO_4/H_2O_2 and oxide removal by HF dip, to remove damaged surface layers and avoid ammonium salt contaminations. Carefully prepared wet-chemical oxides can also be used as a starting point for the subsequent H-termination by a short HF dip even after long time storage in air, because their electronic properties do not degrade and the morphology of atomically flat Si(111) terraces on the initially H-terminated surface can be re-established. Best results were obtained using the hot water pre-treatment.

The wet-chemical H-termination subsequent to a hot water oxidation can be integrated easily in the common wet-chemical pre-treatment technology as batch process. The H-terminated surface itself is not completely stable with respect to re-oxidation in clean-room air. Therefore, the time between the final oxide removal step and the subsequent layer must be as short as possible. To prevent native oxidation, it is helpful to achieve low initial surface state densities and the pre-treated wafer should be additionally stored in dry nitrogen. Controlling the growth rate of native oxides on silicon surfaces in each of the process steps has recently received increasing attention, because it is of great importance for the preparation of thin silicon hetero-structures with excellent electronic properties.

It was shown that combined SPV-, PL and SE measurements can be use as a very sensitive tool to analyse preparations-induced surface electronic properties during wet-chemical preparation of silicon substrates. In order to control surface conditioning immediately prior to thin-film preparation, however, fast non-destructive methods are required for in situ characterisation of wafers directly during the technological wet-chemistry process. The monitoring and control of industrial manufacturing processes demands very fast, robust in-situ measurement equipment. The changes in the surface recombination behaviours can be monitored by in situ PL measurements during removal of the oxide layers in order to optimise HF treatment times. Since surface treatments affect the reflectance of the surface, also optical measurements can be utilised to measure the changes in the spectral reflectance of Si wafers after wet-chemical treatments. In this way, the effects of roughness and texture are obtained separately by a single measurement in a short time (less than 0.1 s) [82].

From the viewpoint of global environment conservation as well as to improve the economics in silicon heterostructure solar cells manufacturing, the employment of highly-efficient cleaning and passivation methods is essential to minimise the chemical consumption and the number of process steps. Silicon surface conditioning by the application of thin, ultra-clean oxides, smoothing and H-termination by few, simple wet-chemical steps in the technological process can be issues of the future. For this purpose, the development of new efficient wet-chemical treatments is required to reach the similar effects of wet-chemical conditioning, as reported here, with a significant cost reduction. Promising methods might be the wet-chemical oxidation in aqueous ozone (O_3) solutions or single step processes for simultaneous wet-chemical smoothing and H-termination. Taking these points into account, both H-termination procedures and optimised ultrathin oxide layer preparations, as reported here, can be successfully employed to passivate silicon interfaces for heterojunction solar cell applications. However, further improvement of wet-chemical passivation methods for n-type substrates is necessary to obtain the same significant reduction in preparation-induced states and recombination losses as achieved on p-type substrates.

As reported by Tsunomura [83], the world's highest solar cell conversion efficiency of 23% (confirmed by AIST) has been obtained by using a HIT structure. This is the world's first practical-size silicon solar cell that exceeds a conversion efficiency of 22% as a confirmed value. This high efficiency has been achieved mainly due to improvements in the a-Si:H/c-Si heterostructure properties, by a new - not specified - cleaning process for the c-Si surface and a lower-damage-deposition process [83].

Up to now there are still many questions that have to be clarified about the effect of preparation induced surface micro-roughness, surface charge and surface states on the resulting optical and electronic properties of silicon heterojunctions. Future investigations of the microscopic interactions between the crystalline silicon surface and the amorphous layers are necessary for a better understanding of the chemical reactions and charge carrier transport on the Si interface and the resulting recombination losses.

Acknowledgments

The financial support of the European Union through the FP7 project "Heterojunction Solar Cells based on a-Si:H/c-Si" (HETSI), grant no. 211821, and by the Bundesministerium für Bildung und Forschung (FKZ 01SF0012) are gratefully acknowledged.

The authors thanks Dr. W. Henrion and Dr. M. Rebien for measuring the UV-VIS spectra and fruitful discussions; Dr. J.-Th. Zettler and Dr. A. Röseler for valuable discussions on optical measurements; Dr. A. Müller and Dr. F. Müller, Dr. K. Hübener and Dr. J. Hauschild for long year cooperation in Si surface characterisation by AFM, I. Sieber and C. Klimm for SEM investigations, Dr. M. Schmidt, Dr. L. Korte, E. Conrad, Dr. K. v. Maydell, Dr. T.F. Schulze, Dr. F. Wünsch, Dr. M. Kunst, Dr. A. Laades and Dr. U. Stürzebecher for cooperation in solar cell preparation and characterisation; and finally D. Patzek, K. Jacob and A. Scheu for technical assistance.

References

[1] Eades, W.D., Swanson, R.M.: Calculation of surface generation and recombination velocities at the Si-SiO2 interface. J. Appl. Phys. 58, 4267–4276 (1985)
[2] Angermann, H.: Passivation of structured p-type silicon interfaces: Effect of surface morphology and wet-chemical pre-treatment. Appl. Surf. Sci. 254, 8067–8074 (2008)
[3] Chabal, Y.J., Higashi, G.S., Raghavachari, K., Burrows, V.A.: Infrared spectroscopy of Si(111) and Si(100) surfaces after HF treatment: Hydrogen termination and surface morphology. J. Vac. Sci. Technol. A 7, 2104–2109 (1989)
[4] Intelmann, C.M., Hinrichs, K., Syritski, V., Yang, F., Rappich, J.: Recombination behaviour at the ultrathin polypyrrole film/silicon interface investigated by in-situ pulsed photoluminescence. Japanese Journal of Applied Physics, Part I: Regular Papers and Short Notes 47, 554–557 (2008)
[5] Rappich, J., Fahoume, M.: Nonradiative recombination and band bending of p-Si(100) surface. Thin Solid Films (2005)
[6] Aberle, A.G.: Surface passivation of crystalline silicon solar cells: a review. Progress in Photovoltaics: Research and Applications 8, 473–487 (2000)
[7] Nemanick, E.J., Hurley, P.T., Webb, L.J., Knapp, D.W., Michalak, D.J., Brunschwig, B.S., Lewis, N.S.: Chemical and Electrical Passivation of Single-Crystal Silicon(100) Surfaces through a Two-Step Chlorination/Alkylation Process. J. Phys. Chem. B 110, 14770–14778 (2006)
[8] Prise, J.B.: Anisotropic Etching of Silicon with KOH-H2O-Isopropyl Alcohol. In: Burges, R.R. (ed.) Semiconductor Silicon. The Electrochemical Society Proceeding Series, Princeton, NJ (1973)
[9] Munoz, D., Carreras, P., Escarre, J., Ibarz, D., Martin de Nicolas, S., Voc, C., Asensi, J.M., Bertomeu, J.: Optimization of KOH etching process to obtain textured substrates suitable for heterojunction solar cells fabricated by HWCVD. Thin Solid Films 517, 3578 (2009)
[10] Hylton, J.D., Burgers, A.R., Sinke, W.C.: Alkaline etching for reflectance reduction in multicrystalline silicon solar cells. J. Electrochem. Soc. 151, 408 (2004)
[11] Weinreich, W., Acker, J., Gräber, I.: The effect of H2SiF6 on the surface morphology of textured multi-crystalline silicon. Semicond. Sci. Technol. 21, 1278–1286 (2006)
[12] Sievert, W.J., Zimmermann, K.-U., Starzynski, J.S.: Wafer Thinning Products. European Semiconductor 27, 17 (2005)
[13] Kern, W.: The Evolution of Silicon Wafer Cleaning Technology. J. Electrochem. Soc. 137, 1887 (1990)
[14] Angermann, H., Henrion, W., Rebien, M., Fischer, D., Zettler, J.-T., Röseler, A.: H-terminated silicon: spectroscopic ellipsometry measurements correlated to the surface electronic properties. Thin Solid Films 313-314, 552–556 (1998)
[15] Kim, W.-B., Matsomoto, T., Kobayashi, H.: Ultrathin SiO2 layer on atomically flat Si(111) surfaces with excellent electrical characteristics formed by nitric acid oxidation method. Appl. Phys. Lett. 93, 072101–072103 (2008)

[16] Rappich, J., Timoshenko, V.Y., Dittrich, T.: In Situ Monitoring of Electrochemical Processes at the (100) p-Si/Aqueous NH4F Electrolyte Interface by Photoluminescence. J. Electrochem. Soc. 144, 493–496 (1997)

[17] Timoshenko, V.Y., Petrenko, A.B., Stolyarov, M.N., Dittrich, T., Fuessel, W., Rappich, J.: Quantitative analysis of room temperature photoluminescence of c-Si wafers excited by short laser pulses. Journal of Applied Physics 85, 4171–4175 (1999)

[18] Heilig, K.: Experimentelle Technik der Physik 14, 135 (1968)

[19] Dittrich, T., Bitzer, T., Angermann, H., Flietner, H., Lewerenz, H.J.: Surface electronic properties of electrolytically hydrogen terminated Si(111). J. Electrochem. Soc. 141, 3595 (1994)

[20] Angermann, H.: Characterisation of wet-chemically treated silicon interfaces by surface photo-voltage measurements. Anal. Bioanal. Chem. 374, 676 (2002)

[21] Lauer, K., Laades, A., Übensee, H., Metzner, H., Lawerenz, A.: Detailed analysis of the microwavedetected photoconductance decay in crystalline silicon. J. Appl. Phys. 104, 104503 (2008)

[22] Laades, A., Brauer, J., Stürzebecher, U., Neckermann, K., Klimm, K., Blech, M., Lauer, K., Lawerenz, A., Angermann, H.: Wet-chemical treatment of solar grade CZ silicon prior to surface passivation. In: 24th European Solar Conference, Hamburg, Germany (2009) 2CV.2.61

[23] Sinton, R.A., Cuevas, A.: Contactless Determination of Curent-Voltage Characteristics and Minority-Carrier Lifetimes in Semiconductors from Quasi-Steady-State Photoconductance Data. Appl. Phys. Lett. 69, 2510–2512 (1996)

[24] Swiatkowski, C., Sanders, A., Buhre, K.-D., Kunst, M.: Charge-carrier kinetics in semiconductors by microwave conductivity measurements. J. Appl. Phys. 78, 1763 (1995)

[25] Henrion, W., Rebien, M., Angermann, H., Röseler, A.: Spectroscopic Investigations of Hydrogen Termination, Oxide Coverage, Roughness, and Surface State Density of Silicon During Native Oxidation in Air. Appl. Surf. Sci. 202, 199 (2002)

[26] Henrion, W., Röseler, A., Angermann, H., Rebien, M.: Application of UV-VIS and FTIR Spectroscopic Ellipsometry to the Characterization of Wet-Chemically Treated Si Surfaces. Phys. Stat. Sol. (a) 175, 121 (1999)

[27] Heilig, K.: Method for reduction of hysteresis effects in MIS measurements. Solid State Electron. 27, 395–396 (1984)

[28] Lam, Y.W.: Surface-state density and surface potential in MIS capacitors by surface photovoltage measurements. I. J. Phys. D: Appl. Phys. 4, 1370–1375 (1971)

[29] Rappich, J., Zhang, X., Rosu, D.M., Schade, U., Hinrichs, K.: Passivation of Si surfaces investigated by in-situ photoluminescence techniques. Solid State Phenomena 156-158, 363–368 (2010)

[30] Rappich, J., Zhang, X., Chapel, S., Sun, G., Hinrichs, K.: Passivation of Si surfaces by hydrogen and organic molecules investigated by in-situ photoluminescence techniques. Phys. Stat. Solidi. (c) 7, 161 (2010)

[31] Aspnes, D.E.: Optical response of microscopically rough surfaces. Phys. Rev. B 41, 10334–10343 (1990)

[32] Bruggeman, D.A.G.: Berechnung verschiedener physikalischer Konstanten von heterogenen Substanzen. Annalen der Physik 5, 636 (1935)

[33] Yasuda, T., Aspnes, D.E.: Optical-standard surfaces of single-crystal silicon for calibrating ellipsometers and reflectometers. Appl. Opt. 33, 7435–7438 (1994)

[34] Aspnes, D.E.: Studies of surface, thin film and interface properties by automatic spectroscopic ellipsometry. J. Vac. Sci. Technol. 18, 289–295 (1981)

[35] Wünsch, F., Citarella, G., Abdallah, O., Kunst, M.: An inverted a-Si:H/c-Si hetero-junction for solar energy conversion. J. Non-Crystalline Solids 352, 1962–1966 (2006)
[36] Angermann, H., Wünsch, F., Kunst, M., Laades, A., Stürzebecher, U., Conrad, E., Korte, L., Schmidt, M.: Effect of wet-chemical substrate pre-treatment on electronic interface properties and recombination losses of a-Si:H/c-Si and a-SiNx:H/c-Si hetero-interfaces. Accepted by Phys. Stat. Sol. (2011)
[37] Laades, A.: Preparation and Characterization of Amorphous/Crystalline Silicon Hetero-junctions, Fachbereich Physik. Technische Universität Berlin, Berlin (2005)
[38] Angermann, H., Henrion, W., Röseler, A., Rebien, M.: Wet-chemical passivation of Si(111)- and Si(100)-substrates. Materials Science and Engineering B 73, 178–183 (2000)
[39] Angermann, H., Henrion, W., Rebien, M.: Electronic Properties of Wet-Chemically Prepared Oxide Layers. Solid State Phenomena 76-77,181–184 (2001)
[40] Goetzberger, A., Heine, V., Nicollian, E.H.: Surface States in Silicon from Charges in the Oxide Coating. Appl. Phys. Lett. 12, 95–97 (1968)
[41] Poindexter, E.H., Geraldi, G.J., Rueckel, M.E., Caplan, P.J., Johnson, N.M., Biegelsen, D.K.: Electronic Traps and Pb Centers at the Si/SiO2 Interface: Band-gap Energy Distributuion. J. Appl. Phys. 56, 2844 (1984)
[42] Flietner, H.: Passivity and Electronic Properties of the Silicon/Silicondioxide Interface. Mat. Sci. Forum 185-188, 73–82 (1995)
[43] Lenahan, P.M., Dressendorfer, P.V.: Hole traps and trivalent silicon centers in metal/oxide/silicon devices. J. Appl. Phys. 55, 3495–3499 (1984)
[44] Poindexter, E.H., Caplan, P.J., Deal, B.E., Radzouk, R.R.: Identification and properties of Pb-like centers in photoluminescent porous silicon. J. Appl. Phys. 52, 879 (1981)
[45] Angermann, H., Kliefoth, K., Füssel, W., Flietner, H.: Defect generation at silicon surfaces during etching and initial stage of oxidation. Microelectron. Eng. 28, 51–54 (1995)
[46] Angermann, H., Henrion, W., Rebien, M., Röseler, A.: Wet-chemical Preparation and Spectroscopic Characterization of Silicon Interfaces. Appl. Surf. Sci. 235, 322–329 (2004)
[47] Drucker, J., Bandari, A., Burrows, V.A.: Si(100) surface corrosion by NH4F studied using high electron imaging in a spatial resolution secondary UHV-STEM. In: Mater. Res. Soc. Symp. Proc., vol. 315, p. 479 (1993)
[48] Neuwald, U., Hessel, H.E., Feltz, A., Memmert, U., Behm, R.J.: Wet chemical etching of Si(100) surfaces in concentrated NH4F solution: formation of (2x1)H reconstructed Si(100) terraces versus (111) facetting. Surf. Sci. Lett. 296, L8–L14 (1993)
[49] Angermann, H., Rappich, J., Klimm, C.: Wet-chemical treatment and electronic interface properties of silicon solar cell substrates. Central Europ. J. Phys. 7, 363–370 (2009)
[50] Angermann, H., Conrad, E., Korte, L., Rappich, J., Schulze, T.F., Schmidt, M.: Passivation of textured substrates for *a*-Si:H/*c*-Si hetero-junction solar cells: Effect of wet-chemical smoothing and intrinsic a-Si:H interlayer. Mat. Sci. Eng. B 159-160, 219–223 (2009)
[51] Angermann, H., Laades, A., Stürzebecher, U., Conrad, E., Klimm, C., Schulze, T.F., Lawerenz, A., Korte, L.: Wet-chemical preparation of textured silicon solar cell substrates: Surface conditioning and electronic interface properties. In: To be Publish in Solid State Phenomena. Scitech Publ., Zuerich-Uettikon (2010)

[52] Steinert, M., Acker, J., Oswald, S., Wetzing, K.: Study on the mechanism of silicon etching in HNO3-rich HF/HNO3 mixtures. J. Phys. Chem. C 111, 2122–2140 (2007)

[53] Nishimoto, Y., Ishahara, T., Namba, K.: Investigation of Acidic Texturization for Multicrystalline Silicon Solar Cells. J. Electrochem. Soc. 146, 457–461 (1999)

[54] Kulkarni, M.S., Erk, H.F.: Acid based etching of silicon wafers: mass-transfer and kinetic effects. J. Electrochem. Soc. 147, 176–188 (2000)

[55] Xi, Z., Yang, D., Que, D.: Texturization of monocrystalline silicon with tribasic sodium phosphate. Sol. Energ. Mat. Sol. Cells 77, 255–263 (2003)

[56] Sievert, W., Zimmermann, K.-U., Hartmann, B., Klimm, C., Jacob, K., Angermann, H.: Surface texturization and interface passivation of mono-crystalline silicon substrates by wet chemical treatments. Solid State Phenomena 145-146, 223–226 (2009)

[57] Angermann H., Rappich J., Korte L., Sieber I., Conrad E., Schmidt M., Hübener K., Polte J., Hauschild J.: Wet-chemical passivation of atomically flat and structured silicon substrates for solar cell application. Appl. Surf. Sci. 254, 3615–3625 (2008).

[58] Angermann, H., Rappich, J., Sieber, I., Hübener, K., Hauschild, J.: Smoothing and passivation of special Si(111) substrates: studied by SPV, PL, AFM and SEM measurements. J. Anal. Bioanal. Chem. 390, 1463–1470 (2008)

[59] Angermann, H., Korte, L., Rappich, J., Conrad, E., Sieber, I., Schmidt, M., Hübener, K., Hauschild, J.: Optimisation of electronic interface properties of a-Si:H/c-Si hetero-junction solar cells by wet-chemical surface pre-treatment. Thin Solid Films 516, 6775–6781 (2008)

[60] Bitzer, T., Lewerenz, H.J.: In situ preparation of hydrogen-terminated silicon single-crystal surfaces. Surf. Sci. 269/270, 886 (1992)

[61] Okorn-Schmidt, H.F.: Characterization of silicon surface preparation processes for advanced gate dielectrics. IBM J. Res. Develop. 43, 351–366 (1999)

[62] Lewerenz, H.J., Bitzer, T.: Electrolytic hydrogenation of silicon. J. Electrochem. Soc. 139, L21 (1992)

[63] Noguchi, H., Adachi, S.: Chemical treatment effects of silicon surfaces in aqueous KF solution. Appl. Surf. Sci. 246, 139–148 (2005)

[64] Yang, S.K., Peter, S., Takoudis, C.G.: Fundamentals of Two-step Etching Techniques for Ideal Silicon-hydrogen Termination of Silicon (111). J. Appl. Phys. 76, 4107–4112 (1994)

[65] Morita, M., Ohmi, T., Hasegawa, E., Kavakami, M., Suma, K.: Control factor of native oxide growth on silicon in air or in ultrapure water. Appl. Phys. Lett. 55, 562–564 (1989)

[66] Gräf, D., Grundner, M., Schulz, R.: Reaction of water with hydrofluoric acid treated silicon(111) and (100) surfaces. J. Vac. Sci. Technol. A 7, 808–813 (1989)

[67] Kolìbal, M., Čechal, J., Bartošìk, M., Mach, J., Šikola, T.: Stability of hydrogen-terminated vicinal Si(1 1 1) surface under ambient atmosphere. Appl. Surf. Sci. 256, 3423–3426 (2010)

[68] Zhao, L., Zhou, C., Li, H., Diao, H., Wang, W.: Characterization on the Passivation Stability of HF Aqueous Solution Treated Silicon Surfaces for HIT Solar Cell Application by the Effective Minority Carrier Lifetime Measurement. Chin. J. Phys. 48, 392–399 (2010)

[69] Angermann, H., Henrion, W., Rebien, M., Röseler, A.: Wet-chemical passivation and characterization of silicon interfaces for solar cell applications. Sol. Energy Mat. Sol. Cells 83, 331–346 (2004)

[70] Rappich, J.H.P., Nickel, N.H., Sieber, I., Schulze, S., Dittrich, T.: Stable electrochemically passivated Si surfaces by ultra thin benzene-type layers. Microelectronic Engineering 80, 62–65 (2005)
[71] Aureau, D., Rappich, J., Moraillon, A., Allongue, P., Ozanam, F., Chazalviel, J.-N.: In situ monitoring of the electronic properties and the pH stability of grafted Si(111). J. Electroanal. Chem. 646, 33–42 (2010)
[72] Schmidt, H.F., Meuris, M., Mertens, P.W., Rotondaro, A.L.P., Heyns, M.M., Hurd, T.Q., Hachter, Z.: H2O2 Decomposition and Its Impact on Silicon Surface Roughening and Gate Oxide Integrity. Jpn. J. Appl. Phys. 34, 727–731 (1995)
[73] Ohmi, T., Miyashita, M., Itano, M., Imaoka, T., Kawanabe, I.: Dependence of thin-oxide films quality on surface microroughness. IEEE Trans. Electron Devices 39, 537–545 (1992)
[74] Akiyama, K., Naito, N., Nagamori, M., Koya, H., Morita, E., Sassa, K., Suga, H.: Effect of SC1 Process on Silicon Surface Microroughness and Oxide Breakdown Characteristics. Jpn. J. Appl. Phys. 34, L153–L155 (1995)
[75] Allongue, P., de Villeneuvea, C.H., Morin, S., Boukherroub, R., Wayner, D.D.M.: The preparation of flat H-Si(111) surfaces in 40% NH4F revisited. Electrochimica Acta 45, 4591–4598 (2000)
[76] Schmidt, M., Korte, L., Laades, A., Stangl, R., Schubert, C., Angermann, H., Conrad, E., van Maydell, K.: Physical aspects of a-Si:H/c-Si hetero-junction solar cells. Thin Solid Films 515, 7475–7480 (2007)
[77] Becker, J.-P., Pysch, D., Leimenstoll, A., Hermle, M., Glunz, S.W.: Wet-Chemical Pre-Treatment of c-Si Substrates Enhancing the Performance of a-Si:H/c-Si Hetero-Junction Solar Cells. In: 24th European PV Solar Energy Conference and Exhibition, Hamburg, Germany (2009)
[78] Angermann, H., Schulze, T.F., Conrad, E., Rappich, J., Korte, L., Schmidt, M.: Cleaning and passivation of structured n-type Si substrates: preparation and interface properties of a-Si:H/c-Si hetero solar cells. In: Lincot, D. (ed.) 23rd European Photovoltaic Solar Energy Conference, Valencia, Spain, pp. 1422–1426. WIP Renewable Energies, München (2008)
[79] Korte, L., Conrad, E., Angermann, H., Stangl, R., Schmidt, M.: Overview on a-Si:H/c-Si heterojunction solar cells - physics and technology. In: 22nd European Photovoltaic Solar Energy Conference, Milan, Italy, pp. 859–865 (2007)
[80] Taira, S., Yoshimine, Y., Baba, T., Taguchi, M., Kanno, H., Kinoshita, T., Sakata, H., Maruyama, E., Tanaka, M.: Our Approaches for Achieving HIT Solar Cells With More Than 23% Efficiency. In: 22nd European Photovoltaic Solar Energy Conference, Milan, Italy, pp. 932–935 (2007)
[81] Scherff, M.L.D., Froitzheim, A., Ulyashin, A., Schmidt, M., Fahrner, W.R., Fuhs, W.: 16.2% Efficiency for amorphous/crystalline heterojunction solar cells on flat p-type silicon wafers. In: PV in Europe - From PV Technology to Energy Solutions, Rome, Italy, p. 7 (2002)
[82] Angermann, H., Uredat, S., Zettler, J.-T.: Surface Texturization and Interface Passivation of Mono- and Polycrystalline Silicon Substrates: Evaluation of the Wet-Chemical Treatments by UV-NIR-Reflectance. In: 24th European Photovoltaic Solar Energy Conference, Hamburg, Germany, pp. 1954–1957 (2009)
[83] Tsunomura, Y., Yoshimine, Y., Taguchi, M., Baba, T., Kinoshita, T., Kanno, H., Sakata, H., Maruyama, E., Tanaka, M.: Twenty-two percent efficiency HIT solar cell. Sol. Energy Mater. Sol. Cells 93, 670–673 (2009)

Chapter 4
Electrochemical Passivation and Modification of c-Si surfaces

Jörg Rappich

Helmholtz-Zentrum Berlin für Materialien und Energie GmbH,
Institut für Silizium-Photovoltaik,
Kekuléstraße 5, D-12489 Berlin, Germany

This chapter addresses the electrochemical passivation of Si surfaces by hydrogen, small organic molecules and ultra-thin polymeric layers which is not yet a standard technique in Si solar cell preparation. The electrochemical surface conditioning leads to different surface structures compared to the wet-chemical techniques (Chapter 3). The electronic properties of the interface depend strongly on these surface morphologies and consequently it is important to measure and control their changes during the electrochemical processing. Therefore, pulsed photoluminescence (PL) spectroscopy is applied as fast and non-destructive method to monitor in-situ and ex-situ the electronic surface properties during electrochemical oxidation, hydrogenation, and grafting of organic molecules and ultra-thin polymeric layers. The additionally used in-situ surface photovoltage (SPV) provides information on the surface charge during the wet-chemical and electrochemical processing. Unusual low concentration of recombination active defects at Si:H surfaces, Si(111) and Si(100), can be obtained after electropolishing in the current oscillating regime in diluted HF solutions. The passivation by hydrogen is influenced by the applied potential, the current flow, the temperature and the solution composition, where nitrogen bubbling of the solution is an important step to enhance the surface passivation.

PL investigations of the organically modified surfaces show that a slightly higher defect concentration at the interface (typical by a factor of 2) is usually observed. However, organically modified Si surfaces have extremely long time stability versus oxidation in ambient air, especially after grafting of 10-carboxydecyl groups via hydrosilylation which shows an ideal Si surface passivation with respect to interface recombination losses.

4.1 Introduction

The etching of silicon oxide (SiO_2) layers in diluted HF solution is the commonly used conditioning step to obtain H-terminated surfaces with appropriate electronic

W.G.J.H.M. van Sark et al. (Eds.): Physics & Tech. of Amorphous-Crystalline, EM, pp. 95–130.
springerlink.com

properties, e.g. low interface recombination [1-3], for further processing. Therefore, the appropriate solution or electrolyte composition, temperature and ambient condition should be evaluated. For this purpose, in-situ techniques are required to measure sensitively and fast the surface recombination velocity or the surface defect concentration. Such techniques are pulsed photoluminescence (PL) spectroscopy [4, 5] and surface photovoltage (SPV) measurements [6]. Additionally, there is need for processing schemes which can be performed at ambient conditions, i.e. at room temperature. One possible route is electrochemical processing, where oxide formation and H-termination can be controlled via the electrolyte composition, temperature, applied potential, and current flow [7] with and without exchanging of the electrolyte solution. Oxide layers and even organically modified surfaces can be prepared anodically in aqueous and non-aqueous solutions. At certain conditions, electropolishing of Si surfaces [8-10] can be performed leading to Si surfaces with extraordinary electronic properties [11] despite of the undulated surface morphology.

In this chapter, processes are described to prepare electrochemically well passivated Si surfaces with low concentrations of recombination active defects. Additionally, it will be shown that the positive charge distribution in the oxide layer can be determined from in-situ surface photovoltage measurements during the etch-back procedure. Alternative routes are presented to modify Si surfaces by organic molecules and ultra-thin polymeric layers to preserve the good passivation by hydrogen and to protect the Si surfaces against oxidation in ambient air and/or water.

4.2 Experimental Techniques to Characterize Si Surface Passivation

The aim of the passivation of Si surfaces is the reduction in the recombination loss of the light induced charge carriers by surface or interface states. Since the photo voltage and the photo current are mainly influenced by such states only well passivated interfaces in hetero structure solar cells are able to produce a high V_{OC} and high I_{SC}. To control these states, sensitive methods are mandatory to measure the density of electronic states in the range from 10^{13} $cm^{-2}eV^{-1}$ to 10^{10} $cm^{-2}eV^{-1}$, or below. One possible and easy to use method is pulsed photoluminescence (PL) spectroscopy [4, 5, 12, 13], which can be used to determine the correlation between the electrochemical substrate treatments [4, 14, 15], the density of recombination active interface states [15], and the stability of different Si surface passivation in ambient air [16]. The Si surface morphology and the type of surface species are usually determined by Atomic Force Microscopy (AFM) and infrared spectroscopic ellipsometry (IRSE) or infrared Fourier-Transform spectroscopy (FTIR) in the attenuated total reflection (ATR) configuration [9, 10, 24].

4.2.1 In-Situ / Ex-Situ Photoluminescence Spectroscopy

The band gap of crystalline silicon (c-Si), which is an indirect semiconductor, is 1.1 eV. The rate of radiative interband recombination is very low for indirect semiconductors due to the fact that phonons are involved during the transition process. Therefore, the radiative recombination lifetime is very long (up to several ms) and the non-radiative recombination processes such as Shockley-Read-Hall (SRH) recombination are usually much faster. Consequently, the efficiency of the interband luminescence at room temperature is very low due to the high efficiency of non-radiative recombination processes. The recombination is dominated by non-radiative bulk and/or surface recombination processes.

The radiative interband recombination can be measured by photoluminescence (PL) techniques, where the quenching of the PL signal contains information about non-radiative recombination [5, 12]. This situation has been used to examine the change in non-radiative surface recombination during surface treatments and processing of c-Si by taking into account that the bulk life time remains constant [5, 12]. For more details on PL spectroscopy, especially the modulated techniques using continuous wave (cw) lasers, see Chapter 8 in this book.

The interband PL of c-Si can be excited with pulsed or cw lasers. High excitation intensity or cooling of the Si sample to very low temperature is required to increase the low PL intensity of c-Si at room temperature. PL excitation by cw lasers has some disadvantages for use as in-situ technique (i) sample heating for high excitation levels, (ii) distortion of the chemical process at the surface by the high amount of excess carriers, and (iii) cooling below about -10 °C is not suitable for electrochemical or wet-chemical processing. These disadvantages are eliminated by excitation with short laser pulses in the ns range due to high excess carrier concentration for very short periods of time. Therefore, the PL intensity can be measured with a very high temporal resolution by the excitation with single laser pulses during chemical processing.

Figure 4.1 gives an overview of the elementary processes at a semiconductor surface under strong illumination ($\delta n >> n,p$; i.e. the concentration of the excess charge carriers is much higher than the doping level of Si). The relevant processes are light absorption (a), carrier diffusion (b), Auger recombination (f), non-radiative surface (d) and bulk (e) SRH recombination, and bimolecular radiative recombination (c) that leads to the PL signal. The efficiency of the radiative interband recombination is proportional to the product of the excess electron (δn) and hole concentration (δp), while the efficiency of the SRH non-radiative recombination is proportional to the excess electron or hole concentration. Therefore, the PL intensity increases much more strongly with increasing excitation intensity than the non-radiative SRH recombination. The efficiency of the Auger recombination is proportional to $\delta n^2 \delta p$ or $\delta p^2 \delta n$ and non-radiative Auger-recombination limits the PL intensity at high excess carrier concentration due to strong illumination.

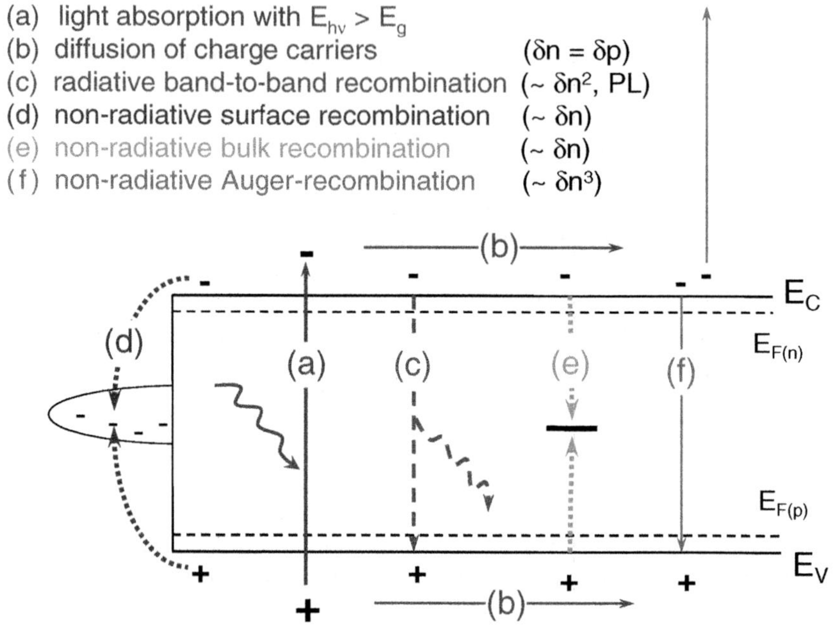

Fig. 4.1 Elementary processes at semiconductor surfaces under strong illumination: (a) excitation of charge carriers, (c) bimolecular radiative band-to-band recombination, (d, e) non-radiative surface and bulk SRH recombination, and (f) Auger recombination.

Figure 4.2 shows a typical setup for in-situ PL and in-situ SPV measurements during chemical processing of Si surfaces. A nitrogen laser (λ=337 nm, τ_{Pulse}=0.6 ns, 220 μJ/pulse) or a tunable dye laser (typically λ=500 nm, τ_{Pulse}=0.5 ns, 60 μJ/pulse) excites the PL which passes an interference filter and is measured by an integrating InGaAs charge coupled detector or a red-shifted avalanche Si-photodiode (time resolution about 3 ns). A laser diode (λ= 905 nm, τ_{pulse}= 150 ns, 200 W) is typically used for SPV measurements (see Chapter 3). The (electro-) chemical cell consists of Teflon with working (Si)-, reference-, and counter-electrode. Fast exchange of the chemical solution can be performed by an inlet and outlet tube. A personal computer (PC) triggers the laser, the analog to digital (AD) conversion process, and reads out the data for further processing. The intensity of the laser is changed over several orders of magnitude by glass plates used as filters. The laser beam is slightly focused on the sample using a quartz lens (spot diameter about 3 to 4 mm). The electrochemical or wet-chemical processes at the Si interfaces are not remarkably influenced by the PL and SPV measurements since the light pulses are so short (0.5 ns and 100 ns) and the repetition rate is about 5 Hz only. An X-Y-table permits recording of PL surface maps.

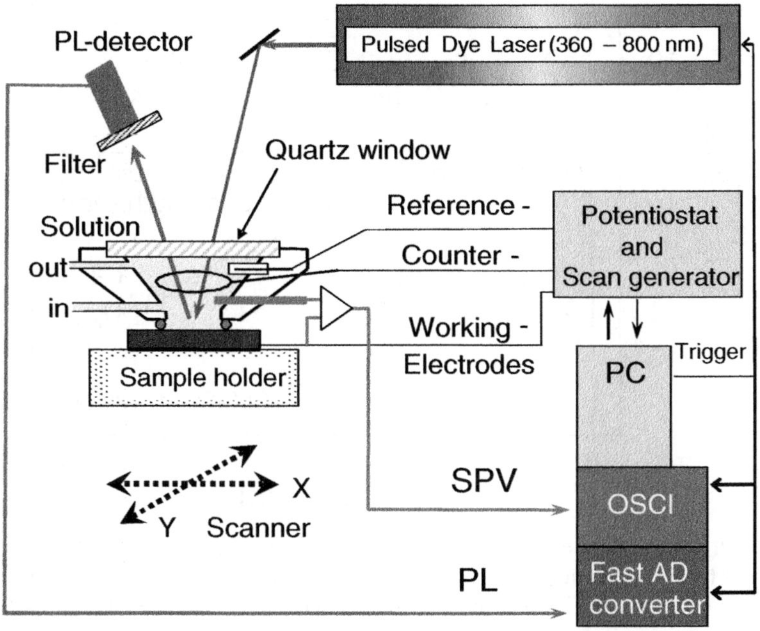

Fig. 4.2 Electrochemical PL and SPV set-up with typical excitation by a nitrogen laser (λ=337 nm, τ_{Pulse}=0.6 ns, 220 µJ/pulse), a tuneable dye laser (λ=500 nm, τ_{Pulse}= 0.5 ns, 60 µJ/pulse) or a laser diode (λ= 905 nm, τ_{Pulse}= 150 ns, 200 W) [17].

Figure 4.3 shows the PL spectrum of an oxidised Si surface measured with the time-integrating InGaAs diode as insert and PL transients of oxidised Si surfaces with different interface state densities (D_{it}) as measured independently by standard high frequency capacitance/voltage (CV) techniques. The higher the defect concentration at the Si/SiO_2 interface, the lower is the integrated PL intensity (I_{PL}) and the faster the PL decay (the lower is τ_{PL}). I_{PL} and τ_{PL} are plotted in Fig. 4.4 as a function of D_{it} as calculated from the CV-measurements. Both show a slope of –1 in the log-log plot what leads to the following relation ship

$$I_{PL} \sim D_{it}^{-1} \tag{4.1}$$

This proportionality enables the calibration of I_{PL} of the sample to measure D_{it} by one independent measurement of the D_{it}, for more details see [5, 7]. Additionally, the integrated PL intensity is inversely proportional to D_{it} in a broad range of excitation levels as outlined in refs. [7, 13].

Additionally, a theoretical calculation of the PL transient allows the determination of D_{it} by either correlation with the experiment or by evaluating τ_{PL} [5, 13]. This makes in-situ PL measurements a comfortable and fast technique [18].

In this chapter, the relative change in I_{PL} is used to demonstrate the changes in D_{it} during the electrochemical or wet-chemical processing and to optimise the respective processes with respect to the reduction in recombination loss.

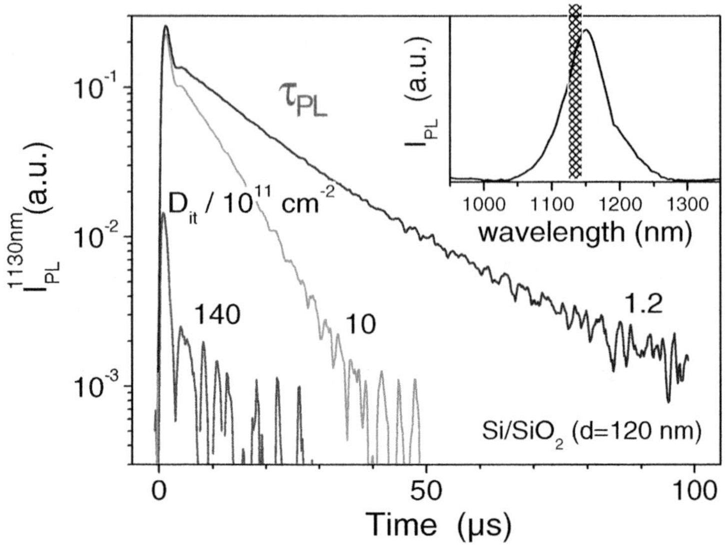

Fig. 4.3 PL transients of oxidised Si surfaces with different interface state densities (D_{it}) as measured by CV-techniques and typical PL spectrum of an oxidised Si surface measured with the time-integrating InGaAs diode (insert). Excitation: $\lambda_{Pulse} = 337$ nm, $\tau_{Pulse}= 0.5$ ns.

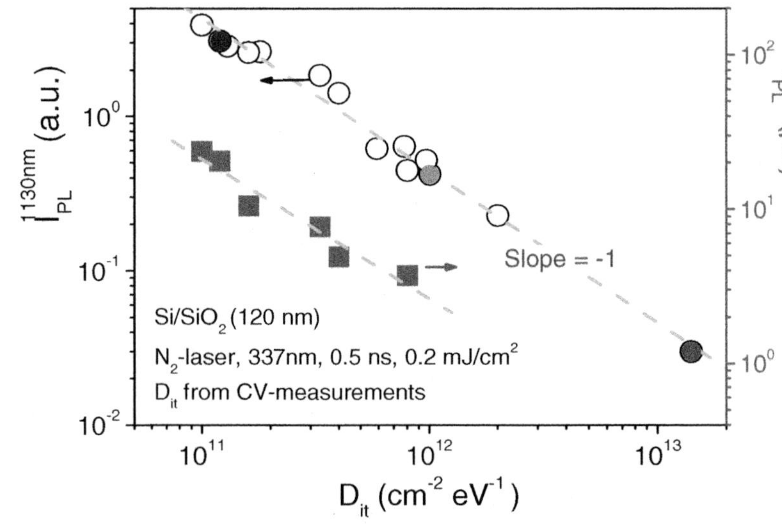

Fig. 4.4 Dependence of the integrated PL intensity and the PL lifetime (τ_{PL}) of Si/SiO_2 samples on D_{it}. Values of D_{it} were obtained by standard high frequency CV measurements.

Figure 4.5 shows an example for in-situ PL measurements where I_{PL} is plotted as a function of time during the etch-back of a wet-chemical oxide on p-Si(111) in 40% NH_4F solution. This process transfers the SiO_2 covered surface into an H-terminated surface. The wet chemically formed oxide has a very low I_{PL} at the beginning. The oxide layer is thinned with ongoing time of etching (see sketch in the upper part of Fig. 4.5) and finally the Si surface is transformed into a flat and H-terminated surface (as is well known from IR spectroscopic measurements [19]). I_{PL} decreases at longer etching times (not shown here) supposedly due to surface roughening and defects which are created at etch-pits steps and kink-sites. In this experiment, the H-termination of the Si(111) surface has an about two order of magnitudes lower surface recombination rate ($I_{PL} \sim D_{it}^{-1}$) compared to a wet-chemically oxidised Si surface.

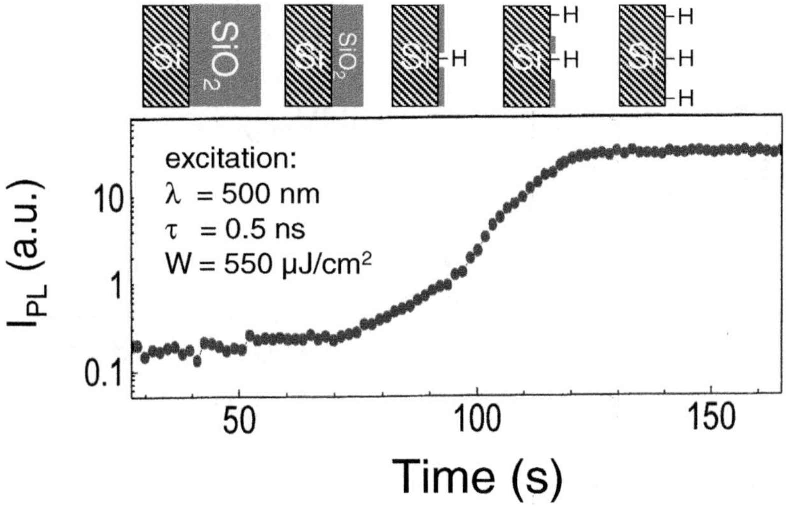

Fig. 4.5 I_{PL} of p-Si(111) during the etch-back of a wet-chemical oxide as a function of time (exc.: 500nm, pulse width 0.5ns, pulse power 550 µJ/cm^2). Top: sketch of the surface change with etching time (transformation from the SiO_2 coverage to the H-termination) [17].

4.2.2 *In-Situ Surface Photovoltage*

The surface photovoltage (SPV) is typically measured by means of a mica spacer used under ex-situ conditions [6]. There is no possibility in electrochemical and wet-chemical processing to apply high electric fields in solutions without any surface reaction, as typically used in standard SPV measurements (see Chapter

3). However, this type of non-standard SPV technique can be applied to electrochemical processing since charges on the Si surface influence the band bending which in turn defines the SPV value but no determination of D_{it} is possible. Here, Si is a probe for surface charges and the signal was capacitively measured via the constant Helmholtz-layer instead of the mica layer. The excitation of the in-situ SPV was performed by a laser diode (λ= 905 nm, τ_{pulse} = 150 ns, 200 W), see Fig. 4.2. A typical SPV transient response is shown in Fig. 6 and only SPV_{max} has been taken for the interpretation of experiment and calculation of the interface / surface related charge. For more details on SPV techniques see Chapter 3.

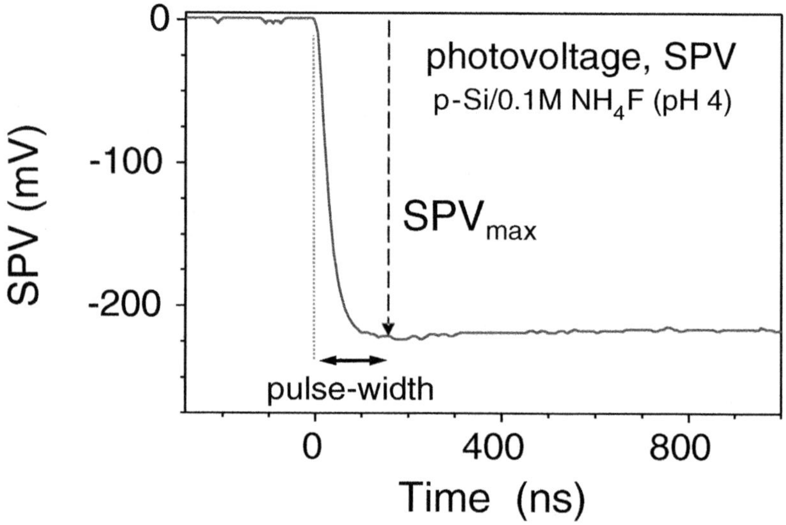

Fig. 4.6 In-situ SPV transient in 0.1 M NH_4F (pH 4) recorded with a pulsed laser diode (λ= 905 nm, τ_{pulse}= 150 ns, 200 W).

According to refs. [2, 6, 7] and citations therein, the correlation of the surface potential, φ_S, and the surface photovoltage, *SPV*, is as follows:

$$SPV = \varphi_S + U_D, \tag{4.2}$$

where U_D is the Dember-voltage. For Si, the value of U_D depends on the doping level of Si and but it is always positive (independent on the type of doping).

4.3 Electrochemical Substrate Preparation Methods

Electrochemical methods can be used to manipulate surfaces in the sub-monolayer regime by appropriate use of the charge flow condition. In Si technology, hydrogenation of Si surfaces takes place whenever an oxide layer on Si is etched-back by an HF containing solution. However, the amount of hydrogen on the Si surface (rough or flat) depends on the solution composition (HF concentration, pH, temperature, solvent, surfactant, etc.) and on the Si oxide / Si interface formed before the etch-back process. This interface can be repeatedly and reproducibly prepared by appropriate choice of the electrochemical conditions. The formation of hydrogenated Si surfaces is one of the most important steps in device manufacturing. HF-dip or buffered NH_4F treatments produce different kinds of surface morphology, i.e. rough or smooth, which is of importance for further processing (deposition, oxide growth, etc.). Four types of hydrogenation of Si surfaces can be distinguished: (i) HF dip (a step in the RCA cleaning process [20, 21]), (ii) treatment in buffered fluoride solutions [22], (iii) electrochemical hydrogenation in diluted fluoride solutions [23, 24], and (iv) formation of porous Si (por-Si) in fluoride solutions [25]. The hydrogenation of the different surfaces that has been produced by these treatments has been extensively investigated by FTIR [1, 19, 22, 24, 26-29] and high resolution electron loss spectroscopy [30-37].

In this section it will be shown that the solution composition and the applied potential during the electrochemical (anodic) oxide formation have a high impact on the final interface recombination loss. Additionally, in-situ SPV measurements have been used to calculate the positive charge in the anodic oxide layer. Furthermore, results on electrochemically prepared H-terminated Si surfaces are presented which have very low defect concentrations, even if they have an undulated surface structure. This behaviour has been recently interpreted as being due to a special kind of reconstruction of surface step facets [7, 38]. Another important part of this chapter addresses the stability of such surfaces in aqueous solutions.

4.3.1 Anodic Oxides: Formation and Etching

The etching of silicon oxide (SiO_2) layers in diluted HF solution is the commonly used processing step to obtain H-terminated surfaces with appropriate electronic conditions. Oxides can be prepared anodically (under positive bias) in aqueous solutions. Using HF containing electrolytes, Si oxides are formed which are simultaneously dissolved by the small amounts of HF present in the electrolyte, leading to the well-known electropolishing behaviour in such solutions [8, 39, 40].

At high anodic potential electrochemical oxidation of Si takes place via:

$$Si + 2\,H_2O + 4\,h^+ \longrightarrow SiO_2 + 4\,H^+ \tag{4.3}$$

Followed by chemical etching of Si oxide according to:

$$SiO_2 + 6\,HF \longrightarrow SiF_2^{2-} + 2\,H^+ + 2\,H_2O \tag{4.4}$$

Figure 4.7 shows typical current-voltage scans of Si in diluted NH_4F solution measured in a set-up as shown in Fig. 4.2, where n-type Si needs additional illumination to ensure a high concentration of light induced holes to force the Si oxidation reaction (eq. 4.3), while p-type Si has holes as majority charge carriers and therefore needs no illumination. At about 0 V there is a strong increase in current for both types of doping, where the Si surface is etched by HF species and a very thin porous Si layer can be formed. At higher potentials, the current decreases and a second broad peak can be observed. This behaviour is due to the formation of an anodic Si oxide over layer whose thickness depends on the applied voltage (eq. 4.3).

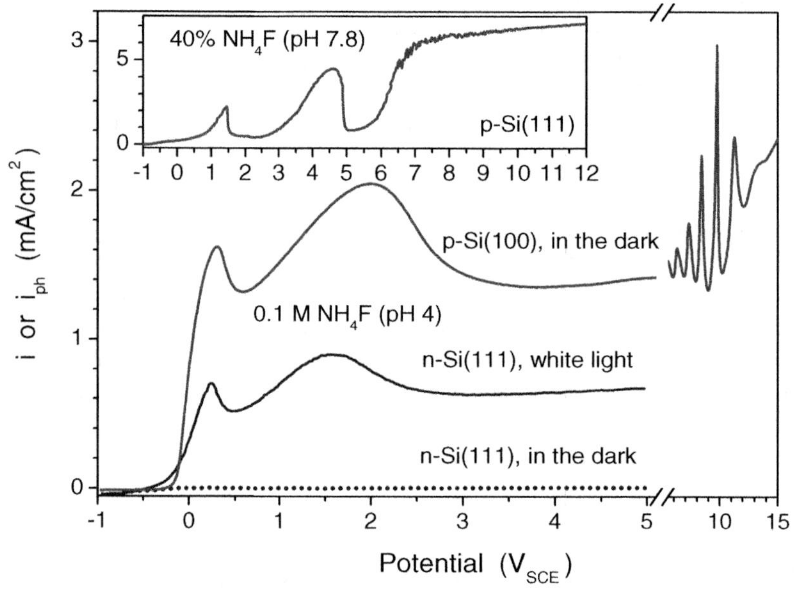

Fig. 4.7 Current (i) and photocurrent (i_{ph}) density for p- and n-type Si surfaces in 0.1 M NH_4F (pH 4) solution, respectively. Insert: potential scan of p-Si(111) in 40% NH_4F solution.

At potentials ranging from approx. +6 to +12 V current oscillations occur due to a competition between the oxide growth (indicated by a current flow) and chemical etch-back of the oxide layer (eqs. 4.3 and 4.4) [41, 42], similarly for n-Si under light illumination (not shown here). The behaviour is slightly different when the potential scan is performed in 40% NH_4F solution [43] as shown in the insert of Fig. 4.7. Here the first and second current peaks occur at higher anodic voltages. Additionally, the voltage breakdowns at +2 and +5 V are much more pronounced and they resemble the typical Flade-potential observed in metal passivation processes [44].

4.3.2 Charge Distribution in the Anodic Oxide Layer

Figure 4.8 shows the time behaviour of (a) the current and (b) I_{PL} after switching the potential from +6 V (anodic oxidation) to -0.4 V (H-termination). Additionally, (c) displays the in-situ measured maximum value of the surface photovoltage (SPV_{max}) measured under open circuit conditions. This reflects the change in the charge of the near surface layer [7].

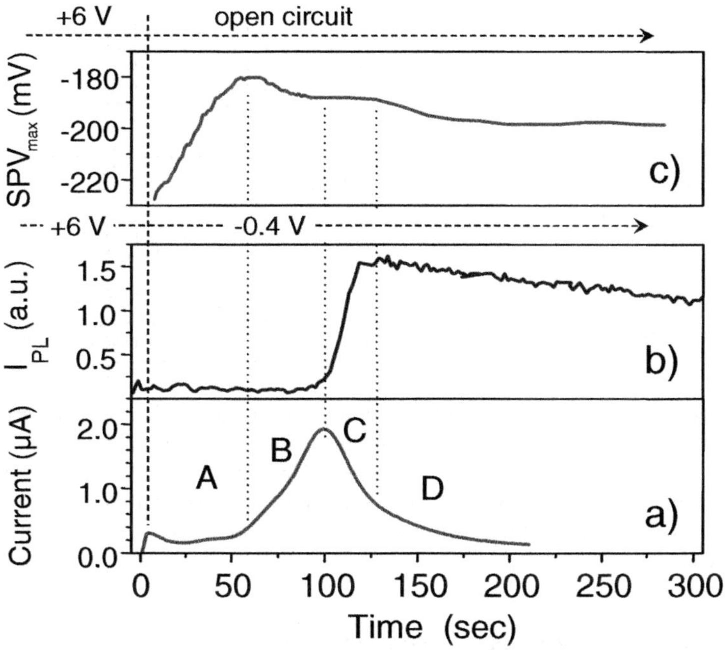

Fig. 4.8 Time behaviour of (a) the current and (b) I_{PL} after switching the potential from +6 V (anodic oxidation) to -0.4 V (H-termination); (c) displays the in-situ measured maximum value of the surface photovoltage (SPV_{max}) measured at open circuit after oxidation at +6 V.

Figure 4.9 depicts the charge distribution in the oxide layer and at the Si surface deduced from the SPV_{max} values in Fig. 4.8. At the origin of the time axis 0 (A) the etch-front starts thinning the anodically prepared oxide layer which has a homogeneously distributed positive charge since SPV_{max} changes linearly with time (decrease in SPV_{max} equals to the reduction in positive charge on the p-Si(111) surface). When the etch-front reaches the Si interface (B and C), SPV_{max} first remains constant and then slowly increases due to an increase in positive charge by surface species which are present in acidic solutions (e.g. H_3O^+). The etching of the Si surface is also reflected by the decrease in I_{PL} due to the formation of recombination active defects at the Si surface (Fig. 4.8b). At last, SPV_{max} is constant at the time when the oxide layer is completely removed and a constant amount of positively charged surface species are adsorbed on the Si surface (D).

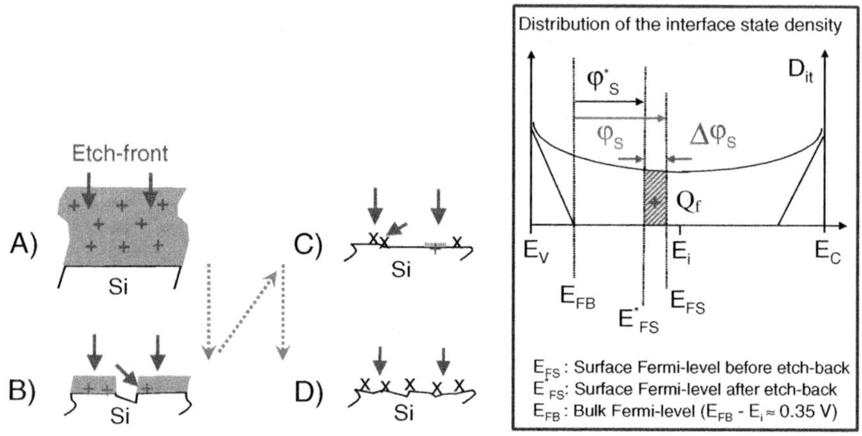

Fig. 4.9 Left: sketch of the charge distribution in the anodic oxide layer and at the Si surface during the etch-back process in HF containing solution (+: positive charge in the oxide layer, x: positively charged interface traps at the Si surface, e.g. H_3O^+). A, B, C, and D correspond to the respective labels in Fig. 4.8. Right: energetic distribution of Dit of the Si surface, the positive charge (Q_f), which has been etched-back, and Si surface potentials (φ), which are correlated to SPV_{max}.

4.3.3 Calculation of the Fixed Oxide Charge

From the above SPV measurement, it is possible to calculate the fixed oxide charge in the SiO_2 layer as follows:

Just after the etch-back of the anodic oxide, the total charge in the space charge region, Q_{SC}, is defined by the interface charge, Q_{it}.

$$Q_{SC} = Q_{SC}^{*} = Q_{it} \tag{4.5}$$

Before etch back, the total charge is influenced by the charged interfaces states, D_{it}, the difference in the surface potentials, and the fixed oxide charge, Q_f.

$$Q_{SC} = (Q_{it} - D_{it} \cdot \Delta\varphi_S) + Q_f \tag{4.6}$$

The charge in the space charge region is defined as:

$$Q_{SC} = \sqrt{2 \cdot \varepsilon \cdot \varepsilon_o \cdot q \cdot p \cdot |\varphi_S|} \quad , \tag{4.7}$$

where q is the elementary charge, p is the donor concentration, and φ_S is the surface potential. The correlation of the surface potential, φ_S, and the surface photovoltage, *SPV*, was already given in eq. (4.2), in which also the Dember-voltage is used.

While $\Delta\varphi_S$ is relatively small (50 mV, see below) and the surface Fermi-level is nearly at midgap (since I_{PL} is very low on Si-H surfaces, see sketch in Fig. 4.9 on the right side), therefore $D_{it} \approx$ const. = D_{it}^{min} and one obtains

$$Q^*_{SC} = 8.3 \cdot 10^{10}\ cm^{-2},$$

$$Q_{SC} = 9.2 \cdot 10^{10}\ cm^{-2}, \text{ and finally}$$

$$Q_f = 1.6 \cdot 10^{10}\ cm^{-2}$$

using the following values: $p = 3 \cdot 10^{15}\ cm^{-3}$ for the used p-Si(100), U_D=+0.04 V, $\varphi_S = SPV_{max} - U_D = -0.19$ V (before etching) and $\varphi^*_S = SPV^*_{max} - U_D = -0.14$ V (just after etching), so that $\Delta\varphi_S$=+0.05 V, and from PL measurements $D_{it} \approx 10^{11}$ $cm^{-2}eV^{-1}$.

Q_f as calculated from ex-situ CV or SPV measurements typically yield $10^{12}\ cm^{-2}$, a value that is much higher than the one obtained here. Obviously, this higher Q_f is a result of adsorbed species from the surrounding which have a stronger influence on Q_f than in solution where the solvent molecules together with the ionic species form a diffuse double layer [45].

4.3.4 Surface Morphology after Etch-Back of Oxides

Figure 4.10 shows AFM images of differently H-terminated Si(111) surfaces after oxide etching in (a) 5% HF dip, (b) after standard smoothing procedure for Si(111) (wet-chemical oxidation in H_2SO_4/H_2O_2 and subsequent etching in 40% NH_4F solution), after electropolishing in 0.1 M NH_4F (pH 4) at (c) +3 V and (d) +8 V with current oscillations, and after oxidation at +8 V in 40% NH_4F. The flattest surface with mono atomic steps was achieved by the standard smoothing process as developed by Chabal et al. [26, 46].

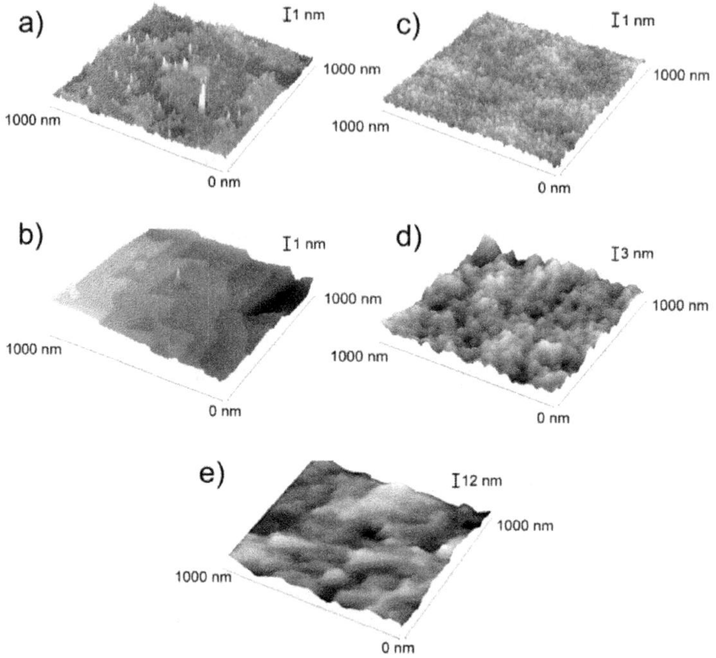

Fig. 4.10 AFM images of a 1x1μm^2 of differently H-terminated Si(111) surfaces after (a) 5% HF dip of a native oxide, (b) after standard smoothing procedure for Si(111) (oxidising in H_2SO_4/H_2O_2 and subsequent etching in 40% NH_4F solution), after electropolishing in 0.1 M NH_4F (pH 4) at (c) +3 V and (d) +8 V with current oscillations, and after oxidation at +8 V in 40% NH_4F. Please note the different z-scaling.

However, all the different types of the surface morphology will have an influence on the electronic properties of H-terminated Si surfaces and Si/SiO_2 interfaces as presented in the following sections.

4.3.5 Electrochemical H-Termination of Si Surfaces

As shown previously in this chapter, the etch-back of oxides on Si can be monitored by PL measurements to find the appropriate point in time to interrupt this process before etch-pits related defects are formed. This is an important advantage as compared to standard wet-chemical treatments that give no indication when the process of H-termination is finished and, finally, etching of the Si surface cannot be effectively suppressed.

On the contrary, the current flow at a certain potential can be used to monitor the transformation from an oxidised to a H-terminated Si surface as presented in

Fig. 4.11, where the end of the oxide etching is accompanied by a peak in the current flow (the so-called dark current transient) and is completed when the current has been decayed on a low constant level (see open circles). Here, the Si suboxide bonds at the interface are further oxidised to SiO_2. This injects some electrons into the Si (measured as a current) and subsequently the oxide is dissolved by HF.

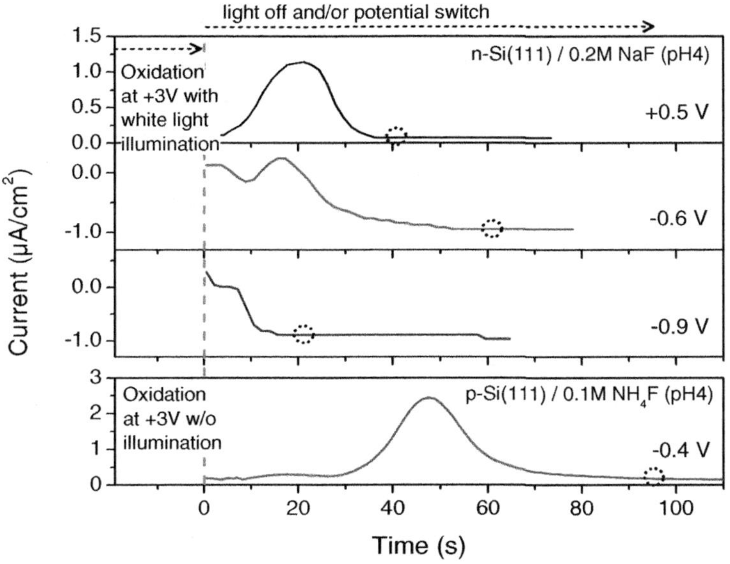

Fig. 4.11 Current transients of Si after switching the potential from +3 V (formation of anodic oxide layer) to lower anodic values (lower band bending), top: n-Si(111) with additional white light illumination on/off, bottom: p-Si(111) [7].

This process of H-termination of the surface was investigated in detail by FTIR spectroscopy and reveals no pronounced dependence on the applied potential during the etch-back of the oxide layer [7]. However, a potential where a current transient is visible is essential for monitoring the process. This requirement limits this technique to small potential windows (i.e. for n-Si: -0.6 to +0.5 V and for p-Si: -0.6 to -0.4 V). The very narrow potential window for p-Si is due to the fact that the majority carriers (holes) are able to oxidise Si surfaces at potentials more positive than -0.4 V and porous structures are created [4, 14].

The higher concentrated solution used for n-Si (0.2 M NaF, pH 4) than for p-Si (0.1 M NH_4F, pH 4) leads to a faster etch-back of the oxide layer and consequently to a faster occurrence of the current transient (see Fig. 4.11).

4.3.5.1 In-Situ PL During Etch-Back of Anodic Oxide

Fig. 4.12 shows the combination of both techniques, dark current transient and I_{PL}, during the etch-back of the oxide layer. At the same time, the ex-situ measured D_{it} value as obtained from ex-situ SPV measurements decreases from approximately 10^{13} down to 10^{11} $eV^{-1}cm^{-2}$. Prior to the SPV measurements, the etching process was interrupted very quickly at specific times of etching, and the Si-sample was rinsed with water and dried by a N_2 stream. I_{PL} has been converted into the number of surface states, N_S ($\cong D_{it}$), which is proportional to $1/I_{PL}$ (eq. 4.1), and is plotted as a function of the etching time. N_S decreases when the oxide layer is etched-back, i.e. when the current transient starts to decrease, and is finished (constant high I_{PL}) as soon as the current transient has completely decayed and the H-termination is reached. Subsequently N_s increases (I_{PL} decreases) with ongoing time of processing pointing to an etching of the Si surface. This behaviour will be discussed in the next paragraphs.

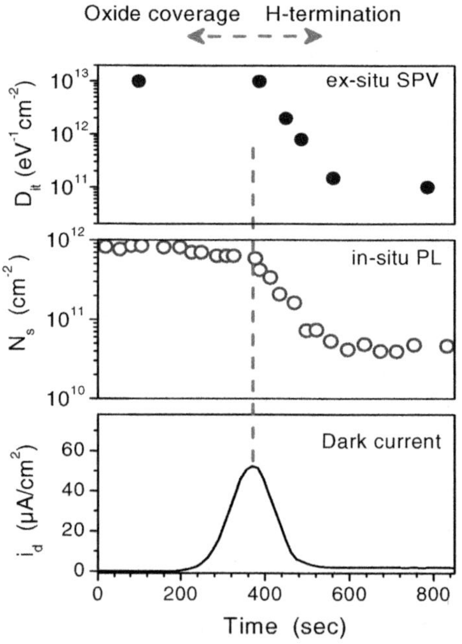

Fig. 4.12 Ex-situ obtained D_{it} at midgap (top) as derived from SPV measurements, N_S as calculated from I_{PL} (middle), and the dark current transient (bottom) as a function of time during the etch-back of an anodic oxide on n-Si(111) in 0.1 M NH_4F (pH 4) solution [7].

4.3.5.2 Oxidation/Hydrogenation Cycles

In electrochemistry, different Si surface orientations behave similar, as can be seen in Fig. 4.13, where I_{PL} of different Si surfaces (111), (100), and (311) during repetitive cycles of oxidation at +3 V and hydrogenation at -0.4 V using 0.1 M NH_4F (pH 4) is presented. The different surfaces were cut from the same ingot, so that the samples have the same bulk properties and changes in I_{PL} are due to changes in the surface recombination velocity only.

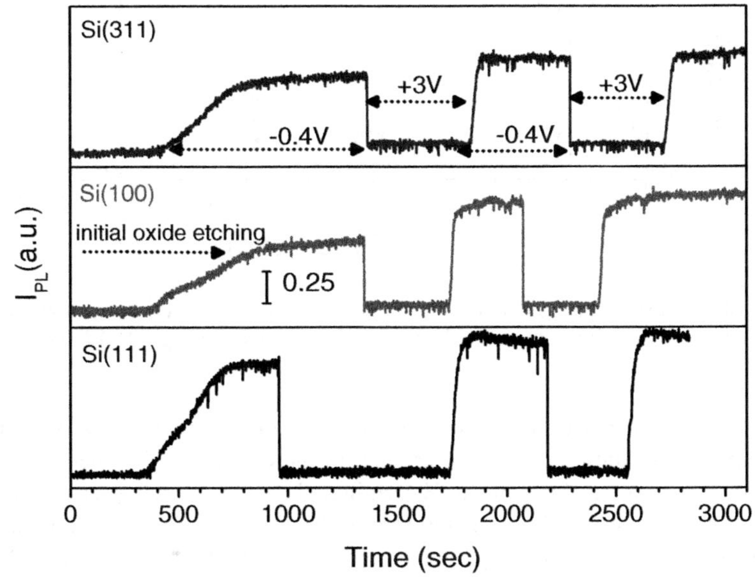

Fig. 4.13 I_{PL} of different Si surfaces orientations, (111), (100), and (311), during repetitive cycles of oxidation at +3 V and hydrogenation at -0.4 V using 0.1 M NH_4F (pH 4). The different surfaces were cut from the same ingot.

In the beginning, the Si samples were covered by a 20 nm thick thermal oxide where the etch-back of the SiO_2 layer is much slower than for the anodically prepared oxide layers, as shown by a slower increase in I_{PL} with etching time. The hydrogenation of the Si surface is reflected by a strong increase in I_{PL} as outlined before.

The increase in the maximum I_{PL} value (I_{PL}^{max}) is much more pronounced when current oscillations are involved in the oxidation process prior to the hydrogenation. Fig. 4.14 shows (b) I_{PL} of the p-Si(100) surface in contact with the same solution, 0.1 M NH_4F (pH 4), but using +8 V as anodisation potential, where (a) current oscillations occur. By switching the potential to -0.4 V the oxidation process is interrupted and hydrogenation sets in as reflected by an increase in I_{PL}.

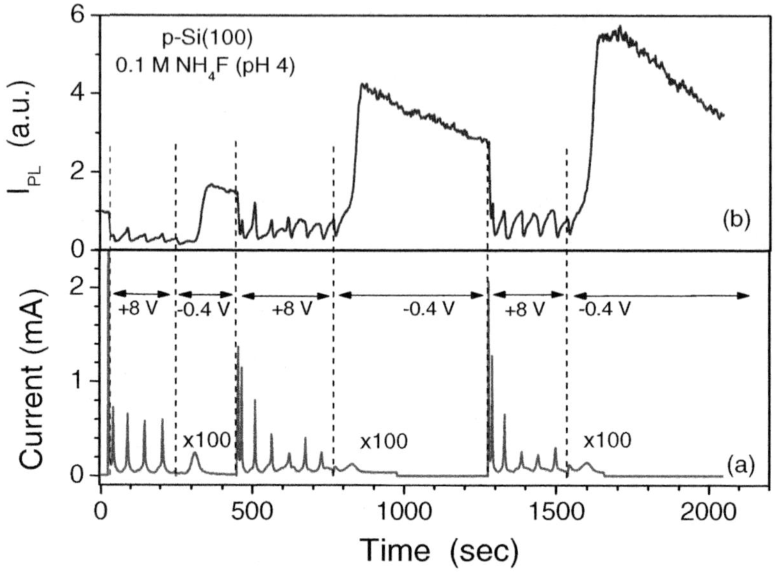

Fig. 4.14 (a) Current and (b) I_{PL} of p-Si(100) during repetitive cycles of oxidation at +8 V, where current oscillations occur, and hydrogenation at -0.4 V using 0.1 M NH_4F (pH 4).

This strong increase in I_{PL}, is correlated to a faster decay of I_{PL} with time the more oxidation/etching cycles are applied to the electrode. This behaviour seems to be due to faster etching of the H-terminated surface which is established after the etch-back of the oxide. The etch rate of the H-terminated surface is increased, with respect to the 1st cycle, by approximately 50% and 300% after the 2nd and 3rd cycle, respectively.

I_{PL}^{max} of the H-terminated Si surface after oxidation at +3 V (see Fig. 4.15a) saturates with the 2nd cycle, and is the highest for Si(111) and the lowest for Si(311). The repetitive oxidation/hydrogenation cycles enhance the interface passivation for all investigated surface orientations by a factor of ≈1.25 for Si(111) and Si(311), and ≈1.6 for Si(100), respectively. I_{PL}^{max} of the H-terminated Si(100) surface after oxidation at +8 V, where current oscillations occur (see Fig. 4.15a), reveals a strong increase and no saturation is reached even after the 3rd cycle; a tendency to saturate seems to be visible. The amount of defects on the H-terminated surface is reduced by a factor of about 10 with respect to the H-termination after oxidation at +3 V as calculated from $I_{PL} \sim D_{it}^{-1}$. This behaviour leads to an unexpected low defect concentration of about 10^{10} cm^{-2} on this Si(100) surface [15] which is typically observed on Si(111) surfaces only [11].

Additionally, I_{PL}^{max} at +3 V (during oxidation) of the Si(111) interface is about half of the I_{PL}^{max} value as measured for the other surface orientations at the same oxidation potential (see Fig. 4.15b). This behaviour points to higher amount of defects due to stress at the Si(111)/SiO_2 interface than for the Si(100)/SiO_2 and Si(311)/SiO_2 interfaces, which can be expected due to better Si-O-Si bond angle alignment at the interface for (100) and (311) orientation of Si as compared to the Si(111) [47-51]. Even during oxidation, I_{PL}^{max} is higher when current oscillations at +8 V occur at the Si(100) surface. I_{PL}^{max} is about 10 up to 50 times higher when the current oscillation peak is at a maximum or a minimum, respectively.

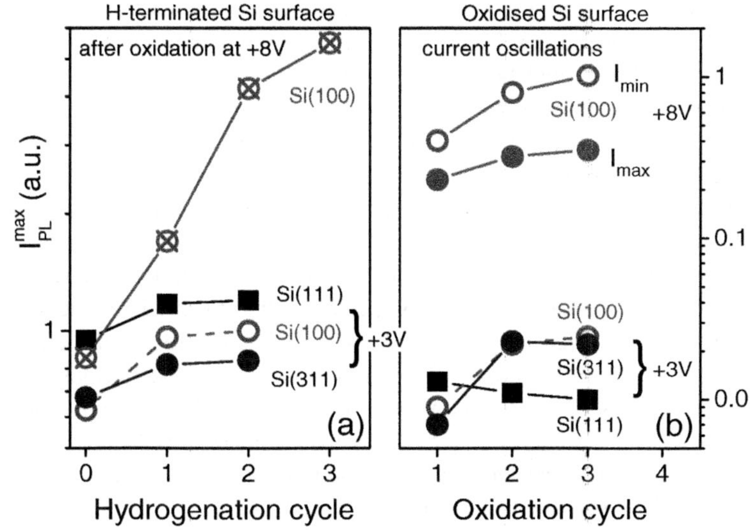

Fig. 4.15 (a) I_{PL}^{max} of H-terminated Si surfaces after anodic oxidation at +3 and +8 V in 0.1 M NH_4F (pH 4) and (b) I_{PL}^{max} of Si surfaces during anodic oxidation at +3 V and +8 V in the same solution, respectively, as a number of the processing cycle.

Electropolishing at +3V (Fig. 4.10c) leads to a smoother surface than after simple 5% HF dip (Fig. 4.10a). Electropolishing at +8 V leads to an undulated surface with about three times higher surface structures. Still, the interface defect density of such surfaces is similar or below that of a flat Si(111) surface shown in Fig. 4.10b [11, 15]. Additionally, this type of surface structure explains the fast decay of I_{PL} after the H-termination process due to 3D-etching of the small hillocks. Obviously, these hillocks (or better the valleys) are the cause for the interface defect concentrations being so low compared the ideally flat Si(111), since hydrogenated Si(111) facets (see Fig. 4.10b) have a relatively large number of surface atoms at steps and corners. At such sites, the Si=H_2 and Si-Si back bonds are weaker and the probability for surface chemical reactions and adsorption of molecules is increased. However, defect sites cannot be avoided by simple chemical treatments on a scale larger than the facets, since the size of facets is thermodynamically limited. The situation is different for electrochemical treatments. Here apparently rounded shapes of Si surfaces can be created at certain conditions, where steps and corners are deactivated by the electrochemical reactions due to the strongly increased oxidation rate at high electric fields at these defect sites, obviously induced by the current boost of an oscillation. However, many Si surface atoms have a similar surface potential from the point of view of reactive surface sites after the electrochemical hydrogenation. This excludes the existence of well oriented facets and terraces on electrochemically hydrogenated Si surfaces. A possible explanation might be that the electrochemical oxidation process with current oscillations is induced by the defects which are than oxidised and etched-back leading to a reduced defect state density.

Figure 4.16 shows infrared (IR) spectra in the Si-H stretching regime of Si(111) surfaces after different treatments to obtain a hydrogenated Si(111) surfaces [26, 52]. The spectra are normalized to the oxidised surface measured in the same environment. The narrowest IR absorption peak can be observed for the smooth and flat Si(111) surface after etching in 40 % NH_4F. This is a result of the flat terraces (see Fig. 4.10) [26, 52]. The IR absorption after current oscillations is also very narrow but exhibits a broad shoulder at lower wavenumbers which is due to Si-H surface species at step facets [53, 54]. The hydrogenation performed with 5 % HF dip leads to a low IR absorption and broadening of Si-H related stretching vibrations as a result of surface roughening (see Fig. 4.10a) [26].

Two possible atomic arrangements of Si-H and Si=H_2 at step facets are shown in Fig. 4.17(a,A), where vicinal Si atoms on a (100) oriented facet (Si=H_2 groups) can be connected as shown in (b) and (B), respectively. This behaviour is similar to a 1x2 reconstruction of a step facet [35, 55, 56], which has a relatively high degree of freedom in the bond angles and bond length, and therefore permits variation in the bonds which can round off corners and steps so that no recombination active defect is formed. This variation in surface bond lengths and angles is reflected by a broadening of the IR absorption in the Si-H stretching vibration regime (Fig. 4.16) by coupled monohydrides [26]. The high electric field with the sudden increase in current during the anodic oxidation in the oscillating regime may be the driving force for such a kind of local reconstruction.

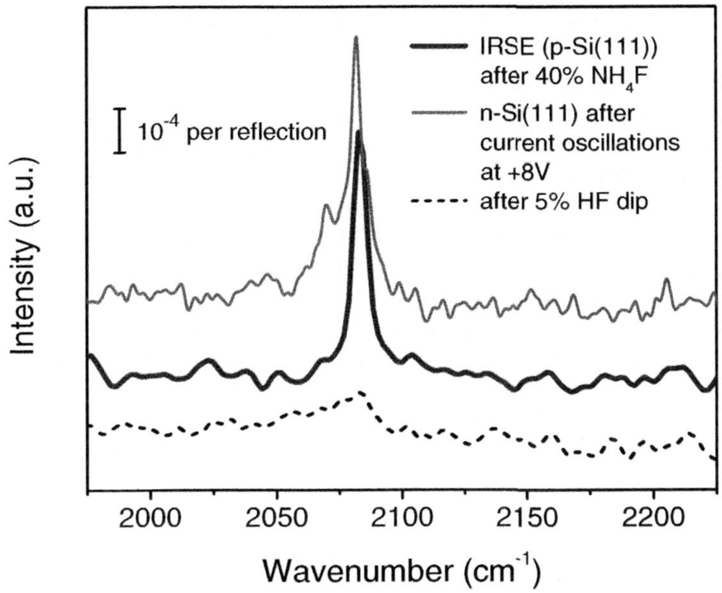

Fig. 4.15 (a) Infrared spectra of Si(111) surface recorded in the Si-H stretching vibration regime after hydrogenation by 40% NH_4F (thick solid line), after current oscillations at +8 V and subsequent hydrogenation at -0.4 V (thin solid line), and after 5% HF dip of an anodically prepared oxide at +4 V (dashed line).

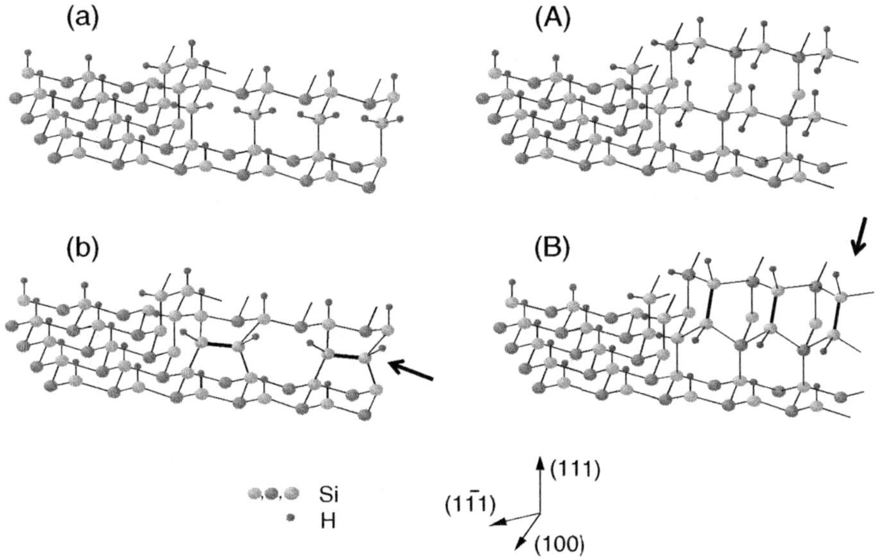

Fig. 4.17 Sketch of the atomic arrangement of hydrogenated Si(111) surfaces with steps along (100) direction (a) and (A), and possible types of respective reconstructions (b) and (B) [7].

4.3.6 Suppression of Defect Formation on H-Terminated Si

The decrease in I_{PL} after the oxide removal (cf. e.g. Fig. 4.13) is a result of etching and defect formation on the Si surface immersed in the HF solution. To suppress these surface reactions, it is essential to purge the solution by nitrogen and to exchange the used solution directly after the hydrogenation process to avoid surface etching as shown in Fig. 4.18. Here purging the solution by N_2 (·····) reduces the deterioration of the Si:H surface (—) in 0.1 M NH_4F (pH 4) slightly, whereas exchanging the HF containing solution by an HF-free solution (e.g. 0.1M K_2SO_4, pH 3) at negative polarisation (----) opens a wide time window of about 200 s before I_{PL} slowly decreases. The decrease in I_{PL} in the non-purged solution is a result of surface etching by the electrochemically induced hydrogen at the Si surface. This hydrogen is induced by the small negative current density of -1 μA/cm^2 which is able to reduce H^+-ions in solution to H-atoms and finally H_2 is formed [57, 58]. Here is some additional potential for optimising procedures.

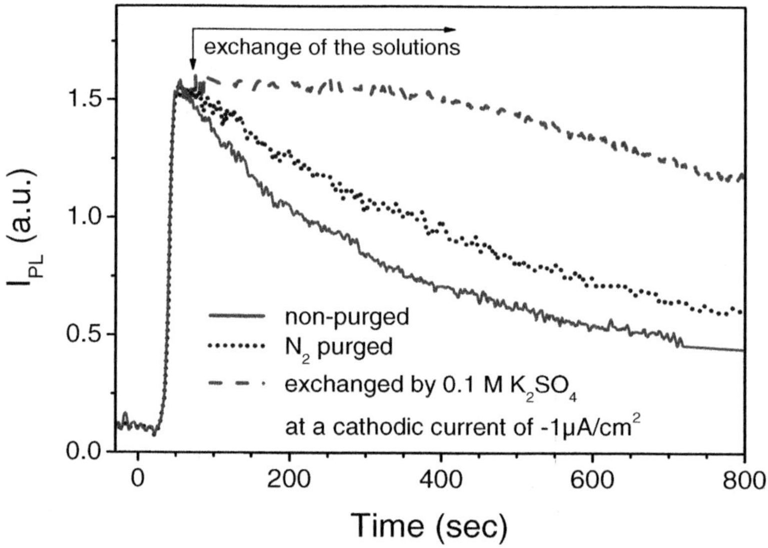

Fig. 4.18 I_{PL} during the hydrogenation process of oxidised p-Si(100) in 0.1 M NH_4F (pH 4) without (—) and with (·····) nitrogen purging, and after exchanging the etching solution by 0.1 M K_2SO_4 (pH 3) at negative polarisation (----) [7].

To investigate the influence of oxidising agents present in the solution/electrolyte on the etch-back of SiO_2 layers on Si(111) in more detail, small amounts of H_2O_2 were added to the etching solution (0.1 M NH_4F, pH 4.2). The time behaviour of I_{PL} during the etching process as a function of the H_2O_2 concentration is shown in Fig. 4.19. The maximum value of I_{PL} directly after the etch-back of the oxide layer is reduced even at a small H_2O_2 concentration of 1 mM/L pointing to a fast attack of the Si-H surface by H_2O_2 molecules to form Si oxides (sub-oxides or peroxides). The oxidation rate is not linear with the concentration of H_2O_2 as can be seen from the relative change in I_{PL}^{max} with respect to the I_{PL}^{max} of the H_2O_2 free solution plotted in Fig. 4.20.

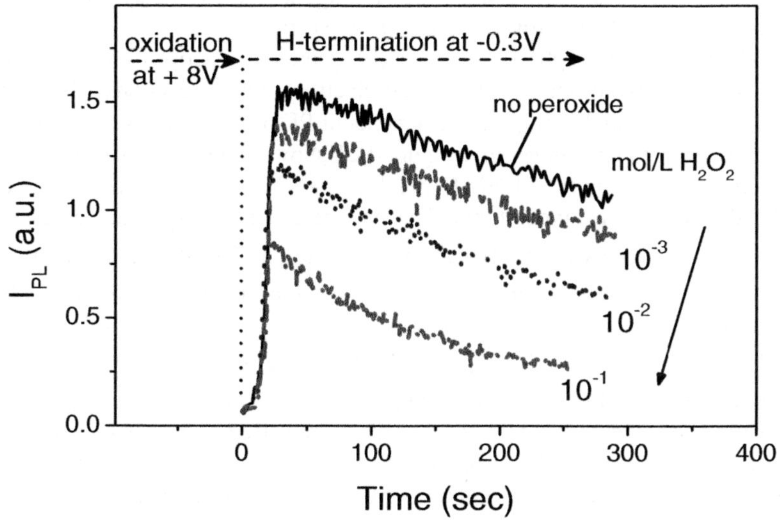

Fig. 4.19 Time dependence of I_{PL} for anodic oxidation of Si(111) in 0.1 M NH_4F (pH 4.2) at +8 V followed by hydrogenation at -0.3 V with different amounts of oxidising agent added to the electrolyte (10^{-3} to 10^{-1} mol/L H_2O_2) [7].

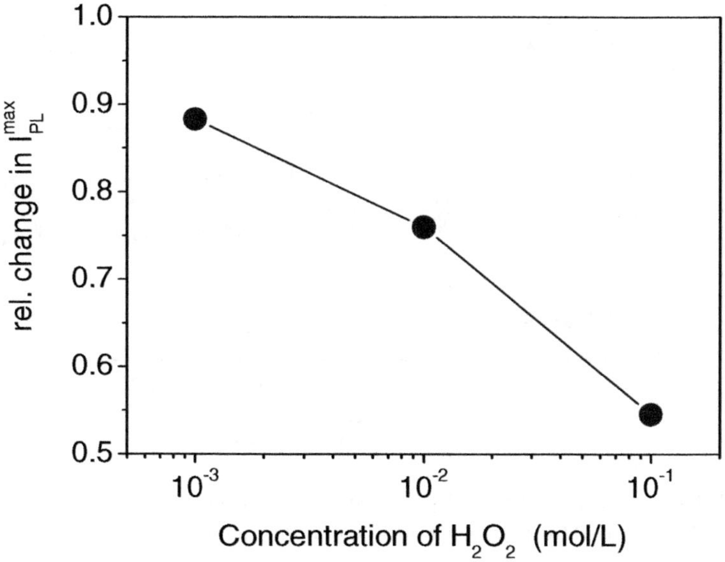

Fig. 4.20 The relative change in I_{PL}^{max} as a function of the H_2O_2 concentration in 0.1M NH_4F (pH 4) normalised to I_{PL}^{max} of the H_2O_2 free solution.

This behaviour suggests that other species in solution enhance the oxidation probability at the Si surface. Therefore, it seems plausible that HF or H_2O molecules support the attack of H_2O_2 species by pre-polarisation of Si-Si back bonds. These oxide species are then additionally etched and a roughening of the surface takes place. This result shows that oxidising agents have to be avoided in the electrolyte to stabilise the hydrogen passivated Si surface.

It is important to emphasize that surface chemical reactions are thermally activated, so that they compete with surface electrochemical reactions which are controlled by the potential. Temperature dependent PL measurements can be performed in a small temperature range only. Figure 4.21 shows the time dependence of I_{PL} for the anodic oxidation in 0.1 M NH_4F (pH 4.2) at +3 V followed by the hydrogenation at -0.4 V at different temperatures. The maximum of I_{PL}, I_{PL}^{max}, decreases strongly with a slightly increased temperature of the solution. The strong change in I_{PL}^{max} is induced by a strong change in the rate of generation of surface defects which act as non-radiative surface recombination defects. I_{PL} decays quite fast for the lowest temperature (8°C) after reaching I_{PL}^{max}, while the decay of I_{PL} is insignificant for the highest temperature (44°C) used. I_{PL} depends only weakly on the temperature after longer times of etching what shows that there are different processes which lead either to the formation of H-terminated Si surfaces or to chemical etching of the H-terminated Si surfaces by additional formation of reactive surface sites.

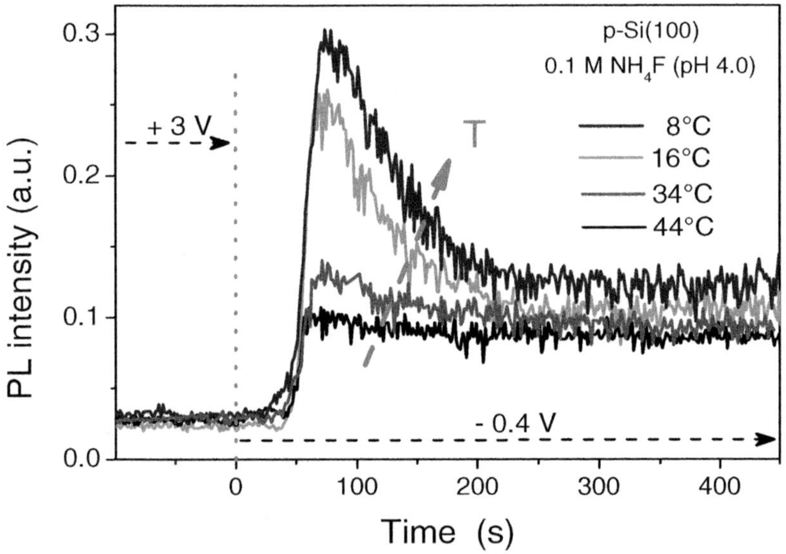

Fig. 4.21 The time dependence of I_{PL} of p-Si(100) for anodic oxidation in 0.1 M NH_4F (pH 4) at +3 V followed by hydrogenation at -0.4 V at different temperatures [7] .

To summarise this section, it can be concluded that low temperatures of the solutions and nitrogen purging or (preferred) exchanging the HF-containing solution by an HF-free solution after the oxide etching step enhances or stabilises the electronic properties of the H-terminated Si surface.

An alternative novel route is to passivate Si surfaces by electrochemical or chemical grafting of organic molecules. The next section describes some results on the passivation of Si surfaces by electrochemical grafting of small molecules and benzene derivatives, and the stabilisation of such surfaces with respect to oxidation in ambient air at room temperature.

4.4 Surface Passivation by Organic Molecules

The passivation of Si surfaces can be performed by electrochemical or chemical grafting of organic molecules. To do this, there are many strategies in literature. The next section will focus on some special grafting techniques and the impact of such surface layers on the recombination losses at Si/organic molecule interfaces and on the long time stability versus ambient air. The following treatments have been applied to Si surfaces.

4.4.1 Grafting Techniques

4.4.1.1 Reduction of Diazonium Ions

The electrochemical grafting process via diazonium ions is sketched in eqs. 4.8a-c and consists of 3 main steps (here for nitrobenzene (NB) as an example):

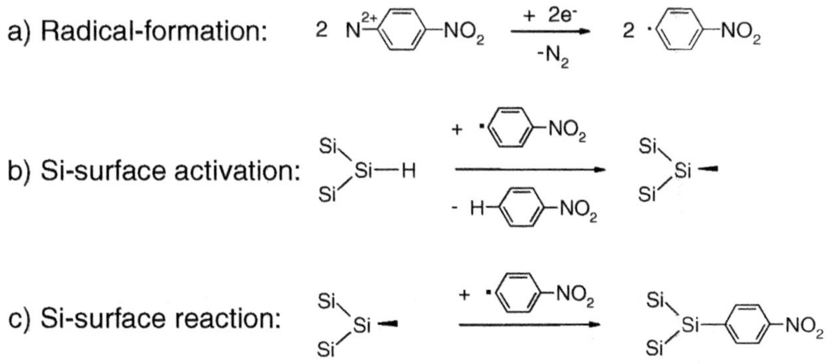

(4.8)

The first step is the (a) radical formation by reduction of the diazonium salt in solution followed by the (b) Si surface activation via abstracting H-atoms, and (c) subsequent surface reaction by binding an aryl radical to the Si dangling bond [59, 60]. The end group in eq. (4.8), NO_2, can be replaced by many other groups leading to different surface dipoles what influences the work function and band bending of Si [61].

4.4.1.2 Oxidation of Grignard Compounds

The initial reaction is the formation of Si dangling bonds on the surface according to eq. (4.9a and 4.9b) followed by bonding of a radical to a dangling bond (eq. (4.9c)) [62].

$$\begin{array}{lll} \text{a)} & \text{R–C≡C–MgX} + h^+ & \rightarrow \text{R–C≡C•} + \text{MgX}^+ \\ \text{b)} & \text{≡Si–H} + \text{•C≡C–R} & \rightarrow \text{≡Si•} + \text{H–C≡C–R} \\ \text{c)} & \text{≡Si•} + h^+ + \text{R–C≡C–MgX} & \rightarrow \text{≡Si–C≡C–R} + \text{MgX}^+ \end{array} \quad (4.9)$$

Several side reactions are reported which lead to halogen incorporation into the layer [63].

4.4.1.3 Polymerisation of Pyrrole and Thiophene

Here, pyrrole (or thiophene, where N is replaced by S) can be anodically polymerised corresponding to:

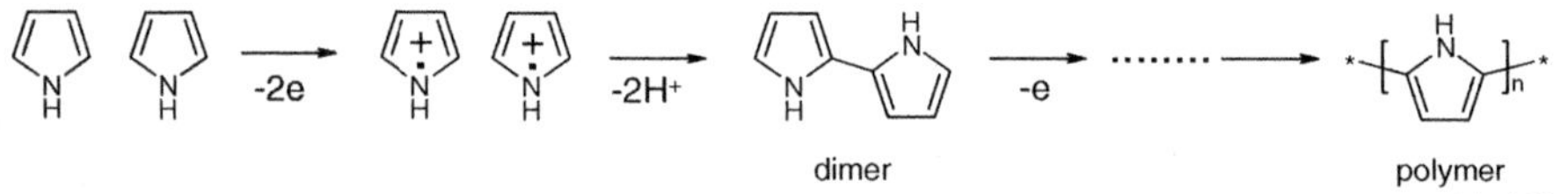

(4.10)

Pyrrole is firstly oxidised to radical cations which recombine to a dimer and additional cation radicals are bonded to the dimer leading finally to a polymeric layer as outlined in eq. (4.10).

4.4.1.4 Thermal Hydrosilylation

This technique is used for aliphatic molecules which have a double or triple bond at the end position (1-alkenes or 1-alkynes) [64]. This double bond can be activated and bound to the Si-H surface according to eq. (4.11):

$$\text{≡Si–H} + \text{H}_2\text{C=CH–R} \xrightarrow{90\text{-}200°\text{C}} \text{≡Si–CH}_2\text{–CH}_2\text{–R} \quad (4.11)$$

The next sections will focus on these types of Si surface modifications with respect to passivation, defect formation and stability of such organic surface layers emphasising their specialities and drawbacks.

4.4.2 Electrochemical Reduction of Diazonium Ions

For this type of surface modification, the Si electrode is cathodically polarised to ensure the electron transfer to the diazonium ions in solution. This has an additional advantage since holes, which are able to oxidise Si to SiO_2, are blocked from the surface and therefore the formation of SiO_2 should be suppressed. However, as will see, this is not completely the case since radical mediated reactions lead to side reactions due to the high oxidation energy of the radicals.

Figure 4.22 shows the time behaviour of (a) I_{PL} and (b) current of Si(111) surfaces during grafting of 4-nitrobenzene (4-NB) from 4-nitrobenzene diazonium tetrafluoroborate (4-NBDT) in aqueous acidic solution (0.01 M H_2SO_4, 0.5 % HF, solid line) or from non-aqueous solution (acetonitrile with tetrabutylamonium hexafluorophosphate, TBAF, as conducting salt, dotted line). The insert shows the time dependence of the current on a smaller time interval. The cathodic current reflects the electron transfer from the Si surface to the ions in solution and the grafting occurs via eq. (4.8) [65].

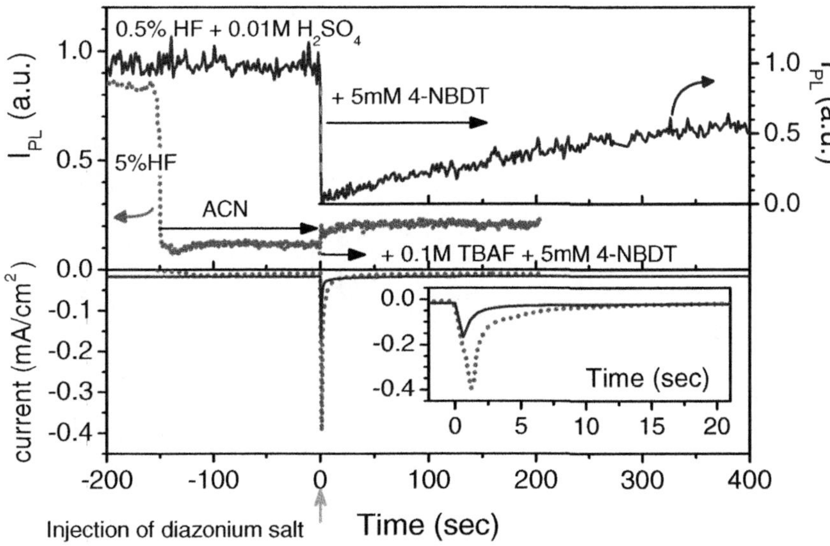

Fig. 4.22 Time behaviour of (a) I_{PL} and (b) current of Si(111) surfaces during grafting of 4-NB from 4-NBDT in aqueous acidic solution (0.01 M H_2SO_4, 0.5 % HF, solid line, -1.2 V) or from non-aqueous solution (acetonitrile, 0.1 M TBAF, dotted line, -2 V). The insert shows the time dependence of the current on a smaller time interval.

Simultaneously with the current flow, I_{PL} drops drastically for the 0.5 % HF + 0.01 M H_2SO_4 solution reflecting the formation of recombination active surface defects which cure out with ongoing time of grafting leading to an I_{PL} half of the starting condition. This means that twice the number of defects as compared to the starting value of D_{it} are formed during the process by side reactions [66, 67], e.g. SiO_2 surface groups or etch pits due to HF etching. The behaviour of I_{PL} is completely different using acetonitrile (CH_3-C≡N) as solvent. Here, the solvent exclusively leads to a strong decrease in I_{PL}, which is partly reduced when 4-NBDT is added to the solution and grafted by electrochemical reduction. There is no further enhancement in I_{PL} as observed for the aqueous solutions. This effect is a result of the stronger interaction of acetonitrile with the hydrophobic Si:H surface which obviously induces some recombination active surface complexes (I_{PL} is reduced by a factor of about 10 which means that 10 times more defects are created) [16]. The small increase in I_{PL} is a result of grafting of 4-NB molecules which replace acetonitrile molecules adsorbed at the surface. However, only small amounts of 4-NB are grafted whereas most of the current is lost by side reactions like dimerisation or reaction with solvent despite of the 2-times higher charge flow compared to the aqueous solution.

Nevertheless, the electrochemical treatment in diluted sulfuric acid leads to well ordered interfaces as can be seen from Fig. 4.23, where a high resolution transmission electron microscopy (HR-TEM) image of a 4-bromobenzene (4-BrB) terminated Si(111) surface is presented. The Si-Si lattice distance of about 3.4 Å and an about 6 Å thick monolayer of 4-bromobenzene at the Si interface, which seems to be highly ordered, can be well seen. This layer completely covers the Si(111) surface [16].

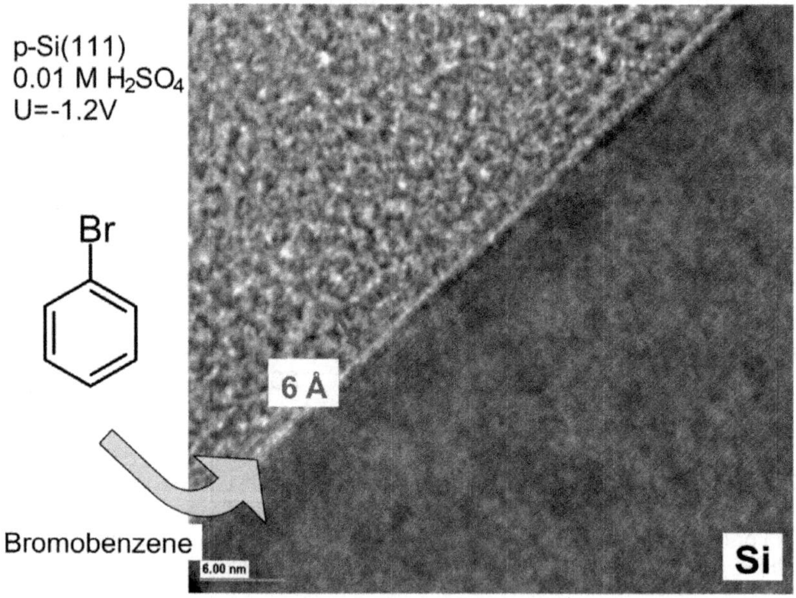

Fig. 4.23 HR-TEM image of the 4-bromobenzene (4-BrB) layer on Si(111) [16].

Figure 4.24 addresses the stability of the modified Si(111) surfaces in ambient air conditions. I_{PL} of the smooth and flat Si(111):H surface (see Fig. 4.24) decreases continuously with time. This behaviour is correlated to the appearance of the SiO_2 related vibrational mode of such surfaces [16].

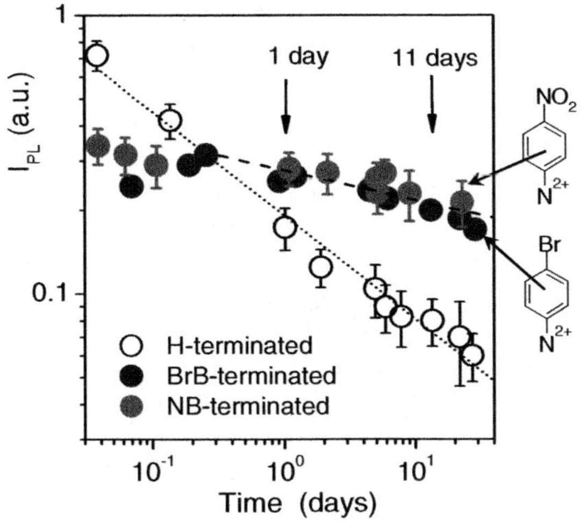

Fig. 4.24 Ex-situ measured I_{PL} at 1130 nm (Si bandgap) of H-terminated, bromobenzene (BrB)- and nitrobenzene (NB)-terminated Si(111) surfaces as a function of time in ambient air (standard lab humidity and temperature) [16].

The decrease in I_{PL} and the appearance of SiO_2 related IR absorption is much less pronounced if the surfaces are covered by organic molecules (e.g. bromobenzene, BrB, and nitrobenzene, NB). Therefore, BrB- and NB-terminated surfaces are relatively stable in ambient air even for months and oxide formation is strongly suppressed. The defect formation rate is about 10 times slower for the organically modified surfaces.

4.4.3 Grafting via Grignard Compounds and Polymerisation of Pyrrole and Thiophene

Figure 4.25 shows PL spectra measured for methyl-groups (CH_3) anodically grafted via electrochemical oxidation of the Grignard compound CH_3MgBr (see eq. (4.9)), and after deposition of ultra-thin layers of polypyrrole and polythiophene according to eq. (4.10). All the grafting processes have been done by scanning the potential from 0 to about +1V [68]. Additionally, the PL spectrum of a NB-terminated Si surface is shown for comparison for a cathodically driven process. These spectra are compared to the PL spectra of freshly H-terminated Si and half a year aged native oxide layers on Si. All Si/organic interfaces show a reduced I_{PL} that is about half of the initial Si-H surface, i.e. the amount of interface states is increased by a factor of 2 only. However, these layers have a quite stable I_{PL} in ambient air whose time behaviour is similar to the results as obtained for benzene layers (Fig. 4.24) [16].

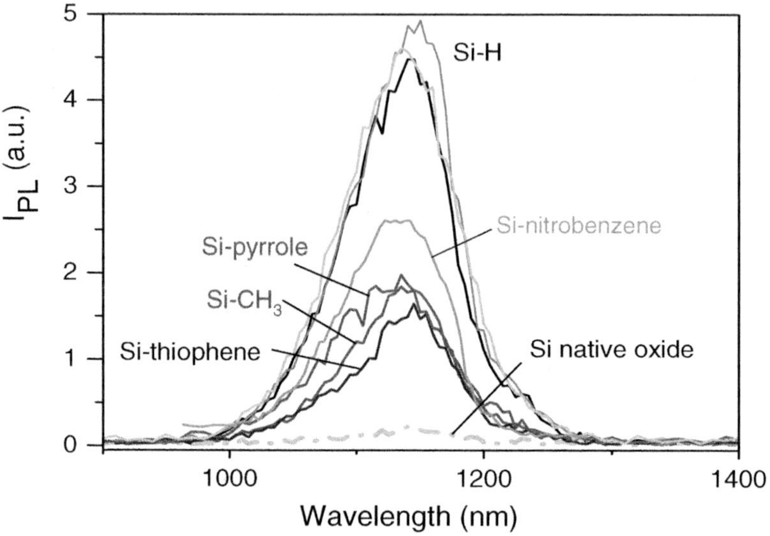

Fig. 4.25 PL spectra of Si(111):H surfaces, native oxidised Si surface, NB-terminated and for anodically grafted methyl-groups ($Si\text{-}CH_3$), and after deposition of a ultra-thin polypyrrole and polythiophene layers.

Figure 4.26 shows SEM images of ultra-thin films of polythiophene or polypyrrole on Si(111) surfaces as deposited from their Grignard compounds which completely cover the Si surface. This thin layer may be partly the cause of the reduced I_{PL} since such material starts to absorb light in the visible regime (here $\lambda_{exc} = 500$ nm).

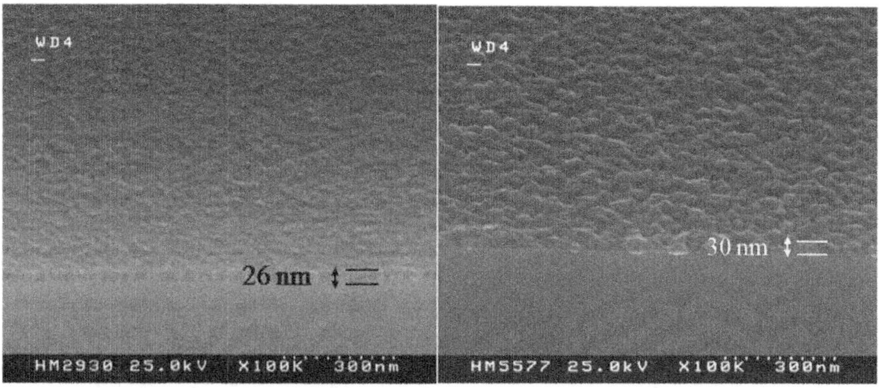

Fig. 4.26 SEM images of polythiophene (top) and polypyrrole (bottom) deposited onto Si(111) surfaces using thiophen-2-yl- or pyrryl magnesium bromide electrolyte in galvanostatic mode (0.1 mA/cm² during 15 minutes).

4.4.4 Thermal Hydrosilylation

A very promising technique is the thermal hydrosilylation of 1-alkenes via eq. (4.11). The thermal hydrosilylation is easy to perform by heating of appropriate compounds in inert gas atmosphere and under dry conditions. Hereby long time stable and very well passivated Si(111) surfaces, nearly similar to H-terminated surfaces, can be prepared [69], which show no change in I_{PL} in ambient air even after a month (Fig. 4.27b), whereas the Si(111):H surface degrades within several hours (Fig. 4.27a) or days depending on the final surface morphology (the flatter the surface the longer can it be handled in ambient air).

This behaviour points to Si surfaces well passivated by Si-carbon bonds, i.e. a Si-C bond does not introduce a recombination active defect. This demonstrates the potential of such surface passivation methods by means of organic molecules.

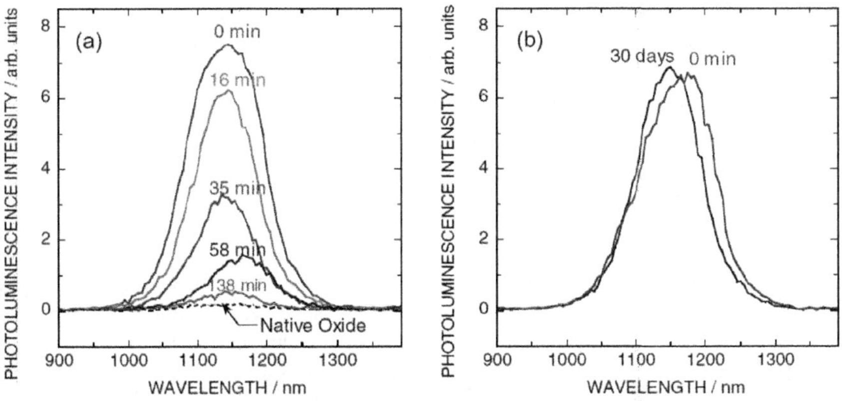

Fig. 4.27 Ex-situ photoluminescence spectra at different times. (a) hydrogenated surface; (b) surface grafted with 10-carboxydecyl groups. Note the huge PL stability upon grafting [69].

This section was dedicated to novel approaches to passivate Si surfaces by means of organic molecules. To conclude, it has been shown that Si-C bonds are very well suitable for passivation of Si surfaces since Si-C is not a recombination active defect. In some cases, low defect concentrations similar to H-terminated surfaces have been reached. The surface passivation can be influenced by the appropriate selection of the solution condition (solvent, pH, etc.). Additionally, these surfaces show low oxidation probability compared to Si-H surfaces and are stable for months. However, such surfaces require low thermal budget processing during further processing, otherwise the surface will be damaged and destroyed.

4.5 Conclusion and Outlook

This chapter complements the preceding one on wet-chemical surface passivation. It shows that electrochemical treatment of Si surfaces leads to completely different surface morphologies with respect to wet-chemical treatments but with similar or even better surface passivation, especially when current oscillations are involved during the oxidative pre-treatment. PL spectroscopy can be applied as sensitive and non-destructive tool to investigate in-situ the development and passivation of the recombination active defects on Si surfaces to optimise the processing steps of oxidation and hydrogenation. The formation of recombination active defects during the etch-back of oxidised Si surfaces can be avoided by using N_2-bubbled electrolytes or solutions, and exchanging the etching solution by a non-etching one or interrupting the etching processes in HF solutions. Additionally, in-situ measured surface photovoltage provides information about the charging of surfaces and interfaces.

However, Si-H surfaces are oxidised in ambient air on a time scale of hours or days depending on the Si surface structure after hydrogen termination. This oxidation can be reduced or even completely suppressed by grafting of organic molecules (benzene, pyrrole, thiophene) onto Si surfaces. Up to now, the simple chemical process of thermal hydrosilylation by 1-alkenes has been proven to be the most promising and leads to excellently passivated Si surfaces.

Acknowledgments

The financial support of the European Union through the EFRE program (ProFIT grant, contract no. 10131870/1 and no. 10136530/1/2), the Senatsverwaltung für Wissenschaft, Forschung und Kultur des Landes Berlin and the Bundesministerium für Bildung, Wissenschaft, Forschung und Technologie are gratefully acknowledged.

The author thanks Dr. K. Hinrichs, Dr. M. Gensch, and Dr. K. Roodenko for measuring the IRSE spectra and fruitful discussions; Dr. T. Dittrich for long years cooperation in Si surface passivation and valuable discussions on SPV measurements; Dr. F. Yang for polymeric layers on Si; Dr. D. Aureau for the samples made by thermal hydrosilylation; and finally X. Zhang and S. Chapel for assistance in the grafting of some diazonium compounds.

References

[1] Yablonovich, E., Allara, D.L., Chang, C.C., Gmitter, T., Bright, T.B.: Unusually low surface-recombination velocity on silicon and germanium surfaces. Phys. Rev. Lett. 57, 249–252 (1986)

[2] Angermann, H., Henrion, W., Rebien, M., Zettler, J.-T., Röseler, A.: Characterization of chemically prepared Si-surfaces by uv-vis and IR spectroscopic ellipsometry and surface photovoltage. Surf. Sci. 388, 15–23 (1997)

[3] Angermann, H., Kliefoth, K., Flietner, H.: Preparation of H-terminated Si surfaces and their characterisation by measuring the surface state density. Appl. Surf. Sci. 104-105, 107–112 (1996)

[4] Rappich, J., Timoshenko, V.Y., Dittrich, T.: In Situ Monitoring of Electrochemical Processes at the (100) p-Si/Aqueous NH4F Electrolyte Interface by Photoluminescence. J. Electrochem. Soc. 144, 493–496 (1997)
[5] Timoshenko, V.Y., Petrenko, A.B., Stolyarov, M.N., Dittrich, T., Fuessel, W., Rappich, J.: Quantitative analysis of room temperature photoluminescence of c-Si wafers excited by short laser pulses. J. Appl. Phys. 85, 4171–4175 (1999)
[6] Dittrich, T., Burke, T., Koch, F., Rappich, J.: Passivation of an anodic oxide / p-Si interface stimulated by electron injection. J. Appl. Phys. 89, 4636–4642 (2001)
[7] Rappich, J., Dittrich, T.: Electrochemical passivation of Si and SiGe. In: Nalwa, H.S. (ed.) Handbook of Thin Film Materials, pp. 1–56. Academic Press, New York (2002)
[8] Turner, D.R.: Electropolishing silicon in hydrofluoric acid solution. J. Electrochem. Soc. 105, 402 (1958)
[9] Ozanam, F., Chazaviel, J.-N., Radi, A., Etman, M.: Current oszillation in the anodic dissolution of silicon in flouride electrolytes. Ber. Bunsenges. Phys. Chem. 95, 98–101 (1991)
[10] Rappich, J., Aggour, M., Rauscher, S., Lewerenz, H.J., Jungblut, H.: Electrochemical surface conditioning of n-Si(111). Surf. Sci. 335, 160–165 (1995)
[11] Rauscher, S., Dittrich, T., Aggour, M., Rappich, J., Flietner, H., Lewerenz, H.J.: Reduced interface state density after photocurrent oscillations and electrochemical hydrogenation of n-Si(111): A surface photovoltage investigation. Appl. Phys. Lett. 66, 3018–3020 (1995)
[12] Timoshenko, V.Y., Petrenko, A.B., Dittrich, T., Füssel, W., Rappich, J.: Photoluminescence characterization of non-radiative defect density on silicon surfaces and interfaces at room temperature. Thin Solid Films 364, 196–199 (2000)
[13] Timoshenko, V.Y., Rappich, J., Dittrich, T.: In Situ Photoluminescence Analysis of Nonradiative Recombination on Silicon Surfaces Treated in Fluoride Solution. Jpn. J. Appl. Phys. 36, L58–L60 (1997)
[14] Rappich, J., Timoshenko, V.Y., Dittrich, T.: Correlation between Surface Non-Radiative Recombination and Current Oscillations at p-Si(100) during Electropolishing in Fluoride Solution. Ber. Bunsenges. Phys. Chem. 101, 139 (1997)
[15] Rappich, J., Timoshenko, V.Y., Würz, R., Dittrich, T.: Is there a Limit for the Passivation of Si Surfaces during Anodic Oxidation in Acidic NH4F Solutions? Electrochim. Acta 45, 4629–4633 (2000)
[16] Rappich, J.H., Hartig, P., Nickel, N.H., Sieber, I., Schulze, S., Dittrich, T.: Stable electrochemically passivated Si surfaces by ultra-thin benzene-type layers. Microelec. Engin. 80, 62–65 (2005)
[17] Rappich, J., Zhang, X., Rosu, D.M., Schade, U., Hinrichs, K.: Passivation of Si surfaces investigated by in-situ photoluminescence techniques. Solid State Phenomena 156-158, 363–368 (2010)
[18] Timoshenko, V.Y., Rappich, J., Dittrich, T.: Express characterization of indirect semiconductors by in-situ photoluminescence during chemical and electrochemical treatments. Appl. Surf. Sci. 123/124, 111–114 (1998)
[19] Higashi, G.S., Chabal, Y.J., Trucks, G.W., Raghavachari, K.: Ideal hydrogen termination of Si(111) surface. App. Phys. Lett. 56, 656–658 (1990)
[20] Kern, W.: Si Cleaning. Semiconductor International 94 (1984)
[21] Kern, W.: The Evolution of Silicon Wafer Cleaning Technology. J. Electrochem. Soc. 137, 1887 (1990)

[22] Jakob, P., Chabal, Y.J.: Chemical etching of vicinal Si(111): Dependence of the surface structure and the hydrogen termination on the pH of the etching solutions. J. Phys. Chem. 95, 2897–2909 (1991)
[23] Lewerenz, H., Blitzer, T., Gruyters, M., Jacobi, K.: Electrolytic Hydrogenation of Silicon. J. Electrochem. Soc. 140, L44–L46 (1993)
[24] Rappich, J., Lewerenz, H.: In Situ Fourier Transform Infrared Investigation on the Electrolytic Hydrogenation of n-Silicon (111). J. Electrochem. Soc. 142, 1233–1237 (1995)
[25] Cullis, A.G., Canham, L.T.: Visible light emission due to quantum size effects in highly porous crystalline silicon. Nature 353, 335 (1991)
[26] Chabal, Y.J., Higashi, G.S., Raghavachari, K., Burrows, V.A.: Infrared spectroscopy of Si(111) and Si(100) surfaces after HF treatment: Hydrogen termination and surface morphology. J. Vac. Sci. Technol. A 7, 2104–2109 (1989)
[27] Peter, L.M., Blackwood, D.J., Pons, S.: In situ characterizazion of the illuminated Silicon-electrolyte interface by Fourier-Transform Infrared Spectroscopy. Phys. Rev. Lett. 62, 308–311 (1989)
[28] Venkateswara Rao, A., Chazaviel, J.-N.: Chemical trends in the electromodulated infrared vibrational spectroscopy of various silicon/electrolyte interfaces. J. Electrochem. Soc. 134, 2777–2783 (1987)
[29] Morita, Y., Tokumoto, H.: Ideal Hydrogen Termination Of Si(001) Surface By Wet-Chemical Preparation. Appl. Phys. Lett. 67, 2654–2656 (1995)
[30] Grundner, M., Schulz, R.: The surface state of silicon (100) and (111) wafers after treatment with hydrofluoric acid. In: AIP Conf. Proc., vol. 167, p. 329 (1988)
[31] Dumas, P., Chabal, Y.J.: Electron-energy-loss characterization of the H-terminated Si (111) and Si (100) surfaces obtained by etching in NH_4F. Chem. Phys. Lett. 181, 537 (1991)
[32] Stuhlmann, C., Bogdány, G., Ibach, H.: Surface phonons of the hydrogen-terminated Si(111) (1x1) surface. Phys. Rev. B 45, 6786–6792 (1992)
[33] Lewerenz, H.J., Bitzer, T., Gruyters, M., Jacobi, K.: Electrolytic Hydrogenation of Silicon: A High Resolution Electron Energy Loss Spectroscopy Investigation. J. Electrochem. Soc. 140, L44–L46 (1993)
[34] Grundner, M., Jacob, H.: Investigations on hydrophilic and hydrophobic silicon (100) wafer surfaces by X-ray photoelectron and high-resolution electron energy loss-spectroscopy. Appl. Phys. A 39, 73 (1986)
[35] Maruno, S., Iwasaki, H., Horioka, K., Li, S.-T., Nakamura, S.: Electronic structures of monohydride (2x1):H and the dihydride (1x1):2H Si(001) surfaces studied by angle-resolved electron-energy-loss spectroscopy. Phys. Rev. B 27, 4110–4116 (1983)
[36] Zaima, S., Kojima, J., Hayashi, M., Ikeda, H., Iwano, H., Yasuda, Y.: Improvements of Electrical Characteristics of Hf/p-Si(100) Interfaces by H-Termination. Jpn. J. Appl. Phys. 34, 741–745 (1995)
[37] Bitzer, T., Gruyters, M., Lewerenz, H.J., Jacobi, K.: Electrochemically prepared Si(111) 1x1-H surface. Appl. Phys. Lett. 63, 397 (1993)
[38] Harrick, N.J.: Internal Reflection Spectroscopy. Wiley Interscience, New York (1967)
[39] Rappich, J., Jungblut, H., Aggour, M., Lewerenz, H.: Electrochemical smoothing of silicon (111). J. Electrochem. Soc. 141, L99–L102 (1994)
[40] Ozanam, F., Chazalviel, J.-N.: In-situ infrared characterization of the electrochemical dissolution of silicon in a fluoride electrolyte. J. Electron Spec. Rel. Phenom. 64/65, 395 (1993)

[41] Lewerenz, H.J., Aggour, M.: On the origin of photocurrent oscillation at Si electrodes. J. Electroanal. Chem. 351, 159–168 (1993)

[42] Aggour, M., Giersig, M., Lewerenz, H.J.: Interface condition of n-Si(111) during photocurrent oscillations in NH4F solutions. J. Electroanal. Chem. 383, 67 (1995)

[43] Yang, F., Roodenko, K., Hinrichs, K., Rappich, J.: Electronic and surface properties during the etch-back of anodic oxides on Si(111) surfaces in 40% NH_4F solution. J. Micromech. Microeng. 17, S56–S70 (2007)

[44] Uhlig, H.H., King, P.F.: The Flade Potential of Iron Passivated by Various Inorganic Corrosion Inhibitors. J. Electrochem. Soc. 106, 1–7 (1959)

[45] Gouy, G.: Sur la constitution de la charge electrique á la surface d'un electrolyte. Compt. Rend. 149, 654–657 (1909)

[46] Chabal, Y.J.: Surface infrared spectroscopy. Surf. Sci. Rep. 8, 211 (1988)

[47] Emoto, T., Akimoto, K., Ishikawa, Y., Ichimiya, A., Tanikawa, A.: Strain near SiO2-Si interface revealed by X-ray diffraction intensity enhancement. Thin Solid Films 369, 281–284 (2000)

[48] Imai, T., Fujimoto, A., Okuyama, M., Hamakawa, Y.: Characterization of Surface Potential and Strain at Ultrathin Oxide/Silicon Interface by photoreflectance Spectroscopy. Jpn. J. Appl. Phys. 35, 1073–1076 (1996)

[49] Neaton, J.B., Muller, D.A., Ashcroft, N.W.: Electronic Properties of the Si/SiO2 Interface from First Principles. Phys. Rev. Lett. 85, 1298–1301 (2000)

[50] Nouwen, B., Stesmans, A.: Dependence of strain at the (111)Si/SiO2 interface on interfacial Si dangling-bond concentration. Mat. Sci. and Engin. A 288, 239–243 (2000)

[51] Stefanov, B.B., Gurevich, A.B., Weldon, M.K., Raghavachari, K., Chabal, Y.J.: Silicon Epoxide: Unexpected Intermediate during Silicon Oxide Formation. Phys. Rev. Lett. 81, 3908–3911 (1998)

[52] Chabal, Y.J., Dumas, P., Guyot-Sionnest, P., Higashi, G.S.: Vibrational dynamics of the ideally H-terminated Si(111) surface. Surf. Sci. 242, 524–530 (1991)

[53] Dumas, P., Chabal, Y.J., Jakob, P.: Morphology of hydrogen-terminated Si(111) and Si(100) surfaces upon etching in HF and buffered-HF solutions. Surf. Sci. 269/270, 867 (1992)

[54] Nakamura, M., Song, M.-B., Ito, M.: Etching processing of Si(111) and Si(100) surfaces in an ammonium fluoride solution investigated by in situ ATR-IR. Electrochim. Acta 41, 681–686 (1996)

[55] Bender, H., Verhaverbeke, S., Caymax, M., Vatel, O., Heyns, M.M.: Surface reconstruction of hydrogen annealed (100) silicon. J. Appl. Phys. 75, 1207 (1994)

[56] Chabal, Y.J., Rafghavachari, K.: Surface infrared study of Si(100)-(2X1)H. Phys. Rev. Lett. 53, 282–285 (1984)

[57] de Mierry, P., Etcheberry, A., Rizk, R., Etchegoin, P., Aucouturier, M.: Defects Induced in p-Type Silicon by Photocathodic Charging of Hydrogen. J. Electrochem. Soc. 141, 1539–1546 (1994)

[58] Rappich, J., Dittrich, T., Timoshenko, Y., Beckers, I., Fuhs, W.: Influence of hydrogen incorporation into Silicon on the room-temperature photoluminescence. In: Mat. Res. Soc. Symp. Proc., vol. 452, pp. 797–802 (1997)

[59] Allongue, P., Delamar, M., Desbat, B., Fagebaume, O., Hitmi, R., Pinson, J., Saveant, J.-M.: Covalent Modification of Carbon Surfaces by Aryl Radicals Generated from the Electrochemical Reduction of Diazonium Salts. J. Am. Chem. Soc. 119, 201–207 (1997)

[60] Allongue, P., de Villeneuve, C.H., Pinson, J., Ozanam, F., Chazalviel, J.N., Wallart, X.: Organic monolayers on Si(111) by electrochemical method. Electrochim. Acta 43, 2791 (1998)
[61] Hartig, P., Rappich, J., Dittrich, T.: Engineering of Si surfaces by electrochemical grafting of p-nitrobenzene molecules. Appl. Phys. Lett. 80, 67–69 (2002)
[62] Teyssot, A., Fidèlis, A., Fellah, S., Ozanam, F., Chazalviel, J.-N.: Anodic grafting of organic groups on the silicon surface. Electrochim. Acta 47, 2565–2571 (2002)
[63] Yang, F., Hunger, R., Roodenko, K., Hinrichs, K., Rademann, K., Rappich, J.: Vibrational and Electronic Characterization of Ethynyl Derivatives Grafted onto Hydrogenated Si(111) Surfaces. Langmuir 25, 9313–9318 (2009)
[64] Linford, M.R., Chidsey, C.E.D.: Alkyl monolayers covalently bonded to silicon surfaces. J. Am. Chem. Soc. 115, 12631–12632 (1993)
[65] Rappich, J., Hinrichs, K.: In situ study of nitrobenzene grafting on Si(1 1 1)-H surfaces by infrared spectroscopic ellipsometry. Electrochem. Commun. 11, 2316–2319 (2009)
[66] Allongue, P., de Villeneuve, C.H., Cherouvrier, G., Cortés, R., Bernard, M.-C.: Phenyl layers on H/Si(111) by electrochemical reduction of diazonium salts: monolayer versus multilayer formation. J. Electroanal. Chem. 550/551, 161–174 (2003)
[67] Rappich, J., Merson, A., Roodenko, K., Dittrich, T., Gensch, M., Hinrichs, K., Shapira, Y.: Electronic properties of Si surfaces and side reactions during electrochemical grafting of phenyl layers. J. Phys. Chem. B 110, 1332–1337 (2006)
[68] Intelmann, C.M., Hinrichs, K., Syritski, V., Yang, F., Rappich, J.: Recombination behaviour at the ultrathin polypyrrole film/silicon interface investigated by in-situ pulsed photoluminescence. Jap. J. Appl. Phys. Part I 47, 554–557 (2008)
[69] Aureau, D., Rappich, J., Moraillon, A., Allongue, P., Ozanam, F., Chazalviel, J.-N.: In situ monitoring of the electronic properties and the pH stability of grafted Si(111). J. Electroanal. Chem. 646, 33–42 (2010)

Chapter 5
Deposition Techniques and Processes Involved in the Growth of Amorphous and Microcrystalline Silicon Thin Films

Pere Roca i Cabarrocas

Laboratoire de Physique des Interfaces et des Couches Minces, CNRS
Ecole Polytechnique, 91128 Palaiseau, France

Abstract. Hydrogenated amorphous and microcrystalline silicon deposition has been a subject of research over the last four decades, supported by its increasing number of applications. Many deposition techniques involving physical (sputtering) or chemical (plasma enhanced chemical vapour deposition) processes have been studied. The choice of the deposition technique may help to favour some type of film precursor, in particular SiH_3 which is often considered as the most suitable to obtain device grade material. However, taking as a general case the growth of μc-Si:H films, we show that the growth process and film properties are mainly controlled by the surface and subsurface reactions. In particular, thanks to in-situ ellipsometry measurements, we demonstrate that there is a growth zone close to the film surface, where cross-linking reactions leading to bulk-like formation take place. In fact, the crystallization front may be located a few tens of nanometers below the surface exposed to the plasma, thus suggesting that the film properties are governed neither by the film precursor, nor by the deposition technique. Finally, we address the issue of the substrate dependence of the growth process, which is fundamental in the case of heterojunction solar cells.

5.1 Introduction

Thin film deposition is a major area of research with industrial applications in various fields: micro-electronics, optics, hard coatings, protective layers, anti-corrosion, large area electronics, etc. To such a variety of applications one can associate a wide range of deposition techniques, which can be divided in two main categories: i) those based on a physical process such as evaporation and sputtering; and ii) those based on chemical processes, known under the generic term of chemical vapour deposition (CVD). Both physical and chemical deposition processes are used in the production of a-Si:H/c-Si heterojunction solar cells, as

W.G.J.H.M. van Sark et al. (Eds.): Physics & Tech. of Amorphous-Crystalline, EM, pp. 131–160.
springerlink.com

schematically shown in Figure 5.1. While evaporation and sputtering are the techniques of choice for the deposition of the metallic contacts and transparent conductive oxide (TCO) layers respectively, hydrogenated amorphous silicon (a-Si:H) thin films are generally produced by plasma enhanced chemical vapour deposition (PECVD). For the heterojunction solar cells addressed in this book, cleaning the c-Si substrate before loading it into the PECVD system is a crucial step in order to achieve a good passivation of the c-Si surface. This is usually achieved by wet chemistry processes [1,2], even though it has been recently demonstrated that an excellent surface passivation can also be achieved by a dry plasma cleaning process using SiF_4 [3].

A plasma is defined as an ionized gas that is macroscopically neutral. The ionization of the gas atoms and molecules is achieved by bringing energy to the system, which can be achieved in various ways: thermally, optically and electrically; the most common for thin film deposition is an electrical discharge between two electrodes. Such a system is under non-equilibrium conditions. Indeed, the species present in the discharge have quite different energies: electrons being the particles with lower mass, they can gain energy from the applied electric field and be heated up to a few tens of eV (1 eV is equivalent to 11.600 K). On the other hand the ions and gas molecules are thermalized, i.e. they are at the temperature of the walls of the reactor (~ 300 K). Charge neutrality can only be achieved in the volume of the plasma; close to the walls space charge regions form, in which positive ions are accelerated towards the surface, leading to ion bombardment. Low temperature plasma processes experience an increasing interest, either for the deposition of thin films or for their etching (about 40% of the processes required to produce an integrated circuit are based on etching processes). As mentioned above plasma deposition of thin films covers a wide range of materials and applications [4,5]. Concerning a-Si:H, the first report goes back to 1879, when Ogier reported on the deposition of hydrogenated amorphous silicon using an electrical discharge in silane [6]. However, it was only hundred years later that the interest on this material really started, with the discovery of the possibility of doping it, in the 1970s by Chittick and Spear and Le Comber [7,8]. This boosted the research on hydrogenated amorphous silicon with thousands of papers studying the relationship between the deposition conditions (pressure in the reactor, substrate temperature, power coupled to the plasma, gas flow rate, reactor geometry) and material properties (optical gap, hydrogen content, structure, defect density,...). An excellent review was published by Street in 1991 [9]. Interestingly, the first report on μc-Si:H due to S. Veprek, also goes back to 1968 [10], however, this is a much more complex material and its industrial application is still limited to tandem solar cells [11]. On the contrary, a-Si:H thin films are the basis of a fast expanding large area electronics industry, based on the fact that this disordered semiconductor can be deposited over large areas and therefore it allows to develop new applications such as thin film transistors for flat panel displays, solar cells, particle detectors, etc. [12].

The combination of crystalline silicon and a-Si:H to produce heterojunction solar cells was recognized as early as 1974 by W. Fuhs [13]. However its real interest appeared when Sanyo demonstrated that efficient solar cells can be produced using this combination of materials [14]. The heterojunction (HJ) solar cell structure shown in Fig. 5.1 looks quite "simple". Still, it took many years to researchers around the world to increase the efficiency of their HJ cells to values close to these of Sanyo and still today, Sanyo has a definite leadership, with record efficiency of 23% at research level and 20% at the module production level [15]. Why is there such gap between Sanyo and the rest of the world? Are we missing something in the "simple" device structure shown in Fig. 5.1, and if so, what is it? This chapter tries to bring an answer to this question by looking at the HJ solar cell from the deposition process point of view.

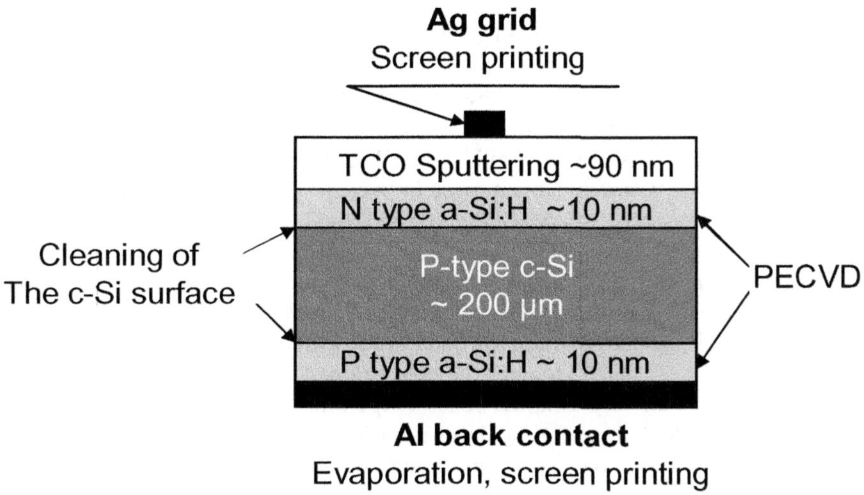

Fig. 5.1 Schematic of a heterojunction solar cell consisting of a p-type c-Si wafer coated with n-type and p-type a-Si:H layers on the front and backside respectively. This is generally achieved by plasma CVD. The cell is completed by the deposition of front transparent conducting oxide (TCO), an Ag grid on the front side, from which the cell is illuminated and an Al back contact. TCO deposition, usually Indium Tin Oxyde (ITO), is achieved by sputtering, while Ag and Al contacts are deposited by screen printing or thermal evaporation.

5.2 Plasma-Based Deposition Techniques

The increasing number of applications of hydrogenated amorphous and microcrystalline silicon thin films has stimulated researchers to develop new deposition techniques. We will come back to the term "deposition" in section 5.5, where we will show that one should think in terms of growth rather than deposition. The development of new techniques is often led by the industrial requirement of increasing the deposition rate, in order to reduce production cost and make the products competitive with respect to existing solutions. This is particularly true in the field of solar energy where equipment cost is often considered as a limiting factor for the development of a-Si:H based solar cells. One could argue that this is not the case of heterojunction solar cells due to the small thickness of the a-Si:H layers to be deposited on each side of the crystalline silicon wafer (see Fig. 5.1). If we consider a production line with a moderate capacity of 100 MW/year of HJ solar cells with 20% efficiency, assuming cells of 200 cm^2, this would imply to process $25x10^6$ cells/year. Now, if we consider that the a-Si:H deposition system is working 80% of the time (292 days/year), a rather optimistic value if we consider that some time will be required for maintenance, cleaning of the reactor, failures, etc, then this means that one has to process ~ 86 thousand wafers per day or 60 wafers/minute (both sides). One can continue these estimations, but the fact is that time matters, and even if only 20 nm are required, one should deposit them fast. Indeed, to achieve the required throughput, one needs to increase either the deposition rate or the number of PECVD reactors. In order to limit the number of PECVD reactors required to achieve a-Si:H deposition on 60 wafers/minute, increasing the deposition rate from the standard 0.1 nm/s up to 1 nm/s is a desirable solution. However, as discussed below, this objective has to be combined with producing a-Si:H films which provide excellent passivation of the c-Si surface. The issue of deposition rate is even more important for amorphous and microcrystalline silicon solar cells, where the thickness of the layers are one or two orders of magnitude higher than for heterojunction solar cells. The same reasons of cost have lead researchers and equipment suppliers to look for deposition techniques that allow increasing the deposition rate: very high frequency PECVD [16], expanding thermal plasma [17], hot-wire CVD [18], micro-wave CVD [19], atmospheric pressure CVD [20], ionized cluster beam deposition [21]. Nevertheless, PECVD in a capacitively coupled reactor operating at 13.56 MHz remains the leading industrial technique for the deposition of a-Si:H and μc-Si:H thin films. Its success is mainly due to the fast development of the active matrix flat panel industry, which has seen up to 10 generations of production systems, allowing today to deposit uniform a-Si:H films on substrate sizes up to 5.7 m^2 [22]. Figure 5.2 shows a schematic diagram of such a parallel-plate PECVD reactor. Besides its simplicity, this type of reactor also benefits from the advantages associated to the plasma processes [4,5], namely:

i) The fact that the dissociation of the gas precursors is produced by collisions with high energy electrons, and therefore deposition is possible even at room temperature.

ii) The fact that ions are accelerated towards the substrate and thus can bring energy to the growth zone, which will generally lead to dense and smooth films.

iii) The wide range of gas precursors which allows to produce a wide range of thin films: a-Si:H, a-SiGe:H, a-SiC:H, a-SiO_x, etc [23]. Moreover these films can be made p-type or n-type by adding either diborane or phosphine to the gas mixture and can be easily stacked on top of each other by changing the flow of gases into the reactor.

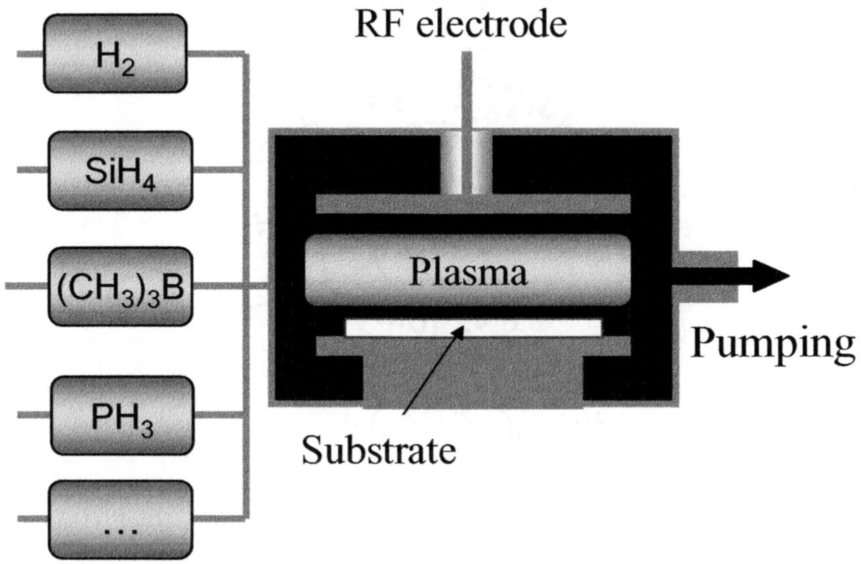

Fig. 5.2 Schematic diagram of a plasma enhanced chemical vapor deposition reactor. It consists of a gas handling system, a vacuum system, a RF power electrode and a substrate holder, which is usually heated to ~ 200 °C. The simple structure of the parallel plate system has made it possible to scale it up to glass plates of up to 5.7 m^2 in industrial systems.

5.3 Gas Phase Reactions

Let us now turn to the processes taking place in the gas phase, which depend on the nature of the gases injected in the reactor. As for the deposition techniques, there is a wide choice of gas precursors for a-Si:H deposition: SiH_4, Si_2H_6, SiF_4, $SiCl_3H$, $SiCl_2H_2$, etc. Let us focus on silane, which is the most studied and widely used gas precursor for a-Si:H deposition.

Once SiH_4 is injected in the reactor and the power is applied to the RF electrode, electrons will be accelerated by the electric field and gain enough energy to dissociate SiH_4. The primary reactions between electrons and silane can be of different types: dissociation, ionization, attachment, etc. The relative importance of each reaction depends on the energy of the electrons, which is a function of the electric field (from which they gain energy) and their mean free path, which depends on the total pressure. Figure 5.3 shows the products of reaction between silane and electrons of 70 eV, as deduced from mass spectrometry measurements [24]. One can see that radicals are the main product, followed by ions, and that within each group of species SiH_3 is the most abundant.

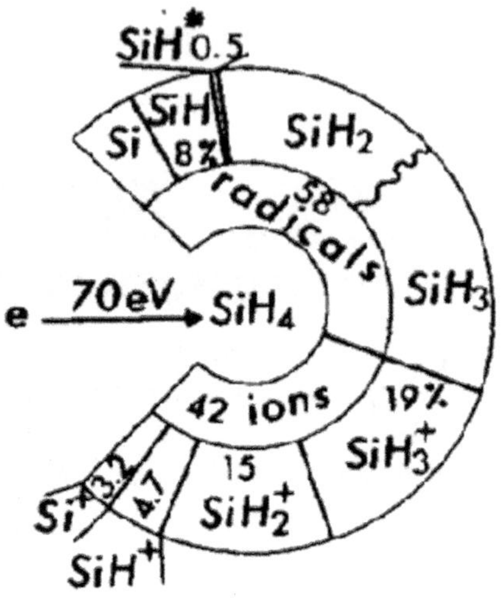

Fig. 5.3 Dissociation pattern of a silane molecule upon impact of a 70 eV electron as deduced from mass spectrometry measurements. Taken from reference [24].

At low pressures (1-10 Pa) the mean free path of the products of silane dissociation is comparable or higher than the inter-electrode distance and thus primary reactions are dominant. However, besides primary reactions, one should also consider secondary ones, that are the reactions which result from the interaction of the products of primary reactions with the most abundant species in the discharge, namely silane molecules. Indeed, the degree of dissociation in most silane plasmas for a-Si:H deposition is typically 10%. When trying to increase silane dissociation, for example by increasing the RF power coupled to the plasma, one also enhances these secondary reactions, which will gain importance and eventually dominate the plasma chemistry [25,26]. The complexity of the plasma processes is schematically illustrated in Figure 5.4 by the rich variety of species present in the plasma. As shown in the figure, the following species will react on the substrate in different amounts depending on the process conditions:

- Silicon radicals (SiH_3, SiH_2, SiH, Si). They will diffuse to the substrate and reach it at thermal energy.
- High energy electrons. Most of the electrons will be confined in the plasma because of the higher potential of the plasma with respect to the substrate (see 5.4). However those having energy higher than the plasma potential will reach the substrate and promote chemical reactions [27].
- Positive ions. Contrary to electrons, all ions arriving at the sheath boundary will be accelerated towards the substrate with energy of the order of the plasma potential (V_p). Depending on their mean free path, they will reach the substrate with energy equal to eV_p or lower if they experience many collisions when crossing the sheath.

- Photons and metastable species will also interact with the growing film. However their effect has been less studied and little is known about their impact on film properties.
- Negative ions as well as large powders (negatively charged) are confined in the plasma. Negative ions are considered as the main candidates for powder formation [28]. Contrary to nanocrystals which can experience charge fluctuations and be positively charged, large powders (diameter > 10 nm) are always negatively charged.
- Clusters and nanocystals formed in the plasma can also interact with the growing film. Their transport to the surface will depend on their charge which is a function of their size. Nanometer size particles experience charge fluctuations and therefore will contribute to deposition when neutral or positively charged [29].

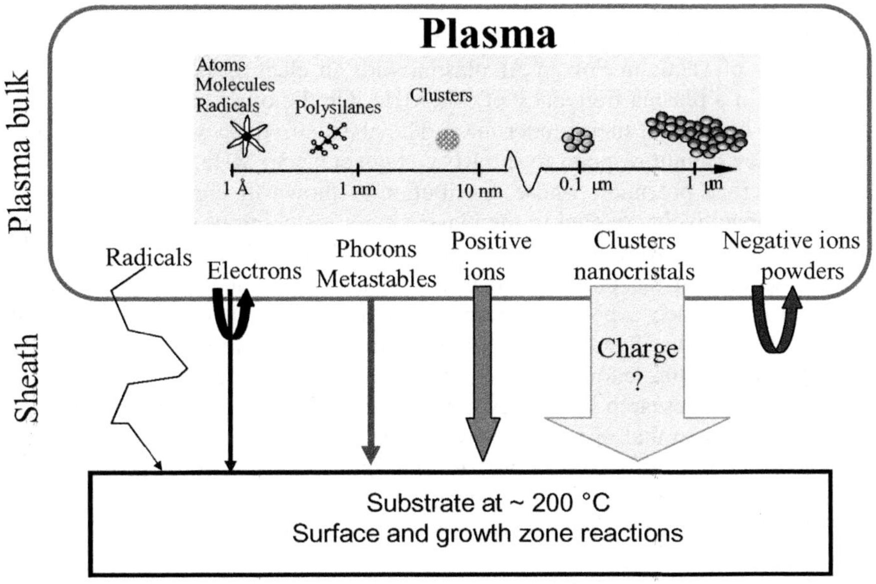

Fig. 5.4 The plasma as a source of reactive species. Their interaction with the substrate depends on their charge due to the presence of a space charge region (sheath) between the macroscopically neutral plasma-bulk and the walls.

The aim of this chapter is not to address the role of each of these film precursors. However, it is important to keep in mind that growth models often correspond to a simplified case of the general situation presented in Figure 5.4. A good understanding of silane plasma processes involves detailed experiments and modelling which are out of the scope of this chapter. The readers interested can look into the abundant literature on both silane plasma modelling [30,31,32] and diagnostics [33,34,35].

5.4 Plasma Surface Interactions

As defined above, plasma is an ionized gas which is macroscopically neutral. This means that in plasmas produced by an electrical discharge between two electrodes as shown in Fig. 2, there must be a transition region (sheath) between the plasma and the reactor walls. Indeed, as the energy of the power supply is mostly coupled to electrons, they are the most energetic species in the discharge and would just move out of the plasma if there was not a potential barrier preventing them from leaving the plasma. Electrons can instantly react to any excitation, in particular to the applied potential at 13.56 MHz. Their plasma frequency (f_{pe}) is given by:

$$f_{pe} = \frac{\omega_{pe}}{2\pi} \quad \text{with} \quad \omega_{pe} = \sqrt{\frac{q^2 n_e}{m_e \varepsilon_0}} \tag{5.1}$$

where q is the charge of the electron, m_e its mass, n_e the electron density and ε_0 the permittivity of vacuum. For an Ar plasma with an electron density of 10^{10} cm^{-3} this results in a plasma frequency of 900 MHz. On the other hand, ions are much less mobile because of their higher mass. The plasma frequency for Ar^+ ions is ~ 1 MHz, so they cannot respond to the RF voltage at 13.56 MHz. They only experience an average potential, whose distribution is shown in Fig. 5.5. Here the RF power is capacitively coupled to the plasma via a matching box, required to minimize the reflected power. This potential distribution implies that negatively charged species in the plasma (electrons, negative ions and negatively charged powders) experience a potential barrier which prevents them from leaving the plasma. On the other hand, positively charged ions and particles will be accelerated to the walls, thus leading to the so-called ion bombardment.

In the case of research laboratory reactors, the area of the grounded surfaces is usually higher than that of the RF powered electrode. Consequently the system is asymmetric and a self-bias (V_B) will develop on the RF electrode (provided there is a capacitive coupling of the RF power). The value of the self-bias depends on the ratio of the grounded versus RF powered areas and is given by the so-called area law [36].

$$\frac{V_1}{V_2} = \left(\frac{A_2}{A_1}\right)^n \quad \text{for } 2 < n < 4 \tag{5.2}$$

Now, in the case of a perfectly symmetric system the two areas are equal and therefore $V_1 = V_2 = V_p$ where the plasma potential will be equal to half the RF voltage (V_{RF}). In the general case where the system is not symmetric and there is capacitive coupling there will be a self bias (V_B) on the RF electrode and the plasma potential is given by [37]:

$$V_p = \frac{1}{2}\left(V_{RF} + V_B\right) \tag{5.3}$$

The energy of the ions will be given by the difference between the plasma potential and the substrate potential. In general the substrate is placed on the grounded electrode and therefore at low pressures, when the ions can travel across the sheath without collisions, the ion energy will be equal to the plasma potential. At a high pressure the ion energy distribution function will be broadened to lower energy. Note that the energy of impinging ions is a crucial parameter which affects the optical and electronic properties of a-Si:H and microcrystalline films [38,39,40].

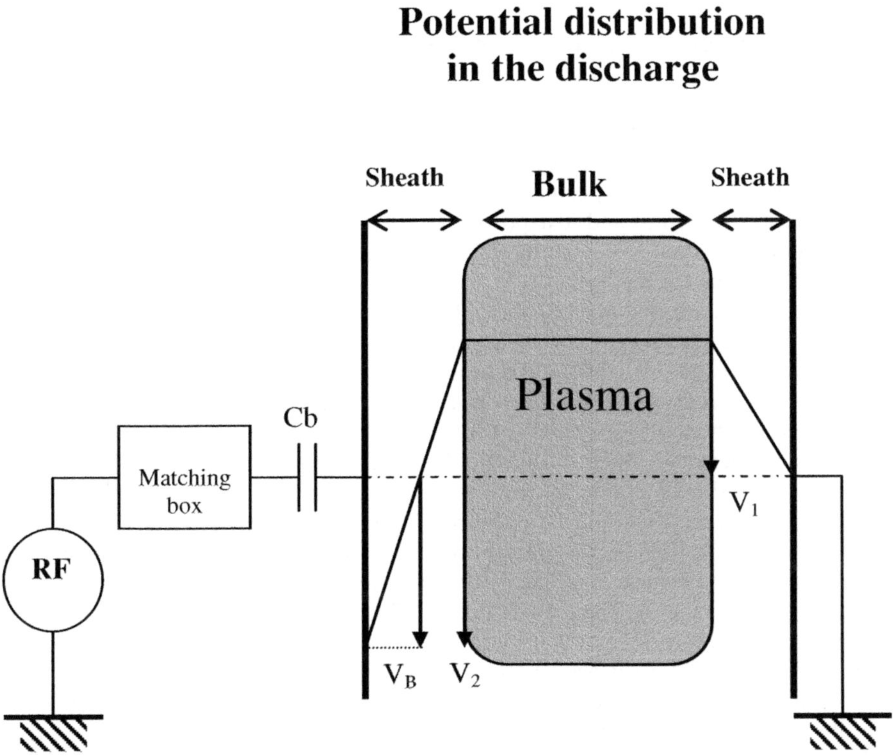

Fig. 5.5 DC potential distribution in a capacitively coupled RF glow discharge reactor. Note that this is the potential experienced by ions. To this DC potential there is a RF potential, which affects electrons.

As the size of the substrates and consequently that of the RF electrode increases, industrial systems become more symmetric and the self bias on the RF electrode tends to zero, while the plasma potential will increase, leading to higher ion energy and possible damage of the crystalline silicon substrate and growing films. This will be even more pronounced when increasing the RF power coupled to the discharge, as is required to increase the deposition rate. To circumvent this effect, various strategies can be adopted such as the use of higher excitation frequency (VHF and microwave plasmas), hot wire CVD, photo-CVD, etc. While these approaches can be adapted to decrease ion energy, the increase of the power coupled to the discharge will bring the desired increase in deposition rate, but also the eventual formation of powders [28]. As a rule, a compromise will have to be found in order to achieve a high deposition rate while keeping low ion energy without powder formation. However this is not an easy task and research is still progressing to find ways to achieve all the requirements in the same reactor.

After all these considerations concerning the plasma processes, let us now turn into the surface processes leading to the deposition of silicon thin films.

5.5 Hydrogenated Amorphous Silicon Deposition Models

It is interesting that despite the richness of silane plasmas presented above, the standard model for a-Si:H deposition is based on SiH_3 as the main film precursor, considered to be the radical leading to "device grade material" [41,42]. Why is this so? Probably because plasma studies [43] have shown that under conditions of low pressure (< 10 Pa) and low power (< 10 mW/cm^2) this is the most abundant radical in the plasma. Note that the fact that a radical is the most abundant species does not necessarily mean that it is the main one contributing to deposition. Indeed, one could argue that SiH_2 and SiH radicals with a sticking coefficient ~ 0.9 are not detected because they are easily lost to the surface, while SiH_3, with a sticking coefficient ~ 0.1 [44] has a higher chance of bouncing back to the plasma and thus be present in the plasma with a much higher density. Nevertheless, SiH_3 is supposed to be the "good" radical for the production of a-Si:H films with a low density of electronic defects, suitable for electronic devices. According to this view, Fig. 5.6 shows a popular cartoon used to explain the growth of a-Si:H films from SiH_3 radicals. The radicals produced in the plasma reach the film surface via diffusion through the sheath and land on a hydrogen terminated silicon surface. This is expected because at the low deposition temperature (~ 200 °C) thermal energy is not sufficient to break Si-H bonds. Thus, once the radical reaches the hydrogen terminated surface, it cannot form a chemical bond, rather it will physisorb and diffuse along the surface of the growing film. During this diffusion process SiH_3 may experience various reactions: i) desorption as SiH_3, with a high probability as its sticking coefficient has been measured to be ~ 0.1; ii) recombination with another SiH_3 to produce a Si_2H_6 molecule; iii) abstraction of a hydrogen atom producing a SiH_4 molecule and leaving a surface dangling bond on which the next SiH_3 can form a chemical bond; and iv) formation of a chemical bond with a silicon dangling bond at the surface, leading to deposition.

The growth model depicted in Figure 5.6 relies on the diffusion of SiH_3 radicals and device quality material is often associated to a high diffusion length. However it is difficult to determine experimentally what is the actual value of the diffusion length, besides the fact that a low sticking coefficient is somehow synonymous to a high diffusion length (a SiH_3 radical can reach a favourable position to be incorporated after some absorption/desorption steps). Modelling can provide a more precise idea about the diffusion length. In the case of a hydrogen terminated (100) c-Si surface it has been reported that the diffusion length is only a few lattice spacings due to fast desorption of SiH_3 radicals [45].

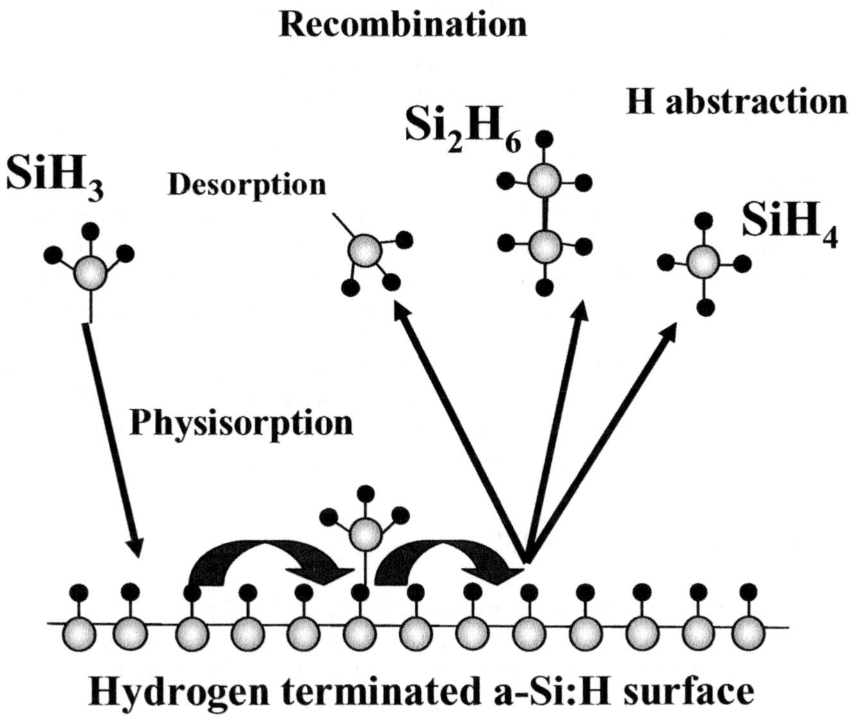

Fig. 5.6 Standard view for the deposition of device quality a-Si:H thin films based on the interaction of SiH_3 radicals with a hydrogen terminated silicon surface.

The growth model depicted in Fig. 5.6 is the most popular to explain a-Si:H and even microcrystalline silicon deposition [46], and it is certainly pertinent in some particular growth conditions of low pressure and low RF power in PECVD processes or in Photo-CVD [44]. However, the deposition conditions used to produce a-Si:H and μc-Si:H films at high deposition rates, as required for industrial applications, will involve more reactive radicals such as SiH_2 and SiH [25,26] as well as silicon clusters and nanocrystals, as schematically shown in Fig. 5.4. On the other hand, besides the simplification of the plasma processes, the surface itself is also highly idealized: a flat and hydrogen terminated silicon surface as shown in Fig. 5.4 hardly corresponds to the case of a-Si:H deposition on glass substrates (interestingly this could apply to the case of deposition on a crystalline silicon wafer for heterojunction solar cells). In practice the surface is not so flat, it rather develops some surface roughness with dangling bonds and strained bonds. To account for the latter aspect, the picture of a-Si:H deposition from SiH_3 has been extended to the case where this radical would insert directly into a strained Si-Si bond [47].

Still, the picture above completely ignores another very important species in the plasma: atomic hydrogen. This is a ubiquitous element in PECVD processes involving SiH_4. As a matter of fact, it took many years to realize that the superior properties of amorphous silicon films produced by PECVD compared to these of the films produced by sputtering from a silicon target were due to the incorporation of hydrogen in the films. Thus, besides surface growth models based on the diffusion of SiH_3, one should also consider chemical reactions taking place in a growth zone [48,49] where cross-linking exothermic reactions involving hydrogen may lead to a rearrangement of the atomic network. This has also been known as chemical annealing [50,51].

The SiH_3 model is thus a quite restrictive approach, which can only apply to a limited number of processes or deposition conditions. Indeed, the need for high deposition rates often leads to the use of deposition conditions such as high pressure and high power, where secondary reactions cannot be ignored. This has been the object of many studies, in particular for the deposition of polymorphous and microcrystalline silicon thin films [52]. But rather than going into complex plasma processes, let us turn our attention to *in situ* studies of the growth of silicon thin films under well controlled plasma conditions where we make sure that only SiH_3 radicals and atomic hydrogen are responsible for the growth.

5.6 Deposition or Growth?

To gain more insight into the processes involved in the deposition or growth of silicon thin films, let us focus on the rather simple case where the growth is achieved in silane-hydrogen plasmas. As a matter of fact the dilution of silane in hydrogen is often related to the growth of microcrystalline silicon films, even though they can also be obtained in pure silane plasmas, provided the dissociation of silane is high enough to produce the atomic hydrogen necessary for μc-Si:H growth. To simplify as much as possible the plasma processes let us consider a growth process where a substrate is alternatively submitted to pure silane and pure hydrogen plasmas under well controlled conditions. We will see how the grown material can change as a function of time (thickness) despite the fact that the "deposition conditions" are kept constant.

5.6.1 Layer by Layer Deposition of Silicon Thin Films

Layer-by-layer (LBL) deposition has been studied by many groups, aiming mostly to the production of microcrystalline silicon thin films [53,54]. In this technique the substrate is alternatively exposed to silane and hydrogen plasmas under well controlled conditions (in general low pressure and low RF power), which minimize secondary reactions in the gas phase so that one can assume that film deposition mostly takes place through the reaction of SiH_3 radicals and atomic hydrogen. Fig. 5.7 shows a schematic diagram of the LBL process in which the two main parameters are the silane plasma exposure time (T_{Si}) and the hydrogen plasma exposure time (T_H). The waiting times (T_{r1} and T_{r2}) can be reduced to zero without affecting the process results. The other process parameters, i.e. RF power, pressure and substrate temperature are kept constant.

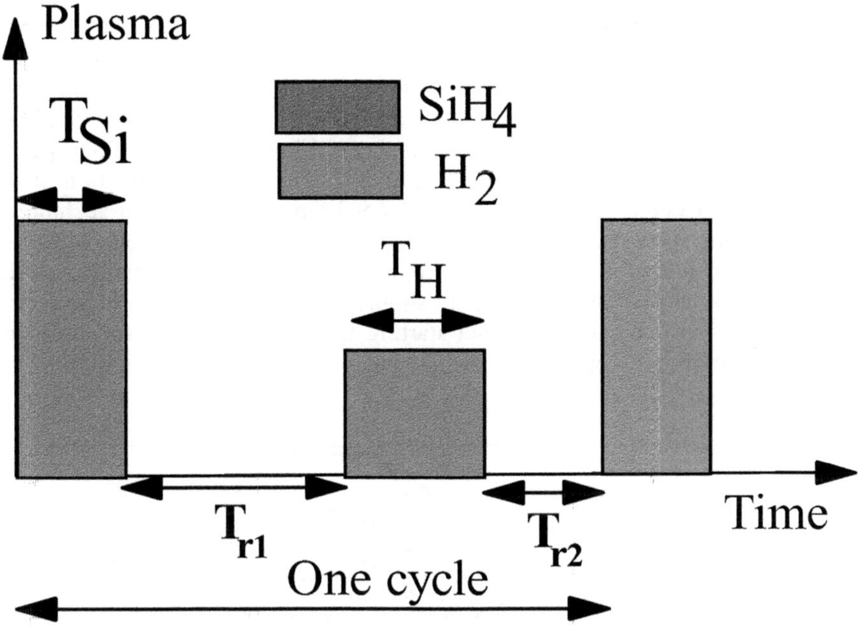

Fig. 5.7 Layer-by-layer process for the deposition of silicon thin films. A substrate is alternatively exposed to silane and hydrogen plasmas. The only process variables are the silane exposure time (T_{Si}) and the hydrogen plasma treatment time (T_H).

Besides the simplification of the plasma chemistry, LBL allows controlling independently the flux of atomic hydrogen and of SiH_3 radicals, and this can be achieved at low ion energy, thus avoiding the damage to the surface and its amorphization. This is not the case in the standard µc-Si:H deposition, where silane is highly diluted in hydrogen, which requires high RF power to dissociate H_2 thereby producing a flux of atomic hydrogen sufficient for crystallization (as discussed below). However increasing the RF power results in an increase of the plasma potential and therefore of the energy of heavy SiH_x^+ ions.

Fig. 5.8 shows the imaginary part of the pseudo-dielectric function ($\langle\varepsilon_2\rangle$), deduced from in-situ UV-visible ellipsometry measurements during a LBL process applied to a glass substrate coated with an a-Si:H layer [55]. One can see that $\langle\varepsilon_2\rangle$ continuously changes with the number of LBL cycles. The amplitude of $\langle\varepsilon_2\rangle$ first decreases up to 15 cycles and continues to decrease with a distinct change in its shape at about 25 cycles. At 35 and 45 cycles the amplitude increases recovering to that of the initial a-Si:H layer, but its shape has completely changed. One can see a plateau in the energy range between 3.5 and 4.1 eV, which is a signature of crystalline material [56]. Further insight into the properties of the film produced by the LBL process can be achieved from modelling of the ellipsometry data. This can be done using the Bruggemann effective Medium Approximation which allows to describe the pseudo-dielectric function of a layer by the dielectric-function of its constituents (amorphous phase, crystalline phase, and voids or porosity) averaged by their fraction. This method can be extended to a layer stack [48,57]. Indeed, modelling the ellipsometry spectra allows revealing the complex structure of the films. In the general case, μc-Si:H films can be described by a four layer model consisting of: i) an interface layer with the substrate, ii) a bulk layer, iii) a subsurface layer, and iv) a surface roughness layer. Thus, modelling of the spectra presented in Fig. 5.8 allows determining the evolution of the composition of the film in a very accurate way, which would not be possible by other techniques such as Raman spectroscopy [58]. Moreover, the in-situ measurements allow monitoring the evolution of the film properties and thus the dynamics of the growth process. Note that modelling is simpler in the case of in-situ measurements, as one can build up the model starting from a well known substrate and increase its complexity step-by-step, with the increase of the film thickness.

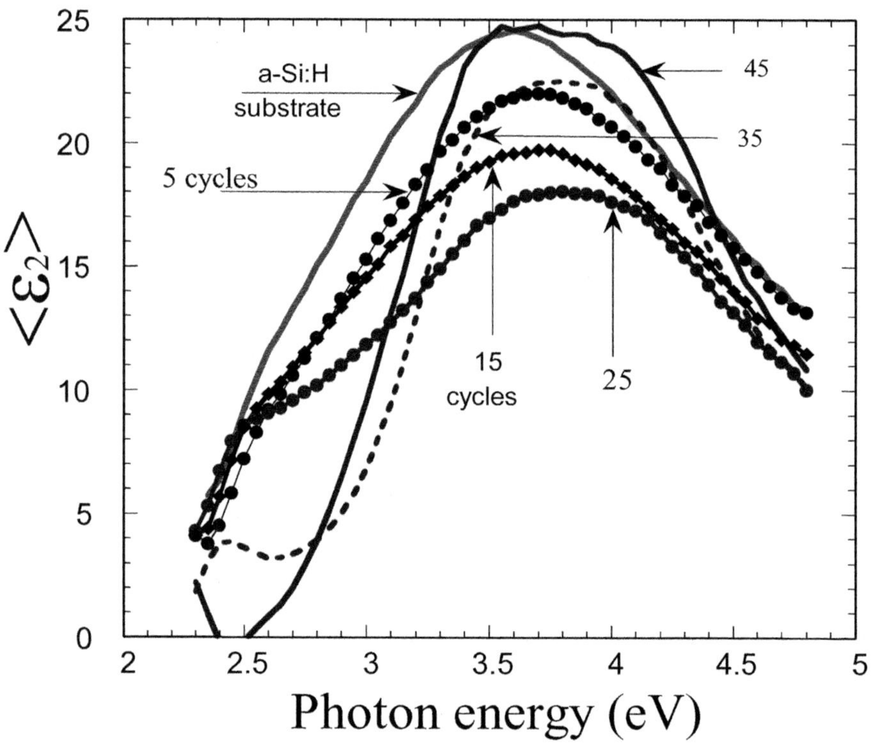

Fig. 5.8 *In situ* ellipsometry measurements of the imaginary part of the pseudo-dielectric function $<\varepsilon_2>$. The LBL process was applied to an a-Si:H film deposited on glass. One can see strong changes in the shape and amplitude of $<\varepsilon_2>$ which reflect the changes in the material from amorphous (initial a-Si:H substrate and up to 5 -15 cycles) to microcrystalline (45 cycles). Taken from [55].

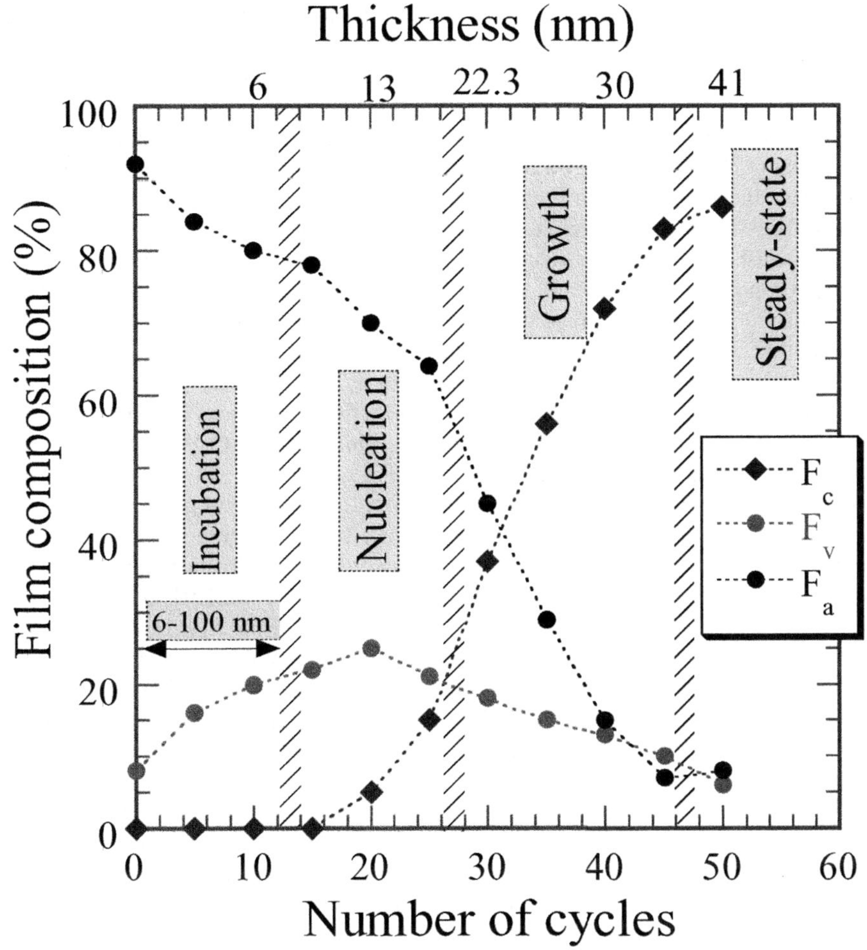

Fig. 5.9 Evolution of the composition of a μc-Si:H film as a function of the number of cycles (film thickness). The μc-Si:H growth process can be divided into four phases: *incubation*, where the amorphous fraction F_a decreases while the void fraction F_v increases; *nucleation*, once a critical void fraction has been reached; *growth*, where the crystalline fraction increases at the expenses of the amorphous phase; and *steady-state*. Note that these four phases are representative of a dynamic process which involves the partial or complete crystallization of the film. In other words, the film may be fully crystallized down to the substrate once it reaches the steady-state phase. Adapted from [55].

The results of modelling the spectra shown in Fig. 5.8 are presented in Fig. 5.9, where we focus on the composition of the layer produced by the LBL process. One can see, that up to 12 cycles the crystalline fraction F_c of the layer is zero (non detectable), while the amorphous fraction F_a steadily decreases to the benefit of the fraction of voids F_v. In other words, during the initial stages of the LBL process the silane plasma produces an a-Si:H layer which becomes more and more porous with increasing the number of cycles, due to the exposure to the hydrogen plasma. This is not surprising as the hydrogen plasma produces atomic hydrogen which can easily diffuse in the a-Si:H layer and modify its properties [59,60]. We denote this as the incubation phase, preceding the nucleation of crystallites, detected by the appearance of a crystalline fraction when the void fraction reaches its maximum (~30%). This can be related to the fact that in the usual PECVD process with SiH_4 and H_2 present in the chamber at the same time, a high hydrogen dilution is necessary to reach such a hydrogen rich and porous material.

Let us discuss the data presented in Fig. 5.9 in the framework of μc-Si:H growth models proposed to explain the growth of μc-Si:H films. They can be grouped into two main families [61]: surface models such as the surface diffusion and the selective etching model; and bulk models such as the chemical annealing model. The data presented in Fig. 5.9 show that nucleation is not a surface process, but that it rather takes place in a porous layer which has a finite thickness (13 nm in the case presented in Fig. 5.9). The data thus supports bulk models. One should pay attention to the fact that the crystalline fraction in Fig. 5.9 relates to the whole thickness and therefore the increase of the crystalline fraction with the number of cycles indicates that the whole layer is crystallizing, the crystalline fraction steadily increasing at the expense of the a-Si:H fraction. In other words, crystallization is not limited to the top surface, but involves the whole layer produced by the LBL process. This is the growth phase in Fig. 5.9, where the crystallization of the a-Si:H phase is mediated by atomic hydrogen. It has been shown that the crystallization is controlled by the diffusion of hydrogen through the a-Si:H layer deposited during T_{Si} [62]. It is interesting to note that the void fraction decreases during this phase and eventually vanishes to 0%. This must be related to the lower solubility of hydrogen in the μc-Si film as compared to that in a-Si:H [63].

Finally, one reaches a steady-state growth where the composition of the film does not change any more and only the film thickness increases. This is a very interesting situation as from now on, the process is similar to that of having changed the substrate. We have the same plasma conditions (assuming reactor walls are not affected) but the substrate has changed from a-Si:H to μc-Si:H. We will come back to substrate effects in section 5.7.

Let us summarize what we have learned by examining this simple LBL process. As schematically shown in Fig. 5.10, and supported by transmission electron microscopy images, the growth (not just deposition) of μc-Si:H involves four phases. The example shown in Fig. 5.10 corresponds to a LBL process for the growth of μc-Si:H on an a-Si:H coated glass substrate (same case as in Fig. 5.8). During the early stages of growth the hydrogen plasma forms a porous phase due to the in-diffusion of hydrogen into the a-Si:H film. Once a critical porosity is reached, some crystallites start to nucleate, indicating that nucleation requires the formation of a disordered phase with high hydrogen content. This shows that atomic hydrogen is the driver for crystallization. The energy to overcome the energy barrier between a-Si:H and μc-Si:H phases is probably provided by recombination of atomic hydrogen in the poorly interconnected silicon network. The combination of both: the porous network and excess hydrogen seem to be the requirements for nucleation to start and are in agreement with a chemical annealing process. It is interesting to note it is this solid phase nucleation which is the limiting step. Indeed, once nucleation has started, then the crystallites grow at the expense of the a-Si:H phase. As shown in Fig. 5.10, this process involves not just the surface, but it takes place in a growth zone. Therefore the process can be considered as a solid phase crystallization one, mediated by atomic hydrogen supplied by the plasma. Interestingly enough, one can see that the porous layer vanishes with increasing crystalline fraction, in other words, hydrogen moves out of the film. This has been proposed as a key for crystallization [64], while our results indicate that it is rather a consequence of the crystallization process. Finally, at steady-state growth conditions one obtains a homogeneous fully crystallized layer on top of the initial a-Si:H substrate. The incubation, nucleation and growth phases cannot be seen anymore, which reveals the extreme importance of the in-situ studies to understand the growth process.

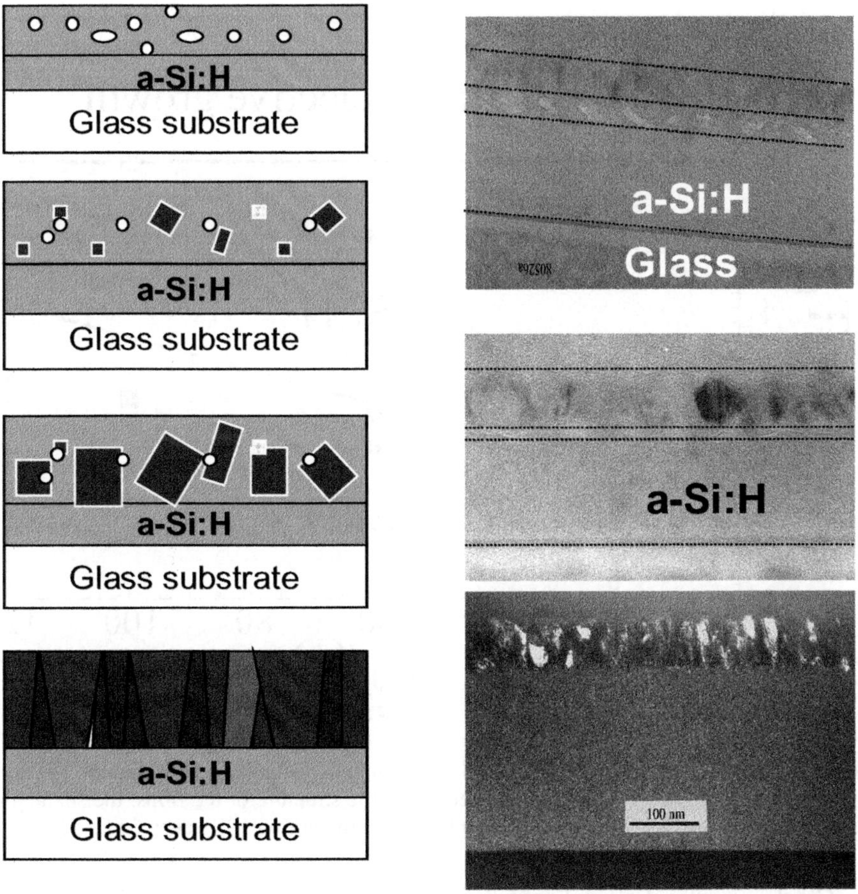

Fig. 5.10 Schematic representation of the evolution of growth of silicon thin films on an a-Si:H coated glass substrate. Here, the four phases described in Fig. 5.9 are shown schematically along with corresponding transmission electron microscopy images.

This approach on glass can be extended to different values of T_{Si} and T_H, to different substrates, deposition temperatures, value of the total pressure, value of the RF power, etc. Indeed, it is amazing that despite of its complexity, μc-Si:H growth can be controlled by just two parameters. Figure 5.11 shows the phase diagram for the growth of silicon thin films on glass at 200 °C. One can see that just by changing the two plasma times one can achieve a wide range of materials: amorphous, μc-Si:H or no film deposition when the hydrogen plasma time is too long and results in etching of the a-Si:H film. Note that this phase diagram will also depend on the material covering the walls of the reactor (chemical transport). It is important to highlight that despite some reports [65], a pure hydrogen plasma will not lead to a crystallization but it will increase the porosity and result in the etching of the film [66]. In other words, we need to reach a critical flux of atomic hydrogen with respect to that of SiH_X radicals in order to achieve nucleation.

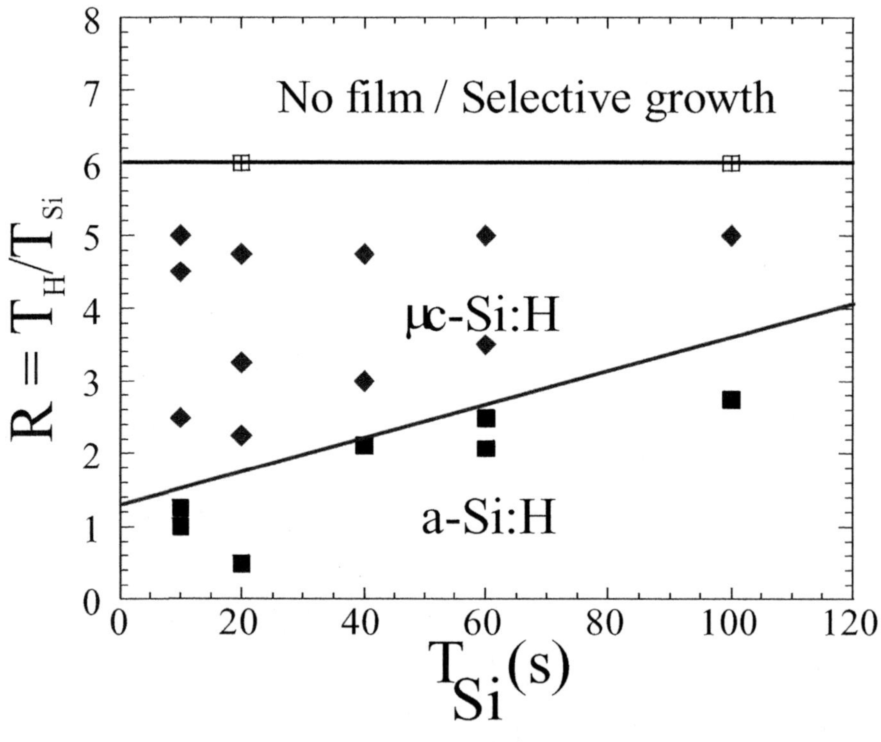

Fig. 5.11 Phase diagram for the growth of silicon thin films on glass substrates by the layer-by-layer process. Only two parameters determine the structure of the films: the silane plasma deposition time T_{Si} and the hydrogen plasma treatment time T_H. Taken from [53].

Combining the data in Fig. 5.9 for a particular deposition condition and the phase diagram in Fig. 5.11, one can conclude that depending on the ratio $R=T_H/T_{Si}$ one can have a growth process which:

i) Remains in the incubation phase. This corresponds to the films which remain in the a-Si:H region of the phase diagram. The differences between these films will be in their void fraction (related to their hydrogen content); the closer to the μc-Si region, the higher the void fraction.
ii) Reaches the steady-state growth. This is the case for the films in the μc-Si:H region of the phase diagram. Again, not all these films are equivalent. In particular, their crystalline fraction increases with the value of R.
iii) The films are completely etched. This corresponds to the case where the hydrogen exposure time is too long and emphasises the fact that a pure hydrogen plasma will not lead to the conversion of a-Si:H into a μc-Si:H film, but to the etching of the a-Si:H.

Thus, an interesting aspect coming forward from the data in Fig. 5.11 is that there is no a sharp transition between the a-Si:H and μc-Si:H phase. As an example, if we follow a vertical axis, say for T_{Si} = 20 seconds, the a-Si:H films will continuously evolve from a dense layer for T_H =0 to a highly porous one for T_H ~30 seconds, before we enter the μc-Si:H region, in which the crystalline fraction (at steady state) continuously increases with T_H from low crystalline fraction at T_H ~40 s to fully crystallized films for T_H ~100 s. Another interesting feature, inherent to these results, is that hydrogen diffusion is always present, even when growing a-Si:H films. Thus one may wonder if when studying a-Si:H deposition one should also consider subsurface effects on the film properties. This will be a critical issue in the case of a-Si:H films for heterojunction solar cells. Before moving to the case of a crystalline substrate, let us consider another ingredient of the deposition of silicon thin films by plasma processes: ion bombardment.

5.6.2 Standard Hydrogen Dilution, Effects on Ion Bombardment

We have seen that the LBL process offers a quite simple situation from the plasma point of view, as well as an excellent control over the growth process. Indeed, only two parameters (the silane plasma time T_{Si} and the hydrogen plasma time T_H) are required to determine the nature of the material: amorphous or microcrystalline. However this technique is often considered as hardly applicable to an industrial process as it seems incompatible with a high deposition rate. The standard process of diluting silane in hydrogen is thus preferred for the growth of μc-Si:H films for solar cells applications, where relatively thick layers (~1 μm) are required. However this brings us back to plasma issues. We have used the same in-situ approach as for the LBL process to analyze μc-Si:H growth from hydrogen dilution and have shown that μc-Si:H growth in standard hydrogen dilution follows the same phases as in the LBL process [67]. In other words, it does not matter so much if hydrogen dilution is performed in time (LBL) or in space (standard hydrogen dilution). This is not surprising, as we are not dealing with a simple deposition process but rather with a growth process where the resulting film properties depend on the chemical reactions induced by atomic hydrogen. However, in the hydrogen dilution case, silane dissociation and atomic hydrogen production are not independent. This imposes a strong limitation in the choice of the plasma parameters. Indeed because the dissociation cross section of hydrogen has a threshold at higher energy than that of silane [68,69], a high RF power is required for an efficient dissociation of hydrogen. This implies to increase the plasma potential and thus the energy of SiH_x ions impinging on the substrate. Thus, the main challenge is to supply enough atomic hydrogen to promote surface and growth-zone cross-linking reactions, but at the same time to avoid energetic ion bombardment by SiH_x^+ ions, which could lead to a decrease in the crystalline fraction and eventually to an amorphization of the growing film. Indeed, submitting a c-Si wafer to a hydrogen plasma results in the amorphization of a region close to its surface [70]. Now the question is over which depth (or layer thickness) this amorphization can take place, and even more important, to which extent atomic hydrogen will be able to heal the defects produced by ion bombardment. Ab-initio molecular dynamics simulations indicate that in the case of bombardment with Ar^+ (similar mass to SiH_x^+ ions) with energies of ~100 eV, typical of the plasma potential in RF PECVD deposition systems [71,72], the thickness of the damaged layer can extend over 1 nm i.e. two to three monolayers [73]. This is in contrast with the LBL results discussed above where we have seen that hydrogen-mediated

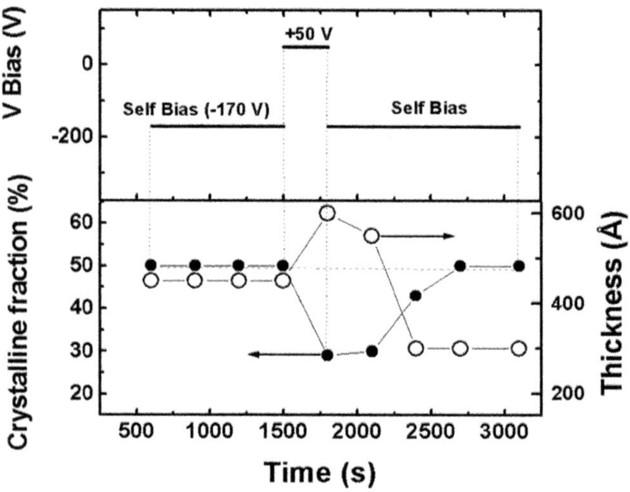

Fig. 5.12 Effect of a DC bias superimposed to the RF voltage applied to the electrode on the thickness and crystalline fraction of the growth zone deduced from *in situ* ellipsometry measurements. During the first 1500 s of deposition, the substrate self-bias on the RF electrode was -170 V and during this time the thickness and crystalline fraction of the growth zone reached steady-state values. After reaching steady-state, a DC bias of + 50 V was applied to the RF electrode, leading to an increase of the plasma potential and as a consequence of the energy of the ions impinging on the grounded substrate. The films are deposited on a Cr coated glass substrate. Taken from [67].

crystallization can extend to at least 30 nm below the surface. Again, in-situ ellipsometry studies have been invaluable to determine the in-depth effect of ion bombardment. Fig. 5.12 shows the evolution of the thickness of the growth-zone measured while superimposing different values of DC bias to the RF voltage applied to the electrode in order to move the plasma potential to higher or lower values [72]. The increase of plasma potential leads to a higher ion energy and, as shown in Fig. 5.12, to an increase in the thickness of the growth zone from 450 Å to 600 Å as well as a decrease of its crystalline fraction from 50% to 30%.

To summarize, even though μc-Si:H growth is a complex process, its main features can be accounted for by three main ingredients: silicon radicals, atomic hydrogen and silicon ions. Depending on the relative contribution of each of these species a rich variety in film composition and structure can be achieved. This is summarized in Fig. 5.13 where we show the layer structure corresponding to the most general case of μc-Si:H growth. It consists of:

1. An amorphous or partly crystallized layer at the interface with the substrate. This layer will be present when the deposition rate is too high compared to the diffusion rate of hydrogen required for crystallization. Thus, a fully crystallized interface can be achieved at the expenses of a low deposition rate.
2. A bulk layer, whose crystalline fraction results from equilibrium between deposition rate and crystallization rate. As for the interface, achieving a high crystalline fraction in this layer requires to have a high hydrogen dilution in order to achieve a high flux of atomic hydrogen compared to that of SiH_x radicals. Indeed, this solid phase crystallization is mediated by hydrogen diffusing through the growth zone
3. Just below the surface, there may be a growth zone where the chemical reactions take place, in particular amorphization by energetic ions. The thickness and crystalline fraction of this layer are controlled by the energy of the ions.
4. A surface roughness layer.

The summary on μc-Si:H growth shown in Figure 5.13 brings two remarks. The first one concerns the nature of the radicals leading to μc-Si:H growth. Indeed, because the crystallization front can lie tens of nanometers below the surface, we suggest that the exact nature of the radicals does not matter, what matters are the cross-linking reactions taking place below the surface. The second one concerns the extension of this general model to the case of a-Si:H deposition. Certainly, as in the case of μc-Si:H, a growing a-Si:H film is exposed to silicon radicals, atomic hydrogen and ions. The fact that the material is amorphous makes it more difficult to distinguish between incubation layer, bulk and growth zone layers. However this does not mean that they are not present. As a matter of fact in-situ studies provide evidence for such rearrangement of the material during growth. This can develop over some thickness 1 to 10 nm [47,48], which will be of relevance in the case of heterojunction solar cells where we aim at growing thin layers on crystalline silicon substrates. This brings us to consider the effect of the substrate on the growth process.

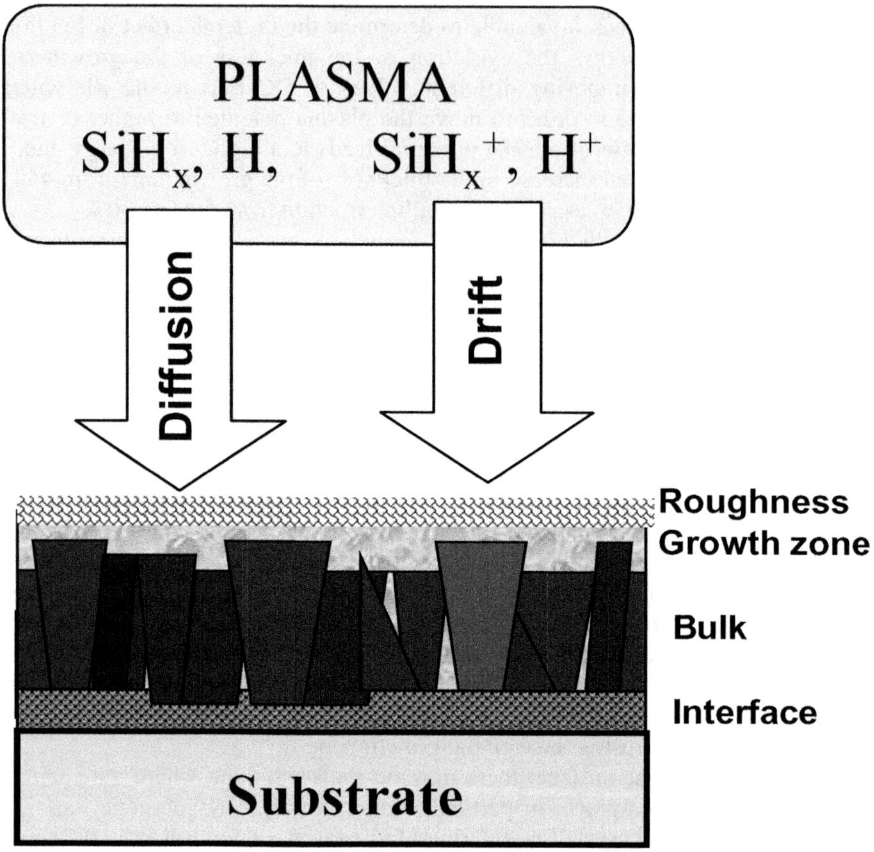

Fig. 5.13 Schematic representation of the main precursors (radicals and ions) involved in the growth of microcrystalline silicon thin films by PECVD and the resulting film structure.

5.7 Substrate Effects

So far we have focussed on the deposition of silicon thin films without paying much attention to the substrate. Hydrogenated amorphous and microcrystalline silicon films are often deposited on glass as this is a convenient substrate for a wide range of optical and electrical characterizations. However, the nature of the substrate can certainly play a role on the growth process [74,75]. This is expected when ions are involved, as the ion energy will be quite different on insulating glass substrates (at floating potential) as compared to conductive substrates (metal, glass coated with a transparent conductive oxide, etc). Moreover, in the case of heterojunction solar cells, the substrate is crystalline silicon. Thus, if deposition conditions are close to those of μc-Si:H growth, one could expect to achieve a steady state growth (see Fig. 5.8) from the beginning of the deposition, in other words, epitaxial-like growth would not be surprising. As a matter of fact, epitaxy has been reported by various groups and has been considered a burden as epitaxial growth cannot provide good surface passivation [76,77]. Indeed, in the case of HJ solar cells the requirement is to achieve "steady-state" a-Si:H growth, with neither subsurface nor interface layers, as the total thickness to be deposited is of the order of 10 nm, i.e. the range over which one can expect hydrogen to diffuse and ion bombardment to strongly impact the film properties.

The effect of the substrate on the growth of silicon thin film deposition is illustrated in Fig. 5.14 where we show the imaginary part of the pseudo-dielectric function of three films co-deposited on i) corning glass, ii) (100) oriented n-type FZ (float-zone) crystalline silicon wafer, and iii) (111) n-type FZ crystalline wafer. Both c-Si wafers had the same resistivity of 1 Ohm.cm. One can see that the spectra of the a-Si:H film deposited on glass and on (111) c-Si are almost the same at high energy, but different at lower energy due to the difference of the substrate. On the contrary, the film deposited on (100) FZ c-Si shows a spectrum very similar to that of crystalline silicon. In other words, the film growth is epitaxial.

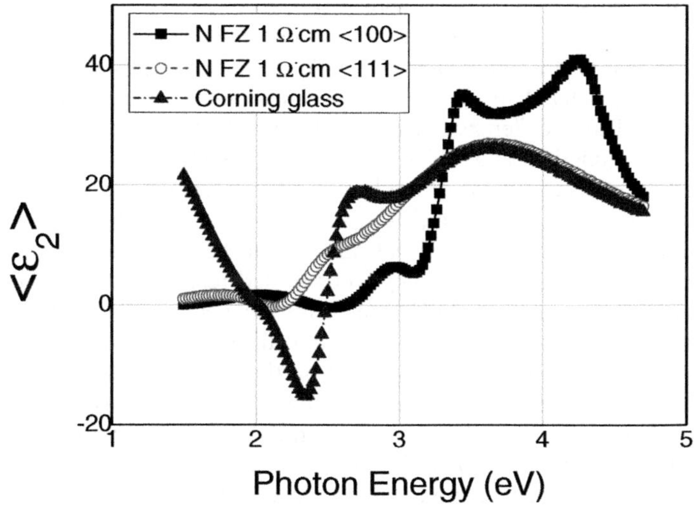

Fig. 5.14 Imaginary part of the pseudo-dielectric function of silicon thin films co-deposited on three different substrates.

The effect of the orientation of the crystalline substrate has been reported by several authors [78,79] and is a critical issue for the application of the heterojunction solar cell concept to multicrystalline silicon wafers, where depending on the grain orientation one can obtain amorphous or crystalline silicon growth and thus a non-homogeneous surface passivation. Besides this striking effect of the orientation of the crystalline substrate it is interesting to note that the optical properties of the a-Si:H films depend on the type of substrate. For that purpose we have modelled the optical data in Fig. 5.14 by using the Tauc-Lorentz dispersion function [80]. Within this approach the imaginary part of the dielectric function is obtained by multiplying the equation of the Lorentz oscillator with the equation of the Tauc joint density of states:

$$\varepsilon_{im\ TL}(E)=\frac{A\cdot E_0\cdot C\cdot\left(E-E_g\right)^2}{\left(E^2-E_g\right)^2+C^2\cdot E^2}\cdot\frac{1}{E}\qquad E>E_g \tag{5.4}$$

$$\varepsilon_{im\ TL}(E)=0\qquad E\leq E_g \tag{5.5}$$

Where E_o is the peak transition energy, E the energy, E_g the Tauc's gap, C a broadening parameter related to the film disorder, and A a parameter related to the film density [81]. The results in Table 5.1 show that the co-deposited films have the same thickness. However the optical properties are significantly different. In particular the optical gap of the a-Si:H film deposited on (111) c-Si is ~ 0.13 eV smaller than that of the film deposited on glass. On the other hand we can see that the density parameter A of the film deposited on glass is higher than that of the film on c-Si. These differences can be of crucial importance when optimizing films for heterojunction solar cells as the properties measured on glass substrates are not representative of those of the films on the actual device. Such substrate effects make it difficult to optimize a-Si:H deposition for heterojuncion solar cells, unless the optimization is carried out directly on the c-Si wafer. Indeed, besides the crystalline orientation, its doping (p or n) and resistivity, may play a role on the quality of the deposited a-Si:H films. This has been studied in the case of μc-Si:H deposition on glass substrates coated with a-Si:H. The growth of μc-Si:H films was found to be strongly dependent on the doping of the a-Si:H layer [82] and related to the dependence of the diffusion coefficient of hydrogen on the type of doping. It is well known that the diffusion coefficient of hydrogen in p-type c-Si is about 100 times larger than that in n-type c-Si [83]. This can play a crucial role on the cross-linking reactions taking place in the growth zone and could be the reason for the lower gap of the a-Si:H film deposited on c-Si compared to that of the film deposited on glass (Table 5.1); hydrogen can diffuse into the crystalline silicon and thus result in a lower hydrogen content and optical gap of the a-Si:H layer deposited on c-Si as compared to the case of a glass substrate. Thus, besides the complexity of the silicon thin film deposition process, one has to consider the strong effect of the growth on c-Si, which can explain why achieving a good surface passivation is not as easy as one could expect from the simple layer stack in Fig. 5.1. We can summarize the boundary conditions of the a-Si:H film for HJ solar cells as follows:

- High deposition rate without surface damage by high energy ions. Ideally one would like to have a high flux of ions (to provide energy to the growing surface) but with low energy per atom (< 10 eV) [84].
- Thin and dense a-Si:H with low surface roughness and a sharp interface to the substrate.
- Low defect density in the a-Si:H films [85] grown at low substrate temperature to minimize thermal budget and reduce thermal stress.
- An excellent surface passivation of the c-Si surface, with a defect density < 10^{11} cm^{-2}.
- Keep all these requirements when a p-type or n-type a-Si:H layer is deposited on top of the intrinsic layer to form the emitter and back-surface field contacts.

Table 5.1 Optical parameters deduced by fitting the ellipsometry data of the films deposited on Corning glass and (111)-oriented crystalline silicon substrate

Substrate	*Thickness (nm)*	*Eg (eV)*	E_0 *(eV)*	*A*	*C*
Corning Glass	74.0	1.74	3.64	231	2.26
c-Si (111)	75.6	1.61	3.58	198	2.27

Thus, in the end, the apparently simple solar cell structure in Fig. 5.1 bears many hidden challenges. The question is whether they can all be met by a single deposition technique or whether a combination of techniques would be more appropriate to meet all these requirements. This will remain an open question of this chapter. Ideally one could argue that Photo-CVD [44] would be an excellent technique to meet all the requirements, except for the deposition rate and possibly film density. In practice one has to make some compromises and try to choose the deposition process which can provide the best trade off among the above requirements.

5.8 Summary and Conclusions

In this chapter we have tried to emphasize the importance of understanding the plasma and surface processes involved in the growth of amorphous and microcrystalline silicon thin films. We have seen that there is a wide variety of deposition techniques and gas precursors to be used. However, rather than looking at deposition techniques, we have focused on the growth process of the silicon films and emphasized the important role of chemical reactions taking place in a growth zone, whose thickness is mostly controlled by hydrogen diffusion and ion bombardment effects. We have illustrated this via the example of μc-Si:H growth, which leads us to suggest that one should consider the process of silicon thin film deposition as a "living" process in which all the parts of the film and not just the surface exposed to the plasma are involved. We have also highlighted the importance of using "soft deposition conditions" and shown that too high ion energies may induce amorphization and damage of the c-Si wafer. This makes it difficult to achieve high deposition rates and low ion bombardment. When it comes to deposition on c-Si one should take care of achieving a sharp interface, i.e. obtaining a steady-state growth of a-Si:H from the first monolayers. Controlling the properties of the thin a-Si:H layer on c-Si substrates is a challenging task which still will require detailed research studies in order to achieve the full potential of heterojunction solar cells.

Acknowledgements

The work reported here is the result of over twenty years of research in the field of deposition of amorphous and microcrystalline silicon thin films. Over the years I have benefited from the collaboration with numerous PhD students and post-doctoral researchers. They will recognize themselves in the list of references. My deep gratitude goes to all of them. Also my particular thanks to the editors of this book who have pushed me to summarize these years of research in (I hope) a clear way.

References

[1] Reinhardt, K.A., Kern, W. (eds.): Handbook of Semiconductor Wafer Cleaning Technology. William Andrew Publisher (2007)
[2] Angermann, H.: Passivation of structured p-type silicon interfaces: Effect of surface morphology and wet-chemical pre-treatment. Appl. Surf. Sci. 254, 8067–8074 (2008)
[3] Moreno, M., Labrune, M., Roca i Cabarrocas, P.: Dry fabrication process for heterojunction solar cells through in-situ plasma cleaning and passivation. Solar Energy Materials and Solar Cells 94, 402–405 (2010)
[4] Chapman, B.: Glow discharge processes, Sputtering and Plasma Etching. Wiley-Interscience, New York (1980)
[5] Lieberman, M.A., Lichtenberg, A.J.: Principles of Plasma Discharges and Materials Processing, 2nd edn. Willey, Chichester (2005)
[6] Ogier, M.J.: Sur la formation thermique de l'hydrogène silicié. Bl. Soc. Chim. Paris T XXXII, 116–118 (1879)
[7] Chittick, R.C., Alexander, J.H., Sterling, H.F.: The preparation and properties of amorphous silicon. J. Electrochem. Soc. 77, 116 (1969)
[8] Spear, W.E., Le Comber, P.G.: Substitutional doping of amorphous silicon. Solid State Commun. 17, 1193 (1975)
[9] Street, R.A.: Hydrogenated Amorphous Silicon. Cambridge University Press, Cambridge (1991)
[10] Vepřek, S., Marecek, V.: The preparation of thin layers of Ge and Si by chemical hydrogen plasma transport. Solid State Electronics 11, 683–684 (1968)
[11] Shah, A., Meier, J., Buechel, A., Kroll, U., Steinhauser, J., Meillaud, F., Schade, H., Domine, D.: Towards very low-cost mass production of thin-film silicon photovoltaic (PV) solar mod-ules on glass. Thin Solid Films 502, 292–299 (2006)
[12] Street, R.A. (ed.): Technology and Applications of Amorphous Silicon. Springer Series in Materials Science. Springer, Berlin (2000)
[13] Fuhs, W., Niemann, K., Stuke, J.: Heterojunctions of amorphous silicon and silicon single crystals. Bull. Am. Phys. Soc. 19, 345–349 (1974)

[14] Sawada, T., Terada, N., Tsuge, S., Baba, T., Takahama, T., Wakisaka, K., Tsuda, S., Nakano, S.: High-efficiency of a-Si/c-Si Heterojunction Solar Cell. In: Proc. of the First IEEE WCPEC, vol. 2, pp. 1219–1226 (1994)
[15] Mishima, T., Taguchi, M., Sakata, H., Maruyama, E.: Development status of high efficiency HIT solar cells. Solar Energy Mat. and Solar Cells 95, 18–21 (2011)
[16] Kroll, U., Shah, A., Keppner, H., Meier, J., Torres, P., Fischer, D.: Potential of VHF-plasmas for low-cost production of a-Si:H solar cells. Sol. Energy Mater. Sol. Cells 48, 343 (1997)
[17] Van de Sanden, M.C.M., Severens, R.J., Kessels, W.M.M., Meulenbroeks, R.F.G., Schram, D.C.: Plasma chemistry aspects of a-Si:H deposition using an expanding thermal plasma. J. Appl. Phys. 84, 2426–2435 (1998)
[18] Matsumura, H.: Study on catalytic chemical vapor deposition method to prepare hydrogenated amorphous silicon. J. Appl. Phys. 65, 4396–4402 (1989)
[19] Guenier-Farret, M.E., Bazin, C., Kleider, J.P., Longeaud, C., Bulkin, P., Daineka, D., Dao, T.H., Roca i Cabarrocas, P., Descamps, P., Kervyn de Meerendre, T., Lempoel, P., Meaudre, M., Meaudre, R.: Device quality a-Si:H deposited from electron cyclotron resonance at very high deposition rates. J. Non Cryst. Solids 352, 1913–1916 (2006)
[20] Kakiuchi, H., Matsumoto, M., Ebata, Y., Ohmi, H., Yasutake, K., Yoshii, K., Mori, Y.: Characterization of intrinsic amorphous silicon layers for solar cells prepared at extremely high rates by atmospheric pressure plasma chemical vapor deposition. Journal of Non Cryst. Solids 351, 741–747 (2005)
[21] Pankkove, J.I. (ed.): Yamada Isao in Semiconductors and Semimetals Part A, ch. 5, vol.21 (1984)
[22] `http://mx.appliedmaterials.com/products/assets/brochures/solar_product_brochure.pdf`
[23] Luft, W., Simon Tsuo, Y. (eds.): Hydrogenated amorphous silicon deposition processes. Marcel Dekker Inc., New York (1993)
[24] Schmitt, J.P.M.: Fundamental mechanisms in silane plasma decompositions and amorphous silicon deposition. J. Non Cryst. Solids 59&60, 649–657 (1983)
[25] Roca i Cabarrocas, P.: Towards high deposition rates of a-Si:H, the limiting parameters. J. Non Cryst. Solids 37, 164–166 (1993)
[26] Amanatides, E., Stamou, S., Mataras, D.: Gas phase and surface kinetics in plasma enhanced chemical vapor deposition of microcrystalline silicon: The combined effect of rf power and hydrogen dilution. J. Appl. Phys. 90, 5786–5798 (2001)
[27] Gillis, H.P., Choutov, D.A., Steiner IV, P.A., Piper, J.D., Crouch, J.H., Dove, P.M., Martin, K.P.: Low energy electron-enhanced etching of Si(100) in hydrogen/helium direct-current plasma. Appl. Phys. Lett. 66, 2476 (1995)
[28] Bouchoule, A.: Dusty Plasmas: Physics, Chemistry, and Technological Impacts in Plasma Processing. Wiley, New York (1999)
[29] Abolmasov, S.N., Kroely, L., Roca i Cabarrocas, P.: Negative corona discharge: application to nanoparticle detection in RF reactors. Plasma Sources Sci. and Technol. 18, 015005 (2009)
[30] Novikova, T., Kalache, B., Hassouni, K., Morscheidt, W., Bulkin, P., Roca i Cabarrocas, P.: Numerical model of hydrogen plasmas: effects of frequency and pressure. J. Appl. Physics 93, 3198–3206 (2003)
[31] Girshick, S.L.: Theory of nucleation from the gas phase by a sequence of reversible chemical reactions. The Journal of Chemical Physics 107, 1948 (1997)

[32] Nienhuis, G.J., Goedheer, W.J., Hamers, E.A.G., van Sark, W.G.J.H.M., Bezemer, J.: A self-consistent fluid model for RF discharges in SiH_4-H_2 compared to experiments. J. Appl. Phys. 82, 2060 (1997)
[33] Gallagher, A., Howling, A.A., Hollenstein, C.: Anion reactions in silane plasma. J. Appl. Phys. 91, 5571–5580 (2002)
[34] Amanatides, E., Mataras, D., Rapakoulias, D.E.: Effect of frequency in the deposition of micro-crystalline silicon from silane discharges. J. Appl. Phys. 90, 5799–5807 (2001)
[35] van Sark, W.G.J.H.M.: Methods of deposition of hydrogenated amorphous silicon for device applications. In: Francombe, M.H. (ed.) Thin Films and Nanostructures. Advances in Plasma-Grown Hydrogenated Films, ch.1, vol. 30, pp. 1–215. Academic Press, San Diego (2002)
[36] Coburn, J.W., Kay, E.: Positive-ion bombardment of substrates in rf diode glow discharge sputtering. J. Appl. Phys. 43, 4965 (1972)
[37] Köhler, K., Coburn, J.W., Horne, D.E., Kay, E.: Plasma potentials of 13. 56 MHz RF argon glow discharge in a planar system. J. Appl. Phys. 57, 59–66 (1985)
[38] Roca i Cabarrocas, P., Morin, P., Chu, V., Conde, J.P., Liu, J.Z., Park, H.R., Wagner, S.: Opto-electronic properties of hydrogenated amorphous silicon films deposited under negative substrate bias. J. Appl. Phys. 69, 2942 (1991)
[39] Gerbi, J.E., Voyles, P.M., Treacy, M.M., Gibson, J.M., Abelson, J.R.: Increasing medium-range order in amorphous silicon with low-energy ion bombardment. Appl. Phys. Lett. 82, 3665–3667 (2003)
[40] Kondo, M., Fukawa, M., Guo, L., Matsuda, A.: High rate growth of microcrystalline silicon at low temperatures. J. Non-Cryst. Solids 266-269, 84–89 (2000)
[41] Perrin, J., Takeda, Y., Hirano, N., Takeuchi, Y., Matsuda, A.: Sticking and recombination of the SiH3 radical on hydrogenated amorphous silicon: The catalytic effect of diborane. Surface Science 210, 114–128 (1989)
[42] Kessels, W.M.M., Smets, A.H.M., Marra, D.C., Aydil, E.S., Schram, D.C., van de Sanden, M.C.M.: On the growth mechanism of a-Si:H. Thin Solid Films 383, 154–160 (2001)
[43] Knights, J.C., Lujan, R.A., Rosenblum, M.P., Street, R.A., Biegelsen, D.K., Reimer, J.A.: Effects of inert gas dilution of silane on plasma deposited a-Si:H films. Appl. Phys. Lett. 38, 331–333 (1981)
[44] Perrin, J., Broekhuizen, T.: Surface reaction and recombination of the SiH_3 radical on a-Si:H. Appl. Phys. Lett. 50, 433–435 (1987)
[45] Ceriotti, M., Bernasconi, M.: Diffusion and desorption of SiH3 on hydrogenated H:Si (100)-(2x1) from first principles. Physical Review B 76, 245301–245309 (2007)
[46] Niikura, C., Itagaki, N., Matsuda, A.: Guiding principles for obtaining high-quality silicon at high growth rates using SiH_4/H_2 glow-discharge plasma. Jpn. J. Appl. Phys. 46, 3052–3058 (2007)
[47] von Keudell, A., Abelson, J.R.: Direct insertion of SiH_3 radicals into strained Si-Si surface bonds during plasma deposition of hydrogenated amorphous silicon films. Phys. Rev. B 59, 5791–5798 (1999)
[48] Winer, K.: Defect formation in a-Si:H. Phys. Rev. B. 41, 12150–12161 (1991)
[49] Roca i Cabarrocas, P., Bouizem, Y., Thèye, M.L.: Defect density and hydrogen bonding in hydrogenated amorphous silicon as functions of substrate temperature and deposition rate. Phil. Mag. B 65, 1025 (1992)
[50] Shirai, H., Hanna, J., Shimizu, I.: Role of atomic hydrogen during growth of hydrogenated amorphous silicon in the chemical annealing. Jpn. J. Appl. Phys. 30, L679–L682 (1991)

[51] Perrin, J.: Plasma and surface reactions during a-Si:H film growth. J. Non Cryst. Solids 137-138, 639–644 (1991)
[52] Roca i Cabarrocas, P., Nguyen-Tran, T., Djeridane, Y., Abramov, A., Johnson, E.V., Patriarche, G.: Synthesis of silicon nanocrystals in silane plasmas for nanoelectronics and large area electronic devices. J. Phys. D: Appl. Phys. 40, 2258–2266 (2007)
[53] Layadi, N., Roca i Cabarrocas, P., Drévillon, B., Solomon, I.: Real-time spectroscopic ellipsometry study of the growth of amorphous and microcrystalline silicon thin films prepared by alternating silicon deposition and hydrogen plasma treatment. Phys. Rev. B 52, 5136–5143 (1995)
[54] Asano, A.: Effect of hydrogen atoms on the network structure of hydrogenated amorphous and microcrystalline silicon thin films. Appl. Phys. Lett. 56, 533–535 (1990)
[55] Hamma, S., Roca i Cabarrocas, P.: In-situ correlation between optical and electrical properties of intrinsic and n-type microcrystalline silicon films. J. Appl. Phys. 81, 7282 (1997)
[56] Drévillon, B.: Progress in crystal growth and characterization of materials, vol. 27, pp. 1–87 (1993)
[57] Losurdo, M., Rizzoli, R., Summonte, C., Cicala, G., Capezzuto, P., Bruno, G.: Anatomy of μc-Si thin films by plasma enhanced chemical vapor deposition: An investigation by spectro-scopic ellipsometry. J. Appl. Phys. 88, 2408–2414 (2000)
[58] Abramov, A., Roca i Cabarrocas, P., Girotra, K., Chen, H., Park, S., Park, K., Hyh, J.M., Choi, J., Kim, C., Souk, J.H.: Jpn. J. Appl. Phys. 47, 7301–7308 (2008)
[59] Fontcuberta i Morral, A., Roca i Cabarrocas, P.: Role of hydrogen diffusion on the growth of polymorphous and microcrystalline silicon thin films. Eur. Phys. J. Appl. Phys. 35, 165 (2006)
[60] Kail, F., Hadjadj, A., Roca i Cabarrocas, P.: Hydrogen diffusion and induced crystallization in intrinsic and doped hydrogenated amorphous silicon films. Thin Solid Films 487, 126–131 (2005)
[61] Roca i Cabarrocas, P.: New approaches for the production of nano-, micro- and polycrystalline silicon thin films. Phys. Status Sol. C 1, 1115–1130 (2004)
[62] Godet, C., Layadi, N., Roca i Cabarrocas, P.: Role of mobile hydrogen in the amorphous silicon recrystallization. Appl. Phys. Lett. 66, 3146 (1995)
[63] Acco, S., Williamson, D.L., Stolk, P.A., Saris, F.W., van den Boogaard, M.J., Sinke, W.C., van der Weg, W.F., Roorda, S., Zalm, P.C.: Hydrogen solubility and network stability in amorphous silicon. Phys. Rev. B 53, 4415 (1996)
[64] Robertson, J.: Thermodynamic model of nucleation and growth of plasma deposited micro-crystalline silicon. J. of Appl. Physics 93, 731–735 (2003)
[65] Sriraman, S., Agarwal, S., Aydil, E.S., Maroudas, D.: Mechanism of hydrogen-induced crystallization of amorphous silicon. Nature 418, 62 (2002)
[66] Hadjadj, A., Pham, N., Roca i Cabarrocas, P., Jbara, O., Djellouli, G.: Ellipsometry investigation of the amorphous-to-microcrystalline transition in a-Si:H under hydrogen plasma treatment. J. Appl. Phys. 107, 083509 (2010)
[67] Kalache, B., Kosarev, A.I., Vanderhaghen, R., Roca i Cabarrocas, P.: Ion bombardement effects on the microcrystalline silicon growth mechanisms and on the film properties. J. Appl. Phys. 93, 1262–1273 (2003)
[68] Perrin, J., Leroy, O., Bordage, M.C.: Cross-sections, rate constants and transport coefficients in silane plasma chemistry. Contrib. Plasma Phys. 36, 3–49 (1996)

[69] Robert Koemtzopoulos, C., Economou, D.J., Pollard, R.: Hydrogen dissociation in a microwave discharge for diamond deposition. Diamond and Related Materials 2, 25–35 (1993)

[70] Neitzert, H.C., Layadi, N., Roca i Cabarrocas, P., Vanderhaghen, R., Kunst, M.: In situ meas-urements of changes in the structure and in the excess charge-carrier kinetics at the silicon surface during hydrogen and helium plasma exposure. J. Appl. Phys. 78, 1438–1445 (1995)

[71] Böhm, C., Perrin, J.: Retarding-field analyzer for measurements of ion energy distributions and secondary electron emission coefficients in low-pressure radio frequency discharges. Rev. Sci. Instrum. 64, 31–44 (1993)

[72] Roca i Cabarrocas, P.: Detailed study of ion bombardment in RF glow discharge deposition systems: the role of He dilution. In: Mat. Res. Soc. Symp. Proc., vol. 149, pp. 33–38 (1989)

[73] Graves, D.B., Humbird, D.: Surface chemistry associated with plasma etching processes. Appl. Surf. Science 192, 72–87 (2002)

[74] Wagner, S.: Amorphous silicon: vehicle and test bed for large-area electronics. Phys. Status Solidi A 207, 501–509 (2010)

[75] Roca i Cabarrocas, P., Layadi, N., Heitz, T., Drévillon, B., Solomon, I.: Substrate selectivity in the formation of microcrystalline silicon: Mechanisms and technological consequences. Appl. Phys. Lett. 66, 3609 (1995)

[76] de Wolf, S., Kondo, M.: Abruptness of a-Si:H/c-Si interface revealed by carrier lifetime measurements. Applied Physics Letters 90, 042111 (2007)

[77] Gielis, J.J.H., Hoex, B., van den Oever, P.J., van de Sanden, M.C.M., Kessels, W.M.M.: Sili-con surface passivation by hot-wire CVD Si thin films studied by in situ surface spectroscopy. Thin Solid Films 517, 3456–3460 (2009)

[78] Das, U.K., Burrows, M.Z., Lu, M., Bowden, S., Birkmire, R.W.: Surface passivation and het-erojunction cells on Si (100) and (111) wafers using dc and rf plasma deposited Si:H thin films. Appl. Phys. Lett. 92, 063504 (2008)

[79] Labrune, M., Moreno, M., Roca i Cabarrocas, P.: Ultra-shallow junctions formed by quasi-epitaxial growth of boron and phosphorous doped films at 175 °C by rf-PECVD. Thin Solid Films 518, 2528–2530 (2009)

[80] Jellison Jr., G.E., Merkulov, V.I., Puretzky, A.A., Geohegan, D.B., Eres, D.B., Lowndes, D.H., Caughman, J.B.: Characterization of thin-film amorphous semiconductors using spectroscopic ellipsometry. Thin Solid Films 377-378, 68–73 (2000)

[81] Fontcuberta i Morral, A., Roca i Cabarrocas, P., Clerc, C.: Structure and hydrogen content of polymorphous silicon thin films studied by spectroscopic ellipsometry and nuclear measurements. Phys. Rev. B 69, 125307 (2004)

[82] Hadjadj, A., Pham, N., Roca i Cabarrocas, P., Jbara, O.: Effect of doping on the amorphous to microcrystalline transition in a hydrogenated amorphous silicon under hydrogen plasma treatment. Appl. Phys. Lett. 94, 061909 (2009)

[83] Hydrogen in Semiconductors. In: Stutzmann, M., Chevallier, J., Frova, A., Tosatti, E. (eds.) Siwth Trieste ICTP-IUPAP Semiconductor Symposium. North Holland, Amsterdam (1990)

[84] Wank, M., van Swaaij, R.A.C.M.M., Kudlacek, P., van de Sanden, M.C.M., Zeman, M.: Hydrogenated amorphous silicon deposited under accurately controlled ion bombardment using pulse-shaped substrate biasing. J. Appl. Phys. 108, 103304–103309 (2011)

[85] Schulze, T.F., Beushausen, H.N., Leendertz, C., Dobrich, A., Rech, B., Korte, L.: Interplay of amorphous silicon disorder and hydrogen content with interface defects in amorphous/crystalline silicon heterojunctions. Applied Physics Letters 96, 252102 (2010)

Chapter 6
Electronic Properties of Ultrathin a-Si:H Layers and the a-Si:H/c-Si Interface

Lars Korte

Helmholtz-Zentrum Berlin GmbH, Department Silicon Photovoltaics, Kekuléstr. 5, D-12489 Berlin, Germany

Abstract. The *a*-Si:H/*c*-Si heterojunction constitutes the core building block of *a*-Si:H/*c*-Si solar cells. In these cells, a key issue to obtain high efficiencies is the minimization of recombination losses at the *a*-Si:H/*c*-Si interfaces: The *a*-Si:H layers induce the band bending at the *p*/*n*-junction, but also passivate the surface of the *c*-Si, by saturation of dangling Si bonds. This is essential to realize the V_{oc} potential > 700 mV of this cell type. High defect densities at the *a*-Si:H/*c*-Si interfaces lead to a pronounced decrease of the cell efficiency, by ~4% absolute at defect densities of 10^{12} cm^{-2} (~ 1 dangling bond per 1000 interface atoms). Thus, it is important to obtain information on recombination-active defects in the ultra-thin *a*-Si:H layer and at the *a*-Si:H/*c*-Si interface. After introducing the basic electronic properties of *a*-Si:H, this chapter discusses the density of occupied valence band and defect states $N_{occ}(E)$ and the position of the Fermi level in the band gap of undoped (so called intrinsic) and of doped ultra-thin *a*-Si:H layers. The measured *a*-Si:H properties are correlated to the band bending in the *c*-Si absorber and to charge carrier recombination at the *a*-Si:H/*c*-Si interface. The connection to solar cell open circuit voltage V_{oc} is made, and the current state-of-the-art of *c*-Si surface passivation by (i)*a*-Si:H is reviewed. Furthermore, the use of temperature-dependent current-voltage measurements on complete *a*-Si:H/*c*-Si solar cells to extract information on recombination and transport is discussed. Finally, the influence of band bending at the TCO/*a*-Si:H interface on cell performance is outlined briefly.

6.1 Introduction

The main difference between conventional silicon homojunction solar cells and the cell type discussed in this book is the use of amorphous/crystalline silicon (*a*-Si:H/*c*-Si) heterojunctions to form the *p*/*n* junction and/or the back surface field (BSF) of the cell. The term 'heterojunction' is generally used for intimate contacts between materials of different chemical composition, physical condition and/or morphology (in our case: crystalline vs. amorphous). The present chapter is

W.G.J.H.M. van Sark et al. (Eds.): Physics & Tech. of Amorphous-Crystalline, EM, pp. 161–221.
springerlink.com

concerned with the specific *electronic* properties of the ultra-thin (~10 nm) *a*-Si:H films used in *a*-Si:H/*c*-Si solar cells and of their junction to the crystalline silicon wafer. Figure 6.1 shows a simplified sketch of this cell type together with a schematic band diagram[1].

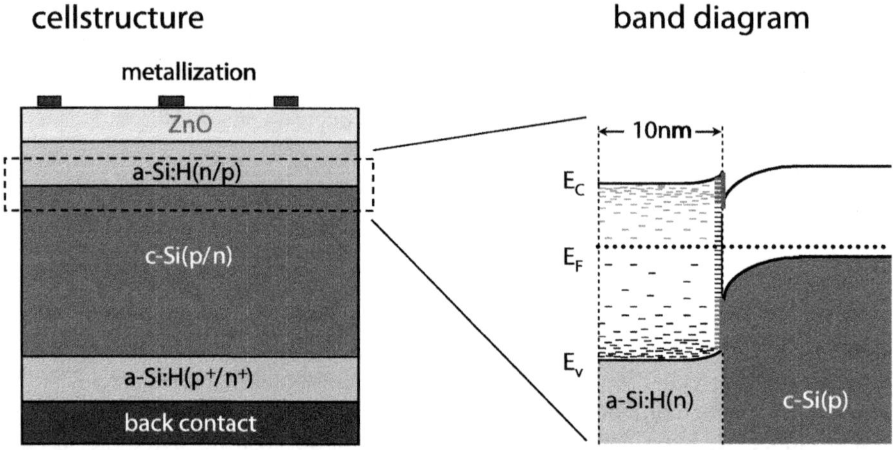

Fig. 6.1 Amorphous/crystalline silicon (*a*-Si:H/*c*-Si) solar cell (schematic) and sketch of the band diagram of the (n)*a*-Si:H/(p)*c*-Si heterojunction. E_C, E_V are the band edges, ΔE_C, ΔE_V the band offsets between the *a*-Si:H and the *c*-Si, E_F the Fermi level and D_{it} the density of defect states at the *a*-Si:H/*c*-Si interface.

The following main differences exist for the heterojunction cell as compared to the equivalent homojunction device:

- *abrupt interfaces:* In conventional silicon solar cells, the *p*/*n* junction and BSF are diffused junctions, the depth distribution of in-diffused dopants is similar to a Gaussian or error function profile. In contrast, the *a*-Si:H/*c*-Si junction is abrupt on the monolayer length scale. Thus the doping profile usually is a box-shaped function.
- *defect states in the a-Si:H band gap:* These are an intrinsic property of over-constrained semiconducting glasses in general and arise from the fact that in order to form the amorphous network, it will be necessary to make bonds between the constituting atoms that are displaced from their equilibrium state, i.e. by changing the bond angle or its length. This gives rise to the so-called Urbach tails, exponential distributions of electronic states reaching from the band edges into the band gap of the material. In addition, if the strain imposed on a silicon-silicon bond by the network configuration becomes too high, the bond breaks,

[1] Note that features such as texturization of the wafer and the so-called intrinsic (actually nominally undoped) *a*-Si:H buffer layer usually present in such cells have been omitted for clarity.

forming so-called dangling bond states deep in the band gap, which can be saturated by the hydrogen present in the *a*-Si:H film.

- *a-Si:H/c-Si interface defects:* Ideally, during *a*-Si:H growth on a *c*-Si wafer, all open (dangling) bonds at the *c*-Si surface should make bonds to Si atoms of the growing film. However, dangling bonds can form at the *a*-Si:H/*c*-Si interface that lead to recombination-active states in the band gap at the interface. Similarly to the situation in the *a*-Si:H bulk, this will happen if the *a*-Si:H network configuration in the vicinity of a *c*-Si surface atom does not allow for the formation of a Si-Si bond. Just like in the *a*-Si:H bulk, also *c*-Si surface dangling bonds can be passivated by hydrogen. Additionally, surface contamination by atoms or molecules adsorbed to the *c*-Si wafer surface prior to *a*-Si:H deposition can lead to extrinsic defect states at the interface. Controlling surface contamination and the initial stages of *a*-Si:H growth are therefore key issues to obtaining low defect density *a*-Si:H/*c*-Si interfaces.
- *Band offsets:* The difference in band gap E_g between *c*-Si ($E_g = 1.12$ eV at room temperature) and *a*-Si:H (E_g=1.6…1.9 eV, depending on deposition conditions) gives rise to discontinuities in the valence and conduction band edges, usually denoted by ΔE_V and ΔE_C, respectively, cf. the schematic band diagram in Fig. 6.1.

Conceptually, the band offsets are the main difference to the *p*/*n* homojunction. They lead to several consequences for charge-carrier transport and recombination in heterojunction devices:

- large band offsets can hinder charge carrier transport over the heterojunction interface, because the charge carrier has to overcome the band offset barrier e.g. by thermionic emission or tunneling processes. Depending on the details of charge distribution and recombination at the heterointerface, this can manifest in the solar cell characteristics as an effect similar to a series resistance, or in an S-shaped *I*-*V* curve ([1, 2] and section 6.5 of this chapter).
- at the same time, the minority charge carriers in the vicinity of the contact are hindered by the band offsets in the heterojunctions to reach the cell contacts, and thereby recombination at the contacts is effectively suppressed. Thus, low reverse diode saturation currents and high open circuit voltages (V_{oc}) can be expected in the cell.

The latter advantage is indeed borne out by experimental results: While the PERL (passivated emitter with rear locally diffused) silicon homojunction cell that currently[2] holds the efficiency record ($\eta = 25.0$ %) for silicon solar cells under AM1.5 illumination has a V_{oc} of 706 mV [3], the best *a*-Si:H/*c*-Si cells reach V_{oc} = 729 mV ($\eta = 23.0\%$) [4].

Highly doped *a*-Si:H layers also influence the optical properties of the cell, as they are parasitic absorbers. This will influence the short circuit current density j_{sc}. The effect is rather small but important and can be put into a useful rule of thumb: Taking the known absorption coefficient α of *a*-Si:H and assuming that all

[2] In November 2010.

absorbed photons are lost for the photocurrent, one finds that every 10 nm of (doped) device grade *a*-Si:H deposited on a *c*-Si absorber decreases j_{sc} by about 1-2 mA/cm² under AM1.5 conditions, see also [5, 6]. For photogenerated electrons and holes in intrinsic *a*-Si:H layers, there is a finite probability of contributing to the photocurrent instead of recombining, thus the decrease in j_{sc} should be less pronounced.

The present chapter focuses on the determination of *a*-Si:H bulk and *a*-Si:H/*c*-Si interface defect densities and recombination properties for the ultrathin *a*-Si:H films (typical thickness 10 nm) as used in *a*-Si:H/*c*-Si cells. After a brief review of results presented in literature, mostly for thicker *a*-Si:H films, investigations using a variant of photoemission excited with near-UV light (NUV-PES) are presented. This technique can be used to measure the density of occupied valence band and defect states $N_{occ}(E)$ and the position of the Fermi level in the band gap. For both intrinsic and doped *a*-Si:H, the measured *a*-Si:H properties are correlated to the band bending in the *c*-Si absorber and to charge carrier recombination time constants. The latter are obtained from photoconductance decay data or by the surface photovoltage method.

From this discussion, the following picture emerges: The microscopic structural and electronic configuration of ultrathin *a*-Si:H layers is governed by the same mechanisms as that of thick *a*-Si:H films. The deposition conditions systematically affect the layer properties, and a combination of low-temperature deposition and a subsequent anneal have shown to yield best results in terms of *a*-Si:H/*c*-Si interface passivation. The as-deposited interface density of states D_{it} (cf. Fig. 6.1) is determined by the local network structure at the interface, which is in a non-equilibrium state for samples deposited at low temperature. The annealing step leads to an equilibration of the interface with the *a*-Si:H bulk and D_{it} decreases to its equilibrium value, defined by the bulk *a*-Si:H network strain reflected in the valence band tail slope. The final D_{it} is in very good agreement with the bulk defect densities of several 100 nm thick *a*-Si:H films. Thus, it appears that the equilibrated a-Si:H/c-Si interface region does not possess unique electronic properties but is determined by the *a*-Si:H bulk defects.

For samples including doped *a*-Si:H layers, the connection between doping, defect generation and parameters of the completed solar cell (i.e. a metal/TCO/*a*-Si:H/*c*-Si/*a*-Si:H/metal structure) is made: For *a*-Si:H/*c*-Si solar cells, an optimum *a*-Si:H doping level exists. For high doping levels, the increase in band bending at the p/n junction is balanced against increasing defect concentration – thus enhanced recombination rate – at the interface and in the *a*-Si:H layers.

In practice, a wealth of information can be obtained from temperature-dependent *I-V* measurements without illumination, as will be discussed in section 6.4: In the high forward bias region, the specifics of a heterojunction (carrier transport across barriers, tunneling etc.) play a minor role, and *a*-Si:H/*c*-Si cells can be described as "one-sided" junctions within the Shockley theory.

Finally, in section 6.5, it will be outlined how the TCO/*a*-Si:H contact influences the band bending at the *a*-Si:H/*c*-Si interface: Depending on the band lineup at the TCO/*a*-Si:H interface an antiparallel diode can be formed in addition to the aSiH/*c*-Si junction, which severely influences the *I-V* curves in solar cells.

6.2 Amorphous Silicon – A Short Overview

The present section is intended as a brief overview on those electronic and optical properties of *a*-Si:H that are important in the context of using thin *a*-Si:H films in heterojunction cells. As in those cells, the *a*-Si:H serves the two main purposes of passivating the *c*-Si surface and inducing a band bending (*p*/*n* junction or BSF), the focus lies on defect densities and doping, whereas such aspects as charge carrier transport and recombination in thick *a*-Si:H films, structural stability and photo-induced effects (Staebler-Wronski effect) etc. are mostly omitted. The reader is referred to the standard literature on this topic, especially the books by Street [7] and Tanaka [8], as well as the collection of reviews in [9].

6.2.1 Unhydrogenated vs. Hydrogenated Amorphous Silicon

Although in the disordered structure of amorphous silicon a long-range order is missing, the local bonding configuration of the atoms is still very similar to that of the corresponding crystalline silicon material. It was shown using tight binding calculations [10], that under these conditions, a similar energetic distribution of the total density of bonding and anti-bonding states should arise as in the crystal, including a band gap of similar width. However, in unhydrogenated *a*-Si, it is found experimentally that the high density of broken silicon-silicon bonds, so-called "dangling bonds", fills the band gap of *a*-Si. When hydrogen is incorporated into the film, most of the dangling bonds are saturated. Usually, the source of hydrogen are the precursor gases SiH_4 and H_2 as well as the dopant precursors used in the standard plasma enhanced chemical vapor deposition (PECVD) processes. Although theoretically, a volume density of 10^{18}-10^{19} cm^{-3} hydrogen would suffice to saturate the dangling bonds, one usually finds around 10-15 at.% hydrogen in device grade (i.e. low defect density) *a*-Si:H. The excess hydrogen is partly incorporated as molecular H_2, covers inner surfaces (voids) of the film or forms higher order silicon hydrides (-SiH_2, SiH_3) [11].

6.2.2 Electronic and Optical Properties

As in amorphous silicon, a long-range periodic structure is missing, it is not possible to describe the electronic properties in the same way as in ideal crystalline materials: In *a*-Si:H, the potential wells formed by the atom cores are non-periodic. Therefore, it is not possible to solve the Schrödinger equation for the charge carriers by a superposition of periodic functions with energy eigenvalues and a wave vector $\boldsymbol{k}$: $\boldsymbol{k}$ is not a "good" quantum number. This means that concepts such as the dispersion relation $E = E(\boldsymbol{k})$, i.e. the band structure, are not applicable to *a*-Si:H (and amorphous materials in general). Still, the concept of an energy-dependent density of electronic states (DOS) $N(E)$ remains valid: It is simply defined as

$$N(E) = \frac{1}{V} \sum_i \delta(E_i - E), \tag{6.1}$$

where the sum extends over all eigenstates of the material with energy E_i (occupied or empty) within the volume V. As many properties of a covalently bonded material depend on the next-neighbour configuration, which is only slightly perturbed in *a*-Si:H as compared to *c*-Si, the general appearance of the *a*-Si:H DOS, is very similar to that of *c*-Si (e.g. [12, 13]).

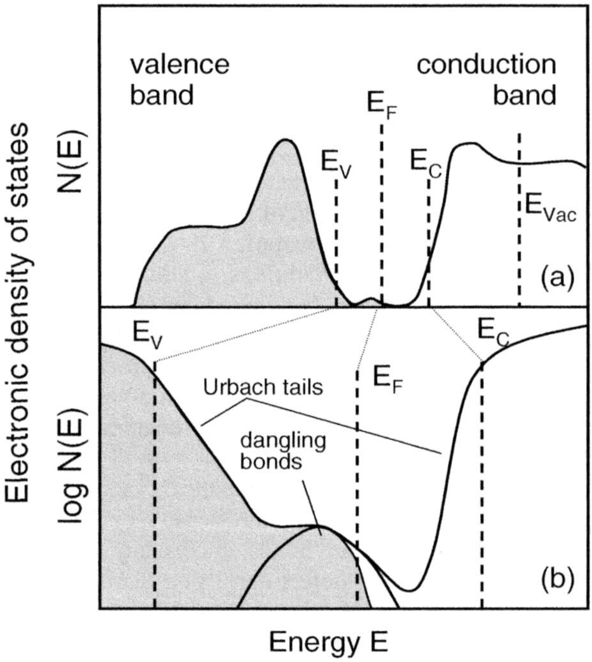

Fig. 6.2 Sketch of the density of electronic states in the valence- and in the conduction band (VB, CB) of hydrogenated amorphous silicon: a) Overview on a linear scale, (b) semi-logarithmic plot of the band gap region, showing the mobility edges E_V and E_C, the valence- and conduction band Urbach tails and the broad distribution of broken silicon bonds ("dangling bonds"). After [7].

Figure 6.2 shows a schematic sketch of the *a*-Si:H DOS. The major difference between *a*-Si:H and *c*-Si becomes apparent in the semi-logarithmic plot of DOS(E): The band edges of *a*-Si:H are not defined by a sharp decrease of the DOS to zero at the band edges, as in *c*-Si, but by a gradual decrease of the DOS from the mobility edges $E_{V\mu}$, $E_{C\mu}$ towards the center of the band gap. These are the so-called Urbach tails. Close to the center of the band gap, a broad distribution of defect states is present, which arise from the remaining dangling bonds that are not saturated by hydrogen.

6.2.2.1 Urbach Tails

The Urbach tails decay exponentially into the band gap of the amorphous silicon, $N(E) \propto \exp(-E/E_0)$, with the Urbach energy E_0 (or E_U) as the slope parameter. Note that E_U often refers to the exponential slope of optical absorption data in the band gap, which arises from the convolution of the valence- with the conduction band-tail DOS, whereas E_{0v} and E_{0c} are used for the individual band tail slopes of valence and conduction band. As the conduction band tail is much steeper than that of the valence band, $E_U \cong E_{0v}$. It is generally accepted that the origin of the band tails is the strain in the *a*-Si:H, i.e. fluctuations in binding angles and -distances in the Si network. The correlation between Urbach slope and bond angle distortion has been investigated e.g. by optical absorption and Raman measurements [14]. The electronic states associated with strongly strained Si-Si bonds appear deeper in the band gap (farther away from the band edge). It can be shown mathematically [15] that a Gaussian (i.e. random) distribution of bond angle and -distance fluctuations around their mean leads to an exponential energy distribution, i.e. the Urbach tail. Consequently, exponentially decaying band tails have been found in many disordered materials, the first being impure AgBr crystals [16].

As the density per unit volume of states in the band tails decreases towards the center of the band gap, the distance in real space between the states correlated to the electronic levels increases. Also, states deeper in the band gap are more localized (their wave function is less extended) than those closer to the band edges. When the distance between states becomes larger than the extent of the wave function, charges in these states cannot move freely anymore and become localized. This extended-to-localized state transition is termed the Anderson transition [17].

A large number of investigations has been devoted to the change of E_U with *a*-Si:H preparation conditions, especially to its dependence on deposition temperature, and on doping. The papers of Stutzmann and coworkers [18, 19] contain both a good overview on experimental findings and some plausible – although not undisputed – suggestions for the physical origin of the observed trends:

For device-grade undoped *a*-Si:H grown by PECVD, the minimum values for the valence band Urbach tail energy are around E_{0v} = 45-50 meV. Their dependence on *a*-Si:H deposition temperature exhibits a broad minimum around 200-300°C, cf. Fig. 6.3.

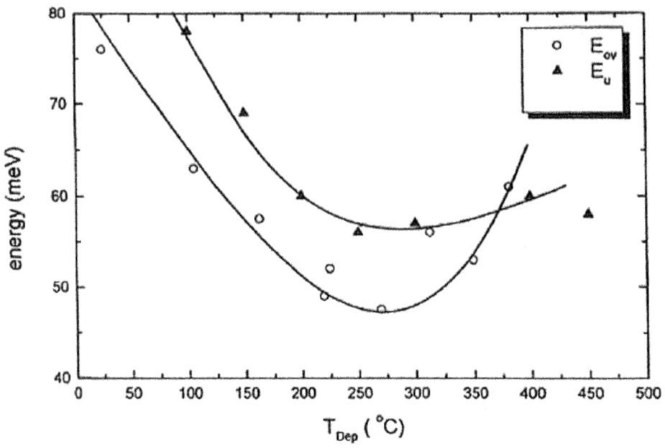

Fig. 6.3 Valence band slope E_{0v} and Urbach energy E_U as determined from photoelectron yield spectroscopy and optical absorption, respectively, vs. *a*-Si:H deposition temperature. From [20] with permission, © 1998: The Institution of Electrical Engineers.

For low-quality or doped films, E_U and E_{0v} lie above these curves. The position of the minimum of the curves has been explained as being due to the fact that the *a*-Si:H network is in thermodynamic equilibrium at a deposition temperature of ~200-300°C, and that the state of the network is "frozen in" when the sample is cooled down after growth. The parabolic shape of the curves can be understood by considering the role of hydrogen for defect passivation and network relaxation within the so-called "hydrogen glass" model [21, 18]. Briefly, within this model, the minimum in E_U is explained by the properties of hydrogen in *a*-Si:H: In the hydrogen glass model, mobile hydrogen is mediating the formation of the *a*-Si:H network by moving through the material, breaking strained bonds and saturating dangling (broken) bonds. At $T < 100$°C, the movement of hydrogen is frozen in, leading to much less ordered (more strained) material, whereas at $T >$ 300-350°C, hydrogen starts to evolve from the film and is thus not available to saturate dangling bonds. While a number of objections have been raised against the hydrogen glass model, based e.g. on annealing experiments of *a*-Si:H films grown at low temperature, it is generally accepted that hydrogen plays a paramount role in the defect equilibration mechanisms in *a*-Si:H.

6.2.2.2 Urbach Tail and Deep Dangling Bond Defects

Measurements of defect densities in *a*-Si:H show a second contribution to the DOS in the band gap, a broad distribution of defects that are usually identified with broken (dangling) silicon bonds. The dangling bond defect in silicon is amphoteric, i.e. it can carry 0, 1 or 2 electrons, corresponding to the charge states Q = +1, 0, -1, respectively. The set of dangling bond states is usually modeled as Gaussian distributions centered around E_d with width σ_d,

$$N_{\rm d}^{Q}(E) = N_{\rm 0d}^{Q} \exp\left(-\frac{({\rm E} - E_{\rm d}^{Q})^2}{2\sigma_{\rm d}} \right) \tag{6.2}$$

In most experimental data where both the Urbach tail slopes $E_{\rm U}$ (or $E_{\rm 0v}$) and the density $N_{\rm d}$ are measured, a correlation between these quantities is found (e.g. [22, 18] and references therein). Figure 6.4 illustrates this finding.

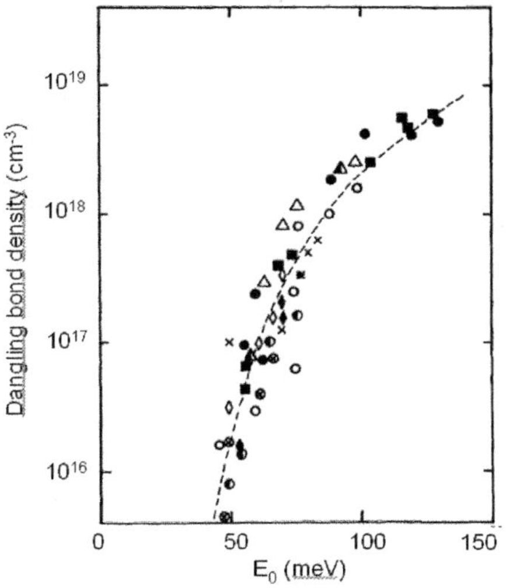

Fig. 6.4 Integrated dangling bond density $N_{\rm d}$ vs. Urbach energy E_0, together with a fit of eq. (6.3) to these data. Reproduced from [18] with permission.

Stutzmann has proposed a simple model to explain this correlation [18]. It relies on the fact that the strain in Si-Si bonds increases with increasing distance of the corresponding energy level in the band tail from the band edge. If the strain becomes too large, the bond breaks. The idea is now to introduce a demarcation energy $E_{\rm db}$, cf. Fig. 6.5: When the electronic level of the strained bond lies above $E_{\rm db}$, the bond breaks and contributes to the dangling bond distribution, i.e. to $N_{\rm d}$. Thus, $N_{\rm d}$ is given by the integral over the part of the tail above $E_{\rm db}$. This leads to a simple equation relating $N_{\rm d}$ and $E_{\rm db}$:

$$N_d = N^* E_0 \exp(-(E_{\rm db} - E^*)/E_0)\,. \tag{3}$$

A fit of this equation to the data is also plotted in Fig. 6.4. Again, this is a rather simple model that does not include e.g. the details of the thermal history of the sample, especially the influence of a time-dependence of defect equilibration. More sophisticated models have been proposed, notably the so-called defect pool model. They take the simple argument by Stutzmann further by defining a "pool" or distribution of defects that could be potentially realized, and then calculating the actual dangling bond distribution by considering the equilibration of elemental chemical reactions involving the separate release and capture of hydrogen [23, 24].

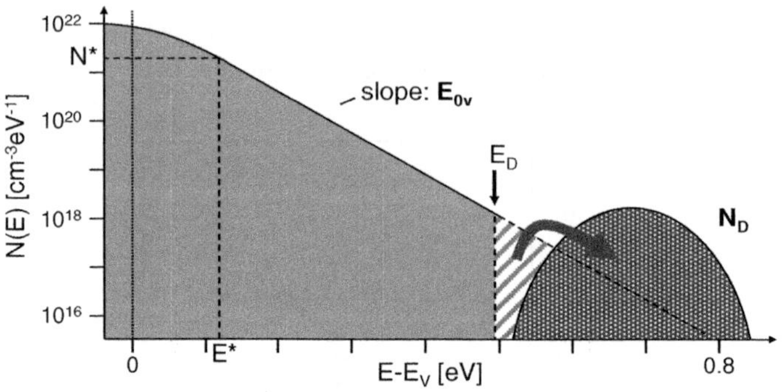

Fig. 6.5 Density of states distribution in the valence band tail and dangling bonds. The quantities used in eq. (6.3) are indicated. Within Stutzmann's model, the strained bonds that would lie in the hatched area are broken and give rise to the deep defect distribution. After [18].

The dynamics of *a*-Si:H defect equilibration is discussed by various authors, in the context of thick *a*-Si:H films notably to account for discrepancies between constant photocurrent mode (CPM) and photothermal deflection spectroscopy (PDS) measurements [25, 26, 27, 28]: While PDS averages over the defect densities in the whole depth of the sample, CPM "sees" mainly the DOS in the low-defect bulk material. This leads to discrepancies that are ascribed to the existence of regions of high defect density close to the *a*-Si:H growth surface, i.e. the *a*-Si:H/vacuum interface, and at the *a*-Si:H/substrate interface. To explain the finding of high defect density close to the surface – that is corroborated by photoelectron spectroscopy, see below – Hata and coworkers [27] have proposed a model that assumes an initially high defect density at the surface of the growing *a*-Si:H film, and a subsequent anneal of the already deposited part of the film while growth proceeds. Depending on the *a*-Si:H deposition conditions – notably, the substrate temperature – the defect density in the final film then shows a depth dependence, from high defect densities at the surface towards an equilibrium density which depends on deposition temperature and Urbach energy (i.e. strain) in the film. This is particularly true for *a*-Si:H deposited at low temperature, where equilibration is hindered by the low thermal energy provided to the amorphous network. Using modulated photocurrent studies, Kleider *et al.* show the inhomogeneity of films grown at low temperature and subsequent homogenization after post-annealing at higher temperature (200°C) [28].

6.2.2.3 Doping of *a*-Si:H and Doping-Induced Defects

Doping of *a*-Si:H is usually done in analogy to that of crystalline silicon, using boron for *p*- and phosphorous for *n*-type doping. Precursors are diborane, B_2H_6, or trimethylboron, $B(CH_3)_3$, and phosphine, PH_3, respectively, which are mixed with SiH_4 in the gas phase. To become electrically active as a donor or acceptor, the dopant atom must be incorporated into the *a*-Si:H network in a four-fold coordinated position. This is usually the case for B and P in *c*-Si. In *a*-Si:H, however, the random network can also easily accommodate the dopants in other configurations. In addition, the hydrogen present in the material can also passivate the dopant atoms, i.e. make them electrically inactive. This leads to rather low doping efficiencies: For a solid phase doping efficiency $\eta_s = N_{act}/N_{sol}$ defined as the ratio between the concentration of activated dopant atoms N_{act} and the total concentration N_{sol} in the film, η_s below 10% (below 1% for device-relevant doping levels) has been reported in literature ([19] and references therein). Another finding is that the probability of incorporation of a dopant atom into the growing film is also dependent on the gas phase dopant concentration N_{gas}. Therefore, the "total doping efficiency" $\eta_{tot} = N_{act}/N_{gas}$ behaves differently than η_s: a square root dependence of η_{tot} on the gas phase dopant concentration is found, cf. Fig. 6.6.

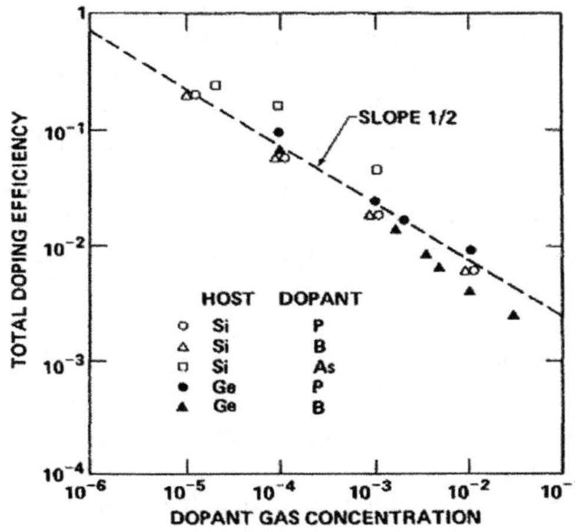

Fig. 6.6 Total doping efficiency vs. dopant concentration in the gas phase. Reproduced with permission from [19], copyright (1987) by The American Physical Society.

This behavior has been tentatively explained by a number of doping models that take into account different chemical reactions at the *a*-Si:H growth surface between dopants and silicon atoms, e.g. [29], as well as the role of hydrogen, e.g. [30]. Tanaka gives a good introduction into the subject matter based on simple thermodynamic considerations ([31], p. 142*ff*).

6.3 Ultrathin *a*-Si:H Layers on Crystalline Silicon

The investigations presented so far focused on thick (at least several 10 nm, typically several 100 nm) *a*-Si:H films. In the following, the properties of ultrathin *a*-Si:H layers with a thickness of the order of 10 nm shall be discussed. It can be expected that this leads to different material properties, as deposition times are short, thus the defect equilibration process outlined above maybe incomplete, and both the film/substrate and the vacuum/film interface regions constitute a significant part of the total sample thickness. The low *a*-Si:H thickness leads to experimental difficulties, as many characterization techniques rely on a certain amount of material to yield a detectable signal.

This section therefore briefly discusses the initial *a*-Si:H growth on SiO_2 and *c*-Si substrates, then presents experimental techniques for its characterization and discusses the properties of ~10 nm thin nominally intrinsic and doped *a*-Si:H films.

6.3.1 *Initial Stages of a-Si:H Film Growth*

In order to passivate the surface of a *c*-Si wafer, it is essential that the ultra-thin *a*-Si:H forms a closed film. The initial stages of *a*-Si:H growth have been the subject of several investigations, e.g. [32, 33, 34]. It was found that usually, two growth phases can be distinguished: a first, fast growth regime that is due to nucleation and growth of separate islands, followed by a second phase after the coalescence of the islands into a compact film. The spectroscopic ellipsometry (SE) and Fourier transform infrared (FTIR) spectroscopy carried out in attenuated total reflection (ATR) geometry reproduced in Fig. 6.7 show such a behavior: After an initial fast increase of the thickness of the surface roughness layer used in the SE model, the thickness of the compact *a*-Si:H bulk layer starts to increase, while the roughness decreases again. This is interpreted as the coalescence of *a*-Si:H islands that have formed during the first ~40 sec of film growth. After ~2 min, the thickness of the surface roughness layer reaches a steady state at a thickness of ~1.5nm, whereas the bulk component of the *a*-Si:H layer is calculated to be ~2.8 nm thick at this point. The absorbance data from FTIR shows a behavior that is compatible with these findings: Due to the enhanced surface during island growth, the respective modes increase quickly in intensity, while the absorption line at 2008 cm^{-1} that is assigned to the Si-H bulk stretching mode here[3] only starts to increase after ~40 sec. Notice, that these data were obtained from *a*-Si:H grown on 3 nm thin native SiO_2 layers on <100> oriented *c*-Si, i.e. not exactly in the configuration (direct *a*-Si:H/*c*-Si interface) used in *a*-Si:H/*c*-Si solar cells.

While these results were obtained with *in-situ* ellipsometry, Fig. 6.8 shows that also *ex-situ*, a similar characterization can be carried out: Laades [35] investigated the evolution of (n)*a*-Si:H[4] thickness by monochromatic *ex-situ* ellipsometry for

---

[3] See [71, 79] for a discussion of the assignment of FTIR modes to Si-H_x bonding configurations.

[4] The deposition conditions are the same as those of the (n)*a*-Si:H films whose electronic properties are discussed below.

a-Si:H grown on oxidized *c*-Si with (100) surface orientation. The figure shows that also here, the growth regime can be divided into two parts:

(i) a fast growth regime for times shorter than 30 s, which may arise in part from a strong dominance of the initial increase in surface roughness. The latter is due to nucleation of separate islands of *a*-Si:H on the oxidized crystalline-Si sub strate. These islands reach a height greater than the nominal film thickness before they begin to coalesce into a compact film. In terms of a two-layer model, this roughness is responsible for the fast increase in the measured film thickness d_{eff} and results in a larger growth rate R_d, the value of which depends on the process conditions. For the present series, R_d amounts to 5.5 Å/s, as inferred from the linear plot in the top left inset[5].

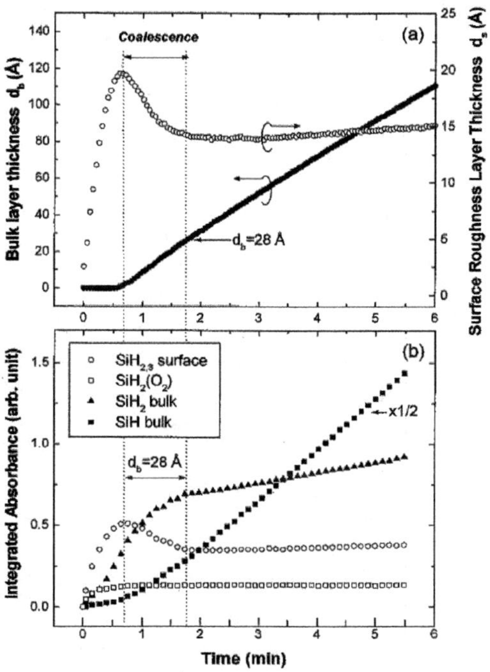

Fig. 6.7 Time evolution of the bulk layer thickness (d_b) and surface roughness layer thickness (d_s) determined by the SE analysis (a), and the integrated absorbance for the SiH_n (n = 1…3) bonds determined by the ATR analysis (b) during a-Si:H deposition on SiO_2/*c*-Si substrate. In (b), the integrated absorbance for the SiH bulk mode (solid square) is scaled down by half for clarity. Reproduced from [33] with permission, copyright (1999) by The American Physical Society.

[5] Note that there is an offset in the linear relation of d_{eff} vs. t_{dep} (inset in Fig. 6.8), i.e. the linear fit does not cross the origin. This is why the log-log plot does not show a straight line at t_{dep} < 60 sec.

(ii) a slow growth regime for times larger than 20 s. The film growth proceeds with a constant R_d of 1.4 Å/s, characteristic of a steady-state growth process, in which the nucleation islands have coalesced to a compact bulk-like film.

It should be noted, however, that the present observation indicates only a gradual transition from the fast regime to the steady state regime. No cut-off time separating the case where the substrate is not covered and the case where the film is dense and pinhole-free can be determined accurately. Also, the determination of d_{eff} is subject to a rather large error of ~50% relative.

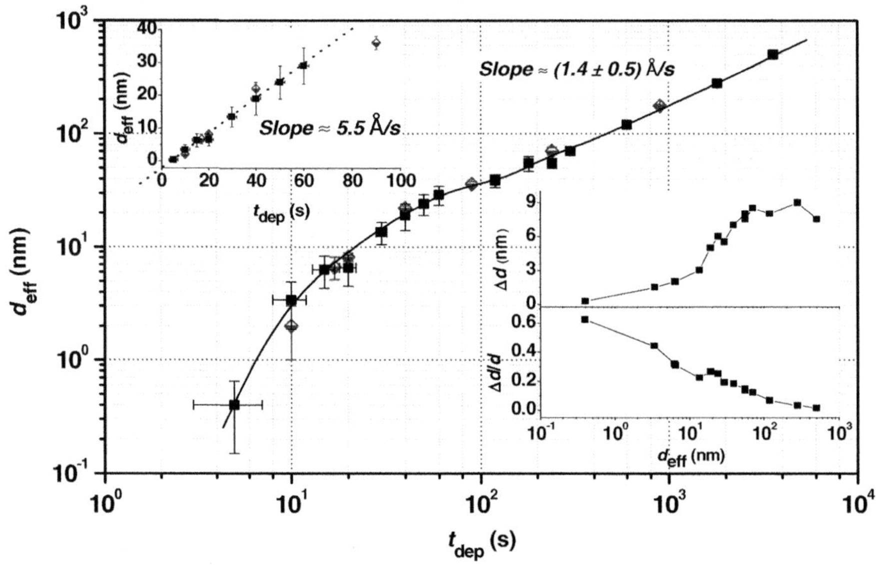

Fig. 6.8 Evolution of *a*-Si:H thickness on thin SiO_2 with deposition time (note the double-log scale). Upper left inset: linear plot of the initial fast growth regime. Lower right: Absolute and relative thickness fluctuations of the films. Different symbols mark different deposition runs. All lines are guides to the eye. From [35] with permission.

As it is to be expected that only closed *a*-Si:H films are able to passivate the *c*-Si surface efficiently, the formation of a closed film should also be apparent in charge carrier lifetime data. Figure 6.9 shows that this is indeed the case: Here, time-resolved surface photovoltage (SPV) measurements are analyzed [35].

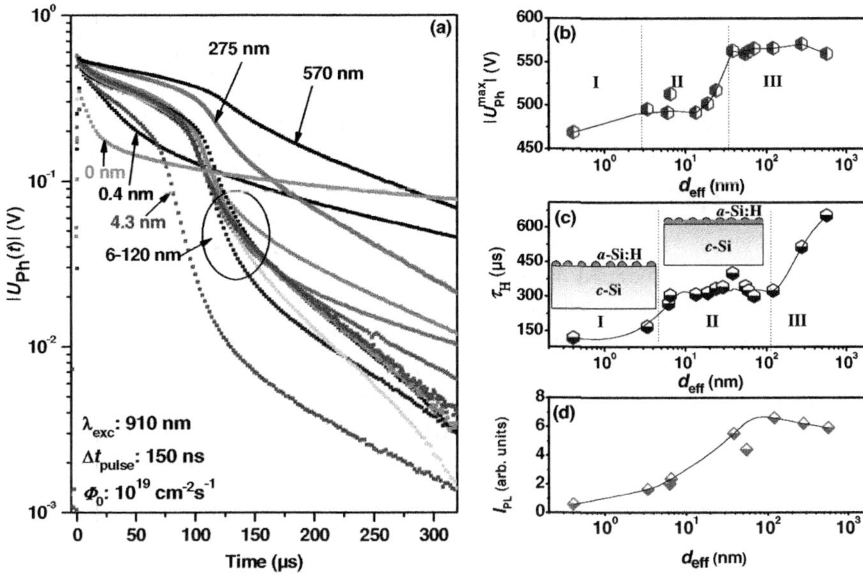

Fig. 6.9 (a) SPV transients of (n)*a*-Si:H/(p)*c*-Si samples with varying *a*-Si:H film thickness d_{eff}; (b) $|U_{\mathrm{SPV}}{}^{\max}| = \mathrm{f}(d_{\mathrm{eff}})$; (c) $\tau_{\mathrm{H}} = \mathrm{f}(d_{\mathrm{eff}})$ calculated from the transients in (a), and (d) photoluminescence intensity $I_{\mathrm{PL}} = \mathrm{f}(d_{\mathrm{eff}})$. From [35] with permission.

Briefly, SPV measures the band bending difference in the *a*-Si:H/*c*-Si system between dark equilibrium and strong illumination [36, 37]. If the illumination is strong enough, the sample will reach high injection and flatband conditions. Then, the measured photovoltage U_{SPV} is (apart from corrections for the so-called Dember voltage) the band bending in the system. After the illumination – provided by a ~150ns laser pulse – ends, the photogenerated charge carriers recombine and the photovoltage decays over time. The time constant of this decay is a measure for the effective carrier lifetime, thus for the surface passivation of the sample. This is very similar to the photoconductance decay discussed below. In Fig. 6.9, the initial photovoltage $U_{\mathrm{SPV}}(t = 0)$ and the time constant of the initial slow decay of surface photovoltage τ_{H} are plotted vs. film thickness. At an effective film thickness $d_{\mathrm{eff}} = 4.3$ nm, the SPV transient shows the typical shape for (n)*a*-Si:H/(p)*c*-Si samples, consisting of a plateau at high photovoltage and subsequent fast decay. Also, $U_{\mathrm{SPV}}(t = 0)$ reaches values that are comparable to those on thicker films, indicating that the *p*/*n* junction has been fully established. Thus, it was concluded, that under the conditions used here, the *a*-Si:H film grown on *c*-Si is closed for $d_{\mathrm{eff}} > 4$-5 nm. For further increasing film thickness, the decay time constant τ_{H} increases, indicating further improving passivation quality. This is corroborated by measurements of the photoluminescence, Fig. 6.9(d). It is likely that this is due to the reduction of recombination-active defects at the *a*-Si:H/*c*-Si interface while the sample stays at elevated temperature for prolonged time spans (see section 6.3.5).

6.3.2 *Measuring a-Si:H Defect States by Near-UV Photoelectron Spectroscopy: Theory and Measurement Set-Up*

For ultra-thin *a*-Si:H films used as buffer, emitter or BSF in *a*-Si:H/*c*-Si solar cells, with film thicknesses around 10 nm, the electronic density of states in the band gap, the position of the Fermi level and the band discontinuities (offsets) to the adjacent *c*-Si and/or TCO and metal layers are difficult to determine: As mentioned above, the growth of the thin *a*-Si:H film, consisting of only a few atomic layers, is influenced by the substrate. Thus, it is important to measure film properties directly on the substrate material that is also used in the solar cell, i.e. crystalline silicon with the same doping and surface preparation as used for the cell. However, if electrical measurements established for the characterization of bulk *a*-Si:H, such as CPM, are carried out on such a sample, the measured quantities are always determined by the contribution from the *c*-Si bulk material; the same holds for techniques such as PDS, where the heat capacity of the ultra-thin *a*-Si:H is not sufficient to produce a detectable signal.

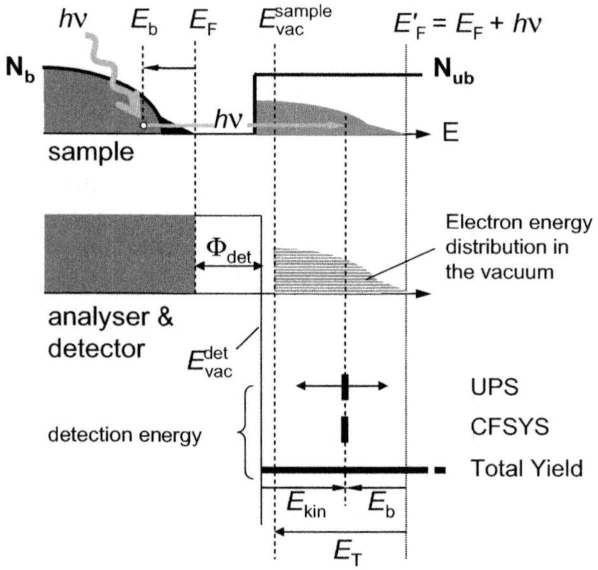

Fig. 6.10 Schematic density of states distribution for an *a*-Si:H sample (upper left), photoemission process (upper right) and photoelectron detection (center and lower right). The binding energy E_b is referenced to the Fermi energy E_F in the sample and is related to the measured quantity E_{kin}, the kinetic energy of the detected photoelectron, by $E_b = E_{kin} - (h\nu - e\,\Phi_{Det})$, where $h\nu$ is the photon energy of the monochromatic light used for excitation of the photoelectrons and Φ_{det} is the work function of the electron detector. The lower part of the sketch shows the differences between the standard UPS mode, Total Yield and Constant Final State Yield Spectroscopy.

To obtain information on the properties of materials on an atomic depth scale, photoelectron spectroscopy (PES) is one of the most widely used methods, e.g. [38]. Briefly, photoelectron spectroscopy works as follows (cf. Fig. 6.10): From an energetic distribution of occupied (valence band) electronic states (occupied DOS, upper left in the figure), photoelectrons are excited, i.e. their energy is increased by the photon energy $h\nu$ of the impinging monochromatic light. This produces an "image" of the occupied DOS (grey area in the upper right) in the unoccupied (conduction band) DOS, here sketched as a step function[6]. The photoelectrons travel to the surface of the sample, and if their kinetic energy and momentum is sufficient to overcome the work function barrier, they are emitted into the vacuum (photoelectron distribution in the middle right), where they are counted by an energy-resolving photoelectron detector (detection window: black bar next to the label "UPS"). The measured count rate vs. kinetic energy distribution is a measure of the occupied density of states in the sample.

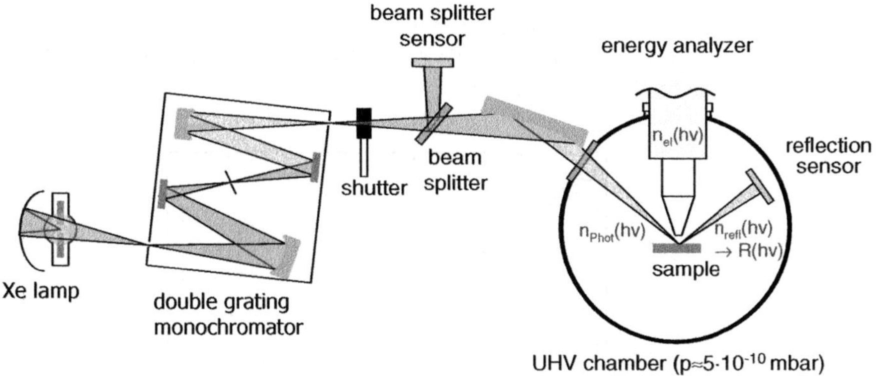

Fig. 6.11 Set-up of the equipment used for photoelectron spectroscopy in the near-UV range (4…7eV), in the Constant Final State Yield (CFSYS) mode.

The standard PES variants, however, are not suitable for the characterization of band gap states in ultrathin *a*-Si:H: PES with excitation by light in the UV range (UPS, typical photon energy $h\nu$=21.2 eV) only probes the first 2-3 monolayers of the material. For X-ray excited PES (XPS, photon energy in the keV range), the excitation cross section of photoelectrons from the band gap region is too small, leading to a very poor signal-to-noise ratio. For a more detailed discussion of these issues, the reader is referred to the work of Winer and Ley, e.g. [39, 40]. To circumvent these problems, a special variant of photoelectron spectroscopy has been developed, which uses light with variable photon energy in the near UV range

[6] Note, the assumption of a step function is not true in general, but justified for the conduction band DOS of *a*-Si:H: Using inverse photoemission (bremsstrahlung isochromate spectroscopy), it has been shown that the conduction band DOS varies by only ~15% in the energy interval $h\nu$ = 4.5…7 eV relevant for near-UV PES [122, 47].

(typically $h\nu$ = 4...8 eV) for the excitation of photoelectrons. Figure 6.11 shows such a set-up: The white light of a high pressure Xe lamp, which provides sufficient flux for photon energies up to ~7.5 eV, is monochromatized using a double grating monochromator, and excites photoelectrons from the sample, which is kept in ultra-high vacuum to prevent contamination. Both the incoming and the reflected photon flux are measured. For these near-UV excitation conditions, the excitation cross section is orders of magnitude higher than with XPS and UPS, and the photoelectron emission depth ("information depth") is of the order of some nm, i.e. ideally suited to characterize ultrathin *a*-Si:H layers. Depending on the type of photoelectron detector used for the experiment, this is termed "Total Yield spectroscopy" (TY) [41, 39, 42, 40] or "Constant Final State Yield Spectroscopy" (CFSYS) [43, 44]: In Total Yield spectroscopy, all photoelectrons that leave the sample are counted, independent of their kinetic energy, while the photon energy $h\nu$ is varied. It can be shown that an analytical expression relates the derivative of the measured count rate vs. $h\nu$ to the density of states, e.g. [42]. In CFSYS, the same energy analyzer as used in UPS serves as detector, and is set to a constant kinetic (final state) energy, hence the name. Both TY and CFSYS can provide similar information on the occupied DOS. In CFSYS, however, also the Fermi level position can be determined easily.

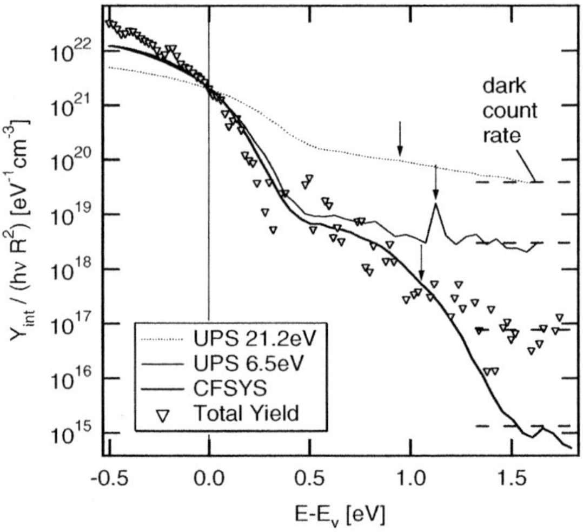

Fig. 6.12 Comparison of different photoelectron spectroscopy modes, applied to the same sample (10 nm (i)*a*-Si:H on *c*-Si): Standard UPS at 21.2 eV and at 6.5 eV, Total Yield and CFSYS. The dashed lines on the right mark the noise levels of the different methods. Reproduced from [45] with permission from Elsevier.

Figure 6.12 compares the sensitivity of the variants of photoelectron spectroscopies discussed above (except XPS, which just shows random noise in the band gap region). It is obvious, that already the lowering of the UPS excitation energy from the standard 21.2 eV to 6.5 eV increases the signal-to-noise ratio (SNR). The reasons are the increase in photoexcitation cross section on one hand, and the low stray light intensity due to the double grating monochromator on the other hand. With Total Yield and CFSYS, further improvements are possible in the SNR, which are due to experimental details beyond the scope of this brief introduction; the interested reader is referred to the literature quoted above. The graph shows that with CFSYS, it is possible to measure the density of occupied states in the gap of amorphous silicon, and in the valence band close to the band edge, over seven orders of magnitude, with a lower detection limit of about 10^{15} states/(cm³eV). In view of the typical defect densities in *a*-Si:H reported above, which are no less than 10^{16}-10^{17} states/(cm³eV) at midgap, it is clear that the Near-UV photoelectron spectroscopy is ideally suited to measure the defect distribution in ultrathin *a*-Si:H layers.

Care has to be taken when evaluating the photoelectron yield data. Photoemission can be described as a three step process [46]: photoionization, transport of the generated photoelectrons to the sample surface and their emission into vacuum before being counted. The energy distribution $J(E,h\nu)$ of electrons after photoionization can be written as [42]

$$J(E,h\nu) \propto h\nu R^2(h\nu) N_{occ}(E) N_{uocc}(E+h\nu)\,, \tag{6.4}$$

where N_{occ} and N_{uocc} are the occupied and unoccupied densities of states, $h\nu$ the photon energy of the monochromatic excitation, and R^2 the dipole matrix element describing the photoexcitation process. It has been shown that R^2 decreases with $h\nu^{-5}$ in *a*-Si:H [47]. This is irrelevant for conventional UPS, where $h\nu$ is fixed, but has to be taken into account in CFSYS and Total Yield spectroscopy, where the excitation energy is varied. Thus, dividing the photoelectron yield Y_{int} by $h\nu R^2$ should directly give the density of occupied states N_{occ}. Accordingly, $Y_{\text{int}}/\, h\nu R^2$ is the quantity plotted in all CFSYS spectra in this chapter. An advantage of CFSYS over UPS is the fixed final state energy: This corresponds to a fixed energetic position within the unoccupied DOS, i.e. for CFSYS, N_{uocc} is a constant factor in eq. (6.4), cf. also the discussion of Fig. 6.10.

Unfortunately, the measured Y_{int} is smeared out by the finite energy resolution of both the light source and photoelectron detector. Mathematically, this can be described by a convolution with the apparatus transfer function, and a deconvolution followed by dividing by $h\nu R^2$ should yield the true DOS. It is found, however, that the deconvolution procedure leads to strong artifacts (oscillations, so-called "ringing" [48], also known from similar approaches in the analysis of CPM and PDS data). Thus, a more robust approach has been adopted [49, 20]: A model density of states consisting of the valence and conduction band DOS, the exponential band tails and a Gaussian defect distribution is multiplied by $h\nu R^2$ and convoluted with the known (energy dependent) apparatus transfer function. The model is then fitted to the measured spectra by varying the parameters (tail slope, valence band edge etc.) of the model DOS. Figure 6.13 gives an example of such a fit. The DOS parameters obtained in this way are plotted in e.g. Fig. 6.17 (see section 6.3.4).

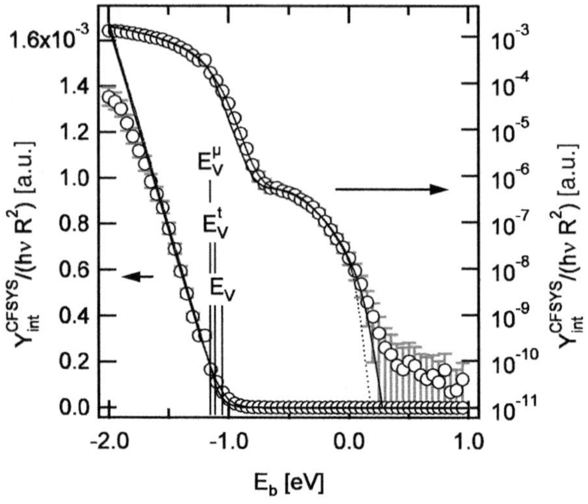

Fig. 6.13 Internal photoelectron yield from a 300 nm (i)*a*-Si:H sample obtained by CFSYS (circles), with error bars. Dotted line: model DOS; full line: model DOS convoluted with the apparatus transfer function.

6.3.3 Defect Densities, Interface Recombination and Solar Cell V_{oc}

In the following, as in most literature on the subject at hand, minority carrier lifetime measurements are used to obtain information on the so-called "implied V_{oc}" of a solar cell [50, 51]. The reasoning behind this is the following: Under open circuit conditions of a solar cell, as no current is drawn externally, the recombination current has to balance the photogenerated current, $j_{ph} = j_{rec}$. Then,

$$j_{ph} = \frac{q\Delta n_{av} W}{\tau_{eff}} . \tag{6.5}$$

Here, a *p*-type wafer is assumed, i.e. the electrons, of concentration *n*, are the minorities. *W* is the thickness of the solar cell, Δn_{av} the average density of minority charge carriers, and τ_{eff} the effective carrier lifetime, which is defined as $\tau_{eff} := \Delta n/G$ (*G* is the generation rate). In fact, eq. (6.5) is just a reformulation of this definition. Now, the implied V_{oc} can be determined from the splitting of the quasi Fermi energies of electrons and holes,

$$V_{oc,impl} = \frac{kT}{q} \ln\left(\frac{\Delta n(0)[N_A + \Delta p(0)]}{n_i^2} \right) , \tag{6.6}$$

where N_a is the concentration of acceptors, n_i the intrinsic carrier concentration, and $\Delta n(0)$, $\Delta p(0)$ the concentration of carriers *at the p/n junction*. In general, care has to be taken because the local carrier concentration at the junction is not necessarily equal to the averaged one, Δn_{av}. For well-passivated surfaces and diffusion lengths exceeding the thickness of the wafer $\Delta n_{av} \approx \Delta n(0)$, though, and eq. (6.5) can be inserted into eq. (6.6) to calculate the implied V_{oc} from τ_{eff}, assuming a given photocurrent j_{ph}.

To measure the carrier lifetime τ_{eff}, photoconductance decay (PCD) measurements, mostly using the commercially available tools by Sinton consulting [50, 51], are widely used. They calculate the dependence $\tau_{eff}(\Delta n)$ of the carrier lifetime on the injection level. Note that similar information can be obtained also by other techniques, such as the transient decay of surface photovoltage (SPV) over time used above. While a quantitative calculation of τ_{eff} from SPV is less straightforward, such data has been used successfully as a qualitative measure of τ_{eff}, and also yields information on the band bending in the system, i.e. the built-in voltage V_{bi} of the cell. Both PCD and SPV can be applied to solar cell "precursor" structures without electrical contacts, which makes them a valuable tool for monitoring the surface passivation quality throughout the solar cell process chain.

The effective carrier lifetime is dependent on recombination both in the bulk of the c-Si wafer and at its interfaces:

$$\frac{1}{\tau_{eff}} = \frac{1}{\tau_b} + \frac{1}{\tau_s}, \tag{6.7}$$

where τ_b is the bulk lifetime and τ_s that at the surface. Following Sproul [52], τ_s can be written for symmetric samples (both wafer surfaces have equal passivation quality) as

$$\tau_s = \frac{W}{2S} + \frac{1}{D}\left(\frac{W}{\pi}\right)^2. \tag{6.8}$$

Here, S is the surface recombination velocity, which is related to the effective interface recombination rate U_{int}: $S = U_{int}/\Delta n$, and D is the minority carrier diffusion constant if low injection conditions are considered, or the ambipolar diffusion coefficient for the high injection case. For the calculation of S, the proper choice of Δn has to be considered: $\Delta n = \Delta n(0)$, i.e. at the recombination active interface, should be chosen to describe the interface. However, in most cases, $\Delta n = \Delta n_{av}$ or $\Delta n = \Delta n(X_s)$, X_s being the limit of the space charge region, is chosen, because these are more easily calculated from experimental data. Then, one gets an effective value of S, S_{eff}.

The recombination rate U is linked to the density of interface states D_{it}:

$$U_{\text{int}} = (n(0)p(0) - n_i^2) \int_{E_V}^{E_C} D_{it}(E) \cdot f(E,n,p,c_n,c_p) \mathrm{d}E\,, \tag{6.9}$$

where f is a function that depends on a number of parameters, like the capture cross sections c_n, c_p of the defects. While for impurity-related defects, the standard Shockley-Read-Hall recombination theory can be applied to calculate U_{int}, at the *a*-Si:H/*c*-Si interface, the amphoteric character of the silicon dangling bonds (see above) has to be considered. This has been done with various degrees of sophistication using analytical and semianalytical models [53, 54, 55, 56] as well as numerical simulations [55].

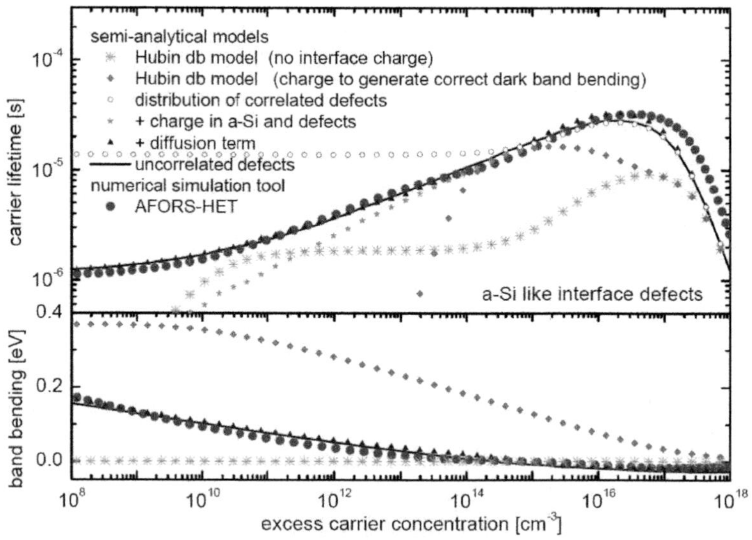

Fig. 6.14 Comparison of effective carrier lifetime and band bending at the (i)*a*-Si:H/(p)*c*-Si interface, as calculated using various semi-analytical models and the numerical simulation tool AFORS-HET developed at Helmholtz-Zentrum Berlin [57, 58]. See [55] for details on the simulation parameters.

Figure 6.14 compares effective carrier lifetime and *a*-Si:H/*c*-Si interface band bending vs. excess charge carrier density calculated with different models (see [55] for details). These simulations show, that it is necessary to consider the re-charging of the *a*-Si:H/*c*-Si interface states and their influence on band bending in the system [55], while it is not possible to simulate the band bending induced by the amorphous silicon layer with a fixed interface charge, as has been done in other models [56]. Furthermore, it becomes clear that depending on the various model parameters, major discrepancies can be found between an elaborate model and simplified ones. Reasons for this can be the diffusion process of carriers to the surface limiting surface recombination velocity and a defect distribution leading to a quasi-Fermi level-dependent recombination.

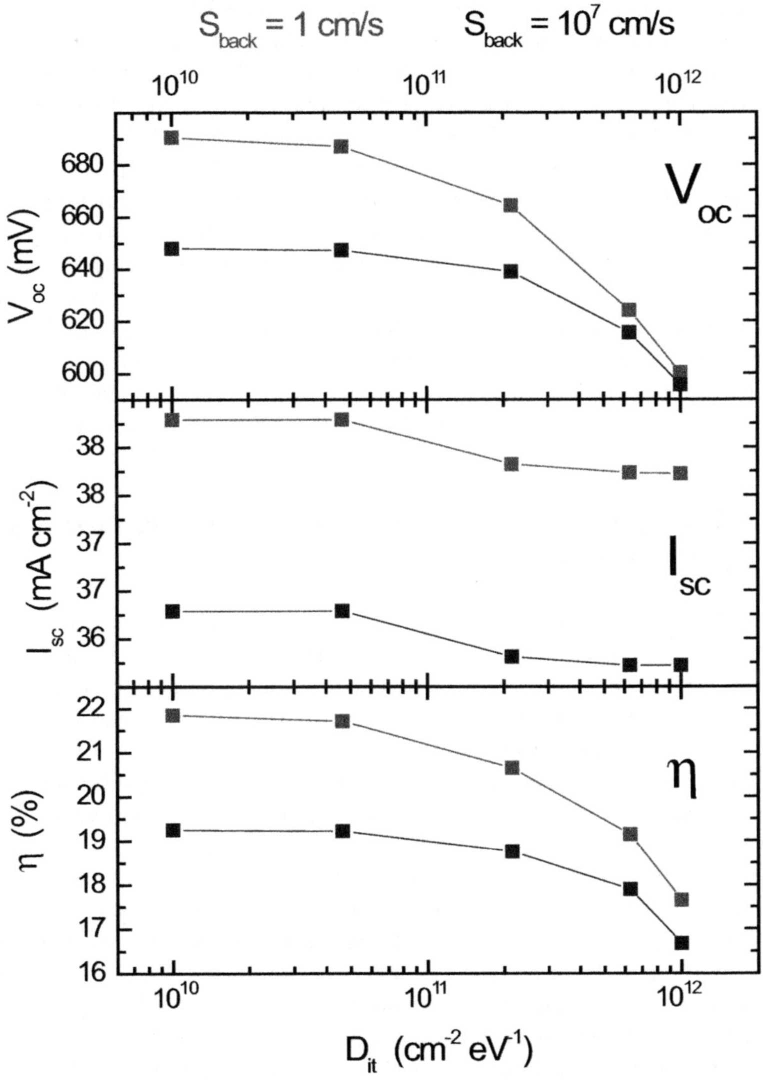

Fig. 6.15 Simulated solar cell parameters of (n)*a*-Si:H/(p)*c*-Si heterojunction solar cells in dependence on the front side interface defect density D_{it}, for two different rear side recombination velocities S_{back} [59].

Using such models, it becomes possible to make use of PCD data not only to monitor passivation quality and expected cell V_{oc}, but to extract key information on the electronic properties of the *a*-Si:H/*c*-Si interface from $\tau_{eff}(\Delta n)$ data. Namely the product of the density of *a*-Si:H/*c*-Si interface states with their capture cross section, $\sigma \cdot D_{it}$, can be extracted, as well as the interface charge, which is influenced by process conditions and deliberate *a*-Si:H doping. With values for σ from literature, D_{it} can be calculated, cf. e.g. Fig. 6.18 (see section 6.3.5).

The question of how solar cell parameters are related to the density of interface states D_{it} can also be approached through simulations: Figure 6.15 shows simulations of solar cell V_{oc}, j_{sc} and efficiency carried out using the numerical device simulator AFORS-HET developed at Helmholtz-Zentrum Berlin [57, 58, and Chapters 13 and 14 in this book]. The simulated cell is a (n)*a*-Si:H/(p)*c*-Si structure (without (i)*a*-Si:H buffer layer), where two different surface recombination velocities S_{back} were assumed for the rear side of the cell, and the density of interface states D_{it} at the front (n)*a*-Si:H/(p)*c*-Si is varied. As expected, mainly V_{oc} is influenced by the varying surface passivation. While for the rather poor rear side passivation with $S_{back} = 10^7$ cm/s, the recombination at the rear side limits the V_{oc} to below 650 mV, close to 700 mV are reached for the well-passivated rear contact if also the front side has a low defect density. According to this simulation, a maximum D_{it} of some 10^{10} cm^{-2} at the *a*-Si:H/*c*-Si interface is allowed if cell efficiencies above 20% are desired[7].

6.3.4 a-Si:H Gap States Distribution in Ultrathin Intrinsic a-Si:H Layers on c-Si

Using Total Yield and CFSYS photoelectron spectroscopy, it has been investigated whether the density of states in the band gap of ultrathin *a*-Si:H layers is different from that in thick films (cf. the first section of this chapter): Based on the arguments in the preceding sections, one might expect increased gap state densities due to the additional constraints imposed on the growing *a*-Si:H network by the vicinity of the *c*-Si surface, and also an incomplete defect equilibration due to the short growth time of the order of one minute for device-relevant ~10 nm *a*-Si:H films. Figure 6.16 shows CFSYS spectra obtained from 11 – 17 nm thin (i)*a*-Si:H layers grown on <111> oriented *c*-Si, with variation of the substrate temperature T_{sub} [49, 60]. It is clear that the general shape of the gap DOS is the same as that obtained for the bulk (using CPM, PDS) and the surface region (from Total Yield) of hundreds of nm thick *a*-Si:H films on glass. Concerning the influence of growth temperature, in the Urbach tail region, clearly a pronounced change is visible, and the steepest slope is found for the 230°C sample. In Fig. 6.17 the DOS parameters extracted from the yield data using the model fit procedure outlined above are plotted.

[7] To be consistent with measured values of D_{it} such as in Fig. 6.18, obviously the same capture cross sections σ have to be assumed in both simulation and evaluation of experimental data.

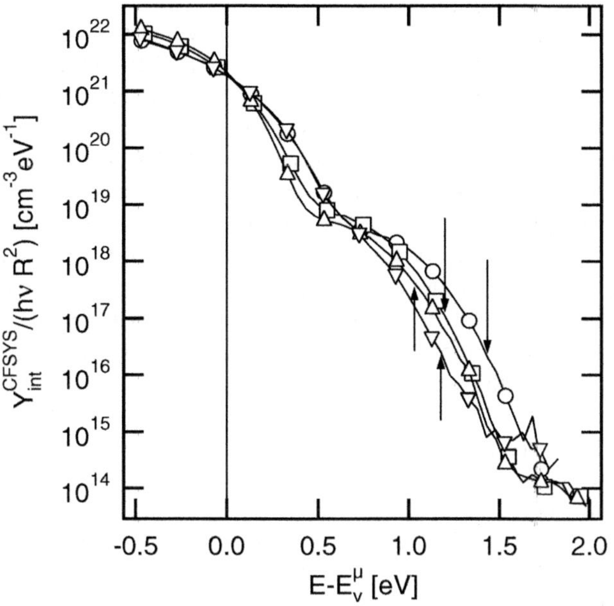

Fig. 6.16 CFSYS spectra from 11-17 nm thin (i)*a*-Si:H films grown on <111> (p)*c*-Si with varying deposition temperature, from 65 to 300°C. Arrows mark E_F. From [49]. E_v^μ is the valence band mobility edge.

The Urbach energy is indeed minimal for T_{sub} = 230°C – in line with findings in thick films, cf. Fig. 6.3 –, and the Fermi level is closest to the valence band edge in this case. Interestingly, the position and density of deep defects show only slight variations, in contrast to what could be expected from findings in thick *a*-Si:H layers. N_d is rather high: for E_{0v} of 60 meV, one finds $N_d \sim 2 \cdot 10^{18}$ cm^{-3}. however, a N_d of some 10^{16}-10^{17} cm^{-3} would be expected in bulk *a*-Si:H, see e.g. Fig. 6.4. This finding is in accordance with Total Yield data [61, 42, 62]: Siebke *et al.* find by Total Yield in a similar (i)*a*-Si:H deposition temperature series, albeit on sapphire substrate and with d_{aSiH} several 100 nm, that N_d is of the order of 10^{18} cm^{-3} and increases monotonously by ~25% for T_{sub} = 100…400°C; a minimal E_{0v} of 63 meV is reported at T_{sub} = 250°C [61]. Bulk defect densities obtained from CPM on the same samples are about an order of magnitude lower, and show a minimum of 10^{17} cm^{-3} at 250°C. Thus, the bulk N_d depends indeed on E_{0v} as expressed by eq. (6.3), while the deep defect states close to the surface behave differently. Note that extrinsic defects due to surface contamination of the PES samples as the source of the increased N_d can be ruled out with good confidence: The measurements reported in the figures were carried out after vacuum transfer from the deposition system into UHV ($p < 10^{-9}$ mbar), and from previous studies, it is known that e.g. adsorbed (atomic) oxygen only starts to change the DOS at exposures of 10 Langmuir (~10^{-5} mbar s) and above [62]. Winer *et al.* conclude that indeed, not only surface states, but an increased defect density close to the surface gives rise to the enhanced N_d. This correlates with an increased concentration of hydrogen – >25% close to the surface *vs.* 10-15% in the *a*-Si:H bulk – in the topmost 0.5-1 nm of *a*-Si:H films grown at optimum conditions, as evidenced by ATR spectroscopy [32, 33, 63] and by photoelectron spectroscopy with varying information depth [64]. Already since the 1980s, it is well known that the *a*-Si:H surface and the interface to the substrate play an important role for the electrical properties of the films. It has been shown for example, that the dark conductivity of *a*-Si:H films doped with 500 ppm phosphorous decreases by a factor of more than 20 when the film thickness is reduced from 500 to 100 nm [65]. Beyer and Overhof have made the connection to a changed concentration of hydrogen in regions close to the surface and the substrate interface, and thereby a different doping efficiency [66].

In the introductory section on *a*-Si:H in this chapter, the argument put forth by Hata *et al.* [27] is explained, that an increased defect density could arise from an incomplete anneal of the topmost *a*-Si:H monolayers that do not stay at elevated deposition temperature long enough to reach their thermodynamic equilibrium. While this argument holds only for the top layers of a thick film, the *a*-Si:H films investigated here consist of only 30-40 monolayers, deposited in ~1 minute. Thus, it is plausible to assume that the whole *a*-Si:H film is incompletely relaxed in this case.

Another aspect to be considered is the vicinity of the *c*-Si surface to the growth surface: In the initial stages of *a*-Si:H growth on *c*-Si, the incoming SiH_x species from the gas phase will bond to the *c*-Si surface dangling bonds, whose positions are fixed by the *c*-Si lattice. This imposes an additional constraint to the film growth. The description of bonding and network relaxation in amorphous solids has been attempted within the framework of *constraint/rigidity theory* ([67] and refs. therein). The influence of the fixed *c*-Si atom positions can be incorporated into this model [68]. As shown by Lucovsky *et al.* [69, 70], this has strong influence on the initial stages of growth of SiO_2 on *c*-Si. Also, the connection to the varying hydrogen content can be made: Fujiwara and Kondo find an *a*-Si:H/*c*-Si interface region with strongly increased hydrogen content, where H is present mainly in the form of SiH_2 (Fig. 6.7). This hints to the existence of a void-rich interface region that extends about ~10Å in the case of this study [32].

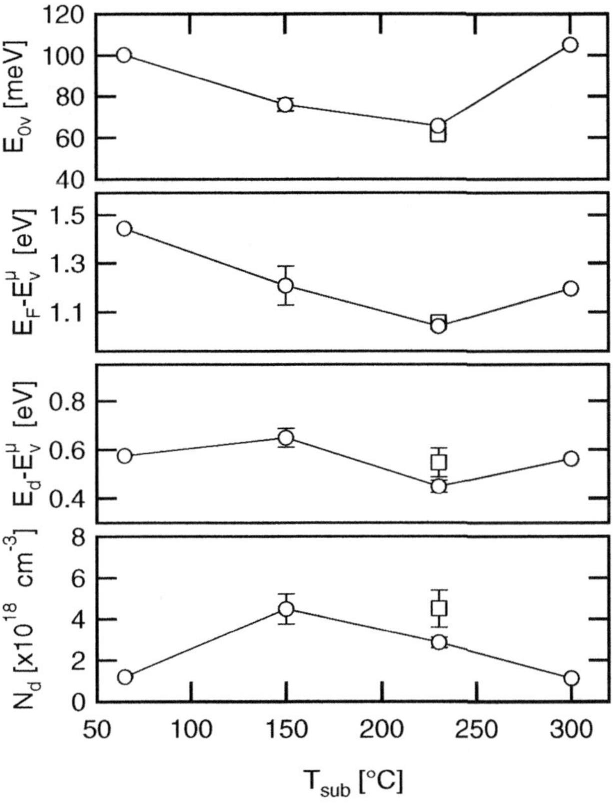

Fig. 6.17 *a*-Si:H DOS parameters extracted from the spectra in Fig. 6.16, plotted vs. substrate temperature: Urbach energy E_{0V}, distance $E_F - E_V$ of the valence band edge from the Fermi level, distance $E_d - E_V$ of the Gaussian defect distribution from E_F, and the density of deep defects N_d integrated over the energy. Squares: fit to the 300 nm (i)*a*-Si:H sample in Fig. 6.13. From [49].

6.3.5 Post-Deposition Anneal of a-Si:H

Recently, post-deposition anneals have been investigated as a means to improve the passivation of *a*-Si:H on *c*-Si [71, 72, 73, 74, 75, 76]. In this context, the effect of post-deposition anneals on network strain as monitored by the Urbach energy E_{0V}, *a*-Si:H/*c*-Si interface defect density D_{it} and hydrogen content in ultra-thin *a*-Si:H on *c*-Si has been investigated [77, 78]. The results are summarized in Fig. 6.18: For all deposition conditions that were investigated – a matrix consisting of three deposition temperatures, 130°C, 170°C and 210°C, and three different deposition regimes termed "low / medium / high pressure regime" – the as-deposited density of interface states D_{it}[8] and the total bulk hydrogen content $C_{H,bulk}$ increase with decreasing deposition temperature for all deposition parameter sets. The excess hydrogen content incorporated upon lowering the deposition temperature is mainly found in the high stretching mode (HSM) of the FTIR spectra while the contribution of the low stretching mode (LSM) is even slightly reduced with decreasing T_{depo}. The two stretching modes are due to different H bonding environments: The LSM is associated to "compactly incorporated" monohydrides SiH, characteristic for compact films, while the HSM arises from clustered monohydrides [79].

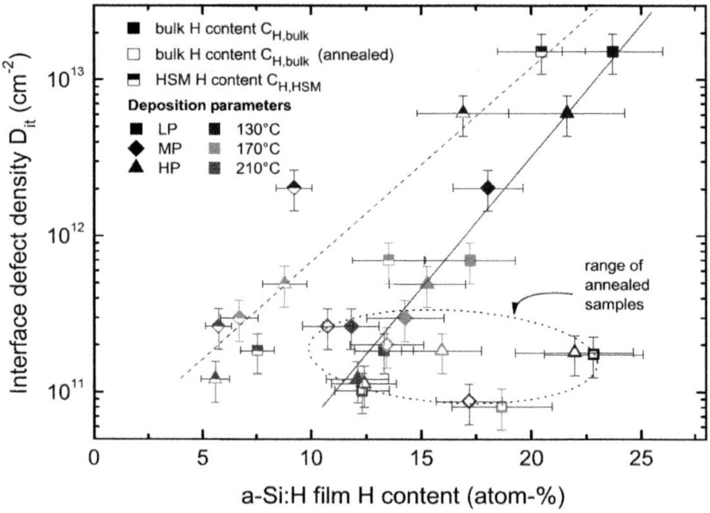

Fig. 6.18 Density of interface states D_{it} for different deposition conditions vs. the H content found in HSM ($C_{H,HSM}$) and the sum of HSM and LSM ($C_{H,bulk}$), as-deposited as well as after annealing (lines are exponential fits). Reprinted with permission from [78]. Copyright 2010, American Institute of Physics.

At a deposition temperature of 210°C, there is no significant difference in hydrogen bonding and recombination between the different parameter sets, and D_{it} reaches state-of-the-art values of the order of 10^{11} cm^{-2}. Note that this is consistent with the decrease of *a*-Si:H defect density shown in Fig. 6.17, where the optimum temperature is also 210°C. The lower T_{depo}, the more pronounced are the differences in H bonding configuration and D_{it} between the three parameter sets.

After annealing for 5 min at 200°C, all the samples reach D_{it} values of around 0.8…3x10^{11} cm^{-2}, which is about the level of the best as-deposited passivation (hot plate (HP), 210 °C) and corresponds to effective Auger-corrected c-Si minority carrier lifetimes of 3…6 ms at an excess charge carrier density $\Delta n = 10^{15}$ cm^{-3}. These final carrier lifetime values are similar to what is reported by other authors (see below).

[8] Obtained by fitting the *a*-Si:H/*c*-Si interface defect model described in [55] and Chapter 13 in this book to carrier lifetime vs. excess charge carrier density, obtained from QSSPC measurements.

Thus, it can be concluded, that in the as-deposited state of *a*-Si:H, the *c*-Si passivation correlates with the bonding structure of thin (i)*a*-Si:H layers, which is defined by the deposition regime, while D_{it} in the annealed state is insensitive to the details of the H bonding and thus the deposition parameters. Further analyzing the data, it is found that the dependence between increasing hydrogen content and decreasing mass density of the *a*-Si:H films is indicative of the presence of microvoids. For a hydrogen content above 10%, a model is proposed where microvoids that are in contact with the *a*-Si:H/*c*-Si interface "depassivate" the *c*-Si surface, thereby increasing D_{it} [78].

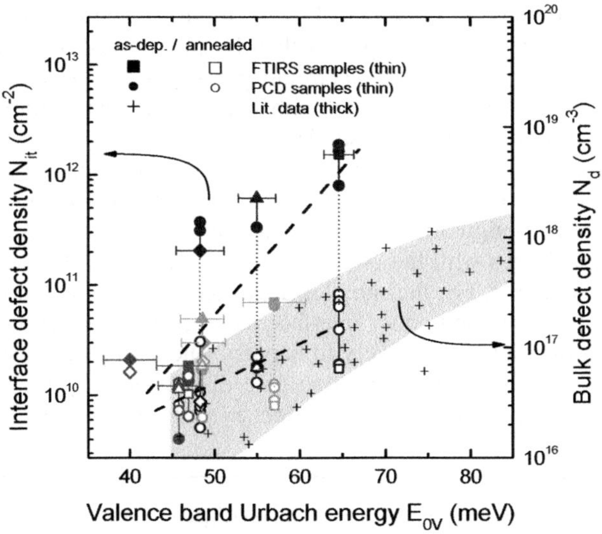

Fig. 6.19 Correlation of the valence band Urbach energy E_{0V} with the interface defect density in the as-deposited and annealed state vs. E_{0V}. The right axis is scaled such that $D_{it} = N_d d_t$, using an effective tunnel length from c-Si into a-Si:H of $d_t = 2.7$nm [123], which allows to map bulk defect densities onto the interface, i.e., to compare a quasi-D_{it} from bulk defect data of thick a-Si:H films (+, data from [19]) with our data. Large symbols: FTIRS samples, small circles: additional PCD samples nominally identical to FTIRS samples. Colors and shapes as in Fig. 6.18.

From Fig. 6.18, it appears also that there is a common lower limit in D_{it}, of the order of 10^{11} cm^{-2}, for all samples after annealing. As shown in Fig. 6.19, this seems to be due to the lower limit of *a*-Si:H bulk defect density set by the strain in the material, just as in bulk *a*-Si:H. The figure shows both the interface defect density at the *a*-Si:H/*c*-Si interface for ultrathin (i)*a*-Si:H passivation layers and the bulk defect density N_d in thick *a*-Si:H films (from [18]) vs. the Urbach energy. The volume density of N_d translates into a surface density via $D_{it} = N_d d_t$, using an effective tunnel length from c-Si into a-Si:H of $d_t = 2.7$nm [123].

As is evident from Fig. 6.19, there is a weak correlation of the as-deposited D_{it} with the Urbach energy. Upon annealing, D_{it} is reduced by up to two orders of magnitude, to 0.8…3x10^{11} cm^{-2} as discussed above. The D_{it} of the samples deposited at 210°C is close to the (two dimensional (2D) projected) defect density in bulk *a*-Si:H already in the as-deposited state, and small changes in D_{it} occur upon annealing. In contrast, the samples deposited at 170°C and more clearly 130°C strongly deviate from the bulk *a*-Si:H data in the as-deposited state. All annealed samples are consistent with the bulk a-Si:H data trend, including a slight increase in D_{it} with the Urbach energy. Thus, two major conclusions can be drawn:

i) At low deposition temperature, the defect density at the *a*-Si:H/*c*-Si interface is strongly enhanced, because the *a*-Si:H network is not able to reach its thermodynamic equilibrium state. Upon annealing, the *a*-Si:H interface defect density decreases to the equilibrium value, which is set by the amount of strain in the film, i.e. the Urbach energy. Thus, it appears that the equilibrated *a*-Si:H/*c*-Si interface region does not possess unique electronic properties but is determined by the a-Si:H bulk defects. This is consistent with studies of recombination and transport in *a*-Si:H/*c*-Si solar cells [78], as also discussed below.

ii) As neither the Urbach energy nor the hydrogen bonding configuration – averaged over the film thickness, from FTIR – change measurably upon annealing, it is clear that no reconfiguration of the bulk *a*-Si:H occurs. Presumably, the heterointerface equilibrates with the a-Si:H bulk by short-range H diffusion and local network reconstruction. For the samples deposited at low *T* that have a large void fraction and an exceptionally poor interface passivation, a partial reconstruction of the network at the interface is likely to happen. The strong decrease in D_{it} cannot be explained by H relocalization alone (see discussion in [78]) and the highly strained Si network is prone to such an effect even starting around 200 °C [80].

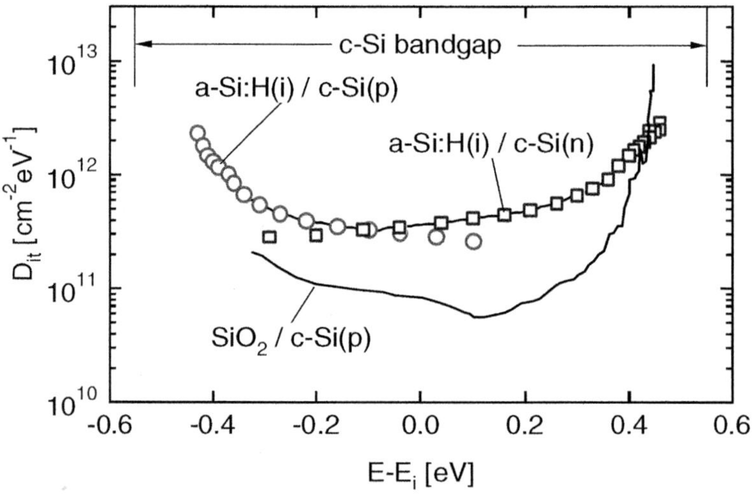

Fig. 6.20 $D_{it}(E)$ of 10 nm (i)*a*-Si:H/(p,n)*c*-Si structures, as calculated from bias voltage dependent surface photovoltage. Measurement temperature: 118 K. For comparison, the D_{it} of a 100 nm thermal oxide (1000 °C) on (n)*c*-Si (100) after forming gas anneal (450 °C, 30 min) is shown. The latter was measured at 295 K [81].

Under certain conditions, it is also possible to measure the D_{it} directly using surface photovoltage measurements under external bias voltage (field-dependent SPV, FD-SPV) [81]. A major limitation is that only undoped *a*-Si:H layers allow the field modulation of band bending in *c*-Si, while doped layers screen the external electric field. Thus, only for (i)*a*-Si:H films on *c*-Si, the D_{it} can be characterized with FD-SPV. Furthermore, it is mandatory that no change of the charge state of the *a*-Si:H and of the interface states takes place during the SPV light pulse. This is known to be the case for the (hydrogen terminated) *c*-Si surface and for the SiO_2/*c*-Si interface. For *a*-Si:H, however, recharging of *a*-Si:H states at or close to the interface is likely to occur. It can be reduced by measuring D_{it} at low temperatures. Accordingly, the $D_{it}(E)$ spectra on (i)*a*-Si:H/*c*-Si depicted in Fig. 6.20 were taken at 120 K. Still, the recharging effect cannot be completely excluded, so that the measured $D_{it}(E)$ spectrum gives an upper limit of the true interface state distribution. Figure 6.20 shows a comparison of the D_{it} obtained on two (i)*a*-Si:H/*c*-Si samples at 120 K to that of a SiO_2/*c*-Si sample (oxide thermally grown at 1100°C in dry O_2 followed by forming gas annealing), measured at room temperature. The two halves of the *a*-Si:H/*c*-Si D_{it} had to be measured on a (n)- and a (p)*c*-Si sample, because it is not possible to drive the *a*-Si:H/*c*-Si interface into inversion at 120 K. The measurements indicate a well passivated *a*-Si:H/*c*-Si interface. This is in good agreement with measured interface recombination velocities below 10 cm/s (see below). On the other hand, the fact that recharging effects are apparent in the data indicates that *a*-Si:H itself may contribute to the recombination via a-Si:H gap-states.

6.3.5.1 Accelerated Interface Defect Reduction by Microwave Annealing

The annealing experiments reported so far were all carried out on conventional hotplates, usually under nitrogen or another inert gas, but sometimes also simply in air. Recently, an alternative means to conduct the post-*a*-Si:H-deposition anneal could be established: The use of a conventional microwave oven (2.45 GHz magnetron, 700 W output power [82]). Figure 6.21 summarizes these results for two samples annealed step-by-step using repeated temperature cycles either on a hotplate (HP) or in the microwave (MW). Each temperature cycle consists of heating the sample up to the desired peak temperature by putting it on the pre-heated hotplate/switching on the microwave, and then letting it cool down radiatively in ambient atmosphere. The thermal budget was kept as similar as possible (same peak temperature and cooling transient; faster heating of the MW sample due to experimental constraints). After each cycle, the effective carrier lifetime τ_{eff} (inset in Fig. 6.21) was measured and the surface recombination velocity S was calculated using eq. (6.8).

The graph shows clearly that additional effects must play a role in the microwave anneal as compared to the hotplate: While the former reaches its lowest S of < 6 cm/s after only two annealing cycles, the hotplate anneal needs more than 20 cycles to reach the same passivation quality. For the hotplate anneal, it has been shown that the evolution of τ_{eff} over time can be fitted with a stretched exponential (e.g. [83] and Chapter 7 on *a*-Si:H/*c*-Si interface passivation in this book). This behavior is often found for the relaxation of disordered systems toward equilibrium, in the case of *a*-Si:H probably also linked to the release of hydrogen from trap sites, cf. the discussion in [83]. These models are based on thermodynamic equilibrium processes. In case of the microwave anneal, although there is no resonant absorption process involved, it is to be expected that in *a*-Si:H the high density of polar Si-H bonds contributes significantly to MW absorption as compared to *c*-Si. This is corroborated by the finding that the MW heating coefficient is reduced by 10–20% in the "naked" *c*-Si substrate as compared to *a*-Si:H/*c*-Si samples. The MW absorption mechanism in *a*-Si:H thus apparently involves exactly the type of bonds that need to be broken in order to mobilize bound hydrogen for *a*-Si:H/*c*-Si interface passivation to happen. Thus, it can be argued that a nonthermal contribution to the vibrational energy of the Si-H bond is imposed by the intense electric field of the microwave, effectively lowering the barrier for bond breaking.

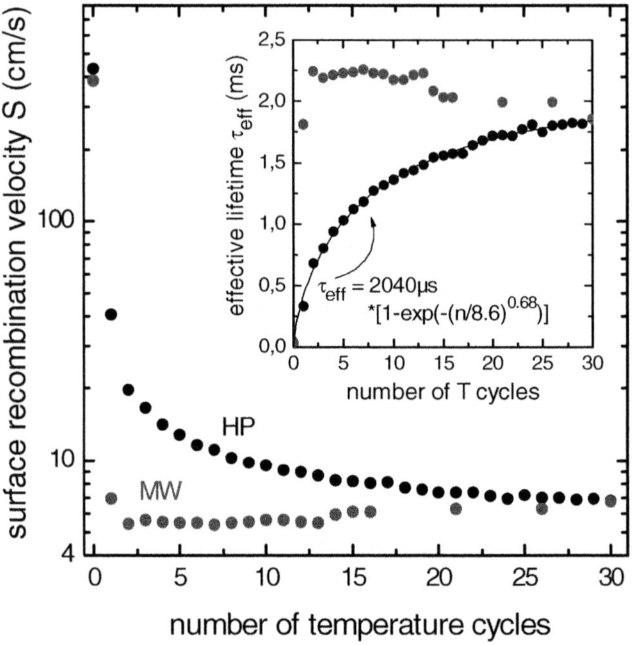

Fig. 6.21. Passivation quality of nominally identical samples that were pulse-annealed with similar temperature profiles. Black: surface recombination velocity *vs.* no. of temperature cycles ("pulses"), each one consisting of putting the sample on a pre-heated hotplate for some sec. followed by radiative cooling in ambient atmosphere. Red: the same, for annealing using a microwave. Inset: Linear plot of τ_{eff} *vs.* no. of temperature cycles. A stretched exponential function was fitted to the hotplate anneal data. Reprinted with permission from [82]. Copyright 2010, American Institute of Physics.

6.3.5.2 *a*-Si:H/*c*-Si Interface Passivation – Current State of the Art

To conclude the section on undoped (so-called intrinsic) *a*-Si:H buffer layers, the best results of several groups for *c*-Si interface passivation by (i)*a*-Si:H as reported in October 2009 [82] are shown in Fig. 6.22.

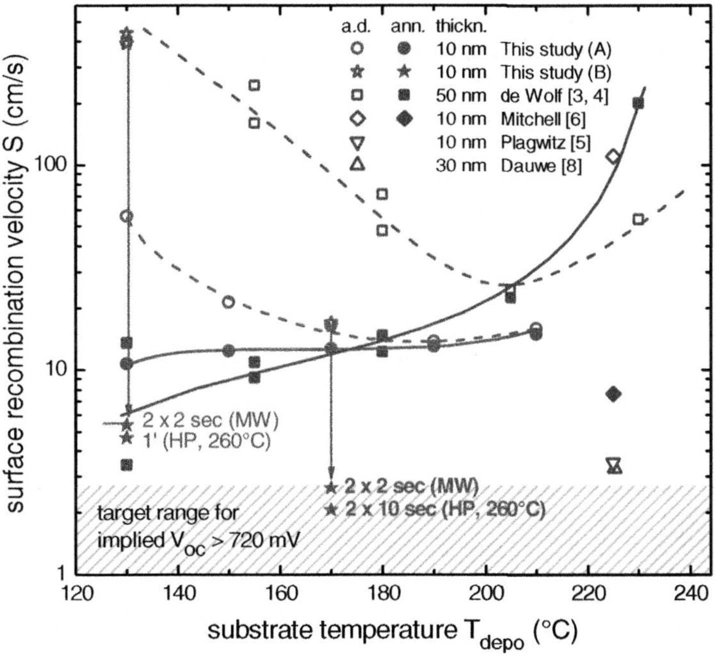

Fig. 6.22. Surface recombination velocity S from this work and previous studies, shown vs PECVD substrate temperature T_{depo}. As deposited passivation (a.d.) typically displays a minimum (open symbols, dashed lines as guides to the eye), with the optimum T_{depo} being deposition parameter dependent. Post deposition annealing (ann., full symbols and lines) increases the passivation – unless the growth regime is epitaxial for high T_{depo} –, the beneficial effect being more pronounced the lower T_{depo}. Numbers state the annealing times needed for the shown improvement in minutes or seconds. The dashed region marks implied open-circuit voltages of $V_{\mathrm{oc}} \sim 720$ mV for 3-5 Ωcm (p)*c*-Si substrate evaluated at 1 sun illumination. Reprinted with permission from [82]. Copyright 2010, American Institute of Physics.

In this figure the surface recombination velocity S is plotted vs. the *c*-Si substrate temperature during *a*-Si:H deposition. S of most samples is shown twice: Before (open symbols) and after (closed symbols) anneal. Most samples were annealed on a conventional hotplate, except the two labeled stars at $T_{\mathrm{dep}} = 130$ and 170°C, where a microwave was used.

While the pre-anneal data shows the characteristic behavior already seen in other parameters of thick and ultra-thin *a*-Si:H samples (notably the Urbach energy, cf. Fig. 6.3, Fig. 6.17) with a minimum around T_{dep} = 180-220°C, it is obvious also from this plot, that the low-*T* deposited *a*-Si:H samples show the greatest potential of reducing *S* by annealing, with minimum values well below 10 cm/s. There is only one notable exception reported in literature: Unusually low *S* for unannealed samples, around 3 cm/s, have been reported by the Institut für Solarenergieforschung Hameln (ISFH) [72, 76]. The samples in both publications stem from the same deposition system, and are also unusual in two other respects: A high RF power density was used (about 10 times higher than usually reported), and the samples were introduced directly into the PECVD system without using a loadlock, i.e. the deposition chamber was exposed to ambient atmosphere after each deposition. This indicates that there are probably other *a*-Si:H deposition conditions – unlike the standard recipe "low ion bombardment, low contamination, low deposition temperature + anneal" – that can also produce excellent surface passivation; also, oxidation of the *c*-Si surface might have occurred during the pump-down phase.

Furthermore, it should be noted, that device-relevant thicknesses for the use of (i)*a*-Si:H as buffer layers are ≤ 10 nm. As the passivation quality decreases strongly below 10 nm, as compared to thicker films (see e.g. the V_{oc} data in [84] and the SPV lifetime data in Fig. 6.9), this represents an additional challenge.

In Fig. 6.22, also a target range is given for the interface recombination velocity: As can be calculated easily using eqns. (6.5-6.8), $S < 3$ cm/s is needed for a $V_{oc} > 720$ mV in ~3-5 Ωcm (p)*c*-Si.

6.3.6 Electronic Film and Interface Properties for n- and p-type Doping

In *a*-Si:H/*c*-Si solar cells, *n*- and *p*-doped *a*-Si:H layers are used as emitter and back surface field layers. As discussed above, due to the short growth time and additional geometrical constraints to the Si-Si bonds from the vicinity of the interface, higher defect densities in the *a*-Si:H band gap can be expected in ultrathin *a*-Si:H layers, as compared to thick *a*-Si:H films. If the Fermi level is lying in a region of high defect density, it becomes more difficult to move it around by either adding dopants to the *a*-Si:H film or by an external electric field, because some of the defects (to first order: those lying between the initial and the final position of E_F) will have to be recharged. For extremely high defect densities, E_F is pinned at a position determined by the charge neutrality condition, and cannot be altered by doping. This is the case e.g. for unhydrogenated *a*-Si, or at poorly passivated surfaces and interfaces. Doping can exacerbate this problem: In thick *a*-Si:H films, it has been shown that the incorporation of dopant atoms into the amorphous network increases the defect density [19]. Thus, by measuring the defect density and Fermi level position in ultra-thin doped *a*-Si:H films, information is obtained that is useful to find the optimum growth conditions and doping level for films to be used as functional layers in *a*-Si:H/*c*-Si cells.

CFSYS measurements on a series of ~10 nm thin (*n*)*a*-Si:H films with different doping concentrations are shown in Fig. 6.23. The films were grown on (*p*)*c*-Si substrates. Again, the spectra have been rescaled using the energy dependence of $h\nu R^2$ to obtain an approximation of the occupied DOS and are plotted *vs.* $E_F - E_V$. The defect parameters, cf. eq. (6.2), obtained from fitting the *a*-Si:H model function to these spectra are plotted against the gas phase doping level [PH3]/[SiH4] in Fig. 6.24.

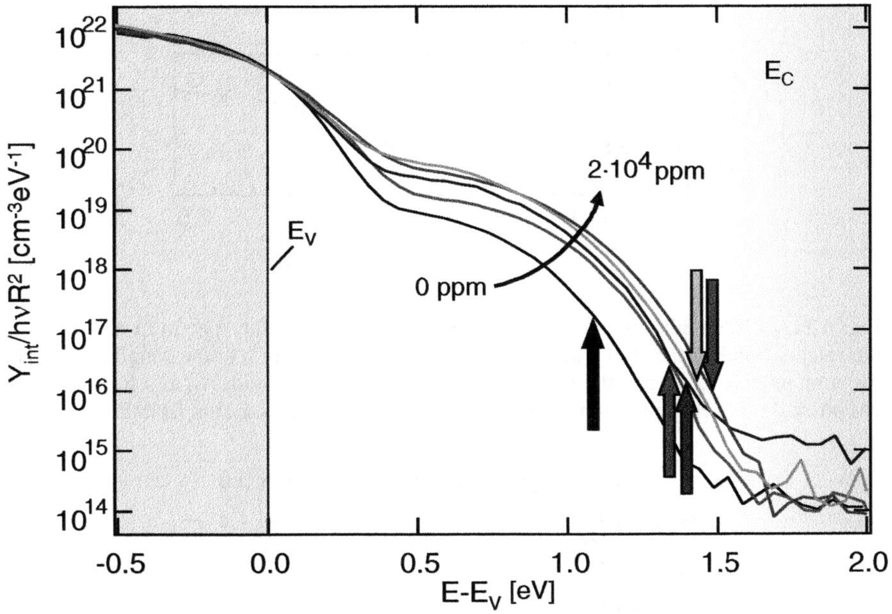

Fig. 6.23. CFSYS measurements, normalized to an approximate density of states, on *a*-Si:H layers with varying gas phase doping $[PH_3]/[SiH_4]$ = 0, 0.1, 0.3, 1 and 2×10^4 ppm. *a*-Si:H film thickness 10 nm, T_{dep} = 170°C. Arrows mark E_F, the vertical line is the valence band mobility edge. After [45].

For thicker films (d_{aSi} ~ 100 nm) deposited on glass using the same deposition conditions, the activation energy $E_C - E_F$ was determined from coplanar conductance measurements. In Fig. 6.25, this quantity is shown together with $E_F - E_V$ obtained from Fig. 6.23. It is obvious, that the change of Fermi level position with doping is identical (but the sign is inverted, due to the change in reference energy) for both the thicker and the ultra-thin layers. With these data, an approximate band gap can be calculated: Taking $E_F - E_V$ from PES and $E_C - E_F$ from conductivity, $E_g = E_C - E_V = (E_C - E_F) + (E_F - E_V)$, which is shown in the lower panel of Fig. 6.25. The calculated band gap is 1.70(1) eV, independent of doping, and deviates only slightly from the optical gap E_g^{opt} = 1.74 eV obtained from a Tauc plot of the optical absorption on a 300 nm thick (i)*a*-Si:H layer.

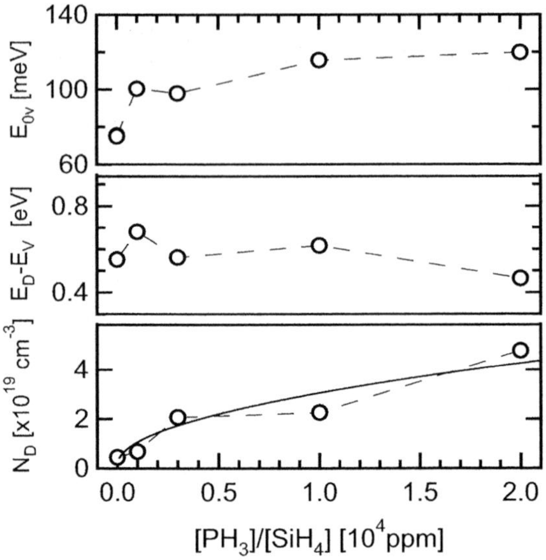

Fig. 6.24. *a*-Si:H defect parameters obtained from evaluating the spectra in 4. Plotted are the valence band Urbach tail parameter E_{0v}, the center position E_D of the dangling bond distribution relative to E_V and the integrated density of dangling bonds N_D as a function of the gas phase doping concentration. Reproduced from [45] with permission from Elsevier.

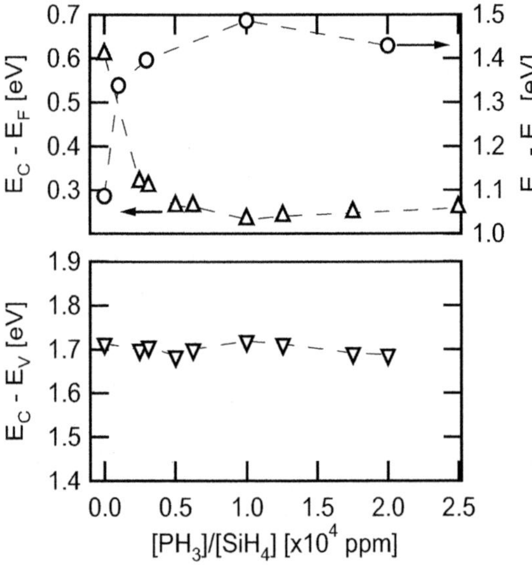

Fig. 6.25. Dependence of $E_F - E_V$ (○, from Fig. 6.23), of $E_F - E_C$ (△, from measurements of the dark conductivity on ~ 100 nm thick films), and of the calculated mobility gap $E_V - E_C$ on gas phase doping. Note that both ordinates show identical intervals of 0.5 eV. Reproduced from [45] with permission from Elsevier.

Finally, in Fig. 6.26, the estimated *c*-Si band bending in the dark, ϕ_0, is plotted vs. *a*-Si:H film doping. ϕ_0 is calculated under the assumption of flat bands in the *a*-Si:H (see sketch of the band diagram in Fig. 6.26),

$$e\phi_0 = E_V^{aSiH} - \Delta E_V - E_V^{cSi}, \quad (10)$$

and with a valence band offset of $\Delta E_V = 0.458$ eV, cf. next section. Note that the assumption of flat bands in the *a*-Si:H is likely to be a good approximation in highly doped *a*-Si:H due to its high defect density (short Debye length), while in low-defect *a*-Si:H (e.g. in an (i)*a*-Si:H buffer layer), the total band bending ϕ_0 would be distributed amongst the *c*-Si and the *a*-Si:H part of the junction. Indeed, in the absence of contributions from other junctions, ϕ_0 can be identified with the built-in voltage of the cell.

For comparison, Fig. 6.26 also shows the open circuit voltage: V_{oc} of a series of (*n*)*a*-Si:H/(*p*)*c*-Si solar cells with the same doping variation of the *a*-Si:H emitter layer [35].

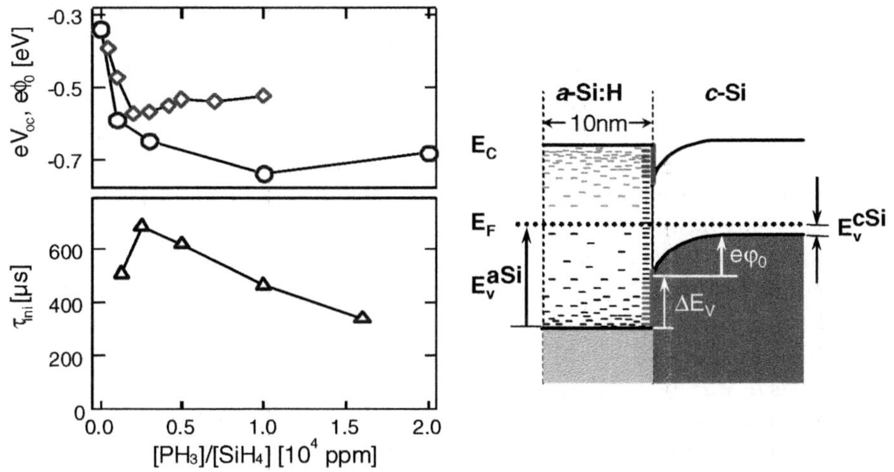

Fig. 6.26. Upper left: built-in potential $e\phi_0$ (circles) and solar cell V_{oc} (lozenges, rescaled to eV units). Lower left: surface photovoltage decay time constant τ_{ini}. The built-in potential $e\phi_0$ was calculated from the Fermi level position $E_F - E_V$, Fig. 6.25, assuming $\Delta E_V =$ 0.458 eV, cf. next section.

For low doping levels, up to ~2500 ppm phosphine in the gas phase, the solar cell V_{oc} follows the band bending closely, i.e. the maximum V_{oc} given by the built-in potential[9] is fully realized. At higher doping level, the band bending further

[9] Under the assumption that the movement of the quasi Fermi levels of the majority charge carriers can be neglected, i.e. under low injection conditions or not too far away from these.

increases up to 10 000 ppm, but the V_{oc} does not follow this trend: It decreases again because the recombination at the *a*-Si:H/*c*-Si interface, which is highly defective at high *a*-Si:H doping levels, limits the charge carrier lifetime and thus the V_{oc}. From these investigations, it was concluded that the optimum doping for cells with direct (n)*a*-Si:H/(p)*c*-Si junctions (without (i)*a*-Si:H buffer) is around 2500 ppm; the V_{oc} of the best solar cell obtained with this junction is 629 mV [85], which clearly demonstrates the drawback of omitting the (i)*a*-Si:H layer.

6.3.7 a-Si:H/c-Si Band Offsets

As outlined in the introduction, the offsets ΔE_V, ΔE_C between the band edges of amorphous and crystalline silicon at the *a*-Si:H/*c*-Si junction (cf. Fig. 6.1) influence strongly the charge carrier transport across the interface; their precise determination is also indispensable for a reliable device modeling. The reader is also referred to Chapter 12 in this book for a discussion of the subject based on electrical conductivity measurements rather than the photoemission experiments discussed in the following.

Theoretical calculations based on first-principle pseudopotential calculations and the so-called *model-solid theory* find a valence band offset of -0.25 eV[10] for unhydrogenated *a*-Si [86]. According to the same work, ΔE_V changes strongly with hydrogen content, by up to 0.04 eV per percent of hydrogen present in the film. As discussed in the previous sections, the hydrogen content varies with *a*-Si:H deposition conditions, and an interfacial layer with enhanced hydrogen concentration has been reported [33]. Taking a hydrogen content of 11% as in device-grade bulk *a*-Si:H would lead to $\Delta E_V = +0.20$ eV, and an enhanced hydrogen concentration of 15% at the *a*-Si:H/*c*-Si interface based on the considerations above gives $\Delta E_V = +0.35$ eV. However, it is highly probable that microvoids will form for films with high hydrogen content and/or at the *a*-Si:H/*c*-Si interface, cf. the section on thin (i)*a*-Si:H films. The calculations in [86], which assume a homogeneous *a*-Si:H, do not describe this case.

Experimentally, from techniques such as capacitance- and admittance spectroscopy or internal photoemission the whole range of possible values for ΔE_V from 0 to 0.7 eV (with the band gaps $E_{g,cSi} = 1.12$ eV and $E_{g,a\text{-Si:H}} = 1.7\ldots1.8$ eV) has been reported, cf. Table 1 and the review in [87].

---

[10] The negative sign means that $E_{V,a\text{-Si:H}}$ lies "above" $E_{V,c\text{-Si}}$, that is, the band lineup is inverted to the case shown i.e. in the sketch in Fig. 6.1, Fig. 6.26.

Table 6.1 Overview of valence and conduction band offsets, ΔE_V and ΔE_C, reported in literature for the *a*-Si:H/*c*-Si heterointerface. Measurement techniques are: IPE – internal photoemission spectroscopy; C-V – capacitance voltage measurements; SR – spectral response; CFSYS – constant final state yield spectroscopy. Adapted and extended from [87].

ΔE_V [eV]	ΔE_C [eV]	deposition technique	measurement technique	comment	Ref.
0	-	Sputtered	IPE	-	[88, 89]
0	-	PECVD	IPE	-	[90]
0.20	0.385	PECVD	I-V	-	[91]
0.58	0.05	PECVD	VFP	-	[92]
0.67	0.01	PECVD	C-V	[H] = 12%	[92]
0.65	0.13	PECVD (H_2 dilution)		[H] = 14%	[93]
0.71	0.09	PECVD	IPE	-	[93]
0.49	0.175	PECVD	SR simulation	independent substrate temperature	[94]
0.44	0.16	PECVD	CFSYS	In situ deposition	[44]
-	0.06	PECVD	C-V	-	[95]
-	0.35	PECVD	I-V, CV	n/p and p/n structures	[96]
-0.06	0.24	PECVD	IPE	(p) a-Si:H	[97]
-	0.15	PECVD	Coplanar conductance	-	[98]
0.458	-	PECVD	CFSYS	(p,i,n) a-Si:H, in-situ	[99]

One of the most widely used techniques to measure ΔE_V in *a*-Si:H/*c*-Si and other heterostructures is photoelectron spectroscopy (PES). For *a*-Si:H/*c*-Si, the variants of PES using near-UV excitation, as introduced above, have proven especially useful due to their high dynamic range and an information depth of up to several nm [44, 42]. Sebastiani *et al.* were the first to apply this technique to measure ΔE_V in (i)*a*-Si:H/*c*-Si [44]. In CFSYS, for film thicknesses up to several nm, both the film and the substrate valence band edges are visible in the same energy-resolved spectrum, and ΔE_V can thus be determined with a minimum of assumptions. The CFSYS technique was later applied to determine ΔE_V in other Si-based heterojunctions such as (i)*a*-Si_xC_{1-x}:H/*c*-Si [100] and the epitaxial Ge/Si(100) heterostructure [101].

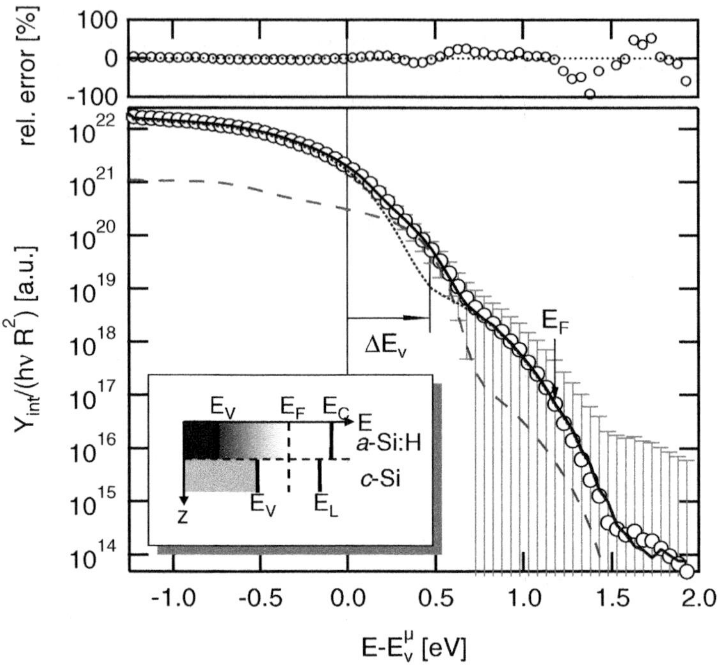

Fig. 6.27 Fit of the CFSYS spectrum obtained from a 2.9 nm (i)*a*-Si:H layer on *c*-Si (circles, with error bars) with the scaled and shifted spectra of a thick (104 nm) (i)*a*-Si:H film (dotted line) and a hydrogen terminated *c*-Si wafer (dashed line). ΔE_V denotes the *a*-Si:H/*c*-Si valence band offset. Also shown is the relative residual error [99].

Figure 6.27 depicts the CFS Yield data from a 2.9 nm (i)*a*-Si:H layer on *c*-Si vs. the energy distance to the valence band edge. By comparing this to the spectrum obtained from a 104 nm (i)*a*-Si:H film deposited under the same conditions (dotted line), it is evident that an additional shoulder has appeared in the spectrum at around 0.3…0.5 eV above the valence band edge. This is due to the contributions from the *c*-Si valence band that "shine through" the *a*-Si:H layer, which is semitransparent to the low-energy photoelectrons. The dashed curve in Fig. 6.27 is the CFSYS measurement on a cleaned and hydrogen terminated *c*-Si sample. The data from the 2.9 nm film can be fitted as the sum of the spectrum from the 104 nm film, Y_{int}^{aSi}, and the *c*-Si spectrum, Y_{int}^{cSi},

$$Y_{int}(E) = C_{aSi} Y_{int}^{aSi}(E) + C_{cSi} Y_{int}^{cSi}(E - \Delta E_V), \qquad (6.11)$$

where both spectra are weighted for their relative contributions ($C_{aSi} \sim 1$ and $C_{cSi} \sim 0.1$ in the example), and the *c*-Si spectrum is shifted along the energy axis by the quantity of interest, the valence band offset ΔE_V. The upper graph in Fig. 6.27 shows the relative error σ_{rel} of the fit, i.e. $\sigma_{rel} = (Y_{int}^{measured} - Y_{int}^{fit}) / Y_{int}^{measured}$. It is obvious, that the procedure yields excellent fits with a relative error well below 50% of the CFSYS yield over six orders of magnitude.

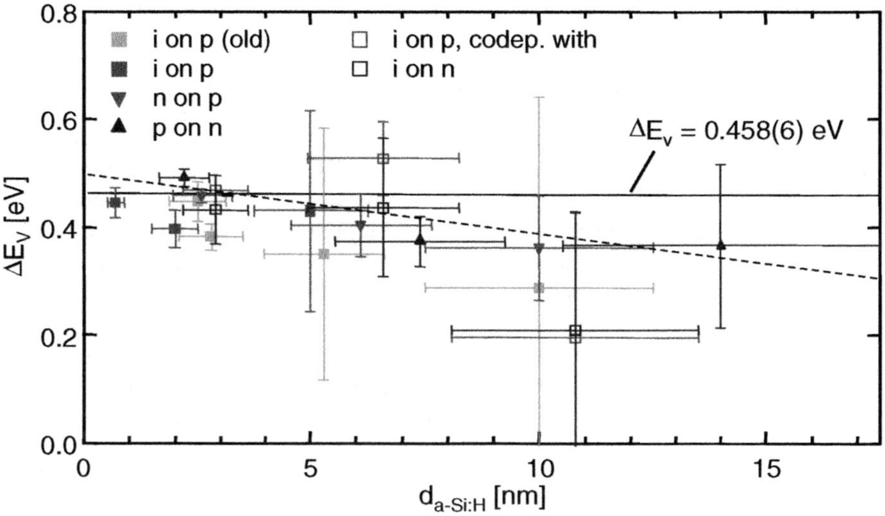

Fig. 6.28 Band offset vs. *a*-Si:H film thickness. Full line: Average d_{aSiH}-independent band offset. Dashed line: Linear fit for a thickness-dependent $\Delta E_V(d_{aSiH})$. Filled symbols denote (p)*c*-Si, open symbols (n)*c*-Si substrates. Symbol shapes: □, Δ, and ∇ indicate (i), (p) and (n)*a*-Si:H films, respectively [99].

Recently, CFSYS has been used to investigate the dependence of ΔE_V on the doping type of both the *a*-Si:H film and the *c*-Si substrate [99]. From theoretical considerations, the band offset should not depend on the position of the Fermi level E_F at both sides of the heterojunction. However, a change in E_g upon changing doping level has been found especially in (p)*a*-Si:H and was linked to structural changes and a change in H content [102, 103]. Consistently, systematic studies carried out for the system *a*-$Si_{1-x}C_x$:H/*a*-Si:H have found a decrease of ΔE_V and the band gap E_g with increasing *p*-type doping of the *a*-Si:H [104, 105].

Figure 6.28 summarizes the results reported in [99]: Each data point represents an *a*-Si:H/*c*-Si sample with different combinations of wafer and film dopings, and varying film thickness d_{aSiH}. For the investigated device-quality *a*-Si:H/*c*-Si heterojunctions, no dependence of ΔE_V on the doping level of either the *c*-Si substrate or the *a*-Si:H film was found. This is consistent with the fact that the same films show no dependence of E_g on doping type or level. Calculating the weighted average $\overline{\Delta E_V}$ over all measurements in Fig. 6.28 yields $\overline{\Delta E_V} = 0.458(6)$ eV. Note, however, that the systematic error is larger by 50-60 meV, i.e. about an order of magnitude than the error in the average. It is introduced by the determination of the mobility edge, a parameter related to electronic transport, from a DOS spectrum, which necessitates cross comparisons between e.g. CFSYS and measurements of the activation energy of electrical conductance.

Surprisingly, ΔE_V is weakly dependent on the *a*-Si:H film thickness. Fitting a linear thickness dependence,

$$\Delta E_V(d_{aSiH}) = 0.49(2)\,\text{eV} - 0.010(5)\,\text{eV/nm} \times d_{aSiH} \qquad (6.12)$$

is found and tentatively explained as being caused by changes in the *a*-Si:H/*c*-Si interface dipole due to changing hydrogen concentration and *c*-Si dangling bond saturation: As discussed above, the interface between *a*-Si:H and *c*-Si consists of intact Si-Si bonds, a large number (~10 %) of silicon-hydrogen bonds, and a comparatively low number of silicon dangling bonds (db). 10^{12} db/cm² can be assumed as an upper bound in well-passivated *a*-Si:H/*c*-Si heterojunctions ([1, 60] and Fig. 6.15). In most modern theories for heterojunction band offsets, the contributions to ΔE_V are separated into contributions from the involved materials themselves and those from additional extrinsic dipoles, e.g. [106]. For example, Tersoff's charge neutrality point theory [107] applied to the *a*-Si:H/*c*-Si heterojunction leads to

$$\Delta E_V = \Phi_{CNL}^{aSiH} - \Phi_{\text{CNL}}^{\text{cSi}} + \Delta_{EN} + \Delta_{ex}, \qquad (6.13)$$

where Φ_{CNL}^{i} are the charge neutrality levels (*i=aSiH, cSi*), referred to the respective valence band edge, Δ_{EN} the potential difference due to differences in electronegativity, and Δ_{ex} the potential difference due to (extrinsic) interface dipoles. For *a*-Si:H/*c*-Si $\Delta_{\text{EN}} = 0$. The *a*-Si:H investigated here contains about 12-17% hydrogen, thus a contribution to the interface dipole Δ_{ex} can be expected to stem from silicon-hydrogen bonds at the *a*-Si:H/*c*-Si interface. In addition, a varying density of dangling bond defects are present in the *a*-Si:H and at the *a*-Si:H/*c*-Si interface. As discussed e.g. in Chapters 7 and 13 in this book, these are amphoteric defects and can thus carry 0, 1 or 2 elementary charges. In the *a*-Si:H, the Si-H- and the dangling bonds are randomly oriented and average to zero net dipole. At the *c*-Si crystal surface, however, both Si-H and charged dangling bonds have a preferential orientation due to the crystal structure (i.e. the hydrogen or the dangling bond is pointing away from the crystal surface) and can thus contribute to the dipole. Note that due to the low growth temperature, it is highly unlikely that after the *a*-Si:H growth has proceeded to a point where a closed *a*-Si:H film has been formed, changes in $\Delta E_V(d_{\text{aSiH}})$ could be induced by diffusing species other than hydrogen segregating at the hetero-interface. For the silicon-vacuum interface, the additional potential step due to a complete coverage of the Si(111) surface with hydrogen was calculated to 1.1 eV [108]. The negative charge is on the vacuum side of the dipole. For the *a*-Si:H/*c*-Si system, the value has to be corrected with the effective relative dielectric constant ε_{eff} of the heterojunction materials, which yields a dipole $\Delta_{\text{Si-H}} \sim 0.09$ eV. Furthermore, the Si surface is not perfectly flat, but has a microroughness due to the HF etching step. This leads, on the one hand, to Si-H bonds also parallel to the wafer surface, i.e. not contributing to an average interface dipole perpendicular to the surface, and, on the other hand, to higher hydrides at the interface (Si-H_2 and Si-H_3) [71, 82, 109] that can increase the dipole. Still, the rough estimate given above should give the correct order of magnitude for the expected dipole. Thus, a change in the Si-H bond density at the interface alone cannot explain the large variation seen in Fig. 6.28.

Another contribution to the dipole will arise from dangling bond defects at the interface. For the amphoteric dangling bond, the contribution p to the total interface dipole from a singly occupied dangling bond Si^0 can be calculated following [108] as $p_{Si0} = \Delta q/\varepsilon_{eff}\, e\, d_{dip}$ assuming total charge transfer, $\Delta q = 1$, of one elementary charge e along the dipole axis of length $d_{dip} = r_{Si}$, where $r_{Si} = 1.1$ Å is the covalent radius of silicon. This yields $\Delta_{Si0} = 1.1$ eV. Again, the negative charge is pointing away from the Si atom, towards the *a*-Si:H. A positively charged dangling bond should have a dipole close to zero (neglecting interactions with neighboring Si-Si bonds) and a layer of Si^- will have a dipole above that of Si^0. These considerations should be seen as rough estimates, as $\Delta q = 1$ is overestimating the charge transfer (experimental and theoretical values of about 0.15 and 0.05 have been reported [110, 108], and on the other hand, the local effective dielectric constant could be reduced due to a weaker screening of bound charges. Still, also here, the order of magnitude of the dipole is expected to be correct, as all values enter linearly into the derivation.

For dangling bond densities of $10^{11}...10^{12}$ cm^{-2}, i.e. about one dangling bond in $10^4...10^3$ interfacial Si atoms, the average dipole potential is reduced to 0.1-1 meV. However, the samples investigated in [99] are similar to those investigated by Laades [35], who has shown that the passivation of *c*-Si surfaces by these *a*-Si:H films is strongly reduced for layer thicknesses below 5 nm (section 6.3.1, see especially Fig. 6.9). This indicates that for low film thicknesses the density of dangling bonds at the *a*-Si:H/*c*-Si interface is strongly increased. Taking $\Delta Si^0 \sim 1$ eV for an initial full monolayer ($\sim 10^{15}$ cm^{-2}) of Si dangling bonds present at the *c*-Si surface after ignition of the PECVD plasma, the reduction in ΔE_V of about 10 meV per nm of *a*-Si:H overlayer may be interpreted as being due to the saturation of the order of 1% or $\sim 10^{13}$ cm^{-2} dangling bonds at the *a*-Si:H/*c*-Si interface per nm of *a*-Si:H overlayer thickness. Therefore, the decrease of ΔE_V with increasing *a*-Si:H film thickness is tentatively explained as follows: During the first stages of *a*-Si:H film growth, the initially high density of *a*-Si:H/*c*-Si interface states leads to an interface dipole due to charged unsaturated ("dangling") bonds of the *c*-Si surface. During further film growth, these dangling bond defects are saturated by hydrogen forming Si-H bonds that lead to a much smaller dipole and thereby a decrease of ΔE_V.

It is interesting to consider the influence of this band offset on charge carrier transport in heterojunction cells: On the one hand, the barrier is essential in reducing recombination at the interface. On the other hand, if it becomes too high, it might limit carrier transport, showing as series resistance or reduced fill factor in the solar cell. A rough estimate of the permissible barrier height has been proposed by Rau and Schock [111]: If one approximates the (p)*a*-Si:H/(n)*c*-Si heterojunction as a Schottky contact with barrier height $\Phi_b^p = \Delta E_V + \Delta E_{Fp}$, where ΔE_{Fp} is the distance of the valence band edge to the Fermi level at the interface, the conductance over this barrier is

$$G = qA^*T/k\exp\left(\frac{-\Phi_b^p}{kT}\right). \tag{6.14}$$

With the effective Richardson constant $A^* = 264$ A/(cm²K²) for <111> oriented (n)*c*-Si [112], and $\Phi_b^p = \Delta E_V + \Delta E_{Fp} = 0.5$ eV (almost ideal case: the interface Fermi level lies within 50 meV of the *c*-Si valence band edge), this yields $G =$ 3.7 S/cm², i.e. a series resistance of 0.27 Ωcm at $T = 300$ K. This is an acceptable value for high efficiency cells, where series resistances below 1 Ωcm are required. Due to the exponential dependence on the barrier height, however, the series resistance increases to 2 Ωcm² already for an increase of the barrier height by only 50 meV. Such an increase is easily conceivable, e.g. by increasing the band gap of the *a*-Si:H, or by moving the interface Fermi level away from the valence band edge by incorporating an (i)*a*-Si:H buffer layer, over which part of the total built in voltage drops. Note that these considerations give only a very rough approximation of transport across the *a*-Si:H/*c*-Si junction, as e.g. additional carrier transport through tunneling processes (see below) is not considered. Still, more advanced numerical simulations show that the critical barrier height for minority carriers at the p/n junction is indeed of the order of 0.5 eV ([113] for *a*-Si:H/*c*-Si cells, see also below).

Using numerical simulations, the sensitivity of the solar cell V_{oc} on interface defects can be investigated. Figure 6.29 shows the result of an AFORS-HET simulation of V_{oc} vs. the band offset for the minority charge carriers, where in addition, the influence of D_{it} on V_{oc} is considered [113, 114]. For the (n)*a*-Si:H/(p)*c*-Si cell, electrons are the minorities, so the relevant band offset is $\Delta E_C \sim 150$ meV. For (p)*a*-Si:H/(n)*c*-Si, $\Delta E_V \sim 450$ meV is the relevant offset.

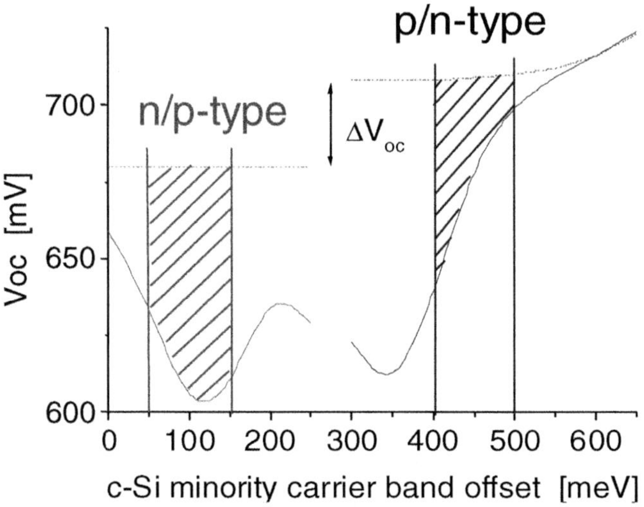

Fig. 6.29. Simulated open circuit voltage of a TCO/(n)*a*-Si:H/(p)*c*-Si/Al (left) and a TCO/(p)*a*-Si:H/(n)*c*-Si/Al solar cell (right). The band offset of the minority charge carriers is varied. The hatched area marks the range of V_{oc}s to be expected for interface defect densities between negligibly small $D_{it} < 10^{10}$ cm^{-2} and $D_{it} = 10^{12}$ cm^{-2} [113, 114].

Figure 6.29 shows simulated V_{oc} data for two different interface defect densities: $D_{it} < 10^{10}$ cm^{-2}, i.e. negligibly small, and 10^{12} cm^{-2}. Note that no (i)*a*-Si:H layer is included in the simulation. From the graph, two major conclusions can be drawn: (i) in (p)*a*-Si:H/(n)*c*-Si, the influence of the D_{it} on V_{oc} is less severe than in the inverse structure, i.e. the cell V_{oc} is more robust towards low quality/degraded *a*-Si:H/*c*-Si interfaces; (ii) for perfect *a*-Si:H/*c*-Si interfaces (negligible D_{it}), (p)*a*-Si:H/(n)*c*-Si cells have a higher V_{oc} potential than (n)*a*-Si:H/(p)*c*-Si ones. A more detailed investigation [113] reveals that this is due to the different minority carrier mobilities in the *c*-Si.

For the simulation in Fig. 6.29, the parameters were chosen such that the only basic differences are carrier mobility and ΔE. For a realistic simulation of actual solar cells, effects like the changing defect distribution in the *a*-Si:H gap have to be considered. This was done for the simulation in Fig. 6.30, which shows how the influence of D_{it} on the different cell parameters combines in the efficiency η [113, 114]. The same basic trend as in the V_{oc} data can be seen: For D_{it}s of ~10^{10} cm^{-2} and below, the cell is not sensitive to D_{it} any more (cf. also Fig. 6.15). An increasing minority carrier band offset leads to increases in cell efficiency that are more pronounced for high D_{it}s, but still substantial also for the high quality interfaces with $D_{it} < 10^{10}$ cm^{-2}. This is because, as discussed above, in the investigated range of up to 350 meV, the ΔE does not act as a barrier to charge transport, but provides the beneficial recombination reduction of a heterojunction.

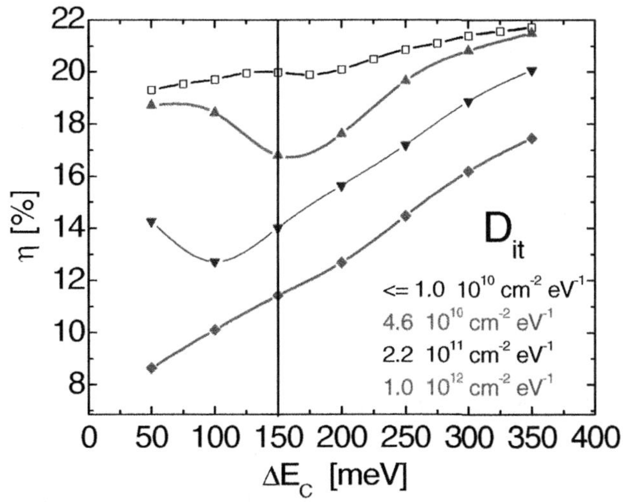

Fig. 6.30 Simulated efficiency of a TCO/(n)*a*-Si:H/(p)*c*-Si/p+)*a*-Si:H/Al solar cell in dependence of the conduction band offset ΔE_C for different *a*-Si:H/*c*-Si interface defect densities D_{it} [113, 114].

Similar investigations have been carried out by various authors, e.g. [115, 116]. The latter find that for (p)*a*-Si:H/(n)*c*-Si cells, the defect density on the front face of the wafer strongly affects cell performance, while D_{it} at the rear has little influence. For (p)*a*-Si:H/(n)*c*-Si (Sanyo HIT type) cells, the D_{it} at the front side interface strongly influences the open-circuit voltage and to some extent the fill factor, while an increased D_{it} at the rear interface reduces mainly the current. Also, it is found that if no tunneling across the heterojunction is considered, the illuminated *I-V* curves of (p)*a*-Si:H/(n)*c*-Si cells show the influence of a barrier (S-shaped *I-V* curves with low fill factor) at $\Delta E_V > 0.5\ldots0.6$ eV, in keeping with the simple considerations above.

6.4 Extraction of *a*-Si:H/*c*-Si Heterojunction Recombination and Transport Parameters from Solar Cell I-V Curves Using an Equivalent Circuit Model

So far, only solar cell parameters obtained under open circuit conditions, i.e. with no external current flow, have been discussed. In the following, it will be outlined briefly how information not only on recombination at the *a*-Si:H/*c*-Si interface but also on charge carrier transport across the interface can be obtained from temperature-dependent current-voltage (*I-V*) measurements.

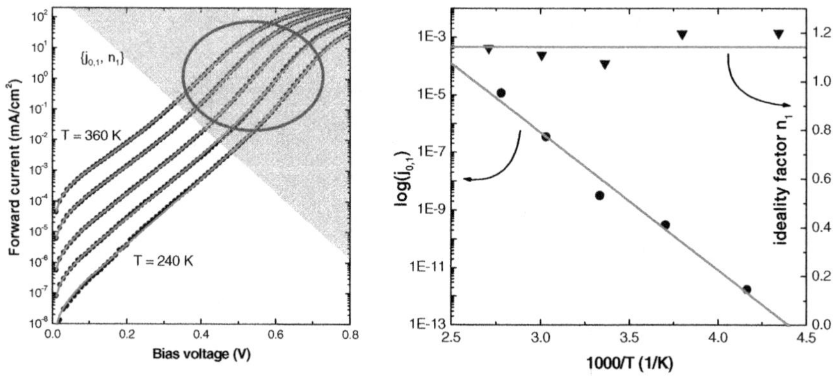

Fig. 6.31 Left: Semilogarithmic plot of I-V curves in forward direction obtained from a 1 cm² *a*-Si:H/*c*-Si solar cell at different measurement temperatures (circles), and fits of the 2-diode-model (lines). Right: reverse saturation current and ideality factor obtained from the 2-diode-model fits. Adapted from [109].

Figure 6.31 shows the semi-logarithmic plot of such *I-V* measurements in forward direction obtained on an *a*-Si:H/*c*-Si solar cell without illumination [109]. It is possible to fit the forward current density j_f with a simple two-diode-model including series and parallel resistance, and to extract the reverse saturation current $j_{0,1}$ and the diode ideality factor n_1 for the diode that fits the high forward bias region (shaded in Fig. 6.31, left). While the ideality factor is close to 1 and temperature-independent, $j_{0,1}$ shows an activated behavior with an activation energy $E_{a,j01}$ close to the *c*-Si band gap. Figure 6.32 reproduces the $E_{a,j01}$ and n_1 data from

[109] *vs*. the solar cell's V_{oc} under standard test conditions (each data point is extracted from *I-V*(*T*) of one solar cell). It is apparent, that for high quality *a*-Si:H/*c*-Si cells (high cell V_{oc}s), the ideality factor deviates systematically from 1.

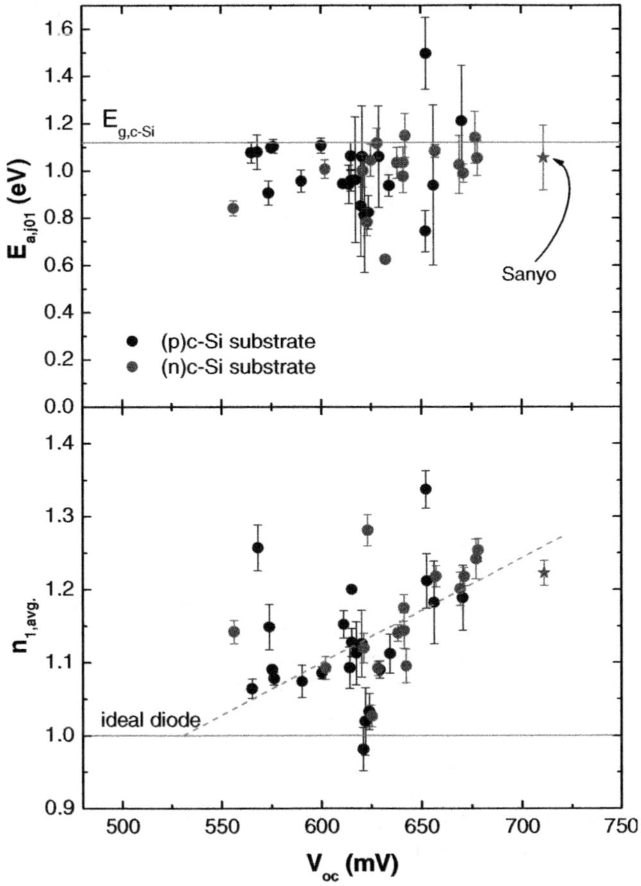

Fig. 6.32. Activation energy of the reverse saturation current $j_{0,1}$ and of the ideality factor for the same diode, averaged over temperature, vs. the V_{oc} of the solar cell under standard test conditions. Adapted from [109].

Comparing the findings for n_1 and $E_{a,j01}$ with theory, it is concluded, that under high forward bias – i.e. at voltages that are close to the working conditions (maximum power point) of the solar cell – the description of an *a*-Si:H/*c*-Si cell in terms of the simple Shockley diffusion model [117] is possible, just as for (ideal) homojunction solar cells. In particular, it is not necessary to invoke tunnel- or tunnel-(hopping-)transport and recombination mechanisms to explain the behavior of the cell under these conditions, although a look at the band diagram of the *a*-Si:H/*c*-Si p/n junction (Fig. 6.33) might suggest that these mechanisms could be of importance (and are indeed for lower forward bias and in the reverse direction as well as for other device structures, cf. [109] and the references therein). The analysis of the $n_1 E_{a,j01}$ product carried out in [109] shows that the heterojunction aspects of carrier transport across the p/n junction become more pronounced with enhanced interface passivation, and are responsible for the deviation from $n_1 = 1$.

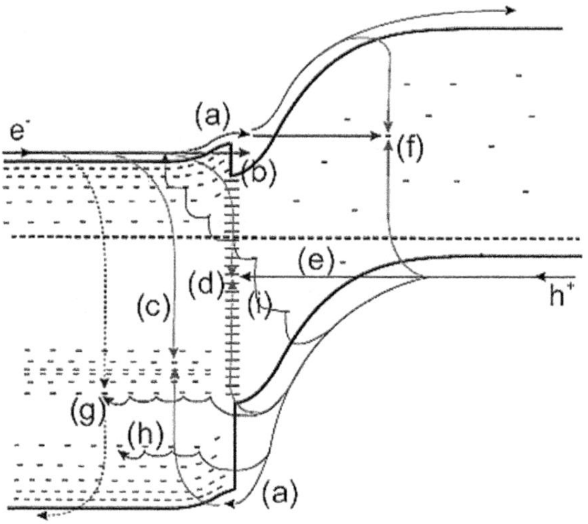

Fig. 6.33. Equilibrium band lineup of a (n)*a*-Si:H/(p)*c*-Si heterojunction solar cell structure (x scale is logarithmic to show the thin *a*-Si:H layer) and possible transport paths for moderate forward bias. Black: emission processes, red: recombination, and blue: tunneling. (a) Emission of carriers across the barriers at the heterojunction imposed by band offsets and spikes in the band. (b) Tunneling through a band spike. (c) Recombination via a-Si:H gap states. (d) Recombination via interface states. (e) Tunneling into interface states and successive recombination. (f) Recombination via deep defects in the c-Si. (g) Multitunneling in the a-Si:H with successive recombination through carrier capture or reemission into the band ("MTCE")[118] (h) Tunnel hopping in the a-Si:H band tail. (i) Band-to-band multitunneling process. Adapted from [109].

As a corollary, it can be shown that temperature-dependent I-V measurements without illumination can serve as a tool to obtain an estimate of V_{oc} under AM 1.5 illumination with good precision. This is obvious from considering only the "high forward bias" diode from the 2-diode-model and setting $j_F = 0$ under open circuit conditions:

$$j_F \approx j_{0,1}\left(e^{\frac{eV_{oc}}{n_1 kT}} - 1\right) - j_{phot} = 0 \tag{6.15}$$

$$\Leftrightarrow eV_{oc} = n_1 kT \ln\left(\frac{j_{phot}}{j_{0,1}}\right). \tag{6.16}$$

In Fig. 6.34, the V_{oc} measured on solar cells under standard test conditions (the same set of cells as in Fig. 6.32) are compared to the estimated V_{oc} calculated from eq. (6.13) taking $j_{0,1}$ and n_1 from the 2-diode fit to the *I-V* data without illumination and the actual photocurrent of the cells j_{phot} as obtained from the illuminated *I-V*. It is obvious, that the V_{oc}s estimated from the dark I-V have a discrepancy of typically less than 2% to the V_{oc}s extracted from the illuminated *I-V* data.

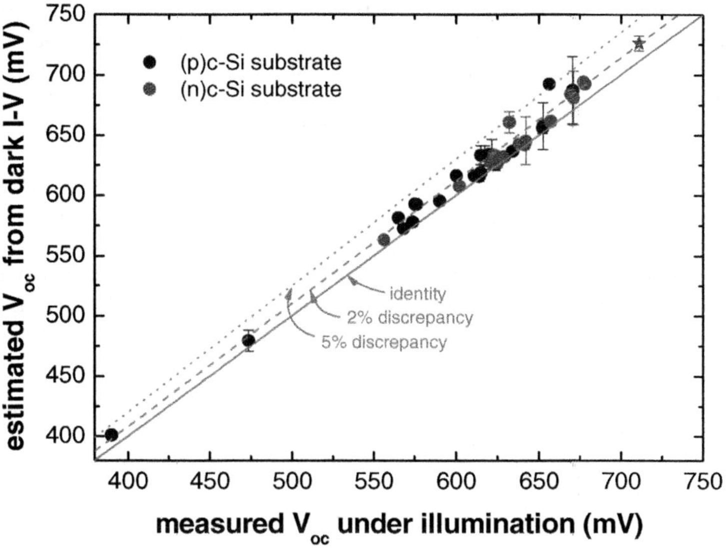

Fig. 6.34 Device V_{oc} as implied from the absolute value of the high-forward-bias saturation current density $j_{0,1}$ obtained from dark I-V curve fits compared to actual V_{oc} as measured under illumination. Adapted from [109].

Furthermore, the Shockley theory [117] can be applied, considering only the effective minority carrier diffusion length L_{eff} on the c-Si side of the *a*-Si:H/*c*-Si junction, i.e.

$$j_{0,1} = \frac{eDN_C N_V}{L_{eff} N_{dop}} e^{-\frac{E_g}{kT}}, \tag{6.17}$$

to calculate an average $j_{0,1}$ based on effective lifetime measurements in *a*-Si:H/*c*-Si structures. Considering only the *c*-Si parameters (effective densities of states in the bands N_C, N_V, doping level N_{dop} and diffusion constant D, band gap E_g) means that the *a*-Si:H/*c*-Si junction is actually considered as a "one-sided" junction. With this approximation, one calculates from eq. (6.17) a $j_{0,1}$ of 9.5×10^{-10} mA/cm^2 for the (p)*a*-Si:H/(n)*c*-Si/(n$^+$)*a*-Si:H samples and $j_{0,1}$ of 5.4×10^{-10} mA/cm^2 for the (p,i)*a*-Si:H/(n)*c*-Si/(i,n^{+})*a*-Si:H samples. These estimated values compare favorably with the measured $j_{0,1}$ of 8.9×10^{-10} mA/cm^2 for the final (p)*a*-Si:H/(n)*c*-Si/(n^{+})*a*-Si:H solar cell and 1.7×10^{-10} mA/cm^2 for the (p,i)*a*-Si:H/(n)*c*-Si/(i,n^{+})*a*-Si:H cell, highlighting the viability of the proposed approach [109].

6.5 Beyond *a*-Si:H/*c*-Si – The Influence of the TCO

To come to an understanding of the whole *a*-Si:H/*c*-Si solar cell, also the influence of the transparent conductive oxides (TCOs) on the front and probably on the rear side of the solar cell have to be considered. In Fig. 6.35, the complete band structure across a TCO/*a*-Si:H/*c*-Si junction is sketched: The TCO – typically InO doped with Sn (ITO) or ZnO doped with Al – is a wide band gap *n*-type semiconductor with degenerate doping, i.e. the Fermi level lies in the TCO conduction band. Due to the high doping in the TCO, it behaves electronically like a metal (with rather poor charge carrier mobility), and the electronic behavior of the TCO/*a*-Si:H junction is usually assumed as similar to a metal-semiconductor junction.

A critical parameter of the TCO/*a*-Si:H junction is the band line-up. For the case sketched in Fig. 6.35, the energetic position of the Fermi level in the TCO leads to a band bending in the *a*-Si:H, i.e. the formation of an (unwanted) space charge region and a rectifying junction in addition to the *a*-Si:H/*c*-Si junction. Furthermore, as the *a*-Si:H emitter and BSF layers are very thin, of the order of only 10 nm, it can be expected that the space charge regions (SCRs) of the TCO/*a*-Si:H and the

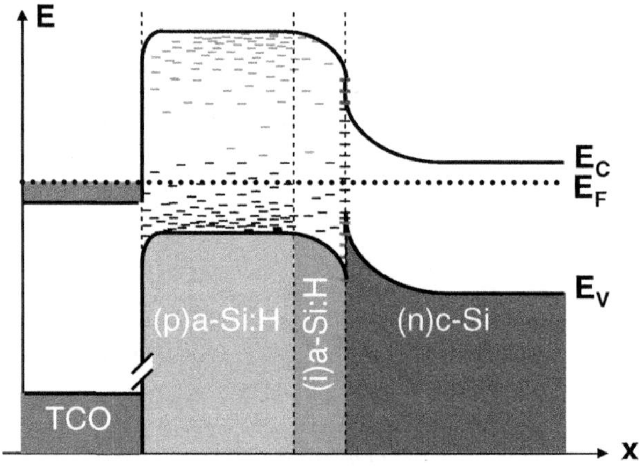

Fig. 6.35 Sketch of the band diagram of a TCO/(p,i)*a*-Si:H/(n)*c*-Si heterojunction. Due to the low work function of the TCO, the band alignment at the TCO/*a*-Si:H interface leads to a Schottky-like junction, i.e. a "reverse" diode with respect to the *a*-Si:H/*c*-Si diode.

a-Si:H/*c*-Si junctions overlap and interact. In Fig. 6.35, the most extreme case would be (for very thin *a*-Si:H) that due to the unfavorable work function of the TCO, the two SCRs could "merge" and effectively deplete the *a*-Si:H of free carriers, thus strongly reducing the built-in voltage and thereby the V_{oc} and fill factor in the cell.

A few experimental and simulation studies have been published that aim at elucidating the role of the TCO in *a*-Si:H/*c*-Si cells [2, 119, 120, 5, 6, 121]. Figure 6.36 sums up the results obtained by Froitzheim and coworkers [2]: Here, the illuminated *I-V* behavior of an (n,i)*a*-Si:H/(p)*c*-Si solar cell is simulated for flatband conditions at the TOC/*a*-Si:H interface (dotted curves) and for an "upwards" band bending of 350 meV[11] (full lines). As an additional parameter, the *a*-Si:H/*c*-Si interface defect density D_{it} was varied by incorporating a 5nm thin defective c-Si layer with volume defect density N_{it} (D_{it} = 5x10^{-7} cm x N_{it} can be assumed for comparisons to e.g. Fig. 6.15). Even for negligible $D_{it} < 10^8$ cm^{-2} ($N_{it} = 10^{14}$ cm^{-3}), the decrease notably in the fill factor is appreciable for the case of band bending at the TCO/*a*-Si:H junction (rightmost full curve). Furthermore, while the simulated cell is rather insensitive to increased *a*-Si:H/*c*-Si D_{it} up to 10^{12} cm^{-2} if the TCO/*a*-Si:H junction is in flatband condition (dashed curves), the cell with the 350 meV band bending at the TCO/*a*-Si:H interface shows S-shaped *I-V* curves with a strong drop in both V_{oc} and fill factor.

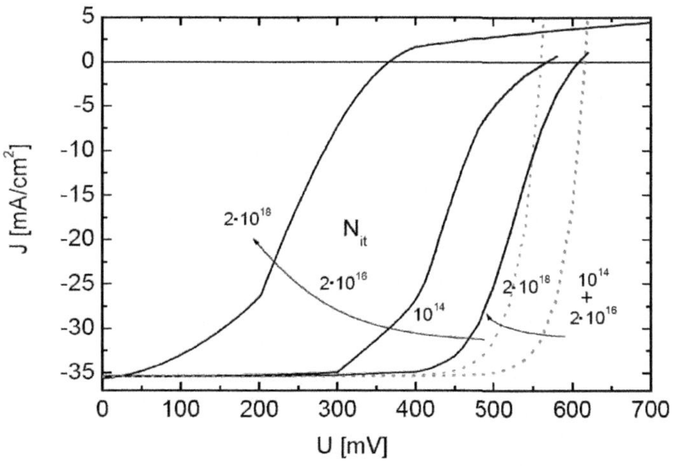

Fig. 6.36 Current voltage characteristics for a n/i/p-type structure with an ohmic front contact (flatband, work function $W = 4.05$ eV, dotted lines) and a Schottky front contact ($W = 4.4$ eV, straight lines). The density of the *a*-Si:H/*c*-Si interface states D_{it} = 5x10^{-7} cm x N_{it} is varied [2].

[11] As a (n)*a*-Si:H/(p)*c*-Si structure is simulated, the "upward" band bending at the TCO/*a*-Si:H interface leads to the antiparallel diode, as does the "downward" band bending at that interface in the (p/n) structure of Fig. 6.35.

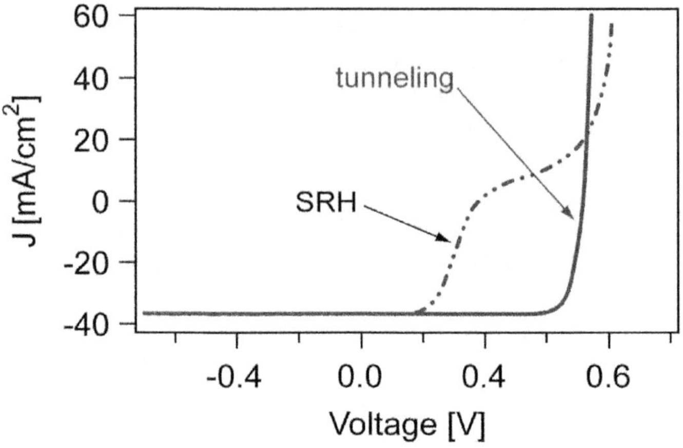

Fig. 6.37 (p)*a*-Si:H/(n)*c*-Si light current-voltage plots showing the influence of tunneling at the *a*-Si:H/ITO interface: The recombination velocity is set to $S = 10^7$ cm/s at the a-Si:H/ITO interface (dashed curve); band-to-band tunneling is included across the a-Si:H/ITO interface (solid curve). Reprinted with permission from [119]. Copyright 2009, American Institute of Physics.

These results are supported by more recent simulation studies [119, 6, 121]. The work by Kanevce and Metzger [119] adds another aspect to the picture: While the other calculations assume only a metal/semiconductor Schottky contact with constant interface recombination velocity *S* at the TCO/*a*-Si:H interface, Kanevce *et al.* include tunneling processes at this junction. Figure 6.37 shows that this alleviates the barrier problem at the junction: While the Schottky contact shows the same S-shape *I*-*V* curve as in Fig. 6.36, the additional charge carrier transport pathway provided by the band-to-band tunneling leads to a "well-behaved" simulated *I*-*V* curve.

From Fig. 6.38, it becomes even more clear that the inclusion of band line-up and charge transport at the TCO/*a*-Si:H interface is necessary to provide a realistic model of *a*-Si:H/*c*-Si solar cells: Here, the doping of the (p)*a*-Si:H layer and the *a*-Si:H layer thickness is varied (upper and lower graph, respectively). Both simulations were carried out twice: Without a TCO, and including a heavily n-type TCO. While the upper graph shows, that only at very high doping levels of ~2×10^{19} cm^{-3} the presence of the TCO plays no role any more for a 10 nm *a*-Si:H layer, the lower graph indicates that the problem outlined above, i.e. the depletion of the *a*-Si:H layer by the oppositely doped TCO, becomes especially severe at low *a*-Si:H thickness (<5 nm), presumably due to the beginning overlap of the *a*-Si:H/*c*-Si space charge region with that of the TCO/*a*-Si:H junction. (Note, however, that the given thicknesses, doping levels etc. should probably be considered as rough indications only, as the simulations were carried out at an illumination density of 5 mW/cm² only).

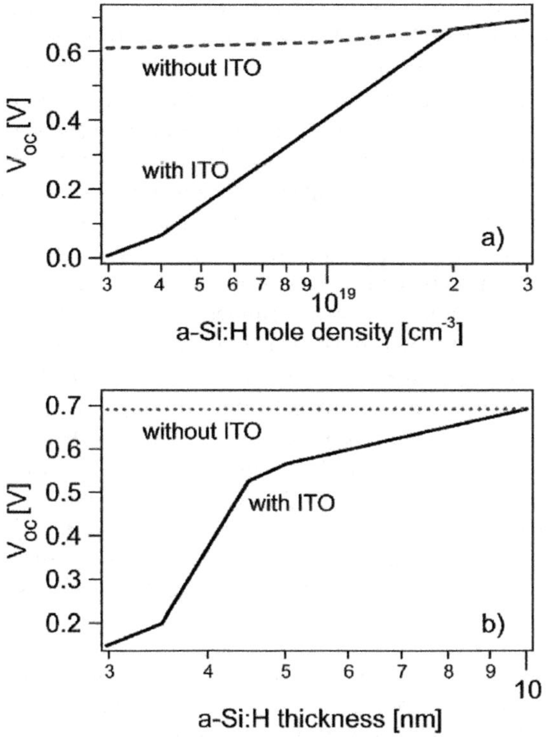

Fig. 6.38 Impact of the (a) *a*-Si:H emitter doping and (b) emitter thickness on the V_{oc} when ITO is included in analysis and when it is not. The entire *a*-Si:H layer is doped, and the simulations are done with an illumination of 5 mWcm². The *a*-Si:H layer thickness in (a) is 10 nm, and the *a*-Si:H carrier density in (b) is 3×10^{19} cm^{-3}. Reproduced from [119] with permission.

6.6 Conclusions

In this chapter, the electronic properties of *a*-Si:H/*c*-Si heterojunctions have been discussed. The electronic properties of the amorphous/crystalline silicon hetero-interface, notably the density of defects in the band gap, govern recombination in *a*-Si:H/*c*-Si based high efficiency solar cells. Thus, the realization of extremely low interface defect densities, allowing for effective surface recombination velocities below 5 cm/s for typical (n)*c*-Si material, is mandatory to achieve V_{oc}s above 700 mV and thereby realize the potential that is inherent in using hetero-contacts for solar cells. Such a low interface recombination velocity corresponds to extremely low interface defect densities, of the order of $10^{10}\ldots10^{11}$ defects/cm^{-2}, according to simulation studies.

The electronic properties of ultrathin (~10 nm) *a*-Si:H layers used for *a*-Si:H/*c*-Si cells can be made comparable to those of thick *a*-Si:H films. However, it appears that unlike in (several 100 nm) thick *a*-Si:H films, this necessitates an additional post-deposition annealing step at moderate temperatures (around 200°C for several minutes, or some seconds with microwave anneal). Microscopically, this can be explained as follows: The ~10 nm *a*-Si:H films are grown in short periods of time (typically 1 minute) on a *c*-Si substrate that constrains the degrees of freedom for the amorphous network growth. Therefore, it appears plausible that the amorphous network is frozen in in a non-equilibrium state with higher density of broken Si-Si bonds than expected for the degree of disorder in the material. The thermal energy provided by the annealing step leads to a marked decrease in defect density, presumably mediated by hydrogen diffusion, similar to the mechanisms proposed in thick *a*-Si:H films.

From near-UV photoelectron spectroscopy, the valence band offset in *a*-Si:H/*c*-Si heterojunctions can be determined for *a*-Si:H layer thicknesses of up to 10 nm, it amounts to $\Delta E_V \approx$ 0.45-0.46 eV for device-grade *a*-Si:H, leaving 0.1…0.2 eV for ΔE_V (assuming $E_{g,aSiH}$ in the range 1.7…1.8 eV).

Finally, a brief look has been taken at the influence of transparent conductive oxides (TCOs) on the overall band structure TCO/*a*-Si:H/*c*-Si: At the (n-type) TCO/(p)*a*-Si:H interface, a rectifying contact is likely to form, that will be antiparallel to the (p)*a*-Si:H/(n)*c*-Si junction and therefore influence charge carrier transport. As shown by simulations, for low *a*-Si:H emitter doping and/or low *a*-Si:H thickness, this might lead to an S-shaped *I-V* curve and a poor fill factor as well as an increased sensitivity to high defect densities at the *a*-Si:H/*c*-Si interface.

Acknowledgments. I would like to thank E. Conrad, K. Jacob and D. Patzek for sample preparations and measurements and R. Stangl, T.F. Schulze and C. Leendertz for valuable discussions. Many of the CFSYS samples were deposited by A. Laades during our joint PhD work at the Helmholtz-Zentrum Berlin (former Hahn-Meitner-Institut). The sections on *a*-Si:H/*c*-Si annealing and *I-V* characterization of solar cells rely on the work carried out by T.F. Schulze during his PhD thesis. The continuous work on thin *a*-Si:H films and *a*-Si:H/*c*-Si solar cells carried out at the Helmholtz-Zentrum Berlin that provided the background for this chapter would not have been possible without the strong support of the whole institute of Silicon Photovoltaics. I would like to thank specifically Manfred Schmidt and Walther Fuhs who have guided this research scientifically and secured the necessary funding over the years. They also supervised my PhD thesis and thereby laid the ground for the sections on near-UV photoelectron spectroscopy. I would like to thank J.-P. Kleider, R. Stangl and S. de Wolf for careful proofreading of the manuscript. This work has been partially funded by the European Commission through the FP7 project "Heterojunction Solar Cells based on a-Si:H/c-Si" (HETSI), grant no. 211821, and by the Bundesministerium für Bildung und Forschung (FKZ 01SF0012).

References

[1] Froitzheim, A., Brendel, K., Elstner, L., Fuhs, W., Kliefoth, K., Schmidt, M.: Interface recombination in heterojunctions of amorphous and crystalline silicon. J. Non-Cryst. Sol. 302, 663–667 (2002)

[2] Froitzheim, A., Stangl, R., Elstner, L., Schmidt, M., Fuhs, W.: Interface recombination in amorphous/crystalline silicon solar cells, a simulation study. In: Conf. Record 29th IEEE Photovoltaic Specialists Conf., pp. 1238–1241. IEEE Operations Center, San Diego (2002)

[3] Zhao, J., Wang, A., Green, M.A., Ferrazza, F.: 19.8% efficient "honeycomb" textured multicrystalline and 24.4% monocrystalline silicon solar cells. Appl. Phys. Lett. 73, 1991 (1998)
[4] Mishima, T., Taguchi, M., Sakata, H., Maruyama, E.: Development status of high-efficiency HIT solar cells. Sol. En. Mat. Sol. Cells 95, 18–21 (2011)
[5] Stangl, R., Froitzheim, A., Schmidt, M., Fuhs, W.: Design Criteria for Amorphous/Crystalline Silicon Heterojunction Solar Cells - a Simulation Study. In: Proc. 3rd World Conf. in Photovoltaic Energy Conversion, art.no. 4P–A8–45 (2003)
[6] Zhao, L., Zhou, C., Li, H., Diao, H., Wang, W.: Design optimization of bifacial HIT solar cells on p-type silicon substrates by simulation. Sol. En. Mat. Sol. Cells 92, 673–681 (2008)
[7] Street, R.A.: Hydrogenated Amorphous Silicon. Cambridge University Press, Cambridge (1991)
[8] Tanaka, K., Maruyama, E., Shimada, T., Okamoto, H.: Amorphous Silicon. John Wiley & Sons, Chichester (1999)
[9] Searle, T. (ed.): Properties of Amorphous Silicon and its Alloys. Emis Datareviews. INSPEC, vol. 19. The Institution of Electrical Engineers, London (1998)
[10] Weaire, D., Thorpe, M.F.: Electronic Properties of an Amorphous Solid. I. A Simple Tight-Binding Theory. Physical Review B 4, 2508 (1971)
[11] Smets, A.H.M., Kessels, W.M.M., van de Sanden, M.C.M.: Appl. Phys. Lett. 82, 865–867 (2003)
[12] Böhmer, E., Lüth, H.: Photoelectron spectroscopy studies of microcrystalline/amorphous silicon interfaces. J. Non-Cryst. Sol. 269, 1038–1043 (2000)
[13] Ley, L.: Photoemission and Optical Properties. In: Joannopoulos, J.D., Lucovsky, G. (eds.) The Physics of Hydrogenated Amorphous Silicon II - Electronic and Vibrational Properties, pp. 61–168. Springer, Berlin (1984)
[14] Müller, G., Krötz, G.: Structural equilibration in pure and hydrogenated amorphous-silicon. In: Mater. Res. Soc. Symp. Proc., vol. 297, pp. 237–248 (1993)
[15] Martin, R.M.: Elastic Properties of ZnS Structure Semiconductors. Phys. Rev. B 1, 4005–4011 (1970)
[16] Urbach, F.: The Long-Wavelength Edge of Photographic Sensitivity and of the Electronic Absorption of Solids. Phys. Rev. 92, 1324 (1953)
[17] Anderson, P.W.: Absence of Diffusion in Certain Random Lattices. Phys. Rev. 109, 1492–1505 (1958)
[18] Stutzmann, M.: The defect density in amorphous silicon. Phil. Mag. B 60, 531–546 (1989)
[19] Stutzmann, M., Biegelsen, D., Street, R.: Detailed investigation of doping in hydrogenated amorphous silicon and germanium. Phys. Rev. B 35, 5666–5701 (1987)
[20] Ley, L.: Band tails of a-Si:H: photoemission and absorption data. In: Searle [9], pp. 113–138
[21] Street, R.A., Kakalios, J., Tsai, C.C., Hayes, T.M.: Thermal-equilibrium processes in amorphous silicon. Phys. Rev. B 35, 1316–1333 (1987)
[22] Smith, Z.E., Wagner, S.: Band Tails, Entropy, and Equilibrium Defects in Hydrogenated Amorphous Silicon. Phys. Rev. Lett. 59, 688–691 (1987)
[23] Powell, M.J., Deane, S.C.: Improved defect-pool model for charged defects in amorphous silicon. Phys. Rev. B 48, 10815–10827 (1993)
[24] Powell, M.J., Deane, S.C.: Defect-pool model and the hydrogen density of states in hydrogenated amorphous silicon. Phys. Rev. B 53, 10121–10132 (1996)

[25] Chahed, L., Thèye, M.L., Fournier, D., Roger, J.P., Boccara, A.C., Li, Y.M., Turner, W.A., Paul, W.: Surface effects in hydrogenated amorphous silicon studied by photothermal-deflection experiments. Phys. Rev. B 43, 14488–14497 (1991)
[26] Curtins, H., Favre, M.: Surface and bulk states determined by photothermal deflection spectroscopy. In: Fritzsche, H. (ed.) Amorphous Silicon and Related Materials, pp. 329–363. World Scientific Publishing Co., Singapore (1988)
[27] Hata, N., Wagner, S., Roca i Cabarrocas, P., Favre, M.: Deposition-induced defect profiles in amorphous hydrogenated silicon. Appl. Phys. Lett. 56, 2448–2450 (1990)
[28] Kleider, J.P., Longeaud, C., Roca i Cabarrocas, P.: Experimental evidence for the annealing of surface defects in a-Si:H during deposition. J. Appl. Phys. 72, 4727–4731 (1992)
[29] Winer, K., Street, R.A., Johnson, N.M., Walker, J.: Impurity incorporation and doping efficiency in a-Si:H. Phys. Rev. B 42, 3120–3128 (1990)
[30] Pierz, K., Fuhs, W., Mell, H.: On the mechanism of doping and defect formation in a-Si:H. Phil. Mag. B 63, 123–141 (1991)
[31] Tanaka, M., Taguchi, M., Matsuyama, T., Sawada, T., Tsuda, S., Nakano, S., Hanafusa, H., Kuwano, Y.: Development of new a-Si/c-Si heterojunction solar cells: ACJ-HIT (Artificially Constructed Junction-Heterojunction with Intrinsic Thin-Layer). Jpn. J. Appl. Phys. 31, 3518–3522 (1992)
[32] Fujiwara, H., Kondo, M.: Real-time monitoring and process control in amorphous/crystalline silicon heterojunction solar cells by spectroscopic ellipsometry and infrared spectroscopy. Appl. Phys. Lett. 86, 032112 (2005)
[33] Fujiwara, H., Toyoshima, Y., Kondo, M., Matsuda, A.: Interface-layer formation mechanism in a-Si:H thin-film growth studied by real-time spectroscopic ellipsometry and infrared spectroscopy. Phys. Rev. B 60, 13598–13604 (1999)
[34] Li, Y.M., An, I., Nguyen, H.V., Wronski, C.R., Collins, R.W.: Thin-film coalescence in hydrogenated amorphous silicon probed by spectroscopic ellipsometry with millisecond-scale resolution. Phys. Rev. Lett. 68, 2814 (1992)
[35] Laades, A.: Preparation and Characterization of a-Si:H/c-Si Heterojunctions. PhD thesis, Technische Universität Berlin. Mensch & Buch Verlag, Berlin (2005)
[36] Heiblum, M., Nathan, M.I., Eizenberg, M.: Energy band discontinuities in heterojunctions measured by internal photoemission. Appl. Phys. Lett. 47, 503–505 (1985)
[37] Lam, Y.W.: Surface-state density and surface potential in MIS capacitors by surface photovoltage measurements. J. Phys. D: Appl. Phys. 4, 1370–1375 (1971)
[38] Hüfner, S.: Photoelectron Spectroscopy - Principles and Applications, 3rd edn. Springer, Berlin (2003)
[39] Ley, L.: Valence and conduction band density of states in a-Si:H: photoemission, inverse photoemission and core level absorption spectroscopy. In: Searle [9], pp. 93–137
[40] Winer, K., Ley, L.: Near-Surface Electronic Properties of a-Si:H. In: Fritzsche, H. (ed.) Amorphous Silicon and Related Materials, pp. 365–408. World Scientific Publishing Co., Singapore (1988)
[41] Griep, S., Ley, L.: Direct spectroscopic determination of the distribution of occupied gap states in a-Si:H. J. Non-Cryst. Sol. 59-60, 253–256 (1983)
[42] Winer, K., Ley, L.: Surface states and the exponential valence-band tail in a-Si:H. Phys. Rev. B 36, 6072–6078 (1987)
[43] Schmidt, M., Schoepke, A., Korte, L., Milch, O., Fuhs, W.: Density distribution of gap states in extremely thin a-Si:H layers on crystalline silicon wafers. J. Non-Cryst. Solids 338-340, 211–214 (2004)

[44] Sebastiani, M., Di Gaspare, L., Capellini, G., Bittencourt, C., Evangelisti, F.: Low-energy yield spectroscopy as a novel technique for determining band offsets: Application to the c-Si(100)/a-Si:H heterostructure. Phys. Rev. Lett. 75, 3352–3355 (1995)
[45] Korte, L., Schmidt, M.: Investigation of gap states in phosphorous-doped ultra-thin a-Si:H by near-UV photoelectron spectroscopy. J. Non-Cryst. Sol. 354, 2138–2143 (2008)
[46] Spicer, W.: Photoemissive, photoconductive, and optical absorption studies of alkali-antimony compounds. Phys. Rev. 112, 114–122 (1958)
[47] Jackson, W.B., Kelso, S.M., Tsai, C.C., Allen, J.W., Oh, S.J.: Energy dependence of the optical matrix element in hydrogenated amorphous and crystalline silicon. Phys. Rev. B 31, 5187–5198 (1985)
[48] Schafer, R., Mersereau, R., Richards, M.: Constrained Iterative Restoration Algorithms. Proc. IEEE 69, 432–450 (1981)
[49] Korte, L.: Die elektronische Struktur des amorph-kristallinen Silizium-Heterostruktur-Kontakts. PhD thesis, Philipps-Universität, Marburg/Lahn. Berichte des Hahn-Meitner-Instituts HMI-B 611 (2006)
[50] Cuevas, A., Sinton, R.A.: Prediction of the open-circuit voltage of solar cells from the steady-state photoconductance. Prog. Photovol.: Res. & Appl. 5, 79–90 (1997)
[51] Sinton, R.A., Cuevas, A.: Contactless determination of current–voltage characteristics and minority-carrier lifetimes in semiconductors from quasi-steady-state photoconductance data. Appl. Phys. Lett. 69, 2510–2512 (1996)
[52] Sproul, A.B.: Dimensionless solution of the equation describing the effect of surface recombination on carrier decay in semiconductors. J. Appl. Phys. 76, 2851–2854 (1994)
[53] Garin, M., Rau, U., Brendle, W., Martin, I., Alcubilla, R.: Characterization of a-Si:H/c-Si interfaces by effective-lifetime measurements. J. Appl. Phys. 98, 093711–093719 (2005)
[54] Hubin, J., Shah, A.V., Sauvain, E.: Effects of dangling bonds on the recombination function in amorphous semiconductors. Phil. Mag. Lett. 66, 115 (1992)
[55] Leendertz, C., Stangl, R., Schulze, T.F., Schmidt, M., Korte, L.: A recombination model for a-Si:H/c-Si heterostructures. Phys. Stat. Sol. (c) 7, 1005–1010 (2010)
[56] Olibet, S., Vallat-Sauvain, E., Ballif, C., Korte, L., Fesquet, L.: Silicon solar cell passivation using heterostructures. In: Proceedings of NREL's 17th Workshop on Crystalline Silicon Solar Cells & Modules: Materials and Processes (2007)
[57] Stangl, R., Kriegel, M., Schmidt, M.: AFORS-HET, Version 2.2, a numerical computer program for simulation of heterojunction solar cells and measurements. In: Record IEEE 4th World Conf. on Photovoltaic Energy Conversion, pp. 1350–1353. IEEE Operation Center, San Diego (2006)
[58] Stangl, R., Leendertz, C., Haschke, J.: Numerical Simulation of Solar Cells and Solar Cell Characterization Methods: the open-source on demand program AFORS-HET. In: Rugescu, R.D. (ed.) Solar Energy, IN-TECH (2009), `http://www.intechopen.com/download/pdf/pdfs_id/8561` (accessed July 20, 2011)
[59] Conrad, E., Korte, L., von Maydell, K., Angermann, H., Schubert, C., Stangl, R., Schmidt, M.: Development and optimization of a-Si:H/c-Si heterojunction solar cells completely processed at low temperatures. In: Hoffmann, W., et al. (eds.) Proc. 21st European Photovoltaic Solar Energy Conf., pp. 784–787. WIP, ETA, Munich, Florence (2006)
[60] Korte, L., Laades, A., Schmidt, M.: Electronic states in a-Si:H/c-Si heterostructures. J. Non-Cryst. Sol. 352, 1217–1220 (2006)

[61] Siebke, F., Beyer, W., Herion, J., Wagner, H.: Surface and bulk density of states of a-Si:H determined by CPM and total-yield photoelectron spectroscopy. J. Non-Cryst. Sol. 137-138, 339–342 (1991)
[62] Winer, K., Hirabayashi, I., Ley, L.: Exponential conduction-band tail in P-doped a-Si:H. Phys. Rev. Lett. 60, 2697–2700 (1988)
[63] Fujiwara, H., Kondo, M., Matsuda, A.: Real-time studies of amorphous and microcrystalline Si:H growth by spectroscopic ellipsometry and infrared spectroscopy. Thin Solid Films 455-456, 670–674 (2004)
[64] Reichardt, J., Ley, L., Johnson, R.L.: Band edges, Fermi level position, and hydrogen concentration in surface near regions of a-Si:H. J. Non-Cryst. Sol. 59-60, 329–332 (1983)
[65] Ast, D.G., Brodsky, M.H.: Thickness and temperature dependence of the conductivity of phosphorus-doped hydrogenated amorphous silicon. Philos. Mag. B 41, 273–285 (1980)
[66] Beyer, W., Overhof, H.: Doping Effects in a-Si: H. In: Pankove, J.I. (ed.) Hydrogenated Amorphous Silicon - Electronic and Transport Properties. Semiconductors and Semimetals, vol. 21(3), p. 385. Academic Press, Orlando (1984)
[67] Thorpe, M.F., Jacobs, D.J., Chubynsky, N.V., Rader, A.J.: Generic rigidity of network glasses. In: Thorpe, M.F., Duxbury, M. (eds.) Rigidity Rheory and Applications, pp. 239–277. Kluwer Academic/Plenum Publishers, New York (1999)
[68] Phillips, J.C.: Constraint theory and defect densities at (nanometer SiO2-based dielectric)/Si interfaces. J. Vac. Sci. Technol. B 17, 1803–1805 (1999)
[69] Lucovsky, G., Wu, Y., Niimi, H., Misra, V., Phillips, J.C.: Bonding constraint-induced defect formation at Si-dielectric interfaces and internal interfaces in dual-layer gate dielectrics. J. Vac. Sci. Technol. B 17, 1806–1812 (1999)
[70] Lucovsky, G., Wu, Y., Niimi, H., Misra, V., Phillips, J.C.: Bonding constraints and defect formation at interfaces between crystalline silicon and advanced single layer and composite gate dielectrics. Appl. Phys. Lett. 74, 2005–2007 (1999)
[71] Burrows, M.Z., Das, U.K., Opila, R.L., Wolf, S.D., Birkmire, R.W.: Role of hydrogen bonding environment in a-Si:H films for c-Si surface passivation. J. Vac. Sci. Technol. A 26, 683–687 (2008)
[72] Dauwe, S., Schmidt, J., Hezel, R.: Very low surface recombination velocities on p- and n-type silicon wafers passivated with hydrogenated amorphous silicon films. In: Schmidt, J., et al. (eds.) Conf. Record 29th IEEE Photovoltaic Specialists Conf., pp. 1246–1249. IEEE Operations Center, San Diego (2002)
[73] de Wolf, S., Kondo, M.: Abruptness of a-Si:H/c-Si interface revealed by carrier lifetime measurements. Appl. Phys. Lett. 90, 042111 (2007)
[74] de Wolf, S., Olibet, S., Ballif, C.: Stretched-exponential a-Si:H/c-Si interface recombination decay. Appl. Phys. Lett. 93, 032101–032103 (2008)
[75] Mitchell, J., Macdonald, D., Cuevas, A.: Thermal activation energy for the passivation of the n-type crystalline silicon surface by hydrogenated amorphous silicon. Appl. Phys. Lett. 94, 162102–162103 (2009)
[76] Plagwitz, H., Terheiden, B., Brendel, R.: Staebler–Wronski-like formation of defects at the amorphous-silicon–crystalline silicon interface during illumination. J. Appl. Phys. 103, 094506–4 (2008)
[77] Schulze, T.F., Beushausen, H.N., Leendertz, C., Dobrich, A., Hannappel, T., Korte, L., Rech, B.: Impact of a-Si:H structural properties on the annealing behavior of a-Si:H/c-Si heterostructures used as precursors for high-efficiency solar cells. In: MRS Symp. Proc., vol. 1268, contrib. no. EE01-07 (2010)

[78] Schulze, T.F., Beushausen, H.N., Leendertz, C., Dobrich, A., Rech, B., Korte, L.: Interplay of amorphous silicon disorder and hydrogen content with interface defects in amorphous/crystalline silicon heterojunctions. Appl. Phys. Lett. 96, 252102 (2010)
[79] Smets, A.H.M., van de Sanden, M.C.M.: Relation of the Si-H stretching frequency to the nanostructural Si-H bulk environment. Phys. Rev. B 76, 073202 (2007)
[80] Beyer, W., Mell, H., Overhof, H.: Electronic transport in hydrogenated amorphous silicon. J. Phys. (Paris) Coll. 42, 103–106 (1983)
[81] Schmidt, M., Angermann, H., Conrad, E., Korte, L., Laades, A., von Maydell, K., Schubert, C., Stangl, R.: Physical and technological aspects of a-Si:H/c-Si heterojunction solar cells. In: Conf. Record 4th IEEE World Conf. on Photovoltaic Energy Conversion, pp. 1433–1438. IEEE Operations Center, San Diego (2006)
[82] Schulze, T.F., Beushausen, H.N., Hansmann, T., Korte, L., Rech, B.: Accelerated interface defect removal in amorphous/crystalline silicon heterostructures using pulsed annealing and microwave heating. Appl. Phys. Lett. 95, 182108 (2009)
[83] de Wolf, S., Fujiwara, H., Kondo, M.: Impact of annealing on passivation of a-Si:H / c-Si heterostructures. In: Proc. 33rd IEEE Photovoltaic Specialists Conf. IEEE Operations Center, San Diego (2008)
[84] Fujiwara, H., Kondo, M.: Effects of a-Si:H layer thicknesses on the performance of a-Si:H/c-Si heterojunction solar cells. J. Appl. Phys. 101, 054516–054519 (2007)
[85] Korte, L., Conrad, E., Angermann, H., Stangl, R., Schmidt, M.: Advances in a-Si:H/c-Si Heterojunction Solar Cell Fabrication and Characterization. Sol. En. Mat. Sol. Cells 93, 905–910 (2009)
[86] van de Walle, C., Yang, L.H.: Band Discontinuities at Heterojunctions Between Crystalline and Amorphous Silicon. J. Vac. Sci. Technol. B 13, 1635–1638 (1995)
[87] Bittencourt, C., Alvarez, F.: Valence Band Offsets of a-Si1-xCx on C-Si and a-Si:H. In: Searle [9], pp. 174–179 (1997)
[88] Cuniot, M., Marfaing, Y.: Study of the band discontinuities at the a-SiH/c-Si interface by internal photoemission. J. Non-Crys. Sol. 77-78, 987–990 (1985)
[89] Cuniot, M., Marfaing, Y.: Energy-Band Diagram of the a-Si:H/c-Si Interface as Determined by Internal Photoemission. Philos. Mag. B 57, 291–300 (1988)
[90] Lequeux, N., Cuniot, M.: Internal photoemission measurements on a - Si1-xGex: H/c - Si heterojunctions. J. Non-Crys. Sol. 114, 555–557 (1989)
[91] Cuniot, M., Lequeux, N.: Determination of the energy band diagram for a-Si1-xYx:H/c-Si (Y = C or Ge) heterojunctions: Analysis of transport properties. Phil. Mag. B 64, 723 (1991)
[92] Essick, J.M., Cohen, J.D.: Band offsets and deep defect distribution in hydrogenated amorphous silicon-crystalline silicon heterostructures. Appl. Phys. Lett. 55, 1232–1234 (1989)
[93] Mimura, H., Hatanaka, Y.: Energy-band discontinuities in a heterojunction of amorphous hydrogenated Si and crystalline Si measured by internal photoemission. Appl. Phys. Lett. 50, 326 (1987)
[94] Eschrich, H., Bruns, J., Elstner, L., Swiatkowski, C.: The dependence of a-Si:H/c-Si solar cell generator and spectral response characteristics on heterojunction band discontinuities. J. Non-Cryst. Sol. 164-166, 717–720 (1993)
[95] Gall, S., Hirschauer, R., Kolter, M., Bräunig, D.: Admittance of a-Si:H/c-Si Schottky Diodes. In: MRS Symp. Proc., vol. 420, p. 245 (1996)
[96] Unold, T., Rösch, M., Bauer, G.H.: Defects and transport in a-Si:H/c-Si heterojunctions. J. Non-Cryst. Sol. 266-269, 1033–1037 (2000)

[97] Sakata, I., Yamanaka, M., Shimokawa, R.: Band lineup at the interface between boron-doped p-type hydrogenated amorphous silicon and crystalline silicon studied by internal photoemission. Jap. J. Appl. Phys. 43, 954–956 (2004)
[98] Kleider, J.P., Gudovskikh, A.S., Roca i Cabarrocas, P.: Determination of the conduction band offset between hydrogenated amorphous silicon and crystalline silicon from surface inversion layer conductance measurements. Appl. Phys. Lett. 92, 162101 (2008)
[99] Korte, L., Schmidt, M.: Doping and thickness dependence of band offsets at the amorphous/crystalline silicon heterojunction. J. Appl. Phys. (2010) (submitted)
[100] Brown, T.M., Bittencourt, C., Sebastiani, M., Evangelisti, F.: Electronic states and band lineups in c-Si(100)/a-Si1-xCx:H heterojunctions. Phys. Rev. B 55, 9904–9909 (1997)
[101] Di Gaspare, L., Capellini, G., Chudoba, C., Sebastiani, M., Evangelisti, F.: Low-energy yield spectroscopy determination of band offsets: application to the epitaxial Ge/Si(100) heterostructure. Appl. Surf. Sci. 104-105, 595–600 (1996)
[102] Chacorn, V., Haneman, D.: Thickness and doping dependence of the optical gap in amorphous hydrogenated silicon films. Solid State Comm. 65, 609–611 (1988)
[103] Wagner, I., Stasiewski, H., Abeles, B., Lanford, W.A.: Surface states in P- and B-doped amorphous hydrogenated silicon. Phys. Rev. B 28, 7080–7086 (1983)
[104] Hayashi, T., Hazama, Y., Miyazaki, S., Hirose, M.: Band offset in boron-doped amorphous silicon heterostructures. Jpn. J. Appl. Phys. 27, L986–L988 (1988)
[105] Hirai, M., Kusaka, M., Iwami, M., Akao, F., Yoshimoto, M., Matsunami, H.: X-ray photoelectron spectroscopy study of band offsets in a-Si0·2C0·8:H/p-(a-Si:H) and a-Si0·2C0·8:H/n-(a-Si:H) heterojunctions. Phil. Mag. B 60, 51–60 (1989)
[106] Mönch, W.: Electronic Properties of Semiconductor Interfaces. Springer, Berlin (2004)
[107] Tersoff, J.: Theory of semiconductor heterojunctions: The role of quantum dipoles. Phys. Rev. B 30, 4874–4877 (1984)
[108] Mönch, W.: Semiconductor Surfaces and Interfaces, 2nd edn. Springer, Heidelberg (1995)
[109] Schulze, T.F., Korte, L., Conrad, E., Schmidt, M., Rech, B.: Electrical transport mechanisms in a-Si:H/c-Si heterojunction solar cells. J. Appl. Phys. 107, 023711 (2010)
[110] Ley, L., Reichardt, J., Johnson, R.L.: Static charge fluctuations in amorphous silicon. Phys. Rev. Lett. 49, 1664–1667 (1982)
[111] Rau, U., Schock, H.: Electronic properties of Cu(In,Ga)Se2 heterojunction solar cells – recent achievements, current understanding, and future challenges. Appl. Phys. A 69, 131–147 (1999)
[112] Sze, S.M.: Physics of Semiconductor Devices, 2nd edn. Wiley Interscience, Hoboken (1981)
[113] Froitzheim, A.: Hetero-Solarzellen aus amorphem und kristallinem Silizium. PhD thesis, Philipps-Universität Marburg (2003)
[114] Stangl, R., Froitzheim, A., Elstner, L., Fuhs, W.: Amorphous/crystalline silicon heterojunction solar cells, a simulation study. In: Proc. 17th European Photovoltaic Solar Energy Conf., p. 1383 (2001)
[115] Datta, A., Rahmouni, M., Nath, M., Boubekri, R., Roca i Cabarrocas, P., Chatterjee, P.: Insights gained from computer modeling of heterojunction with instrinsic thin layer "HIT" solar cells. Sol. En. Mat. Sol. Cells 94, 1457–1462 (2010)
[116] Tucci, M., della Noce, M., Bobeico, E., Roca, F., de Cesare, G., Palma, F.: Comparison of amorphous/crystalline heterojunction solar cells based on n- and p-type crystalline silicon. Thin Solid Films 451-452, 355–360 (2004)

[117] Shockley, W.: The Theory of p-n Junctions in Semiconductors and p-n Junction Transistors. Bell Syst. Tech. J. 28, 435 (1949)
[118] Matsuura, H., Okuno, T., Okushi, H., Tanaka, K.: Electrical properties of n-amorphous/p-crystalline silicon heterojunctions. J. Appl. Phys. 55, 1012–1019 (1984)
[119] Kanevce, A., Metzger, W.K.: The role of amorphous silicon and tunneling in heterojunction with intrinsic thin layer (HIT) solar cells. J. Appl. Phys. 105, 094507–094507 (2009)
[120] Schmidt, M., Froitzheim, A., Stangl, R., Elstner, L., Kliefoth, K., Füssel, W., Fuhs, W.: Photocurrent analysis in TCO/a-Si:H/c-Si solarcell structures. In: Proc. 17th European Photovoltaic Solar Energy Conf., pp. 1383–1386 (2001)
[121] Zhao, L., Zhou, C.L., Li, H.L., Diao, H.W., Wang, W.J.: Role of the work function of transparent conductive oxide on the performance of amorphous/crystalline silicon heterojunction solar cells studied by computer simulation. Phys. Stat. Sol (a) 205, 1215–1221 (2008)
[122] Jackson, W.B., Oh, S.J., Tsai, C.C., Allen, J.W.: Conduction-band density of states in hydrogenated amorphous silicon determined by inverse photoemission. Phys. Rev. Lett. 53, 1481–1484 (1984)
[123] Schulze, T.F.: Structural, electronic and transport properties of amorphous/crystalline silicon heterojunctions. PhD thesis, TU Berlin/Helmholtz-Zentrum Berlin. Reports of the Helmholtz Centre Berlin HZB-B 15 (2011)

Chapter 7
Intrinsic and Doped *a*-Si:H/*c*-Si Interface Passivation

Stefaan De Wolf

École Polytechnique Fédérale de Lausanne (EPFL), Institute of Microengineering (IMT), Photovoltaics and thin-film electronics laboratory (PVlab), Breguet 2, 2000 Neuchâtel, Switzerland

Abstract. The performance of crystalline silicon (*c*-Si) heterojunction (SHJ) solar cells critically depends on the properties of the deposited hydrogenated amorphous silicon (*a*-Si:H) films. Surface passivation is an important role they need to fulfill. Additionally, the *a*-Si:H films should also act as efficient emitter and back surface field (BSF). In this chapter, we focus on the electronic passivation properties of the *a*-Si:H/*c*-Si interface. First, relevant literature on *c*-Si surfaces is briefly reviewed, including the effect of hydrogenation of surface states. This is followed by a discussion of how electronic surface recombination is calculated and measured. Recombination is mainly determined by electronic gap-states. The precise nature of these states is discussed both for the *c*-Si surface and for the *a*-Si:H bulk. Next, the physical passivation mechanism of intrinsic *a*-Si:H is elucidated. It is concluded that it stems from chemical surface state passivation by hydrogen, similar to defect passivation in the *a*-Si:H bulk. For these films, it is also argued how epitaxial growth may detrimentally influence the passivation quality. For heterojunction devices this has its importance, as the deposition of device-grade *a*-Si:H is often very close to the transition to epitaxial growth. A following section focuses on the effect of doping of the amorphous films. Doping is principally expected to improve the passivation quality further, as it should give rise to additional field-effect passivation. Here, it is discussed why this is not necessarily the case, as doping is also linked to Fermi-level dependent Si–H bond rupture in the films. A compromise between doping and surface-passivation may be obtained by employing an intrinsic buffer layer between the doped film and the wafer. By using intrinsic buffer layers, values for the energy conversion efficiency as high as 23% were reported to date for SHJ devices.

W.G.J.H.M. van Sark et al. (Eds.): Physics & Tech. of Amorphous-Crystalline, EM, pp. 223–259.
springerlink.com

7.1 Introduction

Electronic passivation of semiconductor interfaces is of critical importance in the performance of many electronic and photonic devices. Recent examples where such passivation turned out to be crucial include GaAs/AlGaAs heterostructure bipolar transistors [1], AlGaN/GaN high electron mobility transistors [2]. AlGaN/GaN heterojunction field-effect transistors [3] and germanium metal–oxide–semiconductor (MOS) capacitors [4]. In nanoscale devices, electronic control of interfaces is becoming more important as well due to their increased surface area [5]. Passivation is for similar reasons equally important in crystalline silicon (*c*-Si) based solar cells [6,7,8]. For *c*-Si wafers, the lowest reported effective surface recombination velocity, S_{eff}, was achieved using hydrofluoric acid (HF) solutions [9]. For devices, a variety of passivating layers have become available as well. Historically, most of these films were developed for gate dielectrics in microelectronics. Among these, arguably the best known is thermally grown silicon-dioxide (SiO_2), where the electronic properties of its interface with the underlying *c*-Si form the basis of metal-oxide-semiconductor (MOS) transistor operation [10,11]. Perhaps not surprisingly, the *c*-Si based solar cells with the highest energy-conversion efficiency reported to date [25% under a standard air mass 1.5 global (AM 1.5 G) 1-sun spectrum] featured SiO_2 films as well [7,12]. The excellent passivation properties of such layers was also evidenced by efficient silicon-based light emission with similar technology [13]. In addition to enabling record-efficiencies, progress on surface passivation also plays its role in driving down the cost of photovoltaics (PV), as it lowers the amount of expensive, highly refined, silicon required to generate PV power [14]. In microelectronics, scaling-limits dictate the search for alternatives to SiO_2 [15]. In PV, a similar quest exists, although it is motivated rather by the (too) high processing temperature such oxides require [16]. Wet-thermal oxides are grown at reduced temperatures [17] and have proven their use in solar cells [18]. Other PV-suitable dielectrics include amorphous silicon-nitride (*a*-SiN_x:H) [19,20], SiO_2/*a*-SiN_x:H stacks [16,21], or aluminum-oxide (Al_2O_3) films [22,23,24]. Despite this, the presence of highly recombination-active metal contacts remains an important efficiency-limiting factor for such dielectrically passivated solar cells. At best, ignoring cost-issues, a trade-off between total contact area and surface passivation is made by locally opening the dielectric films. Recombination can then further be reduced by defining a local BSF underneath the metal contacts by dopant diffusion [7].

A more elegant solution is obtained by using semiconducting hetero-structure contacts, which simultaneously fulfill the passivation and contacting roles. A key point of such contacts is their displacement of the highly recombination-active (ohmic) contacts from the silicon surface by insertion of a film with wide bandgap [25]. To reach their full potential, the interface state density should be minimal [25]. Practically, *a*-Si:H films are appealing candidates for this: their bandgap is wider than that of *c*-Si and, when intrinsic, such films can reduce the *c*-Si surface state density by hydrogenation [26]. In addition, these films can be made relatively easily either *n*- or *p*-type by substitutional doping [27], allowing the fabrication of electronically abrupt *p*-*n* and low-high hetero junctions [28].

In this chapter we discuss the physical origins of the excellent passivation of intrinisic *a*-Si:H films. For the reduction of gap states associated with either *a*-Si:H or the *c*-Si surface, hydrogen is known to play a benign role, as it can passivate Si dangling bonds. For the passivation of the interface states between these two materials, hydrogen is arguably as important. Next, the impact of doping on the properties of the interface is discussed. Doping of these films may be expected to yield a built-in electrical field, repelling either electrons or holes from the surface states. In principle, this could suppress the *a*-Si:H/*c*-Si interface recombination further, in a similar way as, *e.g.*, in back-surface-field homojunction solar cells [29]. Experimentally, however, such layers were often found to result in poorer electronic passivation of *c*-Si surfaces than their intrinsic counterparts [30,31,32]. We explain this phenomenon as originating from Fermi level (E_F) dependent Si-H bond rupture within such films. The latter phenomenon is attributed to the position of E_F within the bandgap, influencing the formation of (native) defects in the semiconductor. The poor passivation of doped films also explains why, typically, a few nanometer thin intrinsic buffer layer is inserted between the *c*-Si surface and the doped *a*-Si:H films for device fabrication [33]. For heterojunction solar cells featuring such stacked film structures, impressive large area (> 100 cm^2) energy conversion efficiencies (23% to date) were reported [34,35].

7.2 Crystalline Silicon Surfaces

The well-defined band-structure of crystalline semiconductors principally stems from their lattice periodicity. For bulk *c*-Si, the diamond-crystal structure results from the formation of strongly directional covalent sp^3 hybrid bond orbitals. These originate from a linear combination of *s*- and *p*-orbitals. Overlapping sp^3 orbitals on neighbouring tetrahedrally coordinated sites produce bonding and anti-bonding levels, which ultimately broaden into the semiconductor valence band (VB) and conduction band (CB) [36]. To create a surface, bonds need to be broken. These hybrid orbitals appear now as so-called dangling bond (DB) orbitals. Each such orbital is nominally half occupied, since each side of a broken bond can accept one of the electrons of the previously unbroken covalent bond. This locally different electronic structure gives rise to surface states within the bandgap, the so-called Tamm or Shockley states [37,38]. In addition, such local change in chemical bonding may often give rise to surface relaxation and reconstruction, depending on the crystal orientation and preparation conditions [39].

7.2.1 The (111) and (100) Surfaces

For most semiconductor devices, the Si(100) surface is of greatest interest as its surface-state density was observed to be about a factor 3 lower than for Si(111) surfaces (after dry O_2 oxidation) [40]. For many *c*-Si solar cells, including SHJ devices, the Si(111) surface has its importance too as such solar cells are usually fabricated from anisotropically-etched Si(100) substrates. Due to bond-density dependent crystal dissolution, such etching exposes Si(111) faceted pyramids [41,42], which is useful to lower the external optical reflectivity and improve the internal reflection of the device [43,44].

The formation of the ideally truncated (111) surface creates one half-filled surface DB orbital (T_3^0) per surface atom, sticking out perpendicular from the surface and backbonded to three Si atoms. The subscript in the T_3^0 notation gives the coordination number, the superscript the charge state, whereas the T points at the tetrahedrally coordinated site, following a similar notation as for amorphous semiconductors [45]. Electronically, this surface features a single band of states within the forbidden gap, which is split off from the bulk sp^3-states forming the VB and CB. Due to the bond-breaking this band is half-filled and lying at midgap, for an intrinsic semiconductor [36]. As such, this band may be considered to be metallic, were it not for the fact that the band consists of *localized* states. As these DBs belong to atoms which are second-nearest neighbors in the bulk, it is difficult for them to lose energy by forming bonds [46]. Nevertheless, controlled (long range) reconstruction into the ideal Si(111)-(7×7) structure may take place in vacuum at high temperatures, as *e.g.* observed in real space by scanning tunneling microscopy [47]. Wet-chemical etching of these surfaces, such as in NH_4F solutions, will rather yield the [monohydride (MH) terminated] Si(111)-(1×1) structure [48], without reconstruction [49].

Conversely, the unreconstructed Si(100) surface has two broken bonds (T_2) per atom, tilted with respect to another, which is usually denoted as Si(100)-(1×1). Electronically, these two dangling-bond orbitals interact strongly with each other and form principally two gap-state bands, spread out over a wide energy range. This surface proves to be unstable however: Dehybridization of neighbouring sp^3 DBs can occur, yielding a dangling sp_z orbital and a bridging $p_{x,y}$ orbital, giving rise to the Si(100)-(2×1) dimer structure [46]. In this structure, for the two dangling bonds associated with each Si(100) surface atom, one forms a bond with its adjacent surface atom in a σ-bond (the dimer bond), while the other is usually linked in a weak π-bond, forming thus an unoccupied dimer [50]. In the presence of hydrogen, the doubly-occupied Si(100)-(2×1):H dimer structure can be created from such a structure, as it is energetically favourable for hydrogen to occupy both atoms on the dimer [51].

Generally, for silicon surfaces, additional localized surface levels (discrete or continuously distributed) may arise from the presence of impurity atoms and structural imperfections. As an example, in small-sized porous silicon quantum dots, surface states associated with oxidation lead to red shifting of the photoluminescence spectrum [52]. Freshly cleaved silicon exposed to air will rapidly become covered with a few monolayers of oxide as well, saturating at a thickness of about 40 Å [11]. Contrasting to thermal SiO_2 , this native oxide is poorly defined and does not lead to surface passivation. Conversely, other chemisorbed atoms such as hydrogen, have quite a beneficial effect for the electronic properties of such surfaces by removing surface-states from the bandgap.

For a more comprehensive description of silicon surfaces, the reader is kindly referred to, *e.g.*, the books by Bechstedt [46] and Lüth [39].

7.2.2 Surface Hydrogenation

Prior to film deposition, extremely well-controlled surfaces are required to obtain high-quality passivation. Hydrofluoric acid etching is known in the semi-conductor industry for producing silicon surfaces which are contamination-free and chemically stable for subsequent processing [53]. Usually, such etching makes part of a cleaning procedure consisting of sequential oxidation (*e.g.* by peroxide solutions) followed by oxide removal in HF solutions (so-called RCA cleaning) [54]. In between these steps the samples are rinsed in de-ionised water. The purpose of the oxidation step is to grow a layer on the wafer surface which encapsulates present contaminants. In the reduction step, the oxide is etched from the surfaces, taking away these impurities. At the same time, surface states are being hydrogenated [55].

By immersion of the Si(111) surface in HF-based solutions, extremely well-passivated surfaces were demonstrated with values for the surface recombination velocity down to 0.25 $cm.s^{-1}$ [9]. The remarkable chemical stability of such surfaces was initially explained in terms of F-passivation [56]. Experimentally, the presence of SiF_x species in solution may indicate that the mechanism of oxide removal leads to F termination of the surface [57]. In fact, this surface termination was shown to be unstable due to the high polarity of the Si-F bond [58,59]. Passivtion is achieved because HF etching removes the surface silicon atom as SiF_x and leaves the exposed Si terminated by hydrogen [59,60], protecting the surface from further chemical attack. Alternatively, NH_4F-etching is often used for hydrogenation purposes but rather leads to atomically flat (111) surfaces by anisotropic etching due to its basic nature. Such surfaces feature a single dangling bond per surface atom, and this yields the ideal Si(111)-(1×1):H MH termination in such solutions [48]. This termination was observed by attenuated total reflection Fourier transform infrared (ATR-FTIR) spectroscopy of the surface Si-H stretching vibrations [61], which showed very narrow absorption lines at 2083.7 cm^{-1}. This surface features Si-H bonding states far below the VB maximum (about -3eV below E_F) , and Si-H anti-bonding states well above the CB minimum (about +3eV above E_F) [62,63]. Such a large gap explains the excellent electronic passivation by hydrogen of these surfaces.

HF-etched Si(100) surfaces are usually atomically rough. This was attributed to slow oxidation of the H-passivated surface followed by a fast removal of the surface oxide by HF [59]. Surface termination following HF-etching is often non-ideal, featuring mono (MH) and higher hydride states [64,65], likely due to the random nature of this roughness. The different possible structures for the Si(100) surface are depicted in Fig. 7.1. The variety in hydride species on these surfaces has also been observed by thermal desorption spectroscopy (TDS) [66]. For this, usually the sample is heated up to 1000°C, with a linear temperature ramp of *e.g.* 20 $K.min^{-1}$ in ultra-high vacuum (< 1.0×10^{-9} Torr). The species effusing away are detected with a quadrupole mass spectrometer. Typically, such measurements reveal at least two H_2 peaks, as *e.g.* seen in Fig. 7.2. The low-temperature β_2 state is associated with dihydride (DH) and trihydride (TH) species [67], where annealing may transform (adjacent) Si(100)-(1×1):2H structures to doubly-occupied Si(100)-(2×1):H dimers by simultaneous rupture of two Si-H bonds, forming a H_2 molecule in the process [68]. The high-temperature β_1 state is rather linked to MH rupture, such as by the concerted desorption of two hydrogen atoms paired on the same Si(100)-(2×1):H dimer, following first-order desorption kinetics [69]. ATR-FTIR spectroscopy data indicate that such a high temperature state is indeed a

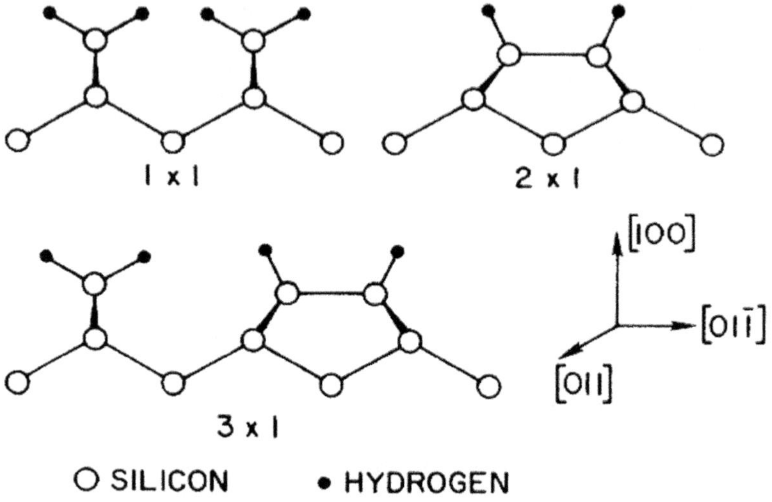

Fig. 7.1 Schematic of the various possible hydrogenated structures present on the Si(100) surface. Reprinted with permission from [65]. © 1990 The American Physical Society.

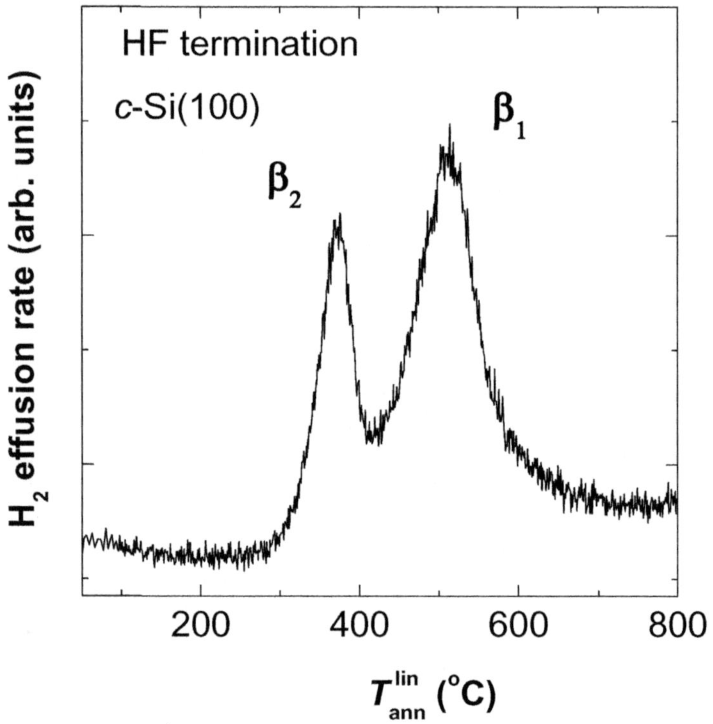

Fig. 7.2 Typical H_2 effusion spectrum from an HF terminated Si(100) surface. Measurement done with a linear temperature ramp of 20 K.min^{-1}. Unpublished data.

signature for monohydride species [70]. Another reported phase on the Si(100) surface consists of alternating rows of Si(100)-(2×1):H and Si(100)-(1×1):2H structures, denoted as Si(100)-(3×1) [70]. TDS experiments on hydrogenated Si(111) surfaces usually reveal only one H_2 peak at higher temperatures, as is expected from the microscopic Si(111)-(1×1):H structure. Noteworthy, etching of Si(100) surfaces in NH_4F-solutions yields microscopically rough surfaces as well, but of a different nature, as such solutions etch anisotropically the Si(100) structure [48]. Microscopically, this roughness is much better defined, consisting of Si(100) terraces separated by evenly distributed square pyramids with (111) facets [71,72]. These facets can again be ideally monohydride (MH) terminated in the form of the Si(111)-(1×1):H structure. The terraces can feature their own ideal Si(100)-(2×1):H structure too [71], even though the Si(100)-(1×1):2H structure may be present as well [72].

7.2.3 Surface Recombination

At a semiconductor surface, principally, the same recombination mechanisms apply to generated charge carriers as in the bulk. However, as a considerable amount of deep-level surface states are present, recombination through such states usually dominates over radiative band-to-band or non-radiative Auger recombination. As recombination centers are tightly coupled to the lattice, the energy and momentum initially belonging to the electron and hole are converted into phonons, with little or no electromagnetic energy being emitted (*i.e.* in a non-radiative fashion) [73]. For bulk recombination through defects, the Shockley-Read-Hall (SRH) theory [74,75] is often applied. This theory considers defects as discrete levels in the band gap that can have two states. Surface-states are however more accurately described as DBs which are amphoteric in nature: at equilibrium and according to the position of the Fermi level, E_F, this defect is either positively (T_3^+), neutrally (T_3^0) or negatively (T_3^-) charged, accommodating respectively 0, 1 and 2 electrons. Consequently, surface recombination is more accurately described using the statistics of correlated electrons [76].

7.2.3.1 Calculating the Surface Recombination

The calculation of electronic recombination at the interface between the *c*-Si surface and a passivating film is based on the so-called extended SRH formalism. This relies on the theory of a surface space-charge layer under non-equilibrium conditions and takes into account the impact of illumination (giving rise to excess carrier densities Δn and Δp), the fixed-charge density, Q_f, the carrier capture cross sections, σ_n and σ_p, in their different charge states, and the interface-state density D_{it} [77,78]. In case a gate electrode is present, as shown in Fig. 7.3, its work-function φ_m and the applied voltage V_G need to be considered too. All charges present at the interface need to balance with the charge density in the silicon, Q_{Si}, to satisfy charge neutrality. This induces a (band bending) surface potential ψ_s in the semiconductor, yielding a change in carrier densities at the surface, n_s and p_s, according to:

$$n_s = (n_{p0} + \Delta n)\exp\left(\frac{q\psi_s}{kT}\right) = n_{p0}\exp\left(\frac{q(\psi_s - \varphi_n)}{kT}\right), \tag{7.1}$$

$$p_s = (p_{p0} + \Delta p)\exp\left(-\frac{q\psi_s}{kT}\right) = p_{p0}\exp\left(-\frac{q(\psi_s - \varphi_p)}{kT}\right), \tag{7.2}$$

with n_{p0} and p_{p0} the equilibrium carrier densities in the bulk, φ_n and φ_p are the separations of the respective quasi-Fermi levels from thermal equilibrium, and k is Boltzmann's constant. Usually, the quasi-Fermi levels in the surface space-charge regions are considered to be flat [79]. Considering this, an algorithm to calculate ψ_s was derived by Girisch *et al.* [80], assuming localized SRH-like interface states [81]. The knowledge of the surface carrier densities and nature of the interface states allows then for the calculation of the surface recombination velocity S_{eff}, which is defined as

$$S_{eff} = \frac{U_s}{\Delta n_{dsc}}, \tag{7.3}$$

and is an experimentally accessible parameter. In this expression, Δn_{dsc} is the excess carrier density at the edge of the space-charge region, as shown in Fig. 7.3, and U_s is the surface recombination rate. Such calculations were *e.g.* used by Aberle *et al.* for SiO_2/*c*-Si interface studies [77]. However, surface states may be more accurately described by the statistics of correlated electrons, rather than by SRH levels [82,83]. Electronically, such defects are characterized by their electron and hole capture cross-sections, respectively in the neutral and charged state, their distribution within the bandgap, but also by their correlation energy U. The (positive) correlation energy stems from the fact that due to Coulomb repulsion usually it requires more energy to place a second electron on a dangling bond than was necessary to do so for the first. A simplified formalism to describe these defects was developed by Hubin *et al.*, yielding a closed-form expression for the recombination rate [84]. No thermal re-emission of charge carriers to defect levels was considered, and the distribution of dangling-bond states is reduced to a single discrete level, featuring the three possible charge states. The validity of these assumptions was discussed by Li *et al.* [85]. We also refer the reader to Chapter 13 by Stangl and Leendertz in this book for a more detailed description of such calculations.

In both the SRH and the (simplified) DB case, the recombination rate is linearly dependent on the density of interface states D_{it} [82]. Thus, it is readily understood that one way to obtain improved passivation is by neutralizing the interface states. This yields lowered values of S_{eff} independent from Δn, which can also be seen in Fig. 7.4, showing some of such calculations. Comparing curves d and c shows the effect of a lowered value for D_{it}. Curve c represents a calculated fit to measured

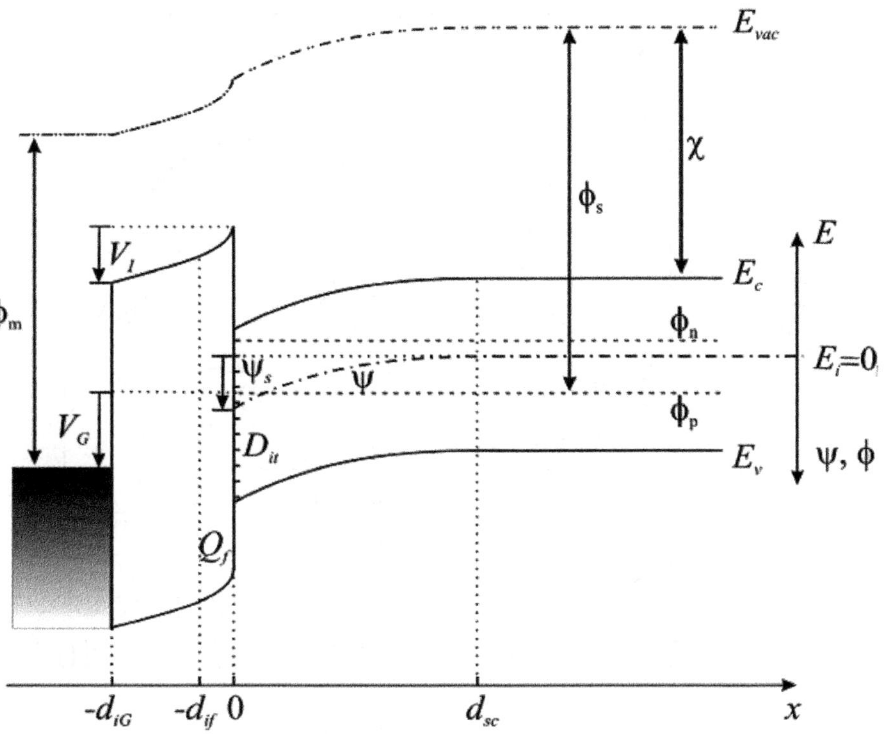

Fig. 7.3 Metal-insulator-semiconductor structure with *p*-type semiconductor under illumination, including a given gate potential V_G. The fixed charges in the dielectric are supposed to be confined within a thin region with thickness d_{if}, in which their concentration is assumed constant [78]. Dimensions are not drawn to scale.

carrier-lifetime data of a wafer passivated with intrinsic *a*-Si:H. Such data are discussed in greater detail in section 7.3.1.2. Alternatively, good passivation can be obtained by appropriately changing the surface potential ψ_s, yielding electrical field effect passivation. This can be accomplished either by the presence of (negative or positive) fixed charges Q_f in a dielectric layer, or by applying an external gate bias V_G. Usually only Q_f is used as a fitting parameter, however, in which case it can be regarded as an 'equivalent' fixed charge, representing all other band bending parameters (such as band offsets, work functions, gate voltages, *etc.*) In case of a surface potential, one type of (excess) carriers is repelled from the surface (as understood from equations (1) and (2)), reducing thus the recombination probability. The effect of such passivation will be strongly dependent on the excess carrier density, with the largest improvement at low injection, which can also be seen in Fig. 7.4, *e.g.* by comparing curves d and b. This figure shows that either a sufficiently large negative (curve a) or positive (curve b) fixed charge yield comparable results. At high injection, the surface minority-carrier density can overcome the potential barrier ψ_s, however, resulting again in recombination.

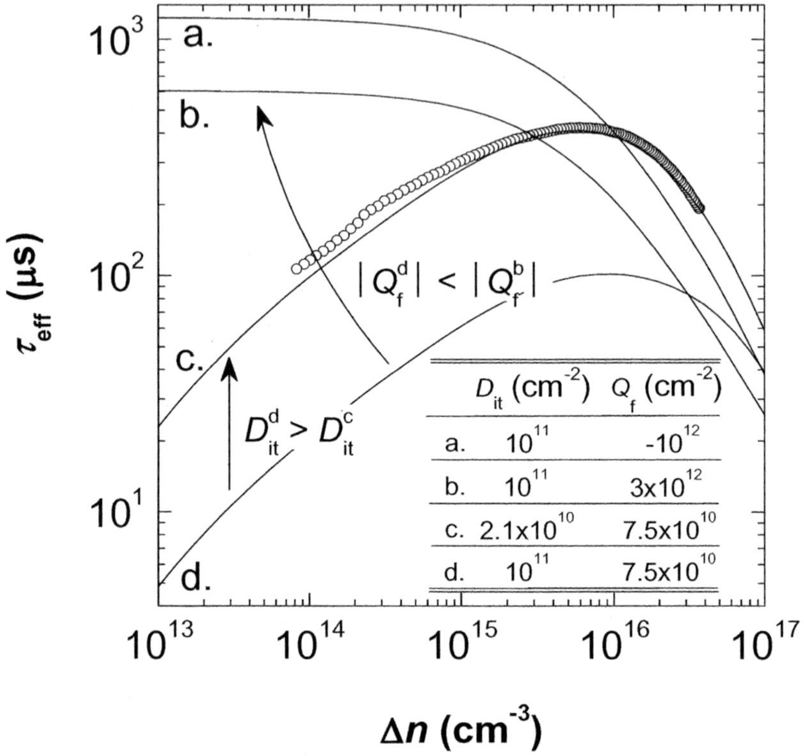

Fig. 7.4 Calculated surface recombination fits, according to the model described in reference [82]. The effect of changing D_{it} and Q_f is highlighted. The data show values for τ_{eff}, taking into account Auger recombination and radiative band-to-band recombination in the bulk. For reference, the symbols represent measured carrier-lifetime data of a wafer passivated with intrinsic *a*-Si:H films.

7.2.3.2 Measuring the Surface Recombination

The most straightforward way to experimentally obtain surface-recombination velocities is to measure the effective carrier lifetime (τ_{eff}) of samples featuring identical layers on both surfaces. Such measurements are known to be extremely sensitive, allowing for detection of bulk defect densities as low as 10^9–10^{11} cm^{-3} in a simple, contactless technique at room temperature [86].

Taking the bulk lifetime (τ_{bulk}) and identical surface recombination velocities at both surfaces (S_{eff}) of a wafer of given width W into account, the excess charge carrier density (Δn) will decay upon generation G(t) according to [87]:

$$\tau_{eff} = \frac{\Delta n}{G(t) - \frac{d\Delta n}{dt}}. \tag{7.4}$$

Steady-state and transient limits are obtained for respectively $G(t) >> d\Delta n/dt$ and $G(t) << d\Delta n/dt$, which are both experimentally accessible by photo-conductance measurements [88]. The effective minority-carrier lifetime τ_{eff} is determined by recombination in the bulk and at the surfaces, but principally also carrier diffusion from bulk to surfaces can play a role, according to [86]:

$$\frac{1}{\tau_{eff}} = \frac{1}{\tau_{bulk}} + D\gamma^2, \tag{7.5}$$

with D being the minority-carrier diffusion constant in the substrate and γ the first real root of [89]:

$$\tan(\gamma W) = \frac{2\gamma D S_{eff}}{\gamma^2 D^2 - S_{eff}^2}. \tag{7.6}$$

For relatively well-passivated surfaces ($S_{eff} << \pi^2 D/W$), this equation simplifies to:

$$\frac{1}{\tau_{eff}} = \frac{1}{\tau_{bulk}} + 2\frac{S_{eff}}{W}. \tag{7.7}$$

As the measurement of τ_{eff} gives access to a global recombination rate that is affected by several physical mechanisms, care has to be taken with its interpretation. For an accurate estimation of passivation properties of high-quality films, obviously high-quality (FZ)-Si wafers must be used. In addition, the effects of radiative band-to-band and Auger recombination in the wafer (especially in high-injection conditions) need to be accounted for by using *e.g.* the Auger recombination parameterization of Kerr and Cuevas [90]. The upper limit for S_{eff} is determined by the thermal velocity of free carriers to [91]:

$$S_{eff} = \sqrt{\frac{kT}{2\pi m^*}}, \tag{7.8}$$

with m^* the minority-carrier effective mass.

7.3 Surface Passivation by *a*-Si:H Films

7.3.1 Intrinsic a-Si:H Passivation

Intrinsic *a*-Si:H(*i*) films are known already for some decades to yield good *c*-Si surface passivation [26,92,93]. Experimentally, such films are mostly prepared by plasma-enhanced chemical vapor deposition (PECVD) with SiH_4 as precursor gas, possibly diluted in H_2. For the plasma-excitation frequency, 13.56 MHz is often used [94,95,96,97], although the successful use of very high frequencies (VHF, *e.g.*, 40 MHz [98] or 70 MHz [82,99]) was reported too. For device-grade films, usually the deposition temperature is about 200°C, and the system is operated at a relatively low pressure (0.1-1 Torr). Other techniques reported to give good results are direct-current PECVD [100] and hot-wire (or catalytic) CVD [101,102]. For a more detailed discussion on deposition technology, we refer the reader to Chapter

5 in this book by Roca i Cabarrocas. As the interface quality is of critical importance, wafers are usually cleaned prior to film deposition by methods such as the RCA cleaning technique. More details on sample cleaning can be found in Chapters 3 and 4 in this book by Angermann and Rappich. Subsequently, fast transfer to the deposition-system is necessary to avoid reoxidation.

Before the attention is turned to the passivation of the interface between an *a*-Si:H film and the *c*-Si surface, a brief discussion is given on some relevant electronic properties of the gap states in both materials, and their physical origins. It is shown that surface states are closely related to the occurrence of Fermi-level pinning. In Si-metal contacts, the position of the pinning position determines the Schottky-barrier height [103]. For heterojunction devices, the absence of such states is equally important to assure the rectifying behavior of the *a*-Si:H/*c*-Si heterojunction.

7.3.1.1 Gap States in the *a*-Si:H Film and at the *c*-Si Surface

In defect-rich *a*-Si:H bulk material, the Fermi level can be pinned in a narrow band of levels near midgap [104,105,106]. For the intrinsic *a*-Si:H(*i*) bulk, two peaks in the distribution of gap states were identified, with one peak located above and the other one below E_F. Based on the similarity in energy levels, these states were speculated to be associated with the divacancy defect present in (bulk) *c*-Si [107]. The divacancy is the lowest-order stable defect in *c*-Si and pairs six DBs. In an amorphous network, defects containing a smaller number of DBs are principally possible as well, including the mono vacancy (featuring four DBs). Even single DBs may exist in this material, as a consequence of structural randomness [108]. As argued earlier, these defects are amphoteric due to covalent bond rupture and are often denoted by T_3^0 in their neutral state [45]. The double nature of the bands in the bandgap originates from another aspect of the amphoteric character of T_3, namely its correlation energy which will be discussed below.

In the *c*-Si bulk, dangling bonds can only exist at a dislocation line; an isolated DB is precluded by crystallographic constraints [109]. For the *c*-Si surface, however, it may constitute the dominant defect, such as at the (unreconstructed) Si(111) surface. Here, pinning of the Fermi level is also a well-known phenomenon [110]. Microscopically, this was linked to the amphoteric nature of the surface states [111]. In this case the pinning usually depends not much on the bulk doping as a space-charge layer is built up at the surface preventing unlimited flow of free (majority) carriers from the *c*-Si bulk to these surface states [36].

A strong electronic similarity between the *c*-Si surface and the bulk of *a*-Si:H may thus exist [112]. In both cases, the gap states originate from an amphoteric defect that can lead to pinning of E_F. For simplicity, we discuss the defect causing the gap states, both for the *c*-Si surface and *a*-Si:H bulk, in terms of a single DB. The true microscopic nature of these defects in *a*-Si:H is however still under debate. Additional T_3 states may be spontaneously created, depending on the position of the Fermi level. This forms the foundations of the *a*-Si:H defect-pool model [113,114]. Such defect formation, and its relevance to surface passivation will be discussed in section 7.3.2. In the current section, the description is limited to the electronic nature of the T_3 state.

In amorphous semiconductors, there is often a strong tendency for electrons to be paired in bonding configurations, where the Coulomb repulsion at the same site is outweighed by a negative term in the energy due to strong electron-phonon (polaron) coupling which leads to configuration changes (*i.e.* relaxation) in the local atomic structure. The energy required to add a second electron to the DB orbital is expressed by the energetic difference between the +/0 and 0/- transition levels, and is called the correlation or Hubbard energy U [115]. This energy is negligible for extended states, but for localized states it may amount to a few tenths of an eV [116] and explains thus why *two* levels (separated by U in the bandgap) may be associated with *one* amphoteric defect. For most amorphous semiconductors, this energy is expected to be negative (exothermic reaction), with the consequence that the gap is essentially free of one-electron states [117]. For *a*-Si:H, however, $U = \varepsilon(0/-) - \varepsilon(+/0) \approx +0.38$ eV [118]. Hence, for this material it costs energy (endothermic reaction) to place a second electron in the T_3 orbital. Defects with respectively a positive and negative U are depicted in Fig. 7.5. It is worth noting that it is thanks to T_3's positive U that *a*-Si:H can be intentionally doped relatively well; negative-U defects pin E_F [119,120].

Each of the two energetic peaks associated with T_3 may be further broadened by disorder (such as bond-angle fluctuations) [109]. Lattice stiffness may make the reaction less exothermic due to steric hindrance from the surrounding network, and hence reduce its (negative) contribution to the Hubbard energy [108]. Note that the surface-state associated correlation energy has its relevance for precise interface recombination calculations too (see section 7.2.3.1).

7.3.1.2 Chemical Interface Passivation

For amorphous silicon films, the incorporation of hydrogen passivates bulk DBs, which removes gap states associated with them [121,122,123]. For the electronic properties of *c*-Si surfaces, hydrogen is known to be as benign. This is evidenced by the surfaces with the lowest *in-situ* measured value for S_{eff} ever reported [9], which was accomplished by HF etching. Such etching yields complete Si-H surface termination [60]. As a consequence, hydrogen can be expected to play an important role for the *a*-Si:H(*i*)/*c*-Si interface passivation as well.

For bulk *a*-Si:H films, low-temperature post-deposition annealing treatments are well-known to be beneficial for defect reduction [124,125]. To study the microscopic passivation mechanism of *a*-Si:H films deposited on *c*-Si surfaces, such treatments were proven to be a valuable tool as well: in a straightforward way they give a single experimental parameter to vary both electronic and structural properties of the studied samples. If the *a*-Si:H(*i*)/*c*-Si interface is atomically sharp, such treatment was seen to improve its electronic interface passivation further [94]. Depending on the deposition conditions, these changes can be quite drastic as shown *e.g.* in Fig. 7.6. This figure depicts the case of a FZ-Si(100) wafer that was bifacially passivated with relatively thick (~50nm) *a*-Si:H(*i*) films, deposited at a rather low temperature (130 °C). The shown data are snapshots over time during low-temperature isothermal annealing at 180 °C in air. For as-deposited films, the carrier lifetime (evaluated at an excess carrier density of 1.0×10^{15} cm^{-3}) is only about 30 μs. However, after long annealing this value exceeds 4 ms. The solid lines represent fits of the (simplified) DB recombination formalism [99] to the data. The uppermost solid curve represents the maximum bulk lifetime based on the Auger recombination parameterization by Kerr and Cuevas [90].

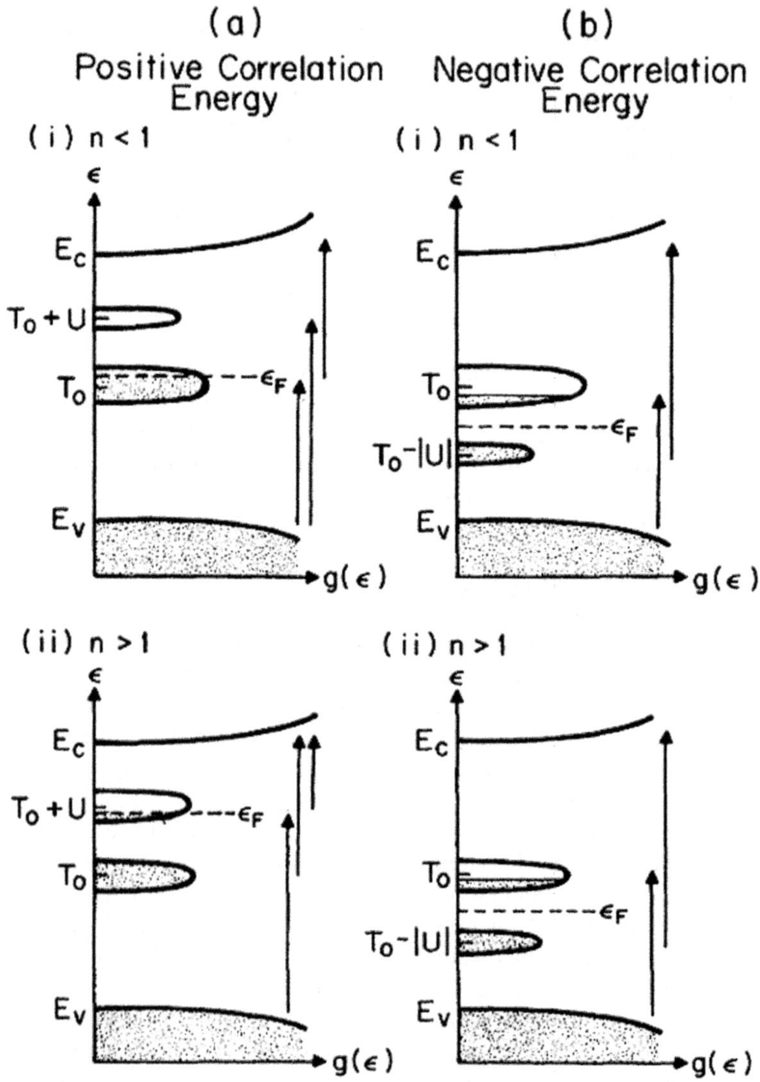

Fig. 7.5 Density of states for a semiconductor with localized states in the gap, relating to a single defect T_3^0: (a) positive correlation energy, *U*; (b) negative correlation energy, -*U*. The average number of electrons per defect is $n = N/N_0$, with N_0 the number of defects and *N* the total number of electrons associated with the defect state. Reprinted from [120]. © 1976 The American Physical Society.

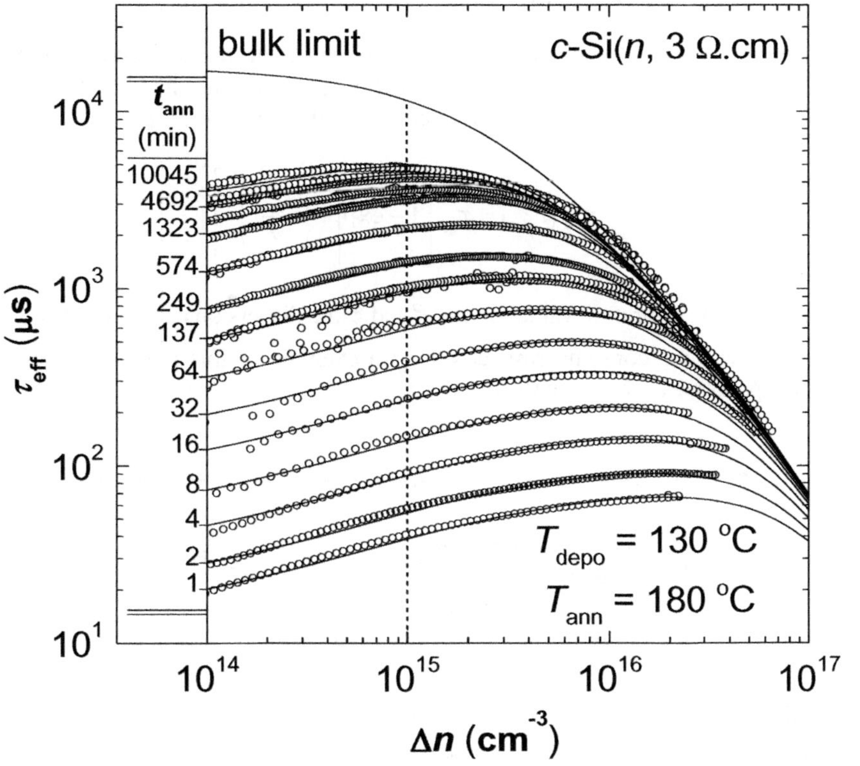

Fig. 7.6. Measured values for τ_{eff} as function of the carrier injection level for films deposited at 130 °C. The different curves show data after different annealing times, t_{ann}. The annealing temperature was fixed at 180 °C. Symbols represent measured data, whereas solid lines show calculated recombination-model fits. The uppermost curve shows the bulk-limited value for τ_{eff}. Reprinted with permission from [99]. © 2008 American Institute of Physics.

Interestingly, the only parameter that needs to be changed to yield a good fit for all cases is the density of interface states, D_{it}. The fixed-charge density, Q_{f}, can be kept constant at a value of -2.2×10^{10} cm^{-2}. As discussed earlier, at low injection all curves remain parallel to each other due to the linear dependence of the recombination rate on D_{it} (see also Fig. 7.4). Physically, the fact that Q_{f} does not change by annealing implies that the surface potential does not change either. Consequently the outstanding passivation obtained by such treatment is not caused by a field effect. Chemical passivation of surface DBs by hydrogen is a more likely explanation for the observed improvement. For ultra thin *a*-Si:H(*i*) films (*i.e.* only a few nm), typically higher DB densities are measured, compared to their thicker counterparts [126]. This may explain why often, when using such thin films, the *a*-Si:H(*i*)/*c*-Si passivation quality is slightly inferior compared to results obtained with thicker (50 nm) films [99].

7.3.1.3 Passivation Kinetics

Closer inspection of the changes in measured value for τ_{eff} (at a chosen injection level, *e.g.* $\Delta n = 1.0\times10^{15}$ cm^{-3}) over annealing time reveals that the data can be well fitted by so-called stretched exponentials of the form [96,99]:

$$\tau_{eff}\left(t_{ann}\right) = \tau_{eff}^{SS}\left[1-\exp\left[-\left(\frac{t_{ann}}{\tau}\right)^{\beta}\right]\right], \tag{7.9}$$

where τ_{eff}^{SS} is the saturation value for τ_{eff}, β and τ the dispersion parameter ($0 < \beta < 1$) and effective time constant, respectively. An example of this is given in Fig. 7.7, with the parameters shown in the inset.

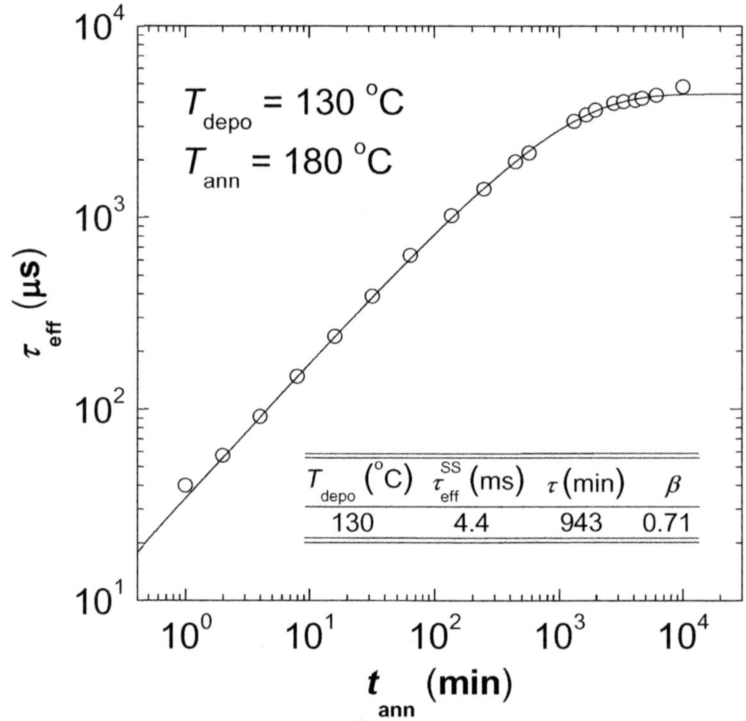

Fig. 7.7 Measured values for τ_{eff} as function of annealing time, t_{ann}, for an *n*-type ~3.0 Ω.cm FZ-Si wafer bifacially passivated by ~50 nm *a*-Si:H(*i*) films. Evaluation was performed at Δn=1.0×10^{15} cm^{-3}. The annealing temperature was fixed at 180 °C. Symbols represent measured data. The solid line represents a stretched-exponential fit to the data. Values for the fitting parameters are given in the inset table. Reprinted with permission from [99]. © 2008 American Institute of Physics.

Stretched-exponential decay is a characteristic phenomenon that often describes the relaxation of disordered systems towards equilibrium [127,128,129]. Mostly, it is used in fitting the dynamical behavior of glasses [130,131], although it may be well applicable to much more diverse phenomena in nature and society [132,133]. For bulk *a*-Si:H, relaxation is often governed by release of atomic hydrogen from trap sites [134]. Dispersive (*i.e.* time-dependent) diffusion of hydrogen, arising from a distribution of energies for trap states and barrier heights, was argued to yield the observed stretched-exponential relaxation [134]. An alternative explanation for such decay was given by Van de Walle [135]. In his interpretation, hydrogen is assumed to be able to reside in three different configurations: either it is in a *trap* state (at relatively high energy), or in a *reservoir* state (at lower energy). The third possibility is that it is in an interstitial state, in transition between a trap and a reservoir state. A schematic representation of this model is given in Fig. 7.8. This model, considering the fact that interstitial hydrogen may also be re-trapped at the same or at another trap site, was shown to yield a functional form behaving much like the stretched-exponential [135]. Practically, the hydrogen reservoir could be assumed to exist in a monohydride form as hydrogenated DBs. This is the hydrogen configuration that can be electronically probed, *e.g.* by a decreased electron spin-resonance signal for annealed as compared to as-deposited samples. Conversely, the trapped hydrogen can be speculated rather to be present in the form of a higher silicon-hydride configuration. This interpretation of stretched-exponential behavior is also quite valuable to understand the annealing-induced passivation changes of the *a*-Si:H(*i*)/*c*-Si interface, as outlined below.

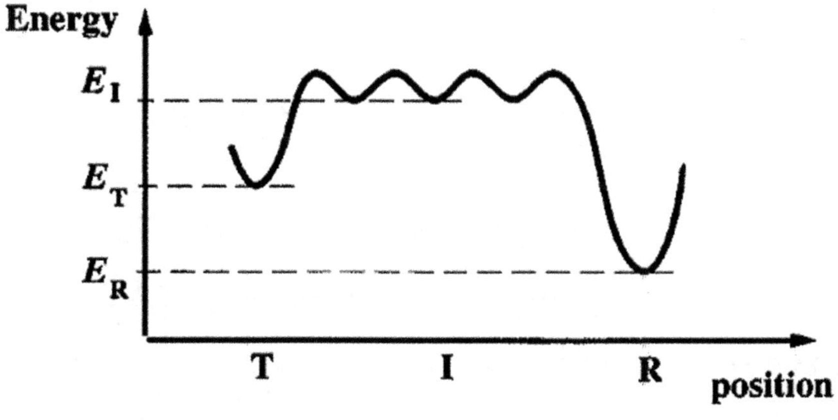

Fig. 7.8 Schematic diagram depicting energy as a function of position of hydrogen in the material. As explained in the text, *R* refers to the ground state (''reservoir''), *T* to the trap state, and *I* to the interstitial hydrogen which diffuses through the crystal. Reprinted with permission from [135]. © 1996 The American Physical Society.

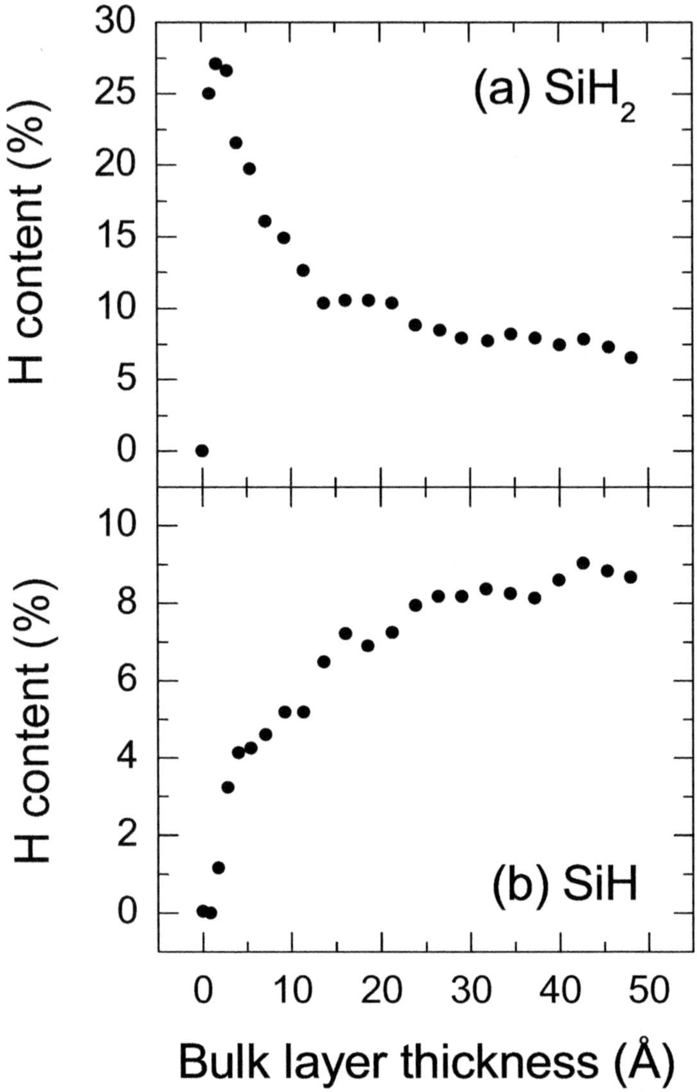

Fig. 7.9 Depth profiles of (a) SiH_2 and (b) SiH hydrogen content in as-deposited *a*-Si:H(*i*) layers formed on a *c*-Si substrate measured by real-time spectroscopic ellipsometry. Reprinted with permission from [95]. © 2005 American Institute of Physics.

For as-deposited *a*-Si:H(*i*) material close to the *c*-Si interface, the hydride modes at higher (lower) stretching frequencies are more (less) dominant than several nanometer into the *a*-Si:H film. This was verified by real-time ATR-FTIR spectroscopy [136]. Such data are represented in Fig. 7.9, with lower stretching frequencies corresponding to mono-hydride configurations [137], and higher stretching frequencies corresponding to di-hydride configurations in the bulk [138], and possibly clustered SiH in microvoids [139]. The presence of a relatively higher density of higher hydrides closer to the interface was confirmed also by H_2 effusion measurements [140]. In terms of the model in Fig. 8, this suggests the presence of a sufficient amount of trapped hydrogen close to the interface of which a fraction may be easily excited to a mobile state by annealing. Such mobile hydrogen may then subsequently be transferred to a reservoir state, in this case a c-Si monohydride state. The latter phenomenon has been studied by differential FTIR measurements on very thin a-Si:H films at the Brewster angle [141]; these measurements suggest that annealing of a-Si:H/c-Si structures yields indeed a transfer of hydrogen to a mono-hydride c-Si surface (reservoir) state. This was seen by a net-increase of the signal around 2075-2080 cm^{-1} [141,96]. For bare c-Si, the mode at 2083.7 cm^{-1} is characteristic of the ideally terminated (111) terraces [48]. The difference between these values may be attributed to their difference in dielectric surrounding: the hydride stretching mode appears at relatively higher/lower frequencies when the surrounding medium's dielectric constant is reduced/increased [142]. As such, this value could thus indeed be interpreted as being a signature for mono-hydride termination of the c-Si surface. This makes a direct link for the a-Si:H(i)/c-Si interface between the electronic passivation and hydrogen termination of the surface.

7.3.1.4 Influence of Epitaxial Growth

A necessary condition for the improvement of the intrinsic a-Si:H(i)/c-Si interface passivation under annealing is that the interface is atomically sharp. In other words: no epitaxial film should be present at the interface. This link between the phase of the interface and its electronic properties is shown in Fig. 7.10. In the graph, the upper panel shows passivation data for films before and after annealing, whereas the lower panel shows spectroscopic ellipsometry (SE) data for similar films in their initial deposition stages. It can be observed that the transition deposition temperature, where annealing starts to lead to passivation losses, corresponds well with the deposition temperature where very thin films are not distinguishable from the c-Si by SE. The latter indicates the occurrence of epitaxial growth. A typical example of such growth is shown in Fig. 7.11.

Epitaxial material grown at such low temperatures is known to be defective at the interface, but also in its bulk. On the one hand, clustered H-related defects such as H-platelets [143] may be present with increasing density for decreasing T_{depo} at the interface [144]. On the other hand, for T_{depo} < 550 °C, breakdown of the epi-Si into an amorphous phase often occurs after a certain thickness of epi-Si growth, h_{epi}, which depends on the precise deposition conditions [145]. At the brink of epi-breakdown, these films are usually very defective [146]. At breakdown itself, a mixed-phase transition region consisting of a-Si:H cones embedded in a c-Si matrix exists [147]. Even for h_{epi} = 0 nm, a mixed-phase layer may exist [148]. The detrimental effect of low-temperature annealing on the passivation quality is likely closely related to the low hydrogen content of the interfacial epi-Si layer. Indeed, epi-Si contains about 30-100 times less hydrogen than a-Si:H [144]. During annealing, hydrogen is exited from higher hydride states, but likely a fraction of interface states may become dehydrogenated too. In case of the *a*-Si:H/*c*-Si interface, a sufficiently large hydrogen source is present in the *a*-Si:H film (in the form of higher hydrides) to guarantee re-passivation of these states. For an epi-Si interface however the hydrogen source is likely too small for such re-passivation.

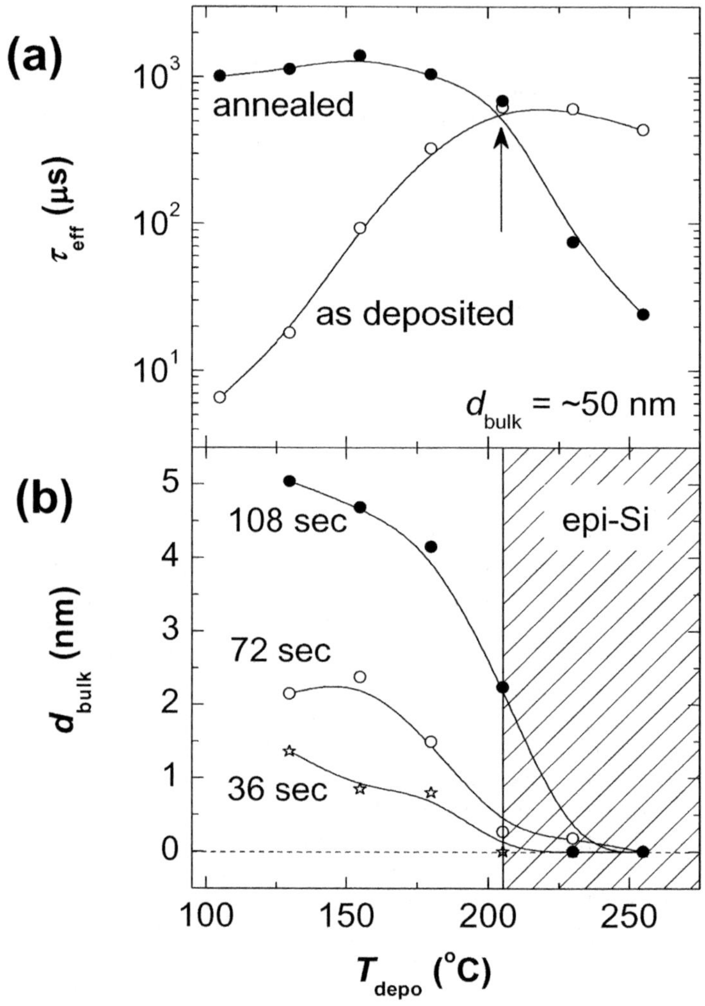

Fig. 7.10 Surface passivation quality of *a*-Si:H films as function of T_{depo}. All films are about 50 nm thick. Results before (open symbols) and after a low-temperature (up to 260 °C) annealing cycle (closed symbols) are shown. (b) Film thickness d_{bulk} at the initial deposition stages, obtained from fitting of SE measurements as function of T_{depo}, given for several values of t_{depo}. All lines are guides for the eye. Reprinted with permission from [94]. © 2007 American Institute of Physics.

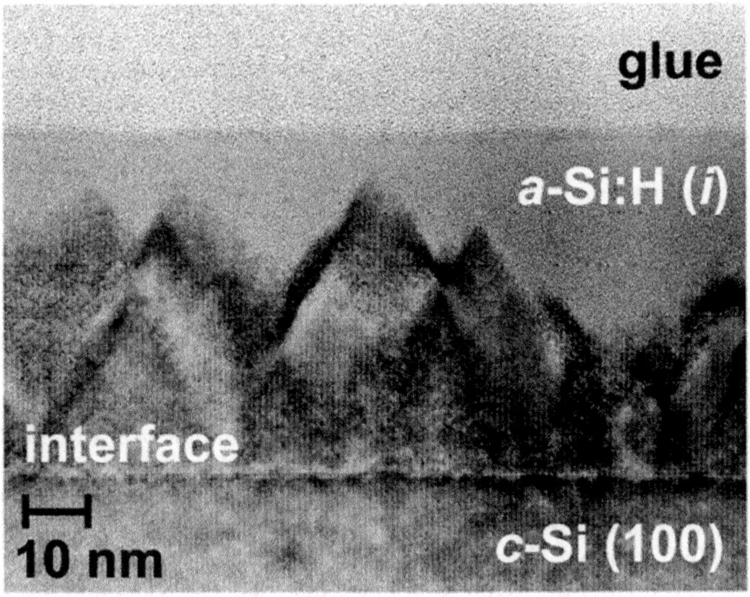

Fig. 7.11 High-resolution transmission electron microscopy image of the *a*-Si:H/*c*-Si interface, with the film deposited at 230 °C. as indicated in Fig. 7.10. The PECVD *a*-Si:H(*i*) film was deposited on a mirror polished *c*-Si(100) surface. Reprinted with permission from [94]. © 2007 American Institute of Physics.

Fundamentally, three parameters play a role in epitaxial growth: the precise nature of the wafer-surface morphology and its chemistry, the deposition temperature, and the silane-depletion fraction in the PECVD plasma. Firstly, it is well-established that epitaxial growth is initiated more easily on (100) than on (111) surfaces [149,141]. Buffered HF-terminated Si(100) surfaces may also lead more easily to such undesired epitaxial growth, compared to HF-terminated surfaces [101]. Both these phenomena may be closely related to the microscopic nature of the surface [140]. As an example, for low-temperature epitaxial growth, strings of Si(100)-(2×1) dimers are known to play an essential role [150,151]. Epitaxial growth occurs then as long dimer strings, perpendicular to the present surface strings. Such dimers are unique to the Si(100) surface. Moreover, their presence on wet-chemically etched surfaces may critically depend on whether NH_4F or HF solutions were used [71]. Secondly, the deposition temperature critically determines the mean free path length of the adatoms on the surface during film growth, and thus the microstructure. Finally, the transition from amorphous to epitaxial silicon can be linked fundamentally to the silane-depletion fraction in the plasma, rather than to any other parameter such as film deposition rate or the occurrence of ion bombardment [152]. Practically, the silane-depletion fraction can be monitored in a straight-forward way by *in-situ* infrared spectroscopy of the plasma [153], using *e.g.* a quantum-cascade laser for measurements with unprecedented resolution [154]. Notably, silane-depletion measurements have provided additional useful insights, including the origin of higher deposition rates when using VHF

[155], the determination of the transition region between *a*-Si:H and *μc*-Si:H depositions [156], the optimization of reactor configurations for fast equilibration at ignition [157], and the link between high-quality interface passivation and highly depleted plasmas, close to the microcrystalline silicon (*μc*-Si) growth regime [98]. The latter observation underlines that even though epi-Si is undesired, the film-growth conditions that guarantee high quality passivation are close to those yielding epi-Si interfaces.

7.3.2. Doped a-Si:H Passivation

For thin-film *p-i-n* *a*-Si:H solar cells, the *a*-Si:H(*p*) layer often was argued to limit device performance. On the one hand, exposure of surfaces simultaneously to B_2H_6 and SiH_4 (before or after PECVD) may give rise to CVD growth of highly defective *a*-SiB_x:H layers, even at very low temperatures, resulting in poor *p-i* interfaces [158,159]. On the other hand, recombination in the *a*-Si:H(p^+) bulk itself was already for the first thin-film *p-i-n* devices recognized to hamper device performance [160]. Doping of such films occurs by incorporation of substitutional impurities [27]. However, such doping may induce additional localized states in the film [161]. Here we discuss how defects in doped layers are critically dependent on the position of the Fermi level in the material. The relation between defect formation and intentional doping turns out to be of a fundamental nature, applying to any type of semiconductor. For the *a*-Si:H/*c*-Si interface, this phenomenon can lead to reduced passivation.

7.3.2.1 Fermi-Level Induced Defect Formation

For *a*-Si:H films, it was observed many years ago that hydrogen is released from doped films at lower temperatures as compared to their intrinsic counterparts. The similarity in behavior of undoped and compensated films indicates that Si-H bond rupture (and thus defect formation) in such films depends on the position of E_F rather than on the physical nature of the present dopants [162,163]. Originally this argument was used to explain doping-dependent hydrogen diffusion phenomena in *a*-Si:H films [164]. The hydrogen diffusion energy E_D^*, defined by $D_H = D_0^* \exp(-E_D^*/kT)$, in *a*-Si:H and *μc*-Si is shown as a function of E_F in Fig. 7.12. The data are reprinted from Ref [165]. The diffusion coefficient, D_H, describes the motion of hydrogen in the silicon matrix, where $D_0^* = 10^{-3}$ cm^{2}s^{-1} is the theoretical diffusion coefficient unaffected by traps [166]. The diffusion activation energy E_D^* equals E_S - μ_H, where E_S is the saddle point (see *e.g.* also Fig. 7.8) for interstitial H migration and μ_H the chemical potential of H atoms [167,168]. These energies are depicted in Fig. 7.13(a), including the quantity $E_{Si\text{-}H}$, which is the energy of a Si-H bond, and E_{BC}, which is the energy of an isolated interstitial H atom at its most stable site, which is taken to be the bond-center site [169]. Figure 7.13 (b) shows the expected broadening of these levels in *a*-Si:H due to the atomic disorder [167]. The occupancy of hydrogen in its different states is determined by the value of μ_H: states with energies above μ_H are mostly empty, those with energies below μ_H are mostly occupied with hydrogen [168]. As observed in Fig. 7.12, the diffusion activation energy for intrinsic material (*i.e.*, where E_F - E_M = ~0 eV) is about 1.5 eV. For doped material it is often lower. An asymmetry with reference to the bandgap can be observed: When E_F is brought closer to the valence-band maximum (VBM, *i.e.* by *p*-type doping) this rapidly results in decreasing values for E_D^* (see region with label (1)). For *n*-type doping, E_F must be brought relatively closer to the conduction-band minimum (CBM) to yield a similar drop (region with label (3)). For *a*-Si:H/*c*-Si interface passivation, doping was observed to yield very similar trends: post-deposition annealing of *p*-type passivation layers yielded losses in passivation at much lower temperatures already, compared to their *n*-type counterparts [170].

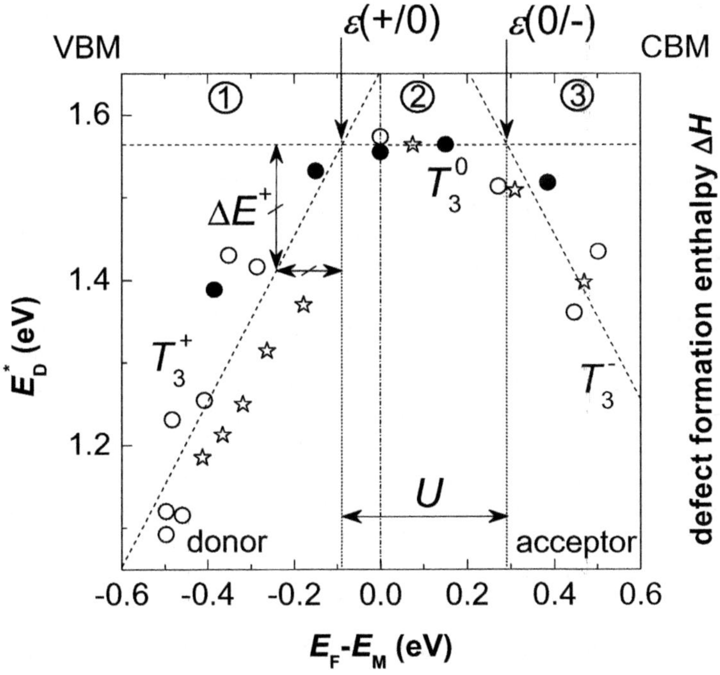

Fig. 7.12 H diffusion energy E_D^* in *a*-Si:H (stars), *μc*-Si:H (open circles) or *c*-Si:H (closed circles) as a function of E_F, at T_{ann} = 350°C. All data are reprinted from Ref.165. The superposed straight lines represent the dependence of the formation enthalpy ΔH of defect α in the respective charge states q = +, 0 and – on E_F (adapted after Ref. 174). E_M represents the middle of the bandgap. Reprinted with permission from [170]. © 2009 American Institute of Physics.

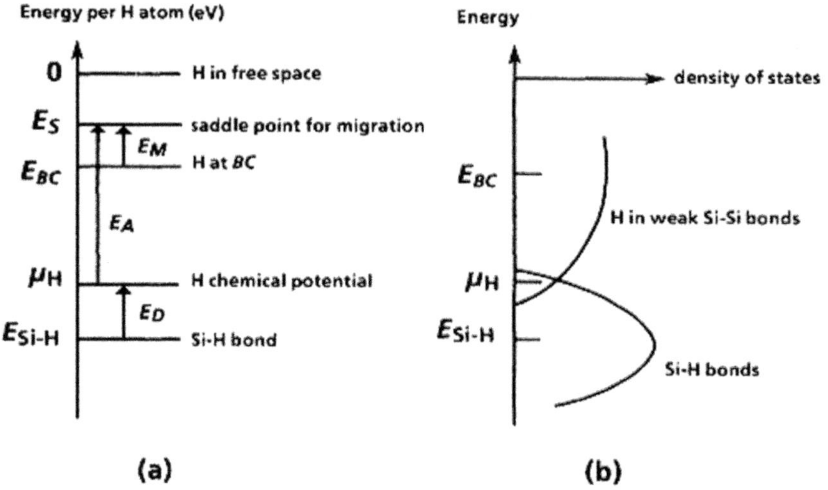

Fig. 7.13 (a) Energies relevant for H interactions with silicon. (b) Schematic hydrogen density of states distribution in *a*-Si:H, corresponding to the level scheme in (a). The broadening of the levels is due to disorder in the amorphous network. Reprinted with permission from [168]. © 1995 The American Physical Society.

Fundamentally, the stability of electrically active defects is known to depend critically on the position of E_F [171]. Moreover, as the bandgap of a semiconductor increases, it often becomes increasingly difficult to dope it in a symmetric way (both *n*- and *p*-type). On the one hand, unintentional conductivity may originate from the presence of extrinsic impurities such as, *e.g.*, is the case for H in ZnO [172,173, and Chapter 9 by Ruske in this book]. Native defects, on the other hand, can act as (self-) compensating centers that counteract intentional doping. Their formation depends on the position of E_F according to the relation [174,175]:

$$\Delta H\left(D^q\right) = qE_F + n_D\left(\mu_D - \mu_{SH}\right) + \Delta E_b\,. \tag{7.10}$$

This relation describes the formation enthalpy of dopant D^q of charge state q (where q = -1, 0 or +1) in the semiconducting host. Here, μ_D and μ_{SH} are the chemical potentials of the dopants and host, n_D the number of dopants, ΔE_b = E(host+defect) – E(host), and E is the total energy. The transition level $\varepsilon(q + 1 / q)$ between charge states q + 1 and q is defined as the position of E_F for which the formation energies of these charge states are equal [176]: Deliberate *p*-type (*n*-type) doping of the material by acceptors (donors) will shift E_F towards the VBM (CBM) and decrease the formation energy of native donors (acceptors) to a point where they are created spontaneously. Usually, E_F can not be brought beyond a certain point, the so-called the *n*- or *p*-type pinning energy $E_{pin}^{(n\ or\ p)}$, due to the occuring electronic compensation. These two processes set an upper and lower bound to the value for E_F [177]:

$$E_{pin}^{(p)} \le E_F \le E_{pin}^{(n)}. \tag{7.11}$$

Practically, such doping-induced Fermi-level pinning by self-compensation forms a powerful phenomenological model that can be used for systemization and prediction of the doping limits in, *e.g.*, II-VI and I-III-VI_2 compound semiconductors [177] and III-nitrides [178].

In *a*-Si:H, it can be speculated that the described (asymmetric) E_F dependent defect formation is related to self-compensation as well. Interestingly, equation (10) can be sketched in Fig. 7.12 by the *ad hoc* superposed straight lines, approximately following the same trends as the experimental data [170]. Here, the most likely formed defect by either type of doping is the amphoteric T_3 defect, *i.e.*, the silicon dangling bond via Si-H rupture [179]. When E_F is located either in the [VBM, $\varepsilon(+/0)$] or [$\varepsilon(0/-)$, CBM] area of the bandgap, T_3 will behave as a compensating center, counteracting the intentional doping. In these cases, the formation energy of a DB, compared to its neutral state, is reduced by an amount respectively equal to $\Delta E^+(E_F) = (\varepsilon(+/0) - E_F)$ and $\Delta E^-(E_F) = (E_F - \varepsilon(0/-))$ [180,181], as indicated in Fig. 7.11. Since dangling bonds are created by breaking of the Si-H bond, the hydrogen diffusion activation energy E_D^* is reduced by similar amounts. The formation energy of T_3^0 is independent of the position of E_F. The asymmetry of the defect-formation enthalpy is closely related to the Hubbard energy U associated with amphoteric defects. As argued earlier, due to Coulomb repulsion, it costs more energy to add a second electron to a dangling bond than was necessary to place the first one. Consequently, to create a negatively charged dangling bond will cost an extra amount of energy U with respect to the formation of the neutral dangling bond.

For the *a*-Si:H/*c*-Si interface the link between doping induced defects and lowered electronic passivation was established by H_2 effusion experiments [182]. Spectroscopic ellipsometry measurements on passivating films similarly indicate that such a link may well exist [170]. Such data are shown in Fig. 7.14. For all three cases (intrinsic, *p*-type and *n*-type doping), the onset of passivation degradation (if present) seems to coincide with a collapse of the optical bandgap. This may point at defect formation in the amorphous host matrix, and was also confirmed by thermal desorption spectroscopy measurements [170]. Interestingly, a similar doping related asymmetry can be observed in this figure as discussed for the T_3 defect. Annealing of *p*-type films leads to degradation of both the passivation quality and the optical bandgap at relatively lower temperatures than for *n*-type films. The decrease of the optical bandgap points at defect formation by Si-H bond rupture, as suggested by H_2 effusion measurements [170]. These arguments lead to the conclusion that the doping asymmetry of *a*-Si:H films, and as a consequence also their difference in *c*-Si surface passivation quality, originate from the asymmetrical location of the T_3^+ and T_3^- states in the *a*-Si:H bandgap, due to the positive value of U for this defect.

In the concept of co-doping, donors are introduced along with acceptors during film growth to overcome the poor doping efficiency of wide-bandgap semiconductors [183]. The argument for doing so is that the value of E_F is then shifted towards midgap, yielding thus reduced compensation center densities. For this approach to work, however, the donors (acceptors) need to be removed from the *p*-type (*n*-type) layer after growth [178], which is obviously not an easy task. In a more straight-forward way, for *p*-type (*n*-type) material, the doping efficiency may be improved by bringing the VBM (CBM) closer to (further away from) the vacuum-level [174]. Fulfilling both conditions simultaneously yields a reduced bandgap material. For *a*-Si:H films, this can be accomplished by optimizing deposition conditions towards a lowered bonded hydrogen content, as the bandgap is mainly determined by the bonded hydrogen content of the films [184]. This corresponds to denser material [185,186]. Reducing the bandgap of the material may also explain why the surface passivation properties of μc-Si(p^+) films (deposited on *a*-Si:H(*i*) buffer layers) appear to be superior, compared to their

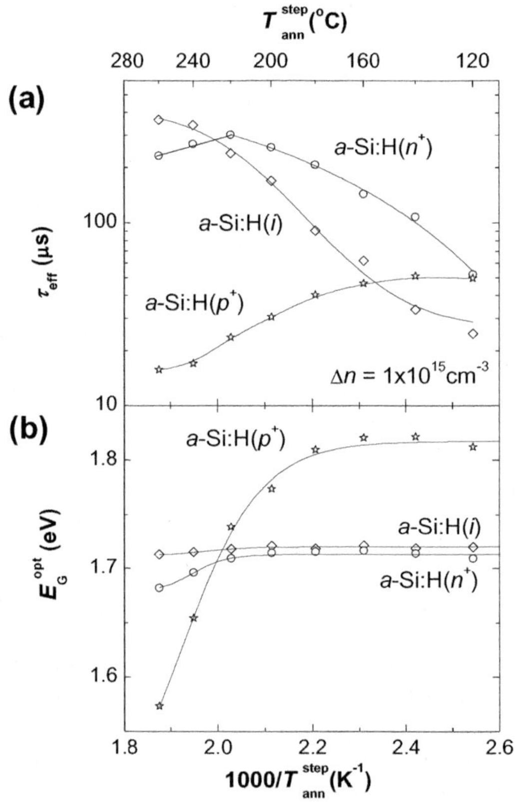

Fig. 7.14 (a) Influence of stepwise annealing on the interface passivation quality of a few nm thin single film *a*-Si:H/*c*-Si, expressed by $\tau_{\rm eff}$ (at $\Delta n = \Delta p = 1.0\times10^{15}$ cm^{-3}) and extracted from QSSPC measurements. No intrinsic buffer layers are present underneath the doped films. (b) Influence of stepwise annealing treatment on the optical bandgap $E_{\rm G}^{\rm opt}$ of the same films as in (a), extracted from SE measurements. Symbols represent experimental data, the lines are guides for the eye. Reprinted with permission from [170]. © 2009 American Institute of Physics.

wider bandgap *a*-Si:H(p^+) counterparts [187,188]. Moreover, the use of μc-Si(p^+) films may also resolve possible contact problems between transparent conductive oxide and the *p*-type film, during device fabrication [188,189]. Care has to be taken to deposit such denser material for the intrinsic buffer layer, however, as it will easily result in undesired epitaxial growth, as argued earlier.

7.3.2.2 Role of the Intrinsic Buffer Layer

In the previous sections it was argued that excellent interface passivation can be obtained by using intrinsic *a*-Si:H(*i*) films featuring atomically sharp interfaces. Doping of these films is needed to create built-in fields for device fabrication, but may lead to important passivation losses. These conflicting issues were first resolved by Sanyo by developing intrinsic/doped *a*-Si:H stacks for emitter and BSF formation with good interface passivation properties [8,33,34]. The intrinsic buffer layer in hetero-junction solar cells is an order or two thinner than the intrinsic absorber layers in thin-film *p*-*i*-*n* *a*-Si:H solar cells [190,191]. Nevertheless, as this layer is also subject to internal fields induced by the doped layers in between which it is sandwiched, similar defect-equilibrium thermodynamics apply to this very thin buffer layer too. The possible defect formation in the intrinsic buffer layer, in case a doped overlayer is present, is the subject of this final section.

For thin-film *p*-*i*-*n* structures, based on the band bending in the device, the energetic position of the Fermi level will deviate from midgap throughout most of the *i*-layer. As noted before, the defect formation in a semiconductor depends on the energetic position of its Fermi level rather than on the physical nature of the dopants. Consequently, internal fields yield a position-dependent change in dangling bond formation enthalpy in the intrinsic absorber layer [192]. As close to the doped interfaces the deviation of E_F from midgap is the largest (at the *p*/*i* interface it is near to the VBM, at the *i*/*n* interface it is near to the CBM), at these locations larger (charged) T_3 densities may be expected, which can be calculated self-consistently [189]. The same Hubbard-energy related asymmetry typical for charged T_3 defects applies in these intrinsic layers. From this, it must be concluded that for a shift of E_F towards the VBM, relatively larger T_3^+ concentrations will be generated compared to the T_3^- concentrations for a similar shift towards CBM (see also Fig. 7.12). For *a*-Si:H(p^+)/*a*-Si:H(*i*)/*c*-Si structures with nanometer thin buffer layers, these findings were confirmed experimentally by effusion of hydrogen out of the buffer layer at relatively lower temperatures, compared to the case without doped overlayer [182]. This phenomenon points to the mediating role the doped layer plays in defect formation in an intrinsic film underneath. With these results, the relatively poor passivation of *a*-Si:H(p^+)/*a*-Si:H(*i*) stacks was explained as well, linking thus Si-H bond rupture to the electronic properties of the interface [31,182]. No enhanced defect formation was observed for an *n*-type overlayer [170], likely because the shift of E_F in the buffer layer towards CBM was insufficient to overcome the energy barrier U (see Fig. 7.11). Likewise, *i*/*n* stacks showed at least as good passivation as intrinsic layers without buffer layer [170]. These results suggest that the use of a buffer layer is indeed beneficial, but even then care has to be taken in the preparation of the doped films (especially for the *p*-type case) to assure good interface passivation.

7.4 Conclusions

In this chapter, a comprehensive overview was given of factors that determine high-quality *a*-Si:H/*c*-Si interface passivation. Hydrogen was seen to play an as crucial role for *c*-Si surface termination as it does for amorphous silicon bulk defect passivation. Annealing kinetics confirm that the passivation of the interfaces is achieved chemically, by mono-hydride termination of dangling bonds. The electronic passivation thus obtained is on par with the best dielectric films. However, it is critical to suppress the growth of low-quality epitaxial material during film deposition. Next, doping of the amorphous films may result in Fermi-level dependent defect formation in the layer, detrimentally affecting its passivation properties. This phenomenon was placed in the broader framework of compensating defect formation in semiconductors. Specifically for the *a*-Si:H case, due to the positive dangling-bond correlation energy, one consequence is that *p*-type films passivate worse than their n-type counterparts. To a certain degree, decoupling of passivating and doping properties is possible by using stacked films, consisting of an intrinsic buffer layer and a doped overlayer. Even so, *p*-type films may still generate defects in the intrinsic buffer layers underneath. For optimal device performance, a careful assessment of the deposition conditions of such films is thus warranted.

Acknowledgements

The author wishes to thank Christophe Ballif, Richard Bartlome, Corsin Battaglia, Harold Dekkers, Bénédicte Demaurex, Antoine Descoeudres, Franz-Josef Haug and Michio Kondo for useful discussions and support. This work was partially funded by Axpo AG, Switzerland in the frame of the Axpo Naturstrom Fonds.

References

[1] Sandroff, C.J., Nottenburg, R.N., Bischoff, J.C., Bhat, R.: Dramatic enhancement in the gain of a GaAs/AlGaAs heterostructure bipolar transistor by surface chemical passivation. Appl. Phys. Lett. 51, 33 (1987)
[2] Green, B.M., Chu, K.K., Chumbes, E.M., Smart, J.A., Shearly, J.R., Eastman, L.F.: The effect of surface passivation on the microwave characteristics of undoped AlGaN/GaN HEMTs. IEEE Electron. Dev. Lett. 21, 268 (2000)
[3] Vetury, R., Zhang, N.Q.Q., Keller, S., Mishra, U.K.: The impact of surface states on the DC and RF characteristics of AlGaN/GaN HFETs. IEEE Trans. Electron. Dev. 48, 560 (2001)
[4] Chui, C.O., Ramanathan, S., Triplett, B.B., McIntyre, P.C., Saraswat, K.C.: Germanium MOS capacitors incorporating ultrathin high-κ gate dielectric. IEEE Electron. Dev. Lett. 23, 473 (2002)
[5] Lauhon, L.J., Gudiksen, M.S., Wang, D., Lieber, C.M.: Epitaxial core–shell and core–multishell nanowire heterostructures. Nature 420, 57 (2002)
[6] Sinton, R.A., Kwark, Y., Gan, J.Y., Swanson, R.M.: 27.5-percent silicon concentrator solar cells. IEEE Electron. Dev. Lett. 7, 567 (1986)

[7] Zhao, J.H., Wang, A.H., Green, M.A.: 24.5% efficiency silicon PERT cells on MCZ sub-strates and 24.7% efficiency PERL cells on FZ substrates. Prog. Photovoltaics 7, 471 (1999)
[8] Taguchi, M., Kawamoto, K., Tsuge, S., Baba, T., Sakata, H., Morizane, M., Uchihashi, K., Nakamura, N., Kiyama, S., Oota, O.: HITTM cells – High-efficiency crystalline Si cells with novel structure. Prog. Photovoltaics 8, 503 (2000)
[9] Yablonovitch, E., Allara, D.L., Chang, C.C., Gmitter, T., Bright, T.B.: Unusually low surface-recombination velocity on silicon and germanium surfaces. Phys. Rev. Lett. 57, 249 (1986)
[10] Nicollian, E.H., Goetzberger, A.: The Si-Si0$_2$ interface electrical properties as determined by the MIS conductance technique. Bell Syst. Tech. J. 46, 1055 (1967)
[11] Ghandhi, S.K.: VLSI fabrication principles. John Wiley & Sons, New York (1982)
[12] Green, M.A.: The path to 25% silicon solar cell efficiency: History of silicon cell evolution. Prog. Photovoltaics 17, 183 (2009)
[13] Green, M.A., Zhao, J., Wang, A., Reece, P.J., Gal, M.: Efficient silicon light-emitting di-odes. Nature 412, 805 (2001)
[14] Swanson, R.M.: Photovoltaics power up. Science 324, 891 (2009)
[15] Kingon, A.I., Maria, J.P., Streiffer, S.K.: Alternative dielectrics to silicon dioxide for memory and logic devices. Nature 406, 1032 (2000)
[16] Aberle, A.G., Hezel, R.: Progress in low-temperature surface passivation of silicon solar cells using remote-plasma silicon nitride. Prog. Photovoltaics 5, 29 (1997)
[17] Deal, B.E., Grove, A.S.: General relationship for the thermal oxidation of silicon. J. Appl. Phys. 36, 3770 (1965)
[18] Schultz, O., Glunz, S.W., Willeke, G.P.: Multicrystalline silicon solar cells exceeding 20% efficiency. Prog. Photovoltaics 12, 553 (2004)
[19] Lanford, W.A., Rand, M.J.: The hydrogen content of plasma-deposited silicon nitride. J. Appl. Phys. 49, 2473 (1978)
[20] Lauinger, T., Schmidt, J., Aberle, A.G., Hezel, R.: Record low surface recombination velocities on 1 Ω.cm p-silicon using remote plasma silicon nitride passivation. Appl. Phys. Lett. 86, 1232 (1996)
[21] Chen, Z., Sana, P., Salami, J., Rohatgi, A.: A novel and effective PECVD SiO2/SiN antireflection coating for Si solar cells. IEEE Trans. Electron Devices 40, 1161 (1993)
[22] Hezel, R., Jaeger, K.: Low-temperature surface passivation of silicon for solar cells. J. Electrochem Soc. 136, 518 (1989)
[23] Agostinelli, G., Delabie, A., Vitanov, P., Alexieva, Z., Dekkers, H.F.W., De Wolf, S., Beaucarne, G.: Very low surface recombination velocities on p-type silicon wafers passivated with a dielectric with fixed negative charge. Sol. Energy Mater. Sol. Cells 90, 3438 (2006)
[24] Hoex, B., Heil, S.B.S., Langereis, E., van de Sanden, M.C.M., Kessels, W.M.M.: Ultralow surface recombination of c-Si substrates passivated by plasma-assisted atomic layer deposited Al_2O_3. Appl. Phys. Lett. 89, 042112 (2006)
[25] Yablonovitch, E., Gmitter, T., Swanson, R.M., Kwark, Y.H.: A 720 mV open circuit voltage SiO_x:c-Si:SiO_x double heterostructure solar cell. Appl. Phys. Lett. 47, 1211 (1986)
[26] Pankove, J.I., Tarng, M.L.: Amorphous silicon as a passivant for crystalline silicon. Appl. Phys. Lett. 34, 156 (1979)
[27] Spear, W.E., LeComber, P.G.: Substitutional doping of amorphous silicon. Solid State Comm. 17, 1193 (1975)

[28] Matsuura, H., Okuno, T., Okushi, H., Tanaka, K.: Electrical properties of n-amorphous/p-crystalline silicon heterojunctions. J. Appl. Phys. 55, 1012 (1984)

[29] Fossum, J.G.: Physical operation of back-surface-field silicon solar cells. IEEE Trans. Electron. Devices 24, 322 (1977)

[30] Taguchi, M., Terakawa, A., Maruyama, E., Tanaka, M.: Obtaining a higher V_{oc} in HIT cells. Prog. Photovoltaics 13, 481 (2005)

[31] De Wolf, S., Beaucarne, G.: Surface passivation properties of boron-doped plasma-enhanced chemical vapor deposited hydrogenated amorphous silicon films on p-type crystal-line Si substrates. Appl. Phys. Lett. 88, 022104 (2006)

[32] Korte, L., Schmidt, M.: Investigation of gap states in phosphorous-doped ultra-thin a-Si:H by near-UV photoelectron spectroscopy. J. Non-Cryst. Solids 354, 2138 (2008)

[33] Tanaka, M., Taguchi, M., Matsuyama, T., Sawada, T., Tsuda, S., Nakano, S., Hanafusa, H., Kuwano, Y.: Development of new a-Si/c-Si heterojunction solar cells: ACJ-HIT (artificially constructed junction-heterojunction with intrinsic thin-layer). Jpn. J. Appl. Phys. 31, 3518 (1992)

[34] Tsunomura, Y., Yoshimine, Y., Taguchi, M., Baba, T., Kinoshita, T., Kanno, H., Sakata, H., Maruyama, E., Tanaka, M.: Twenty-two percent efficiency HIT solar cell. Sol. Energy Mater. Sol. Cells 93, 670 (2009)

[35] Green, M.A., Emery, K., Hishikawa, Y., Warta, W.: Solar cell efficiency tables (version 36). Prog. Photovoltaics 18, 346 (2010)

[36] Zangwill, A.: Physics at surfaces. Cambridge University Press, Cambridge (1988)

[37] Tamm, I.: Uber eine mögliche art der elektronenbindung an kristalloberflächen. Phys. Z. Sowjetunion 1, 733 (1932)

[38] Shockley, W.: On the surface states associated with a periodic potential. Phys. Rev. 56, 317 (1939)

[39] Lüth, H.: Solid surfaces, interfaces and thin films. Springer, Berlin (2001)

[40] Pointdexter, E.H., Caplan, P.J., Deal, B.E., Razouk, R.R.: Interface states and electron spin resonance centers in thermally oxidized (111) and (100) silicon wafers. J. Appl. Phys. 52, 879 (1981)

[41] Bean, K.E.: Anisotropic etching of silicon. IEEE Trans. Electron. Dev. 25, 1185 (1978)

[42] Seidel, H., Csepregi, L., Heuberger, A., Baumgärtel, H.: Anisotropic etching of crystalline silicon in alkaline solutions. J. Electrochem. Soc. 137, 3612 (1990)

[43] Yablonovitch, E., Cody, G.D.: Intensity enhancement in textured optical sheets for solar cells. IEEE Trans. Electron. Dev. 29, 300 (1982)

[44] Campbell, P., Green, M.A.: Light trapping properties of pyramidally textured surfaces. J. Appl. Phys. 62, 243 (1987)

[45] Adler, D.: Density of states in the gap of tetrahedrally bonded amorphous semiconductors. Phys. Rev. Lett. 41, 1755 (1978)

[46] Bechstedt, F.: Principles of surface physics. Springer, Berlin (2003)

[47] Binnig, G., Rohrer, H., Gerber, C., Weibel, W.: 7×7 reconstruction on Si(111) resolved in real space. Phys. Rev. Lett. 50, 120 (1983)

[48] Higashi, G.S., Chabal, Y.J., Trucks, G.W., Ragavachari, K.: Ideal hydrogen termination of the Si (111) surface. Appl. Phys. Lett. 56, 656 (1990)

[49] Hricovini, K., Günther, R., Thiry, P., Taleb-Ibrahimi, A., Indlekofer, G., Bonnet, J.E., Dumas, P., Petroff, Y., Blase, X., Zhu, X., Louie, S.G., Chabal, Y.J., Thiry, P.: Electronic structure and its dependence on local order for H/Si(111)-(1×1) surfaces. Phys. Rev. Lett. 70, 1992 (1993)

[50] Appelbaum, J.A., Baraff, G.A., Hamann, D.R.: The Si(100) surface. III. Surface reconstruction. Phys. Rev. 14, 588 (1976)
[51] D'Evelyn, M.P., Yang, Y.L., Sutcu, L.F.: π-bonded dimers, preferential pairing, and first-order desorption kinetics of hydrogen on Si(100)-(2x1). J. Chem. Phys. 96, 852 (1992)
[52] Wolkin, M.V., Jorne, J., Fauchet, P.M., Allan, G., Delerue, C.: Electronic states and luminescence in porous silicon quantum dots: the role of oxygen. Phys. Rev. Lett. 82, 197 (1999)
[53] Grundner, M., Jacob, H.: Investigations on hydrophilic and hydrophobic silicon (100) wa-fer surfaces by X-ray photoelectron and high-resolution electron energy loss-spectroscopy. Appl. Phys. A: Solids Surf. 39, 73 (1986)
[54] Kern, W., Puotinen, D.A.: Cleaning solutions based on hydrogen peroxide for use in silicon semiconductor technology. RCA Rev. 31, 187 (1970)
[55] Fenner, D.B., Biegelsen, D.K., Bringans, R.D.: Silicon surface passivation by hydrogen termination: A comparative study of preparation methods. J. Appl. Phys. 66, 419 (1989)
[56] Weinberger, B.R., Peterson, G.G., Eschrich, T.C., Krasinski, H.A.: Surface chemistry of HF passivated silicon: X-ray photoelectron and ion scattering spectroscopy results. J. Appl. Phys. 60, 3232 (1986)
[57] Judge, J.S.: A study of the dissolution of SiO_2 in acidic fluoride solutions. J. Electrochem. Soc. 118, 1772 (1971)
[58] Trucks, G.W., Raghavachari, K., Higashi, G.S., Chabal, Y.J.: Mechanism of HF etching of silicon surfaces: A theoretical understanding of hydrogen passivation. Phys. Rev. Lett. 65, 504 (1990)
[59] Ubara, H., Imura, T., Hiraki, A.: Formation of Si-H bonds on the surface of microcrystal-line silicon covered with SiOx by HF treatment. Solid State Commun. 50, 673 (1984)
[60] Burrows, V.A., Chabal, Y.J., Higashi, G.S., Raghavachari, K., Christman, S.B.: Infrared spectroscopy of Si(111) surfaces after HF treatment: Hydrogen termination and surface morphology. Appl. Phys. Lett. 53, 998 (1988)
[61] Chabal, Y.J.: Surface infrared spectroscopy. Surf. Sci. Reports 8, 211 (1988)
[62] Schluter, M., Cohen, M.L.: Nature of conduction-band surface resonances for Si(111) surfaces with and without chemisorbed overlayers. Phys. Rev. B 17, 716 (1977)
[63] Becker, R.S., Higashi, G.S., Chabal, Y.J., Becker, A.J.: Atomic-scale conversion of clean Si-(111):H-1×1 to Si(111)-2×1 by electron-stimulated desorption. Phys. Rev. Lett. 65, 1917 (1990)
[64] Chabal, Y.J., Higashi, G.S., Raghavachari, K., Burrows, V.A.: Infrared spectroscopy of Si(111) and Si(100) surfaces after HF treatment: Hydrogen termination and surface morphology. J. Vac. Sci. Technol. A 7, 2104 (1989)
[65] Boland, J.J.: Structure of the H-saturated Si(100) surface. Phys. Rev. Lett. 65, 3325 (1990)
[66] Flowers, M.C., Jonathan, N.B.H., Liu, Y., Morris, A.: Temperature programmed desorption of molecular hydrogen from a Si(111) surface: Theory and experiment. J. Chem. Phys. 99, 7038 (1993)
[67] Gupta, P., Colvin, V.L., George, S.M.: Hydrogen desorption kinetics from monohydride and dihydride species on silicon surfaces. Phys. Rev. B 37, 8234 (1988)
[68] Ciraci, S., Batra, I.P.: Theory of transition from the dihydride to the monohydride phase on the Si(001) surface. Surf. Sci. 178, 80 (1986)

[69] Wise, M.L., Koehler, B.G., Gupta, P., Coon, P.A., George, S.M.: Comparison of hydrogen desorption kinetics from Si (111) 7×7 and Si (100) 2×1. Surf. Sci. 258, 166 (1991)
[70] Chabal, Y.J., Raghavachari, K.: New ordered structure for the H-saturated Si(100) surface: the (3×1) phase. Phys. Rev. Lett. 54, 1055 (1985)
[71] Neuwald, U., Hessel, H.E., Feltz, A., Memmert, U., Behm, R.J.: Wet chemical etching of Si (100) surfaces in concentrated NH_4F solution: formation of (2×1)H reconstructed Si(100) terraces versus (111) faceting. Surf. Sci. Lett. 296, 8 (1993)
[72] Le Tanh, V., Bouchier, D., Hincelin, G.: Low-temperature formation of Si(001)2x1 surfaces from wet chemical cleaning in NH_4F solution. J. Appl. Phys. 87, 3700 (2000)
[73] McKelvey, J.P.: Solid state and semiconductor physics. Harper & Row, New York (1966)
[74] Shockley, W., Read, W.T.: Statistics of the recombinations of holes and electrons. Phys. Rev. 87, 835 (1952)
[75] Hall, R.N.: Electron-Hole Recombination in Germanium. Phys. Rev. 87, 387 (1952)
[76] Vaillant, F., Jousse, D.: Recombination at dangling bonds and steady-state photoconductivity in a-Si:H. Phys. Rev. 34, 4088 (1986)
[77] Aberle, A.G., Glunz, S., Warta, W.: Impact of illumination level and oxide parameters on Shockley–Read–Hall recombination at the Si-SiO_2 interface. J. Appl. Phys. 71, 4422 (1992)
[78] Aberle, A.G.: Crystalline silicon solar cells: advanced surface passivation and analyses. University of New South Wales, Sydney (1999)
[79] Sze, S.M.: Physics of semiconductor devices. John Wiley & Sons, New York (1981)
[80] Girisch, R.B.M., Mertens, R.P., de Keersmaecker, R.F.: Determination of Si-SiO_2 interface recombination parameters using a gate-controlled pointjunction diode under illumination. IEEE Trans. Electron Devices 35, 203 (1988)
[81] Eades, W.D., Swanson, R.M.: Calculation of surface generation and recombination velocities at the Si-SiO_2 interface. J. Appl. Phys. 58, 4267 (1985)
[82] Olibet, S., Vallat-Sauvain, E., Ballif, C.: Model for a-Si:H/c-Si interface recombination based on the amphoteric nature of silicon dangling bonds. Phys. Rev. B 76, 035326 (2007)
[83] Olibet, S., Vallat-Sauvain, E., Fesquet, L., Monachon, C., Hessler-Wyser, A., Damon-Lacoste, J., De Wolf, S., Ballif, C.: Properties of interfaces in amorphous/crystalline silicon heterojunctions. Phys. Stat. Sol: (a) 207, 651 (2010)
[84] Hubin, J., Shah, A.V., Sauvain, E.: Effects of dangling bonds on the recombination function in amorphous-semiconductors. Philos. Mag. Lett. 66, 115 (1992)
[85] Li, T.T.A., McIntosh, K.R., Cuevas, A.: Limitations of a simplified dangling bond recombination model for a-Si:H. J. Appl. Phys. 104, 113718 (2008)
[86] Schroder, D.K.: Semiconductor Material and Device Characterization, 3rd edn. Wiley-Interscience, Hoboken (2006)
[87] Nagel, H., Berge, C., Aberle, A.G.: Generalized analysis of quasi-steady-state and quasitransient measurements of carrier lifetimes in semiconductors. J. Appl. Phys. 86, 6218 (1999)
[88] Sinton, R.A., Cuevas, A.: Contactless determination of current–voltage characteristics and minority-carrier lifetimes in semiconductors from quasi-steady-state photoconductance data. Appl. Phys. Lett. 69, 2510 (1996)

[89] Luke, K.L., Cheng, L.J.: Analysis of the interaction of a laser pulse with a silicon wafer: Determination of bulk lifetime and surface recombination velocity. J. Appl. Phys. 61, 2282 (1987)
[90] Kerr, M.J., Cuevas, A.: General parameterization of Auger recombination in crystalline silicon. J. Appl. Phys. 91, 2473 (2002)
[91] Spenke, E.: Electronic semiconductors. McGraw-Hill, New York (1958)
[92] Tarng, M.L., Pankove, J.I.: Passivation of p-n junction in crystalline silicon by amorphous silicon. IEEE Trans. Electron. Dev. 26, 1728 (1979)
[93] Weitzel, I., Primig, R., Kempter, K.: Preparation of glow discharge amorphous silicon for passivation layers. Thin Solid Films 75, 143 (1982)
[94] De Wolf, S., Kondo, M.: Abruptness of a-Si:H/c-Si interface revealed by carrier lifetime measurements. Appl. Phys. Lett. 90, 042111 (2007)
[95] Fujiwara, H., Kondo, M.: Impact of epitaxial growth at the heterointerface of a-Si:H/c-Si solar cells. Appl. Phys. Lett. 86, 032112 (2005)
[96] Schulze, T.F., Beushausen, H.N., Hansmann, T., Korte, L., Rech, B.: Accelerated interface defect removal in amorphous/crystalline silicon heterostructures using pulsed annealing and microwave heating. Appl. Phys. Lett. 95, 182108 (2009)
[97] Strahm, B., Andrault, Y., Bätzner, D., Lachenal, D., Guérin, C., Kobas, M., Mai, J., Mendes, B., Schulze, T., Wahli, G., Buechel, A.: Uniformity and quality of monocrystalline silicon passivation by thin intrinsic amorphous silicon in a new generation plasma-enhanced chemical vapor deposition reactor. Mater. Res. Soc. Symp. Proc. 1245, A01–A04 (2010)
[98] Descoeudres, A., Barraud, L., Bartlome, R., Choong, G., De Wolf, S., Zicarelli, F., Ballif, C.: The silane depletion fraction as an indicator for the amorphous/crystalline silicon interface passivation quality. Appl. Phys Lett. 97, 183505 (2010)
[99] De Wolf, S., Olibet, S., Ballif, C.: Stretched-exponential a-Si:H/c-Si interface recombination decay. Appl. Phys. Lett. 93, 032101 (2008)
[100] Das, U.K., Burrows, M.Z., Lu, M., Bowden, S., Birkmire, R.W.: Surface passivation and heterojunction cells on Si(100) and (111) wafers using dc and rf plasma deposited Si:H thin films. Appl. Phys. Lett. 92, 063504 (2008)
[101] Gielis, J.J.H., van den Oever, P.J., Hoex, B., van de Sanden, M.C.M., Kessels, W.M.M.: Real-time study of a-Si:H/c-Si heterointerface formation and epitaxial Si growth by spectroscopic ellipsometry, infrared spectroscopy, and second-harmonic generation. Phys. Rev. B 77, 205329 (2008)
[102] Wang, Q., Page, M.R., Iwaniczko, E., Xu, Y., Roybal, L., Bauer, R., To, B., Yuan, H.C., Duda, A., Hasoon, F., Yan, Y.F., Levi, D., Meier, D., Branz, H.M., Wang, T.H.: Efficient heterojunction solar cells on p-type crystal silicon wafers. Appl. Phys. Lett. 96, 013507 (2010)
[103] Himpsel, F.J., Hollinger, G., Pollak, R.A.: Determination of the Fermi-level pinning position at Si (111) surfaces. Phys. Rev. B 28, 7014 (1983)
[104] Davis, E.A., Mott, N.F.: Conduction in non-crystalline systems. 5. Conductivity, optical absorption and photoconductivity in amorphous semiconductors. Philos. Mag. 22, 903 (1970)
[105] Marshall, J.M., Owen, A.E.: Drift mobility studies in vitreous arsenic triselenide. Philos. Mag. 24, 1281 (1971)
[106] Mott, N.F.: Introductory talk; Conduction in non-crystalline materials. J. Non-Cryst. Solids 8-10, 1 (1972)

[107] Spear, W.E., LeComber, P.G.: Electronic properties of substituted doped amorphous Si and Ge. Philos. Mag. 33, 935 (1976)
[108] Davis, E.A.: Amorphous Semiconductors. In: Brodsky, M.H. (ed.), Springer, Berlin (1979)
[109] Elliott, S.R.: Physics of amorphous materials. Longman, London (1983)
[110] Tersoff, J.: Schottky barrier heights and the continuum of gap states. Phys. Rev. Lett. 52, 465 (1984)
[111] Walukiewicz, W.: Mechanism of Schottky barrier formation: The role of amphoteric native defects. J. Vac. Sci Technol. B 5, 1062 (1987)
[112] Biegelsen, D.K., Johnson, N.M., Stutzmann, M., Poindexter, E.H., Caplan, P.J.: Native defects at the Si/SiO_2 interface-amorphous silicon revisited. Appl. Surf. Sci. 22/23, 879 (1985)
[113] Winer, K.: Chemical-equilibrium description of the gap-state distribution in a-Si:H. Phys. Rev. Lett. 63, 1487 (1989)
[114] Powell, M.J., Deane, S.C.: Improved defect-pool model for charged defects in amorphous silicon. Phys. Rev. 48, 10815 (1994)
[115] Hubbard, J.: Electron correlations in narrow energy bands. Proc. Roy. Soc. London A 267, 238 (1963)
[116] Madan, A., Shaw, M.P.: The physics and applications of amorphous semiconductors. Academic Press, San Diego (1988)
[117] Anderson, P.W.: Model for the electronic structure of amorphous semiconductors. Phys. Rev. Lett. 34, 953 (1975)
[118] Jackson, W.B.: The correlation-energy of the dangling silicon bond in a-Si:H. Solid State Comm. 44, 477 (1982)
[119] Street, R.A.: Hydrogenated amorphous silicon. Cambridge University Press, Cambridge (1991)
[120] Adler, D., Yoffa, E.J.: Electronic structure of amorphous semiconductors. Phys. Rev. Lett. 36, 1197 (1976)
[121] Connell, G.A.N., Pawlik, J.R.: Use of hydrogenation in structural and electronic studies of gap states in amorphous germanium. Phys. Rev. B 13, 787 (1976)
[122] Pankove, J.I., Lampert, M.A., Tarng, M.L.: Hydrogenation and dehydrogenation of amorphous and crystalline silicon. Appl. Phys. Lett. 32, 439 (1978)
[123] Knights, J.C., Lucovsky, G., Nemanich, R.J.: Defects in plasma-deposited a-Si:H. J. Non-Cryst. Solids 32, 393 (1979)
[124] Street, R.A., Knights, J.C., Biegelsen, D.K.: Luminescence studies of plasma-deposited hydrogenated silicon. Phys. Rev. B 18, 1880 (1978)
[125] Biegelsen, D.K., Street, R.A., Tsai, C.C., Knights, J.C.: Hydrogen evolution and defect creation in amorphous Si:H alloys. Phys. Rev. B 20, 4839 (1979)
[126] Korte, L., Laades, A., Schmidt, M.: Electronic states in a-Si:H/c-Si heterostructures. J. Non-Cryst. Solids 352, 1217 (2006)
[127] Kohlrausch, R.: Theorie des elektrischen rückstandes in der leidener flasche. Pogg. Ann. Phys. Chem. 91, 179 (1854)
[128] Williams, G., Watts, D.C.: Non-symmetrical dielectric relaxation behaviour arising from a simple empirical decay function. Trans. Farad. Soc. 66, 80 (1970)
[129] Cardona, M., Chamberlin, R.V., Marx, W.: The history of the stretched exponential function. Ann. Phys (Leipzig) 16, 842 (2007)
[130] Gotze, W., Sjogren, L.: Relaxation processes in supercooled liquids. Rep. Prog. Phys. 55, 241 (1992)

[131] Phillips, J.C.: Stretched exponential relaxation in molecular and electronic glasses. Rep. Prog. Phys. 59, 1133 (1996)
[132] Scher, H., Shlesinger, M.F., Bendler, J.T.: Time-scale invariance in transport and relaxation. Phys. Today 44, 26 (1991)
[133] Laherrere, J., Sornette, D.: Stretched exponential distributions in nature and economy: "fat tails" with characteristic scales. Eur. Phys. J. B 2, 525 (1998)
[134] Kakalios, J., Street, R.A., Jackson, W.B.: Stretched-exponential relaxation arising from dispersive diffusion of hydrogen in amorphous silicon. Phys. Rev. Lett. 59, 1037 (1987)
[135] Van de Walle, C.G.: Stretched-exponential relaxation modeled without invoking statistical distributions. Phys. Rev. B 53, 11292 (1996)
[136] Fujiwara, H., Kondo, M.: Real-time monitoring and process control in amorphous/crystalline silicon heterojunction solar cells by spectroscopic ellipsometry and infrared spectroscopy. Appl. Phys. Lett. 86, 032112 (2005)
[137] Brodsky, M.H., Cardona, M., Cuomo, J.J.: Infrared and Raman spectra of the silicon-hydrogen bonds in amorphous silicon prepared by glow discharge and sputtering. Phys. Rev. B 16, 3556 (1977)
[138] Lucovsky, G., Nemanich, R.J., Knights, J.C.: Structural interpretation of the vibrational spectra of a-Si: H alloys. Phys. Rev. B 19, 2064 (1979)
[139] Langford, A.A., Fleet, M.L., Nelson, B.P., Lanford, W.A., Maley, N.: Infrared absorption strength and hydrogen content of hydrogenated amorphous silicon. Phys. Rev. B 45, 13367 (1992)
[140] De Wolf, S.: Unpublished data
[141] Burrows, M.Z., Das, U.K., Opila, R.L., De Wolf, S., Birkmire, R.W.: Role of hydrogen bonding environment in a-Si:H films for c-Si surface passivation. J. Vac. Sci. Technol. A 26, 683 (2008)
[142] Smets, A.H.M., van de Sanden, M.C.M.: Relation of the Si-H stretching frequency to the nanostructural Si-H bulk environment. Phys. Rev. B 76, 073202 (2007)
[143] Johnson, N.M., Ponce, F.A., Street, R.A., Nemanich, R.J.: Defects in single-crystal silicon induced by hydrogenation. Phys. Rev. B 35, 4166 (1987)
[144] Tsai, C.C., Anderson, G.B., Thompson, R.: Low temperature growth of epitaxial and amorphous silicon in a hydrogen-diluted silane plasma. J. Non-Cryst. Solids 137, 673 (1991)
[145] Eaglesham, J., Gossmann, H.J., Cerullo, M.: Limiting thickness h_{epi} for epitaxial growth and room-temperature Si growth on Si(100). Phys. Rev. Lett. 65, 1227 (1990)
[146] Schwarzkopf, J., Selle, B., Bohne, W., Röhrich, J., Sieber, I., Fuhs, W.: Disorder in silicon films grown epitaxially at low temperature. J. Appl. Phys. 93, 5215 (2003)
[147] Teplin, C.W., Iwancziko, E., To, B., Moutinho, H., Stradins, P., Branz, H.M.: Breakdown physics of low-temperature silicon epitaxy grown from silane radicals. Phys. Rev. B 74, 235428 (2006)
[148] Gielis, J.J.H., van den Oever, P.J., van de Sanden, M.C.M., Kessels, W.M.M.: a-Si:H/c-Si heterointerface formation and epitaxial growth studied by real time optical probes. Appl. Phys. Lett. 90, 202108 (2007)
[149] Levi, D.H., Teplin, C.W., Iwaniczko, E., Yan, Y., Wang, T.H., Branz, H.M.: Real-time spectroscopic ellipsometry studies of the growth of amorphous and epitaxial silicon for photovoltaic applications. J. Vac. Sci. Technol. A 24, 1676 (2006)
[150] Hamers, R.J., Köhler, U.K., Demuth, J.E.: Epitaxial growth of silicon on Si(001) by scanning tunneling microscopy. J. Vac. Sci. Technol. A 8, 195 (1990)

[151] Metiu, H., Lu, Y.T., Zhang, Z.: Epitaxial growth and the art of computer simulations. Science 255, 1088 (1992)
[152] Bartlome, R.: Unpublished data
[153] Bartlome, R., Feltrin, A., Ballif, C.: Infrared laser-based monitoring of the silane dissociation during deposition of silicon thin films. Appl. Phys. Lett. 94, 201–501 (2009)
[154] Faist, J., Capasso, F., Sivco, D.L., Sirtori, L., Hutchinson, A.L., Cho, A.Y.: Quantum cascade laser. Science 264, 553 (1994)
[155] Sansonnens, L., Howling, A., Hollenstein, C.: Degree of dissociation measured by FTIR absorption spectroscopy applied to VHF silane plasmas. Plasma Sources Sci. Technol. 7, 114 (1998)
[156] Strahm, B., Howling, A., Sansonnens, L., Hollenstein, C.: Plasma silane concentration as a determining factor for the transition from amorphous to microcrystalline silicon in SiH4/H2 discharges. Plasma Sources Sci. Technol. 16, 80 (2007)
[157] Howling, A., Strahm, B., Colsters, P., Sansonnens, L., Hollenstein, C.: Fast equilibration of silane/hydrogen plasmas in large area RF capacitive reactors monitored by optical emission spectroscopy. Plasma Sources Sci. Technol. 16, 679 (2007)
[158] Catalano, A., Wood, G.: A method for improved short-wavelength response in hydrogenated amorphous silicon-based solar cells. J. Appl. Phys. 63, 1220 (1988)
[159] Collins, R.W.: In situ study of p-type amorphous silicon growth from diborane-silane mixtures: surface reactivity and interface effects. Appl. Phys. Lett. 53, 1086 (1988)
[160] Carlson, D.E., Wronski, C.R.: Amorphous silicon solar cell. Appl. Phys. Lett. 28, 671 (1976)
[161] Street, R.A.: Localized states in doped amorphous-silicon. J. Non-Cryst. Solids 77&78, 1 (1985)
[162] Beyer, W., Herion, J., Wagner, H.: Fermi energy dependence of surface desorption and diffusion of hydrogen in a-Si:H. J. Non-Cryst. Solids 114, 217 (1989)
[163] Beyer, W.: Hydrogen effusion – a probe for surface desorption and diffusion. Physica B 170, 105 (1991)
[164] Street, R.A., Tsai, C.C., Kakalios, J., Jackson, W.B.: Hydrogen diffusion in amorphous-silicon. Philos. Mag. B 56, 305 (1987)
[165] Beyer, W., Zastrow, U.: Dependence of H diffusion in hydrogenated silicon on doping and the fermi level. In: Mat. Res. Soc. Symp. Proc. vol. 4, p. A20.4.1 (2000)
[166] Beyer, W.: Hydrogen phenomena in hydrogenated amorphous silicon. In: Hydrogen in Semiconductors II Book Series: Semicond. and Semimetals, vol. 61, p. 165. Academic Press, San Diego (1999)
[167] Street, R.A.: Hydrogen diffusion and electronic metastability in amorphous-silicon. Physica B 170, 69 (1991)
[168] Van de Walle, C.G., Street, R.A.: Silicon-hydrogen bonding and hydrogen diffusion in amorphous silicon. Phys. Rev. B 51, 10615 (1995)
[169] Van de Walle, C.G., Denteneer, P.J.H., Bar-Yam, Y., Pantelides, S.T.: Theory of hydrogen diffusion and reactions in crystalline silicon. Phys. Rev. B 39, 10791 (1989)
[170] De Wolf, S., Kondo, M.: Nature of doped a-Si:H/c-Si interface recombination. J. Appl. Phys. 105, 103707 (2009)
[171] Shockley, W., Moll, J.L.: Solubility of flaws in heavily-doped semiconductors. Phys. Rev. 119, 1480 (1960)

[172] Van de Walle, C.G.: Hydrogen as a cause of doping in zinc oxide. Phys Rev. Lett. 85, 1012 (2000)

[173] Van de Walle, C.G., Neugebauer, J.G.: Universal alignment of hydrogen levels in semiconductors, insulators and solutions. Nature 423, 626 (2003)

[174] Baraff, G.A., Schluter, M.: Electronic structure, total energies, and abundances of the elementary point defects in GaAs. Phys. Rev. Lett. 55, 1327 (1985)

[175] Zunger, A.: Practical doping principles. Appl. Phys. Lett. 83, 57 (2003)

[176] Van de Walle, C.G.: Strategies for controlling the conductivity of wide-band-gap semiconductors. Phys. Stat. Sol. (b) 229, 221 (2002)

[177] Zhang, S.B., Wei, S.H., Zunger, A.: A phenomenological model for systematization and prediction of doping limits in II–VI and I–III–VI compounds. J. Appl. Phys. 83, 3192 (1998)

[178] Van de Walle, C.G., Neugebauer, J.G.: First-principles calculations for defects and impurities: Applications to III-nitrides. J. Appl. Phys. 95, 3851 (2004)

[179] Street, R.A., Biegelsen, D.K., Knights, J.C.: Defect states in doped and compensated a-Si: H. Phys. Rev. B 24, 969 (1981)

[180] Bar-Yam, Y., Adler, D., Joannopoulos, J.D.: Structure and electronic states in disordered systems. Phys. Rev. Lett. 57, 467 (1986)

[181] Branz, H.M.: Dangling bonds in doped amorphous silicon: Equilibrium, relaxation, and transition energies. Phys. Rev. B 39, 5107 (1989)

[182] De Wolf, S., Kondo, M.: Boron-doped a-Si:H/c-Si interface passivation: Degradation mechanism. Appl. Phys. Lett. 91, 112109 (2007)

[183] Yamamoto, T., Katayama-Yoshida, H.: Materials design for the fabrication of low-resistivity p-type GaN using a codoping method. Jpn. J. Appl. Phys. 36, L180 (1997)

[184] Matsuda, A., Matsumura, M., Yamasaki, S., Yamamoto, H., Imura, T., Okushi, H., Izima, S., Tanaka, K.: Boron doping of hydrogenated silicon thin-films. Jpn. J. Appl. Phys. 183, L183 (1981)

[185] Fritzsche, H., Tanielian, M., Tsai, C.C., Gaczi, P.J.: Hydrogen content and density of plasma-deposited amorphous silicon-hydrogen. J. Appl. Phys. 50, 3366 (1979)

[186] Smets, A.H.M., Kessels, W.M.M., van de Sanden, M.C.M.: Vacancies and voids in hydrogenated amorphous silicon. Appl. Phys. Lett. 82, 1547 (2003)

[187] Rostan, P.J., Rau, U., Nguyen, V.X., Kirchartz, T., Schubert, M.B., Werner, J.H.: Low-temperature a-Si:H/ZnO/Al back contacts for high-efficiency silicon solar cells. Sol. Energy Mater. Sol. Cells 90, 1345 (2006)

[188] Einsele, F., Rostan, P.J., Schubert, M.B., Rau, U.: Recombination and resistive losses at ZnO/a-Si:H/c-Si interfaces in heterojunction back contacts for Si solar cells. J. Appl. Phys. 102, 094507 (2007)

[189] Stiebig, H., Siebke, F., Beyer, W., Beneking, C., Rech, B., Wagner, H.: Interfaces in a-Si:H solar cell structures. Sol. Energy and Sol. Mater. 48, 351 (1997)

[190] Shah, A., Torres, P., Tscharner, R., Wyrsch, N., Keppner, H.: Photovoltaic technology: the case for thin-film solar cells. Science 285, 692 (1999)

[191] Shah, A.: Thin-film silicon solar cells. EPFL Press, Lausanne (2010)

[192] Branz, H.M., Crandall, R.S.: Defect equilibrium thermodynamics in hydrogenated amorphous silicon: consequences for solar cells. Solar Cells 27, 159 (1989)

Chapter 8
Photoluminescence and Electroluminescence from Amorphous Silicon/Crystalline Silicon Heterostructures and Solar Cells

Rudolf Brüggemann

Institut für Physik, Carl von Ossietzky Universität Oldenburg, 26111 Oldenburg, Germany

Abstract. Photoluminescence and electroluminescence from amorphous silicon / crystalline silicon heterostructures is a measure of the radiative band-to-band recombination, described in terms of the quasi-Fermi level splitting according to Planck's generalised law, and the experiments thus probe the excess carrier densities in the crystalline silicon. Depending on the layer structure of the investigated sample, the contact-less photoluminescence experiment allows the characterisation of precursor structures for solar cell optimisation and for the study of related physical aspects like interface recombination. Both photoluminescence and electroluminescence experiments can be applied to solar cells for which the luminescence yield, or more precisely the deduced quasi-Fermi level splitting, can be related to the open-circuit voltage of the device which itself is limited by factors like the interface recombination rate. The coverage of luminescence techniques is complemented here by an account of modulated photoluminescence, a variant of the experiment, which may be used for the lifetime determination in wafer structures. Numerical modelling provides additional insight into the physics of interface recombination and its impact on the quasi-Fermi level splitting and thus the luminescence yield and the open-circuit voltage.

8.1 Introduction

Photoluminescence (PL) and electroluminescence (EL) probe the excess carrier density in semiconductor structures. While the electroluminescence measurement requires electrical contacts to the structure, the photoluminescence is induced by photogeneration from a suitable light source. In the two cases, the emitted radiation represents spectral information on the splitting of the quasi-Fermi levels and the out-coupling of luminescence photons [1,2].

The luminescence spectrum with its shape and resulting signal height provides information on the loss terms as the quasi-Fermi level splitting or the excess carrier density in the c-Si wafer depends on the bulk lifetime and surface recombination

W.G.J.H.M. van Sark et al. (Eds.): Physics & Tech. of Amorphous-Crystalline, EM, pp. 261–299.
springerlink.com

and in passivated wafers on interface recombination at the interface between passivation layers and the wafers. Efficient hydrogenated amorphous silicon (a-Si:H) / crystalline silicon (c-Si) heterojunction solar cells require a low density of interface defects [3-5] because even moderate interface defect densities reduce the open-circuit voltage. Efficient passivation of interface defects is thus crucial for efficient a-Si:H/c-Si solar cells. Pre-treatment of the c-Si wafer prior to the a-Si:H deposition and the parameters for a-Si:H deposition determine the density of interface defects to a large degree – as described by various articles in this book. Therefore, luminescence techniques have been applied to study wafer passivation and lifetimes in passivated crystalline silicon wafers [6-8].

The focus of this chapter is on the application of photoluminescence and electroluminescence with respect to c-Si wafers, passivated with amorphous silicon. After a review of the basic physical relations, namely Planck's generalised law, which describes the luminescence emission, some conclusions are drawn on what can be determined from the experiment. The internal quasi-Fermi level splitting will be related to the open-circuit voltage of solar cells, the luminescence yield and the interface defect density. Numerical modelling can provide additional insight in relation with the interface defect densities. The modulated photoluminescence (MPL) experiment is presented as a relatively simple tool, in which an effective lifetime of minority carriers is determined from the frequency response of the MPL signal. In this way, the influence of interface defects on the recombination of excess carriers in wafers with different passivation schemes can be studied.

8.2 Theoretical Background

8.2.1 Planck's Generalised Law

According to Lasher and Stern [9], the spontaneous band-to-band radiative recombination rate per energy interval for recombination between conduction-band electrons and valence-band holes is given by

$$\mathrm{d}r_{\mathrm{spont}}(\hbar\omega) = \frac{n_\mathrm{r}^2(\hbar\omega)^2}{\pi^2\hbar^3c^2}\frac{\alpha(\hbar\omega)\mathrm{d}\hbar\omega}{\exp\left(\dfrac{\hbar\omega-(E_\mathrm{Fn}-E_\mathrm{Fp})}{kT}\right)-1} \tag{8.1}$$

where n_r is the refractive index, c the vacuum light velocity, α is the band-to-band absorption coefficient, k is the Boltzmann constant, T is the temperature and the quasi-Fermi levels of the conduction-band electrons and the valence-band holes are E_Fn and E_Fp defined by

$$\begin{aligned} n &= N_\mathrm{c}\exp\left(-\frac{E_\mathrm{c}-E_\mathrm{Fn}}{kT}\right), \\ p &= N_\mathrm{v}\exp\left(-\frac{E_\mathrm{Fp}-E_\mathrm{v}}{kT}\right), \end{aligned} \tag{8.2}$$

where n and p are the densities of free electrons and holes and N_c and N_v are the effective densities of states in the conduction and valence band with the edges at the energies E_c and E_v.

Neglecting the "-1" in the denominator because $\hbar\omega >> (E_{Fn} - E_{Fp})$ for the silicon absorbers, the total radiative band-to-band recombination rate R_{BB} can be described in terms of these densities n and p according to [10]

$$R_{BB} = \underline{B}np\,. \tag{8.3}$$

Based on eq. (8.1), $\underline{B}$ is the integrated coefficient that is calculated from integrating $B(\hbar\omega,T)$, given by

$$B(\hbar\omega,T) = \frac{(\hbar\omega)^2 n_r^2}{\pi^2\hbar^3c^2}\frac{\alpha(\hbar\omega,T)}{n_i^2(T)}\exp\left(-\frac{\hbar\omega}{kT}\right) \tag{8.4}$$

over $\hbar\omega$. Here the relation $np = n_i^2 \exp\left[\left(E_{Fn} - E_{Fp}\right)/(kT)\right]$ with the intrinsic carrier density n_i has been inserted in eq. (8.1). For device simulation, eq. (8.3) is employed for the radiative channel in the recombination-rate terms with the spatial variation $n(z)$ and $p(z)$ where z runs between the front and the back of the structure.

The emitted photon-flux density from a sample with a given thickness into one hemisphere can be calculated, starting from the volume-element rate of eq. (8.1) and by accounting for the out-coupling of emitted photons across the sample/air interfaces. The emitted photon-flux density is then given by the quasi-Fermi level splitting and the absorptivity A of the sample by [11]

$$dj_\gamma(\hbar\omega) = \frac{(\hbar\omega)^2}{4\pi^2\hbar^3c^2}A(\hbar\omega)\exp\left(-\frac{\hbar\omega}{kT}\right)\exp\left(\frac{E_{Fn}-E_{Fp}}{kT}\right)d(\hbar\omega) \tag{8.5}$$

where the "–1" in the Bose-term has been neglected.

Equation (8.5) with its approximation of constant quasi-Fermi level splitting is crucial for the analysis of the luminescence spectra. The term $\exp\left[\left(E_{Fn} - E_{Fp}\right)/(kT)\right]$, i.e., the np/n_i^2 term, independent of photon energy, scales the whole spectrum on an absolute scale. For the PL measurements on solar cells, performed at open circuit, the PL yield is a measure of the open-circuit voltage, which is determined by the quasi-Fermi level splitting that is probed by the PL experiment. In addition, absorber layers that may be incorporated into solar cells can be analyzed by PL measurements and from the determination of the quasi-Fermi level splitting the prospective open-circuit voltage can be estimated [12].

It is noted that the spectral variation of the emitted luminescence spectrum is also determined by the spectral variation of A, which is a temperature-dependent quantity. For a symmetrical wafer and normal incidence, A is given by

$$A(\hbar\omega) = \frac{\left(1 - R_{\text{int}}(\hbar\omega)\right)\left(1 - e^{-\alpha(\hbar\omega)d}\right)}{1 - R_{\text{int}}(\hbar\omega)e^{-\alpha(\hbar\omega)d}} \quad (8.6)$$

with the reflection coefficient of the air/wafer interface R_{int} and the sample thickness d. Multiple reflections are taken into account here.

The term for A is modified for a solar cell with a rear-side mirror-like contact. With the front-reflection coefficient R_{f} and the back-reflection coefficient R_{b} the absorptivity reads

$$A(\hbar\omega) = \frac{(1 - R_{\text{f}})\left[\left(1 - R_{\text{b}}e^{-2\alpha d} - (1 - R_{\text{b}})e^{-\alpha d}\right)\right]}{1 - R_{\text{f}}R_{\text{b}}e^{-2\alpha d}}. \quad (8.7)$$

The variation in the absorption of a crystalline silicon wafer with identical optical interface properties at the front and back interface $R_{\text{f}} = R_{\text{b}} = 0$, represented by $1 - \exp(-\alpha d)$ according to eq. (8.6), and the emitted spectra, calculated with eq. (8.5), is illustrated in Fig. 8.1, shown here for different thicknesses.

For the 300-μm wafer in Fig. 8.1(a), the spectral variation in $A(\hbar\omega)$, calculated with the absorption-coefficient data from [13], becomes constant at higher photon energies. In this case, the spectral variation of high-energy wing of the luminescence spectrum in Fig. 8.1(b) is determined by $(\hbar\omega)^2 \exp\left(-\frac{\hbar\omega}{kT}\right)$, with the absorption in the thicker samples being $A(\hbar\omega) \approx 1 - R_{\text{f}}(\hbar\omega) = 1$. For the thinner wafers, the total luminescence yield is reduced because the emission volume decreases. In addition, the spectral shape changes in the high-energy range as the absorption term of Fig. 8.1(a) is still monotonously increasing with increasing photon energy. In the low-energy range, as $A(\hbar\omega) \approx \alpha(\hbar\omega)d$, A remains a monotonously increasing function with increasing $\hbar\omega$, the photon flux in Fig. 8.1(b) scales with d.

The illustration in Fig. 8.1(b) makes clear that the slope of the high-energy side of the PL or EL spectrum can be determined and compared with the sample temperature. In the cases in which A in Fig. 8.1(a) does not reach a constant value, notably here for the thinner samples, caution is needed. An incorrect temperature and subsequently an incorrect value of the quasi-Fermi level splitting could be determined if the energy-dependent variation of A is neglected.

In the low-energy side of the PL spectra in Fig. 8.1(b) the shoulders correspond to phonon related features in the absorption in Fig. 8.1(a).

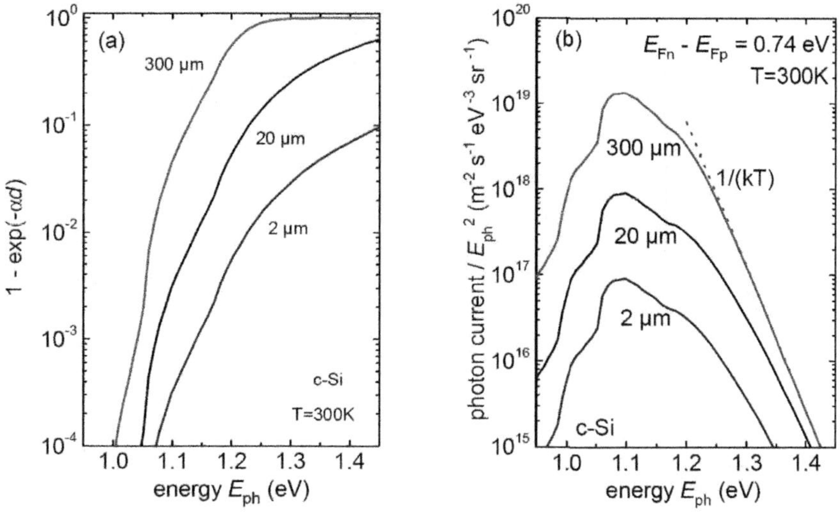

Fig. 8.1 Absorption (a) and spectral luminescence photon flux from identical c-Si wafers with different thicknesses, calculated according to eq. (8.5) with photon energy $E_{\text{ph}} = \hbar\omega$. Note that the ordinate values contain the term E_{ph}^{-2}. In the high-energy range the spectrum is determined by $\exp[-\hbar\omega/(kT)]$ in the range in which the absorption is constant. The dotted line represents the slope $1/(kT)$.

For non-homogeneous carrier distributions and thus a non-constant quasi-Fermi level splitting, a more complicated version of eq. (8.5) involves an integral term to take into account the non-homogeneous generation and propagation of the luminescence photons. It is not possible to split off *A*. For emission through the front interface at $z = d$ according to Fig. 8.2, the photon current density into the hemisphere reads [14,15]

$$\begin{aligned} \mathrm{d}j_\gamma(\hbar\omega) &= \frac{(\hbar\omega)^2}{4\pi^2\hbar^3c^2}\,\frac{\alpha(\hbar\omega)(1-R_\mathrm{f})\mathrm{e}^{-\alpha(\hbar\omega)d}}{1-R_\mathrm{f}R_\mathrm{b}\,\mathrm{e}^{-2\alpha(\hbar\omega)d}} \times \\ &\int_0^d \left[\mathrm{e}^{\alpha(\hbar\omega)z} + R_\mathrm{b}\,\mathrm{e}^{-\alpha(\hbar\omega)z}\right]\mathrm{e}^{-\frac{\hbar\omega}{kT}}\,\mathrm{e}^{\frac{E_\mathrm{Fn}(z)-E_\mathrm{Fp}(z)}{kT}}\,\mathrm{d}z\;\mathrm{d}(\hbar\omega) \end{aligned} \tag{8.8}$$

where again the "−1" in the Bose-term has been neglected. For the emission through the back surface a similar expression holds. For constant quasi-Fermi level splitting, eq. (8.8) reduces to eq. (8.5).

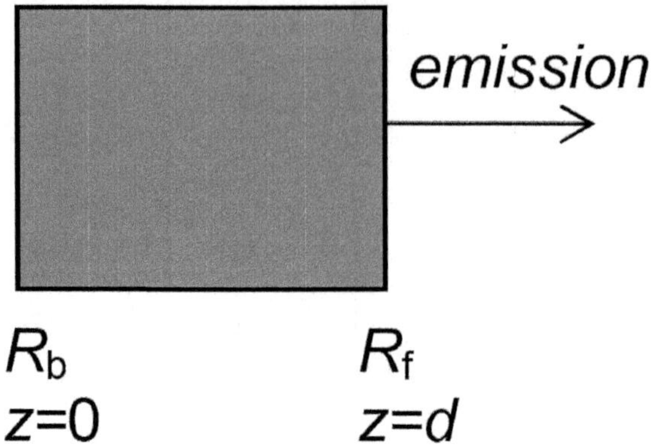

Fig. 8.2 Schematic of the PL emission for eq. (8.8) through the front and back surface with R_b at $z = 0$ and R_f at $z = d$.

A very rigorous account of the analysis of PL emission to either side is by Knabe *et al.* [16] who also take into account coherence effects related with multiple reflection at the air/sample surfaces. Würfel *et al.* [17] exploited the effect of the spatial variation in z-direction of the quasi-Fermi level splitting for short diffusion lengths in Si wafers in order to determine diffusion-length values from PL measurements.

8.2.2 Photoluminescence and Electroluminescence

The equation that describes the radiative recombination rate or the emitted luminescence spectra, eq. (8.5), is identical for PL and EL emission as only the quasi-Fermi level splitting and the optical properties of sample enter. Because of the difference in the continuity equations between EL in the dark and PL under illumination, where the photogeneration rate G enters as a source term, the spatial distributions of the excess carriers may be different, leading to different spectral shape. Under the assumption of constant $E_{\mathrm{Fn}} - E_{\mathrm{Fp}}$ the spectra of PL and EL should be identical.

Typical structures for the PL experiment are symmetrically passivated c-Si wafers to probe the influence of the interfaces or surfaces on the excess minority carrier densities. For further processed samples with electrical contacts, solar cells can be investigated by PL and EL experiments.

8.2.3 Current-Voltage Characteristics and Luminescence

For EL, Würfel [11] noted that for a good luminescent *pn* diode $E_{\mathrm{Fn}} - E_{\mathrm{Fp}} = \mathrm{e}V_{\mathrm{a}}$ where V_{a} is the applied voltage and thus from eq. (8.5) it follows that the luminescent photon flux depends exponentially on the applied voltage according to

$$\mathrm{d}j_{\gamma}(\hbar\omega) = \frac{(\hbar\omega)^2}{4\pi^2\hbar^3c^2} A(\hbar\omega)\exp\left(-\frac{\hbar\omega}{kT}\right)\exp\left(\frac{\mathrm{e}V_{\mathrm{a}}}{kT}\right)\mathrm{d}(\hbar\omega). \tag{8.9}$$

Fuhs *et al.* [18], assuming that $E_{\mathrm{Fn}} - E_{\mathrm{Fp}}$ equals $\mathrm{e}V_{\mathrm{a}}$ and together with the current-voltage diode equation

$$j_d = j_s\left[\exp\left(\frac{\mathrm{e}V_{\mathrm{a}}}{kT}\right) - 1\right], \tag{8.10}$$

formulated

$$\mathrm{d}j_{\gamma}(\hbar\omega) = \frac{j_d}{j_s}\frac{(\hbar\omega)^2}{4\pi^2\hbar^3c^2} A(\hbar\omega)\exp\left(-\frac{\hbar\omega}{kT}\right)\mathrm{d}(\hbar\omega). \tag{8.11}$$

Here, j_{d} is the dark-current density and j_{s} is the saturation-current density. There is thus a linear relationship between the current density and the luminescence yield.

Going a step further, one can substitute j_{s} with the help of the current-voltage characteristics under illumination and eq. (8.11) approximates to

$$\mathrm{d}j_{\gamma}(\hbar\omega) = \frac{j_{\mathrm{d}}}{j_{\mathrm{sc}}}\exp\left(\frac{eV_{\mathrm{oc}}}{kT}\right)\frac{(\hbar\omega)^2}{4\pi^2\hbar^3c^2} A(\hbar\omega)\exp\left(-\frac{\hbar\omega}{kT}\right)\mathrm{d}(\hbar\omega) \tag{8.12}$$

where the link between the open-circuit voltage V_{oc} and the EL flux is obvious: for a given j_{d}, the EL flux depends exponentially on V_{oc}. It is noted that there are other approaches that also establish the link between EL efficiency and open-circuit voltage. Fuyuki included the diode quality factor of the current-voltage characteristics in the analysis [19]. Kirchartz *et al.* [20] apply reciprocity relations.

With a more general form the current-voltage relation reads

$$j(V_{\mathrm{a}}) = j_{\mathrm{s}}\left[\exp\left(\frac{\mathrm{e}V_{\mathrm{a}}}{mkT}\right) - 1\right] - j_{\mathrm{ph}}, \tag{8.13}$$

where *m* is the diode quality factor. In this case, the relation between electroluminescence yield and the current densities becomes

$$\mathrm{d}j_{\gamma}(\hbar\omega) = \left(\frac{j_{\mathrm{d}}}{j_{\mathrm{s}}}\right)^m \frac{(\hbar\omega)^2}{4\pi^2\hbar^3c^2} A(\hbar\omega)\exp\left(-\frac{\hbar\omega}{kT}\right)\mathrm{d}(\hbar\omega). \tag{8.14}$$

This power-law relation was also noted by Fuyuki *et al.* [19] who suggested that the diode quality-factor can be determined from the log-log plot of EL yield vs. dark current density.

A high saturation-current density requires a higher dark-current density to achieve the same EL yield with this power-law relation between EL yield and j_d. In terms of applied voltage, it is noted that the same applied voltage should always lead to the same EL yield unless there is some voltage drop by series resistance [21].

Particularly important for the evaluation of PL measurements is the relation that at open-circuit the quasi-Fermi level splitting can be determined. In this case under illumination $E_\mathrm{Fn} - E_\mathrm{Fp} = \mathrm{e}V_\mathrm{oc}$, where V_oc is the open-circuit voltage. The relative variation in PL yield of different samples or at different lateral positions on a sample can directly be translated into a variation in $E_\mathrm{Fn} - E_\mathrm{Fp}$ by

$$\Delta\left(E_\mathrm{Fn} - E_\mathrm{Fp}\right) = \left(E_\mathrm{Fn} - E_\mathrm{Fp}\right)_1 - \left(E_\mathrm{Fn} - E_\mathrm{Fp}\right)_2 = kT\ln\left(\frac{Y_{PL,1}}{Y_{PL,2}}\right) \tag{8.15}$$

where 1 and 2 denote two different measurement positions or samples. In order to avoid errors from differences in the optical properties, the PL signal should be evaluated in an energy range with high-enough energy where A is only determined by $1-R_\mathrm{f}$. Equation (8.15) is important for the analysis of polycrystalline semiconductors for the study of local fluctuations in the opto-electronic properties [22].

The relations between current density, applied voltage and PL yield may also be used to construct a current-voltage characteristics from the PL measurements. Calibrating the PL yield by a measurement at open circuit and with the open-circuit voltage, the internal voltage V_{PL} curve is constructed by

$$V_{PL} - V_\mathrm{oc} = kT\ln\left(\frac{Y_{PL,V_{OC}}}{Y_{PL,V_a}}\right) \tag{8.16}$$

where $Y_{PL,V_{OC}}$ and Y_{PL,V_a} are the PL yield at open circuit and at an applied voltage V_a. Thus, the relevant voltage is shifted from V_a to the determined internal voltage V_{PL} which is then taken as an input for plotting the measured current density vs. V_{PL} instead of vs. V_a with

$$j(V_a) \rightarrow j(V_{PL})\,. \tag{8.17}$$

It is noted that this approach is only valid when the quasi-Fermi level splitting in the volume of the sample, from which most of the photoluminescence originates, is equal to the quasi-Fermi level splitting at the solar cell junction.

8.2.4 Modulated Photoluminescence

Modulated Photoluminescence (MPL) [23] is measured by illuminating the semiconductor sample with sinusoidally modulated light which is related to a photogeneration rate G according to $G = G_0 + G_1 e^{i\omega t}$ where G_0 represents the bias illumination, G_1 is the amplitude of the modulated generation and ω is the modulation frequency. In steady state with $G = G_0$ a minority carrier lifetime τ_n, taken here to be that of the electrons with excess carrier density Δn_0, can be defined according to $G_0 = \Delta n_0 / \tau_n$. With the Ansatz $\Delta n = \Delta n_0 + \Delta n_1 e^{i\omega t}$ in which Δn_1 is a complex number with amplitude

$$|\Delta n_1| = \frac{\tau_n G_1}{\left(1+\omega^2 \tau_n^2\right)^{1/2}} \tag{8.18}$$

and phase Φ with respect to G_1 one finds

$$\tan(\Phi) = \omega \tau_n \tag{8.19}$$

Fig. 8.3 gives a graphical account for the frequency-dependent amplitude and phase, indicating that 45° phase shift results at $\omega = 1/\tau_n$ This simple analysis follows the standard derivation [24] for the case in which the surface recombination velocity is equal at both surfaces and the bulk lifetime is long enough so that the effective lifetime is mainly determined by surface recombination. The more general case with explicit treatment of two different surface recombination velocities and inhomogeneous excess carrier profile results in more complicated equations for the phase dependence on modulation frequency and is not considered here.

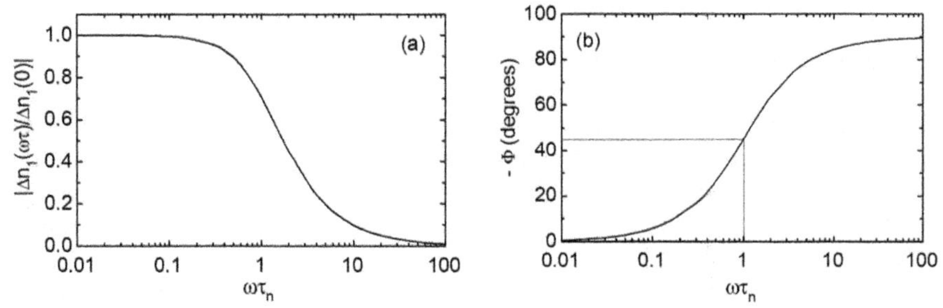

Fig. 8.3 nalytical variation of amplitude (a), normalised to the steady-state value, and phase (b) of the modulated excess carrier density.

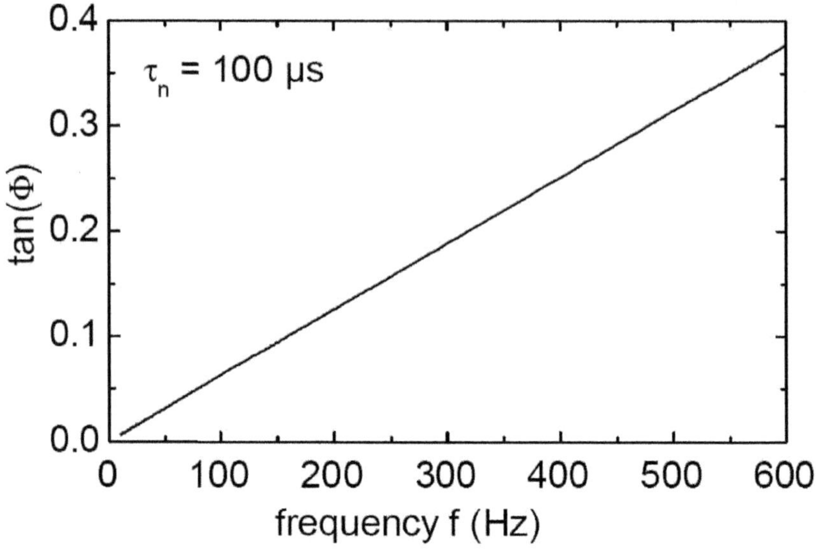

Fig. 8.4 Illustration of the tangent of the phase shift for a lifetime of 100 μs. From the data, the lifetime can be determined from the slope or from any single value by $\tau_n = \tan\left[\Phi(\omega)/\omega\right]$.

The linear variation in $\tan(\Phi)$ with frequency in Fig. 8.4 illustrates that one should check the extrapolation of experimental data to origin. Once having properly calibrated the system, the lifetime can then be determined from the simple relation $\tau_n = \tan\left[\Phi(\omega)/\omega\right]$. The lifetime τ_n determined from MPL is labelled τ_{MPL} in the results section of this article.

8.3 Numerical Modelling

Two aspects are important for the numerical modelling of PL and EL: calculation of the excess-carrier densities and calculation of the spectral photon flux. Basically, the spectral photon flux is calculated by eq. (8.8). This equation is applied to the spatial variation of the quasi-Fermi levels or the electron and hole densities for which separate solvers can be used.

Such numerical device simulation programmes have been developed to solve Poisson's equation and the continuity equations. With respect to a-Si:H/c-Si heterostructures the programmes AFORS-HET from the Helmholtz-Zentrum Berlin [25], the Amorphous Semiconductor Device Modeling Program (ASDMP) adapted for a-Si:H/c-Si structures [26] and SC-Simul from Oldenburg University [27] have been used in recent years for the device modelling of heterojunction solar cells with different layers. The programmes determine all the physical properties of interest, e.g., the quasi-Fermi levels or the free electron and hole densities, the electrostatic potential and other properties like recombination rates.

The input parameters are the semiconductor properties of the different layers with which a structure is built, like the band gap E_g, effective densities of states in the band, free carrier mobilities, electron affinity, capture coefficients for defect states and parameters for defect densities and distributions.

From the numerical solution of the device equations of $E_{Fn}(z)$ and $E_{Fp}(z)$ the luminescence photon flux can be calculated in a next step. For example, eq. (8.8) is applied to calculate the luminescence spectrum. The input for the calculation of the spectral flux requires the temperature-dependent absorption coefficient for calculating the absorption term.

8.4 Experimental

8.4.1 Spectral Photoluminescence

Figure 8.5 shows a typical experimental set-up for PL and EL measurements [28]. The radiation is collected and dispersed by a monochromator and detected by a suitable photodiode. In Fig. 5, the fraction of the photoluminescence is detected which is emitted towards the side from which the samples are excited. The setup can be calibrated with a tungsten band lamp for determining absolute photon fluxes.

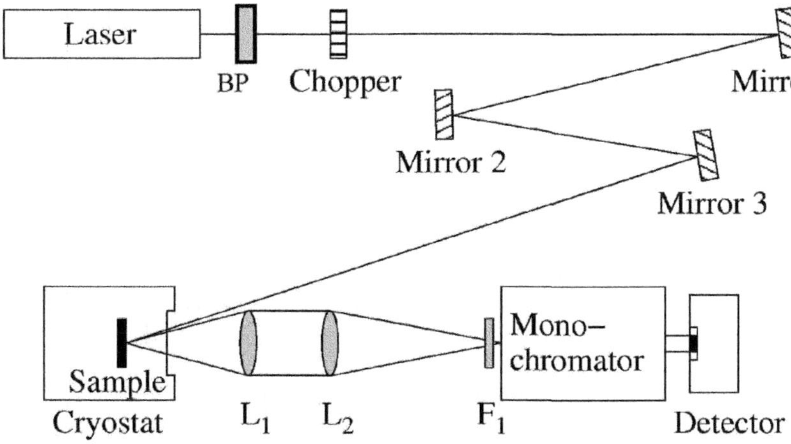

Fig. 8.5 Illustration of a typical set-up for the measurement of photoluminescence. The emitted luminescence is collected by a monochromator where it is dispersed. A bandpass filter (BP) is inserted to block unwanted laser light. The filter F_1 also blocks the laser light. A suitable detector measures the radiation and passes the signal for amplification and signal processing. The set-up can also be arranged with less mirrors. Mirror 3 here adjusts the laser beam to become almost rectangular with respect to the sample.

The laser in Fig. 8.5 is typically a cw laser. Suitable pulsed lasers with high repetition rate can also be employed. In the set-up at the Helmholtz-Zentrum Berlin [29,30] a nitrogen-laser pumped dye laser is employed, where the dye delivers monochromatic light with a suitable wavelength. The PL spectrum is recorded by a detector behind a monochromator which receives the monochromatic PL radiation. For each PL wavelength, the PL transient is integrated in the time range up to a few hundred μs. From the time-integrated PL transients at each measured wavelength the PL spectra are computed. The PL results thus do not represent the steady-state in this case but probe the minority carrier density in a certain time window.

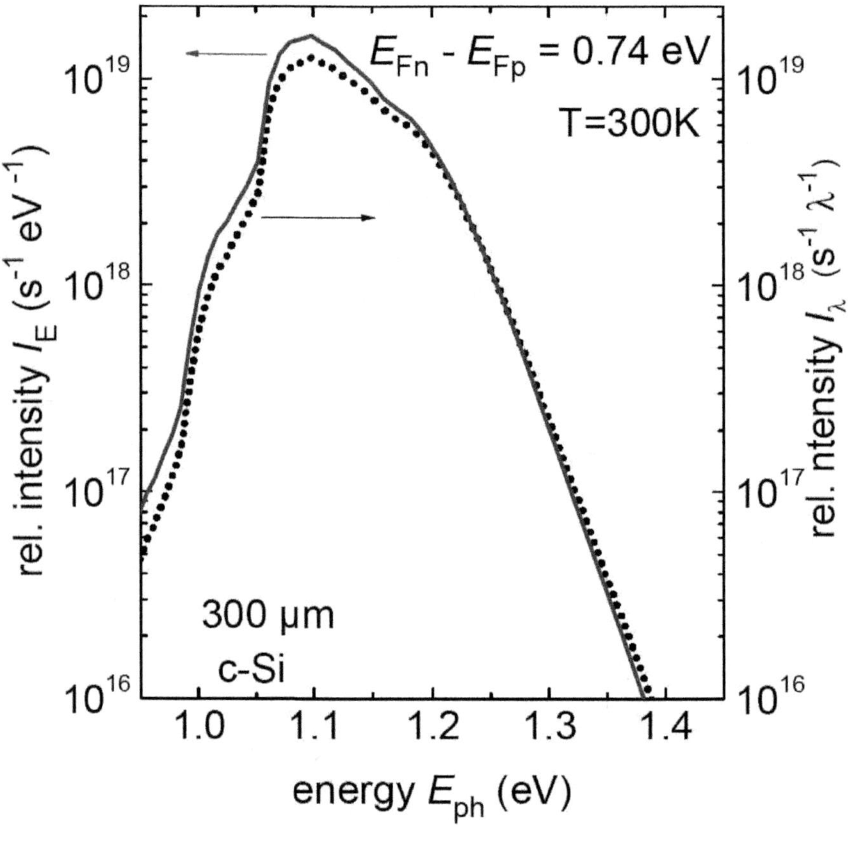

Fig. 8.6 Comparison of PL spectra vs. energy. The dotted line shows the experimental-like data from intensity vs. wavelength when only the abscissa values are converted from wavelength to photon energy. The full line shows the correct conversion and takes into account the energy unit interval with respect to the wavelength unit interval.

PL spectra are often plotted as "intensity" vs. wavelength or vs. energy. It is noted that if one scans with a monochromator linearly in wavelength, the ordinate values of photons per s per wavelength interval need to be changed to photons per s per energy interval [31]. This conversion requires the ordinate values to be multiplied by λ^2 for each data point. Figure 8.6 illustrates the effect if only the abscissa values are converted for a typical c-Si spectrum. Some distortion can be identified, e.g., in the high-energy slope, while the maximum position does not shift. The effect is thus less pronounced here than for much wider spectra like of a black body.

8.4.2 Modulated Photoluminescence

In MPL, phase-sensitive measurements are performed: the sample is illuminated and the illumination is modulated either with a chopper or electronically. In suitable set-ups sketched in Fig. 8.7, a fast photodiode, with an appropriate cut-off filter between the sample and the photodiode, is used for PL detection. In the figure, the PL spectrum weighted with the spectral response of the photodiode is measured.

The PL radiation can be measured in transmission mode (Fig. 8.7, left) or reflection mode (Fig. 8.7, right). For discriminating, the radiation can also be dispersed by a monochromator in a set-up according to Fig. 8.5, giving a smaller signal height in comparison with the full-spectrum measurement. The suitably amplified MPL signal is analyzed by a lock-in amplifier for phase and amplitude determination. Calibration for zero phase is needed for the absolute phase measurement.

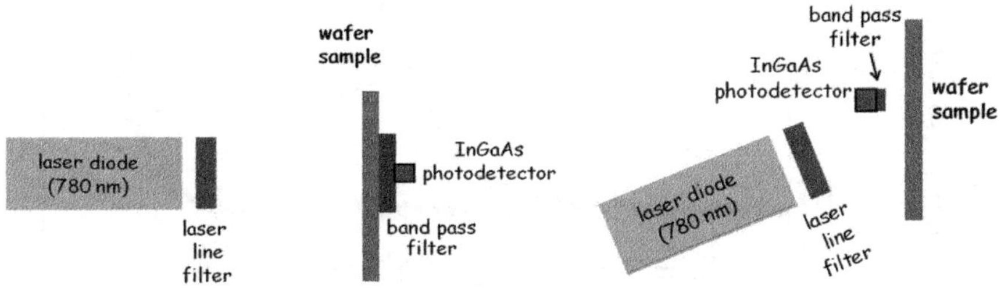

Fig. 8.7 Schematic of two possible set-ups for MPL in transmission mode (left) and reflection mode (right). A band pass filter cuts off unwanted laser light and the photodetector collects all the filtered emitted radiation

8.4.3 Electroluminescence

In the electroluminescence experiment, the luminescent radiation from the solar cell, driven, e.g., by a sourcemeter, is collected. Two modes of driving the solar cell for the electroluminescence experiment can be distinguished. A periodic current is imposed or, alternatively, a stationary dark current can be adjusted and the radiation signal is chopped, with the chopper in front of the monochromator entrance in Fig. 8.5 in the latter case.

8.5 Results

An overview is given on simulation and experimental results and their interpretation based on the luminescence experiments described in 8.4, applied to a-Si:H/c-Si heterostructures.

8.5.1 Numerical Modelling of Interface Recombination and Luminescence Yield

Numerous accounts in the literature identified that interface recombination at the a-Si:H/c-Si interface is detrimental to the solar cell performance, e.g., [6-8]. This relation has been established by numerical simulations in which the defect density in a very thin interfacial layer has been varied. Here, we shall outline the typical features.

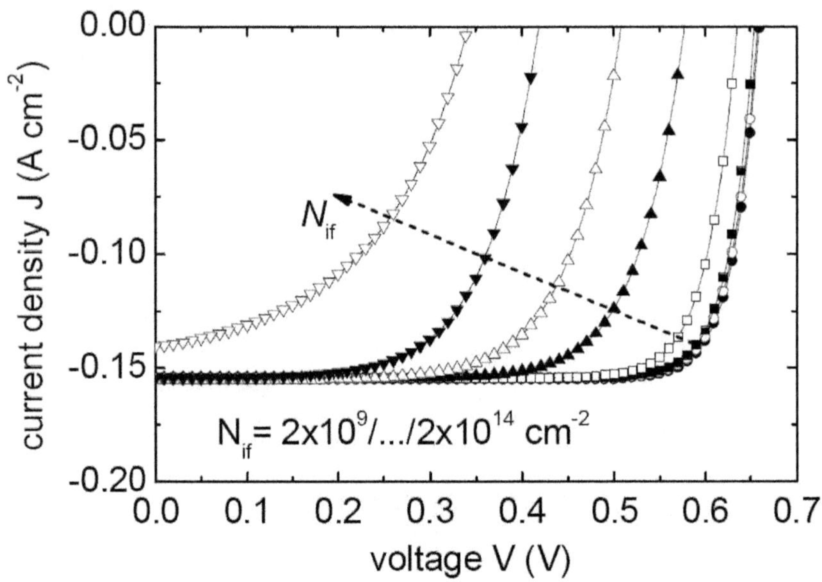

Fig. 8.8 Numerically calculated current-voltage curves of a-Si:H/c-Si heterojunctions for different interface defect densities $N_{if} = 2 \times 10^9, 2 \times 10^{10}, 6 \times 10^{10}, 2 \times 10^{11}, 6 \times 10^{11}, 2 \times 10^{12}, 2 \times 10^{13}, 2 \times 10^{14}$ cm^{-2} at 300 K and with 10^{18}cm^{-2} s^{-1} photon flux (λ = 782 nm); *J-V*-curves of three lowest N_{if} are almost identical, after [7].

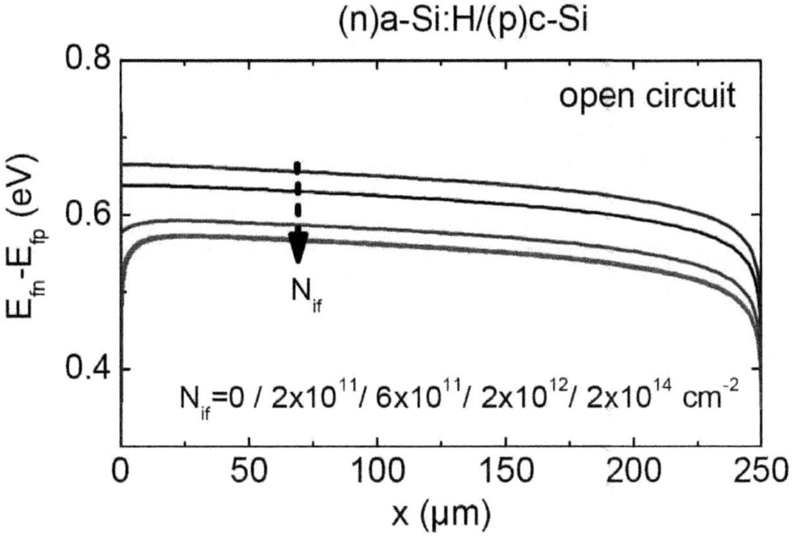

Fig. 8.9 Variation of the quasi-Fermi level splitting for different interface defect densities, simulated at open circuit, after [32]. The a-Si:H/c-Si interface is on the left, an Ohmic back contact on the right.

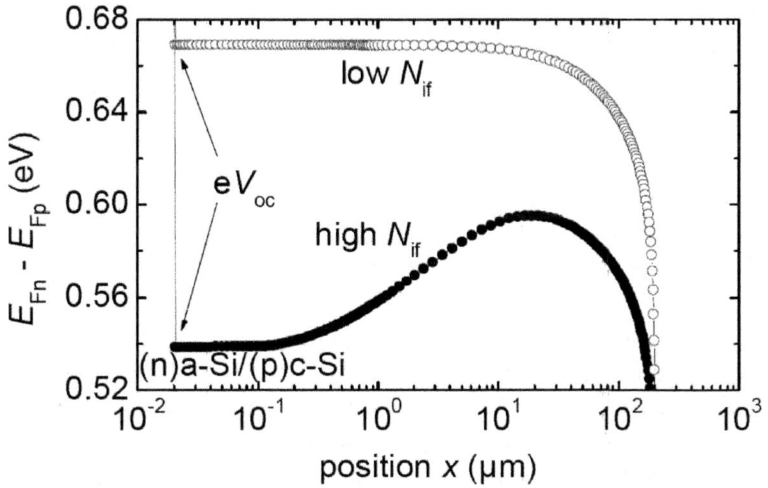

Fig. 8.10 Quasi-Fermi level splitting for two different interface defect densities from Fig. 8.9, indicating that the PL emission for higher defect density is higher than would be expected from the V_{oc} value, after [32].

For an (n)a-Si:H/(p)c-Si solar cell, Fig. 8.8 shows the variation in the current-voltage curves for a variation in the interface defect density N_{if} over 5 orders of magnitude [7,32], simulated here with SC-Simul [27]. These results here under monochromatic illumination at 782 nm wavelength with almost no change in the short-circuit current and a decrease in the open-circuit voltage are similar with other illumination conditions like AM1.5, simulated with AFORS-HET, e.g. [6].

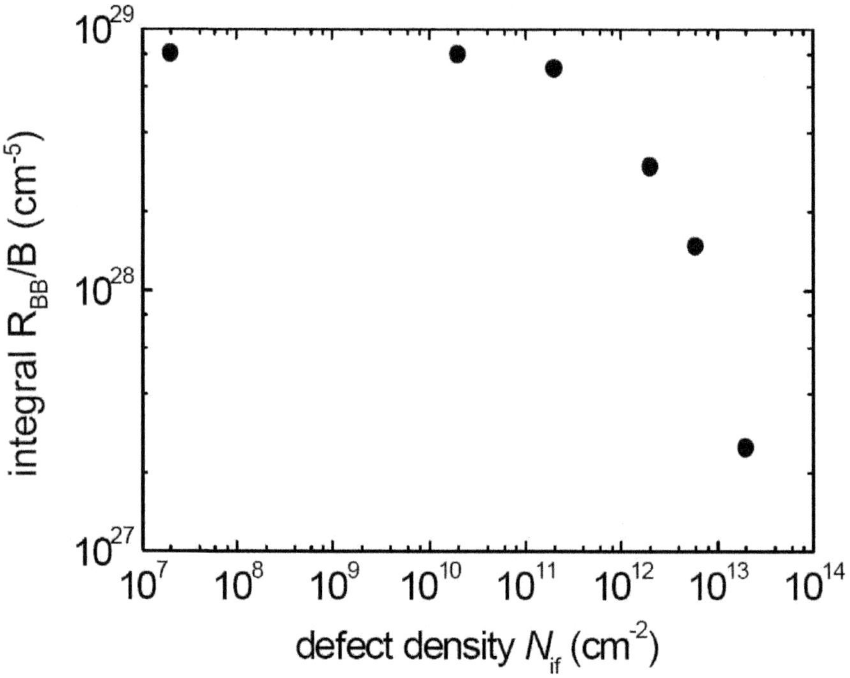

Fig. 8.11 Decrease in the simulated radiative recombination rate with increasing interface defect density, determined at the same dark-current density. Basically, the ordinate is integral of the *np* product according to eq. (8.3) [33].

The relation with the quasi-Fermi level splitting, from which the luminescence intensity is determined, is depicted in Fig. 8.9. The Ohmic back contact on the right leads to carrier loss and a reduction in $E_{Fn} - E_{Fp}$ towards the rear. At the junction on the left side, and with no additional band bending between the contacts and the junction, the quasi-Fermi level splitting there corresponds to the open-circuit voltage for low N_{if}. This is illustrated in Fig. 8.10 which also shows the case for high N_{if}. It is thus possible, as shown in Fig. 8.10, that the PL emission, represented by $E_{Fn} - E_{Fp}$ in the volume of the solar cell, overestimates the

V_{oc} value as for high N_{if} the $E_{Fn} - E_{Fp}$ values in the bulk of the solar cell are much higher than the corresponding eV_{oc}.

Similar to the strong influence on the photoluminescence properties, the interface-induced recombination also results in a strong decrease of electroluminescence with increasing N_{if}. Fig. 8.11 [33] shows the strong decrease of the radiative recombination rate with increasing N_{if}, simulated for constant dark-current density.

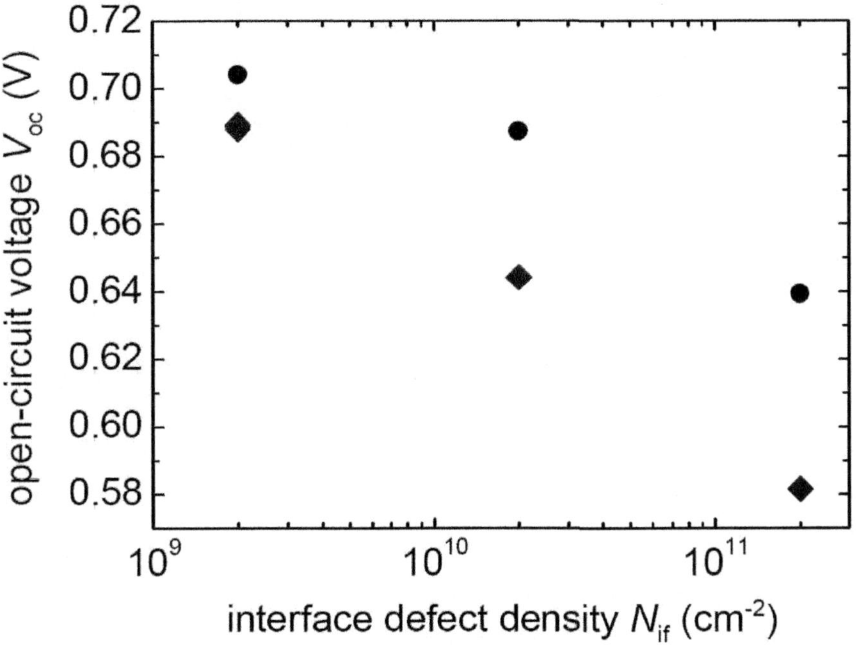

Fig. 8.12 Open-circuit voltage variation for three values of interface defect density and a factor of ten difference in the capture coefficients, respectively. Higher capture coefficients (diamond symbols) reduce the open-circuit voltage.

It is noted that it is not the interface defect density itself but the product of the corresponding capture coefficient and defect density which determines the interface recombination rate. The effect is illustrated in Fig. 8.12 where for one value of N_{if} different open-circuit voltages are achieved by changing the capture coefficients.

It is possible to identify the rate-limiting step in the interface recombination by systematically varying the capture coefficients for electrons and holes. In the dangling-bond model, sketched in Fig. 8.13, there are two channels with four capture-rate terms [34,35]. E.g., path II involves the $D^{0/-}$ transition with both hole capture

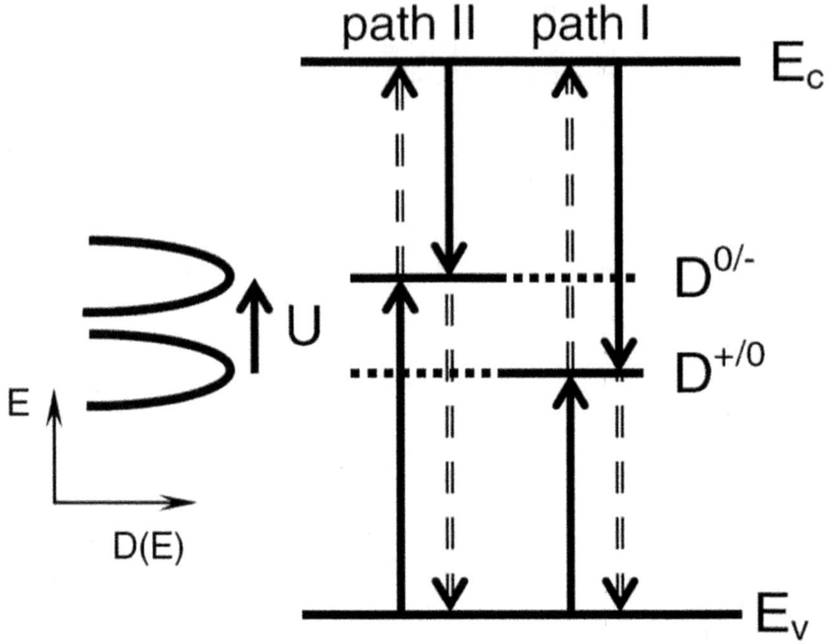

Fig. 8.13 Interface-defect density model with Gaussian distribution of dangling-bond states and correlation energy U [44].

into a negatively charged dangling bond (capture coefficient C_p^-) and electron capture into a neutral dangling bond (capture coefficient C_n^0).

A number of dark and photocurrent-voltage curves in Fig. 8.14 (top) show some variation for three different values of N_{if}. An increase in the saturation current density leads to an increase in the dark current density and a corresponding decrease in the open-circuit voltage. The important point is that the curves are almost identical for the simulations with identical $C_p^- N_{if}$ product values, e.g., the full triangles and open diamonds. This holds for both dark and illuminated curves in Fig. 8.14 (top). The variation in open-circuit voltage leads to a variation in the PL intensity at open circuit, not shown here.

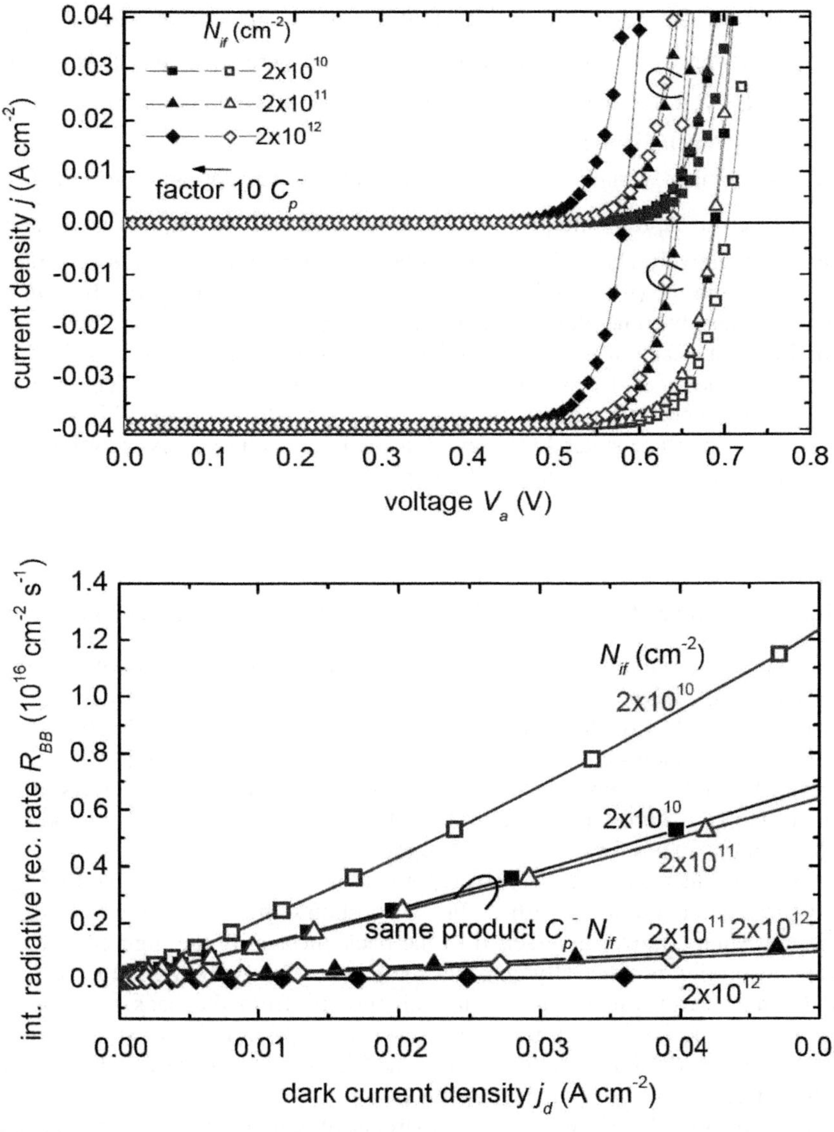

Fig. 8.14 Current-voltage curves upon variation in the capture coefficient. Almost identical values result from the simulations with identical $C_p^- N_{if}$, e.g., full triangles and open diamonds, for both dark and illuminated curves (top). The radiative recombination rate shows a linear dependence on dark current density (bottom). Here, too, the interfaces with identical $C_p^- N_{if}$ value have the same bulk radiative recombination rates. Variation in other capture coefficients had negligible effect.

The EL intensity or the radiative recombination rate in Fig. 8.14 (bottom) shows a linear dependence on dark current density for all combinations of C_p^- and N_{if}, indicating a diode quality factor $m = 1$. And here, too, the interfaces with identical $C_p^- N_{if}$ value have the same bulk radiative recombination rates, e.g., full triangles and open diamonds.

The systematic variation of C_p^- and N_{if} and of other capture coefficients like C_n^+ for electron capture into positively charged states shows that it is the hole capture into negatively charged dangling bonds that determines the interface recombination rate in the (n)a-Si:H/(p)c-Si solar cells. It was noted by Rösch *et al.* [36] that the holes become the minority carriers in the p-type wafer close to the a-Si:H/c-Si interface in which most of the defects are negatively charged. The results by Kleider *et al.* [37,38] are in agreement with these findings. These authors showed that there is an inversion layer at the (p)c-Si interface [37] and using monovalent interface states [38] they identified the capture cross-section of holes determines the recombination rate at the interface.

8.5.2 Photoluminescence

8.5.2.1 Influence of the Optical Properties on the Luminescence Spectrum

From eq. (8.5), it can be seen that the optical properties in terms of absorption coefficient may have a strong influence on the luminescence spectrum. Fig. 8.15 illustrates the effect of PL spectra from a wafer in its polished and textured form where the latter shows enhanced absorption due to optical scattering.

The lifetime determination with modulated photoluminescence shows comparable or slightly shorter MPL-lifetime for the textured wafer but the luminescence yield from the textured wafer is much larger.

A strong effect on the emission output can also be achieved by modifying the back-surface reflection properties. Fig. 8.16 shows the enhancement in the PL emission when a mirror is attached to the back surface. In the high-energy range where absorption is spectrally constant there is no change between the two spectra. In this energy range, any additional photons from the back reflection are absorbed. In the low-energy range, however, absorption is lower so that there is outcoupling of additional photons, here. The PL intensity increases although there is no change in the excess carrier densities.

The two examples show that a comparison of samples may become difficult, unless the outcoupling of photons is identical or can be taken into account.

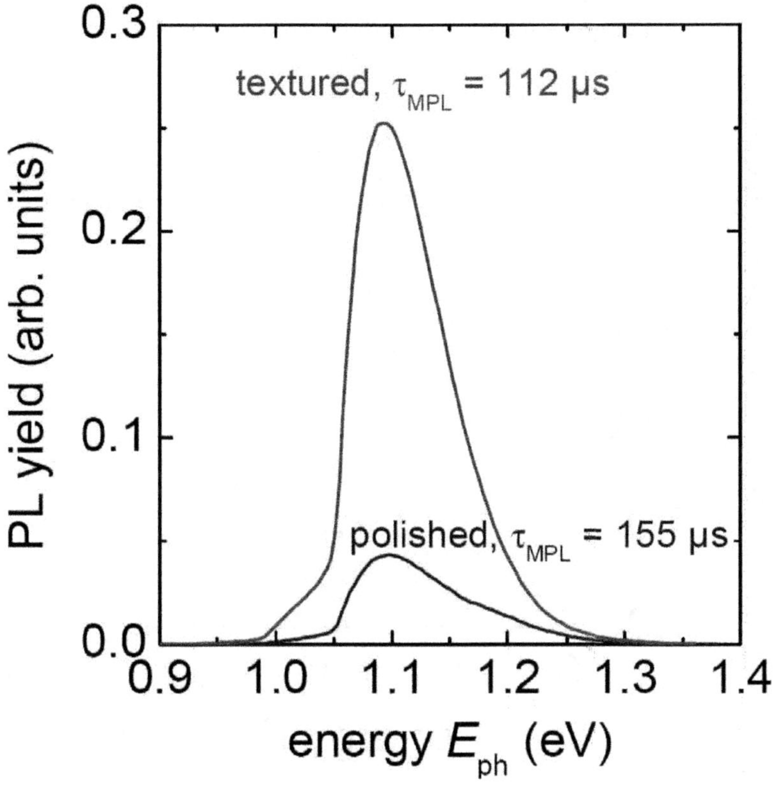

Fig. 8.15 Influence of texture on the emitted room-temperature PL. The minority carrier non-radiative lifetime was determined by the MPL phase shift (after [23]).

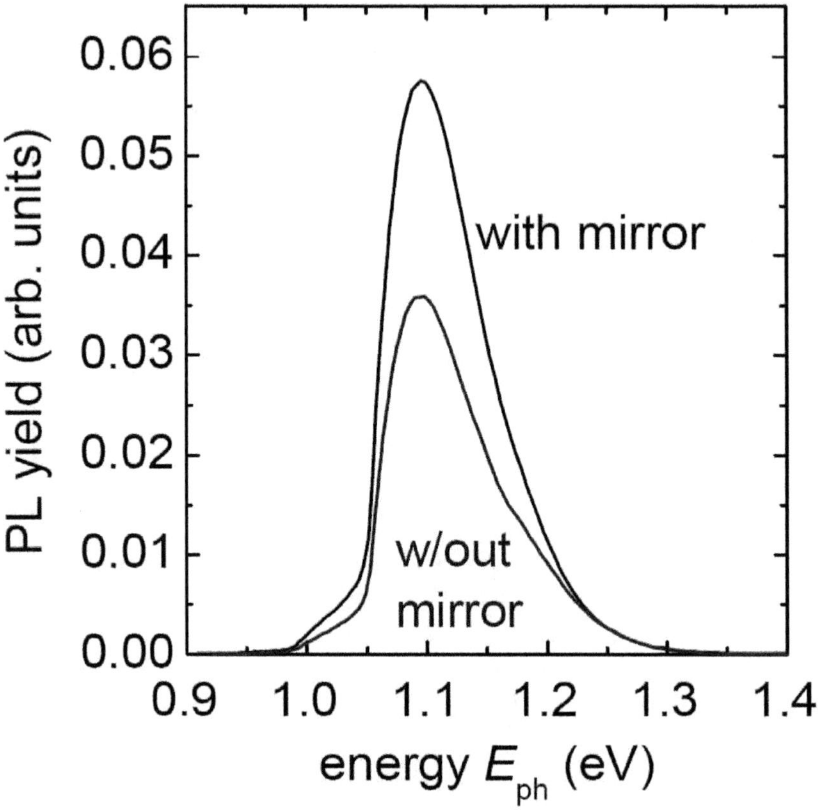

Fig. 8.16 Influence of back-reflection on the measured room-temperature photoluminescence of a c-Si wafer at room temperature. A mirror was attached to the back surface in order to change the absorptivity. Note that at photon energies > 1.25 eV the PL yield is not changed, (after [23]).

8.5.2.2 Spectral Photoluminescence, Quasi-Fermi Levels and Lifetime

PL analysis is useful for characterisation of heterostructures that consist of c-Si wafers passivated by layers of thin-film silicon or silicon alloys. In principle, wafers with different doping and different bulk lifetimes can be compared for optimized choice of the wafer material. For heterostructure optimisation, the same wafer or wafers from the same batch are often used onto which different series of amorphous or microcrystalline layers are deposited. In principle, layer sequences with symmetrical or asymmetrical structures can be investigated. One parameter like the deposition temperature or nominal substrate temperature, the doping-gas ratio in the feed gases or the pressure during thin-film deposition then varies for optimization.

Fig. 8.17 shows the room-temperature relative PL spectra of passivated (p)c-Si wafers for which the substrate temperature T_s for the (n)a-Si:H layer deposition was varied between 65 °C and 300 °C [39]. The PL intensity at T_s = 210 °C is highest. The results indicate that between 140 °C and 300 °C an optimized deposition temperature can be found. Corresponding results for (i)a-Si:H layer deposition showed that optimum conditions were at 230 °C [40].

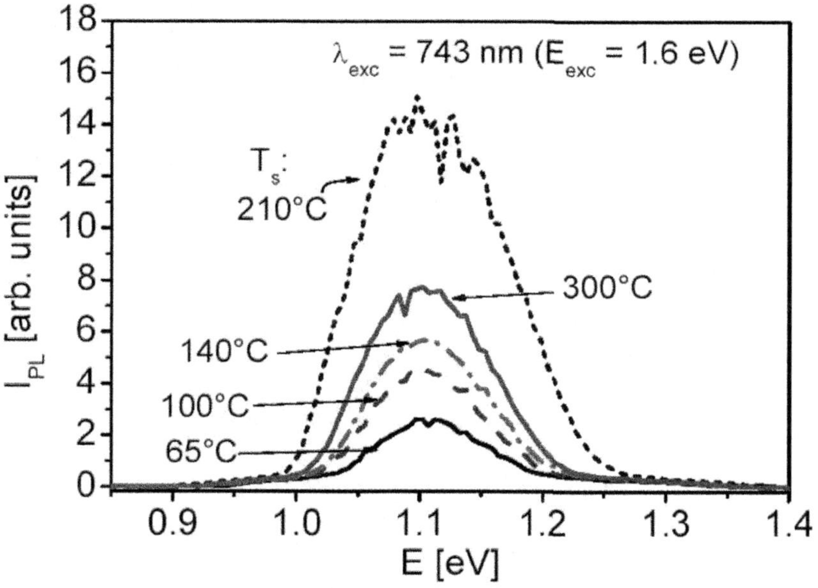

Fig. 8.17 Variation of the relative PL intensity vs. photon energy *E* for different substrate temperatures for the deposition of the (n)a-Si:H layers, from [39]. Reprinted with permission from IEEE, © 2006 IEEE.

The beneficial effect of adding hydrogen to the source gases of the a-Si:H deposition is depicted in Fig. 8.18 [30]. Here, the hydrogen dilution with 10% hydrogen in silane results in a higher PL signal at the maximum positions. The relative increase in the high-energy region where the quasi-Fermi levels are determined is more difficult to see. The authors argued that these PL results correspond with the variation in interface defect density from surface-photovoltage measurements.

The introduction of an intrinsic buffer layer can also be analysed by the contactless PL measurement without the need to prepare full solar cells [41]. The increase in the PL signal upon inserting a 5 nm (i)a-Si:H layer between a (p)c-Si wafer and a 5 nm (n)a-Si:H layer is shown in Fig. 8.19. Here, the moderate increase in the PL signal for the wafer with (i)a-Si:H buffer layer is translated into an increase in expected open-circuit voltage of about 3 mV. This marginal

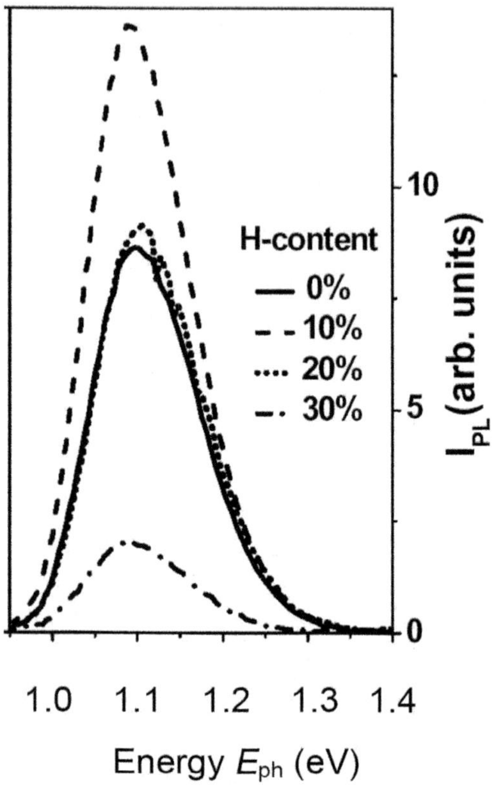

Fig. 8.18 Variation of the relative PL intensity for different hydrogen dilution for the deposition of the (n)a-Si:H layer, after [30].

increase was also found in the open-circuit voltage of solar cells with and without buffer layer. The example may not be sufficient to prove the beneficial effect of a buffer layer in general but it shows that PL results may track the open-circuit voltage of the solar cells prepared with these layers.

The effect of post-processing steps after a-Si:H layer deposition in combination with the deposition of, e.g., TCO can also be analysed by PL measurements [42].

Figure 8.20 shows two sets of PL spectra for a poor and a high-quality front interface with samples where the back surface consisted of a conventional Al BSF. The low-quality front interface was produced by exposing the c-Si surface to a damaging plasma prior to the (n+)a-Si:H deposition. The PL spectra before and after deposition of the front ITO layer, as well as after the final front metallization step show two tendencies: we observe a decrease of the PL signal after the ITO deposition step and the PL signal increases again after the final grid metallization step such as to become even slightly higher than before the ITO deposition. The spectral shape for the high-quality data changes slightly, i.e., the decrease in the PL maximum after ITO deposition is not found in the high-energy spectral range. This may be related to a change in the outcoupling of PL photons.

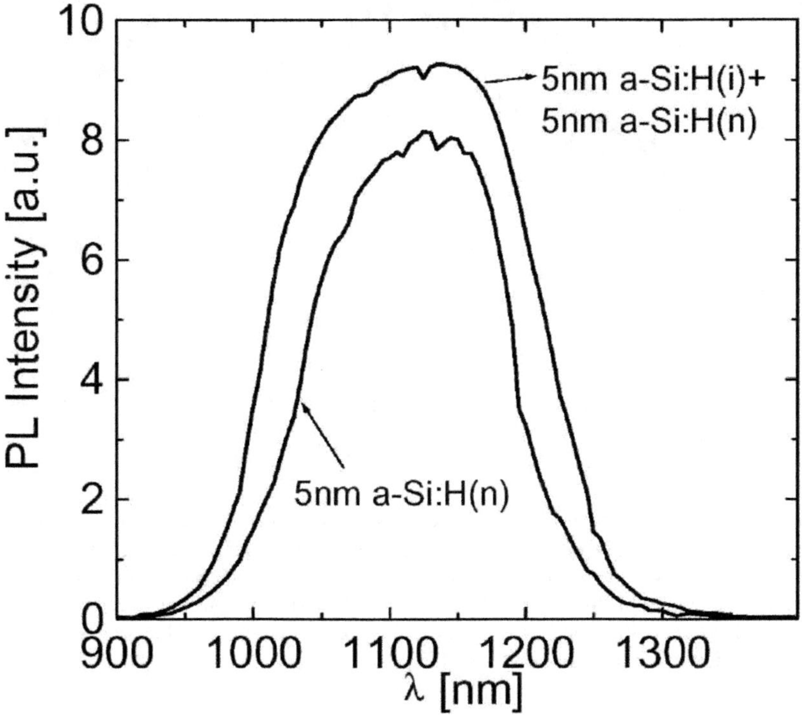

Fig. 8.19 Relative PL intensities vs. wavelength for different hydrogen dilution for the deposition of the (n)a-Si:H layer, from [41]. Reprinted with permission from IEEE, © 2003 IEEE.

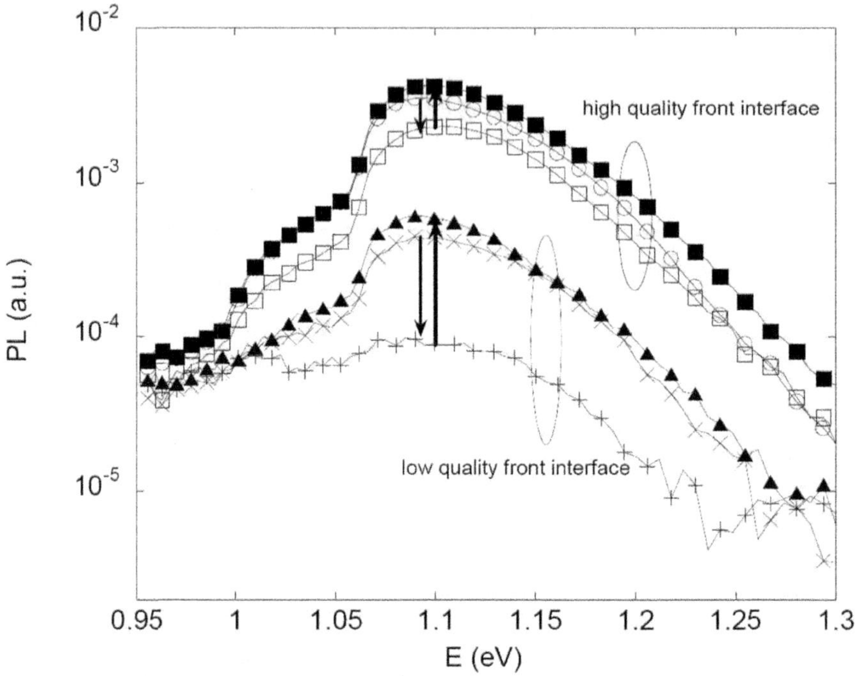

Fig. 8.20 Relative PL spectra *vs.* photon energy E after various steps of the solar cell processing for a high quality front (n+)a-Si:H interface layer (o) before ITO deposition, (□) after ITO deposition, (■) after ITO deposition and subsequent front grid metallization, and for a low quality front (n+)a- Si:H interface (×) before ITO deposition, (+) after ITO deposition, (▲) after ITO deposition and subsequent front-grid metallization. The back-surface conditions were identical in all samples, from [42].

In [42], the variation of the PL was attributed to the creation of interface defects during ITO deposition, and to a defect removal process after front grid metallization. It was mentioned that although only low temperatures ($T < 200°C$) are used in these steps, the front grid metallization includes a thermal annealing step that can lead to a reduction of front interface defects.

It is noted that the relevant quantity for the excess carrier density is not the maximum of the PL signal but its value in the high-energy range in which the absorption is constant. The comparison in the examples above is only allowed because the basis of the investigation is the same wafer for all the samples of a series with similar reflection properties of the air/semiconductor surface. In this case any differences in the outcoupling of photons can be neglected so that the range of maximum PL intensity can be taken as a representative value for comparison.

In a step further, the quantitative rather than the relative PL spectra can be determined and analysed for the corresponding quasi-Fermi level splitting. Fig. 8.21 shows the experimental and analytical room-temperature quantitative PL spectrum from a (p)c-Si wafer [43]. Very good agreement is achieved also for a detail like the shoulder in the low-energy range and the variation in the high-energy range where the quasi-Fermi level splitting of 682 meV was determined.

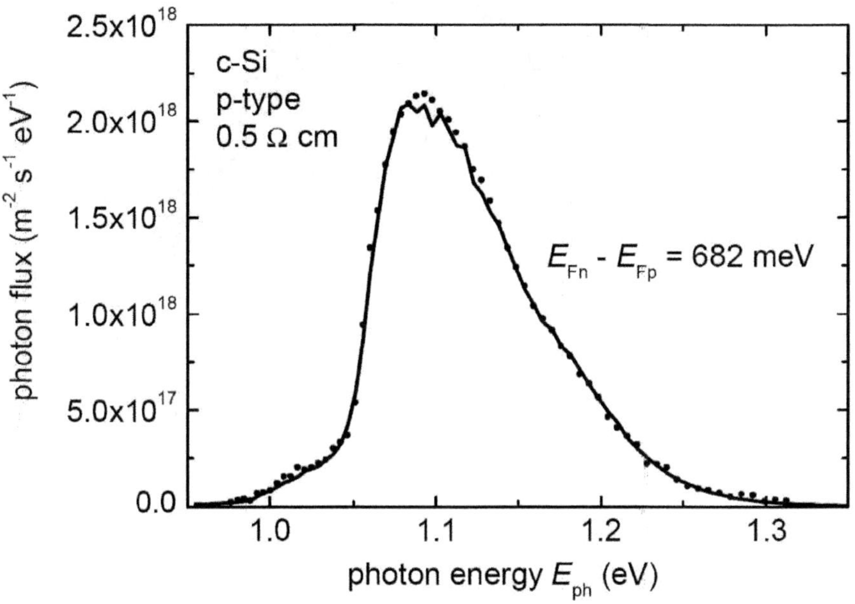

Fig. 8.21 Experimental (dots) and analytical room-temperature PL spectra from a c-Si wafer symmetrically passivated with a-Si:H layers. The full line is calculated with eq. (8.5) with constant quasi-Fermi level splitting and energy-dependent crystalline-silicon absorption coefficient (showing in fact some scatter at the maximum) and energy-independent reflection coefficients were used for the determination of A [43].

Figure 8.22 shows the quantitative PL spectra from a sequence of different layers, deposited on a (p)c-Si wafer [44]. These quantitative PL spectra were measured after different steps of c-Si wafer and a-Si:H/c-Si solar cell processing (full symbols) and of symmetric structures (open symbols). The dotted lines are calculated with the analytical expression of Planck's generalised law [eq. (8.5)] from which the corresponding quasi-Fermi level splittings were calculated. The structures a, b, c labelled in the legend are sketched also in Fig. 8.22.

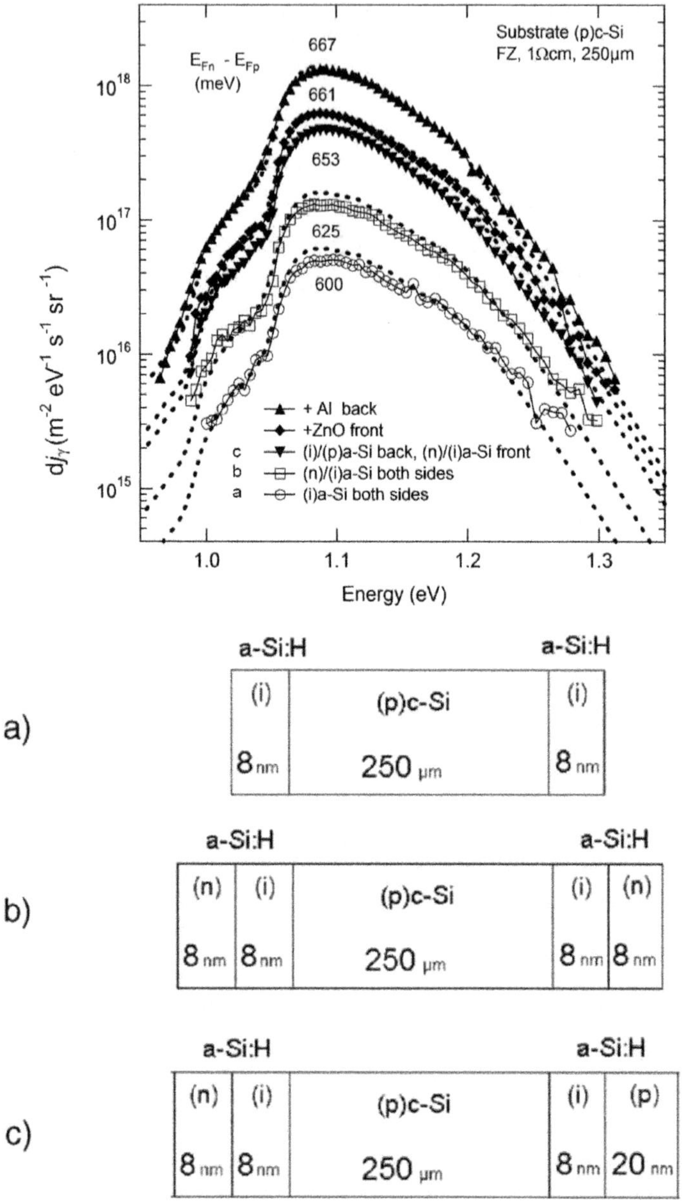

Fig. 8.22 Quantitative PL spectra vs. photon energy E after different steps of c-Si wafer and a-Si:H/c-Si solar cell processing (full symbols) and of symmetric structures (open symbols) with the corresponding quasi-Fermi level splitting (left). The dotted lines are calculated with the analytical expression of Planck's generalised law. The legend is labelled for the structures a, b, c, sketched on the right. The (i) layer is the same with subsequent deposition of the doped layers, from [44].

Especially, the design of these structures a, b, c leads to a pronounced variation in the quasi-Fermi level splitting. Using numerical modelling with SC-Simul, this variation could be tracked by the systematic variation in the dangling-bond like interface defect density, detailed in [44]. Fig. 8.23 compares the quasi-Fermi level splitting from the simulation (full circles) and from different measurement positions on the wafers (open triangles). Simulation of a solar cell with the same interface defect parameters also resulted in comparable open-circuit voltage which corroborates the suggested findings. The plausible defect position from these simulations is 0.7–0.8 eV above the c-Si valence band edge and a Gaussian width in the range of 0.1 eV

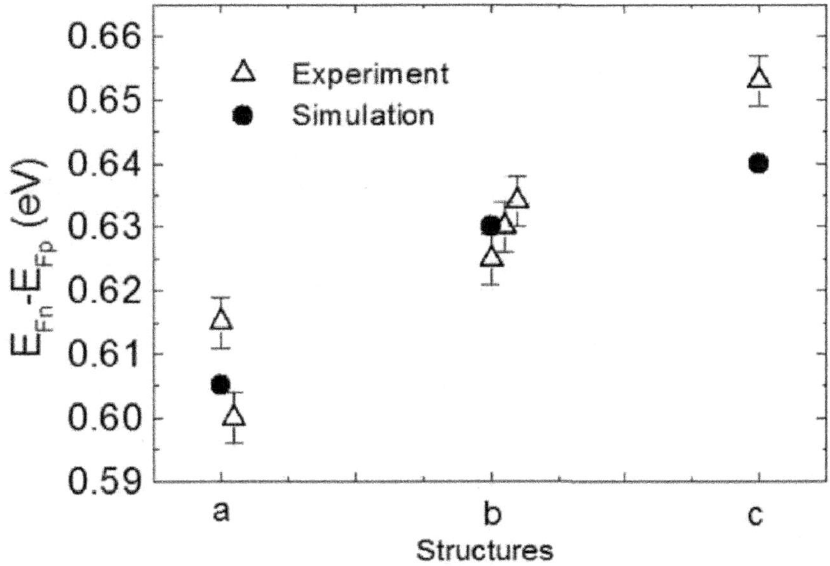

Fig. 8.23 Experimental and simulation results for the quasi-Fermi level splitting for three different a-Si:H/c-Si wafer structures, labelled a, b, c with the different passivation layers.

8.5.2.3 Modulated Photoluminescence

The MPL method is demonstrated for c-Si wafers with different passivation schemes, applied to polished 250 μm thick p-type c-Si wafers from the same batch with nominally identical properties [23]. Symmetric samples were prepared with a-Si:H and SiN_x:H passivation layers on both sides.

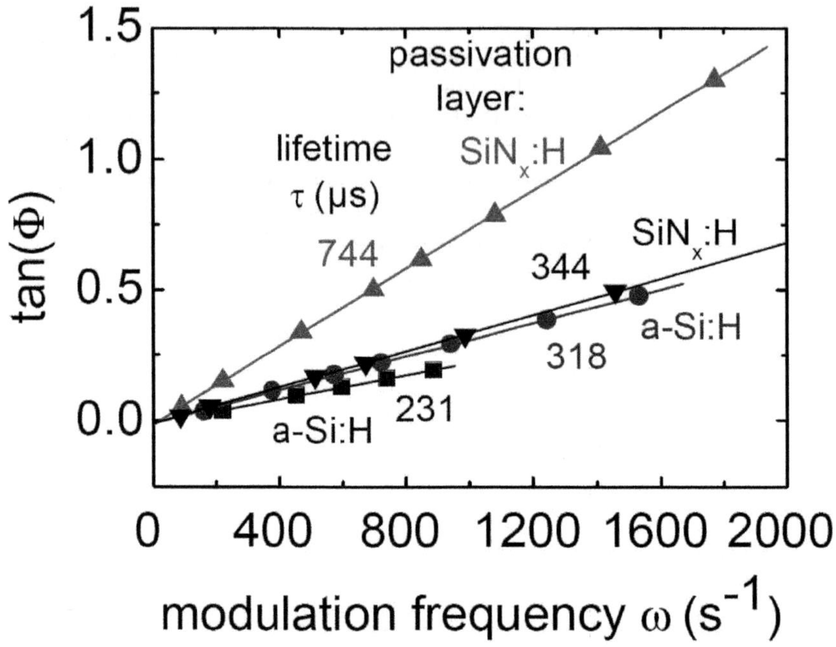

Fig. 8.24 Experimentally determined linear variation of tan **Φ** vs. modulation frequency for two silicon nitride and two amorphous silicon passivation layers deposited on p-type c-Si, from [23].

Figure 8.24 shows the variation of tan(Φ) for a wafer with four different passivation layers. One silicon nitride passivation shows the highest lifetime with some smaller values for the second SiN layer which shows comparable lifetime to a-Si:H layer passivation. It is noted that the functional dependence of experimental tan(Φ) vs. frequency is linear as required by eq. (8.19).

After checking the linear relationship between phase and modulation frequency as in Fig. 8.24, the effective minority-carrier lifetime can be determined from the phase measurement at a single frequency [23]. Such a measurement is evaluated for a chopping frequency of 135 Hz on a polished wafer with natural oxide on both sides in Fig. 8.25. The figure shows the time-dependence of the lifetime, after applying a few drops of iodine/ethanol solution to one side of the wafer. Such a solution has been demonstrated to reduce the density of surface defects [45]. After an increase in lifetime within a few seconds, a sharp drop is observed after about a few tens of seconds.

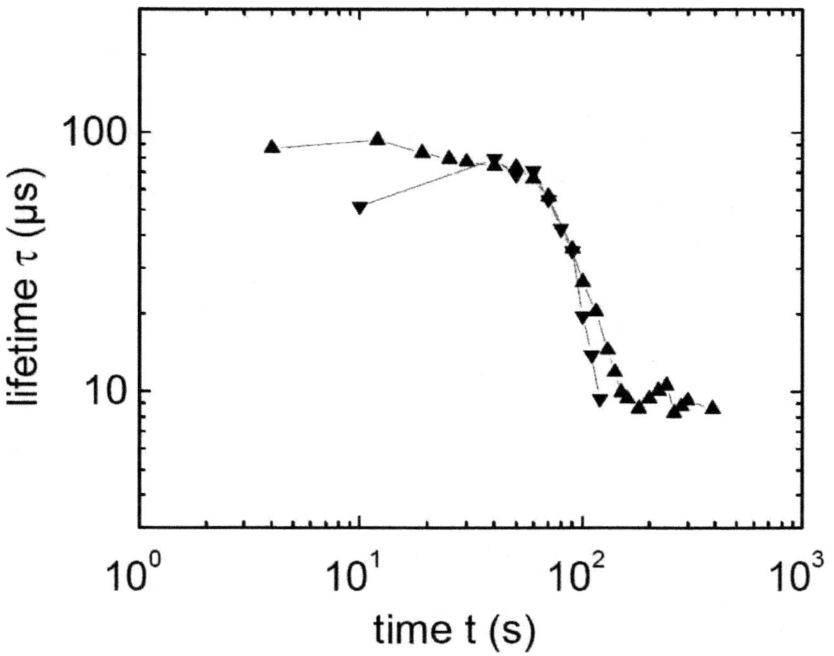

Fig. 8.25 Temporal variation of the MPL lifetime. Two sequences of the time-dependence of the passivation quality of iodine/ethanol solution in air. Upon evaporation the lifetime drops by one order of magnitude.

The MPL method may also be used for investigations of the lateral homogeneity of passivation schemes by performing a line or area scan. In a related way, Fig. 8.26 illustrates the spatial variation of the MPL lifetime from a line scan across a multicrystalline silicon wafer. Here, the MPL amplitude is shown for comparison.

The examples show that the simple MPL set-up, monitoring directly the phase and amplitude of the integrated photoluminescence as detected by a photodiode, is a versatile tool for lifetime characterisation. It can be applied for detailed studies on the link between the effective carrier lifetime and preparation/deposition parameters. Local variations in effective lifetime MPL from inhomogeneity of the passivation properties or in multicrystalline silicon samples are easily detectable. An advantage of the MPL phase measurement may be that the experimental error in comparative measurements is smaller as the phase is less affected by variations in either excitation flux, possible misalignment of samples or local variation in the outcoupling properties of the PL.

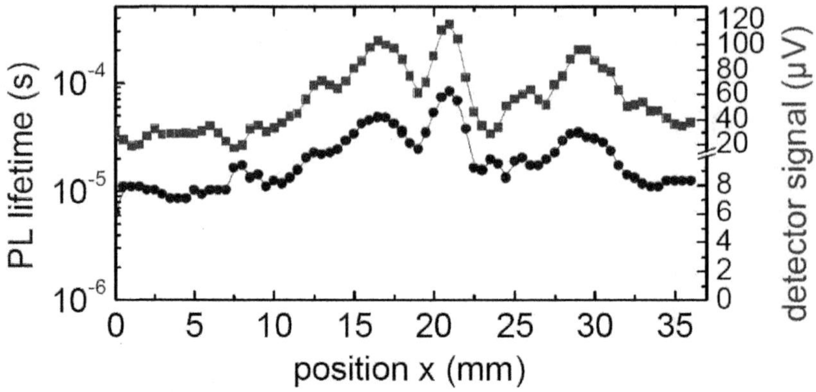

Fig. 8.26 Spatial variation of the MPL lifetime and amplitude for a multicrystalline silicon sample.

8.5.2.4 Comparison of Different Methods

A comparison of the PL results with the lifetime determination from different methods can be helpful to corroborate the respective results. Fig. 8.27 shows the correlation between the effective lifetime τ_{eff} from microwave photoconductive decay (µW-PCD) measurements and the relative PL yield for a number of polished (p)c-Si and n(c)-Si wafers with different a-Si:H passivation layers. The letters in the figure denote the type of doping and the thickness of the a-Si:H layers. The proportionality between decay lifetime and stationary PL-yield indicates that an effective lifetime assumption that includes both bulk and interface recombination processes is suitable to describe both the time-dependent and stationary experiment.

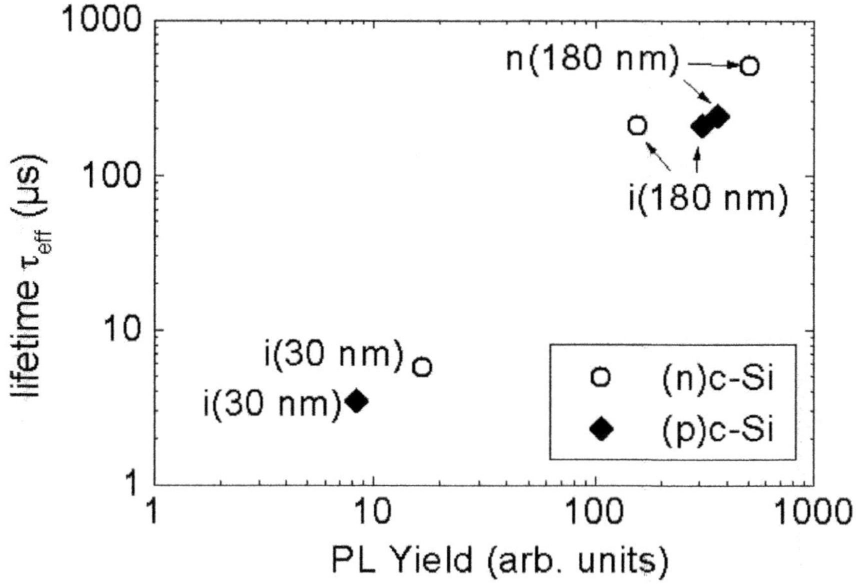

Fig. 8.27 Comparison of the MWPCD lifetime (ordinate) and the PL yield for (p)c-Si and n-c-Si wafers with different a-Si:H passivation layers. The letters in the figure denote the type of doping and thickness of the a-Si:H layer, from [7].

Figure 8.28 shows a comparison between the MPL lifetime and the lifetime from the quasi-steady state photoconductance method (QSSPC). The data points represent measurements on passivated symmetrical wafer structures (polymorphous silicon (pm-Si:H), a-Si:H on nominally identical c-Si wafers) [45]. The straight line indicates a good proportionality between the two different lifetimes. Obviously, there is a discrepancy at short lifetimes. Here, the MPL lifetime is determined to be too low or the QSSPC lifetime too high. More work, taking into account the photogeneration rate profiles may resolve this issue. In addition, numerical modelling would be helpful to assist in identifying any shortcoming of either approach for short minority-carrier lifetimes.

From a pragmatic point of view, characterisation by MPL of high quality interfaces in the long lifetime range is well suited, especially for a-Si:H/c-Si structures which are unsuitable for the QSSPC method because of small lateral size or if metallic contacts are present.

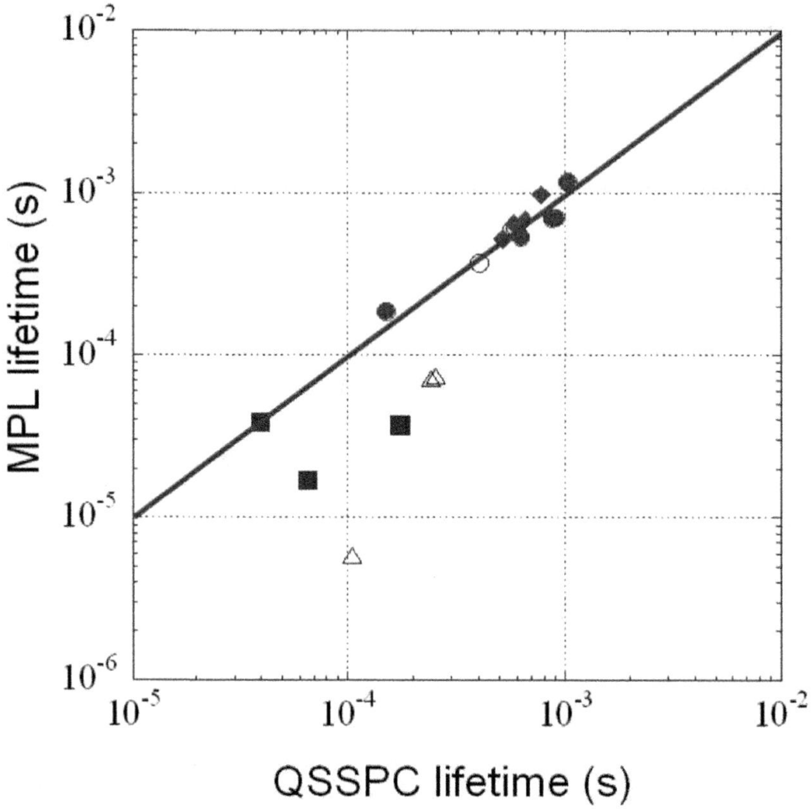

Fig. 8.28 Comparison of the MPL lifetime and the QSSPC lifetime for a number of a-Si:H and pm-Si:H passivated c-Si wafers. The full line is a guide to the eye and indicates the proportionality between the two lifetimes, from [45].

8.5.3 Electroluminescence

8.5.3.1 Temperature-Dependence of the Electroluminescence Yield

Experimental results on the temperature dependence of the EL intensity at constant dark current density show an increase with increasing temperature [18]. For a heterodiode with presumably high interface defect density an almost temperature independent behaviour was determined [16]. One argument for the decrease in the EL intensity or the increase in the interface recombination rate with decreasing temperature is the larger splitting for the quasi-Fermi levels at the a-Si:H/c-Si interface at lower temperature so that interface recombination is enhanced.

8.5.3.2 Electroluminescence Yield and Photovoltaic Properties

From the relation between the open-circuit voltage and the radiative photon flux in eq. (8.12) an exponential increase in the radiative recombination rate with V_{oc} is expected. Figure 8.29 shows that the simulation of different a-Si:H/c-Si heterojunction solar cells confirms this relation. Here, the variation in N_{if} is responsible for the variation in V_{oc}. It is noted that the variation in the thickness of the amorphous emitter layer has almost no effect. The straight line represents a slope of $(kT)^{-1}$ as expected from eq. (8.12).

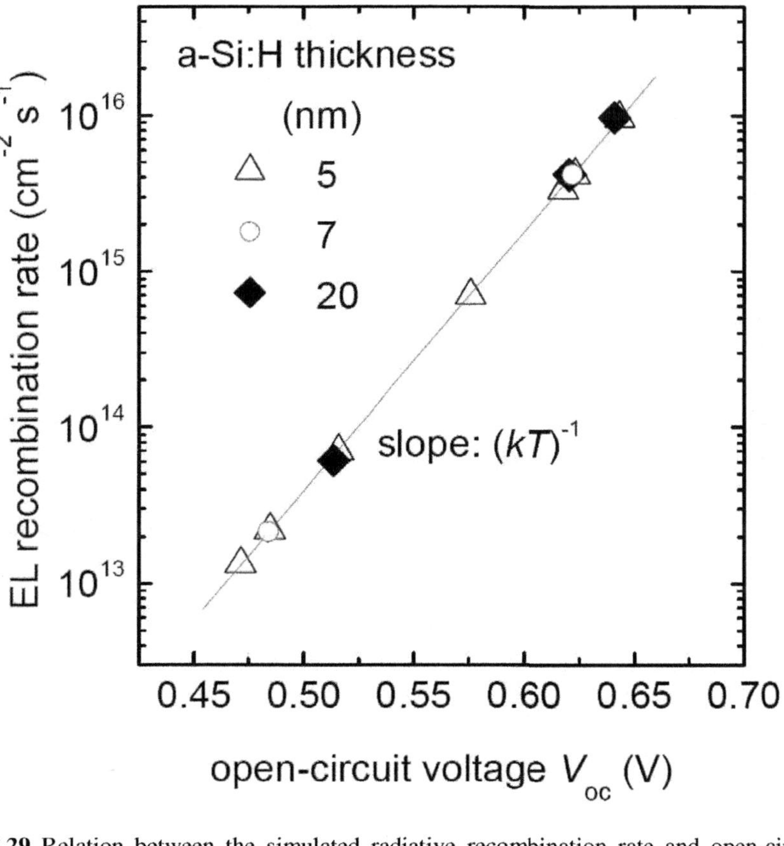

Fig. 8.29 Relation between the simulated radiative recombination rate and open-circuit voltage. The variation in the two quantities is achieved by a variation in the interface defect density. Solar cells with different a-Si:H layer thickness lead to the same EL results.

Heterojunction solar cells with high open-circuit voltage are thus expected to have a high electroluminescence efficiency. For two heterostructure solar cells it was found that the EL intensity at room temperature was higher for the cell with the higher open-circuit voltage [18]. Figure 8.30 shows the experimental EL spectrum in terms of emitted photon flux from a double-heterostructure solar cell with open-circuit voltage around 700 mV from EPFL in Neuchâtel, measured at a current density of 106 mA cm^{-2} [33]. Integration of the spectrum yields the EL quantum efficiency of number of photons per electron, $\eta_{QE} = e\Phi / j_d$ of 0.13%, a high value which agrees with the simulated result for such an open-circuit voltage [33].

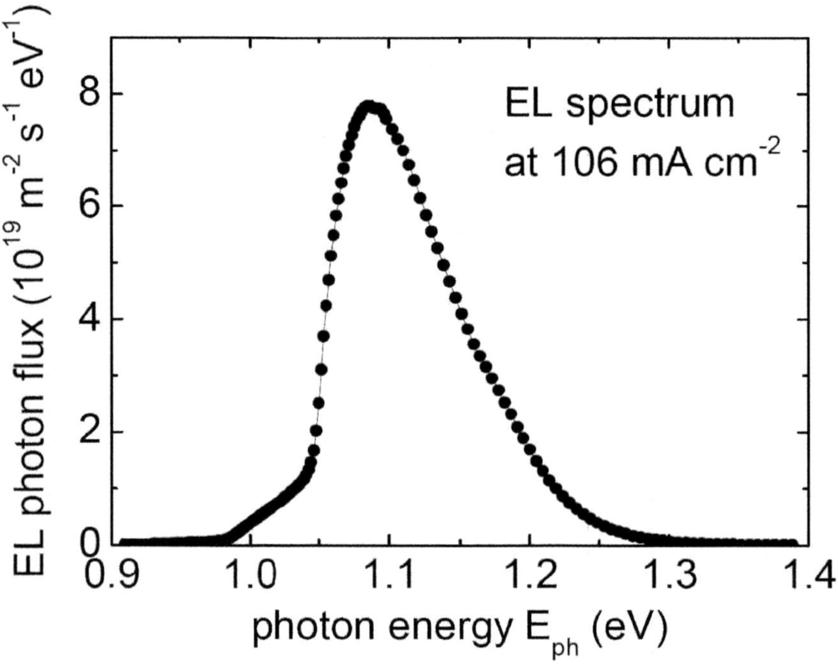

Fig. 8.30 Experimental EL spectrum from a a-Si:H/c-Si solar cell, recorded at a dark current density of 106 mA cm^{-2} [33].

8.6 Conclusion

Photoluminescence techniques are helpful tools for the characterisation of a-Si:H/c-Si heterostructures, allowing a contactless and qualitative comparison between samples of different interface qualities or passivation-layer sequences. Quantitative photoluminescence has been shown to reveal information about interface defect distributions. The capability of the MPL technique for lifetime determination was demonstrated. It is possible to monitor time-dependent changes of surface treatments of c-Si wafers or lateral measurements with a possible variety of further applications. In-line monitoring for quality control of c-Si passivation can easily be achieved by MPL phase measurements. The relation between the open-circuit voltage and electroluminescence yield was outlined and demonstrated by simulation. The highly-efficient EL emission from an a-Si:H/c-Si solar cell was related to the high open-circuit voltage of this device.

Acknowledgements

The author is grateful to the colleagues who contributed to the variants of the PL set-up in Oldenburg (T. Unold, S. Tardon, S. Meier, J. Behrends, F. Heidemann, S. Burdorf, P. Pargmann, G.H. Bauer), which was employed for some of the reported results.

References

[1] Würfel, P.: Physics of Solar Cells. Wiley, Weinheim (2004)

[2] Bauer, G.H., Würfel, P.: Quantum solar energy conversion in organic solar cells. In: Brabec, C.J., Dyakonov, V., Parisi, J. (eds.) Organic Photovoltaics, pp. 118–158. Springer, Berlin (2003)

[3] Taguchi, M., Kawamoto, K., Tsuge, S., Baba, T., Sakata, H., Morizane, M., Uchihashi, K., Nakamura, N., Kiyama, S., Oota, O.: HIT cells—high-efficiency crystalline Si cells with novel structure. Progr. Photovolt. Res. Applic. 8, 503 (2000)

[4] Tanaka, M., Taguchi, M., Matsuyama, T., Sawada, T., Tsuda, S., Nakano, S., Hanafusa, H., Kuwano, Y.: Development of new a-Si/c-Si heterojunction solar cells: ACJ-HIT (Artificially Constructed Junction-Heterojunction with Intrinsic Thin-Layer). Jap. J. Appl. Phys. 31, 3518 (1992)

[5] Scherff, M.L.D., Froitzheim, A., Ulyashin, A., Schmidt, M., Fahrner, W.R., Fuhs, W.: 16.2 % efficiency for amorphous/crystalline silicon heterojunction solar cells on flat p-type silicon wafer. In: Proceedings of the International Conference on PV in Europe - from PV Technology to Energy Solutions, Rome, Italy, p. 216 (2002)

[6] von Maydell, K., Windgassen, H., Nositschka, W.A., Rau, U., Rostan, P.J., Henze, J., Schmidt, J., Scherff, M., Fahrner, W., Borchert, D., Tardon, S., Brüggemann, R., Stiebig, H., Schmidt, M.: Basic electronic properties and technology of TCO/a-Si:H(n)/c-Si(p) heterostructure solar cells: a German network project. In: Proceedings 20th European Photovoltaic Solar Energy Conference, Barcelona, Spain, p. 822 (2005)

[7] Tardon, S., Rösch, M., Brüggemann, R., Unold, T., Bauer, G.H.: Photoluminescence studies of a-Si:H/c-Si heterojunction solar cells. J. Non-Cryst. Solids 338&340, 444 (2004)

[8] Froitzheim, A., Stangl, R., Elstner, L., Schmidt, M., Fuhs, W.: Interface recombination in amorphous/crystalline silicon solar cells, a simulation study. In: Conference Record 25th IEEE Photovoltaic Specialists Conference, New Orleans, USA, p. 1238 (2002)

[9] Lasher, G., Stern, F.: Spontaneous and Stimulated Recombination Radiation in Semiconductors. Phys. Rev. 133, A553 (1964)

[10] Trupke, T., Green, M.A., Würfel, P., Altermatt, P.P., Wang, A., Zhao, J.: Temperature dependence of the radiative recombination coefficient of intrinsic crystalline silicon. J. Appl. Phys. 94, 4930 (2003)

[11] Würfel, P.: The chemical potential of radiation. J. Phys. C 15, 3967 (1982)
[12] Finkbeiner, S., Daub, E., Würfel, P.: Maximum open-circuit voltage for solar cell silicon from absolute intensities of photoluminescence. In: Proceedings of the 11th E.C. Photovoltaic Solar Energy Conference, Montreux, Switzerland, pp. 320–324 (1992)
[13] Green, M.A., Keevers, M.J.: Optical Properties of Intrinsic Silicon at 300 K. Progr. Photovolt. Res. Appl. 3, 189 (1995)
[14] Daub, E.: Photolumineszenz von Silizium, PhD thesis, Universität Karlsruhe (1995) (in German)
[15] Brüggemann, R., Behrends, J., Meier, S., Tardon, S.: Luminescence, quasi-Fermi levels and applied voltage in ideal and real semiconductor structures. J. Optoel. Adv. Materials 9, 77 (2007)
[16] Knabe, S., Langemeyer, M., Heidemann, F., Brüggemann, R., Bauer, G.H.: Study of the effect of excess carrier lifetime depth profiles and depth dependent absorption coefficients on the emitted photoluminescence spectrum of chalcopyrite thin films. Accepted by Prog. Photovolt. Res. Appl. (2011), doi:10.1002/pip.1094
[17] Würfel, P., Trupke, T., Puzzer, T., Schäffer, E., Warta, W., Glunz, S.W.: Diffusion lengths of silicon solar cells from luminescence images. J. Appl. Phys. 101, 123110 (2007)
[18] Fuhs, W., Laades, A., von Maydell, K., Stangl, R., Gusev, O.B., Terukov, E., Kazitsyna-Baranovski, S., Weiser, G.: Band-edge electroluminescence from amorphous/crystalline silicon heterostructure solar cells. J. Non-Cryst. Solids 352, 1884 (2006)
[19] Fuyuki, T., Kondo, H., Kaji, Y., Ogane, A., Takahashi, Y.: Analytic findings in the electroluminescence characterization of crystalline silicon solar cells. J. Appl. Phys. 101, 023711 (2007)
[20] Kirchartz, T., Rau, U., Kurth, M., Mattheis, J., Werner, J.H.: Comparative study of electroluminescence from Cu(In,Ga)Se_2 and Si solar cells. Thin Solid Films 515, 6238 (2007)
[21] Brüggemann, R.: Kirchhoff's generalised law applied to amorphous silicon / crystalline silicon heterostructures. Philos. Mag. 89, 2519 (2009)
[22] Heidemann, F., Brüggemann, R., Bauer, G.H.: The correlation between local defect absorbance and quasi-Fermi level splitting in $CuInS_2$ from photoluminescence. J. Phys. D: Appl. Phys. 43, 145103 (2010)
[23] Brüggemann, R., Reynolds, S.: Modulated photoluminescence studies for lifetime determination in amorphous-silicon passivated crystalline-silicon wafers. J. Non-Cryst. Solids 352, 1888 (2006)
[24] Schroder, D.K.: Semiconductor Material and Device Characterisation, p. 427. Wiley, New York (1998)
[25] AFORS-HET, chapters 13 and 14 in this book
[26] Datta, A., Damon-Lacoste, J., Roca i Cabarrocas, P., Chatterjee, P.: Defect states on the surfaces of a p-type c-Si wafer and how they control the performance of a double heterojunction solar cell. Solar Energy Materials and Solar Cells 92, 1500 (2008)
[27] Brüggemann, R., Rösch, M.: Application of SC-Simul - Numerical simulation for thin-film silicon based devices. J. Optoel. Adv. Materials 7, 65 (2005)
[28] Tardon, S.: Quantitative photoluminescence studies in a-Si:H/c-Si solar cells. PhD thesis, Universität Oldenburg (2005)
[29] Timoshenko, V.Y., Petrenko, A.P., Stolyarov, M.N., Dittrich, T., Füssel, W., Rappich, J.: Quantitative analysis of room temperature photoluminescence of c-Si wafers excited by short laser pulses. J. Appl. Phys. 85, 4174 (1999)

[30] Laades, A., Kliefoth, K., Korte, L., Brendel, K., Stangl, R., Schmidt, M., Fuhs, W.: Surface passivation of crystalline silicon wafers by hydrogenated amorphous silicon probed by time resolved surface photovoltage and photoluminescence spectroscopy. In: Nineteenth European Photovoltaic Solar Energy Conference, Paris, France, p. 1170 (2004)
[31] Lumb, M.D.: Photoluminescence spectroscopy. Academic Press, London (1978)
[32] Bauer, G.H., Brüggemann, R., Rösch, M., Tardon, S., Unold, T.: Numerical modelling as a tool for understanding room temperature photoluminescence in a-Si:H/c-Si heterojunction solar cells. Phys. Status Solidi. (c) 1, 1308 (2004)
[33] Brüggemann, R., Olibet, S.: Analysis of electroluminescence from silicon heterojunction solar cells. Energy Procedia 2, 19 (2010)
[34] Main, C., Berkin, J., Merazga, A.: Photoconductivity in amorphous semiconductors - experiment and computer modelling. In: New Physical Problems In Electronic Materials, p. 55. World Scientific, Singapore (1990)
[35] Okamoto, H., Kida, H., Hamakawa, Y.: Steady-state photoconductivity in amorphous semiconductors containing correlated defects. Philos. Mag. B 49, 231 (1984)
[36] Rösch, M., Brüggemann, R., Bauer, G.H.: Influence of interface defects on current voltage characteristics of amorphous silicon / crystalline silicon heterojunction solar cells. In: 2nd World Conference on Photovoltaic Solar Energy Conversion, Vienna, Austria, p. 946 (1998)
[37] Kleider, J.P., Gudovskikh, A.S., Roca i Cabarrocas, P.: Determination of the conduction band offset between hydrogenated amorphous silicon and crystalline silicon from surface inversion layer conductance measurements. Appl. Phys. Lett. 92, 162101 (2008)
[38] Kleider, J.P., Gudovskikh, A.S.: Characterization of amorphous / crystalline silicon interfaces from electrical measurements. In: MRS Symp. Proc., vol. 1066, p. 75 (2008)
[39] Schmidt, M., Angermann, H., Conrad, E., Korte, L., Laades, A., von Maydell, K., Schubert C., Stangl, R.: Physical and technological aspects of a-Si:H/c-Si heterojunction solar cells. In: 4th World Conference on Photovoltaic Energy Conversion, Hawaii, USA, p. 1433 (2006)
[40] Korte, L., Laades, A., Schmidt, M.: Electronic states in a-Si:H/c-Si heterostructures. J. Non-Cryst. Solids 352, 1217 (2006)
[41] Froitzheim, A.M., Scherff, M.L.D., Ulyashin, A., Milch, O., Schmidt, M., Fahrner, W.R., Fuhs, W.: Amorphous / crystalline silicon heterojunction solar cells with intrinsic buffer layer. In: Proceedings of the 3rd World Conference on Photovoltaic Energy Conversion, Osaka, Japan, pp. 180–183 (2003)
[42] Ribeyron, P.-J., Vandeneynde, A., Damon-Lacoste, J., Eon, D., Roca i Cabarrocas, P., Chouffot, R., Kleider, J.-P., Brüggemann, R.: Polymorphous/Crystalline Silicon Heterojunction Solar Cells: Optimisation on p-type monocrystalline silicon. In: Proceedings 22nd European Photovoltaic Solar Energy Conference, Milan, Italy, p. 1197 (2007)
[43] Brüggemann, R.: Characterisation and optimisation of amorphous silicon crystalline silicon heterojunction solar cells. J. Optoel. Adv. Materials 11, 1072 (2009)
[44] Tardon, S., Brüggemann, R.: Characterization of the interface properties in a-Si:H/c-Si heterostructures by photoluminescence. J. Phys. D: Appl. Phys. 43, 115102 (2010)
[45] Kurata, K., Shingyouji, T.: Low surface recombination velocity on silicon wafer surfaces due to iodine-ethanol treatment. Jap. J. Appl. Phys. 38, 5710 (1999)
[46] Chouffot, R., Brezard-Oudot, A., Kleider, J.P., Brüggemann, R., Labrune, M., Roca i Cabarrocas, P., Ribeyron, P.J.: Modulated photoluminescence as an effective lifetime measurement method: application to a-Si:H/c-Si heterojunction solar cells. Mat. Science and Eng. B 159-60, 186 (2009)

Chapter 9
Deposition and Properties of TCOs

Florian Ruske

Helmholtz-Zentrum Berlin für Materialien und Energie, Institute for Silicon Photovoltaics,
Schwarzschildstraße 3, 12489 Berlin, Germany

a-Si:H/c-Si heterojunction solar cells require different contacting schemes as compared to conventional solar cells with diffused emitters due to the low emitter conductivity. Apart from back-contacted solar cells it is common to use a transparent conducting oxide (TCO) instead of silicon nitride as an anti-reflection (AR) layer. The choice of materials is vast, with materials based on indium oxide and zinc oxide being the most prominent choice. The optical and electrical properties of these films both play a significant role for the solar cell but they are strongly related, meaning that one cannot optimize them independently. Too high carrier concentrations for instance lead to a lower refractive index of the TCO even for light with a wavelength well below 1100 nm, which results in a worsened AR effect. It is therefore advantageous to use materials with moderate carrier concentrations. The challenges for the deposition of these materials are mainly the low thickness required for an optimum AR effect, for which properties are still influenced by inferior film growth during the nucleation phase, and the allowed substrate temperature of around 200 °C which is limited by the thermal stability of the a-Si:H/c-Si interface.

9.1 Introduction

One of the most striking differences of a-Si:H/c-Si heterojunction (HIT) solar cells as compared to conventional wafer-based cells with diffused emitters is the very low conductivity of the emitter. A metal grid alone is not sufficient for current collection from the emitter, which necessitates a different contacting scheme for the solar cell. Possible solutions are back-contacted solar cells [1,2, and Chapters 10 and 16 in this book], which require several additional patterning steps but offer high current potential, or the substitution of the common silicon nitride anti-reflection layer by a conducting and transparent film.

The requirement of simultaneous transparency and conductivity is shared by various applications, especially for optoelectronic devices. Materials offering this combination of properties are thin metal films or transparent conducting oxides, as well as upcoming materials like carbon nanotubes or graphene. Transparent

W.G.J.H.M. van Sark et al. (Eds.): Physics & Tech. of Amorphous-Crystalline, EM, pp. 301–330.
springerlink.com

conducting oxides (TCOs) are mostly used for HIT solar cells, as they exhibit a refractive index close to that of SiN and are therefore a suitable substitute in the optical design of cells with additional electrical conductivity.

Apart from the optical properties the performance of the TCO film is also defined by its electrical performance. The sheet resistance determines the grid layout, with lower sheet resistance enabling a wider grid which reduces shadowing losses. As the thickness of the TCO layer is limited this means that the resistivity of the thin TCO layer has to be minimized. Unfortunately the optical and electrical properties strongly depend on each other, therefore they cannot be optimized independently and a careful balancing of the two has to be found.

Finally a deposition technology is needed, with which high quality films can be deposited onto the emitter without damaging the junction. This mainly requires a deposition technology that can operate at moderate substrate temperatures as the thermal stability of the junction is usually limited to 200 or 250 °C [3,4].

This chapter addresses some aspects of TCO coatings for heterojunction solar cells. Section 9.2 introduces the most important materials and discusses general electrical and optical properties of TCO materials. Section 9.3 discusses the optical layout of heterojunction solar cells and highlights the role of the TCO layer. In section 9.4 deposition technologies for TCO layers are presented, with a focus on magnetron sputtering. Section 9.5 gives a glance at the trends in TCO research that might have an impact for heterojunction solar cells in a not-too-far future. Section 9.6 summarizes this chapter.

9.2 Transparent Conducting Oxides (TCOs)

Generally speaking optical transparency and electrical conductivity are mutually exclusive physical quantities. In most solids electrical current is transported by free electrons, which show a strong interaction with electromagnetic radiation like light. The effect is obvious when comparing a shiny metal surface to an insulator like glass.

The great technological interest in transparent and conducting materials has stimulated a lot of research in that area and nowadays a wide toolbox of materials with unique properties is at hand to choose from. Most transparent conducting materials used today are based on different metal oxides referred to as transparent conducting oxides (TCOs). Suitable oxides have to fulfil three fundamental requirements in order to qualify as a transparent material with high conductivity:

- They should exhibit a large enough bandgap so that absorption due to band-to-band transitions is limited to the UV and will not diminish cell current.
- Suitable extrinsic or intrinsic dopants that form shallow states have to be introducible in a sufficient quantity.
- The material should not be liable to formation of compensating defects upon shifts of the Fermi level (Fermi level pinning)[1].

---

[1] General rules for the dopability of wide bandgap materials can be found in [5].

It is usual to take the minimum of the luminosity function (sensitivity of the human eye) at 380 nm to define the minimum bandgap requirement, which results to around 3.3 eV. This is also a suitable minimum bandgap for solar applications as photons with higher energies contribute little to total cell current.

The materials considered to be TCOs all allow for extremely high doping levels above 10^{20} cm^{-3}, in some cases also above 10^{21} cm^{-3}. At these doping levels the materials are degenerate semiconductors and show a metal-like behaviour. The conductivity can reach values around 10^4 S/cm which is only two orders of magnitude lower than for the best metal conductors like copper or silver. As will be discussed later only this difference allows for the optical transparency of the doped material, see Fig. 9.1.

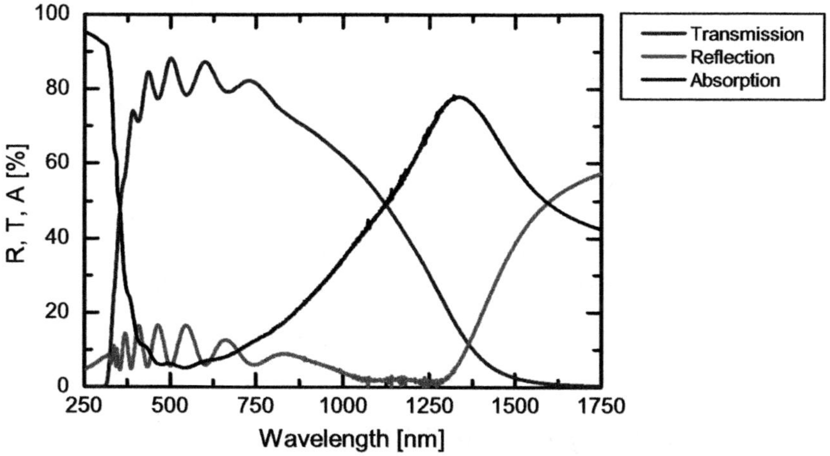

Fig. 9.1 Transmission, reflection and absorption of a 670 nm thick ZnO:Al film on glass. For the spectral region below 400 nm the bandgap absorption limits the transmission. From around 750 nm onwards free carriers strongly influence the optical spectra. For low photon energies the TCO film exhibits a metal-like behaviour with a high optical reflectance. Details on optical properties of TCOs are explained in section 9.2.2.

Finally the formation of compensating intrinsic defects can severely limit the applicability of oxides as semiconductors for transparent, all oxide electronics. It is assumed that this is the main reason for the asymmetry of dopability of most oxides suitable as TCO materials. While most of the materials are easily doped n-type to high levels, the quest for high quality p-type materials is still underway.

Various reviews on TCO materials, their physics and suitable deposition technologies with different focuses exist [6-9]. This huge interest in TCO materials is craven by the vast range of applications of TCO layers as flat panel displays (FPDs), thin film solar cells, energy efficient glazings, gas sensors, touch screens, electrochromic coatings, transparent heaters or radiation shieldings.

From the "classical" TCO materials, doped tin oxide is produced on a large scale on float glass. As the standard deposition technologies of fluorine-doped tin oxide are not compatible to heterojunction solar cells this chapter focuses on the other candidates: indium oxide and zinc oxide. The probably best-known material is tin-doped indium oxide (ITO), which is extensively used in FPD production. Further on zinc oxide based coatings (e.g., see Fig. 9.1) have experienced a high attention in recent years as they offer a cost-efficient substitute for ITO for thin-film solar cells. Some of the basic properties of indium oxide and zinc oxide are summarised in Table 9.1.

Table 9.1 Physical data of indium oxide and zinc oxide. Data taken from [10-13] and references therein.

Host oxide	In_2O_3	ZnO
Crystal structure	bixbyite amorphous	wurzite
Bandgap E_g [eV]	3.75	3.4
static dielectric constant ε_0	≈ 9	$\parallel c \approx 8.4...8.9$ $\perp c \approx 7.4...7.8$
High frequency dielectric constant ε_∞	3.95	$\parallel c \approx 3.7...3.8$ $\perp c \approx 3.6...3.7$
Effective electron mass m^*/m_e	0.35	0.24...0.28
Most important dopants	Sn, Zn H Mo, Ti, Zr, W	Al, Ga, In B H

9.2.1 Electrical Properties of TCOs

In practical applications a transparent conducting oxide thin film will have to exhibit a sheet resistance below a certain maximum value which is defined by the application. Most applications allow for the deposition of thick films (thicker than 500 nm) in order to reach low sheet resistance, but due to the longer deposition times and increased absorption for even thicker films, it is preferable to minimize the resistivity of the transparent conductor. In the case of a-Si/c-Si heterojunction solar cells the situation is even more drastic as the film thickness is fixed due to the fact that the TCO layer is also used as an anti-reflection (AR) coating.

The resistivity of a transparent conducting oxide depends on the concentration of free elecrons N_e and their respective mobility μ:

$$\rho = \frac{1}{e \cdot N_e \cdot \mu}. \tag{9.1}$$

In many cases work on deposition of transparent conducting oxides has focussed solely on the minimization of resistivity. As will be discussed in section 9.2.2 an increase of carrier concentration or mobility will have different consequences on the optical properties of the film. Choosing very high electron concentrations will severely diminish the transparency of the film in the long-wavelength region. Especially for photovoltaics applications it is therefore desirable to obtain films with very high mobilities at defined carrier concentrations. This section will therefore first discuss doping of TCOs and afterwards explain the most important scattering mechanisms of charge carriers that define the electron mobility. Finally the prospects of achieving extremely high mobilities in TCO materials will be briefly discussed.

9.2.1.1 Doping of TCO Materials

Doping of suitable metal oxides, like zinc oxide, indium oxide or tin oxide, can be achieved by a large number of intrinsic and extrinsic defects. Most noticeably, doping can occur through variation of film stoichiometry. For n-type conductivity, as in nearly all TCOs, intrinsic doping can be caused by anion vacancies (oxygen vacancies) or cation interstitials (metal interstitials). The formation of these defects is promoted by oxygen deficient growth conditions.

On the contrary, oxygen rich growth conditions favour the formation of acceptor defects which counteract n-type conductivity, even for extrinsic doping. Therefore in all cases the carrier concentration in films will strongly depend on growth conditions, resulting in small process windows for the deposition of low absorbing but highly doped TCO films.

The importance of intrinsic doping can be readily seen for investigations on films without extrinsic doping. In the case of zinc oxide, carrier concentrations well above 10^{20} cm^{-3} have been achieved for undoped films [14, 15]. Zinc interstitials, oxygen vacancies and also unintentionally introduced hydrogen are among the possible sources for intrinsic n-type doping. In any case the nature of doping causes a reduced thermal stability of the electrical conductivity as compared to extrinsically doped films and annealing in oxidizing environments can severely lower the carrier concentration [16, 17].

For indium oxide the situation is somewhat more complicated, as it can be deposited either in amorphous or crystalline state. It has been suggested, that even for extrinsically doped amorphous films the free carriers are completely generated by intrinsic defects [18] and carrier concentrations up to $5 \cdot 10^{20}$ cm^{-3} are reported [19-21]. Annealing leads to crystallization of the films and for undoped films the carrier concentration is generally lowered. Nevertheless carrier concentrations above 10^{20} cm^{-3} [22, 6] can be reached even without extrinsic doping, although the thermal stability is much lower as compared to doped films [22, 23].

In practice extrinsic doping of indium oxide, tin oxide and zinc oxide is carried out when the materials are to be used as TCO films. The probably best known material is tin-doped indium oxide, known an indium-tin-oxide or ITO. It is usually synthesized from a mixture of indium oxide and 10 weight percent (wt.%) tin oxide. Extensive research has been carried out, especially for applications in flat panel display (FPD) fabrication. A wide range of deposition technologies can be used to produce thin ITO films, among them sputter deposition, pulsed laser deposition, spray pyrolysis, and evaporation technologies (section 9.4). In most cases carrier concentrations above 10^{21} cm^{-3} have been reached, with lower values for the amorphous phase [24, 25].

Other possibilities for extrinsic doping of indium oxide include hydrogen [26], while transitions metals like titanium or molybdenum or rare earth metals like gadolinium have been used to produce TCO films with electron mobilities around or above 100 cm^2/Vs [27].

For zinc oxide TCOs, aluminium is the most prominent dopant and extensively used also in industrial application. The doping level is generally adapted to the application, but carrier concentrations above 10^{21} cm^{-3} have been reported [25, 28]. The same holds for doping with gallium, which is less prominent but has been used for the deposition of high quality films at room temperature [29, 30].

Boron or fluorine are mainly used as dopants in chemical vapour deposition (CVD) of zinc oxide films [31]. Other doping elements have been reviewed by Minami [25].

Tin oxide based TCO layers finally are mostly deposited by CVD or spray pyrolysis, where mainly fluorine is used as a dopant. Antimony has been used for doping of sputtered films [32].

9.2.1.2 Mobility of Free Electrons in TCO Films

For most applications in photovoltaics it is highly desirable to maximize electron mobility instead of carrier concentration (see section 9.2.2). The mobility depends on both the electron effective mass m^* and the relaxation time τ:

$$\mu = \frac{e \cdot \tau}{m^*}. \tag{9.2}$$

The effective mass is a material parameter (see Table 9.1) that cannot readily be altered. Changes may only arise for highly degenerate films in materials which show non-parabolicity of the conduction band [33-35]. It is therefore helpful to study the scattering mechanisms that define the relaxation time τ. According to Matthiesen's rule the relaxation time can be decomposed into the different scattering mechanisms:

$$\frac{1}{\tau_{tot}} = \sum_i \frac{1}{\tau_i}, \tag{9.3}$$

where the index i denotes summation over the different scattering mechanisms. In the case of TCO films at least three different scattering mechanisms have to be taken into account:

- ionized impurity scattering (ii),
- grain boundary scattering (gb) and
- phonon scattering (ps).

Ionized impurity scattering in TCO materials has been investigated by numerous authors and different theories have been tested. A discussion of various different theories and their limitations can be found in [12]. The so-called Brooks-Harring-Dingle expression, describing the scattering of charge carriers in screened potentials of the donor ions, is often applied to TCO materials, as it takes into account that practical TCO films are degenerate[2].

$$\mu_{ii}^{BHD} = \frac{3(\varepsilon_r \varepsilon_0)^2 h^3}{Z^2 m^{*2} e^3} \frac{N_e}{N_i} \frac{1}{F_{ii}} \tag{9.4}$$

In this expression N_e denotes the free electron density and F_{ii} is a screening function. The major parameters determining the scattering at ionized impurities are the static dielectric function ε_r and the effective mass m^* of the host lattice as well as the concentration N_i and charge state Z of the donors. Especially the latter is important, as doping via doubly charged states, like oxygen vacancies, would produce significantly lowered mobility than for singly charged dopants. Further on the BHD expression can be extended to describe the case of partly compensation [36, 12] and a non-parabolic band structure [33].

Phonon scattering has also been discussed in TCO materials [37], but is not affected by the carrier concentration. Therefore it is convenient to take a fixed mobility value that can be derived from measurements on high quality single crystals. Ellmer *et al.* suggested values of 210 cm^2/Vs for zinc oxide and indium oxide and 250 cm^2/Vs for tin oxide, which should be suitable to express the phonon scattering contribution at high carrier concentrations [11].

Finally scattering at grain boundaries in polycrystalline materials is mostly described using Seto's theory [38]. In his rather simple model the mobility is only influenced by the grain size L and the density of assumed acceptor defects Q_t at the grain boundaries. These acceptor defects capture free electrons, leading to a band bending at the grain boundaries. The potential barrier formed finally has to be overcome by thermal excitation of the electrons. For high doping levels ($N_e > Q_t/L$) the potential barrier will be diminished by increasing doping and therefore the mobility μ_{GB} will rise with carrier concentration in this doping range.

Further scattering mechanisms, like neutral impurity scattering [11], will also affect mobility. Nevertheless the density of scattering centers for these additional scattering mechanisms cannot easily be accessed spectroscopically, thus it is sufficient to discuss the above mentioned scattering mechanisms to get an idea of achievable mobility values in polycrystalline TCO films. For very thin films mobility can further be limited by surface scattering [36].

Figure 9.2 shows the maximum mobility for doped zinc oxide. Curve M (green) is the limitation due to ionized impurity scattering and lattice scattering, calculated using equations 9.3 and 9.4, curves A and B (blue) show the additional influence of grain boundary scattering according to Seto's theory. Values can be considerably lower for low quality films. It becomes clear, that the maximum achievable mobility depends heavily on the doping level. For polycrystalline films maximum mobility will normally be achieved for an intermediate doping level, which depends on the defect density Q_t, i.e. the material quality. A low Q_t is crucial to achieve high mobilities at lower doping levels.

---

[2] For In_2O_3 the Mott critical density is about $6 \cdot 10^{18}$ cm^{-3} [39], similar values can be calculated for ZnO.

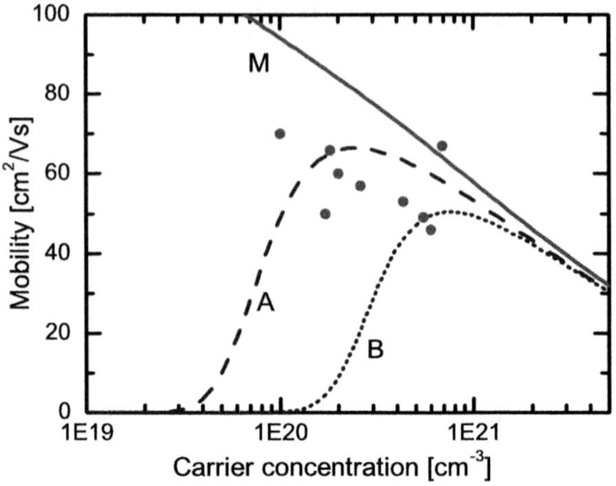

Fig. 9.2 Maximum mobility of doped zinc oxide films. The curve M (green) has been calculated according to the BHD theory (see text) for a non-parabolic band [33, 35] including lattice scattering. The curves A and B (blue) show the influence of grain boundary scattering according to the Seto theory [38] and were calculated assuming a grain size of 50 nm and a defect density Q_t of $1.5 \cdot 10^{13}$ cm^{-2} (A) and $3 \cdot 10^{13}$ cm^{-2} (B). The red data points indicate literature data summarized in Table 2 of Ref. [40].

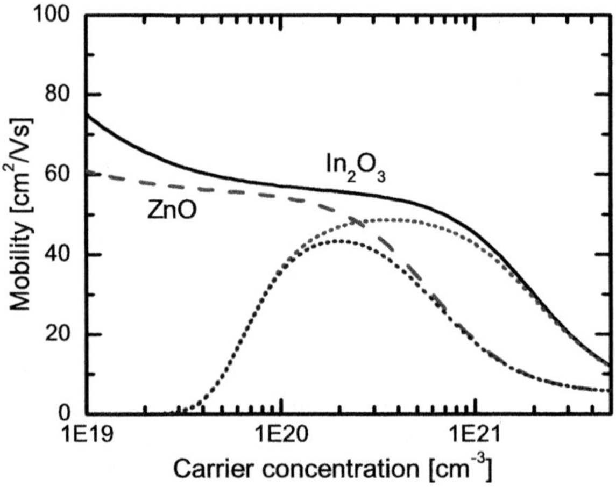

Fig. 9.3 Maximum mobilities in zinc oxide (green) and indium oxide (black) as proposed by Ellmer *et al.* [11] based on values reported in literature. The additional curves indicate the influence of grain boundary scattering (blue: ZnO, red: In_2O_3) and were calculated assuming a grain size of 50 nm and a defect density Q_t of $1.5 \cdot 10^{13}$ cm^{-2}.

In order to achieve a better representation of experimental results and to overcome weaknesses of the BHD theory, Ellmer *et al.* [11] proposed a more phenomenological expression to describe the influence of lattice scattering and ionized impurity scattering. The results are shown in Fig. 9.3 and represent well the fact, that mobilities measured in highly doped, crystalline ITO films are considerably higher than for zinc oxide films with comparable doping level.

It should be noted that the discussion of ionized impurity scattering also holds for amorphous TCO materials. These materials experience attention due to their high mobilities, which can exceed those of amorphous silicon by at least one order of magnitude. Examples include indium oxide and ITO [18, 41] deposited at low substrate temperature or indium-containing alloys like indium zinc oxide (IZO) [42, 43]. While no grain boundary scattering should be active in these materials a similar scattering mechanism limits mobility to lower carrier concentrations [42, 44].

9.2.1.3 Work Function of TCOs

In the actual device the TCO films also has to form a good contact to the underlying a-Si:H layer. The situation is somewhat different, depending whether n-type or p-type wafers are used [45], as charge transfer into the TCO layer will be based on different current paths. For n-type wafers the work function of the TCO layer will determine the contact barrier height at the TCO/(p)-a-Si:H interface [46]. A detailed study based on simulation was carried out by Zhao *et al.* [47], who concluded that the work function of the TCO should be below 4.5 eV for p-type wafers, while the optimum work function for n-type substrates lies between 5.1 and 5.2 eV. A different study for p-type wafers even observed inferior performance for work functions above 4.1 eV for p-type substrates [48].

Despite the uncertainties in exact values, it becomes clear that the work function is a relevant TCO parameter to be studied. This becomes very important considering that the carrier concentration in TCO materials can vary strongly and the work function decreases with increasing doping. This dependence has been studied in ITO [49], while Minami *et al.* investigated the work function for a wide range of ternary oxides [50]. While the work function varies significantly, values below 4.5 eV seem difficult to obtain with indium oxide and zinc oxide. This might be one of the reasons for the difficulty to obtain high efficiencies on p-type substrates.

9.2.2 Optical Properties of TCOs

The optical properties of TCO films are determined by the band structure of the host oxide and the electrical properties. For the application of TCO layers to solar cells it is normally sufficient to analyse the optical properties for the visible and the near infrared spectral range, in the case of silicon based devices down to 1.1 eV. In this spectral range the dielectric function is made up of two major contributions and can be written as

$$\varepsilon(\omega) = 1 + \chi^{VE} + \chi^{FC}, \tag{9.5}$$

where χ^{VE} describes the contribution of valence electrons, and χ^{FC} stands for the susceptibility due to free carriers. Both contributions are complex and frequency-dependant. Contributions of optical phonons can be omitted in the discussion, as the relevant resonances are located in the thermal infrared.

For the relevant spectral range it is convenient to only describe the low energy contribution of the valence electrons, the bandgap contribution, as higher energy transitions only contribute a constant to the dielectric function.

The determination of the bandgap is not straightforward. The main reason is the degeneracy of the TCO films, which means optical transitions will occur into a partially filled conduction band. First this results in a broadening of the onset of intraband absorption [51], causing a deviation from the commonly used square root dependence

$$\alpha(\omega) \propto \left(\hbar\omega - E_g\right)^{1/2}, \tag{9.6}$$

where $\alpha = \dfrac{4\pi \cdot k}{\lambda}$ is the absorption coefficient.

Different expressions have been proposed in order to achieve a better description of the energy dependence of the absorption coefficient α, nevertheless it is concluded that taking the maximum of the derivative $\partial\alpha/\partial(\hbar\omega)$ as a function of $\hbar\omega$ [52] or fitting a Gaussian to it [53] is a suitable way to obtain a good approximation for the bandgap (see Fig. 9.4).

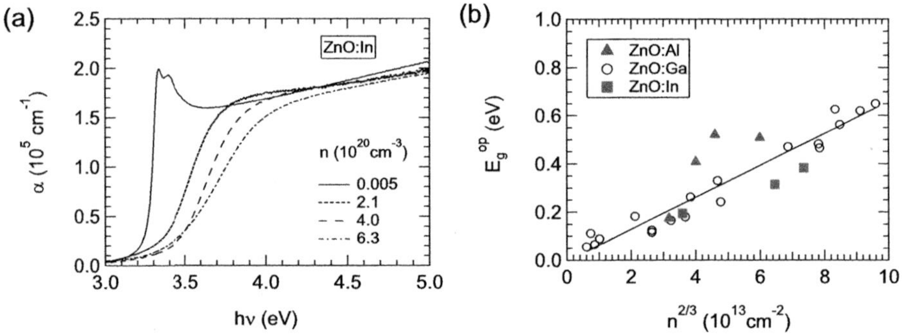

Fig. 9.4 Analysis of bandgap of doped ZnO films for various doping levels. Plotting the absorption coefficient as a function of photon energy a widening of the optical bandgap and a smearing of the transition from transparent to absorbing is observed (a). Actual values for bandgap widening, compared to a reference bandgap of 3.34 eV in the undoped case, approximately scale with $n^{(2/3)}$ with little influence of the chemical nature of the dopant (b). Reprinted with permission from [54]. Copyright (2010) by The American Physical Society.

For modelling of measured optical spectra in practice, various analytical models have been used to describe the dielectric function around the bandgap, including Lorentz [55], Tauc-Lorentz [56] and Leng [57]. The Cauchy expression can be used as an approximation to describe the dispersion close to the bandgap [58].

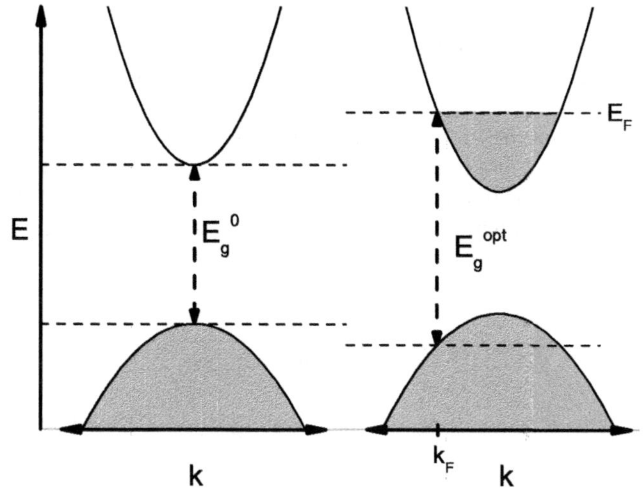

Fig. 9.5 Schematic presentation of the widening of the optical bandgap in heavily doped TCO films. In the undoped case the optical bandgap corresponds to the electronic bandgap. In the degenerate case optical transitions can only occur into unoccupied states of the conduction band and the transitions takes place at the Fermi wave vector k_F. The effect is accompanied by a bandgap narrowing, so that the bandgap widening is somewhat smaller than expected by the Burstein-Moss shift ΔE^{BM} alone.

The most striking feature of bandgap absorption in TCO materials is a strong blueshift of the determined bandgap for increasing doping. In the degenerate case the Fermi level is located above the CBM and will further rise with increasing doping. This causes a blocking of the lowest states in the conduction band for optical transitions and the lowest allowed transition will take place at the Fermi wave vector (Fig. 9.5). The optically determined bandgap therefore increases by the so-called Burstein-Moss shift ΔE^{BM}. Assuming parabolic bands the Burstein-Moss shift will correspond to:

$$\Delta E^{BM} = \left(\frac{1}{m_e^*} + \frac{1}{m_h^*} \right) \frac{\hbar^2}{2} k_F^2 \tag{9.7}$$

This means that the bandgap widening ΔE^{BM} is expected to scale with $n^{-(2/3)}$. In practice this bandgap widening due to band filling is accompanied by a bandgap narrowing that occurs simultaneously [53, 51]. In terms of device optimisation it is therefore convenient to refer to published data on bandgap widening.

For photon energies sufficiently lower than the bandgap energy, the dielectric function can also simplified to

$$\varepsilon(\omega) = \varepsilon_\infty + \chi^{FC}, \tag{9.8}$$

where the contribution of valence electrons is merely a real constant ε_∞, usually referred to as high-frequency dielectric constant. ε_∞ depends on the host lattice. For ZnO a value of 3.85 has been determined [59] while for ITO a value of 4 can be used [60]. Some authors have also reported a drop of ε_∞ with increasing doping [56].

The optical properties towards the long wavelength region are strongly influenced by the free carriers in the material, which show a strong interaction with incident light. The basic interaction is well described by the Drude theory, through which the dielectric function can be calculated as a function of carrier concentration N_e and mobility μ. According to Drude χ^{FC} can be written as:

$$\chi^{FC} = -\frac{\omega_p^2}{\omega^2 + i \cdot \omega\omega_\tau}, \tag{9.9}$$

where the plasma frequency ω_p and the damping ω_τ are defined as:

$$\omega_p^2 = \frac{e^2 \cdot N_e}{\varepsilon_0 \cdot m_e^*} \tag{9.10}$$

$$\omega_\tau = \frac{e}{\mu \cdot m_e^*}. \tag{9.11}$$

The plasma frequency defines the onset of free carrier absorption while the damping is related to the steepness of the transition from transparent to metal-like behaviour. For frequencies below the plasma frequency this means that the light has a very low penetration depth and will be reflected to a high degree.

The Drude theory describes the general optical behaviour of TCO materials quite well, nevertheless deviations have been found. The main reason is the frequency dependence of the scattering effects, which have to be included in an exact analysis. Such approaches have been used to describe the optical properties of ITO [60] and ZnO:Al [59], which also clearly showed that ionized impurity scattering is a dominant scattering mechanism for highly doped TCO materials. Based on this work analytical expressions requiring less computational effort than the full theory have been proposed and applied to different materials [57, 61]. These expressions can be used for accurate modelling of optical spectra.

Figures 9.6 and 9.7 show the effect of a variation of electrical parameters on the index of refraction of a TCO film calculated using a Lorentz oscillator for the description of the bandgap and the Drude theory for the contribution of free electrons. The change of the bandgap with doping was estimated from the data given in Fig. 9.4. While the calculation was carried out for ZnO, the same trends also

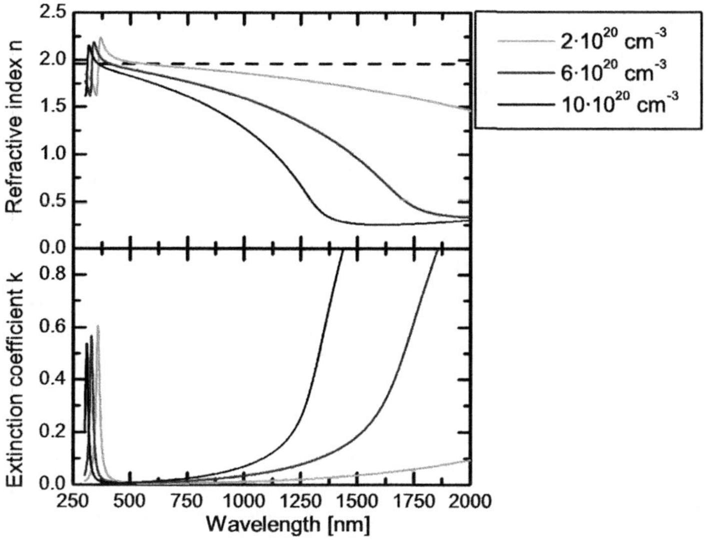

Fig. 9.6 Real (n) and imaginary part (k) of the index of refraction of zinc oxide for various doping levels ranging from $2 \cdot 10^{20}$ cm^{-3} to $1 \cdot 10^{21}$ cm^{-3} calculated according to equation (9.5) using the Drude theory. An electron mobility of 30 cm^2/Vs was assumed. The decrease of n at higher wavelengths causes a deterioration of the AR effect of the thin TCO layer. Further on, absorption can be severe for higher doping levels in the long wavelength region.

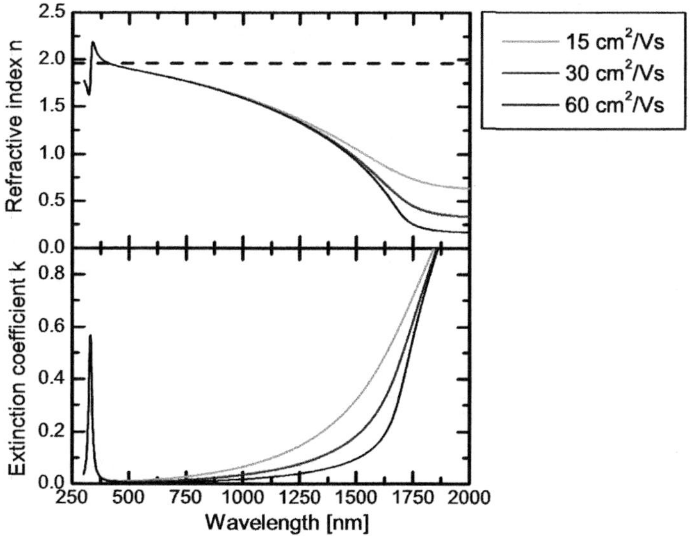

Fig. 9.7 Real (n) and imaginary part (k) of the index of refraction of zinc oxide for various electron mobilities ranging from 15 to 60 cm^2/Vs. A doping level of $6 \cdot 10^{20}$ cm^{-3} was assumed. A variation of mobility does not significantly change the real part n in the spectral range relevant for HIT cells. A significant change of the extinction coefficient is observed close to the plasma frequency.

hold for other TCO materials, where the high-frequency dielectric constant and the effective mass of carriers have to be adopted. The most significant effect of the free carriers is a decrease of the real part of the index of refraction for longer wavelength that significantly changes the cell reflection. The effect gets stronger for increasing doping.

At the same time increasing doping increases the plasma frequency and hence shifts the onset of strong absorption towards the visible spectral range. Doping levels in the order of $10 \cdot 10^{21}$ cm^{-3} can already lead to slight absorption in the relevant spectral range below 1100 nm even for low film thicknesses.

The effect of a variation of mobility (Fig. 9.7) is mostly located around the plasma frequency. The differences in n are negligible below 1100 nm, a change in absorption coefficient will mainly play a role for high doping levels.

It should be noted that additional absorption mechanisms can occur. For low temperature deposition a yellowish appearance is often observed. For sub-stoichiometric films (oxygen deficiency) the films can even appear greyish or black. In any case these parasitic absorption processes will diminish cell performance and have to be avoided by choice of an adequate deposition technology and suitable deposition conditions.

9.3 Optical Design of TCO Front Contact in HIT Cell Structures

The optimization of TCO layers is almost always accompanied by careful weighting of optical and electrical aspects. The previous sections have explained the interaction of electrical properties of TCO layers and their dielectric function.

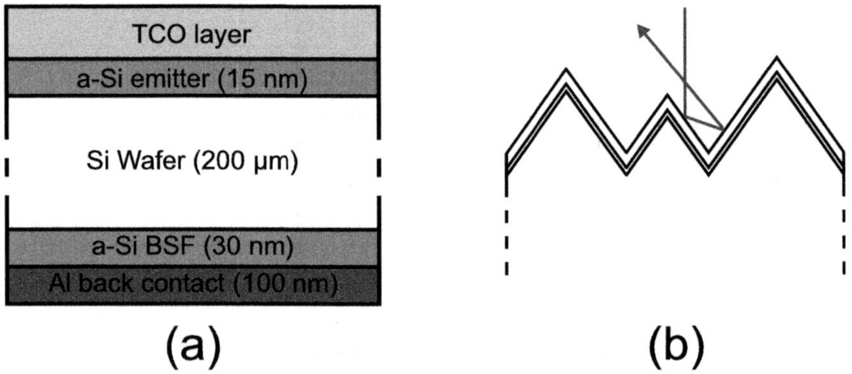

Fig. 9.8 Stack used for simulation of reflectance spectra of HIT cells (a). The total reflectance can be calculated using a transfer matrix formalism. For pyramidal texture, calculation of total reflection is not straightforward. Nevertheless the major part of the reflection is made up of a double reflection indicated in (b). Thus the calculation of this path is a reasonable accurate simplification in the spectral range where the reflection at the wafer backside does not contribute to total reflection.

The presented theories can now be applied to an optimization of the TCO design with respect to HIT solar cells.

While absorption processes in the TCO layer or the emitter layer are best accessed by quantum efficiency measurements, cell reflections can be easily modelled using a matrix formalism as widely used for optical coatings. In this model the whole cell stack using the dielectric properties of all individual layers as shown in Fig. 9.8 has to be considered. The reflectance R can be translated into a maximum cell current $j_{SC}^{\max}$ if a certain internal quantum efficiency IQE is assumed. The maximum cell current is calculated by:

$$j_{SC}^{\max} = e \cdot \int \phi_{AM1.5} \cdot (1-R) \cdot IQE \,\mathrm{d}\lambda \tag{9.12}$$

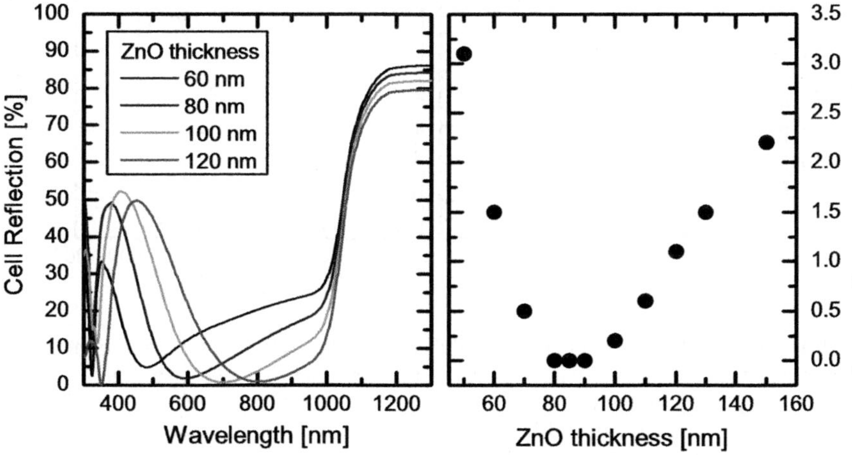

Fig. 9.9 Simulated reflection curves for HIT cells using flat wafers for various TCO film thicknesses (left). Details on the simulation procedure are explained in the text. The right figure shows the current loss, relative to the highest current calculated in this way, due to reflection alone.

Results of the procedure are shown in Fig. 9.9. In the spectral range where no light is reflected from the wafer backside (below 1000 nm in this example) the TCO layer shows the expected behaviour of an anti-reflection (AR) coating. The reflection is minimized for one wavelength given by

$$\lambda = 4 \cdot d \cdot n(\lambda). \tag{9.13}$$

The dispersion of the TCO layer has to be considered, as doping will influence the refractive index already in the visible spectral range (Fig. 9.6). For the shown simulation, zinc oxide with a carrier concentration of $6 \cdot 10^{20}$ cm^{-3} and a carrier mobility of 30 cm^2/Vs has been assumed.

Using an assumed IQE of 1 over the complete spectrum a maximum cell current was calculated according to eq. (9.12). The highest currents were obtained for a TCO thickness between 80 and 90 nm, where the reflection is optimized to the Air Mass (AM) 1.5 spectrum. Figure 9.9 (right) shows the relative loss in the maximum current for deviations in TCO film thickness. The trends shown there are suitable for a general orientation and the optimum film thickness will change for different TCO materials according to eq. (9.13). Absolute values of current loss should be taken with care unless parasitic absorption and the dielectric function of the TCO material to be used are considered.

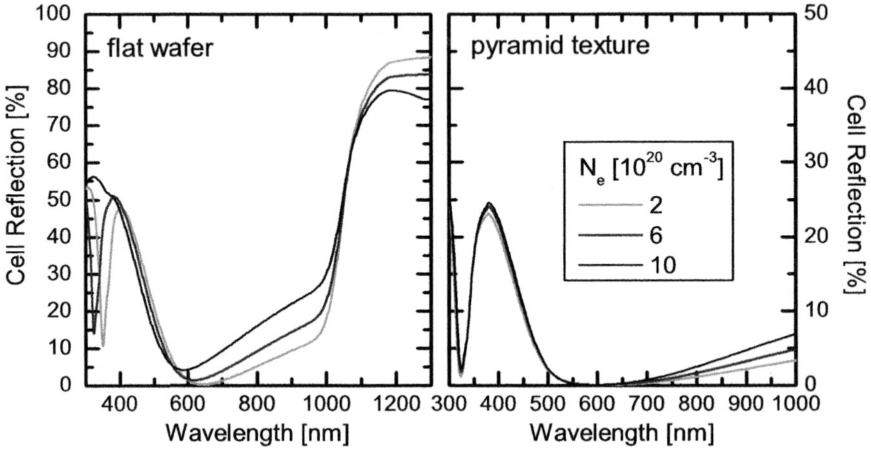

Fig. 9.10 Simulated reflection curves of HIT cells on flat (left) and textured (right) wafers with 85 nm thick TCO layer with varying carrier concentration (see also Fig. 9.5). In the long wavelength region the cell reflection is increased for higher doping due to the lower refractive index of highly doped films. A variation of carrier mobility according to Fig. 9.6 has no significant influence on cell reflection.

The same procedure can also be used to show the influence of various TCO properties on the reflection curves. Figure 9.10 (left) shows calculated reflection curves for cells with 85 nm thick TCO layers with different carrier concentrations as in Fig. 9.6. The lower refractive index for higher doping leads to a worsening of the AR effect and the reflection is increased. Extensive doping will therefore deteriorate the optical properties. Further losses arise from free carrier absorption for the transmitted light in the long wavelength region. It should be noted that the effect of carrier mobility on the cell reflection is negligible for moderate doping levels.

In order to further decrease cell reflection it is common to use textured {100} wafer surfaces, on which random pyramids are formed in alkaline etching procedures. While a complete calculation is best done using ray tracing methods one can also exploit the well-defined angles of the pyramids for a first approximation of the cell reflection [62]. The dominant reflection path of such a surface is shown in Fig. 9.8 (b). The total cell reflection can therefore be approximated by taking the product of the two reflection coefficients for the two defined angles of 54.74°

and 15.70° and averaging the two different polarization directions. This approximation is reasonably good as long as reflection from the wafer backside does not contribute to total reflection. Such a calculation is also shown in Fig. 9.10 (right). From the plot it becomes obvious, that doping has a much less significant influence on total cell reflection on textured wafers as compared to flat ones.

The calculations shown so far offer a significant help to understand the basic design rules for the optical layout of the TCO contact. Nevertheless, not all aspects can be examined. Especially absorption in the films has to be studied in quantum efficiency measurements, where the effect of free carrier absorption can be analysed [63]. Further on alternative optical designs are possible. These include nanostructured materials or multilayer optical coatings [64].

9.4 Deposition of TCO Layers

The choice of a TCO layer for a special application not only consists of picking a certain material and an adequate dopant, there is also a very broad range of deposition technologies available for the deposition of TCO films for various applications. The deposition process not only has to fit the overall product manufacturing line, it also severely influences key material properties, like density, crystallinity and texture, transparency, electrical properties, chemical properties or surface roughness. These differences can be by far larger than any changes that can be achieved by a variation of deposition parameters. Therefore the choice of an adequate deposition process is as important as the choice of the TCO material.

For the deposition of TCO layers for HIT solar cells, the restrictions and requirements are rather challenging. First the deposition has to be carried out on the amorphous silicon emitter, which limits the usable substrate temperature to values around 200 °C. Further on, the most crucial part of the whole device, the a-Si/c-Si interface, is only a few nm below the surface to be coated and any damage to it has to be avoided. Finally deposition will have to be carried out on a textured surface. These requirements are similar to the case of thin film solar cells based on chalcopyrites, where similar restrictions apply. An additional challenge arises from the optical design of the HIT cell, which uses the TCO layer as an antireflective layer with an optimum thickness below 100 nm.

The fact that the total thickness of the layer is restricted is a major challenge to deposition technology as films will often start to grow in a disturbed manner with inferior properties until films reach bulk properties from a certain thickness on.

Due to the wide variety of available deposition technologies an exhaustive review is impossible within the scope of this book. Therefore, after a brief overview, the discussion will focus on the most promising deposition technologies and materials.

9.4.1 Non-vacuum Deposition Processes

Deposition technologies can be divided in vacuum processes and non-vacuum processes. The latter cover sol-gel processes, spray pyrolysis, chemical vapour deposition at atmospheric pressure (APCVD), and electrodeposition. The possibilities are vast and extend to the deposition of nanostructured films, like e.g. zinc oxide based nanorods. These possibilities might lead to an altered device layout, in which the anti-reflection effect is not achieved through choice of a suitable film thickness but by nanostructures [65].

Generally a high crystal quality is required for good electrical transport properties and deposition conditions have to promote high quality crystal growth. This is usually associated with a high mobility of species adsorbed on the substrate surface, which for non-vacuum processes can be raised by a choice of high substrate temperatures, as for spray pyrolysis or APCVD. Other deposition technologies like sol-gel in contrast rely on post-deposition treatments to reach high conductivities [66]. As high temperature treatments cannot be carried out with HIT cells, most of the common non-vacuum processes seem incompatible to cell production at the moment. Tin oxide, for which APCVD is the dominant deposition technology in production, is therefore not used for HIT cells. Nevertheless non-vacuum deposition could in future be a viable option due to the comparably low cost especially when alternative AR concepts are followed and concepts relying on nanostructured precursors continue to evolve rapidly.

9.4.2 Vacuum Based Deposition Technologies

For vacuum deposition technologies the promotion of crystal quality is not limited to high substrate temperatures but ad-atom mobility can also be enhanced by plasma activation. Also, tailored chemical reactions can be used to grow high quality films.

In terms of these aspects the deposition technologies can be divided into chemical vapour deposition (CVD) processes, like plasma enhanced chemical vapour deposition (PECVD), metal-organic CVD (MOCVD), low pressure CVD (LPCVD) or atomic layer deposition (ALD), and physical vapour deposition (PVD) processes such as evaporation, pulsed laser deposition (PLD), ion plating or magnetron sputtering (MS).

CVD type processes, either in the form of LPCVD [31] or plasma-enhanced MOCVD [67], are most commonly reported for ZnO-based TCO films, which are used as front contacts for amorphous/microcrystalline silicon thin film solar cells. The films have excellent optical properties and high deposition rates can be achieved for low substrate temperatures [68]. Nevertheless the films exhibit a very thick incubation layer and crystallites grow in pyramids that increase in size with film thickness. Therefore electrical properties are highly dependant on film thickness. As carrier concentrations are comparably low, the process is not suitable for the deposition of highly conductive thin films. Thicker films could only be applied if the controllable surface features and the accompanying light scattering effects can be used to efficiently couple the incident light into the underlying silicon wafer.

A more attractive option for HIT cells could arise from atomic layer deposition. In this CVD process metal oxide films are deposited by exposing the substrate surface alternately with oxygen or metal precursors. In either step the reaction of the precursor is stopped after a complete monolayer and in principle the films grow in steps of single atomic layers. This procedure can result in high quality films, e.g. for thin barrier layers [69] or buffer layers for thin film solar cells [70]. The process has also been used to deposit TCO layers at moderate substrate temperatures [71] around 150 °C. For slightly higher deposition temperatures of 240 °C an excellent resistivity of $1.35 \cdot 10^{-4}$ Ωcm has been reported [72]. These results indicate that ALD can indeed be an excellent choice for deposition of TCO layers for HIT solar cells, provided the films can be deposited at reasonable cost despite the low deposition rates inherent to the deposition technology.

In the field of physical vapour deposition (PVD) numerous publications on the deposition of various kinds of TCO materials exist. Main drivers have been applications in flat panel display (FPD) fabrication, electrochromics, position sensitive sensors like touch screens, gas sensing, glazings for energy efficient windows, radiation shields, defrosting glazings or thin films photovoltaics. The requirements for the different applications, as well as the restrictions imposed onto the deposition processes, vary strongly. Therefore this review will focus mainly on the investigation of deposition technologies for the deposition of thin films in the order of 100 nm and deposition technologies like magnetron sputtering, that have been so far used to deposit front contacts for HIT solar cells.

A prominent PVD technology in basic science is pulsed laser deposition. The basic setup consists of an excimer laser, which is directed onto a target material. During the intense laser pulse the target material is ablated and ionized to a high degree. While the application of the process in industrial production will not be economic on the short term, PLD is most attractive for material science as its application has led to some of the best material properties achieved so far. Both ITO and Al-doped ZnO have been prepared with resistivities below 10^{-4} Ωcm [73, 28].

PLD is also suitable for deposition of very thin films. Excellent film properties are already reached for film thicknesses well below 100 nm. Figure 9.11 shows an example for ZnO:Al films deposited at a substrate temperature of 260 °C [74]. While this seems too high for deposition on emitter layers of HIT cells, the corresponding publication also shows excellent properties for films deposited at lower substrate temperatures for films thicknesses of only 40 nm. Dong *et al.* [75] have carried out a similar investigation over a wider thickness range. For ITO the thickness dependence was found to be even less pronounced [76], a fact that is also observed for magnetron sputtering.

In summary, TCO films grown by PLD often represent the current performance limit of these materials. They often exhibit the lowest resistivity known for various materials and also show the basic feasibility of high quality films with low film thickness. The values given in Fig. 9.10 would lead to a sheet resistance of only 25 Ω for a film thickness around 85 nm.

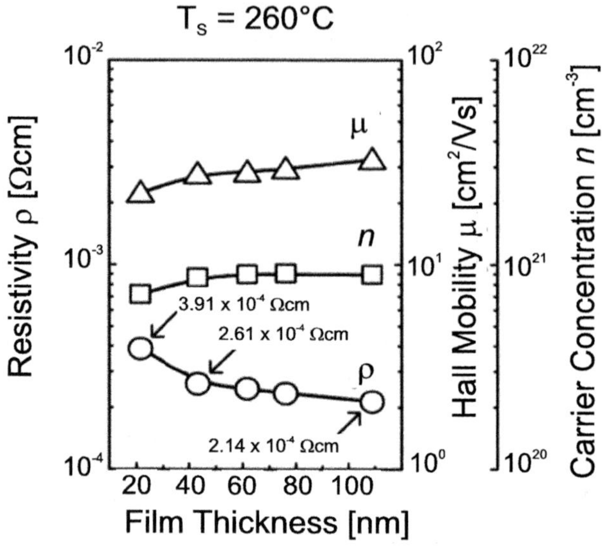

Fig. 9.11 Dependance of resistivity ρ, carrier concentration n and mobility μ on film thickness for ZnO:Al films deposited on glass by pulsed laser deposition at a substrate temperature of 260 °C. Reprinted from [74] with permission of Elsevier.

Nevertheless the most prominent deposition technology for TCO films for HIT solar cells is magnetron sputtering. Literature on the process is extensive due to the large amount of advantages it offers and the vast field of applications. Magnetron sputtering processes are generally easy to scale up, have comparably high deposition rates, the choice of materials is huge and sputtering systems are comparably easy to operate.

The basic setup of a sputtering coater is shown in Fig. 9.12. Deposition of films is achieved by a sputtering effect of positive ions originating from a plasma discharge impinging onto the negatively biased target. Atoms knocked out from the target are collected on the substrate.

In the case of TCO films the target can either consist of metal, e.g. a ZnAl alloy for deposition of ZnO:Al, or a ceramic material. In the case of reactive sputtering of metallic targets oxygen has to be added via the gas phase and the process has to be thoroughly controlled in order to obtain the right film stoichiometry. If ceramic targets are used, oxygen has to be added only in small amounts in order to compensate oxygen from the target lost to the vacuum pumps or oxygen deficiency of the target material. Due to the complexity of the reactive sputtering process ceramic targets are generally favoured in production. Also radio-frequency (RF) sputtering, that is widely used in research, is not a favourable process in production, as deposition rates are significantly lower than for direct current (DC) sputtering and the generation of RF power is associated with higher costs.

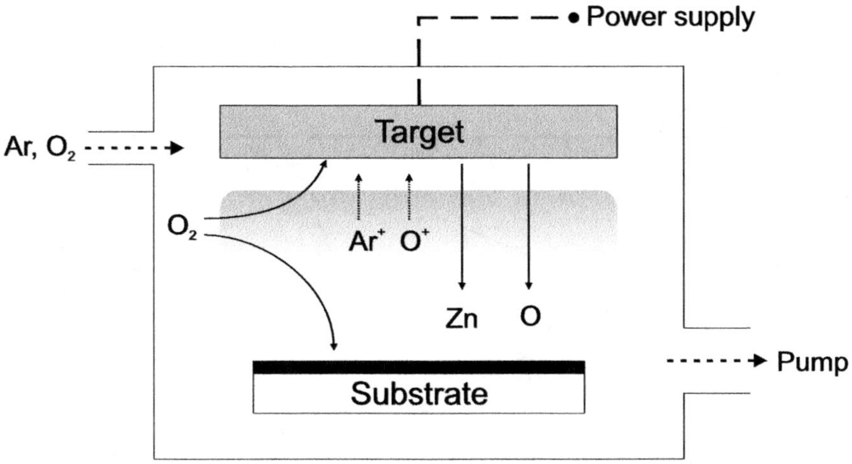

Fig. 9.12 Basic setup of a sputtering process. Either ceramic or metallic target material can be used. They are generally backed by a permanent magnet setup (magnetron, not shown) to increase plasma density in front of the target, resulting in higher deposition rates. The substrate can be heated, temperature up to around 300 °C are common for TCO deposition. Plasma excitation will normally be carried out with DC or pulsed DC power. RF excitation is not favoured in production processes.

The deposition parameters with most impact on the film properties of TCO films are generally substrate temperature and oxygen addition. For ceramic targets the right amount of oxygen can be easily controlled, so substrate temperature is the dominant parameter[3]. For ZnO-based films a substrate temperature of 200 °C is often considered as an important landmark. Above this temperature the vapour pressure of zinc is high enough to allow metallic zinc to evaporate from the substrate surface. In most cases this leads to a much easier control of optical properties of ZnO-based TCO films and greyish or black films are hardly observed [77, 78]. The effect is not as relevant for deposition using ceramic targets, as the oxygen mostly originates from the target materials.

Nevertheless substrate temperature plays a major role for the film quality. For low substrate temperatures the electrical properties will normally improve with substrate temperature [80]. For a target doping of 2 wt.% Al_2O_3 an example of the dependence of electrical transport parameters on substrate temperature is shown in Fig. 9.13. For higher substrate temperatures resistivity was reported to rise again [81, 82], the optimum deposition temperature usually depends on the target doping [83]. Differences occur between DC and RF sputtering due to the additional energy irradiation during RF sputtering which facilitates deposition of high quality films at low substrate temperatures. As results from various publications seem contradictory, it is obvious that other deposition parameters also have significant

[3] It should be noted that other deposition parameters, namely discharge power, have a significant effect on substrate temperature, especially for deposition at low substrate temperatures.

influence, especially in the case of RF sputtering. An elaborate review on magnetron sputtering of ZnO discussing further deposition parameters has been given by Ellmer [84].

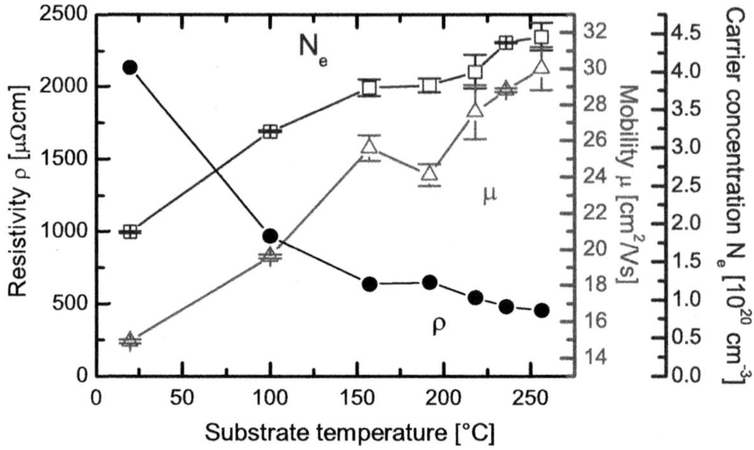

Fig. 9.13 Substrate temperature dependence of electrical transport properties of ZnO:Al films deposited via DC magnetron sputtering using a ceramic ZnO target with an Al_2O_3 doping of 2 weight%. For temperatures below 250 °C the resistivity will normally decrease with substrate temperature, although exact values also depend on coater geometry, target material, substrate material and further deposition conditions. Better resistivities at low substrate temperatures are generally achieved by RF sputtering. Data taken from [79].

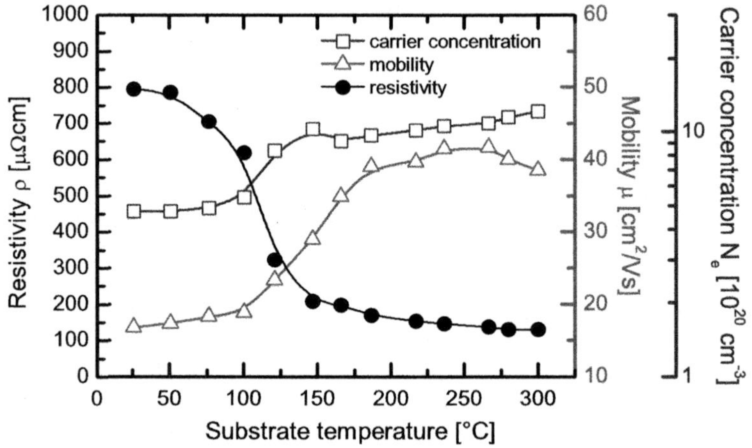

Fig. 9.14 Substrate temperature dependence of electrical transport properties of ITO films deposited by magnetron sputtering. The strong decrease of resistivity at substrate temperatures between 100 and 150 °C is caused by a transition of amorphous film growth at low substrate temperatures to crystalline growth. Reprinted from [85] with permission of Elsevier.

For deposition of ITO films substrate temperature also plays a key role. Figure 9.14 shows the dependence of electrical transport properties on substrate temperature [85]. The main reason for the sharp decrease of the resistivity around 150 °C is a transition to a fully crystalline film growth, while the films can grow partially or fully amorphous at low substrate temperatures. Sometimes the growth of completely amorphous ITO films is desirable, as this phase exhibits a high etching rate and is ideal for patterning processes. Crystallization of ITO can afterwards be carried out by post-deposition thermal treatment and films crystallize at temperatures below 200 °C. The process is therefore compatible to HIT technology.

A newly proposed similar route is the post-deposition crystallization of hydrogen doped indium oxide (In_2O_3:H) [86] which has led to excellent mobilities above 100 cm^2/Vs and already been applied to HIT solar cells [87].

In summary it can be concluded that the substrate temperature limitation for deposition of TCO layers onto HIT cells limits the possibilities, but the problem can be overcome. The challenge is therefore to deposit thin films with thicknesses between 80 and 90 nm. The thickness dependence of electrical transport properties is therefore also very important to study. For deposition of ITO by magnetron sputtering a similar requirement exists for deposition of layers for liquid crystal (LCD) FPDs. High quality is usually achieved by choosing high substrate temperatures above 300 °C. The growth of thin layers at reduced temperature on the contrary is challenging. For room temperature deposition of ITO on poly methyl methacrylate (PMMA) a reduction of resistivity with thickness up to a film thickness of 280 nm is reported [88]. A closer inspection for films with thicknesses below 100 nm deposited on glass has also shown the thickness dependence, but electrical properties seem acceptable over the complete thickness range [89].

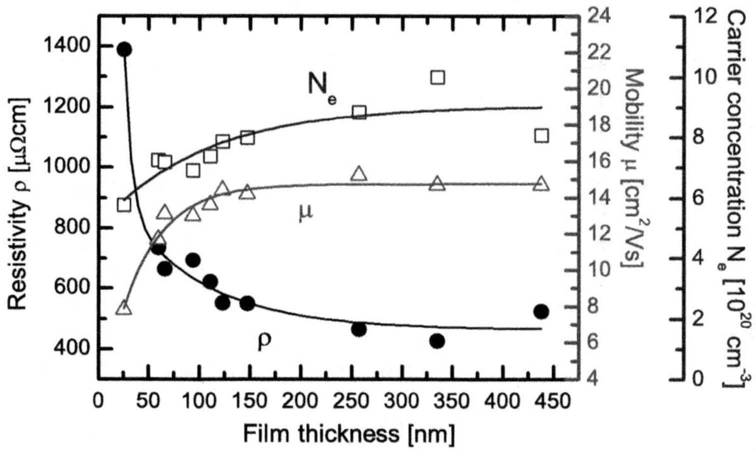

Fig. 9.15 Electrical transport properties as a function of film thickness for reactively sputtered ZnO:Al films on glass [90]. Films have been deposited at a substrate temperature of 180 °C and a sputtering pressure of 410 mPa.

For zinc oxide the trend is even more pronounced. A typical dependence of electrical transport properties on film thickness is shown in Fig. 9.15 for Al-doped ZnO films deposited by reactive DC magnetron sputtering at a substrate temperature of 180 °C [90]. For the same sputtering setup also the influence of substrate temperature and deposition power has been studied [91]. A similar study has e.g. been carried out for RF sputtering of ZnO:Ga from a ceramic target [29]. It becomes clear, that the minimum resistivities obtained for very thick films as well as the dependence of resistivity on thickness are strongly influenced by the deposition conditions. Especially substrate temperature, discharge power and target-to-substrate [29] distance play a major role. Main reason for the low resistivity is a low carrier mobility [29, 91], that is explained by inferior crystal growth in the initial growth stage. Nevertheless also an increase of the carrier concentration is observed with increasing film thickness.

The thickness dependence for other PVD processes has been investigated for vacuum arc plasma evaporation [92], ion beam sputtering [93] and ion plating [94]. For all deposition technologies the same trends are observed, but in some cases excellent properties are achieved for thin films nevertheless.

In summary PVD technologies can fulfil the requirements for the deposition of TCO films for HIT solar cells. Neither the low deposition temperature nor the thin film thickness is a fundamental problem, but the material properties obtained cannot be compared to values for thick films deposited at higher substrate temperatures. The overcoming of this gap is mostly a technological challenge, which demands a constant advancement of established and upcoming deposition technologies.

9.5 Further Trends in TCO Materials Research

Up to this point mainly TCO materials based on indium oxide and zinc oxide, as well as suitable deposition processes, have been discussed. These materials have been researched for more than 50 years. Nevertheless the field of transparent, conducting materials is rapidly evolving and new materials are frequently suggested. Apart from carbon based coatings, namely carbon nanotubes or graphene, two classes of materials have experienced special attention.

While amorphous indium oxide and other amorphous TCOs have been known for quite a while, research on binary or ternary oxides has recently experienced much attention. Especially amorphous indium zinc oxide films have been widely studied. These films can be easily deposited by magnetron sputtering and mobilities around 50 cm^2/Vs at moderate doping levels have been observed for deposition onto unheated substrates [42].

Recent research has also led to the development of TCO materials with mobilities above 100 cm^2/Vs [27]. This leads to excellent optical properties and very low resistivities. The feasibility has already been shown for very thin films [95]. The main drawback is the need for very high substrate temperatures above 300 °C. If this current limitation can be overcome, materials like Mo-doped indium oxide could become very interesting options for HIT cells.

9.6 Conclusion

In summary it can be concluded that the TCO layer is a crucial part of the heterojunction solar cell. Considering its importance it is surprising, that only few detailed studies are known. Performance of solar cells with excellent junctions and passivation properties can still be hampered by poor TCO properties, especially high absorption, high contact barriers or a too high sheet resistance.

If attention is paid to all these aspects, TCO layers for high performing devices and suitable deposition technologies are already available to date. Research should focus on a further decrease of optical absorption, namely free carrier absorption and bandgap absorption, and still lower resistivities for thin films in order to enable wider metal grid layouts. Here future developments in high mobility TCOs or other transparent conducting materials might enable their applicability to HIT solar cells.

Acknowledgements

We would like to thank Lars Korte, Tim Schulze, Robert Rößler, Sonya Calnan, Nicola Mingirulli and Jan Haschke for their contribution in form of various hints, suggestions, explanations and data contributions. Thank to Andreas Pflug for continuous help in using the simulation software RIG-VM.

References

[1] Stangl, R., Haschke, J., Bivour, M., Korte, L., Schmidt, M., Lips, K., Rech, B.: Planar rear emitter back contact silicon heterojunction solar cells. Sol. Energ. Mater. Sol. Cell 93, 1900–1903 (2009)

[2] Tucci, M., Serenelli, L., Salza, E., De Iuliis, S., Geerligs, L.J., Caputo, D., Ceccarelli, M., de Cesare, G.: Back contacted a-Si:H/c-Si heterostructure solar cells. J. Non-Cryst. Solids 354, 2386–2391 (2008)

[3] Zhao, L., Zhou, Z., Peng, H., Cui, R.: Indium tin oxide thin films by bias magnetron rf sput-tering for heterojunction solar cells application. Appl. Surf. Sci. 252, 385–392 (2005)

[4] Dao, V.A., Choi, H., Heo, J., Park, H., Yoon, K., Lee, Y., Kim, Y., Lakshminarayan, N., Yi, J.: rf-Magnetron sputtered ITO thin films for improved heterojunction solar cell applications. Curr. Appl. Phys. 10, S506–S509 (2010)

[5] Zunger, A.: Practical doping principles. Appl. Phys. Lett. 83(1), 57–59 (2003)

[6] Chopra, K.L., Major, S., Pandya, D.K.: Transparent conductors – a status review. Thin Solid Films 102, 1–46 (1983)

[7] Ginley, D.S., Bright, C.: Transparent conducting oxides. MRS Bulletin 25(8) (2000)

[8] Exarhos, G.J., Zhou, X.-D.: Discovery-based design of transparent conducting oxide films. Thin Solid Films 515, 7025–7052 (2007)

[9] Ellmer, K., Klein, A., Rech, B.: Transparent Conductive Zinc Oxide. Springer, Heidelberg (2008)

[10] Pearton, S.J., Norton, D.P., Ip, K., Heo, Y.W., Steiner, T.: Recent advances in process-ing of ZnO. J. Vac. Sci. Technol. B 22(3), 932–948 (2004)

[11] Ellmer, K., Mientus, R.: Carreir transport in polycrystalline transparent conducting ox-ides: A comparative study of zinc oxide and indium oxide. Thin Solid Films 516, 4620–4627 (2008)

[12] Ellmer, K.: Resistivity of polycrystalline zinc oxide films. Current status and physical limit. J. Phys. D: Appl. Phys. 34, 3097–3108 (2001)

[13] Özgür, Ü., Alivov, Y.I., Liu, C., Teke, A., Reshchikov, M.A., Doğan, S., Avrutin, V., Cho, S.-J., Morkoç, H.: A comprehensive review of ZnO materials and devices. J. Appl. Phys. 98, 041301 (2005)

[14] Ito, K., Nakazawa, T.: Transparent and Highly Conductive Films of ZnO Prepared by RF Sputtering. Jap. J. Appl. Phys. 22(4), L245–L247 (1983)

[15] Nanto, H., Minami, T., Shooji, S., Takata, S.: Electrical and optical properties of zinc oxide thin films prepared by RF magnetron sputtering for transparent electrode applications. J. Appl. Phys. 55(4), 1029–1034 (1984)

[16] Minami, T., Nanto, H., Shooji, S., Takata, S.: The Stability of Zinc Oxide Transparent Electrodes Fabricated by R. F. Magnetron Sputtering. Thin Solid Films 111, 167–174 (1984)

[17] Minami, T., Nanto, H., Takata, S.: Highly Conductive and Transparent Aluminum Doped Zinc Oxide Thin Films Prepared by RF Magnetron Sputtering. Jap. J. Appl. Phys. 35(5), L280–L282 (1984)

[18] Bellingham, J.R., Phillips, W.A., Adkins, C.J.: Electrical and optical properties of amorphous indium oxide. J. Phys.: Condens. Matter 2, 6207–6221 (1990)

[19] Lee, B.H., Kim, I.G., Cho, S.W., Lee, S.-H.: Effect of process parameters on the char-acteristics of indium tin oxide thin film for flat panel display application. Thin Solid Films 302, 25–30 (1997)

[20] Shigesato, Y., Hayashi, Y., Haranoh, T.: Doping mechanisms of tin-doped indium ox-ide films. Appl. Phys. Lett. 61(1), 73–75 (1992)

[21] Shigesato, Y., Paine, D.C.: Study of the effect of Sn doping on the electronic transport properties of thin film indium oxide. Appl. Phys. Lett. 62(11), 1268–1270 (1993)

[22] Mizuhashi, M.: Electrical properties of vacuum-deposited indium oxide and indium tin oxide films. Thin Solid Films 70, 91–100 (1980)

[23] Minami, T., Miyata, T., Yamamoto, T.: Stability of transparent conducting oxide films for use at high temperatures. J. Vac. Sci. Technol. A 17(4), 1822–1826 (1999)

[24] Granqvist, C.G., Hultåker, A.: Transparent and conducting ITO films: new developments and applications. Thin Solid Films 411, 1–5 (2002)

[25] Minami, T.: Transparent conducting oxide semiconductors for transparent electrodes. Semicond. Sci. Technol. 20, S35–S44 (2005)

[26] Koida, T., Kondo, M., Tsutsumi, K., Sakaguchi, A., Suzuki, M., Fujiwara, H.: Hydrogen-doped In_2O_3 transparent conducting oxide films prepared by solid-phase crystallization method. J. Appl. Phys. 107, 033514 (2010)

[27] Calnan, S., Tiwari, A.N.: High mobility transparent conducting oxides for thin film so-lar cells. Thin Solid Films 518, 1839–1849 (2010)

[28] Agura, H., Suzuki, A., Matsushita, T., Aoki, T., Okuda, M.: Low resistivity transparent conducting Al-doped ZnO films prepared by pulsed laser deposition. Thin Solid Films 445, 263–267 (2003)

[29] Assunção, V., Fortunato, E., Marques, A., Gonçalves, A., Ferreira, I., Águas, H., Martins, R.: New challanges on gallium-doped zinc oxide films prepared by r.f. magnetron sputtering. Thin Solid Films 442, 102–106 (2003)

[30] Fortunato, E., Assunção, V., Gonçalves, A., Marques, A., Águas, H., Pereira, L., Ferreira, I., Vilarinho, P., Martins, R.: High quality conductive gallium-doped zinc oxide films deposited at room temperature. Thin Solid Films 451-452, 443–447 (2004)

[31] Faÿ, S., Shah, A.: Zinc Oxide Grown by CVD Process as Transparent Contact for Thin Film Solar Cell Applications. In: Ellmer, K., Klein, A., Rech, B. (eds.) Transparent Conductive Zinc Oxide. Springer, Heidelberg (2008)
[32] Stjerna, B., Olsson, E., Granqvist, C.G.: Optical and electrical properties of radio frequency sputtered tin oxide films doped with oxygen vacancies, F, Sb, or Mo. J. Appl. Phys. 76(6), 3797–3817 (1994)
[33] Pisarkiewicz, T., Zakrzewska, K., Leja, E.: Scattering of charge carriers in transparent and conducting thin oxide films with a non-parabolic conduction band. Thin Solid Films 174, 217–223 (1989)
[34] Young, D.L., Coutts, T.J., Kaydanov, V.I., Gilmore, A.S., Mulligan, W.P.: Direct measurement of density-of-states effective mass and scattering parameter in transparent con-ducting oxides using second-order transport phenomena. J. Vac. Sci. Technol. A 18(6), 2978–2985 (2000)
[35] Ruske, F., Pflug, A., Sittinger, V., Szyszka, B., Greiner, D., Rech, B.: Optical modeling of free electron behavior in highly doped ZnO films. Thin Solid Films 518, 1289–1293 (2009)
[36] Look, D.C., Leedy, K.D., Tomich, D.H., Bayraktaroglu, B.: Mobility analysis of highly conducting thin films: Application to ZnO. Appl. Phys. Lett. 96, 062102 (2010)
[37] Chen, M., Pei, Z.L., Wang, X., Yu, Y.H., Liu, X.H., Sun, C., Wen, L.S.: Intrinsic limit of electrical properties of transparent conductive oxide films. J. Phys. D: Appl. Phys. 33, 2538–2548 (2000)
[38] Seto, J.Y.W.: Electrical Properties of Polycrystalline Silicon Films. J. Appl. Phys. 46(12), 5247–5254 (1975)
[39] Edwards, P.P., Porch, A., Jones, M.O., Morgan, D.V., Perks, R.M.: Basic materials physics of transparent conducting oxides. Dalton Trans. 19, 2995–3002 (2004)
[40] Ruske, F., Roczen, M., Lee, K., Wimmer, M., Gall, S., Hüpkes, J., Hrunski, D., Rech, B.: Improved electrical transport in Al-doped zinc oxide by thermal treatment. J. Appl. Phys. 107, 013708 (2010)
[41] Utsumi, K., Iigusa, H., Tokumaru, R., Song, P.K., Shigesato, Y.: Study on In_2O_3-SnO_2 transparent and conductive films prepared by d.c. sputtering using high density ceramic targets. Thin Solid Films 445, 229–234 (2003)
[42] Leenheer, A.J., Perkins, J.D., van Hest, M.F.A.M., Berry, J.J., O'Hayre, R.P., Ginley, D.S.: General mobility and carrier concentration relationship in transparent amorphous in-dium zinc oxide. Phys. Rev. B 77, 115215 (2008)
[43] Tominaga, K., Takada, D., Shimomura, K., Suketa, H., Takita, K., Murai, K., Moriga, T.: Influence of Ga_2O_3 addition on transparent conductive oxide films of In_2O_3-ZnO. Vacuum 83, 561–563 (2009)
[44] Martins, R., Barquinha, P., Pimentel, A., Pereira, L., Fortunato, E.: Transport in high mobility amorphous wide gap indium zinc oxide films. Phys. Stat. Sol. (a) 202(9), R95–R97 (2005)
[45] Kanevce, A., Metzger, W.K.: The role of amorphous silicon and tunneling in heterojunction with intrinsic thin layer (HIT) solar cells. J. Appl. Phys. 105, 094507 (2009)
[46] Rahmouni, M., Datta, A., Chatterjee, P., Damon-Lacoste, J., Ballif, C., Roca i Cabarrocas, P.: Carrier transport and sensitivity issues in heterojunction with intrinsic thin layer solar cells on N-type crystalline silicon: A computer simulation study. J. Appl. Phys. 107, 054521 (2010)
[47] Zhao, L., Zhou, C.L., Li, H.L., Diao, H.W., Wang, W.J.: Role of the work function of transparent conductive oxide on the performance of amorphous/crystalline silicon heterojunction solar cells studied by computer simulation. Phys. Stat. Sol. (a) 205, 1215–1221 (2008)

[48] Vinh Ai Dao, V.A., Heo, J., Choi, H., Kim, Y., Park, S., Jung, S., Lakshminarayan, N., Yi, J.: Simulation and study of the influence of the buffer intrinsic layer, back-surface field, densities of interface defects, resistivity of p-type silicon substrate and transparent conductive oxide on heterojunction with intrinsic thin-layer (HIT) solar cell. Sol. En. 84, 777–783 (2010)
[49] Sato, Y., Ashida, T., Oka, N., Shigesato, Y.: Carrier Density Dependence of Optical Band Gap and Work Function in Sn-Doped In_2O_3 Films. Appl. Phys. Express 3, 061101 (2010)
[50] Minami, T., Miyata, T., Yamamoto, T.: Work function of transparent conducting mul-ticomponent oxide thin films prepared by magnetron sputtering. Surf. Coat. Technol. 108-109, 583–587 (1998)
[51] Hamberg, I., Granqvist, C.G., Berggren, K.F., Sernelius, B.E., Engström, L.: Band-gap widening in heavily Sn-doped I_2O_3. Phys. Rev. B 30, 3240–3249 (1984)
[52] Sernelius, B.E., Berggren, K.F., Jin, Z.C., Hamberg, I., Granqvist, C.G.: Band-Gap Tailoring of Zno by Means of Heavy Al Doping. Phys. Rev. B 37, 10244–10248 (1988)
[53] Sans, J.A., Segura, A., Sánchez-Royo, J.F., Barber, V., Hernández-Fenollosa, M.A., Marí, B.: Correlation between optical and transport properties of Ga-doped ZnO thin films prepared by pulsed laser deposition. Superlattices and Microstructures 39, 282–290 (2006)
[54] Sans, J.A., Sánchez-Royo, J.F., Segura, A., Tobias, G., Canadell, E.: Chemical effects on the optical band-gap of heavily doped ZnO:M_{III} (M=Al,Ga, In): An investigation by means of photoelectron spectroscopy, optical measurements under pressure, and band structure cal-culations. Phys. Rev. B 79, 195105 (2009)
[55] Losurdo, M., Gaingregorio, M., Capezzuto, P., Bruno, G., De Rosa, R., Roca, F., Summonte, C., Plá, J., Rizzoli, R.: Parametrization of optical properties of indium-tin-oxide thin films by spectroscopic ellipsommetry: Substrate interface reactivity. J. Va. Sci. Technol. A 20(1), 37–42 (2002)
[56] Fujiwara, H., Kondo, M.: Effects of carrier concentration on the dielectric function of ZnO:Ga and In_2O_3:Sn studied by spectroscopic ellipsometry: Analysis of free-carrier and band-edge absorption. Phys. Rev. B 71, 75109 (2005)
[57] Pflug, A., Sittinger, V., Ruske, F., Szyszka, B., Dittmar, G.: Optical characterization of aluminium-doped zinc oxide films by advanced dispersion theories. Thin Solid Films 455-456, 201–206 (2004)
[58] Volintiru, I., Creatore, M., van de Sanden, M.C.M.: In situ spectroscopic ellipsometry growth studies on the Al-doped ZnO films deposited by remote plasma-enhanced metalor-ganic chemical vapour deposition. J. Appl. Phys. 103, 033704 (2008)
[59] Jin, Z.-C., Hamberg, I., Granqvist, C.G.: Optical properties of sputter-deposited ZnO:Al thin films. J. Appl. Phys. 64(10), 5117–5131 (1988)
[60] Hamberg, I., Granqvist, C.G.: Evaporated Sn-doped In_2O_3 films: Basic optical properties and applications to energy-efficient windows. J. Appl. Phys. 60(11), R123–R159 (1986)
[61] Mergel, D., Qiao, Z.: Dielectric modelling of optical spectra of thin In_2O_3:Sn films. J. Phys. D: Appl. Phys. 35, 794–801 (2002)
[62] Baker-Finch, S.C., McIntosh, K.R.: Reflection of normally incident light from silicon solar cells with pyramidal texture. Prog. Photovolt: Res. Appl. (2010)
[63] Taira, S., Yoshimine, Y., Baba, T., Taguchi, M., Kanno, H., Kinoshita, T., Sakata, H., Maruyama, E., Tanaka, M.: Our approach for achieving HOT solar cells with more than 23 % efficiency. In: 22nd European Photovoltaic Solar Energy Conference, Milan, Italy, September 3-7, pp. 932–935 (2007)

[64] Wright, D.N., Ulyashin, A., Bentzen, A., Marstein, E.S., Holt, A.: Multi-layer TCO based anti-reflective coatings for silicon solar cells. In: 21st EUPVSEC, Dresden, Germany, September 4-8, pp. 1211–1214 (2006)
[65] Aé, L., Kieven, D., Chen, J., Klenk, R., Rissom, T., Tang, Y., Lux-Steiner, M.C.: ZnO nanorod arrays as an antireflective coating for Cu(In,Ga)Se_2 thin film solar cells. Prog. Photovolt.: Res. Appl. 18, 209–213 (2010)
[66] Schuler, T., Aegerter, M.A.: Optical, electrical and structural properties of sol gel ZnO:Al coatings. Thin Solid Films 351, 125–131 (1999)
[67] Volintiru, I., Creatore, M., Kniknie, B.J., Spee, C.I.M.A., van de Sanden, M.C.M.: Evolution of the electrical and structural properties during the growth of Al doped ZnO films by remote plasma-enhanced metalorganic chemical vapor deposition. J. Appl. Phys. 102, 043709 (2007)
[68] Faÿ, S., Kroll, U., Bucher, C., Vallat-Sauvain, E., Shah, A.: Low pressure chemical vapour deposition of ZnO layers for thin-film solar cells: temperature-induced morphological changes. Sol. Energy Mater. Sol. Cells 86, 385–397 (2005)
[69] Garcia, P.F., McLean, R.S., Groner, M.D., Dameron, A.A., George, S.M.: Gas diffusion ultrabarriers on polymer substrates using Al_2O_3 atomic layer deposition and SiN plasma-enhanced chemical vapor deposition. J. Appl. Phys. 106, 023533 (2009)
[70] Naghavi, N., Abou-Ras, D., Allsop, N., Barreau, N., Bücheler, S., Ennaoui, A., Fischer, C.-H., Guillen, C., Hariskos, D., Herrero, J., Klenk, R., Kushiya, K., Lincot, D., Menner, R., Nakada, T., Platzer-Björkman, C., Spiering, S., Tiwari, A.N., Törndahl, T.: Buffer layers and transparent conducting oxides for chalcopyrite Cu(In,Ga)(S,Se)2 based thin film photovoltaics: present status and current developments. Prog. Photovolt.: Res. Appl. 18, 411–433 (2010)
[71] Banerjee, P., Lee, W.-J., Bae, K.-R., Lee, S.B., Rubloff, G.W.: Structural, electrical, and optical properties of atomic layer deposition Al-doped ZnO films. J. Appl. Phys. 108, 043504 (2010)
[72] An, K.-S., Cho, W., Lee, B.K., Lee, S.S., Kim, C.G.: Atomic Layer Deposition of Un-doped an Al-Doped ZnO Thin Films Using the Zn Alkoxide Precursor Methylzinc Isopropoxide. J. Nanosci. Nanotechnol. 8, 4856–4859 (2008)
[73] Suzuki, A., Matsushita, T., Aoki, T., Yoneyama, Y., Okuda, M.: Pulsed Laser Depo-si-tion of Transparent Conducting Indium Tin Oxide Films in Magnetic Field Perpendicular to Plume. Jpn. J. Appl. Phys. 40, L401–L403 (2001)
[74] Suzuki, A., Nakamura, M., Michihata, R., Aoki, T., Matsushita, T., Okuda, M.: Ultra-thin Al-doped transparent conducting zinc oxide films fabricated by pulsed laser deposition. Thin Solid Films 517, 1478–1481 (2008)
[75] Dong, B.-Z., Fang, G.-J., Wang, J.-F., Guan, W.-J., Zhao, X.-Z.: Effect of thickness on structural, electrical, and optical properties of ZnO:Al films deposited by pulsed laser deposition. J. Appl. Phys. 101, 033713 (2007)
[76] Kim, H., Horwitz, J.S., Kushto, G., Piqué, A., Kafafi, Z.H., Gilmore, C.M., Chrisey, D.B.: Effect of film thickness on the properties of indium tin oxide thin films. J. Appl. Phys. 88(10), 6021–6025 (2000)
[77] Tsuji, N., Komiyama, H., Tanaka, K.: Growth Mechanism of ZnO Film by Reactive Sputtering Method – Significance of Thermodynamics in a Plasma System. Jpn. J. Appl. Phys. 29(5), 835–841 (1990)
[78] Kon, M., Song, P.K., Shigesato, Y., Frach, P., Ohno, S., Suzuki, K.: Impedance control of reactive sputtering process in mid-frequency mode with dual cathodes to deposit Al-doped ZnO films. Jpn. J. Appl. Phys. 42, 263–269 (2003)

[79] Ruske, F.: Magnetronsputtern von hochleitfähigen ZnO:Al-Schichten für die Photovoltaik. PhD Thesis, Fraunhofer IST, Braunschweig and Justus-Liebig-Universität Gießen, Fraunhofer IRB Verlag (2009)

[80] Minami, T., Sato, H., Imamoto, H., Takata, S.: Substrate temperature dependence of transparent conducting Al-doped ZnO thin films prepared by magnetron sputtering. Jpn. J. Appl. Phys. 31, L257–L260 (1992)

[81] Minami, T., Sato, H., Sonoda, T., Nanto, H., Takata, S.: Influence of Substrate and Target Temperatures on Properties of Transparent and Conductive Doped ZnO Thin Films Prepared by RF Magnetron Sputtering. Thin Solid Films 171, 307–311 (1989)

[82] Park, K.C., Ma, D.Y., Kim, K.H.: The physical properties of Al-doped zinc oxide films prepared by RF magnetron sputtering. Thin Solid Films 305, 201–209 (1997)

[83] Berginski, M., Hüpkes, J., Reetz, W., Rech, B., Wuttig, M.: Recent development on surface-textured ZnO:Al films prepared by sputtering for thin-film solar cell application. Thin Solid Films 516, 5836–5841 (2008)

[84] Ellmer, K.: Magnetron sputtering of transparent conductive zinc oxide: relation between the sputtering parameters and the electronic properties. J. Phys. D: Appl. Phys. 33 , R17–R32 (2000)

[85] Lee, B.H., Kim, I.G., Cho, S.W., Lee, S.-H.: Effect of process parameters on the char-acteristics of indium tin oxide thin film for flat panel display application. Thin Solid Films 302, 25–30 (1997)

[86] Koida, T., Kondo, M., Tsutsumi, K., Sakaguchi, A., Suzuki, M., Fujiwara, H.: Hydrogen-doped In_2O_3 transparent conducting oxide films prepared by solid-phase crystallization method. J. Appl. Phys. 107, 033514 (2010)

[87] Koida, T., Fujiwara, H., Kondo, M.: High-mobility hydrogen-doped In_2O_3 transparent-conductive oxide for a-Si:H/c-Si hetero junction solar cells. Sol. Energ. Mater. Sol. Cell 93, 851–854 (2009)

[88] Kim, D.-H., Park, M.-R., Lee, H.-J., Lee, G.-H.: Thickness dependence of electrical properties of ITO film deposited on a plastic substrate by RF magnetron sputtering. Appl. Surf. Sci. 253, 409–411 (2006)

[89] Gao, M.-Z., Job, R., Xue, D.-S., Fahrner, W.R.: Thickness Dependence of Resistivity and Optical Reflectance of ITO Films. Chin. Phys. Lett. 25, 1380–1383 (2008)

[90] Rößler, R., Muske, M.: Unpublished results from Helmholtz-Zentrum, Berlin (2010)

[91] Fenske, F., Brehme, S., Henrion, W., Schmidt, M.: Al-dotierte ZnO-Schichten für a-Si/c-Si Solarzellen. TCO für Dünnschichtsolarzellen II, Jülich, FVS (2002), http://www.fvee.de/fileadmin/publikationen/Workshopbaende/ws2002/ws2002_01_03.pdf (accessed November 15, 2010)

[92] Minami, T., Miyata, T., Honma, Y., Ito, T.: Effect of spatial relationship between arc plasma and substrate on the properties of transparent conducting Ga-doped ZnO thin films prepared by vacuum arc plasma evaporation. Thin Solid Films 517, 6824–6828 (2009)

[93] Qu, Y., Gessert, T.A., Ramanathan, K., Dhere, R.G., Noufi, R., Coutts, T.J.: Electrical and optical properties of ion beam sputtered ZnO:Al films as a function of film thickness. J. Vac. Sci. Technol. A 11, 996–1000 (1993)

[94] Yamada, T., Miyake, A., Kishimoto, S., Makino, H., Yamamoto, N., Yamamoto, T.: Low resistivity Ga-doped ZnO thin films of less than 100 nm thickness prepared by ion plating with direct current arc discharge. Appl. Phys. Lett. 91, 051915 (2007)

[95] Gupta, R.K., Ghosh, K., Patel, R., Kahol, P.K.: Effect of thickness on optoelectrical properties of Mo-doped indium oxide films. Appl. Surf. Sci. 255, 3046–3048 (2008)

Chapter 10
Contact Formation on a-Si:H/c-Si Heterostructure Solar Cells

Mario Tucci[1], Luca Serenelli[1], Simona De Iuliis[1], Massimo Izzi[1], Giampiero de Cesare[2], and Domenico Caputo[2]

[1] ENEA - Research Center Casaccia Via Anguillarese 301 00123 Rome Italy
[2] Department of Electronic Engineering Rome University "Sapienza", Via Eudossiana 18 00139 Italy

Abstract. In this chapter a description of the contact formation in a-Si:H/c-Si heterojunction solar cell is detailed. Firstly the doping of amorphous films is reported together with the possibility to enhance the amorphous film conductivity by using Chromium Silicide formation on top of the doped films. Then a finite difference numerical model is used to describe the a-Si:H/c-Si heterojunction solar cell in which both contacts are made by amorphous films. In particular to evaluate the effect of the bandgap mismatch between amorphous and crystalline silicon at the base contact a detailed investigation is presented comparing experimental current voltage characteristics of heterojunction contacts with the results of a simulation based on numerical model. Subsequently, details about formation and properties of a transparent conductive oxide and a screen printing procedure to form metallic grids are presented as a common way to form the heterojunction solar cell electrodes. Finally three examples of heterojunction solar cells are proposed using different approaches to form the contacts. In particular a double side heterojunction cell fabricated on multicrystalline silicon is presented, a laser fired local contact for the rear side of the cell is shown and finally an interdigitated back contact is described. All the investigations are based on our experience on heterostructure solar cells developed in the past years.

10.1 Introduction

P-type doped silicon is the most common photovoltaics (PV) material, but heterojunction solar cells on p-type c-Si are less popular than on n-type c-Si. On the latter material very impressive results have been obtained by SANYO [1] in the past two decades [2-5]. Many European research groups [6-10] have focused on n-type material, also reaching interesting results. In contrast, up to now, it has been difficult to achieve the same high photovoltaic efficiency on p-type c-Si [11-13], even

W.G.J.H.M. van Sark et al. (Eds.): Physics & Tech. of Amorphous-Crystalline, EM, pp. 331–375.
springerlink.com

considering the fact that the silicon based solar cell is a device in which the main contribution to the photogenerated carriers arises from a diffusion mechanism for which p-type material should be recommended. Several issues limit the heterojunction solar cell efficiency based on p-type doped silicon base. A very relevant one is the lower band gap of n-type doped a-Si:H emitter with respect to the p-type and also the rear side of the device. Indeed, on n-type doped silicon base the heterojunction n-type c-Si/n-type a-Si:H contact works as a barrier for the holes diffusing toward the rear side electrode, thus performing an effective back surface field (BSF) effect. In turn this concept cannot be applied with the same effectiveness on p-type doped silicon base contact as p-type c-Si/p-type a-Si:H, since the band offset at the edge of the two semiconductors represents a critical issue for the majority carrier collection. In fact, it represents a large barrier for majority carrier holes flowing through to the back contact. This chapter is focused on this aspect of the heterojunction cell. To demonstrate the effect of the doping type on the heterojunction contact a detailed investigation is presented comparing experimental heterojunction base contacts with simulations obtained by a numerical model [14]. Indeed, in order to get better insight into the back contact behaviour of the heterojunction solar cell only a numerical model can be used to evaluate the carriers transport mechanism, due to the difficulties in analytical description of the recombination process. The numerical model is based on the solution of the Poisson equation, taking into account the continuity equations and boundary conditions as imposed by bias voltage and/or light exposure. Each layer has been described by a set of parameters imposed to define semiconductor properties: absorption coefficient, energy gap (E_g), electron affinity (χ), thickness, and density of states inside the gap (DOS). The DOS of a-Si:H has been modelled by Gaussian distributions accounting for silicon dangling-bonds, and two tails for weak-bonds within the bandgap. Generation and recombination rates have been taken into account following the Shockley-Read-Hall (SRH) theory.

After an introduction to doping of the amorphous films, the investigation in the subsequent sections is divided in two parts, each of them concerning a different doping type of the silicon base wafer.

10.2 Doping in a-Si:H Film

In order to determine the doping efficiency of each dopant gas on doped a-Si:H layers, several single doped a-Si:H films have been deposited in a three chamber 13.56 MHz Plasma Enhanced Chemical Vapour Deposition (PECVD) system, in which each chamber is devoted to a single doping type of amorphous films, using a RF power density of 28 mW/cm^2, a silane flow of 40 sccm, a gas pressure in the range from 300 mTorr to 700 mTorr, a deposition temperature of 300°C and dopant gas concentrations (phosphine and diborane) ranging from 0 to 1%. The deposition rate has been evaluated to be about 1.5 Å/sec and the sample thickness around 5000 Å.

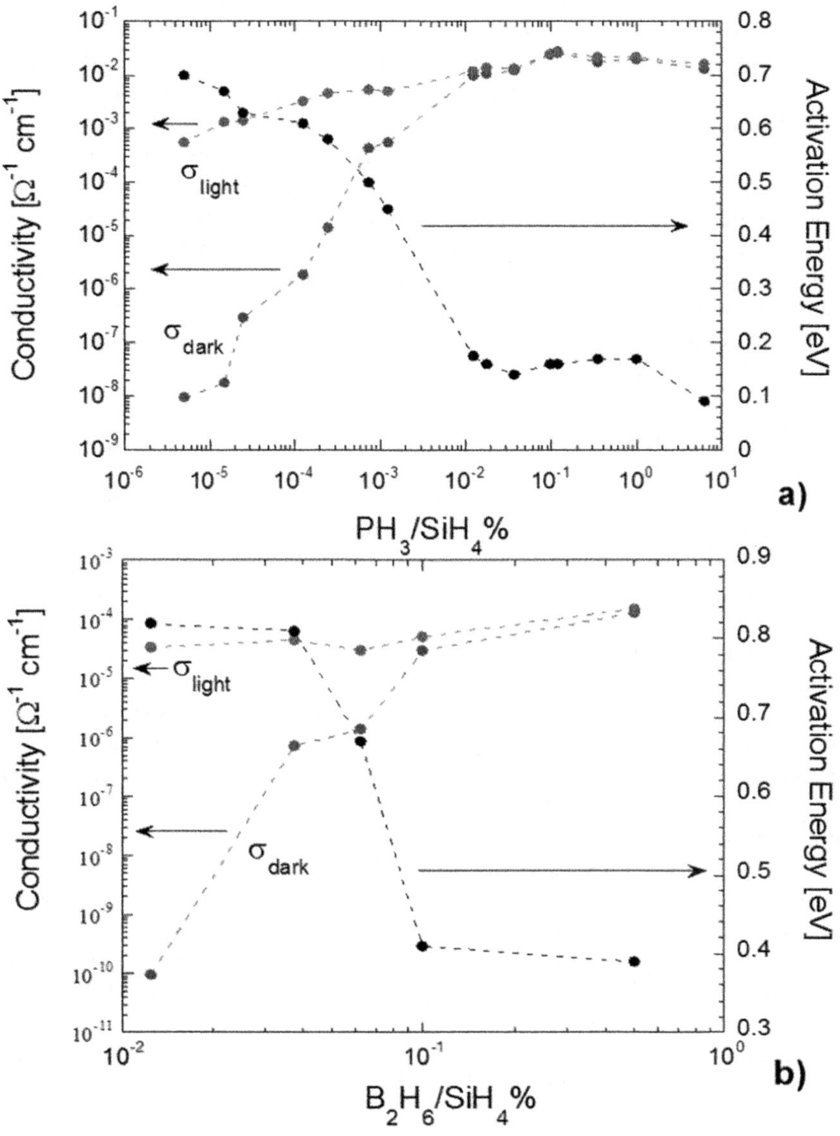

Fig. 10.1 a) Dark and lighted conductivities, activation energy for n-doped films versus phosphine percentage in the gas mixture. b) Dark and lighted conductivities, activation energy for p-doped films versus diborane percentage in the gas mixture.

Electrical and optical characterizations have been performed by dark and light (AM1.5G) conductivity, activation energy and energy gap. For all the samples this latter quantity ranges from 1.68 to 1.72 eV. Dark and light conductivity (σ_{dark}, σ_{light}), as a function of the ratio phosphine/silane and diborane/silane for n-type and p-type doping respectively, are reported in Fig. 10.1. Saturation of the dark conductivity occurs at 10^{-2} $\Omega^{-1}cm^{-1}$ and 10^{-4} $\Omega^{-1}cm^{-1}$ for the n and p single doped films, respectively. These saturated values have been obtained by using a ratio dopant/silane gas around 10^{-2} and 10^{-1} for the n and p material, respectively. The difference in the doping efficiency suggested by these results has been confirmed by the measurements of activation energy, plotted in the same figures, where a saturation value has been obtained for the same gas concentration. While in n-doped material the saturation occurs at 0.2 eV [15, 16], saturation occurs at 0.35 eV in p-doped material [17, 18, 19, 20].

10.2.1 Effect of Chromium Silicide on Doped a-Si:H Films

To improve the electrical properties of doped a-Si:H layer, the formation of a CrSi layer on top of the amorphous film has been suggested [21]. It consists of a Cr thin layer (30 nm) evaporation and subsequent wet chemical removal of the doped amorphous film, which leaves a high conductive CrSi film even at room temperature. To evaluate this effect four differently doped amorphous films have been prepared. Samples #1 and #2 were n-type doped amorphous material deposited at the following conditions: 28 mW/cm^2 of RF Power, a working pressure of 300 mTorr, a temperature of 300°C and a gas mixture of 10 sccm PH_3 and 40 sccm SiH_4. Samples #3 and #4 were p-type doped a-Si:H layers deposited at the following conditions: 28 mW/cm^2 of RF Power, 700 mTorr of working pressure, 280°C of temperature; 4 sccm of B_2H_6 and 40 sccm of SiH_4 as gas mixture. To overcome the difficulty in CrSi formation on the p-type doped a-Si:H layer, a very thin n-type amorphous layer (δn) has been deposited on this films. This thin layer is formed in 3 seconds deposition time at the conditions reported above for the n type doped material. Then, to form the CrSi layer on the top surface of samples #2 and #4 we have evaporated a 30 nm thick Cr layer and subsequently removed it by a wet chemical etch in a solution of 30 g of $Ce(NH_4)(NO_3)_6$, 9 ml CH_3COOH and 200 ml of Deionized Water. After this etching procedure the excess Cr has been removed, leaving the CrSi on the top of the amorphous films. The layer descriptions of each sample are summarized in Table 10.1.

Table 10.1 Structure of the four investigated samples.

sample	structure
#1	n-type a-Si:H
#2	n-type a-Si:H + CrSi
#3	p-type a-Si:H + δn
#4	p-type a-Si:H + δn +CrSi

To verify the effect of Cr treatment in terms of electrical properties the current flowing between two coplanar Cr electrodes evaporated on top of each sample at different temperatures has been measured. The applied external bias voltage has been fixed at 10 V. Arrhenius plots are reported in Fig. 10.2 for samples #1, #2 and samples #3, #4, respectively. In the same figure, the evaluated activation energy (E_{att}) of each sample is shown near to the corresponding experimental curve. For both doping types we observe an increase of the conductivity in Cr treated samples. This increase corresponds to an activation energy reduction of about 0.22 eV. Indeed, in the case of n-type doped a-Si:H the activation energy decreases from 0.24 eV down to 0.017 eV, while for p-type doped the activation energy reduces from 0.36 eV down to 0.14 eV. The conductivities reported in Fig. 10.2 for Cr treated samples arise from the contribution of the currents flowing through both the CrSi layer and the doped amorphous film underneath. The presence of the CrSi layer does not affect the doping type of the amorphous doped film on which it is formed. Since the thickness of the CrSi layer does not exceed 5 nm, an increase of the conductivity cannot be achieved by increasing the thickness of the δn layer otherwise the risk of formation of a p-n junction can occur.

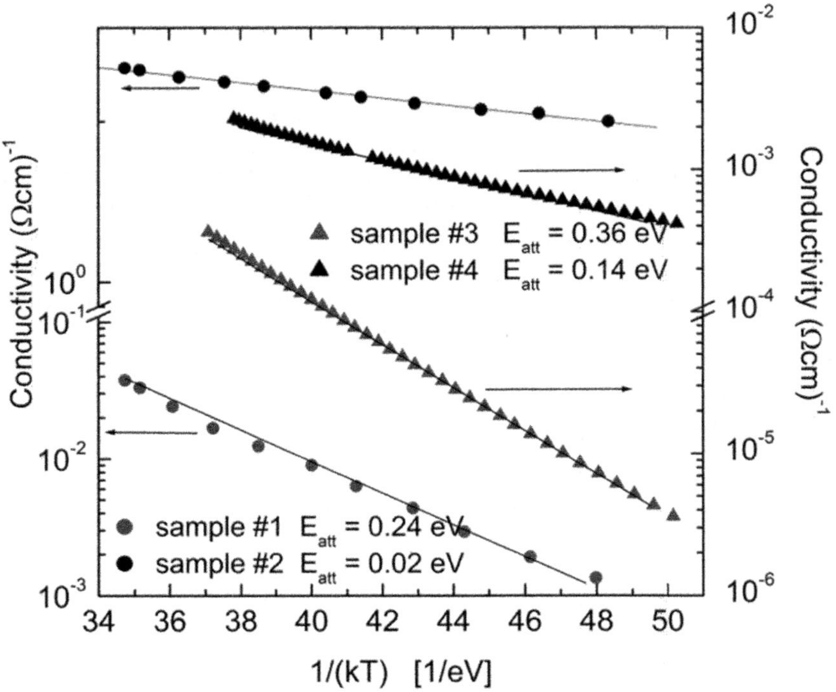

Fig. 10.2 Arrhenius plot and activation energies E_{att} of sample #1 and #2 (left side) and of sample #3 and #4 (right side).

Concerning the doping phase of doped amorphous films, very similar results of E_{att} are reported in literature. Cr treatment is able to form CrSi on top of the film, which reduces the doping E_{att} to lower values for both p and n type doped a-Si:H layers.

10.3 Heterojunction on p-Type Doped c-Si Base

In this section the heterojunction solar cell formed on p-type doped c-Si is firstly considered from the base contact and thereafter from the emitter side. Historically this device configuration has been less investigated than the counterpart based on n-type c-Si wafer for two main reasons: lower bandgap of n type doped a-Si:H emitter with respect to p-type a-Si:H and some difficulties in charge extraction at the rear side. Nevertheless very good efficiencies have been demonstrated by several groups on this device in case of both a mono [11,12,13, 22] and a multi-crystalline silicon base [23, 24]. Up to now the record efficiency can be found in ref [25]. In this paragraph a detailed description of the heterostructure device is reported basing on numerical simulations to clarify the role of front and rear side contact.

10.3.1 Base Contact

The band diagram distribution is the most helpful image to evaluate the transport through heterojunctions between two semiconductors having different band gaps. Looking at the band offsets and conduction and valence band edge alignments at the interface, it is possible to identify any problems related to the carrier collection.

The band bending distributions of heterojunction contact formed on p-type doped c-Si base in Fig. 10.3 a) is considered in dark and short circuit conditions. The simulations have been performed taking into account the E_g, χ and the activation energies (E_{att}) of p-type 1 Ωcm c-Si, the lowest E_{att} that doped a-Si:H films can achieve [6] as reported in Table 10.2 and the DOS within each layer. An intrinsic buffer layer has been introduced between the a-Si:H and c-Si doped materials in order to reduce defect density and thus the recombination at the interface [2].

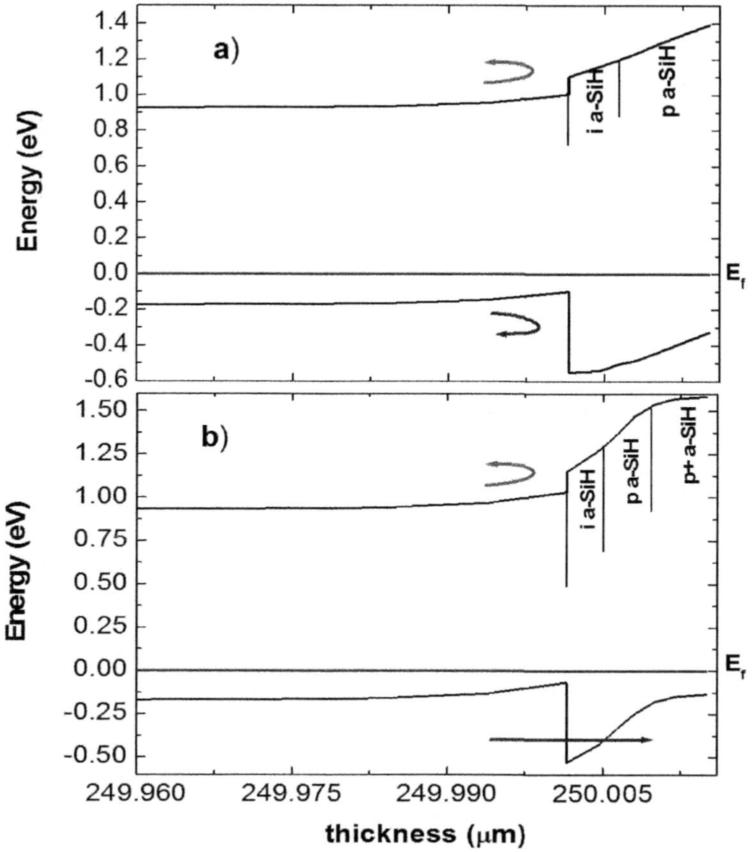

Fig. 10.3 Band bending simulations in dark and 0 V. a) p-type-c-Si/i-a-Si:H/p-aSi:H heterostructure; b) p-type-c-Si/i-a-Si:H/p-aSi:H/ p^+-aSi:H heterostructure.

Table 10.2 Dopant activation energy E_{att}, electron affinity χ, energy gap E_g, and density of states DOS of materials used in the simulations in this chapter.

materials	E_{att}(eV)	χ(eV)	E_g(eV)	DOS (cm^{-3})
n-type c-Si	0.21	4.05	1.12	$1 \cdot 10^{11}$
p-type c-Si	0.19	4.05	1.12	$1 \cdot 10^{11}$
i a-Si:H	-	3.9	1.72	$5 \cdot 10^{15}$
n-type a-Si:H	0.2	3.9	1.67	$1 \cdot 10^{17}$
p-type a-Si:H	0.34	3.9	1.72	$1 \cdot 10^{17}$
n^+-type a-Si:H	0.017	3.9	1.67	$1 \cdot 10^{17}$
p^+-type a-Si:H	0.14	3.9	1.72	$1 \cdot 10^{17}$

These band diagrams can be considered as the rear side contact of a double heterojunction solar cell scheme in which front side emitter and rear side contact are formed using an a-Si:H/c-Si heterojunction. As evident from the band diagrams, the offset at the edge of conduction band at the interface would not represent a serious obstacle if electrons were collected; in turn, the pronounced valence band offset is a critical point for holes flowing toward the metal electrode. In the simulations the carrier collection at the amorphous silicon edge has been supposed not to be influenced by the metal electrode workfunction. Any depletion within the amorphous layer has been neglected due to the high defect density at the metal/ a-Si:H interface that pins the Fermi level.

In order to obtain a good base contact in principle a highly doped layer should be used, but, as previously seen, amorphous doped film cannot be overdoped as for crystalline silicon; therefore the base contact should be carefully considered when amorphous films are used. A way to enhance the doped amorphous film conductivity has been proposed in ref. [11]. In particular we have experimentally found that a CrSi layer formed on top of the a-Si:H doped layer was able to enhance its conductivity. With the aid of numerical model it is possible to verify whether the CrSi layer can be helpful also to enhance the collection at the heterojunction contact. Since the CrSi experimentally forms at the expense of the doped layer, it has been numerically modelled by converting the top part of the a-Si:H doped layer into a heavily doped a-Si:H layer having a reduced E_{att} as reported in Table 10.2. In Fig. 10.3 b) the band diagram refers to a heterojunction similar to the one of Fig. 10.3 a) except for the doped a-Si:H layers that are considered partially converted into a highly doped layer as is achievable by CrSi formation.

To evaluate the role of the heterojunction as a base contact for a c-Si wafer the structures listed in Table 10.3 (p1, p2, p3) have been simulated in terms of current voltage (I-V) behaviour. In particular the sample p1 has been kept as reference for ohmic contacts.

Table 10.3 Modelled structures for p-type doped c-Si base contact.

Sample	structure
p1	Metal /p^+-type c-Si/p-type c-Si/ p^+-type c-Si/ Metal
p2	Metal /p^+-type c-Si/p-type c-Si/i a-Si:H/p-type a-Si:H/ Metal
p3	Metal /p^+-type c-Si/p-type c-Si/i a-Si:H/p-type a-Si:H/ p^+-type a-Si:H / Metal

For each doping type several I-V characteristics have been simulated by varying the thicknesses of both intrinsic buffer and doped a-Si:H layers within the heterostructure. A comparison of samples p1 and p2 in Fig. 10.4 is reported as continuous lines. As evident from the figure, the current voltage behaviour of the p-type doped heterojunction is not linear if compared with the desired ohmic reference contact.

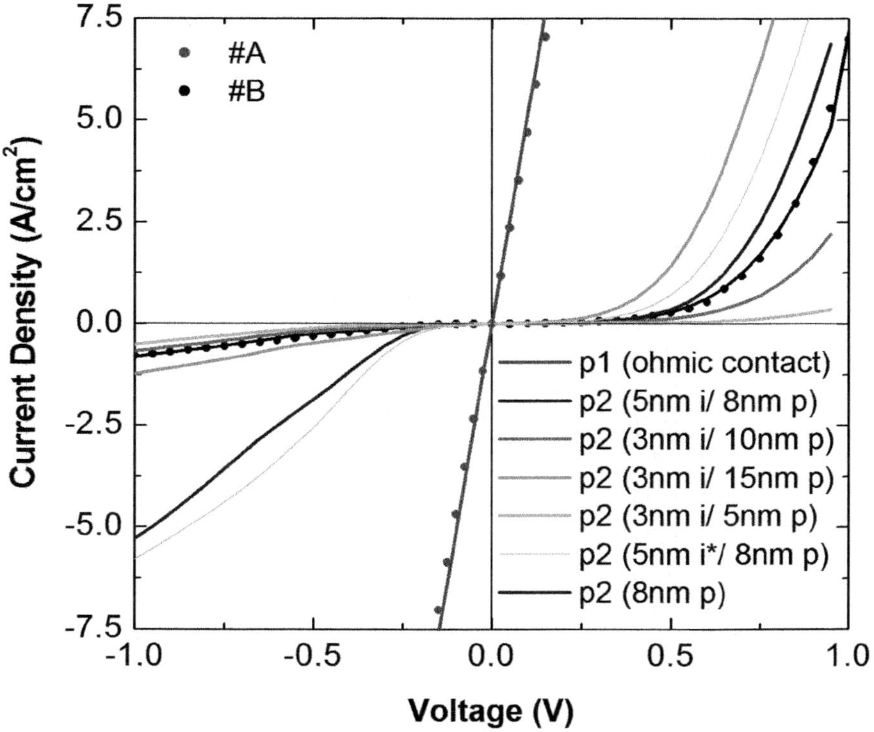

Fig. 10.4 I-V characteristics in dark conditions of p-type doped c-Si/i-a-Si:H/p-a-Si:H contact for several thicknesses of a-Si:H films: simulations (lines) and experiments (dots). i* refers to lower DOS within intrinsic a-Si:H layer ($1 \cdot 10^{15}$ cm^{-3}).

Neither thinning the intrinsic buffer, nor shorting the p-type a-Si:H doped layer has led to an acceptable contact. Reducing the DOS within the intrinsic layer, the unwanted nonlinear effect still influences the I-V curve, as suggested by the simulation reported as grey line in Fig. 10.4 where the intrinsic layer DOS has been reduced down to $1 \cdot 10^{15}$ cm^{-3}. Moreover the p-type doped a-Si:H layer cannot be thinner than 8 nm, otherwise it does not behave as an effective doped layer, also in practical use.

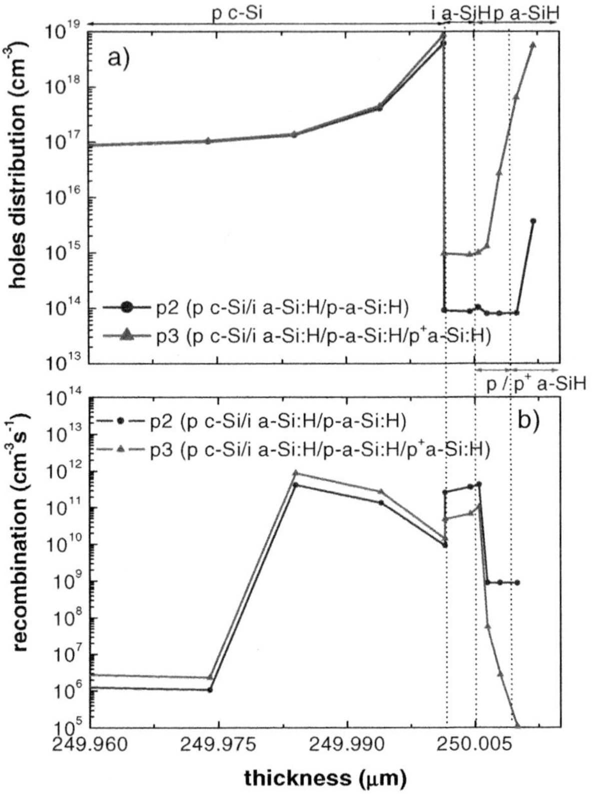

Fig. 10.5 Holes and recombination distributions at the heterojunction at 0.65 V in dark condition.

Completely removing the intrinsic buffer layer the I-V characteristic seems less dominated by a rectifying effect as evident from the simulation p2 (8 nm p) reported in Fig. 10.4. But the removal of the intrinsic layer cannot be suggested in practice, since the growth of a p-type doped a-Si:H layer directly over a c-Si wafer produces a very defected interface, loosing the silicon surface passivation offered by an intrinsic a-Si:H layer [2]. This unwanted effect is mainly due to the valence band offset at the edge of the c-Si wafer that forms a barrier against the carrier collection from the metal electrode as can be deduced from Fig. 10.5. In this figure the hole distribution along the heterojunction (see Fig. 10.5 a), simulated at 0.65 V in case of 5 nm and 10 nm of intrinsic buffer and doped p-type a-Si:H thicknesses respectively, strongly drops at the edge of the a-Si:H/c-Si heterojunction due to recombination, as reported in Fig. 10.5 b). A good improvement of I-V curves has been obtained in the simulations by partially replacing the doped a-Si:H region into higher doped layer as demonstrated in Fig. 10.6 comparing the I-V characteristics of sample p1 and p3. The band bending of this improved heterojunction contact is reported in Fig. 10.3 b) to be easily compared with the previous one.

The effect of a higher doped region is a narrowing of the barrier at the valence band, which promotes a better hole collection with lower recombination at the metal electrode, as can be seen from the hole distribution and recombination of sample p3 reported in Fig. 10.5 and compared with that of sample p2. The recombination at the interface drops by one order of magnitude even using the same amount and distribution of DOS within the band gap at the heterojunction. As a consequence the hole collection at the metal electrode has been strongly enhanced. This enhancement is reflected in a rectifying behaviour reduction of the heterojunction contact as evident from Fig. 10.6 comparing the simulations of sample p3, reported as continuous lines, together with the ohmic reference contact p1. Also in this case several simulations have been performed by varying the intrinsic buffer and the p-type doped layer thicknesses. I-V characteristic very similar to the ohmic contact can be achieved by reducing the intrinsic buffer thickness down to 5 nm.

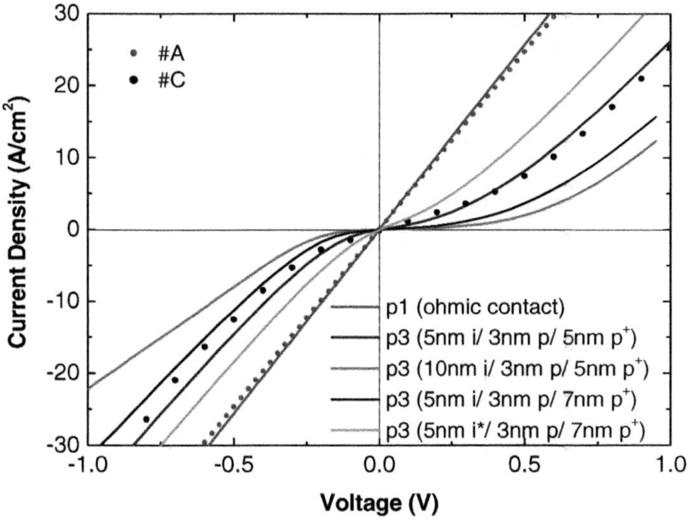

Fig. 10.6 I-V characteristics in dark conditions of p-type doped c-Si/i a-Si:H/p a-Si:H/p^+a-Si:H contact for several thicknesses of a-Si:H films: simulations (lines) and experiments (dots). i* refers to lower DOS within intrinsic a-Si:H layer ($1 \cdot 10^{15}$ cm^{-3}).

To confirm the simulations, a p-type doped c-Si/a-Si:H heterojunction has been fabricated on the RCA cleaned rear side of a 1 Ωcm p-type 200 μm thick CZ silicon wafer. The front side contact has been ensured by 4 μm Al e-beam evaporation followed by a thermal annealing at 700°C to form a p^+ region. On the same kind of wafer, also an Al/p-type c-Si/Al structure has been fabricated and considered as the ohmic contact references. All the amorphous films have been deposited in a three chamber 13.56 MHz RF PECVD system using an RF power density of 28 mW/cm^2, 700 mTorr of working pressure, 200°C of temperature deposition,

40 sccm of SiH_4 for the intrinsic layer and 40 sccm of SiH_4 plus 6 sccm of B_2H_6 for p-type doped films.

Thicknesses of 5 nm and 7 nm for the intrinsic and the p-type doped amorphous films have been chosen, respectively. Finally, Al 4 μm thick circular dots have been e-beam evaporated through a metal mask, useful to limit the metal contact area, as electrode of the heterostructure samples. On similar samples a CrSi layer has been introduced before evaporation of the metal contacts to enhance the p-type doped a-Si:H conductivity. This layer is spontaneously formed at room temperature after Cr evaporation and subsequent removal. All the samples, summarised in Table 10.4, have been characterized by current-voltage (I-V) measurements at room temperature in dark conditions. Their I-V characteristics are reported as dots in Fig. 10.4 and Fig. 10.6 and compared to the related simulations listed in Table 10.3.

Table 10.4 Structure of p-type doped experimental samples.

Sample	structure
#A	Al/p^+-type c-Si/p-type c-Si/ p^+-type c-Si/Al
#B	Al/p^+-type c-Si/p-type c-Si/i a-Si:H/p-type a-Si:H/Al
#C	Al/p^+-type c-Si/p-type c-Si/i a-Si:H/p-type a-Si:H/CrSi/Al

The good agreement between the experimental and simulated data confirms the validity to choose the finite difference model to investigate the transport mechanism at the p-type doped heterojunction contact. From this comparison it is evident that the presence of the higher conductive CrSi layer, with respect to the p-type doped a-Si:H layer, allows the formation of an effective heterojunction contact that can be regarded as a quasi ohmic contact.

10.3.2 Emitter Contact

Selecting the p-type doped c-Si wafer as base for the heterostructure solar cell, an n-type doped a-Si:H layer has to be used as cell emitter. Due to bandgap lowering related to the n-type doping [26], the choice of its thickness becomes relevant to avoid filtering effect of the sunlight incident onto the solar cell. But also in this case the use of CrSi can be helpful to improve the front side contact and, in principle, also the built-in voltage of the a-Si:H/c-Si solar cell [22]. The band diagram distribution of the front side of the heterostructure cell side in Fig. 10.7 is reported for the case of a common n-type doped a-Si:H emitter and for the case of the n-type doped a-Si:H layer partially replaced with a higher conductive CrSi top layer. Both structures contain an intrinsic buffer layer at the heterointerface to reduce the crystalline surface recombination; the simulations use the values reported in Table 10.1 for the n-type doped a-Si:H emitter. Taking into account these results four kinds of solar cells have been simulated. For the first cell (cell-A) a p-c-Si/p^+-c-Si rear contact configuration has been used as generally obtainable using an Al back surface field (BSF). In the other two samples (cell-B and cell-C) the rear side of the cell has been modelled using a p-type doped a-Si:H/c-Si heterojunction. The cell-C rear side has been provided with a p^+-type doped a-Si:H layer. Only in sample D the n-type doped a-Si:H emitter has been considered as would be obtained after a CrSi treatment reducing the doping E_{att} to 0.017 eV from the typical 0.2 eV as reported in Table 10.2.

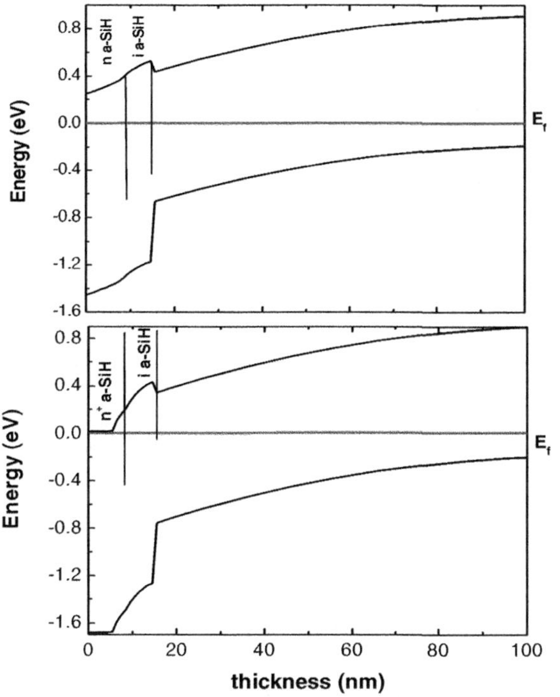

Fig. 10.7 Band bending simulations in dark and 0 V of n-aSi:H/i-a-Si:H/p-c-Si heterostructure without (upper) and with (lower) n^+ a-Si:H layer.

The simulated I-V characteristics under AM1.5G sunlight exposure reported in Fig. 10.8 show the undesired rectifying behaviour at the rear side of the cell strongly affecting the photovoltaic performances of the cell-A. This effect is due to the formation of a small potential barrier at the rear side of the cell that opposes to hole collection at the rear electrode and reduces the built-in potential of the cell. Therefore this effect appears in illuminated I-V characteristics as an “s” shape around the knee of the curve as evident from Fig. 10.8 (cell-A). In turn, the characteristic of cell-C confirms the effectiveness of the p-type high doped top a-Si:H layer to improve the hole collection at the rear side of the cell, also resulting in better photovoltaic performance with respect to the common Al BSF contact of cell-A. Indeed in cell-B a higher open circuit voltage is obtained due to the presence of a heterojunction contact at the both side of the solar cell that improves the cell built in potential. Moreover cell-D shows the possibility to achieve very high open circuit voltage if for both sides of the cell highly doped a-Si:H films would be used.

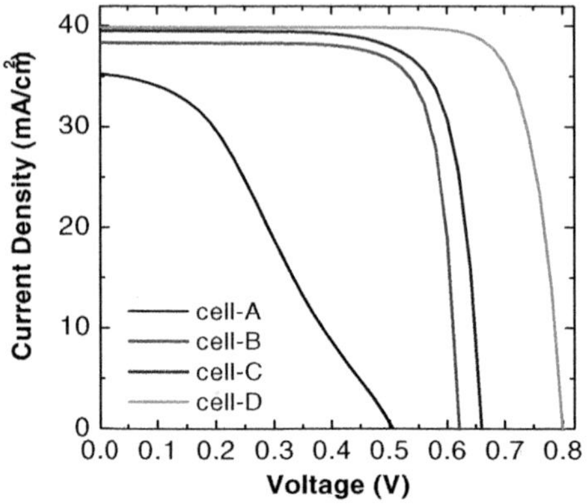

Fig. 10.8 Simulated current voltage characteristics under AM1.5G sunlight conditions.

10.4 Heterojunction on n-Type Doped c-Si Base

In this section the heterojunction solar cell formed on n-type doped c-Si is explained, first regarding the base contact and afterwards the emitter side. Following the very good efficiency obtained by Sanyo during the years [1] on the heterostructure based on n-type doped c-Si wafers, many research groups during the last decades have spent effort to obtain same results on the similar device configuration [6-10]. Up to now none of them succeeded in overtaking Sanyo's cell efficiency, even if some aspects of the device actually are now well addressed and understood [3] with respect to some time before. In this section a detailed analysis of the heterostructure contacts is proposed using numerical simulation to show the role of a-Si:H doping phase in both front and rear side contact.

10.4.1 Base Contact

The band diagram distributions of the heterojunction contact formed on n-type doped c-Si base in Fig. 10.9 a) is reported for dark and short circuit conditions. The simulations has been performed taking into account the E_g, χ and the activation energies (E_{att}) of p-type 1 Ωcm c-Si, the lowest E_{att} that doped a-Si:H films can achieve [6] as reported in Table 10.2 and the DOS within each layer. An intrinsic buffer layer has been introduced between the a-Si:H and c-Si doped materials in order to reduce defect density and thus the recombination at the interface [27]. This band diagrams can be regarded as the rear side contact of a double heterojunction solar cell scheme in which front side emitter and rear side contact are formed using a-Si:H/c-Si heterojunctions. As evident from the band diagrams, the the small offset at the edge of conduction band at the interface does not represent an obstacle for the electron collection at the metal electrode. On the other hand the pronounced valence band offset at the edge of the interface represents a strong barrier for diffusing holes to reach the metal electrode, thus forming a kind of BSF. In the simulations the carrier collection at the amorphous silicon edge has been supposed not to be influenced by the metal electrode workfunction. Any depletion within the amorphous layer has been neglected due to the high defect density at the metal/a-Si:H interface that pins the Fermi level.

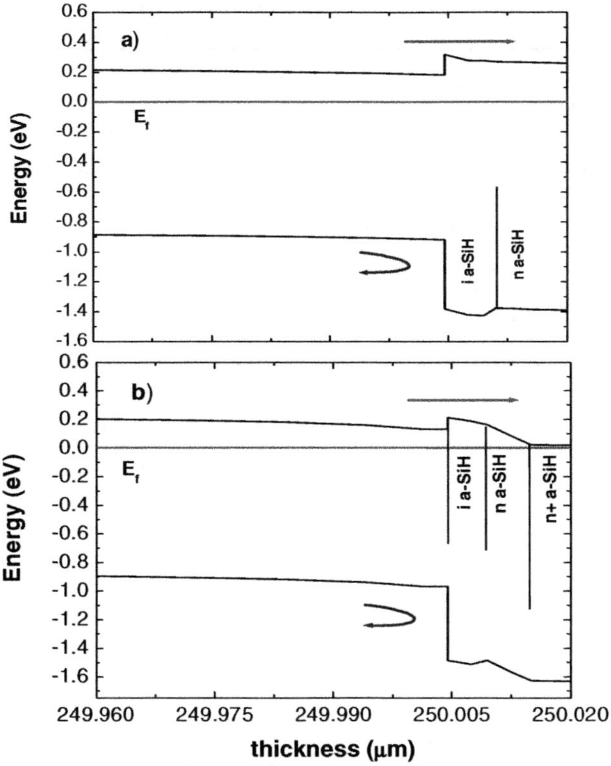

Fig. 10.9 Band bending simulations in dark and 0 V a) n-type-c-Si/i-a-Si:H/n-aSi:H heterostructure; b) n-type-c-Si/i-a-Si:H/n-aSi:H/ n^+-aSi:H heterostructure.

Also in this case the CrSi doped film can be considered to improve the base contact. As previously seen, this layer can be modelled as an overdoped part of the n-type doped a-Si:H layer in which the E_{att} is reduced down to 0.017 eV from the typical 0.22 eV as reported in Table 10.2. The related band bending distribution of the heterojunction base contact as obtained introducing the CrSi concept in the simulation is reported in Fig. 10.9 b). To evaluate the role of the heterojunction as a base contact for c-Si wafer, the structures listed in Table 10.5 have been simulated in terms of current voltage (I-V) behaviour. The structure n1 reported in

Table 5 is the reference ohmic contact as obtainable considering the two sides of the n-type doped c-Si wafer diffused to form n^+ region before metal contact.

The n2 sample is regarded as the base contact of the structure adopted by SANYO in the heterostructure solar cell. The n3 sample is similar to the previous one except for the n^+ doped region introduction at a-Si:H/metal interface. The intrinsic and n-type doped layer thicknesses have been chosen of 5 nm and 10 nm respectively for both n2 and n3 samples. Their DOS are reported in Table 2.

Table 10.5 Modelled structures for n-type doped c-Si base contact.

Sample	structure
n1	Metal /n^+-type c-Si/n-type c-Si/ n^+-type c-Si/ Metal
n2	Metal /n^+-type c-Si/n-type c-Si/i a-Si:H/n-type a-Si:H/ Metal
n3	Metal /n^+-type c-Si/n-type c-Si/i a-Si:H/n-type a-Si:H/ n^+-type a-Si:H / Metal

Table 10.6 Structure of n-type doped experimental samples.

Sample	structure
#D	Ag/n^+-type c-Si/n-type c-Si/i a-Si:H/n-type a-Si:H/Al
#E	Ag/n^+-type c-Si/n-type c-Si/i a-Si:H/n-type a-Si:H/CrSi/Al

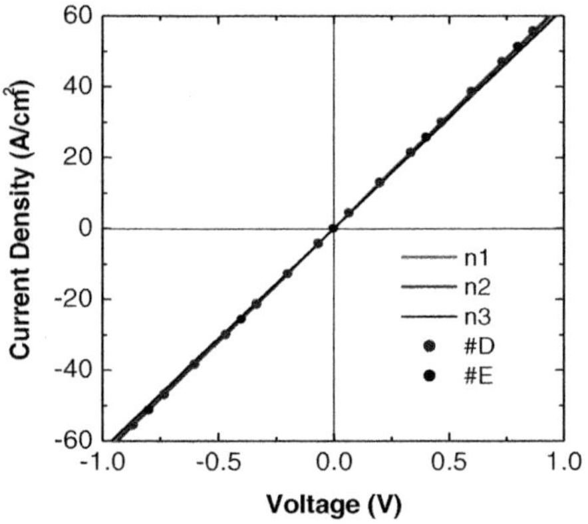

Fig. 10.10 I-V characteristics in dark conditions of n-type doped c-Si/i-a-Si:H/n-a-Si:H contacts: simulations (lines) and experiments (dots). i* refers to lower DOS within intrinsic a-si:H layer ($1 \cdot 10^{15}$ cm^{-3}). Curves of samples n1, n2, n3 are very similar and lie above each other.

The simulated I-V characteristics of the samples listed in Table 10.5 are reported in Fig. 10.10 as continuous line. As evident the three heterostructures always form ohmic contacts. For completely removing the intrinsic buffer layer the I-V characteristic still remains unaffected. But also in this case, the intrinsic removal cannot be suggested in practice, since the growth of a n-type doped a-Si:H layer directly over a c-Si wafer produces a defected interface, loosing the silicon surface passivation offered by an intrinsic a-Si:H layer [28].

Also increasing the thickness of the intrinsic buffer as well as of the n-type doped a-Si:H layer up to 10 nm and 20 nm respectively, the I-V characteristics still remain unaffected.

These simulated I-V characteristics have been compared with experimental I-V data measured on the heterostructure base contact samples listed in Table 10.6. All the amorphous films have been deposited in a three chamber 13.56 MHz RF PECVD system on an RCA cleaned 1 Ωcm n-type doped mono c-Si wafer, using 28 mW/cm^2 of RF power density, 300 mTorr of working pressure, 250 °C of temperature deposition, 40 sccm of SiH_4 for the intrinsic layer and 40 sccm of SiH_4 plus 10 sccm of PH_3 for the n-type doped films. Al 4 μm thick circular dots have been e-beam evaporated through a metal mask, useful to limit the metal contact area, as electrode of the heterostructure samples. On sample #E the CrSi layer has been introduced before metal contacts evaporation to enhance the n-type doped a-Si:H conductivity. The two samples, summarised in Table 10.6, have been characterized by I-V measurements at room temperature in dark conditions. Their I-V characteristics are reported as dots in Fig. 10.10 and compared to the related simulations n2 and n3. The experimental and simulated data are in good agreement and all the curves show a very similar behaviour confirming that the n-type contact is less critical than that on the p-type base. Moreover the CrSi formation, as in the sample #E, does not result in any improvement with respect to sample #D.

The electron distribution and the recombination along the heterostructure base contact are reported in Fig. 10.11 as obtained from simulation of sample n2. Since the simulation of both n1 and n2 samples are very similar only one is reported in Fig. 10.11. As is evident, the conduction band offset at the heterojunction does not affect the electron transfer towards the metal electrode. For clarity the band offset is located at the discontinuity of the two curves reported in the figure.

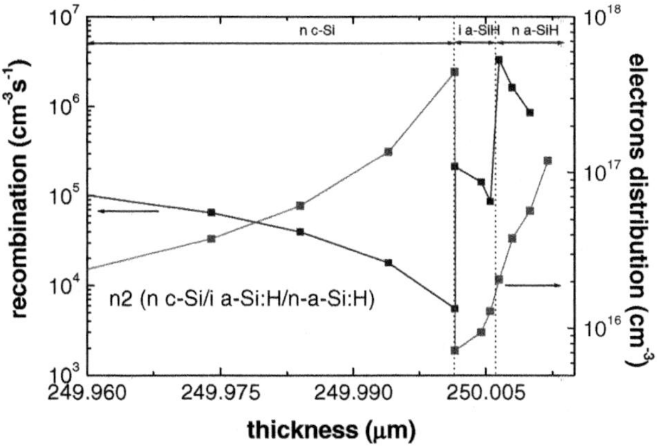

Fig. 10.11 Electron and recombination distributions at the heterojunction at 0.6V in dark condition.

10.4.2 Emitter Contact

Selecting the n-type doped c-Si wafer as base for the heterostructure solar cell, of course a p-type doped a-Si:H layer has to be used as cell emitter. Since the p-type a-Si:H layer has a $E_{gap} = 1.72$ eV the filtering effect of the emitter layer is less pronounced than in n type doped a-Si:H emitter layer. The band diagram distribution of the front side of the heterostructure cell are reported in Fig. 10.12 for the case of common p-type doped a-Si:H emitter and for the case of a p-type doped a-Si:H layer partially replaced with a higher conductive CrSi top layer. Both structures contain an intrinsic buffer layer at the heterointerface to reduce the crystalline surface recombination and the simulations use the values reported in Table 10.2 for the n-type doped a-Si:H emitter. Taking into account these results three kinds of solar cells have been simulated. For the first cell (cell-E) an n-c-Si/n$^+$c-Si rear contact configuration has been used as obtainable using a diffused back surface field (BSF). The rear side of the other two cells (cell-F and cell-G) has been modelled using a n-type doped a-Si:H/c-Si heterojunction. Cell-E and F have been modelled considering a 17 nm thick emitter layer in which 7 nm are composed by intrinsic buffer. In cell-G the p-type doped a-Si:H emitter has been considered as would be obtained after a CrSi treatment reducing the doping E_{att} to 0.14 eV from the typical 0.34 eV as reported in Table 10.2.

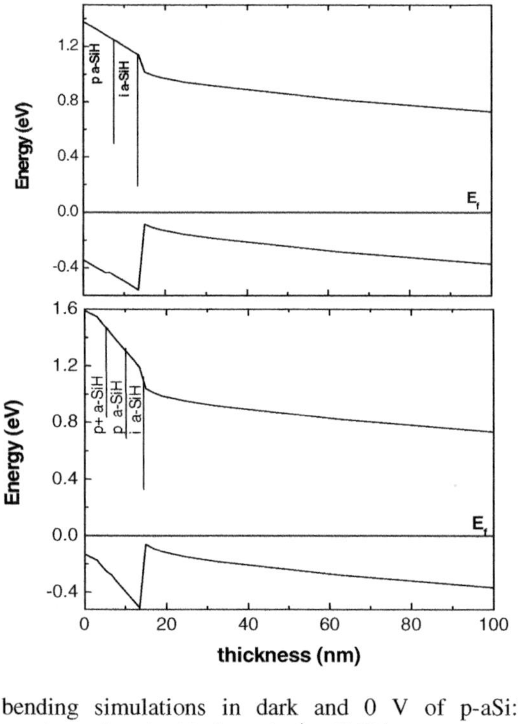

Fig. 10.12 Band bending simulations in dark and 0 V of p-aSi:H/i-a-Si:H/ n-c-Si heterostructure without (upper) and with (lower) p$^+$ a-Si:H layer.

The simulated I-V characteristics under AM1.5G sunlight exposure reported in Fig. 10.13 show the higher photovoltaic performance of the heterostructure cell based on n-type doped c-Si with respect to the p-type c-Si base. Moreover in the cell-F and cell-G higher open circuit voltages have been obtained compared to cell-E, due to the presence of heterojunction contacts at both sides of the solar cell that improve the cell's built-in potential. Finally the presence of a highly doped p-type a-Si:H layer as emitter of cell-G reflects in a very high open circuit voltage of the simulated cell.

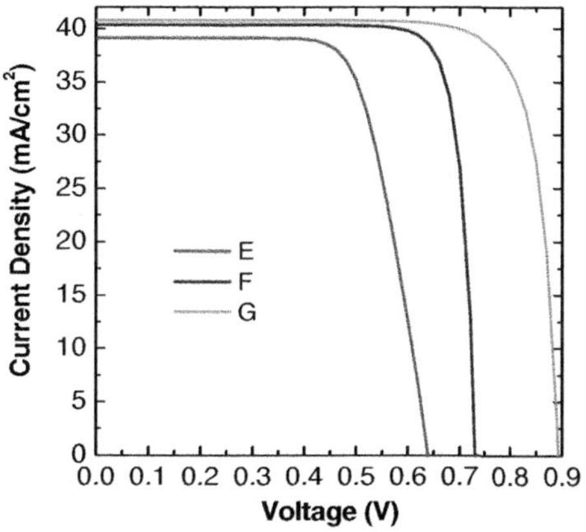

Fig. 10.13 Simulated current voltage characteristics under AM1.5G sunlight conditions.

10.5 ITO Details

Indium-tin oxide (ITO) is the most widely used transparent conducting oxide for flat panel displays, primarily because it has high optical transmittance in the visible region, high electrical conductivity, surface uniformity and process compatibility. ITO films can be prepared by a variety of techniques: dc-sputtering, rf-sputtering, electron beam evaporation, and chemical vapor deposition. The nature of the substrate and experimental conditions like oxygen partial pressure, substrate temperature and post deposition annealing were confirmed to have large effects on the microstructure, electrical and optical properties of the ITO films.

Radiofrequency (RF) magnetron sputtering technique is the most suitable method for ITO deposition on large area thin film optoelectronic devices. In this chapter, RF sputtering conditions appropriate for the deposition of ITO layers on heterojunction silicon solar cells will be extensively investigated.

In silicon heterojunction solar cells the very thin emitter layer exhibits a large sheet resistance and the ITO contact, because of its high conductivity and transparency in the visible range, is necessary to improve the carrier collection in the photovoltaic device. All silicon-based solar cells need an anti-reflecting (AR) coating in order to reduce the large reflectance losses due to the high refractive index of silicon. In the case of heterojunction solar cells, the ITO contact can also be used as AR coating, but an optimization of its layer thickness is essential, taking into account the optical properties of the underlying a-Si:H layer. The electrical and optical properties of ITO films are critically process dependent. Transparency and conductivity of this highly degenerate and wide band-gap oxide semiconductor film can be varied by adjusting the deposition conditions. To obtain the best solar cell performances the ability of depositing highly conductive and transparent ITO films is crucial. However, the optical and electrical properties of this material are inversely related so that a larger conductivity in the ITO film is accompanied by a larger light absorption and opacity. For this reason, a compromise in performance must be made for the intended application.

The ITO front and back contacts in heterojunction solar cells have been optimized by combining experimental and computer simulation techniques, determining also which deviation from the optimal ITO thicknesses can be tolerated. The ITO films useful for heterostructure solar cells have been deposited using a low temperature RF 13.56 MHz magnetron sputtering. The sputtering process temperature has been chosen after several trials varying the temperature from 50°C to 300°C. In Fig. 10.14 are reported, as points, the values of resistivity and effective transmittance from 350 nm to 1200 nm defined as:

$$T_{eff} = \frac{\int_{350nm}^{1200nm} T(\lambda) \cdot AM1.5G(\lambda) d\lambda}{\int_{350nm}^{1200nm} T(\lambda) d\lambda} \tag{10.1}$$

where $T(\lambda)$ is the transmittance and AM1.5G(λ) is the number of photons per unit wavelength (λ).

The deposition process parameters have been optimized in order to obtain a good compromise between the film conductivity and transmittance.

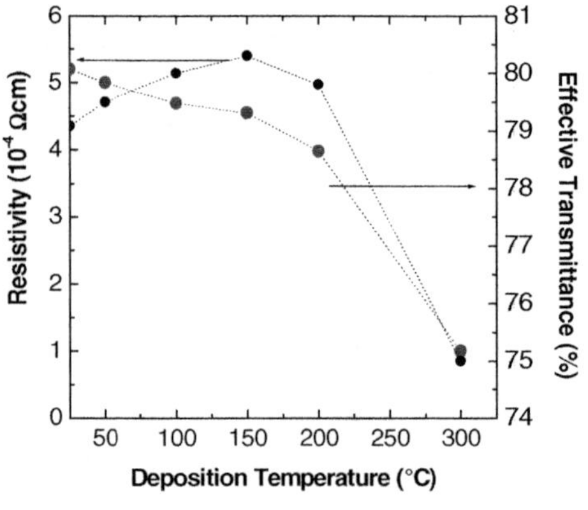

Fig. 10.14 Sputtered ITO resistivity and transmittance as a function of deposition temperature.

In particular, the sputtering conditions have been fixed as follows: the RF power to 200 W, the process pressure to 2.25 mTorr, the Ar flow to 25 sccm, and the process temperature to 150°C. It is worth to note that no oxygen has been used in order to reduce the film resistivity. With these conditions the ITO growth rate equals 4.4 Å/sec, thus the process duration has been set to deposit 62 nm at the front side and 110 nm at the rear of the device. The reflectance profile of the heterostructure device composed by ITO 62 nm / a-Si:H 150 nm / c-Si 250 μm/ a-Si:H 200 nm / ITO 110 nm / Screen printed Ag is reported in Fig. 10.15.

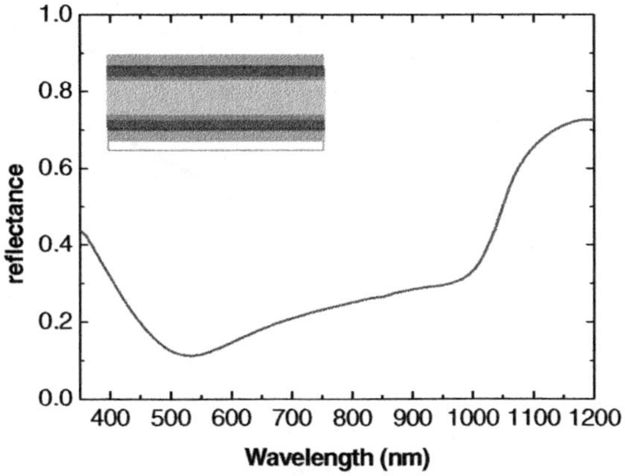

Fig. 10.15 Reflectance of a heterojunction solar cell on a flat wafer with 62 nm ITO and 110 nm ITO sputtered on front and rear side, respectively, as remarked in the inset.

Figure 10.15 shows the reflectance data of a bifacial silicon heterostructure device sample, drawn in the inset, without the front metal grids. The double-sided silicon heterojunction solar cell is the appropriate cell design for n-type crystalline silicon base. In this case the rear contact is formed by an n-c-Si/i-a-Si/n-a-Si/ITO structure.

To quantify the quality of the ITO/a-Si:H contact on both p or n type films four test samples, with and without the CrSi, have been produced and electrically characterized. In particular these samples have been formed depositing two stripes of ITO over a-Si:H doped films and current voltage characteristics have been collected between the two parallel ITO contacts. The measured current density data obtained varying the applied voltage from -2 V to 2 V are illustrated in Fig. 10.16. It is evident that on both base polarities the ITO forms an ohmic contact, that the electrical quality is much better for n-type doped layers and that the CrSi reduces the contact resistance. The coplanar ITO stripes in the insets of Fig. 10.16 are depicted.

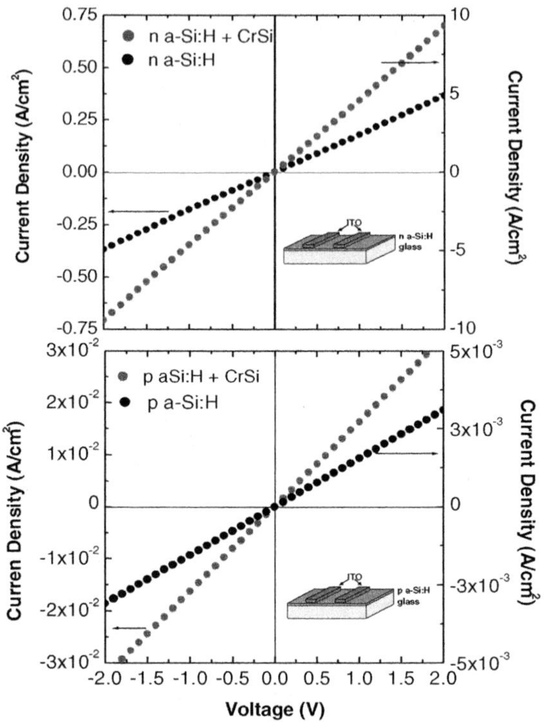

Fig. 10.16 ITO forms ohmic contact on both n- or p-type doped a-Si:H film (upper and lower side respectively).

The physical mechanism for the contact formation at the ITO/p-a-Si:H interface is a tunneling effect, as shown in Fig. 10.16, thus in the design of the ITO layer a crucial parameter to consider is the work function of the transparent conductive layer. In our deposited films this parameter has been evaluated to be 4.3 eV. By modeling the bifacial heterojunction device using DIFFIN [8], the band bending at the amorphous/crystalline silicon interfaces and the electronic affinity distribution can be evaluated, and the minimum value of the ITO work function can be estimated as 5.2 eV for the emitter side of the cell, while 4.3 eV are sufficient for the rear side of the heterostructure cell. Thus we can conclude that to avoid a built-in reduction due to the presence of the ITO/p-a-Si:H contact an ITO with an associated work function more than 5 eV should be designed as suggested by literature [29, 30]. If this rule is neglected and an ITO workfunction lower than 5eV is used for the front side of the heterostructure cell a bandbending at the ITO/p type doped a-Si:H interface can reduce the built-in potential as shown in left side of Fig. 10.17.

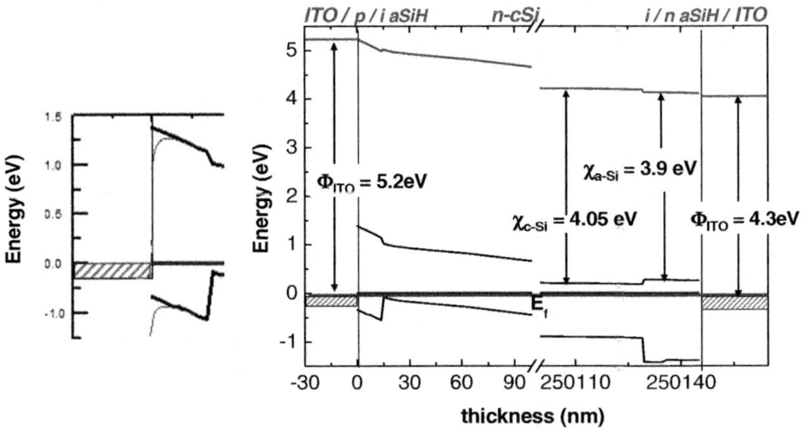

Fig. 10.17 Band bending simulation of the n-type c-Si doped based a-Si:H/c-Si heterojunction cell with two different ITO work functions: 5.2 eV for front and 4.2 eV for the back side. At the left side the band bending at the edge between ITO and p-type doped a-Si:H is magnified to show the undesired band bending distribution when ITO work function lower than 5.2 eV is used for the cell front side.

10.6 Screen-Printing Contact

Actually the screen-printing process for metal deposition is one of the key points for mass production evolution of solar cells, improving the throughput of the metallization processes [31]. It consists in the deposition of metal onto a substrate according to a particular design (usually a grid for the front side contact) with the aid of a mask, through which a screen printable paste containing the desiderate metal is passed. During the printing process, a squeegee moves the paste across the screen. This action causes a decrease in the viscosity of the paste, which in turn allows the paste to pass through the patterned areas onto the substrate. As the squeegee passes, the screen peels off and the paste viscosity returns to normal. Factors that affect the screen peel are the paste and its viscosity, the area of the print, the tension of the screen, the squeegee speed, and the snap-off distance between the sample and the screen. The screen is made of an interwoven mesh kept at high tension, with an organic emulsion layer defining the printing pattern. The effectiveness of screen printed contact formation is related, more than the metal screen printable paste composition, to the thermal process which promotes the paste sintering with the silicon, and influences the bulk resistivity of the metal itself. For diffused silicon solar cells this so-called firing process is usually a rapid thermal annealing involving temperature up to 800°C for a few seconds [32]. In the heterojunction solar cell technology such high temperature steps are detrimental for the cell, due to the low temperature process used to deposit a-Si:H films. However specific silver pastes have been developed from different screen printable paste manufacturers, able to be sintered at temperatures ranging from 120°C to 250°C. Longer thermal annealing time and higher temperatures improve adhesion and bulk resistivity [33, 34]. The actual application of this kind of pastes requires an accurate optimization of printing parameters to obtain a fine-line printing, aspect ratio, non-interrupted fingers. At the same time adequate curing in air atmosphere has to be chosen to obtain good conductivity of front grid but without damaging the cell performances or modifying the TCO properties. Such kinds of pastes are generally rapid in hardening, and the probability to obtain occlusion in the meshes is higher with high mesh count: 350 to 400-mesh stainless steel screens starts to be a limit when 80 mm wide fingers are the target. For clarity the mesh number is the number of steel wires composing the screen per area unit that defines the aperture through which the paste passes during the printing process. The higher this number, the smaller the screen aperture. Therefore the best compromise between resolution, aspect ratio and large area printing quality is represented by 250 to 325 mesh count screen with apertures between 80-100 mm.

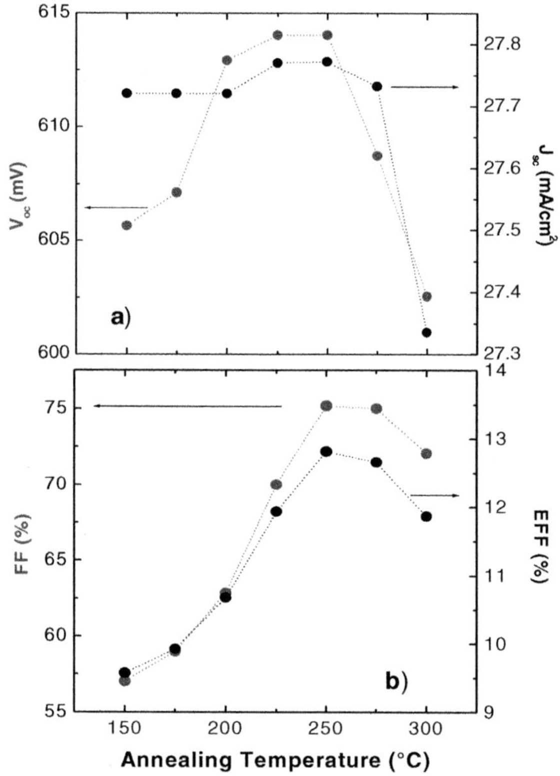

Fig. 10.18 Heterojunction solar cell photovoltaic parameters as a function of annealing temperature used for the screen-printed silver paste process.

Due to the viscosity of the paste (less than 100 poise), a high shear stress is required, therefore when printing at high speed, to get a good resolution, a moderate pressure is required. In this condition also the snap-off distance can be settled as high as possible to obtain higher finger thickness and, at the same time, easiness of paste releasing from the screen. Also a double printing can be optimized in terms of pressure, velocity, snap-off distance to increase the metal thickness. Indeed after the first squeegee pass, the paste deposited onto the substrate is able to recover partially its original viscosity, and so a second pass deposits a sort of second layer, increasing the total finger thickness. Fine line-good aspect ratio is obtainable in this way, like 80 μm wide and 20 μm high fingers. After printing, the curing is the crucial step in which the electrical properties of the metal, the PV parameters of the final device and even the TCO properties are defined. It is easy to understand that the higher the temperature and the longer the time of the curing step, the better the conductivity, the adhesion to the TCO and the final fill factor (FF). However there is a limit, which is fixed by the optimized device, from which on the final overall performances starts to degrade, in spite of this behaviour. A common conveyor dryer can be used for the curing purpose. In Fig. 10.18 the photovoltaic parameters of the heterojunction solar cells, having a sputtered ITO as TCO front side layer, are reported as a function of the temperature used for the curing process ranging from 150°C up to 300°C, keeping the same curing time of 15 minutes for all the samples.

An increase of all photovoltaic parameters is found from 150 to 250°C, obtaining a maximum of V_{oc}, J_{sc}, FF and consequently efficiency. Curing at higher temperatures has resulted in performance reduction, until complete device degradation, especially in terms of V_{oc} and J_{sc}, while FF, partially related to the metal grid condition, is not so much affected. This means that the thermal annealing has to be chosen with more attention to the heterojunction instead of the screen-printed metal itself. The best condition used in the heterojunction cell manufacturing for the screen printing process has been settled as follows: low temperature sintering Ag paste (Dupont PV410); double wet printed through a 350 mesh screen; drying in a conveyor IR belt furnace at 250°C for 15 minutes; 15 micron thick fingers. A resistivity of $2.92 \cdot 10^{-5}$ Ωcm has been determined on screen-printed silver after this curing process. A picture of screen-printed finger of the front side grid shape contact of a heterojunction cell is shown in Fig. 10.19. It has to be mentioned that the ITO sheet resistance increases from 45 Ω/□ up to 60 Ω/□ during the procedure while ITO reflectance and transmittance have been almost unaffected.

Another relevant aspect of the screen-printing process concerns the contact resistance R_c with the layer underneath. Standard Transfer Length Method (TLM) measurements [35] are commonly used for this purpose. In particular after the proposed screen printed process over a 60 Ω/□ sputtered ITO layer a $R_c = 0.444$ Ω with a specific contact resistivity $\rho_c = 3.84$ mΩcm^2 has been found. In Fig. 10.20 an example of a TLM measurement performed on the sample shown in the inset of the same figure is reported.

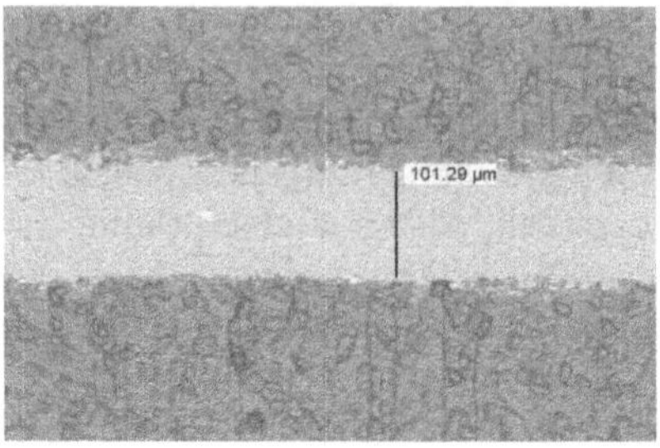

Fig. 10.19 Detail of a screen-printed silver finger of the heterojunction solar cell front side grid contact.

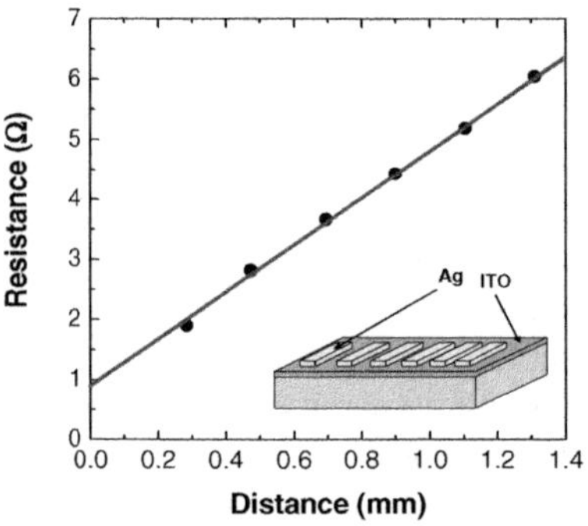

Fig. 10.20 TLM measurement to evaluate R_c between screen-printed silver and ITO.

10.7 Heterojunction Solar Cell on Multicrystalline Silicon

In this section an example of the heterojunction solar cell fabricated on multicrystalline silicon (mc-Si) substrate is described using the concepts introduced in the previous sections.

10.7.1 Sample Fabrication

200 μm thick, 1 Ωcm p-type doped, mc-Si as-cut wafers have been used to fabricate heterostructure solar cells. To remove saw damage and reduce the surface steps between two adjacent grains a CP-4 etching (HF: HNO_3: CH_3COOH=3:5:3) has been adopted. The bulk lifetime of this wafer has been evaluated to be around 50 μs after CP-4 treatment. After a standard RCA cleaning procedure an Ag-grid/ITO/CrSi/n-type a-Si:H/intrinsic a-Si:H structure has been prepared on the front side, and a intrinsic a-Si:H/p-type a-Si:H/CrSi/ITO/full-Ag structure has been used at the back. The a-Si:H layers have been deposited in a three chamber 13.56 MHz RF PECVD reactor using the recipes already reported above. The CrSi layers for both front and rear side of doped a-Si:H layers have been obtained following the suggestion reported in ref. [21]. Then ITO layers have been sputtered

on both sides of the cell in the following conditions: 200 W, 2.25 mTorr, 150 °C, 25 sccm Ar, 4.4 Ω/sec deposition rate. The front side thickness has been chosen to obtain optimal sun antireflection, while the rear side one has been dimensioned to enhance the internal reflectivity back into the silicon wafer. Finally the metal contacts have been formed by low temperature sintering Ag (Dupont PV410) as previously seen in section 10.6. On the front side a grid shaped contact has been adopted and on the rear side a full Ag contact has been preferred. The edge isolation of the cell has been ensured by a dicing machine cut fixing the total area of the cell to 7.5 cm^2. A schematic cross section of the double heterojunction solar cell is depicted not in scale in Fig. 10.21. In the same figure the thicknesses of each layer are reported.

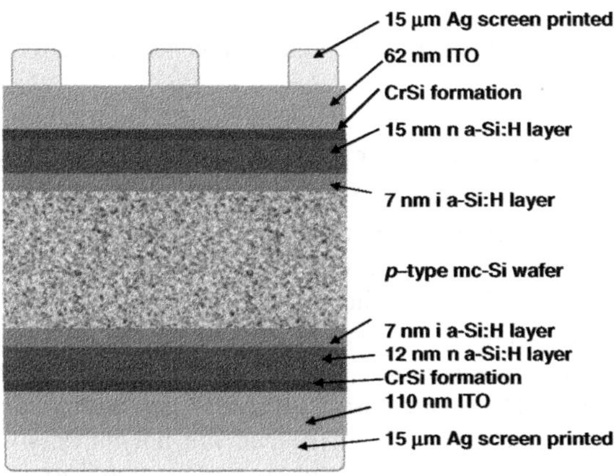

Fig. 10.21 Schematic cross section of double heterojunction solar cell made on p-type doped mc-Si.

10.7.2 Cell Characteristics

One of the main problems in a-Si:H/c-Si heterostructure fabrication is to reduce the a-Si:H emitter thickness in order to avoid absorption in the short wavelength of the sunlight spectrum, while still maintaining a conformal coverage of the silicon surface. On mc-Si surface the conformal coverage is more critical than on mono c-Si since steps between adjacent grains affect the silicon surface. If the surface coverage is not homogeneous, micro-shunts can dominate the junction, strongly reducing the photovoltaic performance of the cell. Concerning this reason the a-Si:H doped layers have been chosen a little thicker than the optimum evaluated above. To evaluate the emitter thickness an optical analysis has been performed measuring the reflectance of the heterojunction before ITO deposition using a numerical simulation in which the optical model of the a-Si:H/c-Si heterostructure has been used [36, 37]. From the good agreement between optical model and measured data, reported in Fig. 10.22 in the wavelength range from 350 nm to 800 nm, a thickness of 22.5 nm of a-Si:H layer has been estimated.

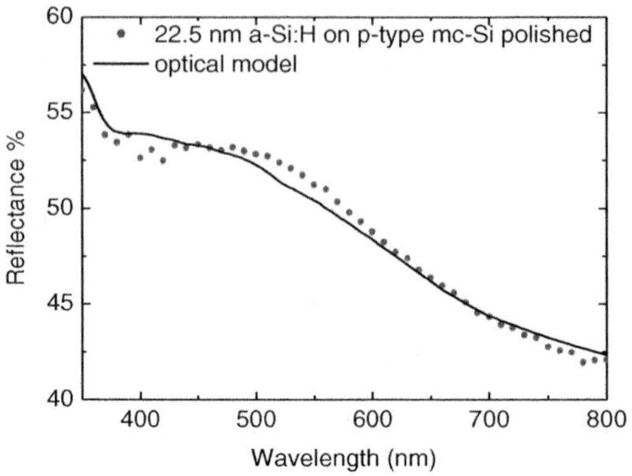

Fig. 10.22 Comparison of reflectance data and numerical model used to evaluate the emitter thickness.

The illuminated I-V characteristic of the solar cells in standard conditions (100 mW/cm^2, AM1.5G, 25°C) is shown in Fig. 10.23, in which also the photovoltaic parameters of the cell are reported. The reflectance as well as the external and internal quantum efficiencies (EQE, IQE) of the cell are reported in Fig. 10.24. The high IQE response around 700 nm indicates the effectiveness of the contact obtained by CrSi formation and the mirroring effect of the ITO/Ag on the backside of the cell. Enhancing the blue response will imply reducing the emitter thickness and optimizing the front ITO, to increase its transparency and conductance. Even though the photovoltaic performances of the cell are not impressive and further investigation is required to find the right mc-Si surface conditioning process before a-Si:H layer deposition, they indicate that the heterojunction technology can be applied also on mc-Si wafer. Indeed very good passivation is obtained on the polished substrate, as indicated by the V_{oc} value, comparable to the state-of-art p-type mc-Si heterostructure solar cells.

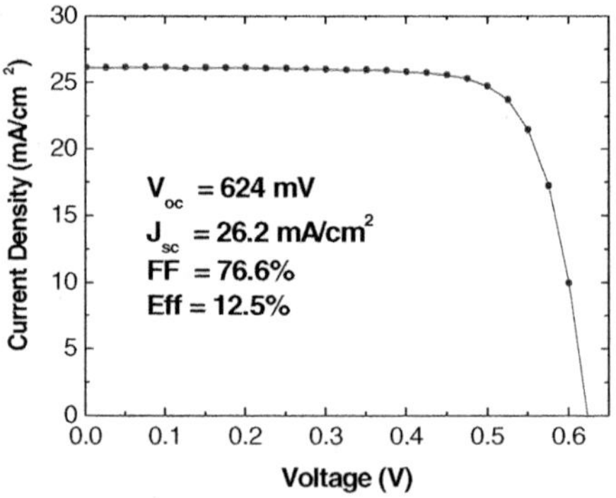

Fig. 10.23 Illuminated I-V characteristic of the double heterojunction solar cell, measured at standard conditions (100 mW/cm^2, AM1.5G, 25 C°).

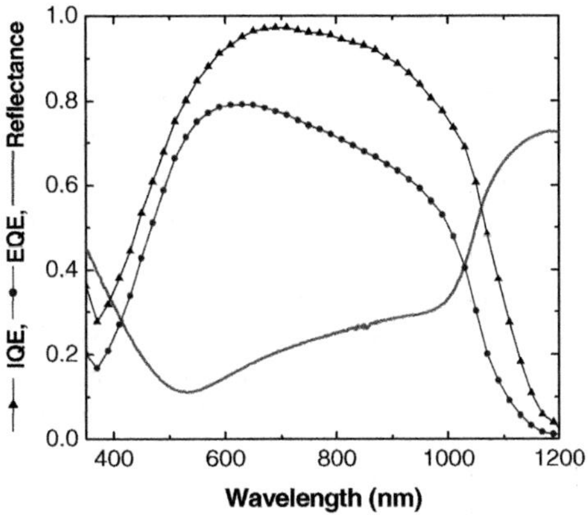

Fig. 10.24 EQE, IQE and Reflectance data of the heterojunction solar cell.

10.8 Laser Fired Local Back Contact

In section 10.3 difficulties in the p-c-Si/i-a-Si:H/p-a-Si:H/ITO structure for the rear contact, due band offset in valence band between a-Si:H and c-Si materials, have been remarked. An alternative approach to overcome this problem and achieve a low temperature BSF formation also on a p-type wafer based heterostructure cell has been proposed using a laser fired local contact onto a well passivated c-Si surface as obtained by a-Si:H/SiN_x layers [38, 39]. Recently, the possibility to merge the technology of laser fired local contact and a-Si:H/c-Si heterojunction to achieve a full low temperature process suitable also for p-type doped thin silicon wafer has been demonstrated [40-41-42-43]. Indeed another aspect should be taken into account. Even if a thinner wafer should results in an easier collection of the diffusion contribution to the total photocurrent, in turn less generation occurs in particular for the near infrared components of the sun spectrum, and light trapping for long-wavelength near-infrared light becomes an issue to be solved, with the adoption of a back mirror. Taking into account the difference in refractive index of a-Si:H and SiN_x layers, the Bragg reflectance concept can be introduced to fabricate a back Dielectric Bragg Reflecting (DBR) mirror composed by couples of these two materials. It is able to enhance the internal reflectance leading to higher cell efficiency.

Here two kinds of rear side silicon surface passivations have been compared. The first has been made by a single couple of a-Si:H/SiN_x layers [44]; the second has been realized introducing the DBR concept. In particular this DBR has been formed by several couples of a-Si:H/SiN_x layers. Moreover, by choosing the right thicknesses of each layer, it is possible to fix the center wavelength of the DBR enhancing the internal reflectance around that center wavelength. From optical simulations, in which the refractive index of each film was used, it has been possible to evaluate that four couples of a-Si:H/SiN_x were sufficient to ensure the 95% internal reflection of the impinging light with a center wavelength around 1000 nm.

10.8.1 Samples Fabrication

a-Si:H / c-Si heterojunction samples have been fabricated starting on 250 μm thick FZ monocrystalline 0.5 Ω cm p-type wafers having one side polished. After the RCA standard cleaning procedure and a 1% HF bath to remove any native oxide, the wafers have been entered in a three chambers 13.56 MHz PECVD system for the emitter formation, consisting in a thin intrinsic (5 nm) followed by a n-type (10 nm) a-SiH layer, deposited following the conditions reported in section 1. Afterwards a 30 nm thick chromium layer has been evaporated on the top of the heterostructure and subsequently wet chemically removed in order to form the highly conductivity CrSi layer [21]. The a-Si:H/SiN_x couple has been formed in a PECVD system on the rear side of the wafer after a 1% HF bath to remove any native oxide. In particular the DBR has been grown repeating four times the following recipes for the film deposition:

- a-Si:H layer deposited at 250 °C, 750 mTorr of working pressure, 31 mW/cm^2 RF power, 120 sccm of 5% SiH_4 diluted in Ar;
- SiN_x layer at the same temperature and pressure conditions, RF power of 264 mW/cm^2 and $\Phi_{NH3}/\Phi_{SiH4} = 1.66$.

Both layers have been deposited avoiding glow discharge interruption in order to reduce the stress at the a-Si/SiN_x interface. Between two subsequent couple deposition, a N_2 purge has been carried out and an adequate vacuum condition has been recovered in order to avoid ammonia residua in a-Si:H layers.

Then an ITO layer has been sputtered over the amorphous emitter following the condition reported in section 10.5. Its thickness has been fixed at 62 nm to obtain an antireflection coating. Afterwards, a Filmtronics B200 Boron spin-on dopant (SOD) film spinned at 2000 rounds/min for 30 sec has been deposited on the DBR and cured at 250 °C for 15 min. The SOD annealing temperature has been chosen as the one sufficient to hard the SOD avoiding at the same time boron contaminations in the amorphous layers underneath. This layer has been introduced in order to help the formation of an overdoped local contact and to relax constrain related to the laser firing process [39]. Subsequently a 2 μm thick Al layer has been e-beam evaporated on the whole rear surface of the mirror.

To form the rear side contact of the solar cells a laser treatment has been locally performed by a Q-Switched Nd-YAG pulsed laser working at a wavelength of 1064 nm in TEM_{00} mode with a power of 320 mW, pulse burst of 100 ms at 1 kHz repetition rate, in order to induce a simultaneous diffusion of Al and B inside the silicon wafer through the amorphous multilayer. Subsequently a silver grid deposited by screen-printing of low temperature sintering DuPont Ag paste has been used to form a grid shape front side electrode (for details see section 10.6). Finally the cell area has been limited to 4 cm^2 by cutting the wafer with a dicing machine. A schematic cross section of the cell with the laser local point contact on the rear side of the cell is reported in Fig. 10.25 and the thickness of each layer within the heterostructure is also indicated. On the rear side of the cell the local contacts have been formed by moving the cell between laser shots with an X-Y stage. Particular

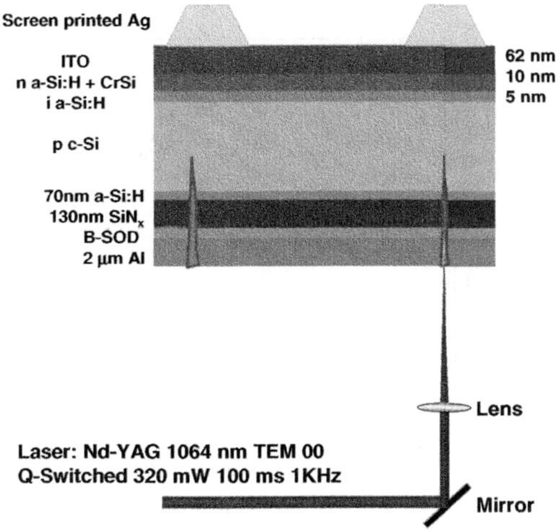

Fig. 10.25 Schematic view of the laser fired local back contact on rear side of a-Si:H/c-Si heterojunction cell.

care should be taken to choose the right distance between two subsequent local contacts to reduce series resistance effects.

10.8.2 Cell Characteristics

To evaluate the behavior of the DBR within the heterostructure solar cell scheme, a comparison of IQE data as a function of wavelength is reported in Fig. 10.26 for the DBR/Al and for the a-Si:H/SiN_x /Al shown as points respectively. As expected the reflectance values of the DBR, in the spectral range from 1000 nm to 1200 nm, are higher then in case of a-Si:H/SiN_x. This effect immediately reflects on the IQE of both samples, leading to a higher IQE value when a DBR is adopted as back reflecting mirror. Moreover a second aspect has to be considered taking into account the properties of an a-Si:H/SiN_x double layer as silicon surface passivation. Higher lifetime values have been estimated if a DBR is used as surface passivation of a FZ silicon wafer instead of the a-Si:H/SiN_x double layer as reported in Table 10.7. This higher value obtained on the DBR treated wafer is mainly due to the higher positive charge content at the a-Si:H/SiN_x interfaces of the stacked structure with respect to the single couple of a-Si:H/SiN_x layers. Indeed the higher the positive charge amount, the stronger will be the field effect close to the a-Si:H/c-Si interface due to inversion layer formation that keeps away the minority carriers from the silicon surface [45]. This results in a better surface passivation. The effect of better surface passivation has been confirmed by the higher V_{oc} value of the heterojunction solar cell having a DBR as reflecting mirror with respect to the similar cell in which only a single couple of a-Si:H/SiN_x is used. A comparison of the illuminated I-V characteristics measured under AM1.5G sunlight class A simulator is reported in Fig. 10.27. In the same figure also the photovoltaic parameters are reported for both cells.

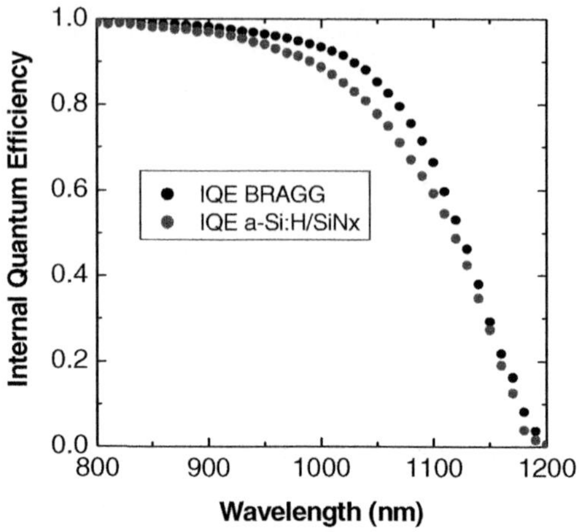

Fig. 10.26 Internal Quantum Efficiency in the near infrared region of a-Si/c-Si heterojunction solar cell with DBR/Al (a) and a-Si/SiN_x/Al (b) back configurations.

Table 10.7 Minority carrier lifetime value of the c-Si surface passivation structures.

Structure	τ_{eff} (μs)
SiN_x/a-Si:H/c-Si/a-Si:H/SiN_x	310
DBR/c-Si/DBR	400

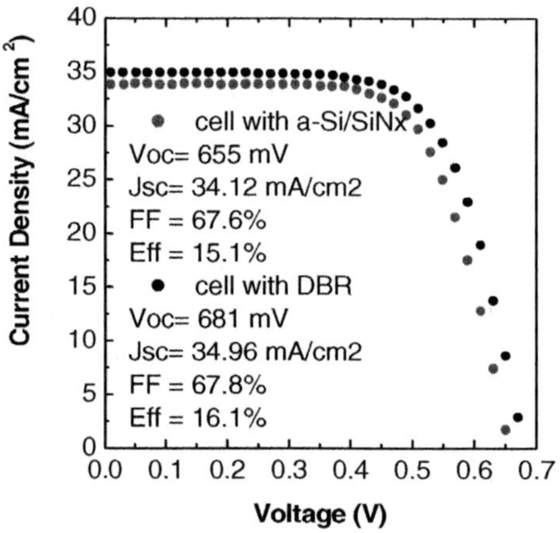

Fig. 10.27 AM1.5G illuminated I-V characteristics for both heterojunction with DBR (black symbols) and a-Si/SiN_x/Al (red symbols) as mirror.

10.9 Heterojunction Cell on n-Type c-Si

Apart from the long experience and actual high efficiency cell achieved by Sanyo group [1] a large interest on n-type c-Si based heterojunction has been actually prompted by industry in order to reduce the cell manufacturing cost and enhance the cell performances. In order to support the industry interest the European Commission has recently invested in a project concerning a-Si:H/c-Si heterojunction technology named HETSI [46]. In this section a brief description of the cell process and steps used to fabricate n-type doped c-Si based heterojunction solar cell is reported.

The heterostructure device has a front side emitter made of p-type a-Si:H layer and a rear base contact formed by n-type a-Si:H layer. Intrinsic a-Si:H layers have been inserted between c-Si wafer and doped a-Si:H layers. ITO layers have been sputtered on both front and rear side of the cell to enhance emitter and base contact conductivity and finally Ag grids have been screen printed on both side of the cell to form the cell electrodes. The resistivity of the ITO deposited without O_2 during the sputtering process was around 6.10^{-4} Ωcm. To enhance the silicon surface passivation particular attention has been paid to wafer alkaline texturing and cleaning. To the first point a sulphuric peroxide mixture (SPM) has been introduced to better remove organic and metal residues left after the texturization step. To the last aim an additional crystalline surface reoxidation and oxide removal has been adopted after standard RCA procedure resulting in a high passivation efficiency and good robustness very helpful at the industrial manufacturing level [47].

The high relevance of crystalline surface passivation in the heterojunction cell manufacturing can be demonstrated by the effective lifetime (τ_{eff}) value higher than 1 ms as evaluated by Quasi Steady State Photoconductance Decay (QSSPC) measurement of a-Si:H passivated c-Si n-type doped wafer reported in Fig. 10.28: an implied V_{oc} of 742 mV has been evaluated on this structure.

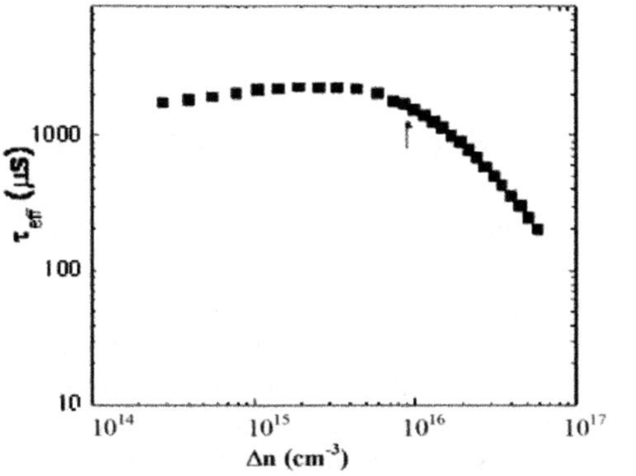

Fig. 10.28 QSSPC performed on 170 mm n-type doped silicon wafer passivated on both side by 50 nm a-Si:H layer [48].

Despite all steps within the heterojunction cell manufacturing process show variability impacting cell performances, the most critical one still remain the crystalline surface passivation. Implied V_{oc} higher than 740 mV and surface recombination velocity lower than 10 cm/s are mandatory to achieve efficiency as high as the cell efficiency of 19.6% obtained on the best sample [48]. In particular the best heterojunction solar cell has been manufactured using a different scheme. Indeed boron doped ZnO has been used for the rear side of the cell followed by an Al foil. This approach is justified by the lower workfunction of the ZnO (around 4.1 eV) with respect to the sputtered ITO (around 5 eV). Indeed a workfunction of 4.1 eV aligns better with the Fermi level of n-type doped a-Si:H layer as indicated in Fig. 10.17. In Fig. 10.29 the illuminatied I-V characteristic of the best cell is reported together with the photovoltaic parameters. Series resistance of 0.95 Ωcm^2 and shunt resistance higher than 1 $M\Omega cm^2$ have been determined. By reducing the edge effects by cutting the cell area to 100 cm^2, the efficiency has been increased up to 20% as reported in Figure 10.29.

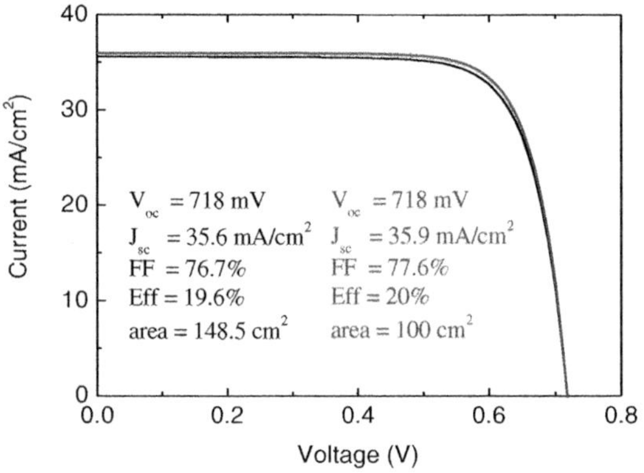

Fig. 10.29 I-V characteristic of the a-Si:H/c-Si heterostructure best cell under AM1.5G solar simulator before and after the area reduction from 148.5 cm^2 to 100 cm^2 [48].

10.10 Interdigitated Back Contact (IBC) Cell

The tendency towards shrinking wafer thickness to reduce the material costs is driving solar cell fabrication to reduce the process temperature. Moreover a way to increase PV conversion efficiency is represented by rear-junction, interdigitated back-contact (IBC) solar cell designs which are able to collect photogenerated carriers entirely from the rear of the cell, avoiding grid shading on the sunward side [49, 50]. Both of these issues can be faced with an amorphous/crystalline silicon heterostructure as firstly demonstrated by the BEHIND Cell concept [51]. The choice of doping type of silicon wafer is relevant to achieve the best efficiency. Several reasons can support both side of choice: p-type is preferable since the minority carriers are the electrons which have higher diffusivity with respect to holes due to their lighter effective mass. Moreover the n-type doped a-Si:H emitter certainly has higher lateral conductivity with respect to the p-type a-Si. In turn the n-type shows no degradation during the sunlight exposure since has no B-O complex is formed [52]. In addition it is easier to passivate with a common SiNx front side layer, which is helpful to form a front surface field due to accumulation induced by the positive charge effect lying within the film. Finally the base contact using n-type a-Si:H/c-Si heterojunction contact is easier to form with respect to the counterpart p-type a-Si:H/c-Si as reported in the previous sections. Notwithstanding these intuitive considerations, the best way to rightly address the problems related to the IBC cell concerning the wafer doping, bulk lifetime and interdigitated contact geometry is a two dimensional simulations of the heterostructure device. To this end very helpful simulations are reported in refs. [53, 54] where p-type doped c-Si wafer is adopted. In Fig. 10.30 the effect of doping concentration on the cell efficiency is reported for different bulk lifetimes taking into account a textured surface for the silicon wafer. From these simulations a doping base of 0.5 Ωcm should be preferred in case of high quality silicon substrate.

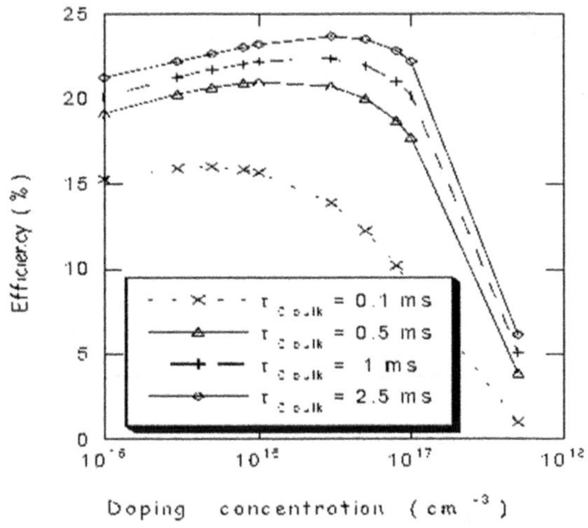

Fig. 10.30 Effect of doping concentration on efficiency for an IBC textured solar cell [53].

In addition the distance between the emitter and base contact is a very relevant parameter since it has to ensure sufficient isolation between n and p-type doped a-Si:H films, but the gap introduced between the two doped regions can strongly reduce cell performance, as reported in Fig. 10.31. Indeed under this region within the crystalline wafer there is no electric field to push minority carriers toward emitter region or to push minority carriers away from the base contact. This reflects in a fill factor strong reduction whether this gap is increased. On the other hand it must fit with the technological aspect needed in the cell manufacturing process, in which photolithographic steps are not industrially appealing. The simulations reported in the Fig. 10.31 are performed on flat silicon wafer taking into account a defect density of 10^{17} cm^{-3} at the heterointerface. If the defect density is taken higher (10^{18} cm^{-3}) a strong degradation of the I-V cell characteristic occurs.

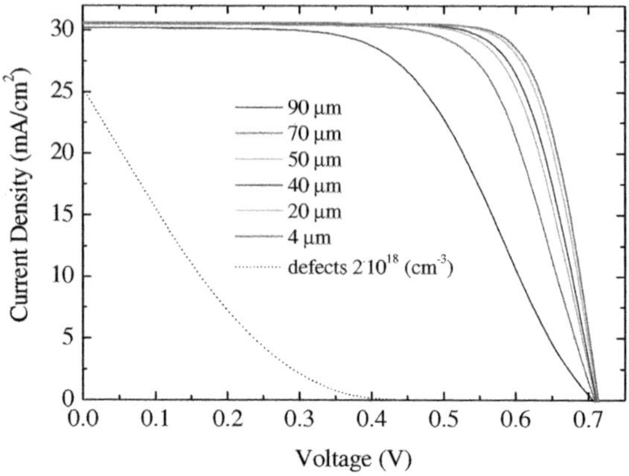

Fig. 10.31 Influence of gap distance between the emitter and base contacts.

Finally the coverage percentage of the emitter region by metal electrode has to be considered to achieve the best efficiency. The higher the emitter coverage, the higher the cell efficiency as reported in Fig. 10.32.

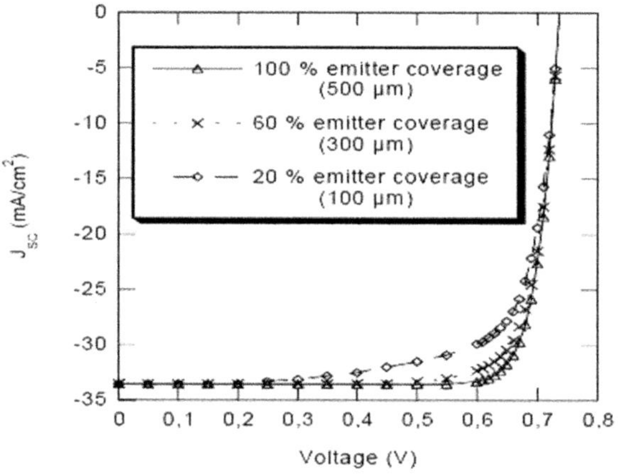

Fig. 10.32 Influence of metal coverage of the emitter contact on the IV characteristics [53].

In the next section the IBC configuration is reviewed on the basis of the BEHIND concept developed at ENEA (Rome Italy) and on the IBC cell fabricated at CEA-INES (France) for p-type and n-type doped c-Si base respectively. Up to now only these two processes based on a-Si:H/c-Si heterojunction are free from photolithographic steps and thus more appealing for industrial application with respect to other IBC approaches reported in literature [55, 56, 57].

10.11 IBC on p-Type Doped c-Si

An actual BEHIND solar cell has been fabricated starting from a 4-inch diameter, 200 μm thick, <100> oriented, 1 Ωcm p-type, one side polished CZ mono-crystalline silicon wafer. After front side alkaline texturing and industrial cleaning, the front side antireflection passivation coating and the rear side emitter and back contact have been deposited in a 13.56 MHz direct Plasma Enhanced Chemical Vapour Deposition (PECVD) system. In particular a double layer stack of a-Si/SiN_x on the sunward side, acting as passivation and anti-reflection layer [38], has been deposited using the conditions reported in previous sections. Then on the whole polished backside of the wafer, after an HF-dipping procedure to remove the native oxide, an intrinsic a-Si:H buffer layer has been deposited followed by a n-type doped a-Si:H one. Over this film a chromium silicide (CrSi) layer has been formed to increase the emitter conductivity [21] by Cr evaporation and wet chemical removal. At this stage a metallic mask has been held and fixed by a particular holder on the rear side of the device. This mask, fabricated from a 100 μm thick Molybdenum foil, has a comb shaped aperture obtained by Nd-YAG laser ablation. A dry etching procedure using NF_3 gas has been performed to remove the n-type a-Si:H portion not covered by the mask, using settings defined on the base of previous experiences [58]. Subsequently, keeping the mask in the same position, the cell base contact has been formed by an intrinsic a-Si:H buffer and a p-type a-Si:H layers, followed by a δn-a-Si:H deposition useful to increase the conductivity of the a-Si layer [21]. Then another comb shaped aperture mask, having narrower fingers with respect to the previous mask, has been held and fixed on the rear side of the device using the previously used holder. The aperture of each finger of this second mask is about the half of the first mask leading to a not difficult mechanical alignment with the pattern underneath. Through these apertures, a 30 nm thick Cr layer followed by 4 μm Al layer have been evaporated. The Cr is used to form the CrSi layer on the p-type a-Si:H layer [21]. Finally, the mask has been rotated 180 degrees and 4 μm of Ag have been evaporated to contact the emitter region, creating the interdigitated shape with respect to the base contact. The total area of the solar cell is 6.25 cm^2. A schematic cross section of the BEHIND cell is depicted in Fig. 10.33.

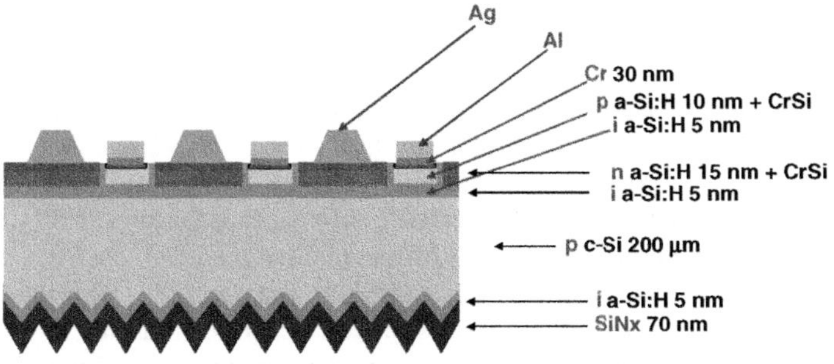

Fig. 10.33 Schematic cross section of the BEHIND cell.

The experimental IV characteristic has been measured at room temperature, under Class A calibrated to 100 mW/cm^2 and AM1.5G illumination and is shown as symbols in Fig. 10.34.

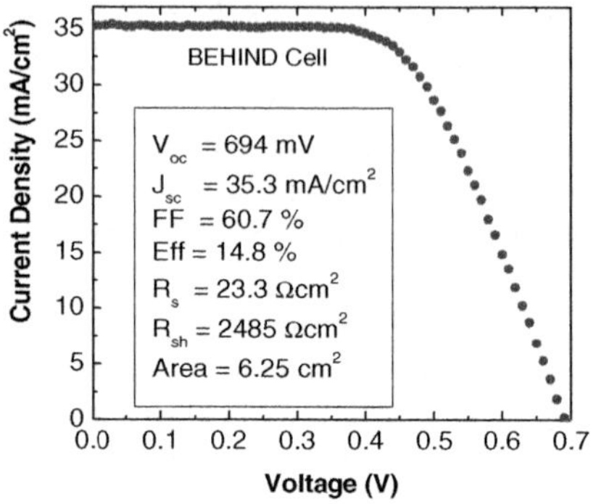

Fig. 10.34 I-V characteristic measured at 25°C, calibrated to 100 mW/cm^2, and AM1.5G illumination in a Class A sun simulator.

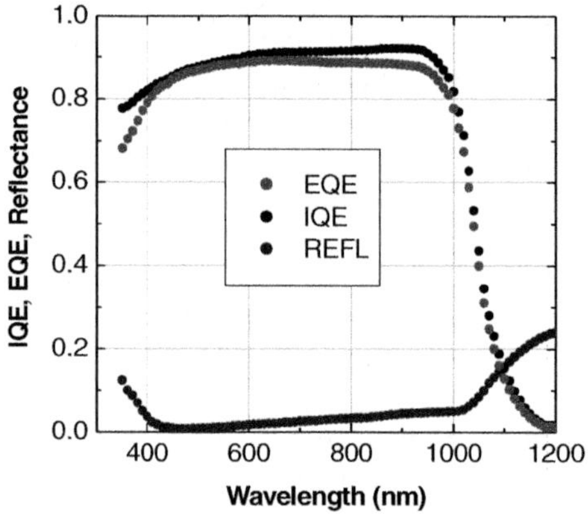

Fig. 10.35 Internal Quantum Efficiency, External Quantum Efficiency and Reflectance of the BEHIND Cell.

Taking into account the best result of a silicon based interdigitated solar cell [2] we can see that the very high open circuit voltage V_{oc} = 694 mV confirms the effectiveness of the a-Si/c-Si heterojunction as the way to improve the silicon based solar cell efficiency. This result also confirms that the uniformity of the deposited amorphous silicon layers is not influenced by the mask-assisted deposition process even when multiple masks are used in the fabrication process. Indeed the alignment between masks and substrate is feasible and the regions where the doped layers can unfortunately overlap are isolated by the intrinsic a-Si:H. The introduction of CrSi on p-type a-Si:H base contact has allowed to overcome the necessity of laser treatment to obtain an effective base contact [21]. The key issues to obtain high J_{sc} depends on diffusion length, L_d, and front surface recombination velocity, $S_{n,p}$. In particular L_d and $S_{n,p}$ have been respectively estimated to be 680 μm and 80 cm/s from a fitting procedure of the IQE [51] reported in Fig. 10.35 together with reflectance and EQE of the cell. Up to now metal thicknesses and interdigitated finger dimension are the main issues that are limiting the series resistance as well as the fill factor of the cell. Low temperature sintering Ag screen-printing process to form the interdigitated metal contact currently is under development [54]. A picture of the screen-printed interdigitated metal electrodes useful to improve the cell efficiency is reported in Fig. 10.36.

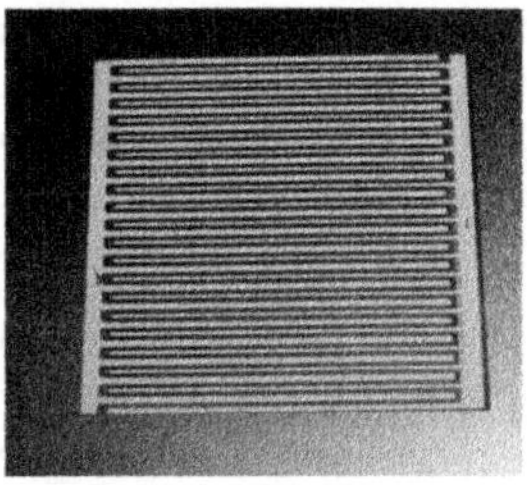

Fig. 10.36 Interdigitated screen-printed silver electrodes.

10.12 IBC on n-Type Doped c-Si

In this case several difficulties have to be considered in the device design essentially due to the poor lateral conductivity of the emitter, since is p-type doped a-Si:H. Therefore in this case the use of high conductive layer on top of amorphous emitter, such as TCO, is mandatory to achieve cell efficiency. In this paragraph a detailed description of the IBC process, as proposed by CEA-INES group, [59] is reported.

The process has been developed starting from n-type FZ 1-5 Ωcm doped 200 μm thick c-Si wafer. After cleaning and HF 2% dipping the wafer has been entered in a RF 13.56 MHz PECVD system at a temperature of 200 °C. Precursor gasses such as SiH_4, PH_3 B_2H_6 and H_2 have been uses to deposit a-Si:H layers. To reduce the problem of poor lateral conductivity of p-type doped a-Si layer an emitter stack consisting in a-Si:H (i) / a-Si:H (p) / ZnO / Al has been chosen for the IBC structure. Then the emitter layer has been patterned in an interdigitated comb shape configuration using a screen-printing etching steps followed by a n-type a-Si:H base contact deposition. Finally a mask assisted process has been used to deposit a metal sheet on top of n-type a-Si:H base contact. This patterning process has not completely reduced the wafer surface passivation. Indeed in Fig. 10.37 the lifetime measurement, as obtained by microwave photoconductance decay (μw-PCD), is reported for the stack heterostructure n a-Si:H/n c-Si/i a-Si:H/p aSi:H interdigitated with n- a-Si:H is reported.

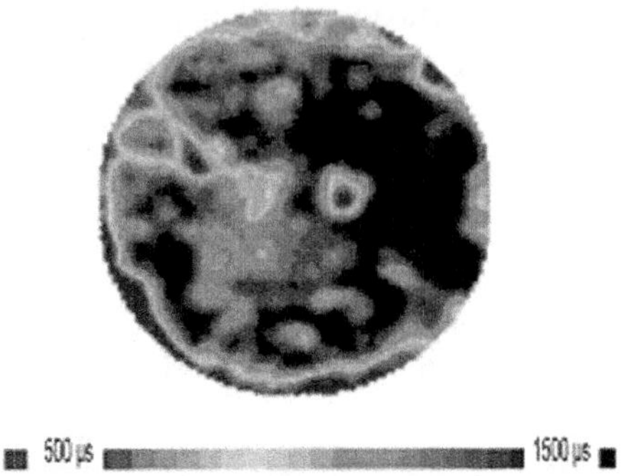

Fig. 10.37 Lifetime mapping of IBC Cell after patterning and emitter deposition (n-type a-Si:H at the front, n- and i/p-types at the rear) [59].

To ensure a good front surface passivation a 80 nm ITO and 10 nm n-type a-Si:H stack is used resulting in an effective *S* below 20 cm/s. On a 25 cm^2 area an efficiency of 15.7 % has been achieved as reported in Fig. 10.38. In the same figure the scheme of the device cross section is reported on the left side, together with the photovoltaic parameters of the IBC.

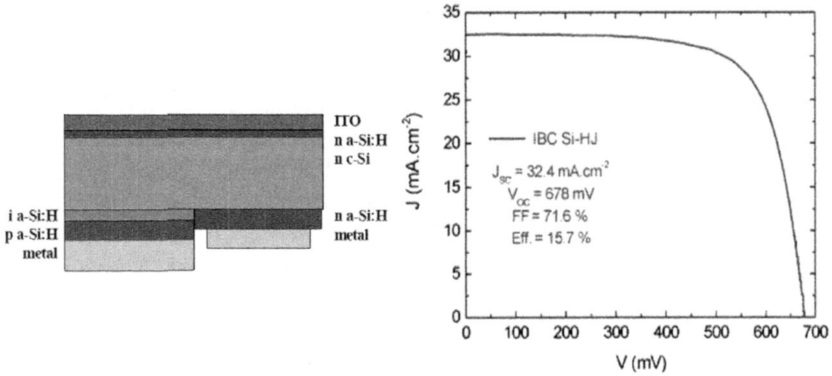

Fig. 10.38 Cross section (left side) and illuminated I-V characteristic of the IBC cell based on n-type c-Si wafer.

10.13 Conclusions

In this chapter a comprehensive description is given of the contact formation in a-Si:H/c-Si heterojunction solar cells. The contact formation is dependant on the doping level of amorphous silicon films. Since the n-type doping efficiency of a-Si:H layer is higher than that of p-type a-Si:H, different approaches have to be considered in device configuration to achieve best performances. To this end a numerical simulator certainly is an helpful tool to determine the right thicknesses of the amorphous film and to evaluate the limit imposed by the band bending distributions at the edge of the a-Si:H/c-Si heterojunctions. Indeed strong differences occur depending on the doping type of silicon base. While n-type c-Si allows a easier base contact formation using an n c-Si/ n aSi:H heterojunction, in turn the emitter suffers from low conductivity being made by p-type a-Si:H. Instead, the p-type doped c-Si base offers a more conductive emitter but a difficulty on the base contact due to the band offset at the edge of valence band between a-Si:H and c-Si. The use of CrSi can be helpful to enhance the conductivity of both n and p –type a-Si:H leading to a better contact formation. An alternative approach, such as laser fired contacts, is considered to overcome this issue on p-type c-Si based heterostructure solar cells. In this case, the use of a dielectric mirror between c-Si wafer and metal contact, results in improved silicon surface passivation as well as in infrared cell response. On the other hand both a-Si:H doped layers need a TCO contact to form a reasonable emitter layer or a low resistivity base contact. Therefore the concern of transparent conductive oxide in terms of electrical and optical properties plays a relevant role to achieve the highest built-in potential at the cell electrodes. Also a TCO is not sufficient to ensure a low resistance cell contact; indeed an Ag grid shaped metal is needed to form the external cell electrodes. Even if Ag low temperature sintering screen printable pastes are actually available in commerce, highly compatible with the low thermal budget process adopted in the a-Si:H/c-Si manufacturing, several aspect have to be carefully considered to avoid fill factor reduction, such as thermal annealing, shadowing and aspect ratio.

All the afore-mentioned concepts play a fundamental rule towards high efficiency solar cells in the interdigitated heterostructure rear side contact concept that, overcoming the shadowing limitation, and realizing an high degree of passivation and an appropriate pitch structure between the contact on the back side, enable the attainment of the high values of J_{sc} and V_{oc} respectively that commonly belong to this structures as the company Sunpower has largely demonstrated.

References

[1] Sakata, H., Tsunomura, Y., Inoue, H., Taira, S., Baba, T., Kanno, H., Kinoshita, T., Taguchi, M., Maruyama, E.: R&D Progress of Next-Generation Very Thin HITtm Solar Cells. In: Proc. of 25th World Conference and EPVSEC, Valencia, Spain, p. 1102 (2010)

[2] Tsunomura, Y., Yoshimine, Y., Taguchi, M., et al.: Twenty-two percent efficiency HIT solar cell. Sol. En. Mat. And Sol. Cell 93(6-7), 670–673 (2009)

[3] Taguchi, M., Terakawa, A., Maruyama, E.: Obtaining a higher Voc in HIT cells. Prog. Photovolt.: Res. Appl. 13(6), 481–488 (2005)

[4] Taguchi, M., Kawamoto, K., Tsuge, S.: HIT (TM) cells - High-efficiency crystalline Si cells with novel structure. Prog. Photovolt. Res. Appl. 8(5), 503–513 (2000)

[5] Tanaka, M., Taguchi, M., Matsuyama, T.: Development of a new a-Si c-Si heterojunction solar cells ACJ-HIT (Artificially Constructed Junction-Heterojunction with Intrinsic Thin layer). J. J. of Appl. Phys. 31(11), 3518–3522 (1992)

[6] Olibet, S., Monachon, C., Hessler-Wyser, A., Vallat-Sauvain, E., De Wolf, S., Fesquet, L., Damon-Lacoste, J., Ballif, C.: Textured Silicon Heterojunction Solar Cells With Over 700 mV Open-Circuit Voltage Studied by Transmission Electron Microscopy. In: Proc. of the 23rd EU PV Conference, Valencia, Spain, p. 1140 (2008)

[7] Korte, L., Conrad, E., Angermann, H., Stangl, R., Schmidt, M.: Overview on a-Si:H/c-Si heterojunction solar cells - physics and technology. In: 22nd EPVSEC, Milan, p. 859 (2007)
[8] Rizzoli, R., Centurioni, E., Plá, J., Summonte, C., Migliori, A., Desalvo, A., Zignani, F.: Open circuit voltage in homojunction and heterojunction silicon solar cells grown by VHF-PECVD. J. Non-Cryst. Solids 299-302, 1203 (2002)
[9] Van Cleef, M.W.M., Rubinelli, F.A., Rizzoli, R., Pinghini, R., Schropp, R.E.I., van der Weg, W.F.: Amorphous silicon carbide/crystalline silicon heterojunction solar cells: A comprehensive study of the photocarrier collection. J. of Appl. Phys. 37(7), 3926–3932 (1998)
[10] Damon-Lacoste, J., Cabarrocas, P.R.I., Chatterjee, P., et al.: About the efficiency limits of heterojunction solar cells. J. of Non-Cryst. Solids 352(9-20), 1928 (2006)
[11] Tucci, M., de Cesare, G.: 17% efficiency heterostructure solar cell based on p-type crystalline silicon. J. of Non-Cryst. Solids 338-340, 663 (2004)
[12] Conrad, E., Korte, L., Maydel, K., Angermann, H., Schubert, C., Stangl, R., Schmidt, M.: Development and Optimization of a-Si:H/c-Si Heterojunction Solar Cells Completely Processed at Low Temperatures. In: 21st EPVSEC, Dresden (D), p. 784 (2006)
[13] Angermann, H., Conrad, E., Korte, L., Rappich, J., Schulze, T.F., Schmidt, M.: Passivation of textured substrates for a-Si:H/c-Si hetero-junction solar cells: Effect of wet-chemical smoothing and intrinsic a-Si:H interlayer. Mat. Sci. Eng. B 159-60, SI 219–SI 223 (2009)
[14] Wang, T.H., Page, M.R., Iwaniczko, E., Wang, Q., Xu, Y., Yan, Y., Roybal, L., Levi, D., Bauer, R., Yelundur, V., Rohatgi, A., Branz, H.M.: 17.5% p-Type Silicon Heterojunction Solar Cells with HWCVD a-Si:H as the Emitter and Back Contact. In: DOE Solar Energy Technologies Program Review Meeting, Denver, Colorado (2005)
[15] Tucci, M., Serenelli, L., Salza, E., De Iuliis, S., Geerligs, L.J., Caputo, D., Ceccarelli, M., de Cesare, G.: Back contacted a-Si:H/c-Si heterostructure solar cells. Journal of Non-Cryst. Solids 354, 2386 (2008)
[16] Caputo, D., Forghieri, U., Palma, F.: Low-temperature admittance measurement in thin film amorphous silicon structures. J. Appl. Phys. 82, 733 (1997)
[17] Stutzmann, M., Biegelsen, D.K., Street, R.A.: Detailed investigation of doping in hydrogenated amorphous-silicon and germanium. Phys. Rev. B 35, 5666 (1987)
[18] Goldie, D.M., Spear, W.E., Liu, E.Z.: Electron and hole transport in compensated amorphous silicon. Philos. Mag. B 62, 509 (1990)
[19] Tucci, M.: Optimization of n-doping in n-type a-Si:H / p-type textured c-Si heterojunction for photovoltaic applications. Sol. En. Mat. and Sol. Cel. 57, 249 (1999)
[20] Caputo, D., de Cesare, G., Palma, F., Tucci, M.: Phosphorus and boron compensation in hydrogenated amorphous silicon films. Solid State Phenomena 1, 61 (2001)
[21] Labrune, M., Roca i Cabarrocas, P., Chouffot, R., Brezard-Oudot, A., Kleider, J.-P., Ribeyron, P.J., Vandeneynde, A., Brúggemann, R.: Polymorphous/crystalline silicon heterojunction solar cells: optimization on monocrystalline silicon. In: 23rd European Photovoltaic Solar Energy Conference, Valencia, Spain, vol. 1696 (2008)
[22] Jackson, W.B., Amer, N.M.: Direct measurement of gap-state absorption in hydrogenated amorphous-silicon by photothermal deflection spectroscopy. Phys. Rev. B 25, 5559 (1982)
[23] Caputo, D., de Cesare, G., Ceccarelli, M., Nascetti, A., Tucci, M., Meda, L., Losurdo, M., Bruno, G.: Characterization of chromium silicide thin layer formed on amorphous silicon films. J. of Non-Cryst. Solids 354, 2171 (2008)

[24] Caputo, D., de Cesare, G., Tucci, M.: Built-in Enhancement in a-Si:H Solar Cell by Chromium Silicide Layer. IEEE El. Dev. Lett. 31(7), 689 (2010)

[25] Tucci, M., Serenelli, L., De Iuliis, S., Caputo, D., Nascetti, A., de Cesare, G.: Amorphous/crystalline silicon heterostructure solar cell based on multi-crystalline silicon. In: Proc. 21st EUPVSEC, Dresden, Germany, p. 902 (2006)

[26] Borchert, D., Rinio, M.: Interaction between process technology and material quality during the processing of multicrystalline silicon solar cells. J. Mat. Sci.: Mat. El. 20-1, 487 (2009)

[27] Angermann, H., Korte, L., Rappich, J., Conrad, E., Sieber, I., Schmidt, M., Hübener, K., Hauschild, J.: Optimisation of electronic interface properties of a-Si: H/c-Si hetero-junction solar cells by wet-chemical surface pre-treatment. Thin Solid Films 516, 6775 (2008)

[28] Street, R.A.: Hydrogenated amorphous silicon. Solid State Science Series. Cambridge University Press, Cambridge (1991)

[29] Taguchi, M., Terakawa, A., Maruyama, E., et al.: Obtaining a higher V-oc in HIT cells. Prog. Photovolt. Res. Appl. 13(6), 481 (2005)

[30] Maydell, K., Conrad, E., Schmidt, M.: Efficient silicon heterojunction solar cells based on p- and n-type substrates processed at temperatures < 220 degrees C. Prog. Photovolt.: Res. Appl. 14, 289 (2006)

[31] Centurioni, E., Iencinella, D.: Role of front contact work function on amorphous silicon/crystalline silicon heterojunction solar cell performance. IEEE El. Dev. Lett. 24-3 (2003)

[32] Rubinelli, F.A., Arch, J.K., Fonash, S.J.: Effect of contact barrier heights on P-I-N detector and solar cell performance. J. Appl. Phys. 72(4), 15, 1621–1630 (1992)

[33] Hosberg, C., Bowden, S.: PV-CDROM, `http://pvcdrom.pveducation.org/`

[34] Hilali, Mohamed, M.: PhD thesis, Georgia Institute of Technology, 59 (2005), `http://hdl.handle.net/1853/7284`

[35] DuPont: `http://www2.dupont.com/Photovoltaics/en_US/news_events/article20090325.html`

[36] Ferro: LF33-series pastes: `http://www.ferro.com/non-cms/ems/Solar_2009/interconnect/LF33-700.pdf,LF33-701.pdf, LF33-750.pdf`

[37] Meier, D.L., Schroder, D.K.: Contact Resistance – its measurement and relative importance to power loss in a solar cell. IEEE Trans. Electron Devices 31, 647 (1984)

[38] `http://www.bo.imm.cnr.it/~centurio/optical.html`

[39] Centurioni, E.: Generalized matrix method for calculation of internal light energy flux in mixed coherent and incoherent multilayers. Appl. Opt. 44, 7532 (2005)

[40] Tucci, M., Talgorn, E., Serenelli, L., Salza, E., Izzi, M., Mangiapane, P.: Laser Fired Back Contact for silicon solar cells. Thin Solid Film 516, 6767 (2008)

[41] Kreinin, L., Bordin, N., Broder, J., Eisemberg, N., Tucci, M., Talgorn, E., De Iuliis, S., Serenelli, L., Izzi, M., Salza, E., Pirozzi, L.: Comparison of two BSF technologies suitable for high efficiency multi-crystalline solar cells. In: Proc. of 21st EUPVSEC, Dresden, Germany, p. 855 (2006)

[42] Tucci, M., Serenelli, L., Salza, E., Pirozzi, L., De Cesare, G., Caputo, D., Ceccarelli, M.: Bragg reflector and laser fired back contact in a-Si:H / c-Si heterostructure solar cell. Mat. Sci. and Eng. B 159, 48 (2009)

[43] Brendle, W., Nguyen, V., Brenner, K., Rostan, P.J., Grohe, A., Schneiderlochener, E., Preu, R., Rau, U., Palfinger, G., Werner, J.H.: Low temperature back contact for high efficiency silicon solar cells. In: Proc. of 20th EUPVSEC, Barcellona Spain, vol. 745 (2005)
[44] Munoz, D., Voz, C., Blanque, S., et al.: Development of laser-fired contacts for amorphous silicon layers obtained by Hot-Wire CVD. Mat. Sci. Eng. B 159-60, SI 23 (2009)
[45] Schneiderlöchner, E., Emanuel, G., Grupp, G., Lautenschlager, H., Leimenstoll, A., Glunz, S.W., Preu, R., Willeke, G.: Silicon solar cells with screen printed-front contact and dielectrically passivated, laser-fired rear electrode. In: 19th EUPVSEC, Paris, France, p. 447 (2004)
[46] Tucci, M., Serenelli, L.: Metastability of SiNx / a-Si:H crystalline silicon surface passivation for PV application. Thin Solid Films 516, 6939 (2008)
[47] Tucci, M., Serenelli, L., De Iuliis, S., Izzi, M.: Characterization of SiNx / a-Si:H crystalline silicon surface passivation under UV light exposure. Thin Solid Films 515, 7625 (2007)
[48] http://www.hetsi.eu/
[49] Danel, A., Hickel, P.E., Souche, F., Salvetrat, T., Letiec, Y., Nolan, T., Ribeyron, P.J.: Control of Texturized Silicon Substrates Surface Passivation for a-Si:H/c-Si Heterojunction Solar Cells. In: 25th EUPVSEC, Valencia, Spain, pp. 1951–1953 (2010)
[50] Muñoz, D., Ozanne, A.S., Harrisson, S., Danel, A., Souche, F., Denis, C., Favier, A., Nguyen, N., Hickel, P.E., Mur, P., Kang, M.S., Kim, K.M., Janin, R., Pesenti, C., Blin, D., Nolan, T., Kashkoush, I., Ribeyron, P.J.: Toward high efficiency on full wafer a-Si:H/c-Si heterjunction solar cells: 19.6% on $148cm^2$. In: Proc. of the 35th PVSC, Hawaï (2010)
[51] Swanson, R.M.: Approaching the 29% limit efficiency of silicon solar cells. In: Proc. of the 31st IEEE PVSEC, Lake Buena Vista, p. 889 (2005)
[52] Mulligan, W.P., Carandang, M.A., Dawson, M., De Ceuster, D.M., Stone, C.N., Swanson, R.M.: Reducing silicon consumption by leveraging cell efficiency. In: Proc. of 21st European Photovoltaic Solar Energy Conference, p. 1301 (2006)
[53] Tucci, M., Serenelli, L., Martufi, P., Salza, E., De Iuliis, S., Geerligs, L.J., de Cesare, G., Caputo, D., Ceccarelli, M.: Back enhanced heterostructure with interdigitated contact – BEHIND - solar cell. In: Proc. of IEEE COMMAD, p. 242 (2008)
[54] Macdonald, D., Rougieux, F., Cuevas, A., Lim, B., Schmidt, J., Di Sabatino, M., Geerligs, L.J.: Light-induced boron-oxygen defect generation in compensated p-type Czochralski silicon. J. Appl. Phys. 105(9), 093704 (2009)
[55] Diouf, D., Kleider, J.P., Desrues, T., Ribeyron, P.-J.: Interdigitated Back Contact a-Si:H/c-Si Heterojunction Solar Cells Modelling: Limiting Parameters Influence on Device Efficiency. In: 23rd EUPVSEC, Valencia, Spain, p. 1949 (2008)
[56] Tucci, M., De Iuliis, S., Serenelli, L., Izzi, M., Salza, E., De Luca, V., Velardi, F., Sanseverino, N., de Cesare, G., Caputo, D.: Behind 2.0: The Back Enhanced Heterostructure with Interdigitated Contact Solar Cell Evolution. In: 25th EUPVSEC, Valencia, Spain, p. 2089 (2010)
[57] Stangl, R., Bivour, M., Conrad, E., Didschuns, I., Korte, L., Lips, K., Schmidt, M.: A novel high efficiency buried grid rear contact amorphous/crystalline silicon heterojunction solar cell concept. In: Proc. of the 22nd EUPVSEC, Milan Italy, p. 870 (2007)

[58] Lu, M.J., Das, U., Bowden, S., Hegedus, S., Birkmire, R.: Optimization of interdigitated back contact silicon heterojunction solar cells by two-dimensional numerical simulation. In: 34th IEEE Photovoltaic Specialist Conference, vol. 1-3, p. 1475 (2009)

[59] Hertanto, A., Liu, H., Yeghikyan, D., Rayaprol, B.G., Kherani, N.P., Zukotynski, S.: 34th Back Amorphous Crystalline silicon Heterojunction (BACH) photovoltaic device. In: IEEE Photovoltaic Specialists Conference (PVSC), p. 1767 (2009)

[60] Tucci, M., Serenelli, L., De Iuliis, S., Salza, E., Pirozzi, L.: Dry NF_3/O_2 plasma texturing of multicrystalline silicon. In: Proc. of 21st EUPVSEC, Dresden Germany, p. 1250 (2006)

[61] Desrues, T., Ribeyron, P.-J., Vandeneynde, A., Ozanne, A.-S., Souche, F., Veschetti, Y., Bettinelli, A., Roca i Cabarrocas, P., Labrune, M., Diouf, D., Kleider, J.-P., Lemiti, M.: Progress in Contacting a-Si:H/c-Si Heterojunction Solar Cells and its Application to Interdigitaded Back Contact Structure. In: 25th EUPVSEC, Valencia, Spain, p. 1673 (2010)

Chapter 11
Electrical Characterization of HIT Type Solar Cells

Jatin K. Rath

Utrecht University, Faculty of Science, Debye Institute of Nanomaterials Science, Section Nanophotonics-Physics of Devices, The Netherlands

Abstract. The silicon heterojunction solar cell (SHJ) has made rapid progress in reaching high efficiency and it is already developed as an industrially viable product. However, much of its progress has come through process development while there is scarce knowledge on the microscopic nature of the functioning of this device. Although this device as a whole can be considered as bulk type, the parts of a SHJ solar cell that control the charge transport behavior are limited to very thin regions, either interface or a very thin layer. This poses problems on accurate determination of the physical quantities, such as defect densities and energetic positions, conductivity, carrier recombination and the overall charge transport behavior. This chapter gives the present understanding of electrical characterization of SHJ solar cells and provides a study of defects in the interesting regions of the device.

11.1 Introduction

Heterojunction solar cells are a very widely used type of solar cells. In addition to the amorphous/crystalline silicon cells considered here, other examples are cadmium telluride (CdTe/CdS) [1] and copper indium (Ga) diselenide (CIGS/CdS) [2]. Heterojunctions are generally encountered in multijunction solar cells where different band gap materials are used for sub-cells. In these cases, for example the high efficiency triple junction GaAs based solar cells [3], even at the buffer layers and tunnel junctions, charge collection faces the band offsets and barriers. Silicon based solar cells are, on the other hand, predominantly homojunction type; the standard crystalline silicon (c-Si) solar cells are homojunctions. Thin film silicon solar cells are strictly speaking not homojunction types, especially the interface of the wide bandgap p-layer (window layer) with the intrinsic absorber layer [4]. However, the silicon heterojunction solar cell (SHJ), combining thin films and crystalline silicon, has been making continuous progress in the last 20 years, starting with laboratory [5, 6] and then at industry [7], mainly at Sanyo [8] that

W.G.J.H.M. van Sark et al. (Eds.): Physics & Tech. of Amorphous-Crystalline, EM, pp. 377–404.
springerlink.com

reported 23% efficiency. Many other groups are trying to catch up with Sanyo [10, 11, 12, 13, 14] and are in the process of finding the technology behind Sanyo's success story.

The silicon heterojunction solar cell can be best characterized as a device structure that marriages thin film and bulk silicon technologies as well as their basic physics of operation. Of particular interest is the heterojunction cell with intrinsic thin layer (HIT) for which Sanyo has achieved a formidable efficiency of 23% [8] which challenges the standard high cost c-Si homojunction cell. The emitter can be amorphous [8] or microcrystalline [9] silicon type. It is claimed by Sanyo [7] and also commonly accepted by others that an excellent surface passivation of c-Si at the c-Si/a-Si interface does the trick. However, this hypothesis is based on indirect measurements, which have to be read with a certain amount of caution. This is a multilayered structure which demands a proper analysis of not only the bulk properties, but also the various interfaces and band offsets at the hetero-junctions. Because of a rather small number of defects at the interfaces, only a few methods such as capacitance-voltage (C-V) [15], electrically detected magnetic resonance (EDMR) [16] and photo emission spectroscopy (PES) [17] can be used for direct estimation of defect density. On the other hand, the interface defect characteristics is generally analyzed by indirect methods, such as various carrier life time measurements; resonance-coupled photoconductive decay (RCPCD) [18], quasi steady state photo conductance (QSSPC) [19], microwave photo conductive decay (μPCD) [20], (modulated) photoluminescence [21] and surface photovoltage (SPV) [22]. The optical properties of the extremely thin emitter and the intrinsic silicon layers can only be characterized by ellipsometry technique [23], whereas the electrical properties of these thin layers are almost impossible to measure by any existing experimental methods. Because of the above mentioned difficulties in obtaining some of the basic physical properties, SHJ or HIT solar cells are predominantly qualitatively characterized by external quantum efficiency (EQE) [24], internal quantum efficiency (IQE) [25] and current-voltage (I-V) [26] measurements in combination with various simulations [27, 28, 29, 30] that are applied to obtain various bulk and interface properties as output parameters. In this chapter we will discuss how to best characterize the HIT cells.

11.2 Basic Charge Transport Process in SHJ Cell

A heterojunction by its very nature contains a very sensitive interface whose characteristics play a big role in any electrical characteristics of this heterojunction or a device containing such a junction, in addition to the bulk characteristics of the individual layers. It is a daunting task to characterize the microscopic nature of the interface states, especially when one or more layers are very thin, in the range of few nm. It is difficult to separate bulk states from surface and interface states. We will discuss later what sorts of possibilities exist for such direct density of states (DOS) characterizations. Before that, we will discuss what the alternatives are. In fact, the common processes of characterizing heterojunctions are experiments in the device structure. Thus, the age old current voltage characterization and spectral response still provide the most valuable information. However, the difficulty lies in the interpretation of such data and the effective ways to use these characterization techniques to derive any conclusion on interface properties.

11.2.1 Current-Voltage Relation

Before we discuss the special case of heterojunction, let us refresh our memory on how a simple solar cell (an ideal p-n junction diode) is characterized. An equivalent circuit as represented in Fig. 11.1a is used as the model to build up the equation of charge transport and the relation between current and voltage. In addition to the usual configuration of an ideal diode, a parallel resistance that represents current leakage through the diode and a series resistance that represents any voltage loss due to contact resistance and internal resistance of the diode are added to the equivalent circuit. The difference in the equivalent circuits of a diode in the light and dark condition lies only in a current source parallel to the diode, but the current in this path is in opposite direction to the forward current through the diode. An ideal diffusion limited solar cell (c-Si solar cell for example, in which the current collection is predominantly diffusion limited) structure can be well represented by such a circuit diagram. The current-voltage equation for such a case is given by,

$$J\,(V) = -J_{ph} + J_0\left(e^{\frac{q(V-JR_S)}{nkT}} - 1\right) + \frac{V-JR_S}{R_P} \tag{11.1}$$

where J_{ph} is the photocurrent density, R_s and R_p (or R_{sh}) are the series and shunt resistance respectively, V is the voltage across the load, J is the current though the load and R_L is the resistance of the load. J_0 and n are the saturation current and diode quality factor, respectively. However, solar cells contain a region where the built-in-field plays a crucial role in the transport of carriers. In a simple p/n junction the transport of carriers across the junction (in the depletion region) is indeed through a drift field. Hence, even a crystalline silicon p/n junction is not completely described with diffusion type of carrier transport and hence a voltage dependence of carrier collection, thus current, exits. However, the diffusion characteristics dominate in the charge transport process and the crystalline silicon type of solar cells can be analyzed by the simple equivalent circuit shown in Fig. 11.1a.

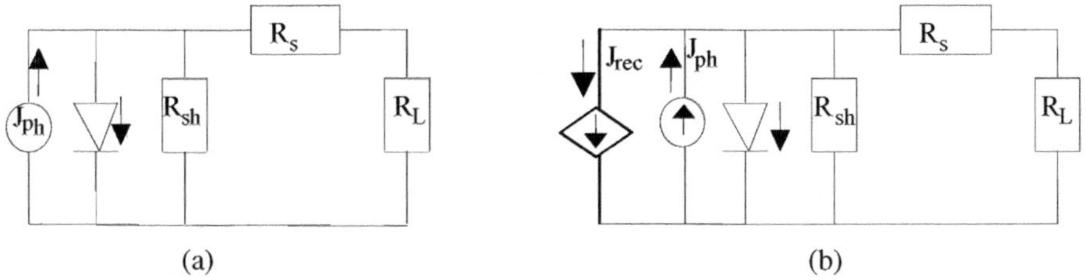

Fig. 11.1 Equivalent circuit of an (a) ideal homojunction solar cell and (b) a solar cell with a significant recombination component in the charge transport. J_{ph}: Photogenerated current, R_s: Series resistance, R_{sh}: Shunt resistance, R_L: Load, J_{rec}: Recombination current. Arrows indicate the current path.

The situation changes substantially when the photocurrent no more remains voltage independent. In that case, the simple constant current source picture of Fig. 11.1a is incorrect. This happens in case of thin film silicon solar cells and heterojunction type of solar cells. We will compare these two cases. In case of thin film solar cells, the drift type of carrier collection due to the field inside the carrier generating bulk region, which is the i-layer inside a p-i-n type of cell, demands a recombination path, in addition to the current generation, to be inserted in the equivalent circuit. This changes the equivalent circuit diagram of Fig. 11.1a to Fig. 11.1b and the current-voltage relation in Eq. (11.1) is also modified to take into account the recombination part.

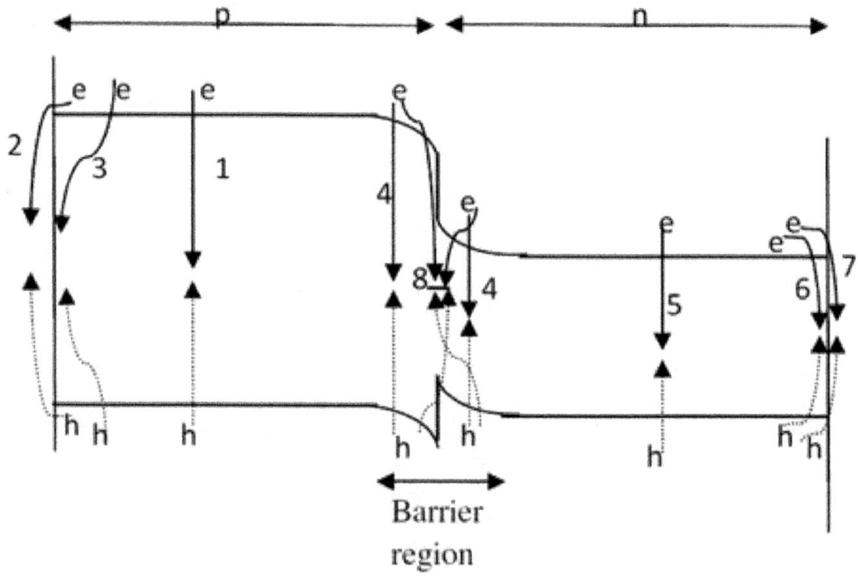

Fig. 11.2 Band diagram, charge transport and recombination pathways of a silicon heterojunction solar cell. Path 2 and 3 are recombination losses at top contact, path 6 and 7 are the recombination losses at the back contact, path 1 and 5 are the bulk recombinations in the emitter and base respectively, path 4 is the recombination in the barrier region, path 8 is the interface recombination path, which is typical for the heterojunction.

Let us look how the carrier transport in a heterojunction [31] differs from that of a simple homojunction p/n solar cell. Figure 11.2 shows the band diagram of a heterojunction cell with carrier transport and recombination pathways. There is a marked difference in the recombination behavior between a heterojunction and a homojunction at the interface between the base and the emitter. Because of the different electron affinities of the layers there is no smooth transition in the bands in case of heterojunction; in fact we end up having kinks or spikes in the energy bands at the conduction band as well as valence band. The spikes, which are caused by band offsets, are a boon as well as a curse to the operation of a heterojunction solar cell. When the light falls on a heterojunction solar cell, first, the light is absorbed by both the top and the bottom parts of the cell (though substantially more in the base than in the emitter). The situation is the same as in a homojunction. Once the electrons and holes are formed in both these parts, according to their generation rate and quantum efficiency, the carriers from both parts flow in the direction dictated by the field and the gradient of charge carriers, by drift and diffusion respectively. The transport of carriers inside the bulk, especially in the base, is basically dominated by diffusion as there is practically no field in this region. The carriers here suffer from recombination loss. This effect is the same as in a homojunction. The minority carriers move towards the junction. The carriers also suffer from recombination losses at the contacts on both ends of the cell. This is also the same as in case of a homojunction. There are two other recombination losses which are typical of heterojunctions: the recombination between the electrons and holes of the same layer in the depletion region and the electrons and holes from the opposite layers. These recombinations are negligible in case of a homojunction. However, in a heterojunction the band offsets and the defective region at the interface change the behavior compared to a homojunction. The presence of a highly defective interface region and charge trapping increase recombination. This recombination is voltage dependent and the equivalent circuit diagram is adapted by taking this recombination into account as an additional current path. The current voltage equation is also modified as [26],

$$\begin{aligned} J &= -P_{1c}P_{2c}P_{1}J_{ph} + P_{1c}J_{r1}\left(e^{\frac{V}{kT}}-1\right) + P_{1c}P_{1}J_{r2}\left(e^{\frac{V}{kT}}-1\right) + P_{1c}P_{2c}P_{1}J_{diff}\left(e^{\frac{V}{kT}}-1\right) \\ &= -P_{1c}P_{2c}P_{1}J_{ph} + \left(P_{1c}J_{r1} + P_{1c}P_{1}J_{r2} + P_{1c}P_{2c}P_{1}J_{diff}\right)\left(e^{\frac{V}{kT}}-1\right) \end{aligned} \tag{11.2}$$

where J_{r1} and J_{r2} are the recombination currents through paths 4 and 8 respectively in Fig. 11.2. P_{2c} is the probability of the carriers to be swept across the band bending region of the absorber against the tendency to diffuse away from the junction and P_{1c} is the similar probability from the window side. P_{1} is a probability term for the carriers to be emitted over the barrier (thermoionic emission) at the interface, against the tendency to recombine with path 8 or back-diffuse away from the interface.

For a homojunction the terms J_{r1} and J_{r2} are set to zero, because the recombination probability at this interface is negligible due to a strong field and no barrier to transport. Also, P_{1c}, P_{2c} and P_{1} are set to 1, for the same reason. The current voltage relation then changes back to Eq. (11.1) without the R_s and R_p terms.

This above Eq. (11.2) is an exact equation, which can be used for computer simulation. However, for I-V fitting using simple numerical modeling, this relation has to be simplified. This situation is similar to the case of thin film silicon (p-i-n type) solar cells, for which the following equation has been proposed [32], which is modified from Eq. (11.1) to take into account recombination J_{rec}

$$J\,(V) = -J_{ph} + J_{ph}\cdot\frac{d_i^2}{(\mu\tau)_{eff}\cdot[V_{bi}-(V-JR_S)]} + J_0\left(e^{\frac{q(V-JR_S)}{nkT}}-1\right) + \frac{V-JR_S}{R_p} \tag{11.3}$$

$$J_{rec} = J_{ph}\cdot\frac{d_i^2}{(\mu\tau)_{eff}\cdot[V_{bi}-(V-JR_s)]}$$

$$(\mu\tau)_{eff} = \frac{2(\mu_n^0\tau_n^0 . \mu_p^0\tau_p^0)}{\mu_n^0\tau_n^0 + \mu_p^0\tau_p^0}$$

where $(\mu\tau)_{eff}$ is the effective mobility, V_{bi} is the built-in potential and d is the thickness of the absorber layer.

As Eq. (11.3) shows, the photoconductivity J_{ph} is reduced by a factor given by the second term in the bracket. This loss in the photocurrent is determined by the recombination through bulk states, which is dependent on the built-in field.

An equation similar to that used for a thin film silicon cell given by Eq. (11.3) can also be used for the heterojunction cell as given below [33],

$$J = -J_{ph}.H(V) + J_0\left(e^{\frac{q(V-JR_S)}{nkT}} - 1\right) + \frac{V-JR_S}{R_{sh}} \tag{11.4}$$

$$H(V) = \frac{1}{1 + \frac{S^*}{\mu E(V)}}$$

where μ is the carrier mobility and S^* (cm/s) is surface-recombination velocity, used as fit parameter for fitting of the I-V curve. The electric field E(V) at the interface is defined as,

$$E(V) = \left[\frac{2q\varepsilon_{S2}N_{A1}N_{D2}}{\varepsilon_{S1}(\varepsilon_{S1}N_{A1}+\varepsilon_{S2}N_{D2})}\right]^{\frac{1}{2}}[V_{bi} - (V - JR_s)]^{\frac{1}{2}} \tag{11.5}$$

where, N_A and N_D are the accepter and donor concentrations, 1 and 2 represent the absorber and emitter respectively for a p-c-Si/n-a-Si solar cell.

The differences between these two equations Eq. (11.3) and Eq. (11.5) are in the way the defects and the recombination in both these cases are taken into account: in case of thin film silicon it is the bulk defects represented as μτ of the intrinsic layer, whereas in case of the heterojunction cell, the interface states are represented by a surface recombination velocity.

One of the consequences of the recombination term is that the cell parameters (resistances) cannot be extracted in a simple way as is done for homojunction solar cells such as crystalline silicon solar cells. The slope of the curve at open circuit voltage (V_{oc}) and short circuit current (J_{sc}) are affected by the recombination effects and R_s and R_p values will be masked by it. Based on the light intensity dependence of I-V characteristics [34] as proposed for p-i-n cell, it is possible to separate the loss terms; R_s, R_p and recombination term.

11.3 Experiments

Now that the fundamental electrical transport behavior in a heterojunction is formulated, the next question is whether these formulations can be used to extract the solar cell parameters. There are two ways of approaching this problem, (i) through experiments and data analysis, and (ii) simulations based on a model such as we defined above. I-V measurements and spectral response measurements provide valuable information on the characteristics of solar cells.

11.3.1 Current Voltage Characteristics

Both dark and light current voltage measurements are done to extract a variety of parameters. Whereas the dark I-V measurements throws light on the diode behavior and the various band offsets at the junctions, the light I-V gives information on the carrier trapping and recombination.

11.3.1.1 Device Characteristics in the Dark

The dark diode equation does not change with recombination effects that we discussed above; hence the relation is a simple one to apply to explain solar cell behavior in dark. This relationship is an exponential function and when current is plotted in natural log against voltage, the electrical transport behavior in the diode will be seen as a straight line with a slope from which the diode quality factor and saturation current can be extracted. The behavior deviates from the straight line at the lower voltage region due to shunting/recombination and at the higher voltage region due to series resistance. In case of a diode structure where multiple types of transport of carriers across the junction take place, the diode characteristics do not show such simple behavior, i.e., a single straight-line is not observed. A double diode equivalent circuit is commonly used to describe such a behavior. The heterojunction solar cell is a good example where the transport of carriers at the interfaces goes through many paths, hence, a complex equivalent circuit is needed to accurately explain the dark diode characteristics. Figure 11.3 shows the adapted equivalent circuit for a double diode behavior in the dark. The part of the equivalent circuit except the photocurrent (J_{ph}) and the recombination of photocarrier (J_{rec}) parts, are relevant for dark conductivity behavior. It is observed that for the dark I-V curve, instead of a single slope (ideal diode), the slope changes with voltage.

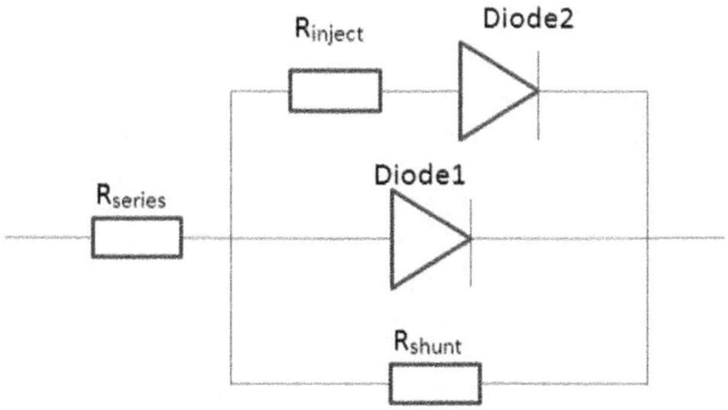

Fig. 11.3 Equivalent circuit diagram for the charge transport in a silicon heterojunction solar cell without barrier to current flow.

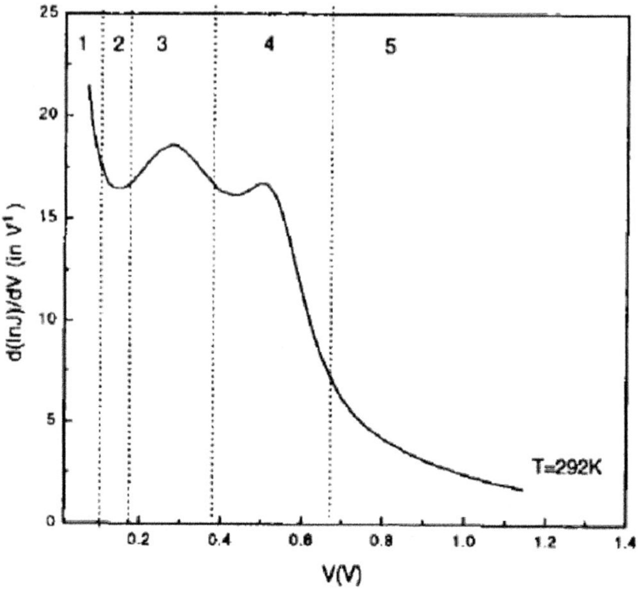

Fig. 11.4 Voltage derivative of the natural logarithm of dark current as a function of applied voltage to a silicon heterojunction solar cell [35]. Region 1 : High slope , region 2: a local minimum, attributed to shunt resistance, region 3 : a local maximum in slope indicating the first diode, region 4 : a local maximum indicating the second diode, region 5 : a rapid decrease of slope, attributed to series resistance.

One of the ways to identify the number of equivalent diodes and the behavior of each of these diodes is to plot the first derivative of the logarithm of the current against voltage (Fig. 11.4), instead of the logarithm of current versus voltage. In such a case one gets a peak value for the electrical transport of each of the diodes [35]. In case of a c-Si/a-SiC:H heterojunction two peaks have been observed, indicating double diode behavior [36]. This result shows that there are two possible transport paths. The temperature dependence of these peaks can provide information on the type of transport occurring in these diodes. Though both of them contribute to the I-V characteristics, each one is dominant at a certain voltage range; at low voltages and low temperature one peak seems to dominate, whereas at high temperature and high voltages the other peak (diode) seems to dominate. Computer simulation has shown that the dark I-V characteristics can be fitted perfectly with such an equivalent circuit (Fig. 11.3), confirming the double diode behavior [36]. Let us consider the general expression of the current voltage relation,

$$j = j_0\,(T)\left\{e^{A(T)V} - 1\right\} \tag{11.6}$$

$$j_0 \sim e^{\frac{E_a}{kT}}$$

where $j_0(T)$ and $A(T)$ are temperature dependent quantities that represent the specific current mechanism one obtains for the behavior of specific diodes. The slopes of the so-called Arrhenius plots of ln j_0 versus 1/T and A (expressed as q/nkT) versus T give the activation energy of current generation and diode quality factors respectively. One observes that the diode quality factor of one of the peaks is temperature independent whereas the other is temperature dependent. The second one, which dominates at low temperature, has a diode quality factor of 2, showing recombination character. The activation energy from ln J_0 confirms that the recombination occurs through the mid gap states of the crystalline part, as this activation energy is found to be almost half of the band gap of the c-Si [36]. The A value of the other peak is temperature independent, showing a tunneling behavior. It is not clear whether this transport is through multistep (trap assisted) tunneling with capture and emission (MTCE) [37] or multiband tunneling through the spike at the bands, as the J_0 can be fitted to both the functions $e^{E/kT}$ and e^{T/T_0}. In summary, the transport is either recomination limited or tunneling limited depending on the temperature of solar cell operation. However, this is not a universal observation. For the case of high efficiency (HIT type) solar cells of Sanyo [38], for example, the dark current at high forward bias is dominated by diffusion current, just as in case of conventional n/p homojunction. Hence, it all depends on the efficiency of passivation of the junction. The diffusion limited transport is also reported by other groups [40]. However, the Sanyo cell still shows tunneling transport at low forward voltages, similar to the observation by Rubinelli [39] for n-a-Si/p-c-Si hetero junction. Sanyo's cell shows a tunneling path (just as in c-Si/a-SiCH) which can be explained by either multistep tunneling capture emission (J_0 varying exponentially with -1/T) or direct tunneling through spike near valence band at the interface (J_0 varying exponentially with T). Computer simulation supports this latter hypothesis [41]. A diode quality factor of ~1.2 at high forward voltage has been obtained, which confirms the diffuson character. The simulation also shows that the current at low forward voltages could be dominated by tunneling. At high forward voltages the field inside the cell is weak and thus the electrons can freely diffuse. On the other hands at low forward voltages, the field is sufficiently high; in this region the electrons can cross the barrier through tunneling. In fact, also under illumination, this dark transport mechanism holds and carriers move in the direction opposite to the field direction. However, the tunneling current is too small compared to the photogenerated current to have any effect on solar cell performance [38] as reported by Sanyo, which is supported by simulation [41]. To sum up, at low bias voltages, the forward current is dominated by tunneling for a SHJ cell with or without i-layer passivation, however the forward current at high bias voltage, will depend on whether the c-Si is well passivated or not, i.e., it will be drift-diffusion limited if a passivating i-layer is present and recombination limited without the i-layer [50]. This demonstrates the importance of the i-layer in a HIT cell.

11.3.1.2 Current Voltage Characteristics under Illumination

The electrical transport characteristics of a heterojunction cell gives information on the barrier and trapping of the photogenerated carriers at the interfaces. The method to probe the barrier characteristics is by looking at the light I-V characteristics with varying measurement conditions, such as temperature and light intensity.

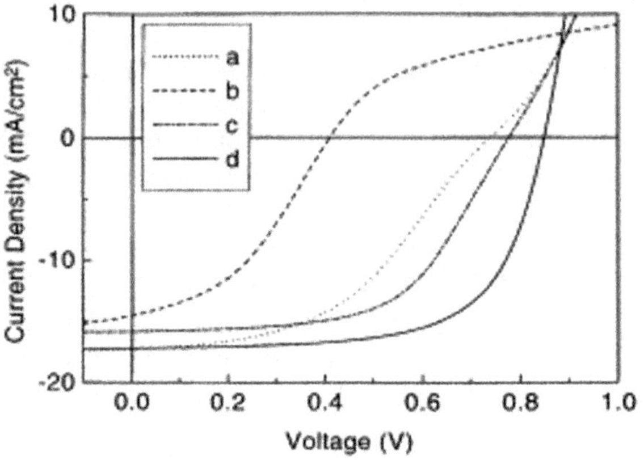

Fig. 11.5 *I–V* characteristics of a-Si p-i-n cells with and without buffers layer at the p/i interface. (a) and (b): no buffer, (c) a-SiC :H buffer, (d) wide band gap a-Si :H buffer.

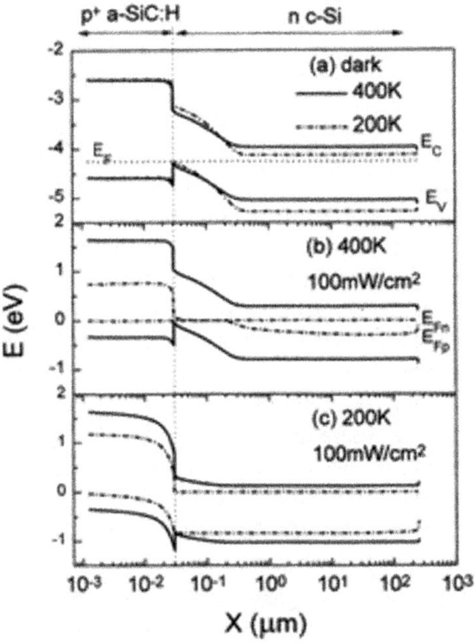

Fig. 11.6 Band diagram of a heterojunction solar cells obtained from device simulation at different light and temperature conditions [42].

The first method to discuss is the temperature dependence. A common observation in light I-V characteristics of an unoptimised cell is an S-type character at the forward voltages, not only in SHJ [42] but also thin film silicon solar cells [43], an example of it is given in Fig. 11.5. This happens whenever there is barrier to current collection, especially at the interfaces. This S-character becomes stronger as the temperature of measurement is lowered. Even a sample whose I-V curve shows a perfect diode behavior at room temperature starts to show S-character at low temperature. One of the useful aspects of low temperature measurements is that for a sample whose I-V curve at room temperature does not clearly show S-character, this may become prominent and easily detectable at low temperatures. In order to understand this S-type character, one has to visualize what is happening at the base/emitter interface. Let us take the case of an a-Si(p)/c-Si(n) interface. Here we have an emitter layer with a high band gap of ~1.7 eV and an absorber of band gap ~1.1 eV. Normally the emitter p-layer is heavily doped ($> 10^{18}$ cm^{-3}) whereas the n-type absorber is nominally doped ($\sim 10^{16}$ cm^{-3}).

There are two situations that could arise. The first case is when the p-layer is sufficiently doped. In that case, due to the difference in the doping of the two sides, the depletion region is predominantly in the n-side of the interface and the p-side is a rather flat band region. This is the ideal case for the functioning of a heterojunction cell, because the carrier generation region experiences a strong field, which allows the holes to drift from the n-side and get collected at the p-side. It has to be pointed out here that the holes face a barrier at the interface; however, if the temperature is high enough the holes will have no difficulty in drift-diffusing across this band offset (see the spike at 0.03 μm in Fig. 11.6). Thus, the band diagram and the depletion region remain unaffected with light condition, as compared to the dark state, as there is no piling or trapping of charges at the interface. This leads to a normal I-V curve. At low temperature, the situation changes considerably. The holes have difficulty in diffusing across the barrier. This region where the hole concentration exceeds the donor concentration behaves as an inversion layer. The accumulation of holes thus changes the field at the junction. The depletion region now shifts to the emitter side of the interface and the absorber side of the interface becomes a weak field region. Due to the weak field the electrons in the absorber layer diffuse to the interface region and recombine with the trapped holes at the interface or through the defects at the interface. This recombination becomes severe with forward bias and a sharp decrease in current occurs that gives an S-type character seen in the I-V characteristics. At negative bias the trapped holes can easily drift-diffuse to the emitter side and the current is fully collected. The lower the temperature, the more pronounced is the accumulation of holes at the absorber-emitter interface and one observes stronger S-character in the I-V characteristics. The main physical characteristics that causes the S-type character is the band offset (ΔE_v for p-a-Si/n-c-Si heterojunction) at the valence band and the temperature dependence of the drift-diffusion across this barrier that exponentially depends on the barrier energy ΔE_v as exp ($-\Delta E_v/kT$). As long as the concentration of accumulated holes at the absorber side of interface is lower than the acceptor ion concentration at the emitter side the field distribution will not deviate much from the dark state and the I-V characteristics will show normal behavior. At low temperature this balance is broken as the trapped holes exceed the acceptor ion concentration in the emitter side and the I-V curves exhibit S-character. A detailed experiment on the temperature dependence will thus indicate how the interface behaves and how delicate the charge carrier transport is for such a structure with band offset.

One of the conclusions from the above discussion is that the behavior of the solar cell concerning the collection of light generated carriers at the heterojunction interface depends on the distribution of the field at the interface. This, in turn, is a combined effect of the field in the dark state and the amount of photogenerated carriers crossing the barrier. If the absorber and emitter materials are fixed, this means that the band offsets at the valence and conduction band remain unchanged, the ability of the carriers to drift-diffuse will solely depend on the field at the absorber side of the interface. This will depend on the relative doping of the emitter with respect to the absorber. This has been confirmed through experiments by using different p-layer doping levels with changing boron concentration (Fig. 11.7) [44]. As long as the doping in the emitter is high enough compared to the absorber layer, the depletion is entirely on the absorber side of the interface and the photogenerated carriers will have no problem in drift-diffusing across the barrier. However, if the p-layer is lowly doped, the depletion region shifts to the emitter side and even the whole of p-layer may be depleted, whereas the depletion region inside c-Si shrinks. This low field on the c-Si side allows the electrons to diffuse to the interface and recombine with accumulated holes at the interface. This effect becomes stronger at forward voltage whereas at a reverse voltage, the field inside the c-Si depletion region recovers and the recombination losses are reduced, resulting in high current. This is the cause of the S-character. A systematic change of the I-V characteristics, starting from normal shape to S-shape by decreasing the emitter doping has been observed even at room temperature [41, 44]. Needless to say, the S-character becomes more severe at low temperature. This means that the threshold of the emitter doping to cause the S-character will systematically shift to higher values with decreasing temperature.

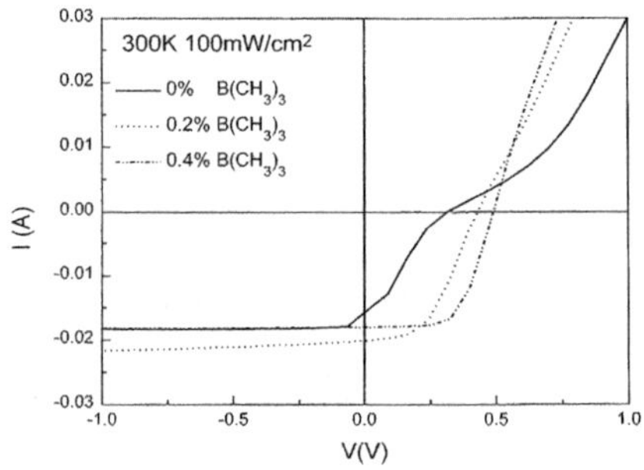

Fig. 11.7 Current-voltage characteristic of an n c-Si/p a-SiC heterojunction solar cell with varying dopant concentration in the emitter layer [44].

Another experiment to characterize a heterojunction cell is variation of light intensity. This is a very important technique in case where recombination plays a big role in limiting the photocurrent generation of a cell.

As was mentioned above, the behavior of charge transport across the depletion region under light depends on the amount of accumulated carriers at the c-Si side of the junction compared to the accepter concentration in the emitter. The accumulation of holes and recombination depends on the amount of photrogenerated carriers and thus the light intensity. A cell that shows S-character at normal light condition, such as AM1.5 light, can show perfect I-V characteristics when the light intensity is decreased below some threshold [44]. For example for a heterojunction cell, the S-character that is seen in I-V characteristics at low temperature or at low boron concentration can be recovered to normal shape at low light illumination. This change in S-character is actually systematic with respect to change in light intensity. This behavior again can be explained by studying the behavior of the depletion at the interface. At high light intensity condition, the accumulated holes in the c-Si absorber at the interface due to the band offset exceeds the acceptor concentration in the emitter and hence, the depletion is redistributed such that a big part of the emitter becomes a depletion region whereas the field in the c-Si is weakened. Due to this effect, electrons generated in c-Si diffuse into the depletion region and recombine with the accumulated holes at the interface. This leads to decrease in current and S-character in the I-V curve. At low light intensity, the accumulated holes are still below the accepter density in the emitter and the depletion is entirely in the c-Si side. This allows a strong field at the interface in the absorber layer side and holes have no problem in drift-diffusing across the valence band offset.

The S-character in the I-V characteristics originates from many reasons. Broadly speaking, this is a consequence of barriers to the flow of charge carriers. We discussed above the case where even for a perfectly passivated and clean interface one gets S-character, due to loss of field in the absorber layer. However, there are other reasons that obstruct flow of carriers, such as 1) an unclean c-Si surface, for example native SiO_x layers on c-Si surface due to unoptimized cleaning, 2) barrier caused by the intrinsic layer between absorber and emitter, 3) barrier at the emitter/TCO interface at the cell's front side etc. In such cases, the barrier behaves as a diode in the opposite direction. Hence, an equivalent circuit in this case (see Fig. 11.8) consists of diodes in series in opposite polarity direction. The I-V characteristics in such a double diode case will tend to show S-character at low temperature, even when the characteristics at room temperature may show normal behavior. This double diode behavior is common to almost all types of solar cells, which means also for non silicon type heterojunction solar cells, for example in case of a $Mo/CuIn_2Se$ interface [45].

11.3.2 Current Collection

One of the advantages of a heterojunction cell is that the emitter is a very thin high band gap layer; hence, the absorber layer receives more light, which should convert to a higher photo current as compared to a homojunction. However, when one compares the best HIT cell current density from Sanyo [46, 48], with the current collected in a passivated emitter with rear locally diffused (PERL) type crystalline silicon cell, one finds that the HIT cell has almost 4 mA/cm^2 lower current compared to the PERL cell, which generates a current of ~43 mA/cm^2 [47]. The Sanyo record cell has a current density of 39.5 mA/cm^2 [48], which

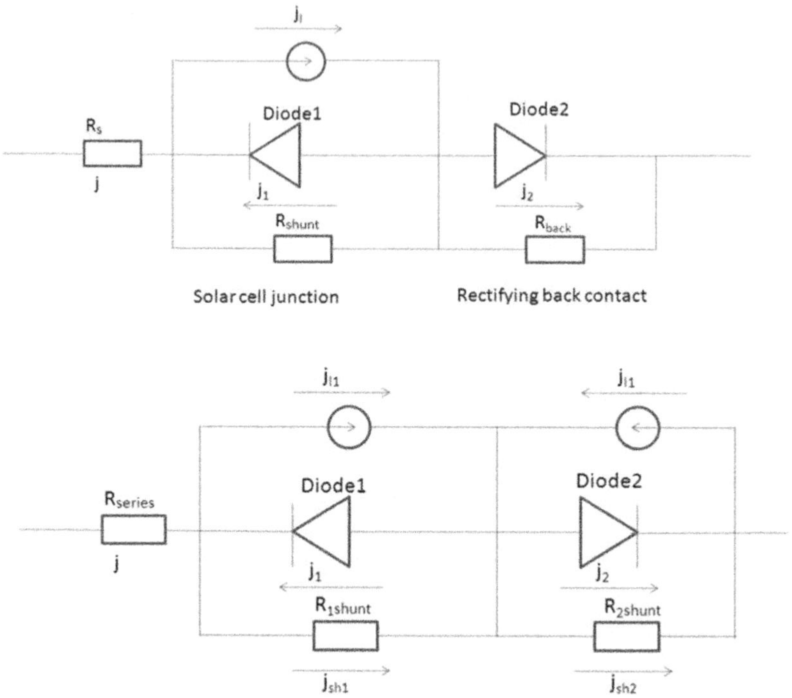

Fig. 11.8 Equivalent circuit for a silicon heterojunction solar cell having barrier to charge transport at the back contact (top) and at the absorber/emitter interface (bottom).

other groups are also approaching [49]. A comparison of spectral response (Fig. 11.9) between the c-Si PERL cell and the HIT cells from Sanyo and NREL shows that whereas the maximum in quantum efficiency values are comparable, there is a substantial loss in the low wavelength (<500nm) and high wavelength region (>900nm) for the HIT cell, which is attributed to absorption loss in the window layer and free carrier absorption in the TCO layer respectively. At present ITO seems to be the best option for the TCO layer. However, research is still ongoing to find a high mobility low carrier concentration layer to have maximum conductivity with low free carrier absorption. As far as photogeneration in the emitter is concerned, it is a dead layer due to the high defect density. Hence, any absorption in this layer is a current loss [50], unlike for c-Si homojunction cells. Increasing the band gap of the emitter of a SHJ cell will lead to an increasing barrier to carrier collection at the heterojunction interface due to an increasing band offset [51]. A band gap of around 1.85 eV is considered to be the best option as emitter window layer for a crystalline silicon/amorphous silicon heterojunction solar cell. Increasing band gap (with a concomitant increasing band offset), for example by adding more and more carbon to the silicon layer (SiC layer), will need higher doping in the emitter layer to help charge carriers to cross the spike in the band at the interface. This process is not desirable because the emitter layer will become too defective.

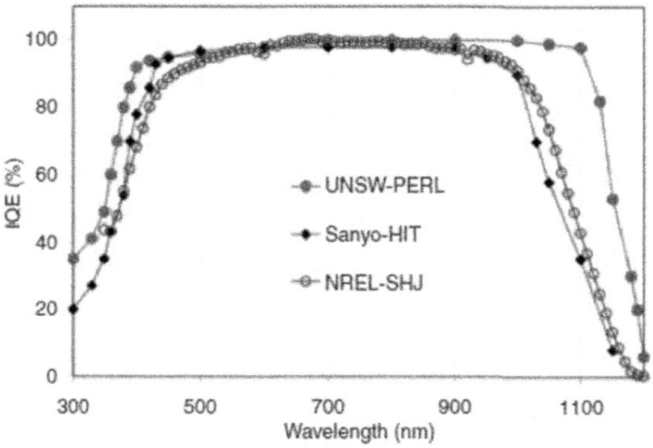

Fig. 11.9 Internal quantum efficiency of silicon heterojunction solar cells from Sanyo, and NREL in comparison with that of a PERL type c-Si homojunction solar cell [49].

Amorphous silicon oxide (a-SiO_x) i-layer and p-layer have been applied to improve the transparency and suppress epitaxial growth of the emitter [52]. It should to be noted that the response in the short wavelength for a SHJ cell with a-SiO_x or a-Si is worse even compared to a microcrystalline silicon cell [53]. This result again emphasizes the importance of optical properties of the emitter layer, and by its optimization significant current can be gained. New types of emitter layers have to be developed to address this challenge. One of the new layers tried for this purpose is n-type microcrystalline cubic silicon carbide (μc-3C-SiC) made by Hot Wire Chemical Vapor Deposition (HWCVD). Although, the quantum efficiency indeed showed improved response in the low wavelength region, reaching almost 80% at a wavelength of 400 nm, it is still lower than the spectral response of the PERL cell at this wavelength [53]. The spectral response with this emitter, however suffers quite a loss in the long wavelength region, which is speculated to be due to the damage to the interface at the back side of the cell. One of the new concepts for heterojunction solar cells would be to use quantum dot (QD) type of emitter layers [54]. These layers can be gas phase deposited by plasma CVD process at low temperature on c-Si absorber, as depicted in Fig. 11.10. Silicon quantum dots embedded in a wideband gap matrix, such as a-SiC, a-SiN, or a-SiO_x can be used for this purpose. This emitter layer will not suffer from epitaxial growth (as is the case for a μc-Si emitter layer) because of the choice of the matrix layer, and it will have a high conductivity due to the array of quantum dots. Moreover, the thin intrinsic layer as used for the HIT cell is not needed, because silicon nitride and silicon oxide can act as good passivation layers. The doping of the QD layer is, however, a challenge considering the self-purification effect. However, recent reports have shown encouraging results on phosphorous doping of silicon quantum dots in the gas phase [55]. Designing a QD array structure that would yield good conductivity due to tunneling and proper QD size to maintain a high transparency will help to improve the current.

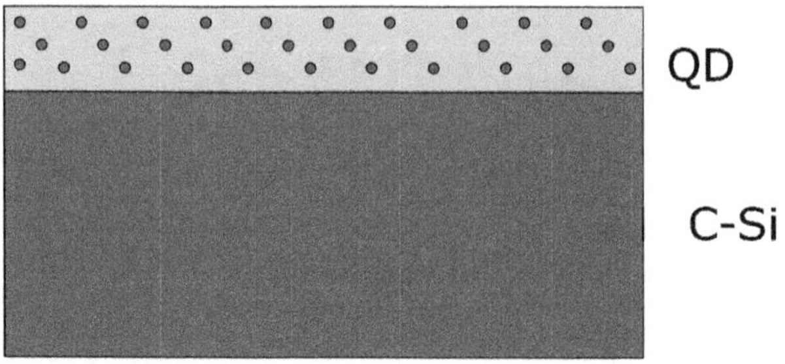

Fig. 11.10 Silicon heterojunction solar cell with quantum dot emitter layer. The host matrix of the emitter can be a-SiC, a-SiO_x, a-SiN_x etc.

11.3.3 Band Offset and Barrier to Charge Collection

One of the main causes for the barrier to the photogenerated carrier collection at the heterojunction interface is the band offset. In the above discussion we had assumed that for p-a-Si/n-c-Si heterojunction, there is a valence band offset between the emitter and the absorber. A positive (defined as conduction band edge of c-Si being below conduction band edge of a-Si) band offset at the conduction band is beneficial to the solar cell operation because the electron back diffusion is suppressed and this is in fact one of the major advantages of a heterojunction [31]. However, a positive band offset at the valence band side (the valence band edge of c-Si being above the valence band edge of a-Si) has a negative effect for the carrier transport at the junction. The best situation could be when the band offset is entirely at the conduction band. Unfortunately that is not the case with most heterojunctions and most probably not for c-Si/a-Si, according to the bulk of experimental data. In case of the c-Si/a-S heterojunction, there are conflicting reports on how much the offsets at the two bands really are. On the one hand, there is a report to suggest that the offset is entirely at the conduction band [56]; there is also another report telling us that the offset is entirely at the valence band [57]; see also the discussions in chapter Korte and chapter deWolf. Presently, the majority view is that there is a substantial offset at the valence band, which causes all the barrier related transport problems as we have discussed above. It is necessary to find an accurate value of this offset for a useful device simulation. It is helpful at this juncture to discuss the experimental methods to obtain these offset values so that fresh ideas for more accurate methods to do such experiments can be found.

One of the popular experiments to achieve this end is internal photoemission (IPE) [58]. This is basically a spectral response measurement followed by a certain way to analyze the data. According to Kane's model [59], the quantum yield of an indirect semiconductor in the vicinity of a threshold energy E_t is,

$$y \sim (h\nu - E_t)^{\frac{5}{2}} \tag{11.7}$$

A plot of the 2/5 root of the quantum yield y versus energy will give a straight-line whose intercept with the energy axis gives the threshold energy. This energy is basically a sum of the band gap of the absorber layer and the offset at one of the bands (depending on the device structure). If the mobility gap of the emitter layer is known, then the offset at the other band can be estimated. This simple concept however needs a carefully made device structure to do experiments with [60, 61]. The main condition to be met is that the photo carrier generating layer should have no band bending, hence should not have a depletion region. This problem is overcome by using an intrinsic layer at the junction with the doped absorber material, which ensures that the depletion region is completely inside the emitter region [61]. However, a doped layer is still needed to make Ohmic contact at the front side. For evaluating the valence band offset, an n-type absorber material is used whereas a p-type absorber is used to evaluate the conduction band offset. A typical structure is ITO/p-a-Si/i-a-Si/n^+c-Si/metal for valence band offset and ITO/n-a-Si/i-a-Si/p^+c-Si/metal for conduction band offset measurements. The second condition that needs to be fulfilled is that no tunneling is taking place across the spike at the valence band interface for the holes to be collected. This problem is solved by using a thick intrinsic layer compared to the typical thickness for the emitter used in solar cell structures. However, the thickness has to meet another demand, namely, that the absorption in this layer is kept low. Computer simulation is very useful in this regard to determine the minimum thickness of this layer. Typically around a few tens of nm is used, which of course depends on the doping of the absorber layer. In such a case, as mentioned above, the band offset is estimated from subtracting the band gap value of the absorber layer from the intercept energy of the plot of 2/5 root of quantum yield versus photon energy. Recent reports [62] have shown that tunneling can also be effectively utilized for extracting other informations. The field dependence of tunneling behavior is utilized for this purpose. By choosing an emitter layer with a certain thickness it is possible to observe tunneling at 0-Volt or reverse bias, whereas there is a blocking behavior at forward voltage. The advantage of this method is that one does not have to assume the band gap of the absorber layer or to find it by any other experiment. At the low voltage condition where the tunneling of minority carriers at the junction is present, the quantum yield plot will straightaway give the band gap value as the threshold voltage value of carrier collection. On the other hand, at high forward voltage, the 2/5 root plot of quantum yield will give, as mentioned earlier, the sum of the band gap of absorber and one of the band offsets. From these two studies one can obtain the relevant band offset. For p^+c-Si/a-Si heterojunction one gets the values of the band gap (E_g) of c-Si and the valence band offset, whereas n^+c-Si/a-Si structure provides the conduction band offset. From these values one can calculate the band gap of the a-Si emitter as $E_g+E_{v\text{-offset}}+E_{c\text{-offset}}$. This has been confirmed from the experiments by comparing with the band gap of a-Si obtained from optical measurements such as ellipsometry [62].

There is a discrepancy between the band offsets reported in various publications, not only from various types of experiments, but within the same experimental category such as IPE. The band offset estimation by IPE can be seriously affected by the presence of charge carrier transport behavior in the high band gap material of the heterojunction, namely; (1) tunneling through the heterojunction spike and (2) hopping transport in the band tail. In other words, IPE measures the effective band discontinuity.

11.3.4 Band Discontinuity and Tunneling

Much of the discussion above is centered on the main characteristics of a heterojunction; the band gap mismatch that manifests itself as band discontinuity. This is the main advantage of the heterojunction i.e., the suppression of back diffusion of majority carriers and more light entering the absorber layer due to a transparent emitter. For a p a-Si/n c-Si heterojunction, the electron back diffusion is reduced, however, the valence band offset limits its device performance. The experimental observation and simulation studies point out that the barrier to hole collection increases systematically with an increase in band offset and above certain band offset, it is difficult for holes to cross the barrier and they accumulate at the interface leading to the S-character of the I-V characteristics [41]. On the other hand, the band offset should not be too low: below a certain value of band offset the voltage of the device decreases substantially [41]. The band offset is an essential feature of a heterojunction cell from which the high voltage in a heterojunction solar cell results. Coming back to the barrier to hole transport, an example may be given on the threshold to band offset for hole collection. The simulation studies have shown for a case (p-a-SiC emitter of doping level $2x10^{19}$ cm^3 and the absorber wafer of doping level of $1x10^{11}$ cm^3) that the I-V characteristics should show barrier behavior to charge transport by drift-diffusion [42, 41]. However, experimentally no such S-characteristics have been observed at room temperature and the I-V characteristics have normal behavior. This leads to the hypothesis that there are other mechanisms operating for the transport of carriers than simple drift-diffusion. Tunneling has been proposed as the possible mechanism to collect the holes across the spike [41, 42]. This is expected as we have seen in the dark electrical transport that such a tunneling is present. In a capacitance study, temperature independence of the release of carriers across the junction has been observed, which points to such a behavior [63]. However, the precise mechanism for the carrier transport has not been established. The temperature dependent capacitance study for a heterojunction structure subject to light pulses shows that the carrier transport is not thickness dependent [63]. This suggests a multistep tunneling to be more appropriate than direct tunneling. An alternative mechanism is that the tunneling occurs at the top of the spike where the thickness of the barrier is small enough, however a field is needed to allow the carriers to reach this top of the spike region. We have discussed this behavior of charge transport in the dark current voltage electrical transport in an SHJ cell, through thermionic emission coupled with tunneling (Fig. 11.11) [35]. For light I-V, a similar field dependent tunneling may be operating. In fact, such electric field dependent tunneling is observed, manifested as the tunneling at reverse bias and low forward voltages and barrier controlled transport at high forward bias [41].

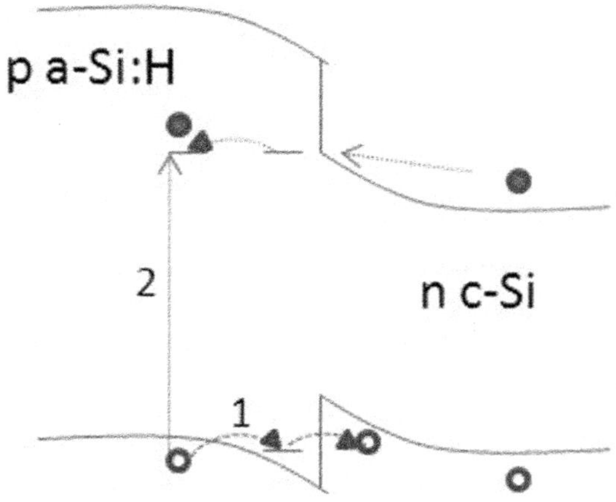

Fig. 11.11 Schematic of the multitunneling capture emission (2) (J_0 varying exponentially with -1/T) or multitunneling through spike (1) near valence band at the interface (J_0 varying exponentially with T) for a heterojunction solar cell in dark condition.

11.3.5 Interface Defects

From the I-V analysis (Eq. (11.4)) and as depicted in the equivalent circuit diagram (Fig. 11.1b), it is clear that the recombination path plays a big role in the functioning of the SHJ solar cell. Hence, to understand the electrical properties of solar cells, such as current-voltage and spectral response, and to carry out device simulation, defect characteristics, especially at the heterojunction interface, are important. Unfortunately, there are not many experimental tools that can be used to extract the density of states (DOS) for very thin layers such as that used for emitter and interface defects. The most sensitive technique to extract information on the interface defects and the nature of defects is electrically detected magnetic resonance (EDMR) [64], also called spin dependent photoconductivity (SDPC) [65] when dealing only with the photoconductivity aspect. These techniques combine electron spin resonance (ESR) and electrical conductivity. Hier, the ESR technique is used to understand the transport properties based on the fact that in a system having electrons, holes and dangling bond defects, each of these are paramagnetic in nature. The dangling bond can be paramagnetic or not depending on its charge state, as these states are amphoteric in nature, which holds also for the c-Si/a-Si interface defect states [66]. The energy of a paramagnetic center (for example of spin ½) in the presence of a magnetic field H_o splits into two energy levels (m=-1/2 and m=+1/2); this is called Zeeman splitting. Transition by a photon (microwave energy quantum) from the ground state (-1/2) to an upper state (+1/2) occurs at the spin resonance condition when a magnetic field at a frequency $g\mu_B H_o/h$, perpendicular to the static magnetic field is applied. The terms g and μ_B are the Landé factor and the Bohr magneton respectively, and h is Planck's constant. To sustain an ESR condition, it is essential that the spin lattice relaxation time $T_1 \ll \tau_{SF}$, the spin flip time, given by $\tau_{SF} \approx 1/\gamma H_1$ where $\gamma = g\mu_B/h$ is the gyromagnetic ratio. This condition is necessary to maintain the spin polarization by thermodynamic relaxation that suppresses the spin system to reach saturation in resonance transition. The EDMR experiment, on the other hand, is a technique that relies on the spin dependent electronic transition between states; here the necessary condition for the lifetimes is very different, as explained below. The spin states of electrons, holes and midgap defects can be either in spin parallel or anti parallel state. If we consider electron recombination or tunneling to the midgap state, only the pair with spins antiparallel to each other can participate in the transition and recombine (Fig. 11.12). Thus, only the singlet states are involved. Therefore, in the thermal equilibrium state triplet states (spins parallel to each other) outnumber the amount of singlet states. However, if $T_1 << \tau_{SF}$, then during resonance condition there will be equalization of singlet and triplet states due to saturation of spins and hence, the recombination and tunneling transitions will increase. This will show up as a quenching signal of the photo-conductivity given by $\Delta\sigma/\sigma$, where σ is the electrical conductivity. In an EDMR/SDPC experiment, however, both quenching and enhancing signals are observed depending on the type of transitions involved.

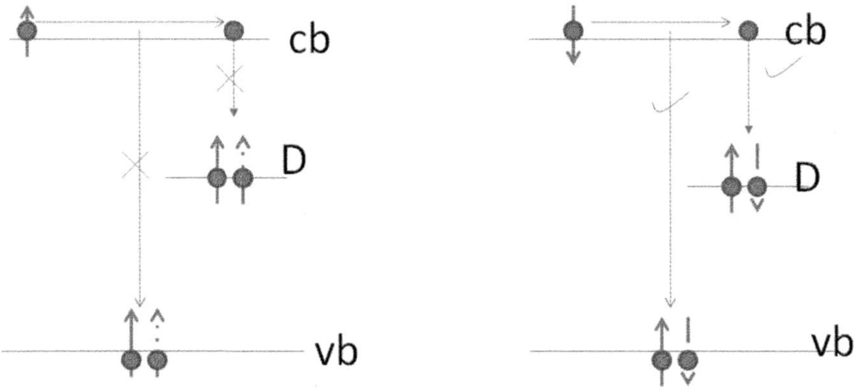

Fig. 11.12 Mechanism of recombination limited spin dependent conductivity.

In principle ESR is a very sensitive technique to extract information on the defects, however the total number of defects in a sample is decisive in detecting the concerned ESR signal. Moreover, ESR can, not only tell us what the orientation of spins (defects) with respect to the reference structure (surface) is, but also provide information on the environment, the reactions with surrounding electron spins (for example dangling bond exchange pairing [67]) and atoms (nuclear spins), and the movement of adjacent atoms such as hydrogen [68] by looking at and comparing the hyperfine interactions with neighboring atoms in a series of temperature dependent ESR measurements. This latter effect is very useful for silicon materials, especially for the c-Si/a-Si interface where the electron spins (dangling bond defects) are surrounded by hydrogen to passivate these defects. Moreover, interactions with other impurity atoms, such as oxygen, phosphorous etc. can be detected. However, the sensitivity limit of ESR is ~10^{11}/G, where G stands for unit of magnetic field in Gauss. Hence it cannot detect the interface defects signals.

EDMR, on the hand, has a 7 orders of magnitude larger sensitivity. An EDMR signal of 10^5 spins has been detected in a thin film transistor (TFT) structure whereas for ESR the minimum detection limit is ~ 10^{12} spins. The high sensitivity of EDMR/SDPC is attributed to the optical transition energies which are in the range of ~1V, whereas for the case of ESR it is in the range of a few μeVs between the Zeeman states. Basically, ESR can be coupled with a range of other phenomena to derive additional information about the defects: optically detected magnetic resonance (ODMR) that couples luminescence with resonance; electric detected magnetic resonance (EDMR)/Spin dependent photoconductivity (SDPC) that couples electrical conductivity with resonance; capacitance detected magnetic resonance (CDMR) that couples capacitance with resonance; noise detected magnetic resonance (NDMR) that couples noise spectroscopy with resonance etc. Out of these, EDMR/SDPC, is the most promising technique to characterize interface states, while the capacitance technique still needs further refining of data analysis. It is difficult to separate the capacitance effect from EDMR due to the presence of a leakage current and this calls for finding an experimental condition where the EDMR signal is minimized to study the CDMR accurately.

The spin dependent transport (SDT) is a complex phenomenon that has to satisfy conditions of lifetimes given by $T_1 > \tau_{SF} > \tau_{trans}$, where τ_{trans} is the transition time (recombination time constant). This complexity comes from the nature of this measurement. The usefulness of the SDT measurement is quite clear; the sensitivity of the method allows one to detect a very small number defects involved in the transport phenomena. There is also another advantage in that only the defects/spins involved in the transport path are recorded and the contaminations in the microwave cavity or the measurement tube etc. do not contribute to the signal. Another advantage is that, unlike ESR, the sensitivity of EDMR does not decrease with decreasing magnetic field; hence it is possible to do the measurements at a low static magnetic field. In case of a-Si:H, a static magnetic field as low as possible is used to suppress the inhomogeneous broadening of the signal. This allows proton-hyperfine structure to be resolved and the coordination and interaction of the spin (defect) with hydrogen can be evaluated. This is very important for interface characterization, especially for the SHJ type of cells where the passivation of interface states/dangling bonds by hydrogen is very crucial.

The issue that is still not well resolved is how to quantify the number of defects from the SDT measurement. This is because a rather small effect of spin resonance on the conductivity is expected. This does not mean that only a small part of paramagnetic states are involved in the spin dependent processes. Detailed calculations based on the spin polarization effect give a quantitative dependence of the rate of change of conductivity σ signal at the resonance on the defects as;

$$-\frac{\Delta\sigma}{\sigma} \leq \left(\frac{\Delta N}{N}\right)^2 = \left(\frac{h\nu_o}{2kT}\right)^2 \tag{11.8}$$

where ΔN is the occupation difference between the excited and ground states ($N_- - N_+$) in thermal equilibrium and N is the total number of spins ($N_- + N_+$), ν_o is the resonance frequency. The term ΔN/N represents the spin polarization that occurs when the system is in thermal equilibrium given by a Boltzmann distribution. From this term it is obvious that the relative change of spin dependent photoconductivity (for example) does not depend on the absolute number of defects directly, rather a small part of it, termed as spin polarization ($\Delta N/N$). This is, however, similar to the case of ESR in which case also one measures only this spin polarization part of the total defect density and a calibration material (such as a Tempol solution or Mn^{2+} doped MgO) is used to get the absolute number of spins or defect density. This concept holds also for SDT, however, there are also other effects that affect the signal. One of the models [69] proposes that the spins form a correlated bound pair before the transition takes place and the spin pair has a characteristic lifetime τ_{Diss} within which the transition has to take place. Hence, the lifetime condition for SDT to take place is modified from the earlier relation as; τ_{Diss}, $T_1 > \tau_{SF} > \tau_{trans}$. This effect leads to a much larger SDT signal which is generally observed compared to what the Eq. (11.7) would predict. All these effects make the accurate determination of the absolute number of defects difficult to obtain.

Notwithstanding the abovementioned ongoing issues, this experimental technique promises to be the best option to detect the defects at the interfaces and in very thin materials and understand its nature. So far, a lot of informations has been gathered by this technique, some of which are mentioned here. EDMR experiments on HIT type of cells show that carrier transport is midgap defect (recombination) limited, whereas for high efficiency cells the mobility of electrons and holes in the band tail is the dominant factor [70]. P_b type centers (dangling bond back bonded to three silicon atoms), which are generally present at Si/SiO_2 interface, have been detected, for example on the top surface of the epitaxial thin film silicon emitter of a HIT cell [71]. Angular variation studies are easily made on such spin centers to identify their location and orientation (Fig. 11.13). Deliberate passivation of the top surface of the emitter with Si_3N_4 or SiO_2 decreases the density of P_b centers, in addition to the intended functioning of these layers to improve antireflection. In case of TFTs, the leakage current through the nitrogen dangling bonds in the Si_3N_4 dielectric layer has been detected (Fig. 11.14) [72]. The beauty of spin dependent processes is that, just as in ESR, the various interactions show up such as hyperfine spitting in the signal. Hence, from the signal, one can derive information on the specific type of atoms or molecules to which the spin belongs and the other types of species surrounding it. The identification of nitrogen dangling bonds mentioned above from the EDMR study of TFTs was possible due to the hyperfine structures from the interaction of spin with the nitrogen nucleus. A second example is the detection of hyperfine structures for the electron transport through hopping in the band tails of n-a-Si emitter layers in a p-c-Si/n-a-Si heterojunction structure [73]. In this case, the lines are attributed to the P-donor hyperfine structure. Further improvement on the SDT measurements are being made, for example using pulsed mode, called pEDMR, by which coupling strengths of the spins and coherence times as well as transition probabilities for the electronic processes can be investigated [74].

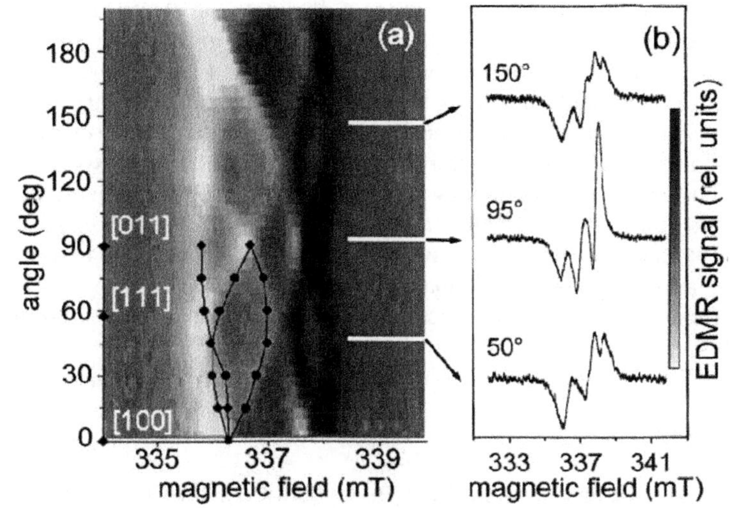

Fig. 11.13 (a) 2D plot of the angular dependence of the EDMR signal of a silicon heterojunction solar cell measured with short circuit condition. The gray scale in (b) indicates the signal amplitudes of the 2D plot. The symbols represent literature data of the P_{bo} center. (b) 1D spectra at some selected angles. [71]

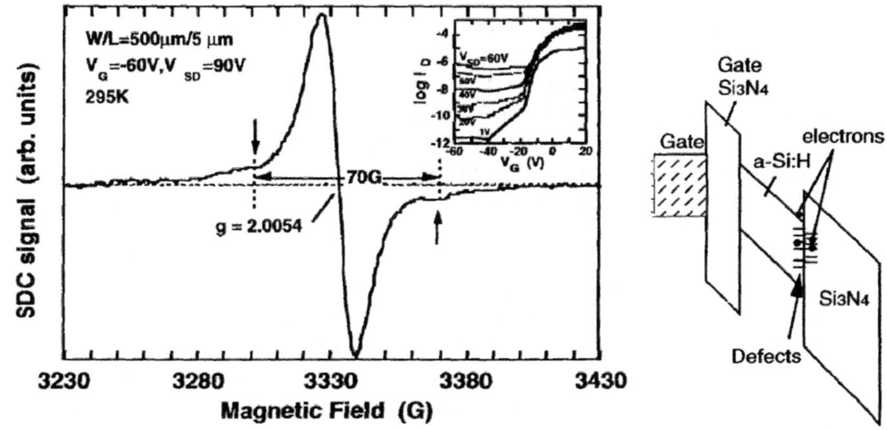

Fig. 11.14 Spin dependent conductivity spectrum in dark of a TFT (left). The band diagram and the interface defect states of the TFT (right) [72].

11.4 Conclusion

Silicon heterojunction cells are yet to be understood well. The electrical properties of a heterojunction cell differ substantially from that of a homojunction cell. An equivalent circuit and a current voltage relation specific to the heterojunction cell have been developed. The interface states and charge carrier collection across the band offset and the recombination process dominate the electrical properties both in dark and light current voltage characteristics. The origin of the S-type character in the I-V characteristics and its origin and solutions to overcome this have been discussed. Experimental methods to determine the band offset and the tunneling behavior at the spikes in the bands have been described. Determining interface states is a difficult task to perform and there are not many experimental tools available to probe interface states. One of the promising techniques, namely spin dependent transport or EDMR, has been discussed and some of the interesting results obtained with this technique have been presented. This review makes it clear that a lot has to be done to make microscopic characterization of the different parts of the cells, most specifically the defects and charge transport at the interfaces.

References

[1] Repins, I., Contreras, M.A., Egaas, B., de Hart, C., Scharf, J., Perkins, C.L., To, B., Noufi, R.: 19·9%-efficient ZnO/CdS/CuInGaSe_2 solar cell with 81·2% fill factor. Prog. Photovolt. Res. Appl. 16, 235 (2008)

[2] Wu, X.: High-efficiency polycrystalline CdTe thin-film solar cells. Sol. Energy 77, 803–814 (2004)

[3] Guter, W., Schöne, J., Philipps, S.P., Steiner, M., Siefer, G., Wekkeli, A., Welser, E., Oliva, E., Bett, A.W., Dimroth, F.: Current-matched triple-junction solar cell reaching 41.1% conversion efficiency under concentrated sunlight. App. Phys. Lett. 94, 223504 (2009)

[4] Niemann, E., Fischer, R., Leidich, D., Linhart, E.: Electrical properties of a-SiC_x /a-Si heterojunctions. J. Non-Cryst. Sol. 77-78, 991–994 (1985)

[5] Hattori, Y., Kruangam, D., Katoh, K., Nitta, Y., Okamoto, H., Hamakawa, Y.: High-conductive wide band gap p-type a-SiC:H prepared by ECR CVD and its application to high efficiency a-Si basis solar cells. In: Proc. 19th IEEE Phtovoltaic Specialists Conf., p. 686 (1987) (a-Si/c-Si)

[6] Han, M.-K., Mastsumoto, Y., Hirata, G., Okamoto, H., Hamakawa, Y.: Characterization of boron doped μc-SiC/c-Si heterojunction solar cells. Journal of Non-Crystalline Solids 115, 195–197 (1989) (μc-SiC/c-Si)

[7] Tanaka, M., Taguchi, M., Matsuyama, T., Sawada, T., Tsuda, S., Nakano, S., Hanafusa, H., Kuwano, Y.: Development of new a-Si/c-Si heterojunction solar cells: ACJ-HIT (Artificially Constructed Junction-Heterojunction with Intrinsic Thin-layer). Jpn. J. Appl. Phys. 31, 3518–3522 (1992)

[8] Taguchi, M., Tsunomura, Y., Inoue, H., Taira, S., Nakashima, T., Baba, T., Sakata, H., Maruyama, E.: High Efficiency HIT Solar Cell on Thin (<100 μm) Silicon Wafer. In: Proceedings of the 24th European Photovoltaic Solar Energy Conference, Hamburg, Germany, p. 1690, WIP-Munich, Germany, (2009)

[9] van Cleef, M.W.M., Rath, J.K., Rubinelli, F.A., van der Werf, C.H.M., Schropp, R.E.I., van der Weg, W.F.: Performance of heterojunction p+ microcrystalline silicon n crystalline silicon solar cells. J. Appl. Phys. 82, 6089 (1997)

[10] Descoeudres, A., Barraud, L., Bartlome, R., Choong, G., de Wolf, S., Zicarelli, F., Ballif, C.: The silane depletion fraction as an indicator for the amorphous/crystalline silicon interface passivation quality. Appl. Phys. Lett. 97, 183505 (2010)

[11] Lachenal, D., Andrault, Y., Bätzner, D., Guerin, C., Kobas, M., Mendes, B., Strahm, B., Tesfai, M., Wahli, G., Buechel, A.: High efficiency silicon heterojunction solar cell activities in neuchatel, switzerland. In: 25th European Photovoltaic Solar Energy Conference and Exhibition / 5th World Conference on Photovoltaic Energy Conversion, Valencia, Spain, p. 1272 (2010)

[12] Daniel, A., Hickel, P.E., Souche, F., Salvetrat, T., Letiec, Y., Nolan, T., Ribeyron, P.J.: Control of texturized silicon substrates surface passivation for a-Si:H/c-Si heterojunction solar cells. In: 25th European Photovoltaic Solar Energy Conference and Exhibition /5th World Conference on Photovoltaic Energy Conversion, Valencia, Spain, p. 1951 (2010)

[13] Conrad, E., Korte, L., Maydell, K.V., Angermann, H., Schubert, C., Stangl, R., Schmidt, M.: Development and optimization of a-Si:H/c-Si heterojunction solar cells completely processed at low temperatures. In: 21st European Photovoltaic Solar Energy Conference and Exhibition, Dresden, Germany, p. 784 (2006)

[14] Wang, Q., Page, M.R., Iwaniczko, E., Xu, Y., Roybal, L., Bauer, R., To, B., Yuan, H.-C., Duda, A., Hasoon, F., Yan, Y.F., Levi, D., Meier, D., Branz, H.M., Wang, T.H.: Efficient heterojunction solar cells on p-type crystal silicon wafers. Applied Physics Letters 96, 013507 (2010)

[15] Gudovskikh, A.S., Kleider, J.P., Froizheim, A., Fuhs, W., Terukov, E.I.: Investigation of a-Si:H/c-Si heterojunction solar cells interface properties. Thin Solid Films, 451-452, 345–349 (2004)

[16] Poindexter, E.H., Rong, F.C., Buchwald, W.R., Gerardi, G.J., Keeble, D.J., Warren, W.L.: Electrically-detected magnetic resonance near the p-doped/n-doped interface of Si junction diodes. Colloids and Surfaces A: Physicochemical and Engineering Aspects 72, 119–125 (1993)

[17] Korte, L., Schmidt, M.: Journal of Non-Crystalline Solids 354, 2138–2143 (2008)

[18] Ahrenkiel, R.K., Johnston, S.W.: An advanced technique for measuring minority-carrier parameters and defect properties of semiconductors. Mater. Sci. Eng. B 102, 161–172 (2003)

[19] Sinton, R., Cuevas, A., Stuckings, M.: Quasi-steady-state photoconductance, a new method for solar cell material and device characterization In: Proceedings of the 25th IEEE Photovoltaic Specialists Conference, pp. 457–460 (1996); De Wolf, S., Kondo, M.: Nature of doped a-Si:H/c-Si interface recombination. Journal of Applied Physics 105, 103707 (2009)

[20] Klein, D., Wuensch, F., Kunst, M.: The determination of charge-carrier lifetime in silicon. Physica Status Solidi (B)—Basic Solid State Physics 245, 1865–1876 (2008); Zhao, L., Diao, H., Zeng, X., Zhou, C., Li, H., Wang, W.: Comparative study of the surface passivation on crystalline silicon by silicon thin films with different structures. Physica B 405, 61–64 (2010)

[21] Bruggemann, R., Reynolds, S.: Modulated photoluminescence studies for lifetime determination in amorphous-silicon passivated crystalline-silicon wafers. Journal of Non-Crystalline Solids 352, 1888–1891 (2006)

[22] Lam, Y.W.: Surface-state density and surface potential in MIS capacitors by surface photovoltage measurements. J. Phys. D: Appl. Phys. 4, 1370–1375 (1971) ; Korte, L., Conrad, E., Angermann, H., Stangl, R., Schmidt, M.: Advances in a-Si:H/c-Si heterojunction solar cell fabrication and characterization. Solar Energy Materials & Solar Cells 93, 905–910 (2009); Maydell, K.V., Korte, L., Laades, A., Stangl, R., Conrad, E., Lange, F., Schmidt, M.: Characterization and optimization of the interface quality in amorphous/crystalline silicon heterojunction solar cells. Journal of Non-Crystalline Solids 352, 1958–1961 (2006)

[23] Fujiwara, H., Kondo, M.: Impact of epitaxial growth at the heterointerface of a-Si:H/c-Si solar cells. Applied Physics Letters 90, 013503 (2007)
[24] Hartman, S., Lind, M.A.: Spectral response measurements for solar cells. Solar Cells 7, 147–157 (1982-1983)
[25] Bordin, N., Kreinin, L., Eisenberg, N.: The use of the self-calibration method for accurate determination of silicon solar cell internal quantum efficiency. Solar Energy Materials & Solar Cells 63, 247–257 (2000)
[26] Schulze, T.F., Korte, L., Conrad, E., Schmidt, M., Rech, B.: Electrical transport mechanisms in a-Si:H/c-Si heterojunction solar cells. J. Appl. Phys. 107, 023711 (2010)
[27] Diouf, D., Kleider, J.P., Desrues, T., Ribeyron, P.-J.: Study of interdigitated back contact silicon heterojunctions solar cells by two-dimensional numerical simulations. Materials Science and Engineering B, 159-160, 291–294 (2009) (ATLAS)
[28] Froitzheim, A., Stangl, R., Elstner, L., Kriegel, M., Fuhs, W.: AFORS-HET: a computer-program for the simulation of heterojunction solar cells to be distributed for public use. In: Proceedings of the 3rd World Conference on Photovoltaic Energy Conversion, vol. A, pp. 279–282 (2003) (AFORS-HET)
[29] Hernández-Como, N., Morales-Acevedo, A.: Simulation of hetero-junction silicon solar cells with AMPS-1D. Solar Energy Materials & Solar Cells 94, 62–67 (2010) (AMPS-1D)
[30] Clugston, D., Basore, P.: PC1D version 5: 32-bit solar cell modeling on personal computers. In: Proceeding of the 26th IEEE Photovoltaic Specialists Conference, Anaheim, CA, pp. 207–210 (1997); Lien, S.-Y., Wuu, D.-S.: Simulation and Fabrication of Heterojunction Silicon Solar Cells from Numerical Computer and Hot-Wire CVD. Prog. Photovolt: Res. Appl. 17, 489–501 (2009) (Pc1D)
[31] Fonash, S.J.: Solar Cell Device Physics. Academic Press Inc., London (1981)
[32] Merten, J., Asensi, J.M., Voz, C., Shah, A., Platz, R., Andreu, J.: Improved Equivalent Circuit and Analytical Model for Amorphous Silicon Solar Cells and Modules. IEEE Trans. Electron Devices 45, 423–429 (1998)
[33] Barrio, R., Gandıa, J.J., Carabe, J., Gonzalez, N., Torres, I., Munoz, D., Voz, C.: Surface recombination analysis in silicon-heterojunction solar cells. Solar Energy Materials & Solar Cells 94, 282–286 (2010)
[34] Merten, J., Voz, C., Munoz, A., Asensi, J.M., Andreu, J.: The role of the buffer layer in the light of a new equivalent circuit for amorphous silicon solar cells. Solar Energy Materials & Solar Cells 57, 153–165 (1999)
[35] van Cleef, M.W.M., Philippens, M.W.H., Rubinelli, F.A., Kolter, M., Schropp, R.E.I.: Electrical transport mechanisms in p+a-SiC:H/n c-Si heterojunctions: dark J-V-T characteristics. In: Mat. Res. Soc. Symp. Proc., vol. 420, p. 239 (1996)
[36] van Cleef, M.W.M.: Amorphous crystalline silicon heterostructures and solar cells, Ph.D Thesis, Utrecht University (1998)
[37] Matsuura, H., Okuno, T., Okushi, H., Tanaka, K.: Electrical properties of n-amorphous/p-crystalline silicon heterojunctions. J. Appl. Phys. 55, 1012 (1984)
[38] Taguchi, M., Maruyama, E., Tanaka, M.: Temperature Dependence of Amorphous/Crystalline Silicon Heterojunction, Solar Cells. Jap. J. Appl. Phys. 47, 814–818 (2008)
[39] Jensen, N., Rau, U., Hausner, R.M., Uppal, S., Oberbeck, L., Bergmann, R.B., Werner, J.H.: Recombination mechanisms in amorphous silicon/crystalline silicon heterojunction solar cells. J. Appl. Phys. 87, 2639 (2000)

[40] Rubinelli, F., Albournoz, S., Buitrago, R.: Amorphous-crystalline silicon heterojunction: Theoretical evaluation of the current terms. Solid-state Electron. 32, 1055 (1989)
[41] Rahmouni, M., Datta, A., Chatterjee, P., Damon-Lacoste, J., Ballif, C., Roca i Cabarrocas, P.: Carrier transport and sensitivity issues in heterojunction with intrinsic thin layer solar cells on N-type crystalline silicon: A computer simulation study. J. Appl. Phys. 107, 054521 (2010)
[42] van Cleef, M.W.M., Rubinelli, F.A., Rizzoli, R., Pinghini, R., Schropp, R.E.I., van der Weg, W.F.: Amorphous silicon carbide/crystalline silicon heterojunction solar cell: A comprehensive study of the photocarrier collection. Jap. J. App. Phys. 37, 3926–3932 (1998)
[43] Rath, J.K., Schropp, R.E.I.: Incorporation of p-type microcrystalline silicon films in amorphous silicon based solar cells in a superstrate structure. Solar Energy Materials and Solar Cells 53, 189–203 (1998)
[44] van Cleef, M.W.M., Rubinelli, F.A., Rath, J.K., Schropp, R.E.I., van der Weg, W.F., Rizzoli, R., Summonte, C., Pinghini, R., Centurioni, E., Galloni, R.: Photocarrier collection in a-SiC:H/c-Si heterojunction solar cells. J. Non-Cryst. Sol. 227-230, 1291 (1998)
[45] Bowron, J.W., Damaskinos, S.D., Dixon, A.E.: Characterization of the anomalous second junction in Mo/CuInSe2/(CdZn)S/ITO solar cells. Solar Cells 31, 159–169 (1991)
[46] Tsunomura, Y., Yoshimine, Y., Taguchi, M., Baba, T., Kinoshita, T., Kanno, H., Sakata, H., Maruyama, E., Tanaka, M.: Twenty-two percent efficiency HIT solar cell. Solar Energy Materials & Solar Cells 93, 670–673 (2009)
[47] Wang, A., Zhao, J., Green, M.A.: 24 % efficient silicon solar cells. Appl. Phys. Lett. 57, 602 (1990)
[48] Maruyama, E., Terakawa, A., Taguchi, M., Yoshimine, Y., Ide, D., Baba, T., Shima, M., Sakata, H., Tanaka, M.: Sanyo's challenges to the development of high-efficiency HIT solar cells and the expansion of HIT business. In: Proceedings of WCPEC-4, Hawaii, vol. 3, pp. 1455–1460 (2006)
[49] Wang, Q.: High-efficiency hydrogenated amorphous/crystalline Si heterojunction solar cells. Phil. Mag. 89, 2587–2598 (2009)
[50] Dao, V.A., Lee, Y., Kim, S., Kim, Y., Lakshminarayan, N., Yi, J.: Interface characterization and electrical transport mechanisms in a-Si:H/c-Si heterojunction solar cells. Journal of The Electrochemical Society 158(3), H312–H317 (2011)
[51] Datta, A., Rahmouni, M., Nath, M., Boubekri, R., Roca i Cabarrocas, P., Chatterjee, P.: Insights gained from computer modeling of heterojunction with instrinsic thin layer "HIT" solar cells. Solar Energy Materials & Solar Cells 94, 1457–1462 (2010)
[52] Fujiwara, H., Sai, H., Kondo, M.: Japanese Journal of Applied Physics 48, 064506 (2009)
[53] Banerjee, C., Narayanan, K.L., Haga, K., Sritharathikhun, J., Miyajima, S., Yamada, A., Konagai, M.: Fabrication of microcrystalline cubic silicon carbide/crystalline silicon heterojunction solar cell by hot wire chemical vapor deposition. Japanese Journal of Applied Physics 46, 1–6 (2007)
[54] Park, S., Cho, E., Song, D., Conibeer, G., Green, M.A.: n-Type silicon quantum dots and p-type crystalline silicon heteroface solar cells. Solar Energy Materials & Solar Cells 93, 684–690 (2009)
[55] Stegner, A.R., Pereira, R.N., Klein, K., Lechner, R., Dietmueller, R., Brandt, M.S., Stutzmann, M., Wiggers, H.: Electronic Transport in Phosphorus-Doped Silicon Nanocrystal Networks. Phys. Rev. Lett. 100, 026803 (2008)

[56] Cuniot, M., Marfaing, Y.: Energy band diagram of the a-Si:H/c-Si interface as determined by internal photoemission. Philos. Mag. B 57, 291 (1988)
[57] Mimura, H., Hatanaka, Y.: Energy-band discontinuities in a heterojunction of amorphous hydrogenated Si and crystalline Si measured by internal photoemission. Appl. Phys. Lett. 50, 326 (1987)
[58] Wronski, C.R.: Review of direct measurements of mobility gaps in a-Si:H using internal photoemission. J. Non-Cryst. Solids 141, 16–23 (1992)
[59] Kane, E.O.: Theory of Photoelectric Emission from Semiconductors. Phys. Rev. 127, 131 (1962)
[60] van Cleef, M.W.M., Schropp, R.E.I., Rubinelli, F.A.: Significance of tunneling in p+ amorphous silicon carbide n crystalline silicon heterojunction solar cells. Appl. Phys. Lett. 73, 2609 (1998)
[61] Xu, X., Yang, J., Banerjee, A., Guha, S.: Band edge discontinuities between microcrystalline and amorphous hydrogenated silicon alloys and their effect on solar cell performance. Appl. Phys. Lett. 67, 2323 (1995)
[62] Sakata, I., Yamanaka, M., Kawanami, H.: Characterization of heterojunctions in crystalline-silicon-based solar cells by internal photoemission. Solar Energy Mater. Sol. Cell 93, 737–741 (2009)
[63] Page, M.R., Iwaniczko, E., Xu, Y.-Q., Roybal, L., Hasoon, F., Wang, Q., Crandall, R.S.: Amorphous/crystalline silicon heterojunction solar cells with varying i-layer thickness. Thin Solid Films 519, 4527–4530 (2011)
[64] Stutzmann, M., Brandt, M., Beyer, M.W.: Spin-dependent processes in amorphous and microcrystalline silicon: a survey. J. Non-Cryst. Solids 266-269, 1–22 (2000)
[65] Stuke, J.: Recent results on hydrogenated amorphous silicon. Ann. Rev. Mater. Sci. 15, 79–102 (1985)
[66] Olibet, S., Vallat-Sauvain, E., Ballif, C.: Model for a-Si:H/c-Si interface recombination based on the amphoteric nature of silicon dangling bonds. Physical Review B 76, 035326 (2007)
[67] Rath, J.K., Barbon, A., Schropp, R.E.I.: Clustered defects in hot wire chemical vapor deposited poly-silicon films. Journal of Non-Crystalline Solids 266-269, 548–552 (2000)
[68] Rath, J.K., Radhakrishna, S.: EPR studies of MoO centres in KDP single crystals. Phys. Status Solidi. (a) 100, 593 (1987)
[69] Kaplan, D., Solomon, I., Mott, N.F.: Explanation of the large spin-dependent recombination effect in semiconductors. J. Phys. Paris 51, L51 (1978)
[70] Lips, K., Muller, R., Kanschat, P., Finger, F., Fuhs, W.: Spin-dependent processes in thin-film silicon solar cells. In: Mat. Res. Soc. Symp. Proc., vol. 609, A18.2.1 (2000)
[71] Muller, R., Kanschat, P., Von Aichberger, S., Lips, K., Fuhs, W.: Identification of transport and recombination paths in homo and heterojunction silicon solar cells by electrically detected magnetic resonance. J. Non-cryst. Solids 266-269, 1124–1128 (2000)
[72] Kawachi, G., Graeff, C.F.O., Brandt, M.S., Stutzmann, M.: Spin-dependent transport in Si thin-film transistors. In: Mat. Res. Soc. Symp. Proc., vol. 467, p. 851 (1997)
[73] Boehme, C., Behrends, J., von Maydell, K., Schmidt, M., Lips, K.: Investigation of hopping transport in n-a-Si:H/c-Si solar cells with pulsed electrically detected magnetic resonance. Journal of Non-Crystalline Solids 352, 1113–1116 (2006)
[74] Boehme, C., Lips, K.: Theory of time-domain measurement of spin-dependent recombination with pulsed electrically detected magnetic resonance. Phys. Rev. B 68, 245105 (2003)

Chapter 12
Band Lineup Theories and the Determination of Band Offsets from Electrical Measurements

Jean-Paul Kleider

Laboratoire de Génie Electrique de Paris
CNRS UMR8507; SUPELEC; Univ. Paris-Sud; UPMC Univ. Paris 06;
11, Rue Joliot-Curie, Plateau de Moulon, 91192 Gif-sur-Yvette Cedex, France

Abstract. Semiconductor heterojunctions have been used in the last decades to build devices with enhanced electrical or optoelectrical properties compared to those of equivalent homojunction devices. Examples of heterojunction devices are encountered in laser applications using band gap engineering possibilities in crystalline III-V compounds, and in bipolar transistors in crystalline silicon based electronics. More recently, heterojunctions formed between hydrogenated amorphous silicon (a-Si:H) and crystalline silicon (c-Si) were introduced for the fabrication of silicon solar cells. The main advantage of heterojunctions over homojunctions is due to the band offsets that can provide selective barriers for one type of carriers. The determination of band offsets and band lineup at the interface is thus of crucial importance. A lot of theoretical work has been devoted to this issue. In parallel, various characterization techniques have been developed to provide experimental insight into band offsets. In this chapter the principal models for band lineup at interfaces are recalled, with particular emphasis on Anderson's electron affinity rule and Tersoff's branch-point energy alignment theory. The application to the a-Si:H/c-Si system is discussed. Then, the principal electrical characterization tools based on capacitance and admittance measurements are presented. After a general overview of the widely used capacitance versus bias voltage technique (so-called C-V or $1/C^2$ method), the main potential problems and sources of uncertainty when applying this technique to the a-Si:H/c-Si system are addressed. Some features specific to the a-Si:H/c-Si interface are identified and illustrated using both numerical simulations and experimental data. These features are related to the amorphous nature of a-Si:H, e.g. the high density of band gap states, and to the existence of a strong inversion regime at the c-Si surface that can lead to two dimensional electron or hole gases. A simple technique based on the measurement of the planar conductance of a-Si:H/c-Si structures is presented. The determination of band offsets from such measurements and related modelling on both (p) a-Si:H / (n) c-Si and (n) a-Si:H / (p) c-Si structures is discussed. For interfaces used in high efficiency solar cells the band offsets are found to be 0.15 eV for the conduction band and 0.40 eV for the valence band.

W.G.J.H.M. van Sark et al. (Eds.): Physics & Tech. of Amorphous-Crystalline, EM, pp. 405–444.
springerlink.com

12.1 Introduction

There are several key parameters to make a good solar cell. One should avoid recombination of the photogenerated carriers and provide good selective collection of these carriers. Thus, in a usual crystalline silicon (c-Si) solar cell, the silicon material should possess long carrier lifetimes and suitable contacts for carrier collection. With n-type silicon this is generally achieved in the usual one-dimensional (1D) solar cell geometry where carriers are extracted at front and back contacts using a front p-n junction and a back $n\text{-}n^+$ junction forming the so-called back-surface field (BSF). The p-n junction provides a barrier favoring collection of holes and repelling electrons, while the $n\text{-}n^+$ junction provides a barrier at the back electrode favoring collection of electrons and repelling holes, as schematically depicted in Fig. 12.1. Device operation can be greatly improved by using heterojunctions instead of usual homojunctions. Indeed, a semiconductor heterojunction can be used to control selectively the carrier transport across the junction thanks to the band offsets. In heterojunctions formed on c-Si using hydrogenated amorphous silicon (a-Si:H), the larger bandgap of a-Si:H introduces band offsets that can help repelling one type of carriers and minimizing their loss, as depicted in Fig. 12.2. At the (p) a-Si:H / (n) c-Si heterojunction the conduction band offset can decrease the loss of electrons in c-Si that could diffuse into the a-Si:H layer, while the valence band offset can decrease the loss of holes in c-Si at the (n) c-Si / (n) a-Si:H heterojunction. However, the valence band offset could be a drawback for the collection of holes at the (p) a-Si:H / (n) c-Si heterojunction, while the conduction band offset could be a drawback for the collection of electrons at the (n) c-Si / (n) a-Si:H heterojunction. This is why these band offsets should not be too large; additionally, tunneling processes through the spikes generated by the band bending, e.g. in the valence band at the (p) a-Si:H / (n) c-Si interface in Fig. 12.2, could also be involved.

It is thus obvious that a precise determination of band offsets between a-Si:H and c-Si is of major importance both to understand the heterojunction solar cell operation, and to be able to provide a good analytical or numerical modeling. This is also important in order to be able to evaluate the potential of such solar cells in terms of energy conversion efficiency, and to study the influence of material parameters and properties, as can be done for instance in the AFORS-HET simulation tool [1]. In this chapter, a short overview of band lineup theories will be given. Their potential applicability to the a-Si:H/c-Si system will be discussed. Then, the determination of band offsets from electrical characterization techniques will be addressed. The capacitance techniques will be analysed and a powerful technique based on planar conductance measurements will be detailed.

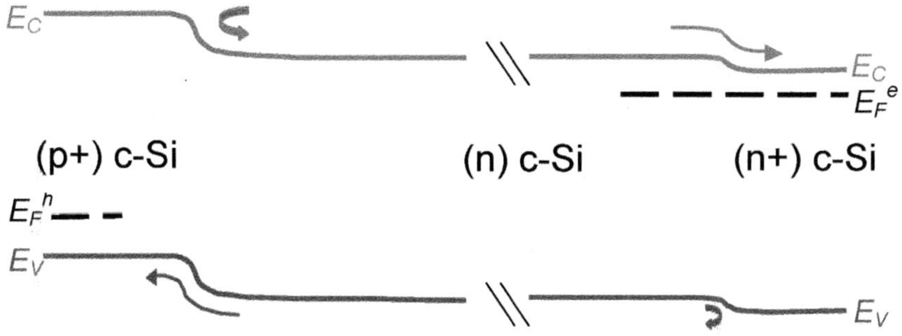

Fig. 12.1 Schematic band diagram of a traditional p$^+$-n-n$^+$ homojunction silicon solar cell under illumination. The top and bottom lines represent the bottom of the conduction band, E_C, and the top of the valence band, E_V, respectively. An indication of the positions of the quasi Fermi levels for free electrons and free holes, $E_F{}^e$ and $E_F{}^h$, respectively, is given at the edges of the system. The arrows are indicative of the repelling or collecting properties for electrons and holes at the interfaces.

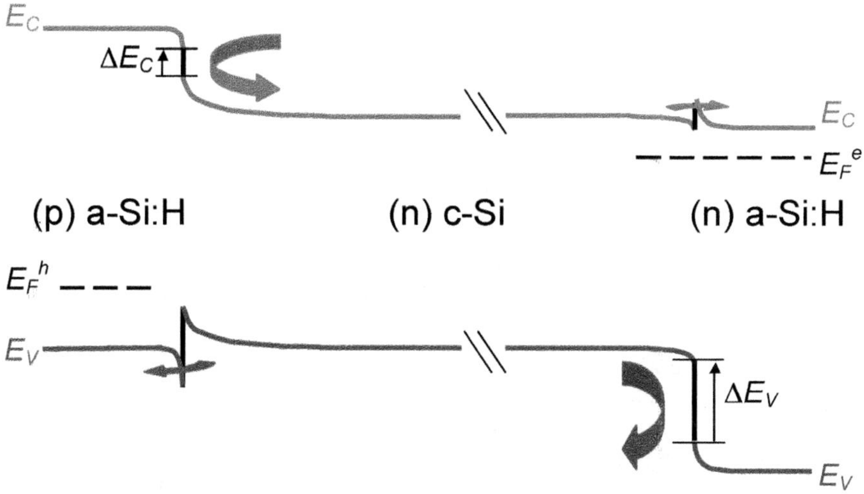

Fig. 12.2 Schematic band diagram of a (p) a-Si:H / (n) c-Si / (n) a-Si:H heterojunction solar cell under illumination. Same indications and notations as in Fig. 12.1. The conduction and valence band offsets at the interfaces are denoted ΔE_C and ΔE_V, respectively.

12.2 Band Lineup Theories and Their Applicability to the a-Si:H/c-Si System

The problem of band lineup is schematically described in Fig. 12.3. Being two semiconductors, SC A and SC B, how can we describe the equilibrium band diagram at the interface? If the system consisting of the two semiconductors in contact is in thermodynamic equilibrium, this implies that the Fermi level should be the same at any point. However, the condition of Fermi level alignment is not sufficient to properly describe how the bands are connected at the interface. The problem of band lineup has been widely discussed in literature. A good review of heterojunction band discontinuities has been published in 1987 in a book edited by Federico Capasso and Giorgio Margaritondo [2], where several chapters address the theoretical and measurement issues. This is why only a very short overview will be given here for the sake of self-consistency.

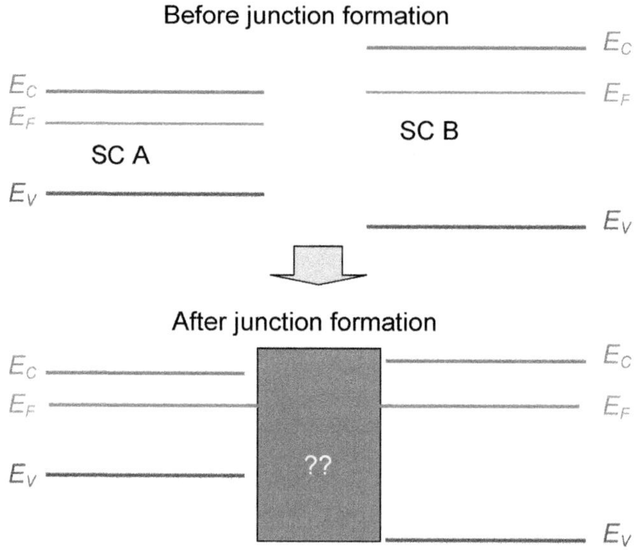

Fig. 12.3 Principle of band lineup between two semiconductors, SC A and SC B. Equilibrium band diagrams are shown before and after formation of the junction.

Band lineup theories always rely on assumptions and simplifications, sometimes on empirical rules. The basic idea of most theories is to introduce a reference level, E_r, in order to put the semiconductors on a common absolute energy scale, and then to align the reference levels in each semiconductor with each other for band lineup. A critical assumption and requirement is that E_r should be a well-defined bulk property, so that the bottom of the conduction band and top of valence band, E_C and E_V, respectively, can be easily described with respect to E_r in the bulk. The reference level approach applied to the SC A/ SC B heterojunction then states that

$$\Delta E_r(A,B) = E_r(B) - E_r(A) = 0. \qquad (12.1)$$

The conduction band offset $\Delta E_C(A,B)$ is thus equal to:

$$\Delta E_C(A,B) = E_C(B) - E_C(A) = \Delta E_C{}^r(A,B), \qquad (12.2)$$

where $E_C{}^r$ is the relative position of E_C referred to the reference level E_r, i.e.

$$E_C{}^r(\mathrm{A}) = E_C(A) - E_r(A). \qquad (12.3)$$

Similar relations hold for the valence band offset. The two main reference levels described in literature are the so-called vacuum level and branch-point energy. These are the starting points of two famous models, namely the electron affinity rule and the branch-point energy (or neutral energy) alignment model.

12.2.1 Electron Affinity Rule

The most familiar reference level is the vacuum level, E_{vac}, leading to the most popular theory, known as the *electron affinity rule*, or the *ionization potential rule*, or also *Anderson's model* [3]. The principle is described in Fig. 12.4.

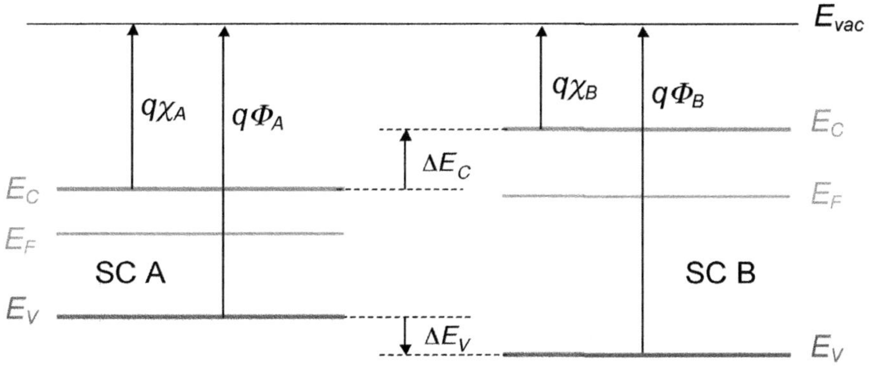

Fig. 12.4 Illustration of the electron affinity rule with the vacuum level being the reference level in calculating the band lineup, for the system consisting of the two semiconductors SC A and SC B.

E_C^r is directly related to the electron affinity, defined as

$$E_C^r(A) = -q\chi_A, \tag{12.4}$$

q being the absolute value of the electron charge. Thus the conduction band offset is simply given by the difference in the electron affinities:

$$\Delta E_C\,(A,B) = E_C(B) - E_C(A) = -q(\chi_B - \chi_A). \tag{12.5}$$

This model is also sometimes referred to as the ionization potential rule [4]. Indeed, the ionization potential of semiconductor A being Φ_A, the valence band maximum referred to the vacuum level is

$$E_V^r(A) = -q\Phi_A, \tag{12.6}$$

and according to eq. (12.2) Anderson's model leads to the valence band offset directly linked to the difference of ionization potentials:

$$\Delta E_V\,(A,B) = E_V(B) - E_V(A) = -q(\Phi_B - \Phi_A), \tag{12.7}$$

Of course, the schematic diagram of Fig. 12.4 is not compatible with a constant equilibrium Fermi level. The consistency with the requirement of a Fermi level being constant everywhere in the material at equilibrium comes with the introduction of band bending related to changes in the electrostatic potential in the vicinity of the junction. The full equilibrium diagram emphasizing the constant Fermi level and the band lineup is shown in Fig. 12.5. The local vacuum level is linked to the electrostatic potential. One can introduce a flat vacuum level $E_{vac\ \infty}$, that would correspond to the energy of an electron being at rest far away from any material – in a region where the electrostatic potential can be taken equal to zero. This can then be taken as the zero energy reference.

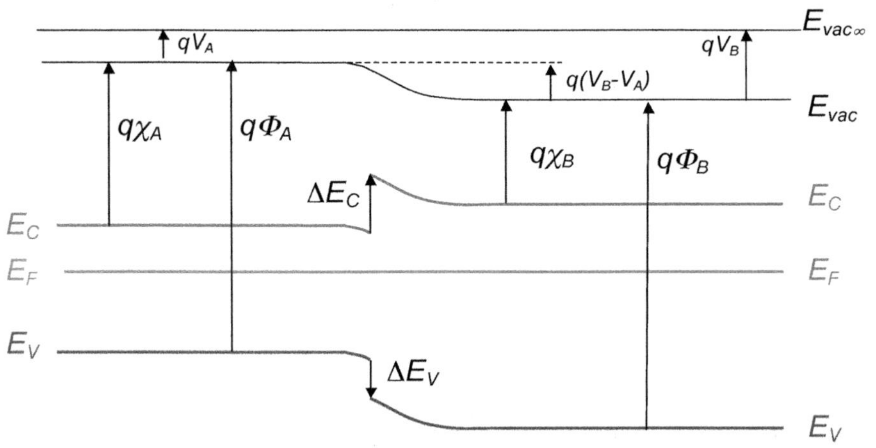

Fig. 12.5 Equilibrium band diagram at SC A/ SC B heterojunction in the framework of Anderson's model (electron affinity rule or ionization potential rule).

The problem with Anderson's model is that neither ionization potential nor electron affinity are true bulk properties. They depend on the structure of the surface, so are they affected by surface relaxation and surface reconstruction. Also, when the interface is formed, there may be charge transfer, microdiffusion of atoms across the interface, interface reconstruction or chemisorption. These phenomena can then induce formation of a dipole layer that is not accounted for in this simple theory. This will add an additional potential drop at the interface within atomic distance that changes the apparent electron affinity or ionization potential.

Also, the electron affinity rule can be regarded as an extension of the Mott-Schottky model of metal-semiconductor contacts [5, 6] that are a particular type of heterojunctions. Indeed, in that case, as can be seen in Fig. 12.6, the so-called Schottky barrier for electrons, $q\Phi_{Bn}$, defined as the difference in energy between the bottom of the conduction band of the semiconductor at the interface and the Fermi level, is given by

$$q\Phi_{Bn} = q(\Phi_M - \chi), \tag{12.8}$$

Φ_M being the metal work function. This is the Schottky-Mott rule. Comparing with eq. (12.5), the Schottky barrier for electrons in a Schottky junction is equivalent to the conduction band offset in a semiconductor heterojunction, where the metal work function plays the role of the electron affinity in the metal. In the same way, the Schottky barrier for holes $q\Phi_{Bp}$ can be defined as the difference in energy between the top of the valence band of the semiconductor at the interface and the Fermi level. In the Schottky-Mott rule, $q\Phi_{Bp}$ can simply be written in terms of the difference between the ionization energy of the semiconductor and the metal work function, yielding a relation similar to eq. (12.7) with the metal work function playing the role of ionization energy in the metal. In other words, the Schottky-Mott rule is the Anderson rule for vanishing bandgap energy.

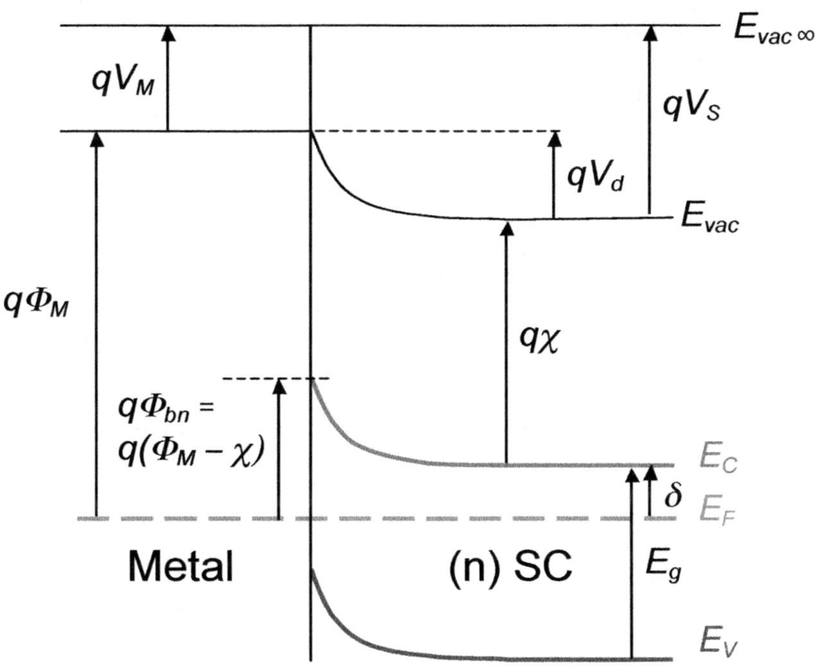

Fig. 12.6 Equilibrium band diagram of a Schottky barrier formed between a metal and an n-type semiconductor in the framework of the Schottky-Mott rule. Compared to the heterojunction between semiconductors the metal work function plays the role of both the electron affinity and the ionization potential.

According to eq. (12.8), for a given semiconductor, the Schottky barrier should vary linearly with the metal work function with a slope $d\Phi_{Bn}/d\Phi_M$ equal to 1. It has been recognized quite early that experimental data do not confirm this model (for a nice historical review, see [7]). In 1947 Bardeen proposed an explanation involving contact interface states [8], the main idea being that the Fermi level could be pinned at such interface states. In such a case, the Schottky barrier becomes independent of the metal work function, as opposed to the Schottky-Mott rule.

12.2.2 Interface Gap States, Branch-Point Energy and Neutral Lineup

A lot of work was devoted to the physical origin and to the explanation of surface or interface states. It was suggested in the early work of Heine [9] that in the energy range of the semiconductor forbidden gap metal wave functions tail into the semiconductor, giving rise to the so-called metal-induced gap states, or MIGS [10]. These tails are described by complex wave vectors (i.e. with an imaginary part) and they thus derive from the complex band structure of the semiconductor [11]. The gap states calculated from the complex band structure are called the virtual induced gap states (VIGS), and they have a density of states D_{VIGS} with singularities at E_C and E_V. Figure 12.7 gives a schematic illustration of the concept of virtual induced gap states that occur at a semiconductor surface or interface. More details on MIGS and VIGS can be found in reference textbooks [7, 12]. The VIGS derive from both conduction and valence bulk states, and they possess a donor or acceptor-like character. States lying in the upper part of the bandgap exhibit a more pronounced acceptor character, while those in the lower part of the gap have a more pronounced donor character. Therefore, if the Fermi level lies close to the conduction band a negative charge arises, while a positive charge arises if the Fermi level is close to the valence band. There does exist a neutrality level E_n, also called crossover energy or branch-point energy, E_{bp}. This energy separates the more acceptor-like from the more donor-like states. In a simple one-dimension model with symmetric bands E_{bp} lies at midgap.

E_{bp} is the central information in the neutral lineup model since it plays the role of the reference level. Calculations of the charge neutrality level of the interface-induced gap states have been performed by several authors and good correlations were found with both Schottky barrier heights and heterojunction band offsets [4, 13-18]. In particular, the ability of this theory to explain Schottky barrier values has been reviewed in detail by Mönch [7, 19].

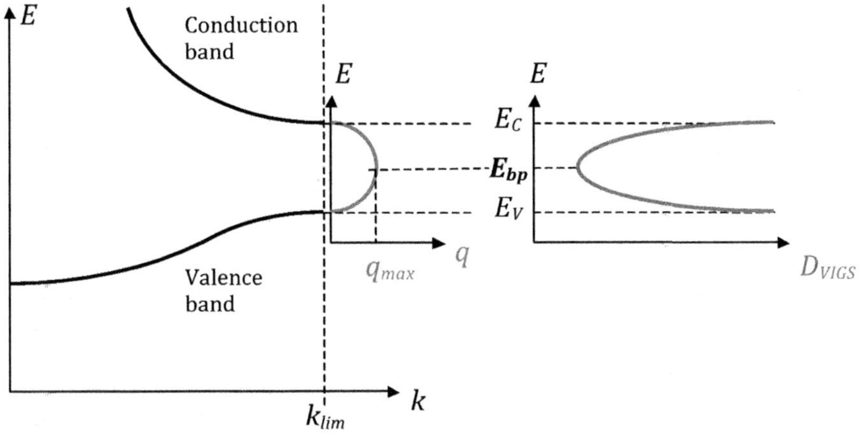

Fig. 12.7 Schematic illustration of the virtual induced gap states (VIGS) at a semiconductor surface or interface. The left side shows the bulk energy dispersion in the semiconductor as a function of wave vector k (in a one-dimensional system). The central part shows the VIGS energy as a function of complex wave vector q, while the right part shows the density of VIGS. There is a maximum value, q_{max}, of the complex wave vector that corresponds to the branch-point energy, E_{bp}.

12.2.3 Reconciling the Electron Affinity Rule and the Neutral Lineup Models

The two models for band lineup can be reconciled by considering interface dipoles and charge transfer effects. As an illustration, Tersoff developed a simple approach including dipole effects. He obtained the following expression for the valence band offset in a semiconductor heterojunction [4]:

$$\Delta E_V = \frac{\alpha}{1+\alpha}\Delta E_V^n + \frac{1}{1+\alpha}\Delta E_V^0, \qquad (12.9)$$

where ΔE_V^0 and ΔE_V^n are the values of the valence band offset predicted in the electron affinity rule and the neutral lineup models, respectively, and α is a dimensionless susceptibility. Obviously, the value predicted by the neutral lineup model prevails if $\alpha \to \infty$, while $\alpha = 0$ corresponds to the electron affinity rule. Tersoff proposed that α is simply related to the usual bulk dielectric susceptibility, ε, by $\alpha = \varepsilon - 1$. Since ε is of the order of 10 in semiconductors, the actual lineup given by eq. (12.9) is very close to that predicted by the neutral lineup, with only 10% contribution from the electron affinity rule. Note that, for (110) surfaces where electronic properties are assumed to have a metal-like behavior, Tejedor and Flores proposed a value for α of 2.5 [13] which gives a stronger impact of the affinity rule.

Interestingly, an interface dipole effect related to interface states was also proposed from a device physicist point of view for a Schottky junction by Cowley and Sze [20]. The preceding equation for band lineup in semiconductor heterojunctions can be compared to the equation derived by Cowley and Sze for the electron Schottky barrier height, Φ_{Bn}. If the small barrier lowering due to image force effect is neglected, Φ_{Bn} can be approximated by:

$$\Phi_{Bn} = \frac{\frac{d_i q^2 N_{SS}}{\varepsilon}}{1 + \frac{d_i q^2 N_{SS}}{\varepsilon}} \Phi_{Bn}^{n} + \frac{1}{1 + \frac{d_i q^2 N_{SS}}{\varepsilon}} (\Phi_M - \chi). \tag{12.10}$$

In this expression, Cowley and Sze took into account the contribution of surface states at the semiconductor surface having a constant density N_{ss}, separated from the metal by a thin interfacial layer of thickness d_i, as shown in Fig. 12.8.

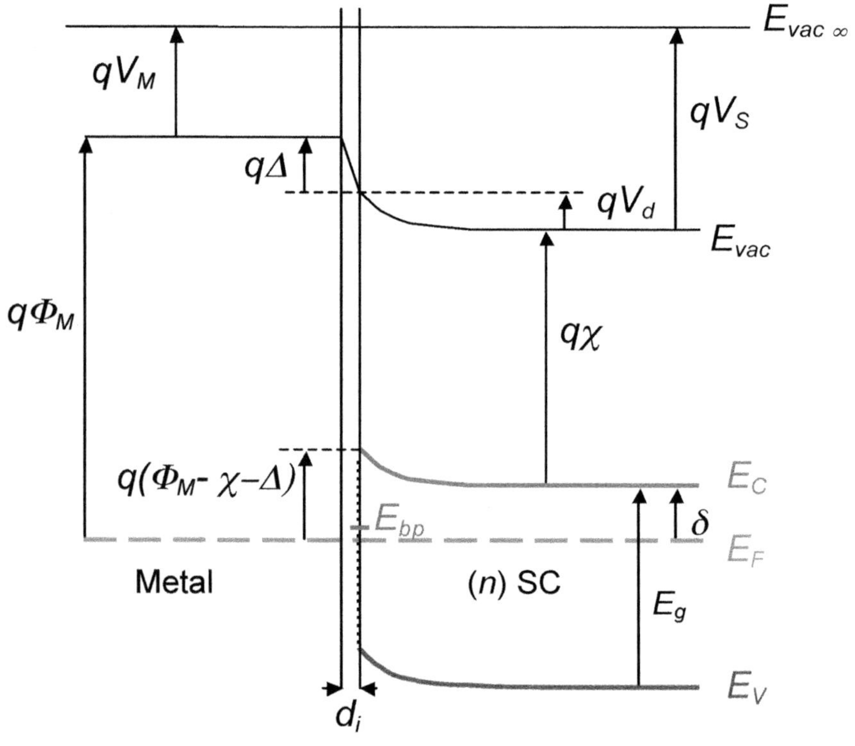

Fig. 12.8 Equilibrium band diagram of a metal-semiconductor contact taking account of surface states and an interfacial dielectric layer, as proposed by Cowley and Sze in 1965.

The Schottky barrier can thus be expressed as a function of the value, Φ_{Bn}^{n}, standing in the neutral lineup theory and of the value ($\Phi_M - \chi$) standing in the affinity rule (Mott-Schottky value). Equations (12.9) and (12.10) have obviously the same form. The dimensionless susceptibility introduced on the basis of theoretical and chemistry arguments by Tersoff is related to the surface state density in the device physicist point of view of Cowley and Sze. Compared to the ideal Mott-Schottky case of Fig. 12.6, the Schottky barrier is modified by the quantity $q\Delta$ linked to the interfacial layer that leads to a dipole effect. Thus one can say that, within the frame of the electron affinity rule, the electron affinity has to be replaced by an effective electron affinity $\chi+\Delta$ that includes the dipole effect.

12.2.4 Application to the a-Si:H/c-Si Interface

For crystalline semiconductors, the branch-point energy concept has been introduced as a neutrality point in the virtual induced gap states. These derive from the complex band structure originating from the breaking of bulk bond ordering at the semiconductor surface. In an amorphous semiconductor, however, there is already lack of ordering within the bulk, which leads to the existence of bulk gap states. Also, in bulk a-Si:H it has been shown that the density of silicon dangling bond states and their energetic distribution strongly depend on hydrogen concentration and on the Fermi level position, according to the defect-pool model [21, 22]. Possible extrapolations of this concept to the a-Si:H surface and to VIGS in a-Si:H, and calculations of branch-point energy values in a-Si:H are still open questions, and I will not go into theoretical calculations here. Some arguments can however be put forward from preceding calculations performed in crystalline semiconductors.

Indeed, it has been argued that the branch-point energy closely corresponds to a midgap energy [16], or to an average hybrid sp^3 energy [18, 23]. It was also suggested that the branch-point energy used for band lineup corresponds to the hybrid orbital energy [24]. Several calculations based on different methods and approximations (empirical tight-binding method, augmented plane-wave method and local-density approximation, ...) have been presented in literature. When looking at the values calculated for a large number of crystalline semiconductors the results obtained from these various calculations are quite close on average (see Table 5.1 in [7] or Table 8.1 in [12]). However, when focusing on a specific semiconductor, significant differences can be observed. For instance, for crystalline silicon, branch-point energies referred to the valence band maximum of 0.36, 0.23 and 0.03 eV were reported by Tersoff [15], Cardona and Christensen [18], and Mönch [7], respectively.

An indicative value of the band offsets in the a-Si:H/c-Si system using the neutral lineup model can be obtained using a branch-point energy in a-Si:H taken at midgap. This is illustrated in Fig. 12.9. With such an assumption regarding the branch-point in a-Si:H, values of 0.14 and 0.54 eV are obtained for the conduction band and valence band offset, respectively, for a bandgap energy of 1.8 eV in a-Si:H. These indicative values may depend on the hydrogen concentration since the bandgap also depends on it.

There are not many reports on theoretical calculations of band offsets at the a-Si:H/c-Si heterojunction. From a tight-binding approximation and a supercell of 4096 atoms, Allan and co-workers found a value of 0.36 eV for the valence band offset, for a hydrogen content of 8% [25], while Van de Walle and Yang obtained a value of 0.2 eV from ab initio calculations, with 11% hydrogen content in a-Si:H [26].

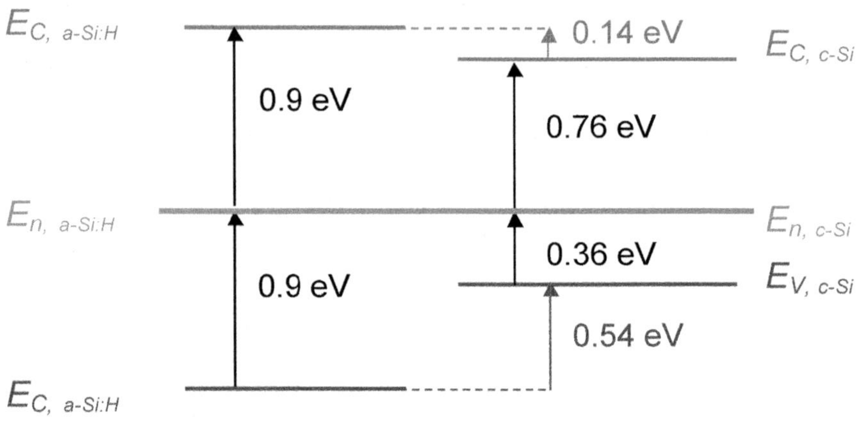

Fig. 12.9 Illustration of the neutral band lineup in the a-Si:H/c-Si system. The value of the branch-point energy in c-Si is taken from Tersoff. The branch-point energy in a-Si:H is taken at midgap for a bandgap energy of 1.8 eV.

Several extrinsic effects may also affect the band offsets. In particular, formation of an ultra-thin over-layer or intra-layer can modify the interface dipole and thus the band lineup. This can be achieved through various mechanisms like formation of interface bonds, saturation of dangling bonds, activation or inhibition of microdiffusion processes, etc. There are numerous examples of such effects in different kinds of surfaces and heterointerfaces. Just to give a few examples, formation of an ultra-thin Al intra-layer in ZnSe/Ge heterostructures can increase the valence band offset by more than 0.2 eV [27, 28]. Hydrogen was shown to play a crucial role at semiconductor surfaces. Hydrogen in diamond is known to produce very strong effects that lead to negative electron affinities [29], and to the decrease of the Schottky barrier height for holes by about 1.4 eV in Schottky diodes formed on p-type diamond [7]. Effects of the opposite sign were reported in Pb-Schottky diodes formed on c-Si(111), where a decrease of 0.35 eV of the Schottky barrier of electrons was reported [7]. The opposite sign of the changes in barrier heights caused by interfacial hydrogen in Pb/H/silicon and metal/H/diamond was explained by the different sign in the ionicities of H-C and H-Si interface molecules [4]. Indeed, the electronegativity of hydrogen (X_H=2.2 Pauling-units) is intermediate between the values of carbon (X_C=2.55) and silicon (X_{Si}=1.9). Thus, H-Si and H-C termination molecules produce opposite dipoles that could explain the opposite shift of Schottky barrier heights [4]. The role of hydrogen and surface reconstruction at the c-Si surface may depend on the orientation and on the preparation conditions of the surface. A review can be found in [30]. Hydrogen was also found to modify the valence band offset by up to 0.5 eV at the Si/SiO_2 interface [31]. Regarding the a-Si:H/c-Si structure, it bas to be noted that hydrogen is naturally present in a-Si:H and it plays a crucial role at the c-Si surface to passivate silicon dangling bonds. The effect on band offsets of surface preparation and deposition conditions of a-Si:H is however still a matter of debate.

Owing to the uncertainties in the theoretical models, the lack of precise knowledge of the branch-point energy in a-Si:H and the extrinsic effects of intra-layers, the reasonable conclusion is that, at the present stage of knowledge, theory is not able to predict the conduction and valence band offsets in the a-Si:H/c-Si system with the precision and reliability that are required to properly model the device operation. This gives an even more important role and motivation for reliable experimental band offset determinations.

12.3 Band Offsets in the a-Si:H/c-Si System

12.3.1 Literature Survey of Experimental Work

There has been a lot of experimental work on the a-Si:H/c-Si system in the past three decades. Several kinds of techniques have been employed in order to study band offsets in this system: photoyield spectroscopy, internal photoemission, spectral response, capacitance and conductance measurements.

Table 12.1. Experimentally determined conduction and valence band offset values, ΔE_C and ΔE_V, reported in the literature. Indicated are: the type of interfaces that were analysed, X/y where X indicates the type of doping of a-Si:H and y the type of doping of c-Si (I stands for the so-called "intrinsic" layer, meaning that the a-Si:H layer is intentionally undoped); the type of experiment that was used: IP: Internal Photoemission; SR: Spectral Response; C-V: Capacitance vs voltage; CS: Capacitance spectroscopy; IV: Current-Voltage characteristics; VFP: Voltage Filling Pulse method; AS: Admittance spectroscopy; PS: Photoyield spectroscopy; PC: Planar Conductance.

ΔE_C (eV)	ΔE_V (eV)	Structure	Method	Reference
0.09	0.71	I/n, I/p	IP	[34]
0.4...0.8	-0.1...+0.15	I/p	IP	[35, 36]
-	0	I/n, I/p	IP	[37]
0.24	-0.06	P/n	IP	[38]
0.14	0.63	I/n	IP	[39]
0.15...0.175	0.46...0.49	P/n, I/n	SR	[40]
0.06	large	I/n	SR	[41]
0.20	-	N/p	C-V	[42]
0.14	-	I/p	C-V	[43]
0.45	-	I/p	C-V	[44]
0.01	0.67	N/p	CS	[45]
0.05	0.58	I/n	CS, VFP	[46]
0.25	-	N/p	C-V	[47]
0.35	-	N/p	CS, IV	[48]
0.05	-	I/n	AS	[49]
-	0.44	I/p	PS	[50]
0.14	0.46	N/p, P/n	PS	[51, 52]
0.15	-	N/p	PC	[53]

Table 12.1 gives a list of reported values for conduction and band offsets in a-Si:H/c-Si heterojunctions (see also Chapter 6 by Korte in this book, [32]). The striking feature when looking at this table is that the values of band offsets spread over a quite large interval, with an average tendency of ΔE_C being smaller than ΔE_V. Such a spread could arise from different preparation conditions of the a-Si:H/c-Si interface, since it has been argued in the preceding section that intra-layers and dipole effects can play a crucial role. However, part of such a spread is probably related to misinterpretations of the experimental techniques. To cite a sentence from a publication of Herbert Kroemer, who did a huge work on heterojunctions: *"There does not exist any experimental technique to determine band offsets that is simultaneously simple, reliable and universally applicable."* [33]. Indeed, one has to take care of the specific situation (type of materials, etc.) and reconsider the analysis of the techniques accordingly.

The use of the photoyield or UV-photoelectron spectroscopy, which is one of the more powerful techniques for band offset determinations, is detailed and illustrated in Chapter 6 of this book [32].

I will focus in the following on some electronic characterization techniques.

12.3.2 Electronic Characterization Techniques

Capacitance and current measurement techniques for the determination of band offsets have been widely used and reviewed [54, 55]. The basic ideas of using such techniques are that (i) band offsets can play a dominant role in determining the current across the junction and the carrier collection (b) band offsets will influence the electrostatic potential drops on both sides of the junction and these may be reflected in some capacitance measurements.

Current across the junction can be related to and determined by the band offsets. More specifically, for solar cells, it has been shown that large band offsets can be the origin of the so-called S-shape in the I-V (current versus voltage) characteristics under illumination [56]. However, a precise determination of band offsets from I-V curves is not possible because of tunneling phenomena. Indeed, tunneling can partly shunt the barrier that carriers should normally overcome. To what extent this tunnel-induced shunting process does affect the current depends on many parameters (the type of tunneling, the density of defects, etc.). So I will not address further the use of current measurements *across* the junction for the determination of band offsets. Instead, I will focus on capacitance measurements and on *planar* conductance measurements.

12.3.2.1 Capacitance Techniques

12.3.2.1.1 Basic Ideas

For band offset determination using capacitance techniques, one can distinguish measurements on isotype heterojunctions (i.e. heterojunctions between either two p-type or two n-type semiconductors) from that on anisotype heterojunctions. The former have been used in combination with an adjacent Schottky barrier. The idea in this case is to vary the width of the Schottky space charge region by varying the applied voltage in such a way that it sweeps across the heterojunction. This technique has been described and used in particular in III-V compounds by Kroemer and co-workers who refined the analysis [57]. To my knowledge this technique has not been used on isotype a-Si:H/c-Si junctions, and I will rather consider the case of anisotype heterojunctions.

The equilibrium band diagram for the heterojunction formed between an n-type wide bandgap semiconductor and a p-type small bandgap semiconductor (N/p junction), that is illustrative of (n) a-Si:H / (p) c-Si, is shown in Fig. 12.10.

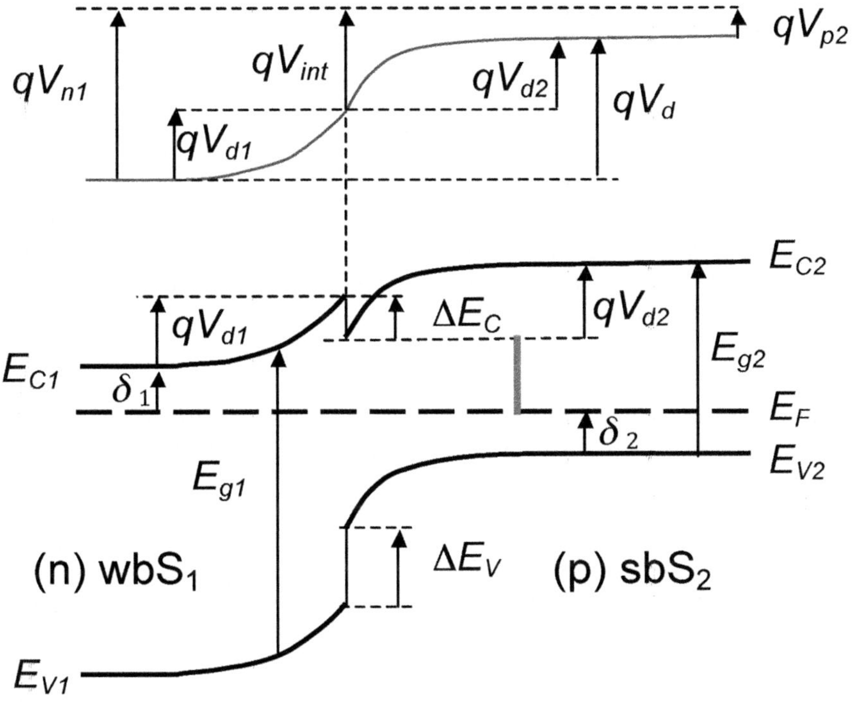

Fig. 12.10 Schematic equilibrium band diagram for the N/p anisotype heterojunction between an n-type wide band gap semiconductor, (n) wbS_1, and a p-type small band gap semiconductor, (p) sbS_2; this is representative of the (n) a-Si:H/ (p) c-Si structure. The top (blue) curve shows the electrostatic potential distribution accross the junction.

One can write the energy difference $E_{C2}–E_F$ at the interface, indicated by the bold (red) bar in Fig. 12.10, in two different ways:

$$\delta_1 + qV_{d1} - \Delta E_C = E_{g2} - \delta_2 - qV_{d2}. \tag{12.11}$$

Here ΔE_C is the conduction band offset (counted positively here from the conduction band of the wide bandgap semiconductor to that of the small bandgap semiconductor, $\Delta E_C = E_{C1}–E_{C2}$ at the interface); $\delta_1 = E_{C1}–E_F$ and $\delta_2 = E_F–E_{V2}$ are the bulk Fermi level positions in each semiconductor referred to the corresponding majority carrier band, which mainly depend on the semiconductor doping; E_{g2} is the bandgap energy of the p-type small bandgap semiconductor; V_{d1} and V_{d2} are the (positive) electrostatic potential drops, so-called diffusion potentials on each side of the junction. With the total diffusion potential accross the junction, $V_d = V_{d1} + V_{d2}$, the conduction band offset can be written

$$\Delta E_C = \delta_1 + \delta_2 - E_{g2} + qV_d. \tag{12.12}$$

Since the values of bandgap and bulk Fermi level positions are generally known parameters, it follows that the conduction band offset can be deduced from eq. (12.12) if the total diffusion potential can be measured.

Symmetrically, the band diagram for the heterojunction formed between a p-type wide bandgap semiconductor and an n-type small bandgap semiconductor (P/n) that is illustrative of (p) a-Si:H / (n) c-Si is shown in Fig. 12.11.

In that case one can easily obtain:

$$\Delta E_V = \delta_1 + \delta_2 - E_{g2} + qV_d, \tag{12.13}$$

where ΔE_V is the valence band offset (which is counted positively here from the valence band of the small bandgap semiconductor to that of the wide bandgap semiconductor, i.e. $\Delta E_V = E_{V2}–E_{V1}$ at the interface); δ_1 and δ_2 still correspond to the bulk Fermi level positions referred to the corresponding majority carrier band, which are now $\delta_1 = E_F - E_{V1}$ and $\delta_2 = E_{C2}–E_F$. Equation (12.13) shows that measurement of the total diffusion potential in a P/n anisotype junction leads to the determination of the valence band offset in a similar way as the determination of the conduction band offset from eq. (12.12) in an N/p heterojunction.

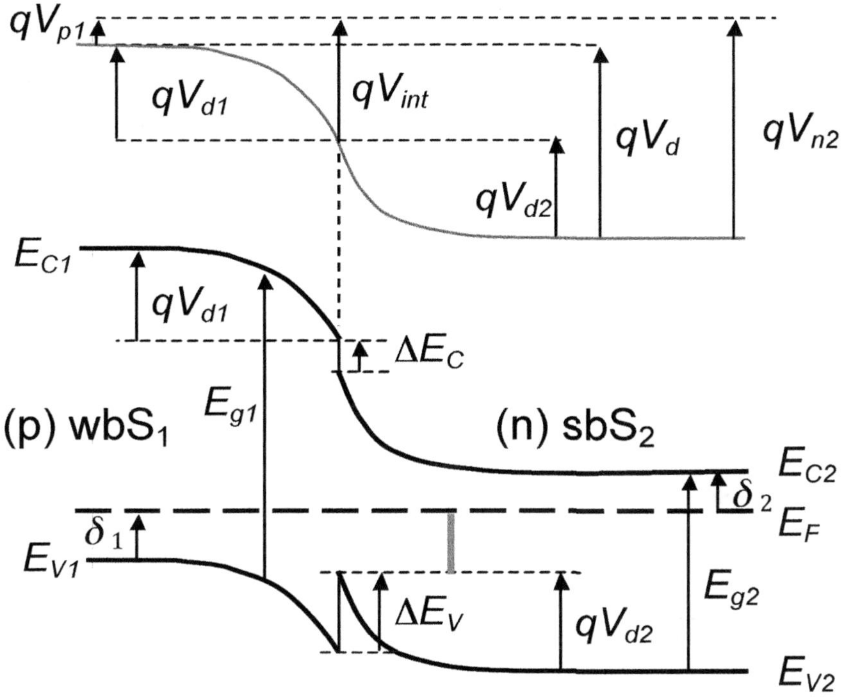

Fig. 12.11 Schematic equilibrium band diagram for the P/n anisotype heterojunction between a p-type wide band gap semiconductor, (p) wbS_1, and an n-type small band gap semiconductor, (n) sbS_2; this is representative of the (p) a-Si:H/ (n) c-Si structure. The top (blue) curve shows the electrostatic potential distribution accross the junction.

In a crystalline p/n junction with acceptor and donor doping densities N_{a1} and N_{d2} on the p- and n- side, respectively, the traditional depletion approximation states that the depletion of free carriers leaves a space charge density that can be considered as consisting in two slabs with constant values, equal to $-qN_{a1}$ within a distance x_{p1} on the p-side, and equal to qN_{d2} within a distance x_{n2} on the n-side, x being the coordinate perpendicular to the junction, the zero abscissa being taken at the interface. In order to measure the junction capacitance a small ac bias is superposed to a dc bias voltage. This ac bias produces variations in the space charge density. These occur at the edges of the depletion region. The capacitance can thus be seen as the series of the capacitances associated to the depletion widths in each semiconductor, $C_p=\varepsilon_1 A/x_{p1}$ and $C_n=\varepsilon_2 A/x_{n2}$, respectively, ε_1 and ε_2 being the corresponding dielectric permittivities and A the diode area. A schematic illustration is given in Fig. 12.12. Expressions of the widths of the space charge density on each side of the junction are easily found within the depletion approximation:

$$x_{1p}^2 = \frac{2}{q} \frac{N_{d2}\varepsilon_1\varepsilon_2\left(V_d - V_{app}\right)}{N_{a1}\left(\varepsilon_1 N_{a1} + \varepsilon_2 N_{d2}\right)}, \tag{12.14}$$

$$x_{2n}^2 = \frac{2}{q}\frac{N_{a1}\varepsilon_1\varepsilon_2\left(V_d - V_{app}\right)}{N_{d2}\left(\varepsilon_1 N_{a1} + \varepsilon_2 N_{d2}\right)}, \tag{12.15}$$

where V_{app} is the applied dc bias voltage. The junction capacitance is thus given by

$$\frac{A}{C} = \sqrt{\frac{2}{q}\frac{\left(\varepsilon_1 N_{a1} + \varepsilon_2 N_{d2}\right)\left(V_d - V_{app}\right)}{\left(\varepsilon_1\varepsilon_2 N_{a1} N_{d2}\right)}}. \tag{12.16}$$

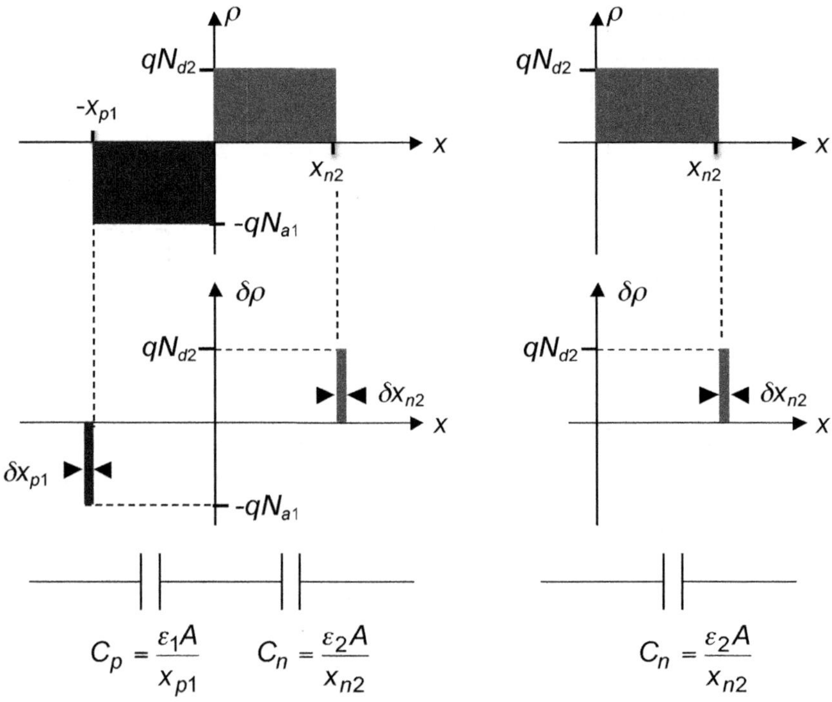

Fig. 12.12 Static space charge density, ρ, ac space charge density, $\delta\rho$, and equivalent capacitance circuit in the depletion approximation for a p-n junction (left) and a Schottky barrier (right) with crystalline semiconductors.

Thus the square of the inverse capacitance varies linearly as a function of the applied voltage and the intercept of the extrapolated value to zero is equal to the diffusion potential. This is the basis of the C-V method [58]. A potential drop of $2k_BT/q$ may be subtracted to V_d (k_BT/q for each side of the junction) to account for the non abrupt variation of the space charge density at the borders of the space charge regions if the contribution of majority free carriers is taken into account [58], but this does not change the fundamentals of the method and it will be neglected here. Since only the space charge junction capacitance has been considered, measurements have to be taken in the dark and at reverse or moderate forward bias in order to avoid a contribution from the diffusion capacitance that varies exponentially with the dc voltage and becomes dominant at higher forward bias. When doping of the p-side is stronger than that of the n-side, the depletion width x_{p1} is smaller than x_{n2} and the capacitance C_p is higher than C_n. Thus the measured capacitance is more sensitive to the side of the junction that has the larger space charge width. For very high space charge density and very small space charge width on the p-side, the situation becomes similar to that of an equivalent Schottky barrier, as depicted in Fig. 12.12, and the inverse capacitance then reads:

$$\frac{A}{C}=\sqrt{\frac{2}{q}\frac{\left(V_d-V_{app}\right)}{\left(\varepsilon_2 N_{d2}\right)}}. \tag{12.17}$$

12.3.2.1.2 Specific a-Si:H Issues

The technique has been widely used for crystalline heterojunctions, especially in III-V compounds, and it is tempting to use it for the a-Si:H/c-Si case. However, two issues have to be considered. First, since a-Si:H is not a crystalline semiconductor, it contains a huge quantity of bulk defects within the bandgap, and the space charge density is determined by the localized gap states around the bulk Fermi level rather than by the doping density. Therefore the simple depletion layer expressions are no longer valid. The second issue, also related to the bandgap states, concerns the dynamic response to an ac voltage. Indeed, in a-Si:H changes in the space charge density are linked to capture/emission of charges into/from defects around the Fermi level. This has been studied in detail by several authors [59-61]. In order to emphasize the changes that have to be taken into account for a-Si:H with respect to a crystalline semiconductor, a simple illustration is given in Fig. 12.13 for a Schottky barrier on undoped or n-type a-Si:H.

Apart from a narrow region close to the interface, the space charge density can be mainly divided into two regions. These are separated by the abscissa x_i^* defined as the point where the bulk Fermi level crosses the level E_i^* that corresponds to the energy where the emission frequency of electrons towards the conduction band, e_n, equals the emission frequency of holes towards the valence band, e_p. This level is close to midgap. For $x > x_i^*$, the gap states occupancy is determined by the bulk Fermi level E_F that corresponds to the quasi-Fermi level for trapped electrons. For $x < x_i^*$, states are essentially full below E_i^* and empty above, and E_i^* plays the role of the quasi-Fermi level with however an occupation function slightly different from the usual Fermi-Dirac form [59-61]. The space charge density is constant for $x < x_i^*$, equal to $\rho_i^*=qN_i^*$, where N_i^* is the integral of the density of states (DOS) between the bulk Fermi level and E_i^*. This is because all gap states lying between E_F and E_i^* are unoccupied in this region, while being occupied in the bulk. For $x > x_i^*$, the space charge density decreases with increasing x, corresponding to less gap states being unoccupied. Note that if the quasi Fermi level for trapped electrons does not cross E_i^*, which may be the case for small band bending, then $x_i^* = 0$.

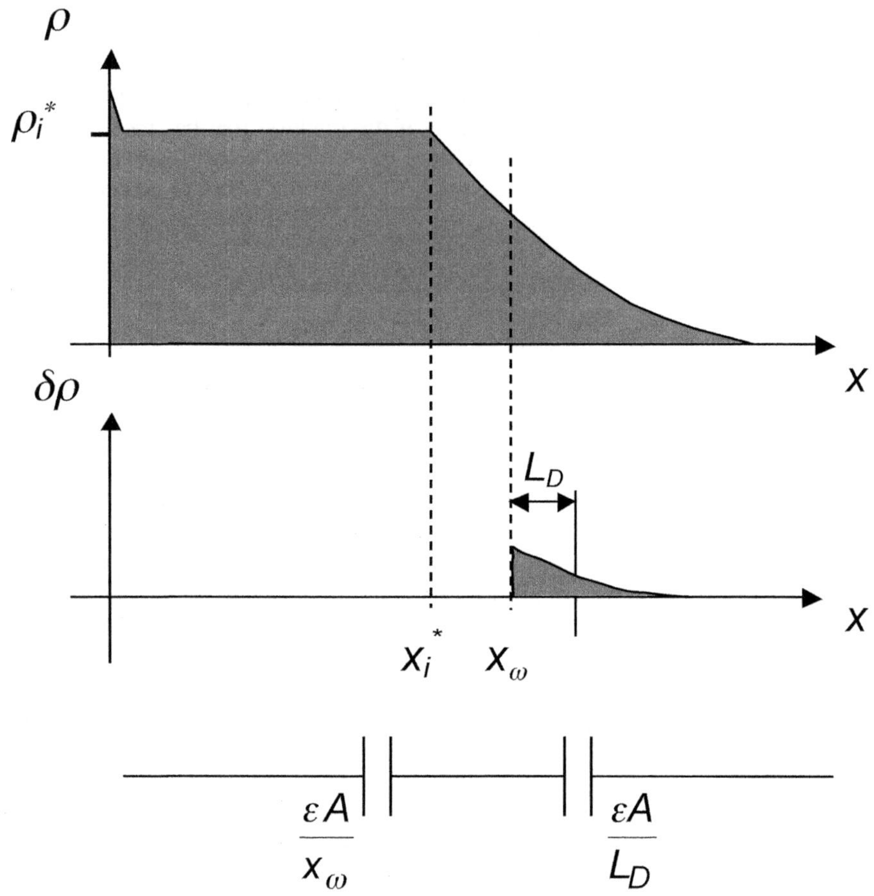

Fig. 12.13 Static space charge density, ρ, ac space charge density, $\delta\rho$, and equivalent capacitance circuit for a Schottky barrier formed on undoped or n-type a-Si:H.

Regarding the ac changes in the space charge density related to the ac voltage, one has to consider the dynamics of gap states. While exchanges of holes can be neglected in the two considered regions since only states at energies $E > E_i^*$ play a role, exchanges of electrons with the conduction band for states at an energy E are characterized by a time constant $\tau(E, x)$ that involves the capture and emission of electrons, such that $\tau(E,x) = 1/(c_n n(x)+e_n(E))$, where c_n is the capture coefficient, $n(x)$ the free electron concentration at x and $e_n(E)$ the emission frequency towards the conduction band. States at the Fermi level thus have a characteristic time given by:

$$\frac{1}{\tau_c(x)} = 2e_n(E_F, x) = 2\nu_n \exp\left(-\frac{E_C(x) - E_F}{k_B T}\right), \tag{12.18}$$

where v_n is the attempt-to-escape frequency, k_B Boltzmann's constant and T the temperature. A cut-off abscissa x_ω for the response of gap states is then defined by

$$\frac{1}{\tau_c(x_\omega)} = \omega, \tag{12.19}$$

where ω is the angular frequency of the ac voltage. Due to the band bending and the strong dependence of $\tau_c(x_\omega)$ on x, for $x > x_\omega$, states can be considered as "fast enough" to follow the ac voltage, while for $x < x_\omega$, they cannot respond to the ac voltage and can be considered as frozen-in. The ac space charge density has a characteristic width equal to the Debye length L_D, defined as $(L_D)^2 = \varepsilon/(q^2 g(E_F))$, where $g(E_F)$ is the DOS at the bulk Fermi level. Within this simple cut-off approach, there is no gap states response for $x < x_\omega$, while there is a full gap states response for $x > x_\omega$ with a characteristic width L_D. Therefore the junction capacitance can be regarded as the series of the capacitance $\varepsilon A/x_\omega$, corresponding to the frozen-in region, and of the capacitance $\varepsilon A/L_D$ expressing the full response of gap states for $x > x_\omega$. This simple cut-off approach is a zero-temperature like approximation that should be corrected for finite temperature effects; also, corrections to the expression of the capacitance should include the finite thickness of the amorphous layer. These are however details that do not need to be included here for the comprehension of the phenomena.

Looking at the temperature dependence of the characteristic response time, it appears that τ_c increases with decreasing temperature, so that the cut-off abscissa x_ω also increases since shallower states become unable to respond on the time scale imposed by the measurement frequency. Thus there exists a turn-on temperature T_{on}, defined by

$$2v_n \exp\left(-\frac{E_{CB} - E_F}{k_B T_{on}}\right) = \omega, \tag{12.20}$$

where E_{CB}–E_F is the value of E_C–E_F in the bulk. For $T < T_{on}$, no states at all can respond in a-Si:H, that then behaves like an insulator with a capacitance equal to the geometric dielectric layer capacitance $C_{diel} = \varepsilon A/d_{a\text{-}Si:H}$, where $d_{a\text{-}Si:H}$ is the thickness of the a-Si:H layer. This is the lowest capacitance of the a-Si:H layer. On the contrary, increasing the temperature leads to a movement of the cut-off abscissa towards the junction and an increase of the capacitance until x_ω reaches x_i^*. At this stage, all the states that could potentially respond are actually responding, and a further increase in temperature does not lead to an increase of the gap states response since the occupancy of gap states for $x < x_i^*$ is independent of the Fermi level and cannot be modulated by the bias modulation.

If one increases the modulation frequency the cut-off abscissa is shifted further into the bulk, and the whole capacitance versus temperature curve will be shifted towards higher temperatures, as schematically illustrated in Fig. 12.14.

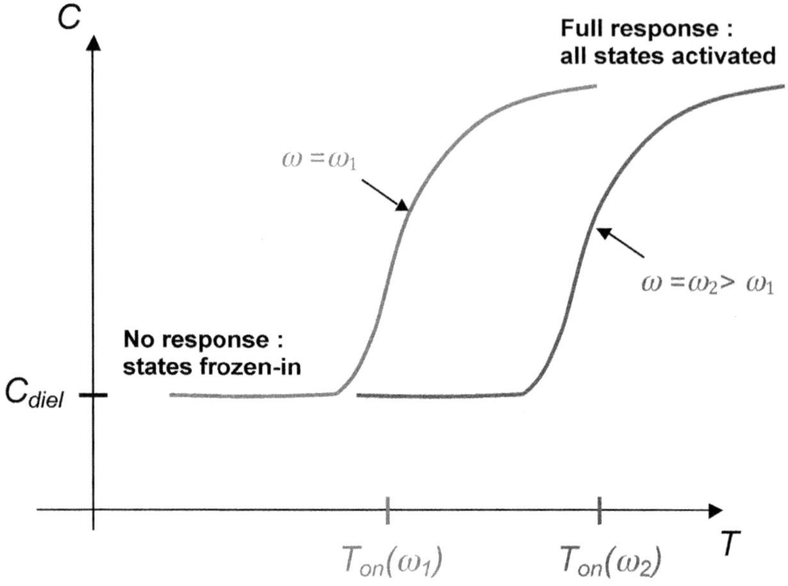

Fig. 12.14 Schematic capacitance variation in an a-Si:H depletion layer as a function of temperature for two angular frequencies.

Note that the above theory assumes that the limitation of the ac response in a-Si:H comes from the capture and emission processes. Another limitation could come from the transport in a-Si:H. In that case, the characteristic time is the dielectric relaxation time rather than the capture and emission time. Expressions of the cut-off abscissa and a-Si:H capacitance would then be different. However, the step-like behaviour of Fig. 12.14 separating a frozen-in region and a full response region would still be valid because the dielectric relaxation time has the same kind of activation energy as τ_c in eq. (12.18).

If one would consider a Schottky barrier on p-type a-Si:H, the same kind of behaviour would be expected, with appropriate changes in the signs and modulation of holes instead of electrons.

12.3.2.1.3 Application to the a-Si:H/c-Si System

Let us now apply the physical background described above for a-Si:H Schottky diodes to the case of a (p) a-Si:H / (n) c-Si heterojunction. The expected simplified space charge densities and the equivalent junction capacitance circuit are illustrated in Fig. 12.15.

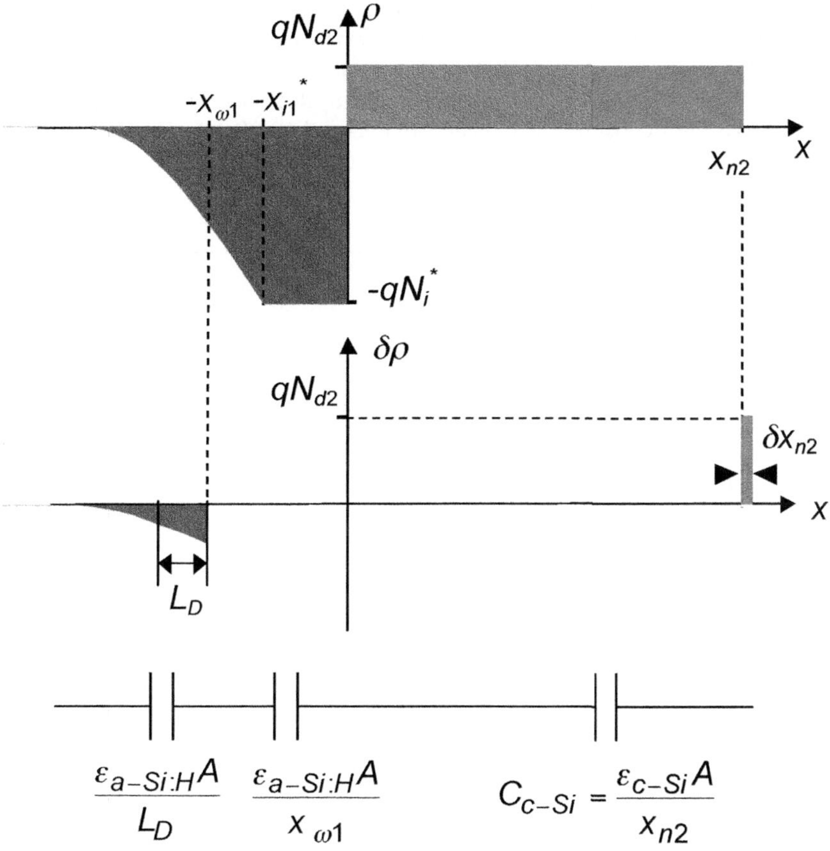

Fig. 12.15 Schematic illustration of the static space charge density, ρ, ac space charge density, $\delta\rho$, and equivalent capacitance circuit in a (p) a-Si:H/ (n) c-Si heterojunction. The dielectric permittivity in a-Si:H can be taken equal to that of c-Si, $\varepsilon_{a\text{-}Si:H} = \varepsilon_{c\text{-}Si} = \varepsilon$.

Compared to the simple case of a pure crystalline p-n heterojunction, two important features are worth noting.

Firstly, the ac charge density in a-Si:H depends on temperature and frequency. As a consequence, we expect a step-like behaviour in the capacitance versus temperature plot, just like for an a-Si:H Schottky diode. The same also applies to a (n) a-Si:H / (p) c-Si heterojunction. Such a capacitance step behaviour has indeed been observed experimentally on silicon heterojunctions formed with undoped a-Si:H [49] as well as with doped a-Si:H, and it has been confirmed in simulations [62]. An example of simulated capacitance versus temperature curves for (p) a-Si:H / (n) c-Si heterojunctions is shown in Fig. 12.16 for two values of the a-Si:H thickness and three frequencies. One can observe the predicted step-like behaviour and shift to higher temperature of the curves when increasing the measurement frequency. Also, one can observe that the low capacitance plateau obtained at low temperature when all states in a-Si:H are frozen-in depends on the a-Si:H thickness, while the high capacitance plateau obtained when all a-Si:H states are responding does not depend on it, as expected from the above physical arguments.

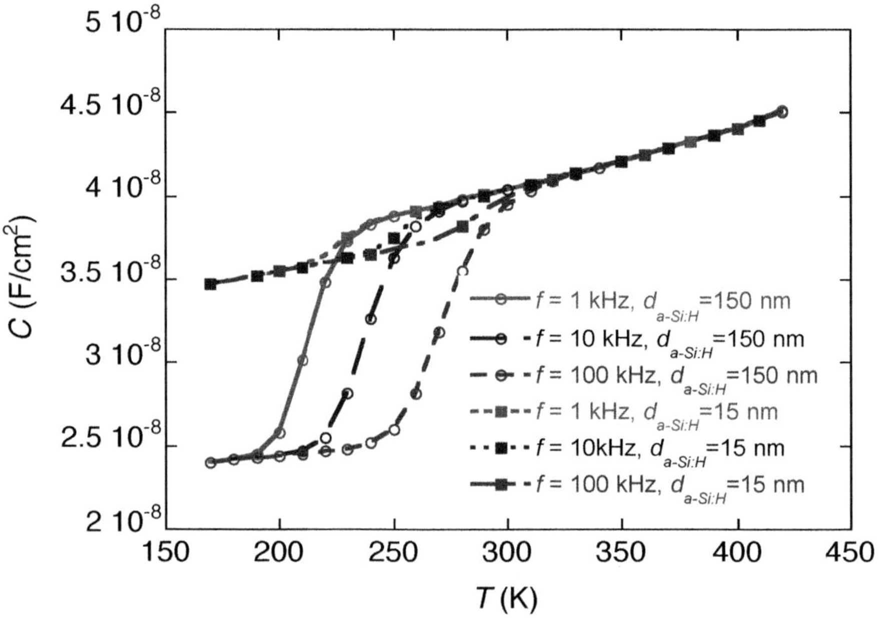

Fig. 12.16 Capacitance of a (p) a-Si:H / (n) c-Si heterojunction calculated in the dark at zero dc bias as a function of temperature, for two values of a-Si:H layer thickness (15 nm and 150 nm), and three values of frequency (1 kHz, 10 kHz, 100 kHz).

Secondly, due to the peculiar charge distribution in a-Si:H, one cannot use the simple eqs. (12.14) and (12.15) to calculate the widths of the space charge regions. Therefore the simple bias dependence of the total capacitance expressed by eq. (12.16) should not apply and one can wonder whether the diffusion potential can be obtained or not from this bias voltage dependence. Due to the temperature and frequency dependence of the capacitance, one also wonders what is the impact of frequency on the bias voltage dependence and hypothetical determination of the diffusion potential from the C-V curve. In order to address this issue let us concentrate on the capacitance calculated numerically using the AFORS-HET software [1] for a (p) a-Si:H / (n) c-Si heterojunction. The temperature dependence at zero dc bias is shown for two very different frequencies (100 Hz and 1 MHz) in Fig. 12.17.

In order to study the dc bias voltage dependence, let us now select the temperature T=250 K, where the low frequency value of the capacitance is in the high plateau regime, while the high frequency value is in the low plateau regime. Thus, at this temperature the two selected frequencies exhibit the extreme behaviours of the a-Si:H gap states response. Calculations were performed with a donor density in c-Si N_d=1.5×10^{15} cm^{-3}, a DOS with absolute values and energy dependence typical of p-type a-Si:H with a Fermi level position in a-Si:H at 0.35 eV above the top of the valence band, and a valence band offset ΔE_V = 0.48 eV. The results are shown in Fig. 12.18.

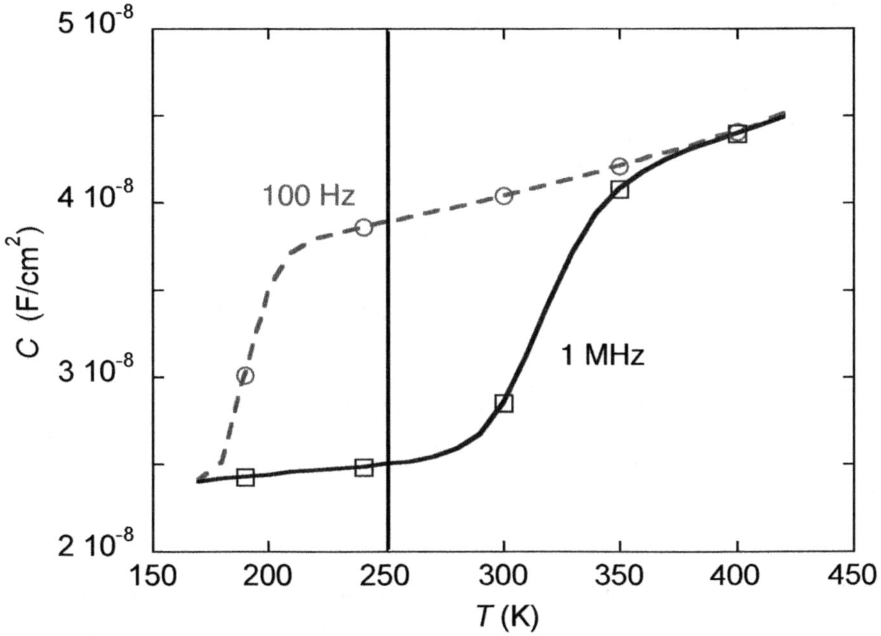

Fig. 12.17 Capacitance versus temperature of a (p) a-Si:H / (n) c-Si heterojunction (a-Si:H thickness of 150 nm) calculated at two very different frequencies (100 Hz and 1 MHz). At 250 K these frequencies express the extreme behaviours for the response in a-Si:H. This temperature was selected for calculating the bias voltage dependence.

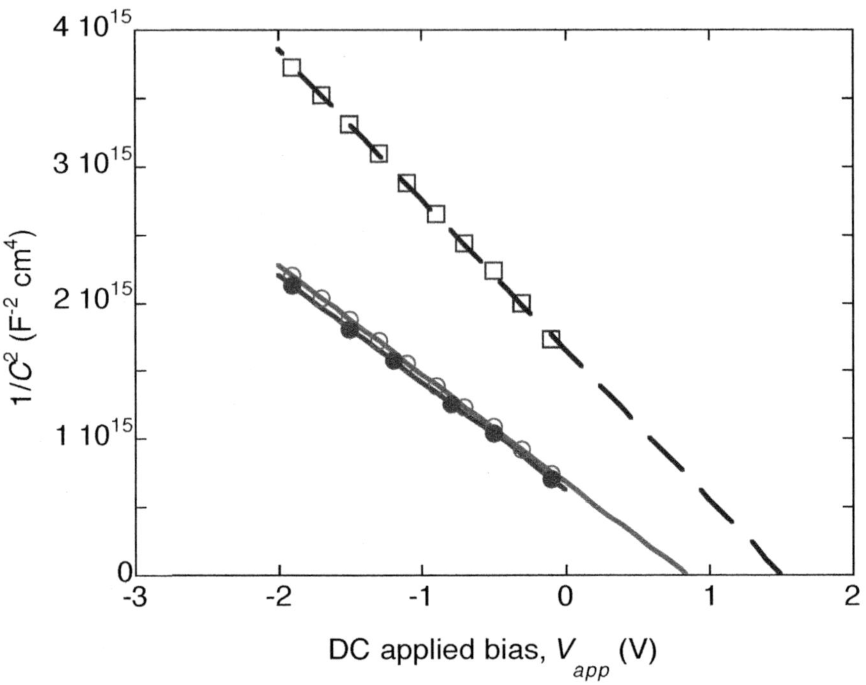

Fig. 12.18 Calculated inverse square capacitance as a function of DC applied bias at the temperature (250 K) and frequencies (100 Hz and 1 MHz) selected from Fig. 12.17 for a (p) a-Si:H / (n) c-Si heterojunction; open blue squares: 1 MHz; open red circles: 100 Hz. Also shown are the data obtained at 1 MHz after correction for the a-Si:H geometric dielectric capacitance as indicated in the text (full green circles). The lines are linear fits to the data in order to extract the intercept voltage.

It can be observed that the square of the inverse capacitance follows a linear dependence on the dc bias for both frequencies. However, the slope and the voltage intercept of the linear fit are very different for the two frequencies. Fitting the data as:

$$\frac{A}{C}=\sqrt{\frac{2}{q}\frac{\left(V_{int}-V_{app}\right)}{\left(\varepsilon_{c-Si}N_{d}^{eff}\right)}}, \tag{12.21}$$

one obtains the values of V_{int} and N_d^{eff} displayed in Table 12.2. Interestingly, exploitation of the capacitance data obtained at low frequency yields the correct value for the doping density introduced in the calculation for c-Si, i.e. $N_d^{eff} = N_d = 1.5\times10^{15}$ cm^{-3}, while the value deduced from the high frequency regime capacitance data are wrong by about 40%. In the high frequency regime the a-Si:H layer behaves like an insulator and the corresponding capacitance is thus simply equal to the geometric dielectric capacitance $\varepsilon A/d_{a\text{-}Si:H}$, as explained above. Since the a-Si:H thickness is known one can extract the capacitance corresponding to the c-Si, C_{Si}, from the total measured capacitance, C_{tot}, using: $1/C_{Si}=1/C_{tot} - d_{a\text{-}Si:H}/\varepsilon A$, and then plot the square of C_{Si} versus applied voltage.

Table 12.2 Values of effective doping density in c-Si, N_d^{eff}, intercept voltage, V_{int}, obtained using eq. (12.21) from the C-V method applied to the simulated data of Fig. 12.17, and corresponding valence band offset ΔE_V obtained from eq. (12.13) assuming that $V_d = V_{int}$. The values introduced in the simulation for the doping density and valence band offset were $N_d = 1.5\times10^{16}$ cm^{-3} and $\Delta E_V = 0.48$ eV.

Frequency (Hz)	N_d^{eff} (10^{16} cm^{-3})	V_{int} (V)	ΔE_V (eV)
100	1.5	0.84	0.28
10^6	1.1	1.50	0.94

This has been done from the calculated 1 MHz data as shown by the full circles in Fig. 12.18. One can see that these 1 MHz C_{Si} data are very close to the data corresponding to the *total* capacitance at 100 Hz. This means that when the gap states in a-Si:H are able to follow the ac signal the corresponding a-Si:H capacitance is very large compared to the c-Si capacitance. This also means that the space charge layer width in a-Si:H is very narrow compared to that in c-Si. From this point of view the system behaves like a p^+/n one-sided junction. In that case, despite the amorphous nature of the a-Si:H, the simple treatment of eq. (12.17) works for the determination of N_d. Regarding the determination of the intercept voltage, V_{int}, the two frequencies lead to very different results. Since the low frequency data yield the correct value of doping density in c-Si, one could expect that these data also yield the correct value of band offset. As can be seen in Table 12.2, assuming that the diffusion potential V_d is equal to the value of V_{int} determined from Fig. 12.18 at low frequency, and using eq. (12.13), one obtains a reconstructed value of ΔE_V of 0.28 eV. This is much less than the actual value of 0.48 eV introduced in the calculations.

This example shows that the C-V intercept method applied to the (p) a-Si:H / (n) c-Si structure leads to a strongly undersestimated value for the valence band offset. In the same way, the method applied to (n) a-Si:H / (p) c-Si structure leads to underestimated values for ΔE_C. For the latter N/p structure, a critical study of the C-V intercept method has been performed [63]. By changing the conduction band offset value in the calculations (all other parameters being constant) it has been shown that there are two reasons that explain the underestimated values obtained for the band offset. First, increasing the band offset leads to an increase of the diffusion potential in the a-Si:H that becomes no longer negligible. This can be seen in Fig. 12.19, where the part of the diffusion potential in c-Si, $V_d^{c\text{-}Si}$ departs from the total diffusion potential V_d for increasing values of ΔE_C. Second, part of the potential drop in c-Si is not reflected in the simple depletion layer approximation because of the existence of a strong inversion layer in c-Si at the interface. In this layer the potential drop does not follow the quadratic dependence as in a depleted region but it has much steeper variations. The intercept voltage of the C-V method reflects the potential drop of the depleted region in the c-Si layer, and does not account for the two additional potential drops in a-Si:H and in the strong inversion layer in c-Si. This is why, when increasing the band offset the intercept voltage of the C-V method saturates at a value close to the potential drop in the depleted region of c-Si and does not increase linearly as does V_d [63].

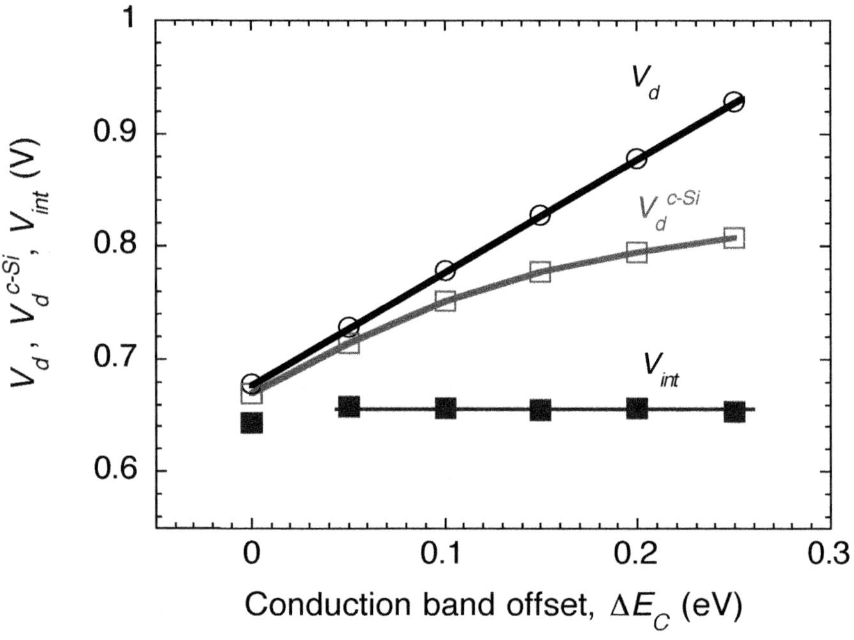

Fig. 12.19 Calculated values of the total diffusion potential, V_d, the part of the diffusion potential in c-Si, $V_d^{c\text{-}Si}$, and the intercept voltage of the inverse square capacitance, V_{int}, in an (n) a-Si:H / (p) c-Si heterojunction ($N_a = 10^{15}$ cm^{-3}), as a function of conduction band offset, ΔE_C.

So far, the theoretical considerations and calculations did not take account of interface states that do exist at the a-Si:H/c-Si interface. When interface states are present, the charge in the interface states also leads to a shift of V_{int} compared to the ideal value. This shift becomes significant when the charge contained in the interface states becomes comparable to the charge in the space charge regions of the two semiconductors. With the typical values of doping densities in c-Si that are used in a-Si:H/c-Si solar cells (of the order of 10^{15}-10^{16} cm^{-3}), the space charge regions have charges of the order $q \times (10^{12}$ cm$^{-2})$. Therefore the effect of interface states becomes relevant for interface states densities above 10^{12} cm^{-2} [64]. Interface states densities have to be minimized in solar cell applications in order to reduce the recombination of photogenerated carriers. In the actual high efficiency solar cells the interface state densities are well below this limit [52], so that the influence of interface states on the band bending and on the determination of band offsets from the C-V method is of minor importance.

The linear dependence of the inverse square capacitance as a function of the applied voltage obtained in numerical calculations as described above has indeed been observed experimentally, as shown for example in Fig. 12.20 for an (n) a-Si:H / (p) c-Si solar cell. For this solar cell, according to conductivity measurements in a-Si:H and to the doping density in c-Si, $N_a = 9\times10^{14}$ cm^{-3}, the parameters δ_1 and δ_2 in eq. (12.13) where equal to 0.2 eV and 0.25 eV, respectively, which leads to $\Delta E_C \approx 0$ [65]. From the above discussion we know that this determination can underestimate the band offset. This has been confirmed from the planar conductance technique, detailed below, which yields a value $\Delta E_C \approx 0.15$ eV [53].

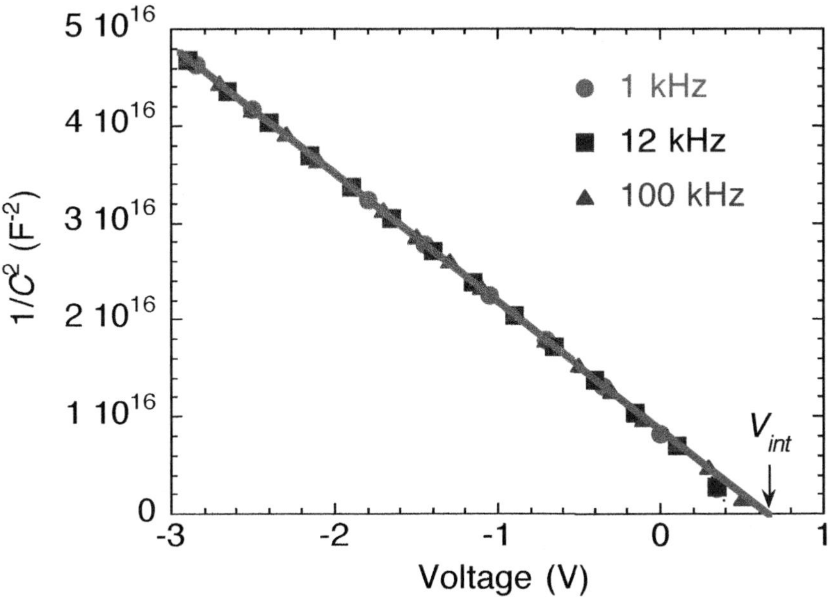

Fig. 12.20 Experimental inverse square capacitance plots for three frequencies from measurements performed at 300 K on an (n) a-Si:H / (p) c-Si solar cell. The plain (red) line is the linear fit for determination of the intercept voltage, $V_{int} = 0.65$ V.

12.3.2.2 Planar Conductance Technique

The capacitance measurements and modeling suggests that in both (n) a-Si:H / (p) c-Si and (p) a-Si:H / (n) c-Si structures there exists a strong inversion layer in c-Si at the interface. This has been confirmed by the planar conductance technique that allowed us to determine values for the band offsets. The schematic equilibrium band diagrams of Fig. 12.21 emphasize the strong inversion layer and also indicate the relative amplitude of band offsets, with the valence band offset being much larger than the conduction band one.

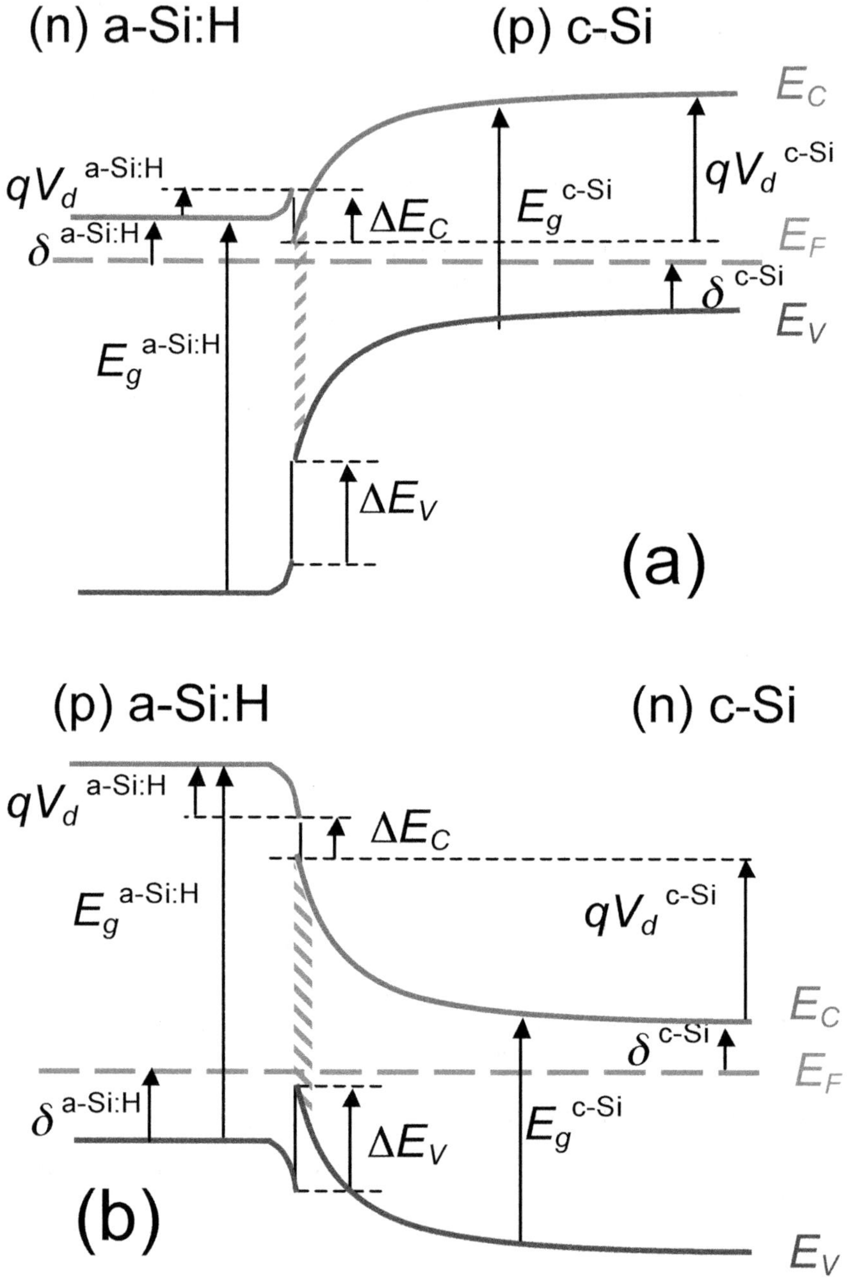

Fig. 12.21 Equilibrium band diagrams for (a): (n) a-Si:H / (p) c-Si; and (b): (p) a-Si:H / (n) c-Si (right) heterojunctions, emphasizing the strong inversion layers in c-Si at the interface (dashed regions) and the larger valence band offset with respect to the conduction band offset.

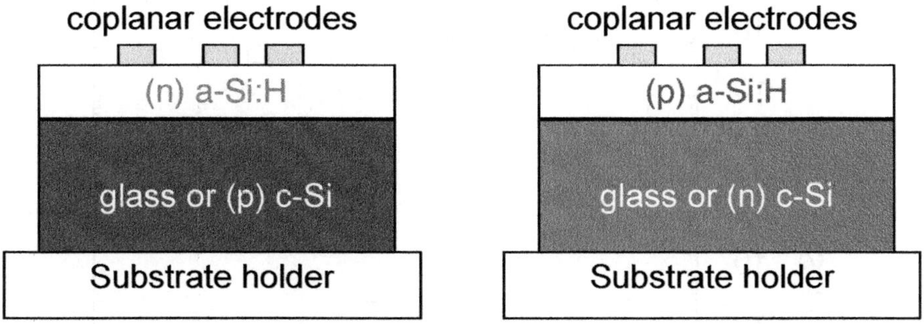

Fig. 12.22 Structures for planar conductance measurements for the determination of the conduction band offset at the (n) a-Si:H / (p) c-Si interface (left) and of the valence band offset at the (p) a-Si:H / (n) c-Si interface (right). Top coplanar electrodes with various gaps like in a TLM configuration can be used to verify the linear dependence of the measured current with the distance between electrodes and to eliminate contact resistance effects.

The structures to be used for planar conductance measurements are depicted in Fig. 12.22. They simply consist of the (p) (or (n)) a-Si:H layer deposited onto (n) (or (p)) c-Si just like for the solar cell fabrication, the structure being fitted with top coplanar electrodes. The a-Si:H layers were also deposited in the same run on a glass substrate in order to measure the conduction properties of the a-Si:H alone. The technique simply consists in measuring the temperature dependence of the dc current, I_{dc}, flowing when applying a dc bias, V_{dc}, between two adjacent top electrodes. The conductance is then defined as the ratio $G = I_{dc}/V_{dc}$. This should be independent of the dc bias in the absence of contact effects. This has been found experimentally for low dc bias.

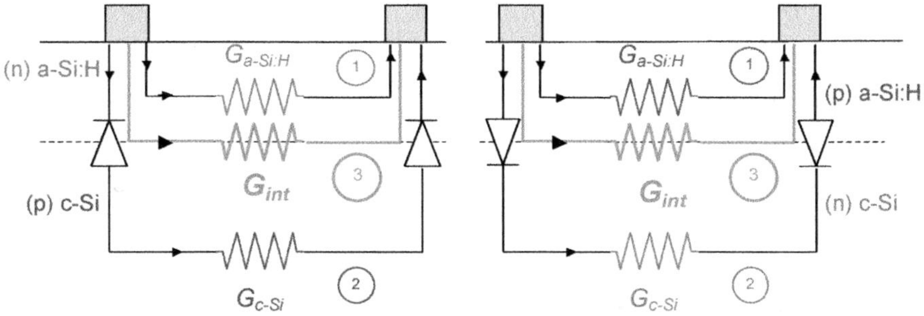

Fig. 12.23 Simplified conduction paths for the current flowing between two top electrodes. The dominant conductance, G_{int}, corresponds to path (3) related to the flow in channel formed by the strong inversion regions.

The idea of using such a technique is physically interesting and very simple. Indeed, if a strong inversion layer does exist in an (n) a-Si:H / (p) c-Si structure, a highly conductive channel of electrons forms in c-Si at the hetero-interface. Similarly, if a strong inversion layer exists in a (p) a-Si:H / (n) c-Si structure, a highly conductive channel of holes forms in c-Si at the hetero-interface. This is the same

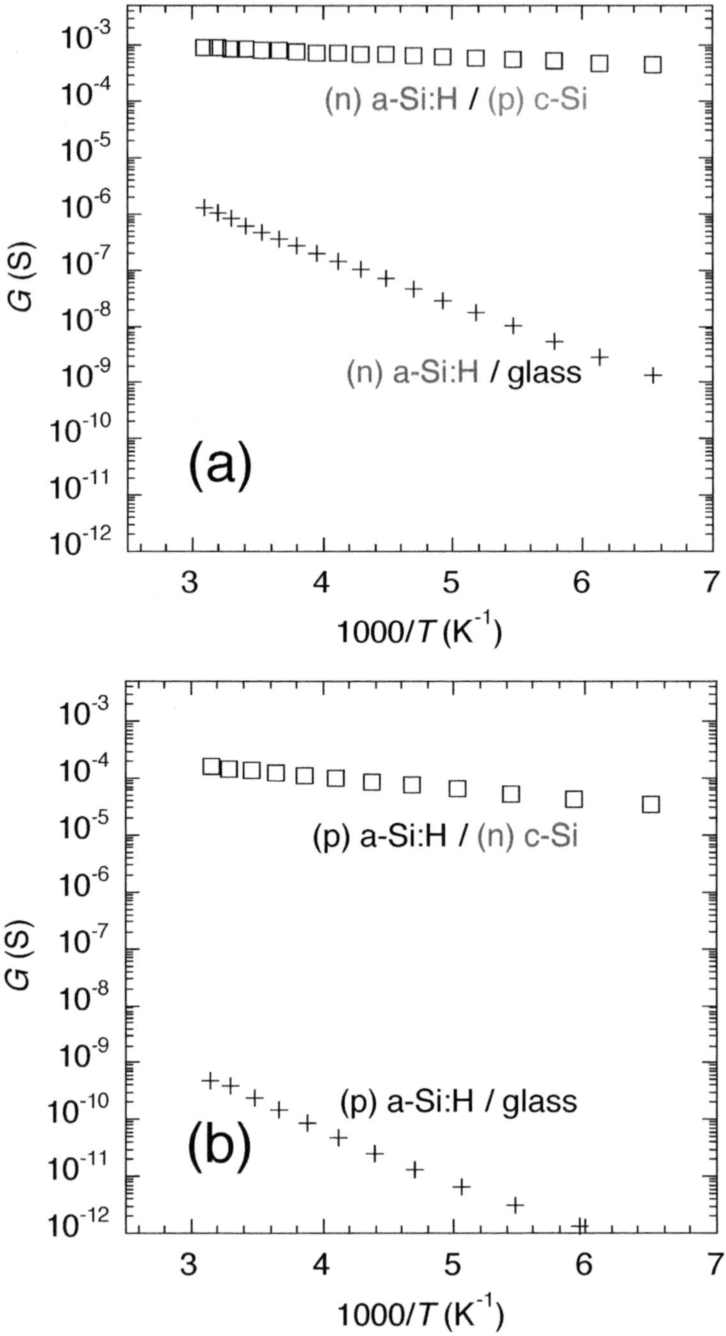

Fig. 12.24 Examples of Arrhenius plots of the conductance measured for (a): (n) a-Si:H; and (b): (p) a-Si:H layers deposited on c-Si of the opposite type and on glass.

type of 2 dimensional electron gas (or hole gas) as can be found in Si/SiO_2 and AlGaAs/GaAs heterostructures. In the a-Si:H/c-Si system, there is a further ingredient that makes measurement of the interface channel conductance more easy, namely that the carrier mobilities are much lower in a-Si:H than in c-Si. As a consequence, from the 3 parallel paths identified in Fig. 12.23 for the current flow, the upper first path consisting in the flow within the a-Si:H layer is negligible.

This has been verified experimentally by comparing the current measured in the a-Si:H/c-Si structure with that measured in the a-Si:H/glass structure [66, 67]. An example is shown in Fig. 12.24, where it is clear that the conductance of the a-Si:H/c-Si structures is much higher than that of the corresponding a-Si:H/glass structures, and that the activation energy is much lower. In both structures of Fig. 12.23, the path (2) crossing the interfaces and flowing through the crystalline silicon has also a negligible contribution since it involves a reverse biased p/n junction in series. The prevalence of the interface conductance related to the strongly inverted layer was verified experimentally by etching the a-Si:H layer between the electrodes. Indeed, after etching, it was found that the current is orders of magnitude lower and it has a much more pronounced temperature dependence. Recently, a direct experimental proof for the existence of a strongly inverted c-Si layer at both (n) a-Si:H / (p) c-Si and (p) a-Si:H / (n) c-Si was obtained from conductive probe atomic force microscopy where a conductive channel was clearly observed by probing cleaved sections of heterojunctions [68].

12.3.2.2.1 (n) a-Si:H / (p) c-Si Structure

For the (n) a-Si:H / (p) c-Si structure, the measured conductance is related to the electron concentration n through

$$G = \frac{qh}{L} \int_0^{d_{c-Si}} \mu_n n(x) dx, \tag{12.22}$$

h being the length of the top electrodes, L the gap between electrodes, μ_n the electron mobility. The local conductivity $q\mu_n n$ is integrated over the c-Si of thickness $d_{c\text{-}Si}$, the origin $x = 0$ being taken at the a-Si:H/c-Si interface. Since the c-Si is of p-type it is clear that the main contribution to this integral comes from the strong inversion region in c-Si at the interface, so that it may be written:

$$G = \frac{qh}{L} \mu_n N_S, \tag{12.23}$$

with

$$N_S = \int_0^{d_{c-Si}} n(x) dx \tag{12.24}$$

being the sheet electron density determined by the inversion layer channel and μ_n the electron mobility in the channel.

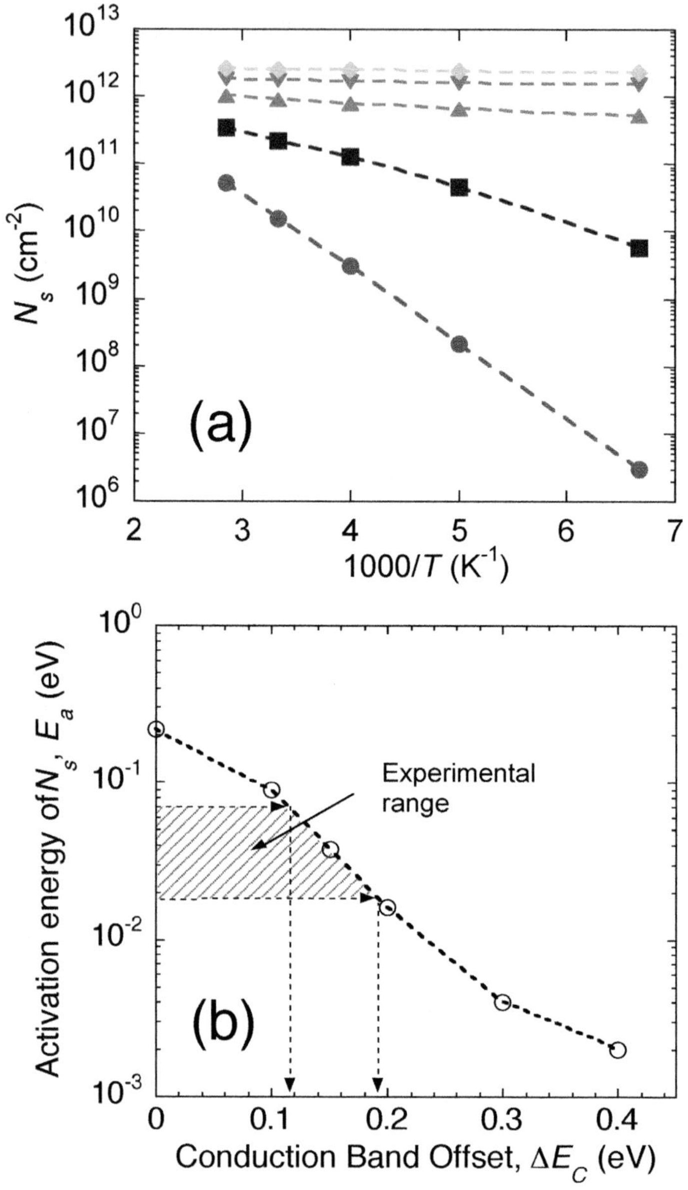

Fig. 12.25 (a): Arrhenius plots of the calculated sheet electron density in the c-Si channel at the (n) a-Si:H/ (p) c-Si interface for various values of the conduction band offset: (●) ΔE_C=0; (■)ΔE_C=0.1 eV; (▲)ΔE_C=0.2 eV; (▼) ΔE_C=0.3 eV; (◆)ΔE_C=0.4 eV; (b): Activation energy of N_s as a function of conduction band offset. Also shown in this graph is the range of experimental values, taking account of the uncertainty on the value and temperature dependence of the electron mobility in the strong c-Si inversion layer and of the dispersion in measurements from a set of several samples deposited on c-Si having various origins, material qualities (CZ, FZ) and different nominal resistivities (1-22 Ω.cm).

N_s can be easily calculated from the knowledge of the band bending in c-Si which in turn depends on the doping density in c-Si, the position of the Fermi level in a-Si:H and the DOS in a-Si:H. However, the most sensitive parameter for the temperature dependence of N_s is the conduction band offset. This can be seen in Fig. 25. Extracting experimental values of N_s from the conductance measurements according to Eq. (23) one can compare the experimental temperature dependence with that obtained from calculations and thus obtain the corresponding conduction band offset. Taking account of several scenarii for the temperature dependence of the electron mobility in the inverted surface layer (this may be different than in bulk silicon due to different diffusion mechanisms) and of the results obtained on various samples, it was possible to deduce a very precise value for the conduction band offset: $\Delta E_C = 0.15$ eV with an uncertainty of about 0.04 eV [53].

12.3.2.2.2 (p) a-Si:H / (n) c-Si Structure

In this structure the planar conductance is related to the hole sheet density

$$P_s = \int_0^{d_{c-Si}} p(x)dx, \tag{12.25}$$

p being the free hole concentration, through

$$G = \frac{qh}{L}\mu_p P_s. \tag{12.26}$$

The same study as for (n) a-Si:H / (p) c-Si can then be performed and compared to experimental data.

Since the theoretical calculation of N_s or P_s from the band bending also depends on a-Si:H defect parameters, a detailed study of the influence of the position of the Fermi level in a-Si:H, the DOS at the Fermi level in a-Si:H, and the shape of the DOS, e.g. the widths of the exponential bandtails, was performed by means of numerical simulations. The temperature dependence of the bandgap energies was also taken into account. This dependence is well known in c-Si, and much less documented in a-Si:H. Also, it is not clear from literature whether changes in the bandgap energies do affect the conduction band or the valence band, or both, so various possible cases were investigated. This complete study will be published elsewhere. The main conclusion from this study is that the value of the band offset has a very strong influence on the activation energy of the sheet carrier densities. The whole set of experimental data available for both (n) a-Si:H / (p) c-Si and (p) a-Si:H / (n) c-Si structures have been shown to be fully described by band offset values $\Delta E_V = 0.4$ eV and $\Delta E_C = 0.15$ eV with less than 0.1 eV absolute error. It is worth pointing that these values are in agreement with that obtained from UV-photoelectron spectroscopy [32, 50-52]. Also, these values are in quite good agreement with the neutral band lineup theory if one takes the branch-point energy in a-Si:H close to midgap (cf. Fig. 12.9).

12.4 Summary and Conclusions

While very high solar cell efficiencies have been obtained in silicon heterojunction solar cells, the detailed physics of the devices is still a matter of research. In particular, band offset values reported in literature for the a-Si:H/c-Si system are very widespread.

The two main band lineup theories, namely the electron affinity rule and the neutral band lineup theory based on the branch-point energy alignment have been reviewed and the application to the a-Si:H/c-Si case has been discussed. Part of the spreading observed in the published values for both conduction and valence band offsets can be due to misinterpretations of experimental techniques when an amorphous semiconductor is involved. This has been emphasized for the C-V intercept method that has been widely used in crystalline semiconductor heterojunctions. Indeed, it has been shown that this method underestimates the band offsets when the c-Si is strongly inverted at the interface. From planar conductance measurements, the existence of strong inversion surface layers has been evidenced. Using the temperature dependence of the planar conductance, the band offsets have been determined as $\Delta E_V = 0.4$ eV and $\Delta E_C = 0.15$ eV from measurements on (p) a-Si:H / (n) c-Si and (n) a-Si:H / (p) c-Si structures, respectively, with less than 0.1 eV absolute error. These are today's most precise and most reliable determinations from electrical techniques. The obtained values are in agreement with recent UV-photoelectron spectroscopy measurements. Comparing with theoretical works, these values are quite in agreement with the branch-point energy alignment if one takes the branch-point energy in a-Si:H at midgap.

Acknowledgments

I would like to thank people of my research group at LGEP who have contributed to increase the amount of data on silicon heterojunctions from either experimental or modeling work: J. Alvarez, E. Blanc, A. Brézard-Oudot, R. Chouffot, F. Dadouche, D. Diouf, W. Favre, A.S. Gudovskikh, M.E. Gueunier-Farret, O. Maslova, C. Pareige and R. Varache. Colleagues at partner laboratories (CNRS-LPICM, CEA/INES, Fern-Universität Hagen, Helmholtz-Zentrum Berlin, Ioffe Institute) who provided samples are also kindly acknowledged.

References

[1] Stangl, R., Leendertz, C.: General principles of solar cell simulation and introduction to AFORS-HET. chapter 13 in this book

[2] Capasso, F., Margaritondo, G. (eds.): Heterojunction band discontinuities, physics and device applications. North Holland, Amsterdam (1987)

[3] Anderson, R.L.: Experiments on Ge-GaAs heterojunctions. Solid-State Electron. 5, 341–351 (1962)

[4] Tersoff, J.: The theory of heterojunction lineups. In: Capasso, F., Margaritondo, G. (eds.) Heterojunction Band Discontinuities – Physics and Device Applications, pp. 3–57. North Holland, Amsterdam (1987)

[5] Mott, N.F.: Note on the contact between a metal and an insulator or semi-conductor. Proc. Camb. Philos. Soc. 34, 568–572 (1938)

[6] Schottky, W.: Discrepencies in Ohm's laws in semiconductors. Physik. Zeitschr. 41, 570–573 (1940)

[7] Mönch, W.: Electronic properties of semiconductor interfaces. Springer, Berlin (2004)

[8] Bardeen, J.: Surfacc States and Rectification at a Metal Semi-Conductor Contact. Phys. Rev. 71, 717–727 (1947)

[9] Heine, V.: Theory of surface states. Phys. Rev. 138, A1689–A1696 (1965)

[10] Louie, S.G., Cohen, M.L.: Electronic structure of a metal-semiconductor interface. Phys. Rev. B 13, 2461–2469 (1976)

[11] Flietner, H.: The *E(k)* relation for a two-band scheme of semiconductors and the application to the Metal-Semiconductor contact. Phys. Stat. Sol. (b) 54, 201–208 (1972)

[12] Lüth, H.: Surfaces and interfaces of solid materials, 3rd edn. Springer, Berlin (1995)

[13] Tejedor, C., Flores, F.: A simple approach to heterojunctions. J. Phys. C: Solid State Phys. 11, L19–L23 (1978)

[14] Flores, F., Tejedor, C.: Energy barriers and interface states at heterojunctions. J. Phys. C: Solid-State Phys. 12, 731–749 (1979)

[15] Tersoff, J.: Schottky barriers heights and the continuum of gap states. Phys. Rev. Lett. 52, 465–468 (1984)

[16] Tersoff, J.: Theory of semiconductor heterojunctions: the role of quantum dipoles. Phys. Rev. B 30, 4874–4877 (1984)

[17] Tersoff, J.: Schottky barriers and semiconductor band structures. Phys. Rev. B 32, 6968–6971 (1985)

[18] Cardona, M., Christensen, N.E.: Acoustic deformation potentials and heterostructure band offsets in semiconductors. Phys. Rev. B 35, 6182–6194 (1987)

[19] Mönch, W.: Barrier heights of real Schottky contacts explained by metal-induced gap states and lateral inhomogeneities. J. Vac. Sci. Technol. B 17, 1867–1876 (1999)

[20] Cowley, A.M., Sze, S.M.: Surface states and barrier height of metal-semiconductor systems. J. Appl. Phys. 36, 3212–3220 (1965)

[21] Deane, S.C., Powell, M.J.: Defect chemical potential and the density of states in amorphous silicon. Phys. Rev. Lett. 70, 1654–1657 (1993)

[22] Powell, M.J., Deane, S.C.: Defect-pool model and the hydrogen density of states in hydrogenated amorphous silicon. Phys. Rev. B 53, 10121–10132 (1996)

[23] Harrison, W.A., Tersoff, J.: Tight binding theory of heterojunction band lineups and interface dipoles. J. Vac. Sci. Technol. B 4, 1068–1073 (1986)

[24] Hasegawa, H., Ohno, Sawada,T.: Hybrid Orbital Energy for Heterojunction Band Lineup. Jpn. J. Appl. Phys. 25, L265-L268 (1986)

[25] Allan, G., Delerue, C., Lannoo, M.: Electronic structure and localized states in a model amorphous silicon. Phys. Rev. B 57, 6933–6936 (1998)

[26] Van de Walle, C.G., Yang, L.H.: Band discontinuities at heterojunctions between crystalline and amorphous silicon. J. Vac. Sci. Technol. B 13, 1635–1638 (1995)

[27] Niles, D.W., Margaritondo, G., Perfetti, P., Quaresima, C., Capozi, M.: Heterojunction band discontinuity control by ultrathin intralayers. Appl. Phys. Lett. 47, 1092–1094 (1985)

[28] Niles, D.W., Margaritondo, G., Colavita, E., Perfetti, P., Quaresima, C., Capozi, M.: Understanding and controlling heterojunction band discontinuities. J. Vac. Sci. Technol. A 4, 962–964 (1986)

[29] Ristein, J.: Diamond surfaces: familiar and amazing. Appl. Phys. A 82, 377–384 (2006)
[30] Dabrowski, J., Müssig, H.-J.: Silicon surfaces and formation of interfaces. World Scientific, Singapore (2000)
[31] Perfetti, P., Quaresima, C., Coluzza, C., Fortunato, C., Margaritondo, G.: Dipole-induced changes of the band discontinuities at the SiO_2-Si interface. Phys. Rev. Lett. 57, 2065–2068 (1986)
[32] Korte, L.: Electronic properties of ultrahin a-Si:H layers and the a-Si:H/c-Si interface. chapter 6 in this book
[33] Kroemer, H.: Heterostructure devices: a device physicist looks at interfaces. Surf. Sci. 132, 543–576 (1983)
[34] Mimura, H., Hatanaka, Y.: Energy band discontinuities in a heterojunction of amorphous hydrogenated Si and crystalline Si measured by internal photoemission. Appl. Phys. Lett. 50, 326–328 (1987)
[35] Cuniot, M., Marfaing, Y.: Study of the band discontinuities at the a-Si:H/c-Si interface by internal photoemission. J. Non-Cryst. Solids 77&78, 987–990 (1985)
[36] Cuniot, M., Marfaing, Y.: Energy band diagram of the a-Si: H/c-Si interface as determined by internal photoemission. Phil. Mag. B 57, 291–300 (1988)
[37] Lequeux, N., Cuniot, M.: Internal photoemission measurements on a-Si1-xGex:H/c-Si heterojunctions. J. Non-Cryst. Solids 114, 555–557 (1989)
[38] Sakata, I., Yamanaka, M., Shimokawa, R.: Band Lineup at the Interface between Boron-Doped P-Type Hydrogenated Amorphous Silicon and Crystalline Silicon Studied by Internal Photoemission. Jpn. J. Appl. Phys. 43, L954–L956 (2004)
[39] Sakata, I., Yamanaka, M., Kawanami, H.: Characterization of heterojunctions in crystalline-silicon-based solar cells by internal photoemission. Sol. Energy Mater. Sol. Cells 93, 737–741 (2009)
[40] Eschrich, H., Bruns, J., Elstner, L., Swiatkowski, C.: The dependence of a-Si:H/c-Si solar cell generator and spectral response characteristics on heterojunction band discontinuities. J. Non-Cryst. Solids 164-166, 717–720 (1993)
[41] Gall, S., Hirschauer, R., Kolter, M., Braünig, D.: Spectral characteristics of a-Si:H/c-Si Heterostructures. Sol. Energy Mater. Sol. Cells 49, 157–162 (1997)
[42] Matsuura, H., Okuno, T., Okushi, H., Tanaka, K.: Electrical properties of n amorphous / p-crystalline silicon heterojunctions. J. Appl. Phys. 55, 1012–1019 (1984)
[43] Matsuura, H.: Hydrogenated amorphous silicon/crystalline-silicon heterojunctions: properties and applications. IEEE Trans. Electron. Dev. 36, 2908–2914 (1989)
[44] Song, Y.J., Park, M.R., Guliants, E., Anderson, W.A.: Influence of defects and band offsets on carrier transport mechanisms in amorphous silicon/crystalline silicon heterojunction solar cells. Sol. Energy Mater. Sol. Cells 64, 225–240 (2000)
[45] Essick, J.M., Nobel, Z., Li, Y.-M., Bennett, M.S.: Conduction- and valence-band offsets at the hydrogenated amorphous silicon-carbon/crystalline silicon interface via capacitance techniques. Phys. Rev. B 54, 4885–4890 (1996)
[46] Essick, J.M., Cohen, J.D.: Band offsets and deep defect distribution in hydrogenated amorphous silicon crystalline silicon heterostructures. Appl. Phys. Lett. 55, 1232–1234 (1989)
[47] Rösch, M., Brüggemann, R., Bauer, G.H.: Influence of interface defects on the current-voltage characteristics of amorphous silicon/crystalline silicon heterojunction solar cells. In: Proc. of the 2nd World Conf. on Photovoltaic Solar Energy Conversion, Vienna, Austria, pp. 964–967 (1998)
[48] Unold, T., Rösch, M., Bauer, G.H.: Defects and transport in a-Si:H/c-Si heterojunctions. J. Non-Cryst. Solids 266-269, 1033–1037 (2000)

[49] Gall, S., Hirschauer, R., Braünig, D.: Admittance measurements at a-Si:H/c-Si heterojunction devices. In: Proc. of the 13th European Photovoltaic Solar Energy Conference, Nice, France, pp. 1264–1267 (1995)
[50] Sebastiani, M., Di Gaspare, L., Capellini, G., Bittencourt, C., Evangelisti, F.: Low-Energy Yield Spectroscopy as a Novel Technique for Determining Band Offsets: Application to the c-Si(100)/a-Si:H Heterostructure. Phys. Rev. Lett. 75, 3352–3355 (1995)
[51] Fuhs, W., Korte, L., Schmidt, M.: Heterojunctions of hydrogenated amorphous silicon and monocrystalline silicon. J. Opt. Adv. Mat. 8, 1989–1995 (2006)
[52] Schmidt, M., Korte, L., Laades, A., Stangl, R., Schubert, C., Angermann, H., Conrad, E., von Maydell, K.: Physical aspects of a-Si:H/c-Si hetero-junction solar cells. Thin Solid Films 515, 7475–7480 (2007)
[53] Kleider, J.P., Gudovskikh, A.S., Roca i Cabarrocas, P.: Determination of the conduction band offset between hydrogenated amorphous silicon and crystalline silicon from surface inversion layer conductance measurements. Appl. Phys. Lett. 92, 162101 (2008)
[54] Forrest, S.R.: Measurement of energy band offsets using capacitance and current measurement techniques. In: Capasso, F., Margaritondo, G. (eds.) Heterojunction Band Discontinuities – Physics and Device Applications, pp. 311–375. North Holland, Amsterdam (1987)
[55] Lang, D.V.: Measurement of energy band offsets by space charge spectroscopy. In: Capasso, F., Margaritondo, G. (eds.) Heterojunction Band Discontinuities – Physics and Device Applications, pp. 377–396. North Holland, Amsterdam (1987)
[56] van Cleef, M.W.M., Rubinelli, F.A., Rizzoli, R., Pinghini, R., Schropp, R.E.I., van der Weeg, W.F.: Amorphous silicon carbide/crystalline silicon heterojunction solar cells: a comprehensive study of the photocarrier collection. Jpn. J. Appl. Phys. 37, 3926–3932 (1998)
[57] Kroemer, H., Chien, W.Y., Harris, J.S., Edwall, D.D.: Measurement of isotype heterojunction barriers by CV profiling. Appl. Phys. Lett. 36, 295–297 (1980)
[58] Sze, S.M.: Physics of semiconductor devices, 2nd edn. John Wiley & Sons, New York (1981)
[59] Cohen, J.D., Lang, D.V.: Calculation of the dynamic response of Schottky barriers with a continuous distribution of gap states. Phys. Rev. B 25, 5321–5350 (1982)
[60] Archibald, I.W., Abram, R.A.: More theory of the admittance of an amorphous silicon Schottky barrier. Philos. Mag. B 54, 421–438 (1986)
[61] Kleider, J.P.: Capacitance techniques for the evaluation of electronic properties and defects in disordered thin film semiconductors. Thin Solid Films 427, 127–132 (2003)
[62] Gudovskikh, A.S., Kleider, J.P., Terukov, E.I.: Characterization of an a-Si:H/c-Si interface by admittance spectroscopy. Semiconductors 39, 904–909 (2005)
[63] Gudovskikh, A.S., Ibrahim, S., Kleider, J.P., Damon-Lacoste, J., Roca i Cabarrocas, P., Veschetti, Y., Ribeyron, P.-J.: Determination of band offsets in a-Si:H/c-Si heterojunctions from capacitance–voltage measurements: Capabilities and limits. Thin Solid Films 515, 7481–7485 (2007)
[64] Gudovskikh, A.S., Kleider, J.P., Damon-Lacoste, J., Roca i Cabarrocas, P., Veschetti, Y., Muller, J.-C., Ribeyron, P.-J., Rolland, E.: Interface properties of a-Si:H/c-Si heterojunction solar cells from admittance spectroscopy. Thin Solid Films 511-512, 385–389 (2006)

[65] Kleider, J.P., Gudovskikh, A.S.: Characterization of amorphous/crystalline silicon interfaces from electrical measurements. In: Mater. Res. Symp. Proc., vol. 1066, pp. 75–86 (2008)
[66] Kleider, J.P., Soro, Y.M., Chouffot, R., Gudovskikh, A.S., Roca i Cabarrocas, P., Damon-Lacoste, J., Eon, D., Ribeyron, P.-J.: High interfacial conductivity at amorphous silicon/crystalline silicon heterojunctions. J. Non-Cryst. Solids 354, 2641–2645 (2008)
[67] Favre, W., Labrune, M., Dadouche, F., Gudovskikh, A.S., Roca i Cabarrocas, P., Kleider, J.P.: Study of the interfacial properties of amorphous silicon/n-type crystalline silicon heterojunction through static planar conductance measurements. Phys. Status Solidi. C 7, 1037–1040 (2010)
[68] Maslova, O.A., Alvarez, J., Gushina, E.V., Favre, W., Gueunier-Farret, M.E., Gudovskikh, A.S., Ankudinov, A.V., Terukov, E.I., Kleider, J.P.: Observation by Conductive-Probe Atomic Force Microscopy of strongly inverted surface layers at the hydrogenated amorphous silicon / crystalline silicon heterojunctions. Appl. Phys. Lett. 97, 252110 (2010)

Chapter 13
General Principles of Solar Cell Simulation and Introduction to AFORS-HET

Rolf Stangl and Caspar Leendertz

Helmholtz-Zentrum für Materialien und Energie, Institut für Silizium-Photovoltaik,
Kekuléstr.5, D-12489 Berlin, Germany

Abstract. The principles of numerical solar cell simulation are described, using AFORS-HET (**a**utomat **for** **s**imulation of **het**erostructures) which is a device simulator program for modelling multi layer homo- or heterojunction solar cells and typical characterization methods in one dimension. The basic equations for the optical and electrical calculations used in AFORS-HET are explained including a detailed description of the equations needed to calculate the recombination via defects in the semiconductor layers.

13.1 Principles of Solar Cell Simulations

Most solar cells on the market today, can be described as a one-dimensional (1D) sequence of different semiconductor layers. If they are uniformly illuminated, one-dimensional solar cell modelling is sufficient (the internal electron/hole current can flow only in one dimension). This is the case for most wafer based silicon solar cells as well as for most thin film solar cells as long as the integrated series connection shall not be explicitly modelled. However, some solar cells use stripe- or point-like metallic contacts, which are embedded in a passivation layer in order to minimize contact recombination. These contacts can either be placed on both sides of the solar cell or favourably only at the rear side of the solar cell, thereby avoiding shadowing due to the contacts. In these cases, the resulting solar cells have to be modelled as two (2D) or even three-dimensional (3D) problems (the internal electron/hole current can flow in 2 or even 3 dimensions). In the current version 2.4 of AFORS-HET only 1D simulations are possible; however, there is a 2D model under development.

The material properties of the semiconductor layers do not necessarily have to be constant. For example, a conventional wafer based silicon solar cell will have a doping profile within the emitter layer. Also thin film solar cells may have non-constant defect state densities, as soon as high temperature processes are involved in the production, which allow defects to migrate within the cell.

W.G.J.H.M. van Sark et al. (Eds.): Physics & Tech. of Amorphous-Crystalline, EM, pp. 445–458.
springerlink.com

In some solar cells, the interface between some semiconductor layers will be also a critical parameter, significantly influencing the solar cell characteristics. A typical example is the amorphous/crystalline silicon heterojunction solar cell, where the interface between the crystalline c-Si wafer and the amorphous a-Si:H emitter layer has to be explicitly modelled. Thus, not only the material properties of the semiconductor layers have to be stated, but also the material properties of the interfaces, like for example an interface defect state density. Furthermore, an appropriate model for the transport of charge carriers across that interface has to be chosen (for example transport through an interface can be considered to take place with or without tunnelling).

Solar cell simulation generally subdivides into two steps: optical and electrical simulation. In most cases they can be performed separately. By optical simulation the local generation rate $G(\mathbf{x},t)$ within the solar cell is calculated, that is the number of excess carriers (electrons and holes) that are created per second and per unit volume at the time t at the position $\mathbf{x}$ within the semiconductor layers of the solar cell due to light absorption. Depending on the optical model chosen for the simulation, effects like external or internal reflections, coherent superposition of the propagating light or light scattering at internal surfaces can be considered.

The local generation rate $G(\mathbf{x},t)$ obtained from the optical simulation is an input needed for the second step: the electrical simulation, in which the local carrier densities $n(\mathbf{x},t)$, $p(\mathbf{x},t)$ and the local electric potential $\varphi(\mathbf{x},t)$ within the semiconductor layers of the solar cell are calculated for different boundary conditions (for example for a solar cell operated under open-circuit conditions). All other internal cell quantities, such as band diagrams, local recombination rates, local cell currents and local phase shifts can then be calculated from the local electron and hole densities and from the local potential. The solution for $n(\mathbf{x},t)$, $p(\mathbf{x},t)$ and $\varphi(\mathbf{x},t)$ is found by numerically solving a set of three coupled differential equations: the Poisson equation and the continuity equations for electrons and holes. This was suggested for the first time in 1964 by Gummel [1] and is described in detail in the book by Selberherr [2]. In order to solve the semiconductor equations the local recombination rate $R(\mathbf{x},t)$ has to be explicitly stated in terms of the unknown variables $n, p, \varphi, R(\mathbf{x},t) = f(n,p,\varphi)$. Depending on the recombination model chosen for the simulation, effects like direct band to band recombination (radiative recombination), indirect band to band recombination (Auger recombination) or recombination via defects (Shockley-Read-Hall recombination, dangling-bond recombination) can be considered.

In order to simulate a real measurement, the optical and electrical simulations are repeatedly calculated while changing a boundary condition of the problem, which is specific to the measurement, like the cell illumination, the external applied cell voltage or the cell current.

To assure a numerical simulation with reliable results, a good model calibration, i.e. a comparison of simulation results to a variety of different characterisation methods is necessary and the solar cell measured under different operation conditions should be compared to the simulations. Furthermore different characterisation methods for the solar cell components, i.e. for the individual semiconductor layers and sub stacks should be tested against simulation. Only then the adequate physics models as well as the corresponding model input parameters can be satisfactorily chosen. Thus it is an important feature of AFORS-HET to be able to simulate the common characterisation methods for solar cells and its components.

13.2 Brief Description of AFORS-HET

AFORS-HET is a 1D simulation program which is able to handle homojunctions as well as heterojunctions, that is an arbitrary stack of different semiconductor layers can be treated. Interface defects can be introduced in order to handle not only bulk recombination $R(\mathbf{x},t)$ within the semiconductor layers but also interface recombination $R^{it}(t)$ at the corresponding interfaces between the layers. Depending on the physical assumption how to describe an electron/hole transport across a heterojunction interface, a distinct interface model can be chosen, such as the drift-diffusion and the thermionic emission interface models. Tunnelling interface models are currently under development. The semiconductor material properties of the layers can be non-constant (linear, exponential, Gaussian, Errorfunction like increasing or decreasing). The two external boundaries of the semiconductor stack (the front and back contact of the solar cell) can be calculated as Schottky boundary, insulator boundary or metal/insulator/semiconductor boundary, so that different experimental configurations can be modelled. The program allows for arbitrary parameter variations and multidimensional parameter fitting in order to match simulated measurements to real measurements.

AFORS-HET, version 2.4, is an open source on demand program. It is distributed free of charge and it can be downloaded from the HZB website (http://www.helmholtz-berlin.de/). A detailed description of all implemented models and their equations can be found in the e-book: "Solar Energy" [3]. In the following we repeat the equations for the basic optical and electrical calculations of AFORS-HET and for the most important recombination models available.

13.2.1 Optical Calculations

In order to describe the generation rate $G_n(x,t), G_p(x,t)$, of electrons and holes due to photon absorption within the bulk of the semiconductor layers a distinction between super-band gap generation (for photons with an energy $E_{photon} = hc/\lambda \geq E_g$) and sub-band gap generation (for photons with an energy $E_{photon} = hc/\lambda \leq E_g$) is made ($\lambda$: photon wavelength h : Planck´s constant, c : velocity of light, E_g : band gap of the semiconductor layer in which the photon absorption takes place). Only the super-band gap generation rate is calculated by optical modeling as it is independent of the local particle densities $n(x,t), p(x,t)$. Sub-band gap generation depends on the local particle densities and must therefore be calculated within the electrical modeling part and will not be discussed further in this book.

The optical super-band gap generation rate is equal for electrons and holes $G(x,t)=G_n(x,t)=G_p(x,t)$. It can either be imported by loading an appropriate file or it can be calculated within AFORS-HET.

So far, two optical models are implemented in AFORS-HET, i.e. the optical model Lambert-Beer absorption and the optical model coherent/incoherent internal multiple reflections. The first one takes textured surfaces and multiple internal boundary reflections into account but neglects coherence effects. It is especially suited to treat wafer based crystalline silicon solar cells. The second takes coherence effects into account, but this is done only for plain surfaces. If coherence effects in thin film solar cells are observable it may be used.

Optical model: Lambert-Beer absorption

Using this model, the absorption within the semiconductor stack will be calculated assuming simple Lambert-Beer absorption, allowing for multiple forward and backward travelling of the incoming light, however disregarding coherent interference. A reflectance and absorptance file of the illuminated contact $R(\lambda)$, $A(\lambda)$ can be loaded or constant values can be used. The incoming spectral photon flux $\Phi_0(\lambda,t)$ is weighted with the contact reflection and absorption, i.e. the photon flux impinging on the first semiconductor layer is given by $\Phi_0(\lambda,t), R(\lambda), A(\lambda)$. To simulate the extended path length caused by a textured surface, the angle of incidence δ of the incoming light can be adjusted. On a textured Si wafer with <111> pyramids, this angle is δ=54.74°, whereas δ=0° equals normal incidence. The angle γ by which the light travels through the layer stack depends on the wavelength of the incoming light and is calculated according to Snellius' law:

$$\gamma(\lambda)=\delta-\arcsin\left(\sin(\delta)\cdot\frac{1}{n(\lambda)}\right), \tag{13.1}$$

where $n(\lambda)$ is the wavelength dependent refraction index of the first semiconductor layer at the illuminated side. Note, that within this model, the change in $\gamma(\lambda)$ is neglected, when the light passes a semiconductor/semiconductor layer interface with two different refraction indices. Thus it is assumed that all photons with a specified wavelength cross the layer stack under a distinct angle γ.

Photon absorption is then calculated from the spectral absorption coefficient $\alpha_x(\lambda)=4\pi\, k(\lambda)/\lambda$ of the semiconductor layer corresponding to the position x within the stack, which is calculated from the provided extinction coefficient $k(\lambda)$ of the layer. The super band gap electron/hole generation rate for one single run through the layer stack (no multiple passes) is then given by:

$$G(x,t)=\int_{\lambda_{min}}^{\lambda_{max}} d\lambda\ \Phi_0(\lambda,t)R(\lambda)A(\lambda)\ \alpha_x(\lambda)\ e^{\frac{-\alpha_x(\lambda)\,x}{\cos(\gamma)}}\ . \tag{13.2}$$

The minimum and maximum wavelengths $\lambda_{\min}, \lambda_{\max}$ for the integration are generally provided by the loaded spectral range of the incoming spectral photon flux, $\Phi_0(\lambda,t)$. However, if necessary, $\lambda_{\max}$ is modified in order to ensure that only super-band gap generation is considered: $\lambda_{\max} \le hc/E_g$.

To simulate the influence of light trapping mechanisms, internal reflections at both contacts can be additionally specified. They can either be set as a constant value or a file can be loaded specifying a wavelength dependent reflection. The light then passes through the layer stack several times as defined by the user. Internal reflections and refractions within the solar cell layers are neglected. The residual flux after the defined number of passes is added to the transmitted flux at the contact, at which the calculation ends.

This model was especially designed to estimate the influence of light trapping on textured wafers and to adapt the optical simulation to optical measurements on real solar cells. Thus, if coherence effects can be neglected (which is the case as long as considering wafer based solar cells, but may be even true for some thin film solar cells) the measured reflectance and transmittance of the solar cell can be directly used to consistently simulate the optical generation rate $G(x,t) = G_n(x,t) = G_p(x,t)$.

13.2.2 Electrical Calculations

In the following, the differential equations and corresponding boundary conditions, which are solved by AFORS-HET under the various conditions, are stated.

An arbitrary stack of semiconductor layers can be modeled. Within each semiconductor layer the Poisson equation and the continuity equations for electrons and holes have to be solved. At each semiconductor/semiconductor interface and at the front and backside boundary of the stack the current through these interfaces/boundaries can be described by different physical models. This leads to a nonlinear coupled system of three differential equations with respect to time and space derivatives. The electron density $n(x,t)$, the hole density $p(x,t)$, and the electric potential $\varphi(x,t)$ are the independent variables, for which this system of differential equations is solved. A numerical discretisation scheme is used as outlined by Selberherr [2] in order to linearize the problem. It can be solved for different calculation modes:

- EQ calculation mode, describing thermodynamic equilibrium at a given temperature,
- DC calculation mode, describing steady-state conditions under an external applied voltage or current and/or illumination,
- AC calculation mode, describing small additional sinusoidal modulations of the external applied voltage/illumination, and
- TR calculation mode, describing transient changes of the system, due to general time dependent changes of the external applied voltage or current and/or illumination.

In case of using the EQ or the DC calculation mode, all time derivatives vanish, resulting in a simplified system of differential equations. The system of differential equations is then solved for time independent, but position dependent functions:

$$\begin{aligned} n(x,t) &= n^{EQ}(x), & n(x,t) &= n^{DC}(x), \\ p(x,t) &= p^{EQ}(x), & p(x,t) &= p^{DC}(x), \\ \varphi(x,t) &= \varphi^{EQ}(x), & \varphi(x,t) &= \varphi^{DC}(x). \end{aligned} \quad (13.3)$$

In case of using the AC calculation mode, it is assumed that all time dependencies can be described by small additional sinusoidal modulations of the steady-state solutions. All time dependent quantities are then modelled with complex numbers (marked by a dash ~), which allows determining the amplitudes and the phase shifts between them, i.e., for the independent variables of the system of differential equations, one gets

$$\begin{aligned} n(x,t) &= n^{DC}(x) + \tilde{n}^{AC}(x)\, e^{i\omega t}, \\ p(x,t) &= p^{DC}(x) + \tilde{p}^{AC}(x)\, e^{i\omega t}, \\ \varphi(x,t) &= \varphi^{DC}(x) + \tilde{\varphi}^{AC}(x)\, e^{i\omega t}. \end{aligned} \quad (13.4)$$

In case of using the TR calculation mode, the description of the system starts with a steady-state (DC-mode) simulation, specifying an external applied voltage or current and/or illumination. An arbitrary evolution in time of the external applied voltage or current and/or illumination can then be specified by loading an appropriate file. Then, the time evolution of the system, i.e. the functions $n(x,t)$, $p(x,t)$, $\varphi(x,t)$ during and after the externally applied changes are calculated.

Within the bulk of each semiconductor layer, the Poisson equation and the continuity equations for electrons and holes are to be solved in one dimension.

Poisson´s equation, which is to be solved at each x, reads:

$$\frac{\varepsilon_0 \varepsilon_r(x)}{q} \frac{\partial^2 \varphi(x,t)}{\partial x^2} = p(x,t) - n(x,t) + N_D(x) - N_A(x) + \sum_{trap} \rho_{trap}(x,t) \quad (13.5)$$

with

$$\rho_{trap}(x,t) = -\int dE \; f_{1,trap}^{SRH}(E,x,t)\, N_{trap}(E) \quad (13.6)$$

in case of acceptor-like defects, and

$$\rho_{trap}(x,t) = +\int dE \; f_{0,trap}^{SRH}(E,x,t)\, N_{trap}(E) \quad (13.7)$$

in case of donor-like defects, q being the electron charge and ε_0, $\varepsilon_r(x)$ being the absolute/relative dielectric constant of the semiconductor layer. $N_D(x)$, $N_A(x)$ are the doping densities of fixed donor/acceptor states at a position x within the layer.

They are assumed to be always completely ionized and are an input parameter of the semiconductor layer. Further input parameters are the defect distributions $N_{trap}(E)$ for each defect within the layer, that is the amount of traps at an energy position E within the band gap. Shockley-Read-Hall traps are either empty or singly occupied by an electron, depending on the local particle densities $n(x,t)$, $p(x,t)$ within in the layer, thus they can be locally charged/uncharged within the system. The defect distributions $N_{trap}(E)$ together with the corresponding defect occupation functions $f_{0,trap}^{SRH}(E,x,t)$, $f_{1,trap}^{SRH}(E,x,t)$, specifying the probability that traps with an energy position E within the band gap are empty or singly occupied with electrons, define the defect density of charged defects $\rho_{trap}(x,t)$, which enters in the Poisson equation. The explicit formula for the defect occupation functions are described in a later section. Dangling bond defects can also be modeled in AFORS-HET, these can be empty, singly or even doubly occupied by electrons, however this is not covered here.

The one-dimensional equations of continuity and transport for electrons are given by:

$$-\frac{1}{q}\frac{\partial j_n(x,t)}{\partial x} = G_n(x,t) - R_n(x,t) \; - \frac{\partial}{\partial t} n(x,t) \qquad (13.8)$$

with

$$j_n(x,t) \;=\; q\;\mu_n\;n(x,t)\;\frac{\partial E_{Fn}(x,t)}{\partial x} \qquad (13.9)$$

$$E_{Fn}(x,t) = E_C(x) + kT\,\ln\frac{n(x,t)}{N_C(x)} = -q\chi(x) + q\varphi(x,t) + kT\,\ln\frac{n(x,t)}{N_C(x)} \qquad (13.10)$$

and for holes:

$$+\frac{1}{q}\frac{\partial j_p(x,t)}{\partial x} = G_p(x,t) - R_p(x,t) \; - \frac{\partial}{\partial t} p(x,t) \qquad (13.11)$$

with

$$j_p(x,t) \;=\; q\;\mu_p\;p(x,t)\;\frac{\partial E_{Fp}(x,t)}{\partial x} \qquad (13.12)$$

$$E_{Fp}(x,t) = E_V(x) - kT\,\ln\frac{p(x,t)}{N_V(x)} = -q\chi(x) + q\varphi(x,t) - E_g(x) - kT\,\ln\frac{p(x,t)}{N_V(x)}. \qquad (13.13)$$

The electron/hole currents $j_n(x,t)$, $j_p(x,t)$ are driven by the gradient of the corresponding quasi Fermi energy $E_{Fn}(x,t)$, $E_{Fp}(x,t)$. Input parameters are the electron and hole mobilities μ_n, μ_p, the electron affinity χ, the band gap E_g, the conduction/valence band energy E_C, E_V and the effective conduction/valence band density of states N_C, N_V of the semiconductor layer, which can be spatial dependant. The electron/hole super-band gap generation rates $G_n(x,t)$, $G_p(x,t)$ have to be determined by optical modeling as discussed in the previous section. The recombination rates $R_n(x,t)$, $R_p(x,t)$ are determined by the recombination models discussed in the next section.

13.2.3 Basic Recombination Models

Recombination from the conduction band into the valence band may occur directly, i.e. via radiative band to band recombination [4], $R_{n,p}^{BB}(x,t)$, or via Auger recombination [5], $R_{n,p}^{A}(x,t)$. It may also occur via defect states located within the band gap of the semiconductor, i.e. via Shockley-Read-Hall recombination [6, 7] $R_{n,p}^{SRH}(x,t)$ or via dangling bond recombination [8], $R_{n,p}^{DB}(x,t)$:

$$R_{n,p}(x,t) = R_{n,p}^{BB}(x,t) + R_{n,p}^{A}(x,t) + R_{n,p}^{SRH}(x,t) + R_{n,p}^{DB}(x,t) \tag{13.14}$$

Radiative recombination

The radiative band to band rate constant r^{BB} has to be specified in order to equate the radiative band to band recombination rates $R_{n,p}^{BB}(x,t)$. The resulting electron and hole recombination rates are always equal with

$$R_{n,p}^{BB}(x,t) = r^{BB}\left\{n(x,t)\,p(x,t) - N_C\,N_V\,e^{-E_g/kT}\right\}. \tag{13.15}$$

In case of using the DC or AC calculation mode and neglecting second order terms in case of the AC calculation mode, this simplifies to

$$\begin{aligned} R_{n,p}^{BB}(x) &= r^{BB}\left\{n^{DC}(x)\,p^{DC}(x) - N_C\,N_V\,e^{-E_g/kT}\right\}, \\ R_{n,p}^{BB}(x,t) &= R_{n,p}^{BB}(x) + \tilde{R}_{n,p}^{BB}(x)\,e^{i\omega t}, \\ \tilde{R}_{n,p}^{BB}(x) &= r^{BB}\,n^{DC}(x)\,\tilde{p}^{AC}(x) + r^{BB}\,p^{DC}(\mathrm{x})\,\tilde{n}^{AC}(x). \end{aligned} \tag{13.16}$$

Auger recombination

The electron/hole Auger rate constants r_n^A, r_p^A have to be specified in order to calculate the Auger recombination rates $R_{n,p}^{A}(x,t)$. Again, the resulting electron and hole recombination rates are always equal:

$$R_{n,p}^{A}(x,t) = \left[r_n^A\,n(x,t) + r_p^A\,p(x,t)\right]\left\{n(x,t)\,p(x,t) - N_C\,N_V\,e^{-E_g/kT}\right\} \tag{13.17}$$

In case of using the DC or AC calculation mode, neglecting second order terms within the AC calculation mode, this simplifies to

$$R_{n,p}^{A}(x)=\left[r_n^A\, n^{DC}(x)+r_p^A\, p^{DC}(x)\right]\left\{n^{DC}(x)\,p^{DC}(x)-N_C\, N_V\, e^{-E_g/kT}\right\},$$
$$R_{n,p}^{A}(x,t)=R_{n,p}^{A}(x)+\tilde{R}_{n,p}^{A}(x)\,e^{i\omega t}, \tag{13.18}$$
$$\tilde{R}_{n,p}^{A}(x)=\left[r_n^A\, n^{DC}(x)^2+2\,r_p^A\, n^{DC}(x)\,p^{DC}(x)\right]\tilde{p}^{AC}(x)$$
$$+\left[r_p^A\, p^{DC}(x)^2+2\,r_n^A\, n^{DC}(x)\,p^{DC}(x)\right]\tilde{n}^{AC}(x).$$

Shockley Read Hall recombination

Shockley-Read-Hall (SRH) recombination requires specifying the character (acceptor-like or donor-like) of each defect, its electron/hole capture cross sections σ_n^{trap}, σ_p^{trap}, and its energetic defect distribution $N_{trap}(E)$ within the band gap of the semiconductor. These are input parameters for each defect chosen. An arbitrary number of defects with either one of the following energetic defect distributions $N_{trap}(E)$ can be chosen:

- point like distributed
- constantly distributed within a specific region within the band gap
- exponentially decaying from the conduction/valence band into the band gap:

$$N_{trap}(E)=N_{trap}^{C,tail}\, e^{-(E_C-E)/E_{trap}^{C,tail}},\quad N_{trap}(E)=N_{trap}^{V,tail}\, e^{-(E-E_V)/E_{trap}^{V,tail}} \tag{13.19}$$

  with $N_{trap}^{C,tail}$, $N_{trap}^{V,tail}$: tail state density at the conduction/valence band, and $E_{trap}^{C,tail}$, $E_{trap}^{V,tail}$: characteristic decay energy.
- Gaussian distributed within the band gap:

$$N_{trap}(E)=\frac{N_{trap}^{db}}{\sigma_{trap}^{db}\sqrt{2\pi}}\,e^{-\frac{\left(E-E_{trap}^{db}\right)^2}{2\sigma_{trap}^{db\,2}}} \tag{13.20}$$

  with N_{trap}^{db}: total dangling bond state density, E_{trap}^{db}: specific energy of the Gaussian dangling bond peak, σ_{trap}^{db}: standard deviation of the Gaussian dangling bond distribution.

For each defect, electron/hole capture coefficients $c_{n,p}^{trap}$ are equated $c_{n,p}^{trap}=v_{n,p}\,\sigma_{n,p}$ from the electron/hole thermal velocity $v_{n,p}$ of the semiconductor (another input parameter of the semiconductor layer). The corresponding electron/hole emission coefficients $e_{n,p}^{trap}(E,x,t)$ are then given by:

$$e_n^{trap}(E,x,t)=c_n^{trap}\,N_C\,e^{-(E_C-E)/kT}$$
$$e_p^{trap}(E,x,t)=c_p^{trap}\,N_V\,e^{-(E-E_V)/kT} \tag{13.21}$$

Finally, the Shockley-Read-Hall recombination rate due to the defects is

$$R_n^{SRH}(x,t)=\sum_{trap}\int dE\left\{c_n^{trap}\,n(x,t)N_{trap}(E)\,f_{0,trap}^{SRH}(E,x,t)-e_n^{trap}(E,x,t)N_{trap}(E)\,f_{1,trap}^{SRH}(E,x,t)\right\}$$

$$R_p^{SRH}(x,t)=\sum_{trap}\int dE\left\{c_p^{trap}\,p(x,t)N_{trap}(E)\,f_{1,trap}^{SRH}(E,x,t)-e_p^{trap}(E,x,t)N_{trap}(E)\,f_{0,trap}^{SRH}(E,x,t)\right\}$$

(13.22)

A positive electron/hole SRH recombination rate means recombination of an electron/hole from the conduction/valence band into the defect state (trap), a negative electron/hole SRH recombination rate means sub-band gap generation of an electron/hole from the defect state (trap) into the conduction/valence band.

13.2.4 Defect Occupation Functions

A Shockley-Read-Hall defect can be either empty or occupied by an electron, thus the two defect occupation functions sum up to unity:

$$f_{0,trap}^{SRH}(E,x,t)+f_{1,trap}^{SRH}(E,x,t)=1 \tag{13.23}$$

The Shockley-Read-Hall defect occupation function $f_{1,trap}^{SRH}(E,x,t)$, which is stating the probability for a trap to be singly occupied with an electron, will be explicitly stated in case of using the EQ, DC, AC or the TR calculation mode. $f_{0,trap}^{SRH}(E,x,t)$ can then directly be equated.

Generally, a local change of the trapped charge stored in SRH defects must be determined by the difference between the local electron and hole SRH recombination rates:

$$\frac{d}{dt}\rho_{\mathrm{trap}}(x,t)=R_p^{SHR}(x,t)\;-\;R_n^{SHR}(x,t) \tag{13.24}$$

This defines for each defect an additional differential equation for its SRH defect occupation function $f_{1,trap}^{SRH}(E,x,t)$ with respect to its time derivative:

$$\frac{d}{dt}f_{1,trap}^{SRH}(E,x,t)=$$
$$\left(c_n^{trap}\,n(x,t)+e_p^{trap}(E,x,t)\right)\left(1-f_{1,trap}^{SRH}(E,x,t)\right)\;-\;\left(c_p^{trap}\,p(x,t)+e_n^{trap}(E,x,t)\right)f_{1,trap}^{SRH}(E,x,t) \tag{13.25}$$

In case of using the EQ or the DC calculation mode, the time derivative vanishes, and an explicit expression for the SRH defect occupation function, $f_{1,trap}^{SRH}(E,x)$, which is no longer time dependant, can be derived:

$$f_{1,trap}^{SRH,DC}(E,x)=\frac{c_n^{trap}\,n^{DC}(x)+e_p^{trap}(E,x)}{c_n^{trap}\,n^{DC}(x)+e_n^{trap}(E,x)+c_p^{trap}\,p^{DC}(x)+e_p^{trap}(E,x)} \tag{13.26}$$

In case of using the AC calculation mode, this differential equation can be explicitly solved, assuming the time dependencies $n(x,t) = n^{DC}(x) + \tilde{n}^{AC}(x)\, e^{i\omega t}$, $p(x,t) = p^{DC}(x) + \tilde{p}^{AC}(x)\, e^{i\omega t}$. Neglecting second order terms, one gets for the SRH defect occupation function in the AC calculation mode, $f_{1,trap}^{SRH,AC}(E,x,t)$:

$$\begin{aligned} f_{1,trap}^{SRH,AC}(E,x,t) &= f_{1,trap}^{SRH,DC}(E,x) + \tilde{f}_{1,trap}^{SRH,AC}(E,x)\, e^{i\omega t} \\ \tilde{f}_{1,trap}^{SRH,AC}(E,x) &= \frac{c_n^{trap}\, f_{0,trap}^{SRH,DC}(E,x)\, \tilde{n}^{AC}(x) \;-\; c_p^{trap}\, f_{1,trap}^{SRH,DC}(E,x)\, \tilde{p}^{AC}(x)}{c_n^{trap}\, n^{DC}(x) + e_n^{trap}(E,x) + c_p^{trap}\, p^{DC}(x) + e_p^{trap}(E,x) + i\omega} \end{aligned} \tag{13.27}$$

In case of using the TR calculation mode, the transient SRH defect occupation function $f_{1,trap}^{SRH,TR}(E,x,t_{i+1})$ at the time step t_{i+1} for an evolution of the system from the time point t_i towards the time point t_{i+1} can be stated by solving the differential equation using a full implicit time discretisation scheme with respect to the particle densities and the emission rates:

$$\begin{aligned} &\frac{d}{dt} f_{1,trap}^{SRH,TR}(E,x,t) \\ &= \left(c_n^{trap}\, n(x,t_{i+1}) + e_p^{trap}(E,x,t_{i+1})\right)\left(1 - f_{1,trap}^{SRH,TR}(E,x,t)\right) \\ &- \left(c_p^{trap}\, p(x,t_{i+1}) + e_n^{trap}(E,x,t_{i+1})\right) f_{1,trap}^{SRH,TR}(E,x,t) \end{aligned} \tag{13.28}$$

An analytical solution of this differential equation leads to:

$$f_{1,trap}^{SRH,TR}(E,x,t_{i+1}) = f_{1,trap}^{SRH,DCtr}(E,x,t_{i+1}) - \frac{f_{1,trap}^{SRH,DCtr}(E,x,t_{i+1}) - f_{1,trap}^{SRH,TR}(E,x,t_i)}{e^{dt\,\left(c_n^{trap}\, n(x,t_{i+1}) + e_n^{trap}(E,x,t_{i+1}) + c_p^{trap}\, p(x,t_{i+1}) + e_p(E,x,t_{i+1})\right)}} \tag{13.29}$$

with

$$f_{1,trap}^{SRH,DCtr}(E,x,t_{i+1}) = \frac{c_n^{trap} n(x,t_{i+1}) + e_p^{trap}(E,x,t_{i+1})}{c_n^{trap} n(x,t_{i+1}) + e_n^{trap}(E,x,t_{i+1}) + e_p^{trap} p(x,t_{i+1}) + e_p^{trap}(E,x,t_{i+1})} \tag{13.30}$$

In the steady-state limit, i.e. for the limit $dt \to \infty$, $dt = t_{i+1} - t_i$, this formula converts to the well-known steady state SRH defect occupation function $f_{1,trap}^{SRH,DC}(E,x) = f_{1,trap}^{SRH,DCtr}(E,x,t_{i+1})$.

13.2.5 Input Parameters of AFORS-HET

In the following, the most important input parameters of AFORS-HET are explained.

Optical parameters

The incoming spectral photon flux $\Phi_0(\lambda,t)$, that is the number of incident photons of wavelength λ at the time t, has to be stated. An optical calculation determines the super-band gap generation rate $G(x,t)$ within the semiconductor stack. The thicknesses L_i and the dielectric properties of the semiconductor layers have to be specified, i.e. the complex refractive indices, $\tilde{n}_i(\lambda)=n_i(\lambda)-\mathrm{i}\,k_i(\lambda)$ with refractive index $n(\lambda)$ and extinction coefficient $k(\lambda)$. If the model Lambert-Beer absorption is chosen a measured reflectivity $R(\lambda)$ of the semiconductor stack can be specified and the resulting absorption $A(\lambda,x,t)$ within the semiconductor stack will be calculated assuming Lambert Beer absorption by using the specified values for $k_i(\lambda)$ only. If the model coherent/incoherent internal multiple reflections is chosen, the reflectivity $R(\lambda)$, the transmisivity $T(\lambda)$ and the absorption $A(\lambda,x,t)$ of the semiconductor stack is calculated from the specified values $n_i(\lambda)$, $k_i(\lambda)$, assuming plain surfaces within the stack but taking coherent internal multiple reflections into account, if desired. For both models, $G(x,t)$ is calculated from $A(\lambda,x,t)$ by integration over all wavelengths of the incident spectrum.

Layer parameters

For each semiconductor layer, the thickness L, the electron/hole mobilities μ_n, μ_p, the effective valence/conduction band densities N_V, N_C, the electron/hole thermal velocities v_n, v_p, the electron affinity χ, the relative dielectric constant ε, the doping profile $N_D(x)$, $N_A(x)$ and the band gap E_g of the semiconductor has to be specified. Up to four different recombination models can be chosen, (1) radiative recombination, (2) Auger recombination, (3) Shockley-Read-Hall recombination,. For radiative recombination, the radiative band to band rate constant r^{bb} has to be specified. For Auger recombination, the electron/hole Auger rate constants r_n^{Aug}, r_p^{Aug} have to be specified. For Shockley-Read-Hall recombination, the defect density distribution within the band gap of the semiconductor $N_{trap}(E)$ and two capture cross sections σ_n, σ_p have to be specified.

Interface parameters

The electron/hole current transport across a semiconductor/semiconductor interface can be described by three different interface models,

(1) drift-diffusion without interface defects ("no interface" in AFORS-HET)
(2) drift-diffusion with interface defects ("drift-diffusion" in AFORS-HET)
(3) thermionic emission interface

If model (1) is chosen, no additional interface defects can be specified. Otherwise, an interface defect distribution $N_{trap}^{it}(E)$ and two electron/hole interface capture cross sections σ_n^{it}, σ_p^{it} can be specified. For both models (1) and (2) an interface thickness L_{it} has to be stated and transport across the semiconductor/semiconductor interface layer is treated like in the bulk of the semiconductor layers (drift-diffusion approximation), assuming that within the interface layer all semiconductor properties vary linearly from their given values located left and right side of the interface. In model (3) the interface is assumed to be infinitely thin. Interface defects are specified by an interface defect distribution $N_{trap}^{it}(E)$ and four capture cross sections $\sigma_{n,I}^{it}$, $\sigma_{n,II}^{it}$, $\sigma_{p,I}^{it}$, $\sigma_{p,II}^{it}$ for electron/hole capture from both sides of the interface. Transport across the interface is treated according to the theory of thermionic emission.

Boundary parameters

The boundaries of the semiconductor stack may either be metallic (usually constituting the contacts of the solar cell) or they may be insulating in order to simulate some specific measurement conditions requiring insulator contacts. Four different boundary models can be chosen:

(1) flatband metal/semiconductor contact
(2) Schottky metal/semiconductor contact
(3) insulator contact
(4) metal/insulator/semiconductor contact

If choosing the flatband metal/semiconductor contact, there will be no band bending induced within the semiconductor due to the contact (flatband contact). The electron/hole surface recombination velocities $S_n^{front/back}$, $S_p^{front/back}$ of the metallic contact have to be specified. If choosing the Schottky metal/semiconductor contact, an additional work function $\phi^{front/back}$ of the metal contact has to be defined which can give rise to a band bending within the semiconductor. If choosing the insulator/semiconductor or the metal/insulator/semiconductor contact, interface states between the insulator and the semiconductor can be defined, that is an interface defect distribution $N_{trap}^{it}(E)$ and interface capture cross sections for electrons and holes σ_n^{it}, σ_p^{it} have to be specified. In case of the metal/insulator/semiconductor contact an additional interface capacitance $C^{front/back}$ has to be specified. The charge in the interface defects can give rise to a band bending within the semiconductor.

Circuit elements

A series resistance R_s, a parallel resistance R_p, a parallel capacitance C_p and in case of a metal/insulator/semiconductor contact also a series capacitance C_s can be specified. If circuit elements are specified, the internal cell voltage V_{int} and the internal cell current I_{int} of the semiconductor stack will differ from the external cell voltage V_{ext} and external cell current I_{ext} of the modeled device, which are the measurable quantities .

External parameters

External parameters in AFORS-HET are parameters that are externally applied to the device. These are the temperature T of the device, the spectral photon flux $\Phi_0(\lambda,t)$ and the external cell voltage $V_{ext}(t)$ or the external cell current $I_{ext}(t)$. The remaining quantity, i.e. the external cell current $I_{ext}(t)$ or the external cell voltage $V_{ext}(t)$, respectively, will be calculated. In order to simulate a characterization method, some external parameters are varied, while some other parameters which are specific for the measurement, are monitored. For example, in order to simulate the IV characteristic of a solar cell, the external cell voltage is varied, and the resulting external cell current is monitored.

References

[1] Gummel, H.: A self-consistent iterative scheme for one-dimensional steady state transistor calculations. IEEE Trans. Electron Devices 11(10), 455–465 (2005)

[2] Selberherr, S.: Analysis and simulation of semiconductor devices. Springer, Vienna (1984)

[3] Stangl, R., Leendertz, C., Haschke, J.: Numerical simulation of solar cells and solar cell characterization methods: the open-source on demand program afors-het. In: Solar Energy, IN-TECH (2010), `http://www.intechopen.com/source/pdfs/8561/InTech-Numerical_simulation_of_solar_cells_and_solar_cell_characterization_methods_the_open_source_on_demand_program_afors_het.pdf`, ISBN: 978-953-307-052-0

[4] Sze, S.M., Kwok, K.N.: Physics of Semiconductor Devices, p. 40. John Wiley & Sons, Inc., Hoboken (2007)

[5] Landsberg, P.: The band-band auger effect in semiconductors. Solid-State Electron 30, 1107–1115 (1987)

[6] Shockley, W., Read, W.T.: Statistics of the recombinations of holes and electrons. Phys. Rev. 87, 835 (1952)

[7] Hall, R.N.: Electron-hole recombination in germanium. Phys. Rev. 87, 387 (1952)

[8] Sah, C.-T., Shockley, W.: Electron-hole recombination statistics in semiconductors through flaws with many charge conditions. Phys. Rev. 109, 1103 (1958)

Chapter 14
Modeling an *a*-Si:H/*c*-Si Solar Cell with AFORS-HET

Caspar Leendertz and Rolf Stangl

Helmholtz-Zentrum für Materialien und Energie, Institut für Silizium-Photovoltaik, Kekuléstr.5, D-12489 Berlin, Germany

Abstract. The physics models and material parameters needed to simulate an *a*-Si:H/*c*-Si solar cell with AFORS-HET are discussed and a simulation study showing solar cell characteristics subject to emitter doping, i-layer thickness and interface quality is presented. The AFORS-HET user interface is introduced so that the interested reader can repeat the simulation study. It is explained in detail how to define a structure and how to simulate a solar cell under different external conditions such as external current, voltage and illumination and how to calculate I-V curves to obtain solar cell characteristics.

14.1 Modeling an *a*-Si:H/*c*-Si Solar Cell

Simulation of an *a*-Si:H/*c*-Si solar cell is a demanding task due to the advanced physics models and the large number of material parameters involved and it has thus been the subject of various publications [1, 2, 3, 4, 5].

While the modeling of crystalline silicon wafers is well understood, the amorphous layer and especially the interface between the *a*-Si:H layer and the crystalline Si are by far more difficult to model. As discussed in Chapter 6 in this book, *a*-Si:H can be described as a semiconductor with a defect distribution consisting of band tail and dangling bond states in the band gap. A realistic simulation of the amorphous layer must therefore allow for the calculation of the recombination via distributed Shockley-Read-Hall (SRH) and dangling bond defects. Furthermore, it is crucial to include the interface explicitly in the simulation since this is one of the most critical parts of the *a*-Si:H/*c*-Si solar cell, which strongly influences the solar cell characteristics. A realistic simulation must therefore account for distributed defects at the interface and comprise a model for carrier transport across a heterojunction. The importance of tunneling effects for the transport across the *a*-Si:H/*c*-Si interface as well as at the TCO/*a*-Si:H interface has been discussed in several publications and is still a matter of debate [1, 2].

A further demanding task concerns the various material parameters that enter into the simulation. Of crucial importance are the density, the distribution as well

W.G.J.H.M. van Sark et al. (Eds.): Physics & Tech. of Amorphous-Crystalline, EM, pp. 459–482.
springerlink.com

as the capture cross sections of the defects in the *a*-Si:H layer and at the interface. Moreover, the band offsets at the interface have a substantial influence on the cell characteristics [5]. Further parameters are the *a*-Si:H mobilities and, for optical calculations, the dielectric function of the *a*-Si:H, i.e. the *n*,*k*-data. All the material parameters mentioned above have in common that they are difficult to measure and that they can vary for *a*-Si:H of different doping levels, deposited under different conditions and possibly even with different thicknesses, as has been discussed in Chapters 6 and 7 in this book. Therefore, the simulation model has to be carefully calibrated with different measurement techniques such as IV-curves, EQE and QSSPC measurements. Additionally, input from techniques such as constant final state yield photoelectron spectroscopy (CFSYS) to determine the defect distribution in the *a*-Si:H layer [6] and planar conductance measurements or CFSYS to determine band offsets [7] are needed.

In the following, simulations of *a*-Si:H/*c*-Si solar cells calculated with AFORS-HET will be presented. The focus will lie on the discussion of the structure based on an n-type wafer as this is the structure with intrinsic advantages with regard to efficiency [5]. It is also the one used in the record HIT-cell produced by SANYO [8]. The basics of the semiconductor simulation, as implemented in AFORS-HET, have been described in Chapter 13 in this book, while in this section the physical effects that are critical for the *a*-Si:H/*c*-Si solar cell are discussed in more detail.

14.1.1 Physics Models for Simulating Crystalline and Amorphous Silicon

The crystalline wafer is modeled following in most respects the approach of the widely used simulation software PC1D whereby the user only specifies the wafer doping and a defect density and all the other material parameters are calculated from this input according to different models [9]. The electronic defect levels in the band gap that arise from the residual impurities of oxygen, carbon or 3d transition metals that typically determine the wafer quality are combined and calculated as one neutral, mid-gap SRH defect with temperature- and doping-dependent recombination. Moreover, models are implemented, which describe the doping dependence of band gap, effective band densities and mobilities [10]. The Auger recombination is calculated dependent on doping and excess carrier concentration following the model of Kerr and Cuevas [11].

To model the amorphous layers accurately, a detailed model to describe the recombination processes and the charge state of the defects is crucial. The tail states of the amorphous layer are modeled by Shockley-Read-Hall (SRH) defects. In order to model the dangling bond states of the amorphous layer, recombination via amphoteric defects according to Sah and Shockley [12] should be used. Since this is not yet fully implemented in AFORS-HET a typical approximation is used, in which one amphoteric defect is modeled by a pair of acceptor-like and donor-like single-electron-states (SRH defects), separated by a certain correlation energy U [13]. In this way the four capture/emission processes characterized by the capture emission coefficients σ and the two energy levels of an amphoteric defect can be described (see Fig. 14.1).

Transport across the *a*-Si:H/*c*-Si heterojunction interface is calculated with the drift-diffusion model as described in Chapter 13, while tunneling effects are neglected. The TCO front and aluminum back contact are modeled as Schottky boundary conditions. For the optical simulation the simple Lambert-Beer model is used.

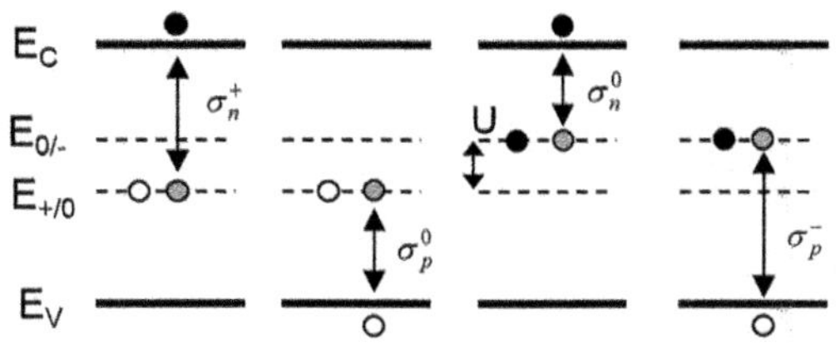

Fig. 14.1 Capture/emission processes and energy levels for recombination via amphoteric defects. The low energy state is donor-like while the high energy state is acceptor-like.

14.1.2 Device Structure and Material Parameters for a-Si:H/c-Si Cell Simulation

The structure used for simulation consists of a crystalline silicon wafer sandwiched between two amorphous layers that act as emitter and back surface field respectively. Additionally, an intrinsic buffer (i-layer) is introduced on the emitter side. The structure and the defect distributions are displayed in Fig. 14.2 and the material parameters are given in Table 14.1.

To model the interface between the emitter and the absorber, a thin layer is added, which is *c*-Si-like with respect to the band gap but comprises Gaussian distributed dangling bond defects. Although a direct characterization of the interface has not been carried out, this is an obvious approximation, as it has to be assumed that open bonds from the crystal reaching into the amorphous network act as recombination centers at the interface. The details of the defect distribution at the interface, for example its exact position and width, however, have not yet been clarified. For the calculation presented here, a mid gap-centered distribution with a width of 180meV and a correlation energy of 150meV is assumed at the interface. The uncharged capture cross sections are defined to be 10^{-18}cm^2 with a ratio of charged to uncharged capture cross sections of 10. The exact value of capture cross sections as well as the ratios of charged and uncharged, electron and hole capture cross sections have also not yet been fully clarified. The parameters for defect distribution in the amorphous layers can be approximated by analyzing CFSYS data of the given layers [6]. In accordance with these measurements, Gaussian distributed dangling bond defects and band tails have been defined. Gaussian distributed dangling bond defects centered at about 0.55meV above the

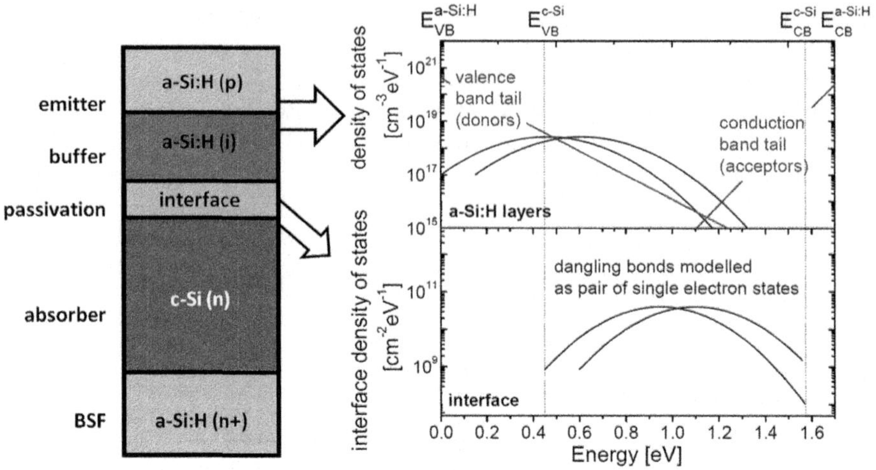

Fig. 14.2 Device structure for simulating a heterojunction solar cell with crystalline absorber and amorphous buffer, emitter and back surface field layer (left). The displayed defect distributions in the *a*-Si:H bulk and at the interface (right) have a strong influence on the solar cell performance and the exact distribution of defects are still a matter of debate.

Table 14.1. *a*-Si:H/*c*-Si material parameters used in the simulation. The integrated dangling bond defect densities are given in the table (see text for details on the distribution).

material parameters					
	emitter	buffer	interface	absorber	BSF
	a-Si:H(p)	*a*-Si:H(i)	p-type	*c*-Si(p)	*a*-Si:H(p+)
thickness [μm]	0.01	varied	0.001	250	30
doping [cm^{-3}]	varied	$1.25x10^{17}$	$1.5x10^{16}$	$1.5x10^{16}$	$9x10^{18}$
band gap [eV]	1.7	1.7	1.124	1.124	1.7
electron mobility [$cm^2V^{-1}s^{-1}$]	5	5	1111	1111	5
hole mobility [$cm^2V^{-1}s^{-1}$]	1	1	421.6	421.6	1
defect type	dangling bonds	dangling bonds	dangling bonds	neutral mid gap	dangling bonds
defect density [cm^{-3}]	$1.5x10^{18}$	$1.3x10^{17}$	varied	$1x10^{10}$	$5.88x10^{18}$

valence band with a width of 160meV are defined for the p-type emitter layer. The band tails typical have defect density of about $10^{21}cm^{-3}eV^{-1}$ at the band edges and for the p-type emitter layer they decay exponentially with a decay rate of

0.08eV for the valence band tail and 0.04eV for the conduction band tail. These values vary for different doping levels and deposition conditions. For the band offsets that result from the electron affinity and band gap given in Table 14.1 we assume 150meV for the conduction band offset and about 430meV for the valence band offset, which is in agreement with CFSYS [6] and planar conductance measurements [7].

14.1.3 Sensitivity Study for an a-Si:H(p)/c-Si(n) Solar Cell

As discussed in Chapters 6 and 7 in this book, the defect density at the *a*-Si:H/*c*-Si interface is a very critical parameter since interface recombination is often the limiting factor for the open circuit voltage the solar cell can achieve. There is an experimentally proven interrelation between emitter doping and defect density in that the defect density rises with increasing emitter doping [14]. On the other hand, a high emitter doping level is also desirable as the open circuit voltage can be increased by means of strong *c*-Si dark band bending. Hence in record cells a thin intrinsic buffer layer (i-layer) has been introduced to passivate the interface. Depending on the i-layer thickness, there will be a major influence of this buffer layer on the overall band structure (e.g. distribution of total band bending between the *c*-Si and the *a*-Si:H side) which in turn will influence the cell performance.

The simulation study presented in Fig. 14.3 elucidates the effects of this interrelation between emitter doping, interface defect density and i-layer thickness on the solar cell performance. Open circuit voltage, fill factor and solar cell efficiency are shown as functions of the emitter Fermi level position, which is determined by the doping and the defect distribution in the *a*-Si:H emitter layer. For this study the defect density of the emitter layer has been fixed (see Table 14.1) and the doping varied between $10^{16}cm^{-3}$ and $10^{20}cm^{-3}$. This study has been completed for four different technologically interesting i-layer thicknesses between 0nm and 15nm, for two typical interface defect densities and for structures with and without back surface layer.

Since some approximations have been made concerning the material parameters and physics models, which may influence the details of the simulation results. we will only qualitatively analyze the strong trends in the simulated data. As has also been shown in earlier simulation studies [3, 15, 5], the interface defect density has a negative effect on the open circuit voltage, but for many sets of parameters the negative influence on the fill factor is even be more important (c.f. Fig. 14.3). Furthermore we can make the following observations by analysis of the data in Fig. 14.3:

- While in experiments typically a higher V_{OC} is observed for thicker i-layers [16, 17] due to reduced interface recombination, simulations show that a thicker buffer layer in turn leads to a decrease in fill factor.

- While it is known from experiments that higher emitter doping gives rise to more interface recombination if no i-layer or a very thin i-layer is used [18, 19, 20, 21], simulations show that a high emitter doping level can, on the other hand, neutralize the negative effects of higher i-layer thickness and high interface defect density.
- A back surface field enhances the maximum possible open circuit voltage and slightly optimizes the fill factor for low emitter doping levels.

The most obvious conclusion for designing solar cells that can be drawn from these simulations is that the emitter doping level should be maximized and the thinnest possible i-layer selected for effective interface passivation. A set of parameters which might not be far from current record cells would suggest a structure with 5nm i-layer, an interface defect density of $5x10^{10}$ cm^{-2}, a Fermi level <350meV above the valence band edge in the emitter and a back surface field. According to our simulation study this cell design could achieve an efficiency of above 22%.

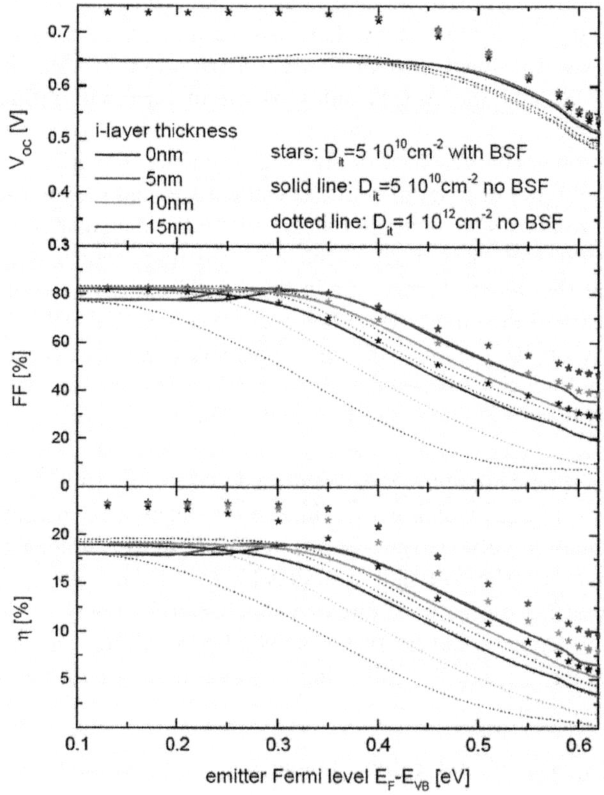

Fig. 14.3 Solar cell characteristics dependent on the emitter Fermi level position determined by the doping level. Simulations are carried out for different i-layer thicknesses, with varying interface defect density and with or without a back surface layer.

14.2 Introduction to the AFORS-HET User Interface

In the previous section the material parameter and physics models needed to simulate a silicon heterojunction solar cell have been discussed and a simulation study was shown. In this subchapter we will now introduce the AFORS-HET user interface so that the reader will be able to do his/her own simulation studies.

To give an impression of the broad scope AFORS-HET has to offer, this section starts out with a brief overview of the AFORS-HET main window. Fig. 14.4 shows the main window when the most advanced calculation mode is activated. In this layout the whole range of simulation options is visible.

To start working with the program, first a structure has to be defined. The button to open the "Define structure" window is to be found in the left-hand section of the main window. Also on the left-hand side, the numerical settings can be found behind the "settings" button and the calculation mode can be set e.g. to steady state (DC) or transient as described in Chapter 13.

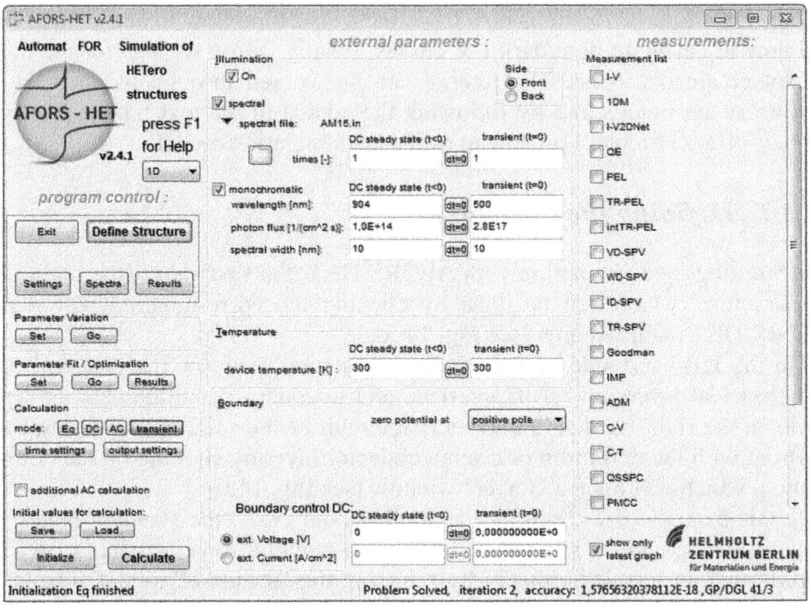

Fig. 14.4 The AFORS-HET main window with all options visible (after defining a structure and activating the transient calculation mode).

After the structure has been defined and the simulation has once been run under equilibrium conditions, a calculation under different conditions can be started with the "calculate" button, or a set of calculations with one parameter varied can be set up and started in the "Parameter Variation" section. It is also possible to start a fit to measured data or the structure can be optimized in the "Parameters Fit / Optimization" section, for example with respect to its efficiency. In the central section of the main window the external parameters: illumination, temperature and voltage or current can be defined. In the illumination section a spectrum or generation file can be loaded. Similarly, a monochromatic illumination of a certain wavelength and photon flux can be defined. For all external parameters the time dependence can be specified if the calculation modes AC or transient are activated. In the right-hand section of the main window one can set up and start the simulation of varied solar cell and material characterization techniques including typical measurements like I-V curve (IV), quantum efficiency (QE) and injection-dependent lifetime (QSSPC) used to characterize *a*-Si:H/*c*-Si solar cells.

To demonstrate how to work with AFORS-HET in more detail, a step-by-step tutorial on how to simulate a crystalline silicon solar cell is presented in the following. This tutorial explains how to define a structure, how to analyze the results and how to calculate light/dark I-V curves. Finally, some settings for the numerical solver are discussed. The tutorial can be worked through independently by looking at the figures and by following the advice in the text boxes. The speech bubbles offer extra ideas on how to refine the simulation model.

14.2.1 Defining the Structure

When setting up a simulation with AFORS-HET, the first step is the definition of the structure, which can be done by clicking the corresponding button in the AFORS-HET main window (see Fig. 14.5).

On the left-hand side of the "Define Structure" window the different optical and electrical layers and their interface and boundary conditions can be defined while on the right-hand side the external circuit of the solar cell can be specified. Start out with the definition of a semiconductor layer by clicking on the "Layer 1" caption, which will open a "Layer" window (see Fig. 14.6).

In this example, by choosing the "bulk model *c*-Si", the standard material parameters for crystalline silicon are selected. It would also be possible to define all material parameters individually by selecting the "standard" model. The layer is defined as p-type with a doping density of 10^{16} cm^{-3} and a thickness of 1μm. For optical calculation the *n,k*-data can be loaded (find sample files for different materials in the "Spectra" folder of your AFORS-HET installation). Enter an appropriate name and confirm your choices by clicking the "OK" button.

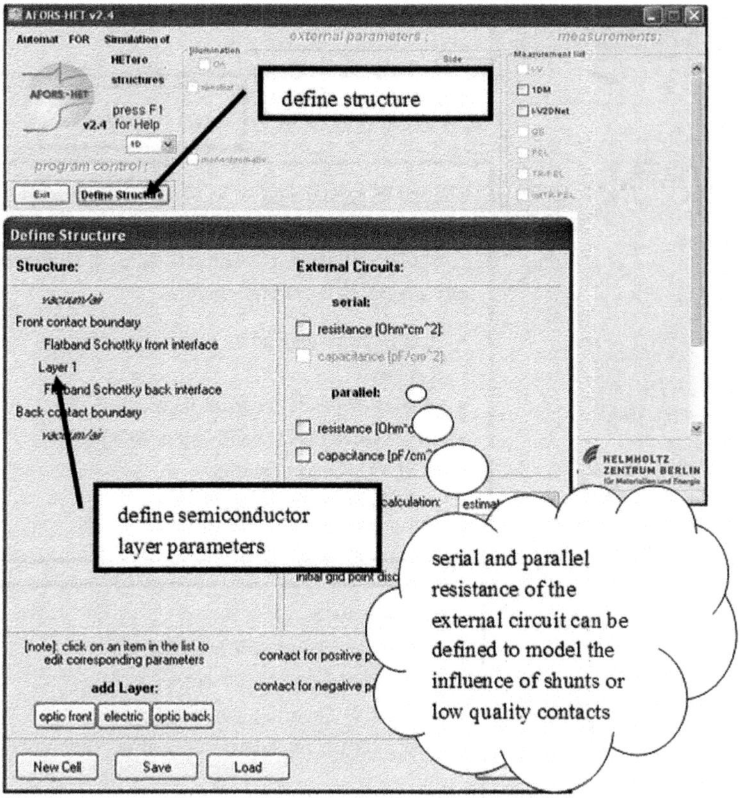

Fig. 14.5 The AFORS-HET main window and "Define Structure" window.

Returning to the "Define Structure" window (see Fig. 14.7), a further electrical layer can be added. This layer is defined as n-type (see Fig. 14.8) so that a p/n-junction can be achieved which works as a solar cell. In addition to the material parameters, deep defects can be defined to model the wafer quality (see Fig. 14.8). Finally, the structure can be saved, the simulation started (see Fig. 14.9) and the results analyzed (see Fig. 14.10).

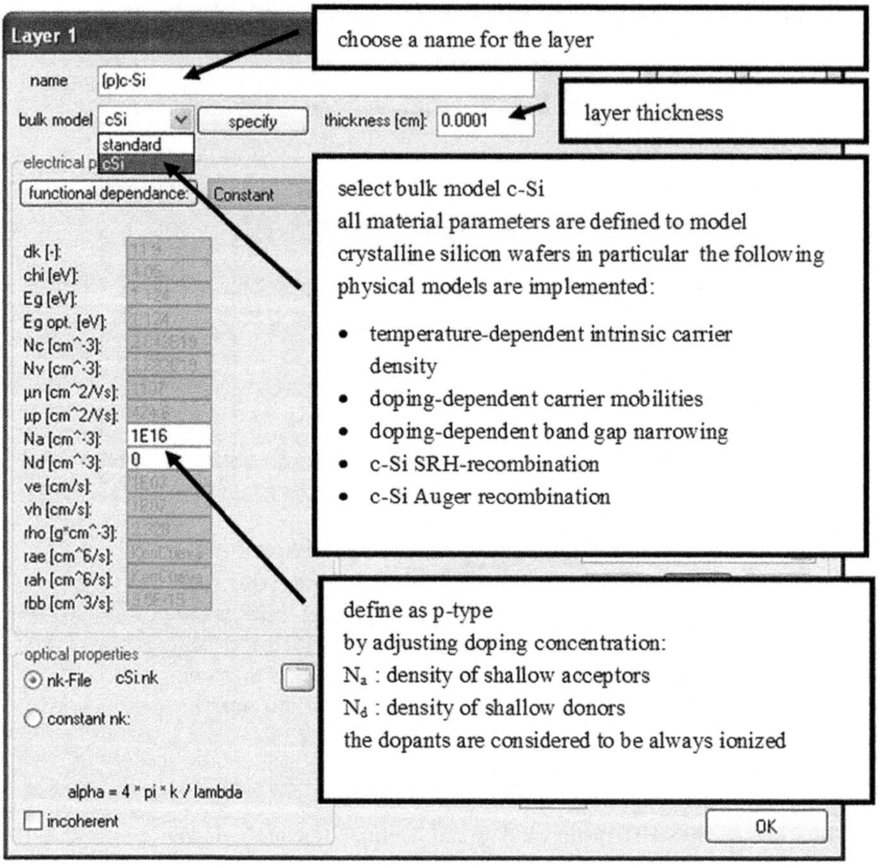

Fig. 14.6. For each layer the electrical and optical material parameters can be defined in the "layer" window.

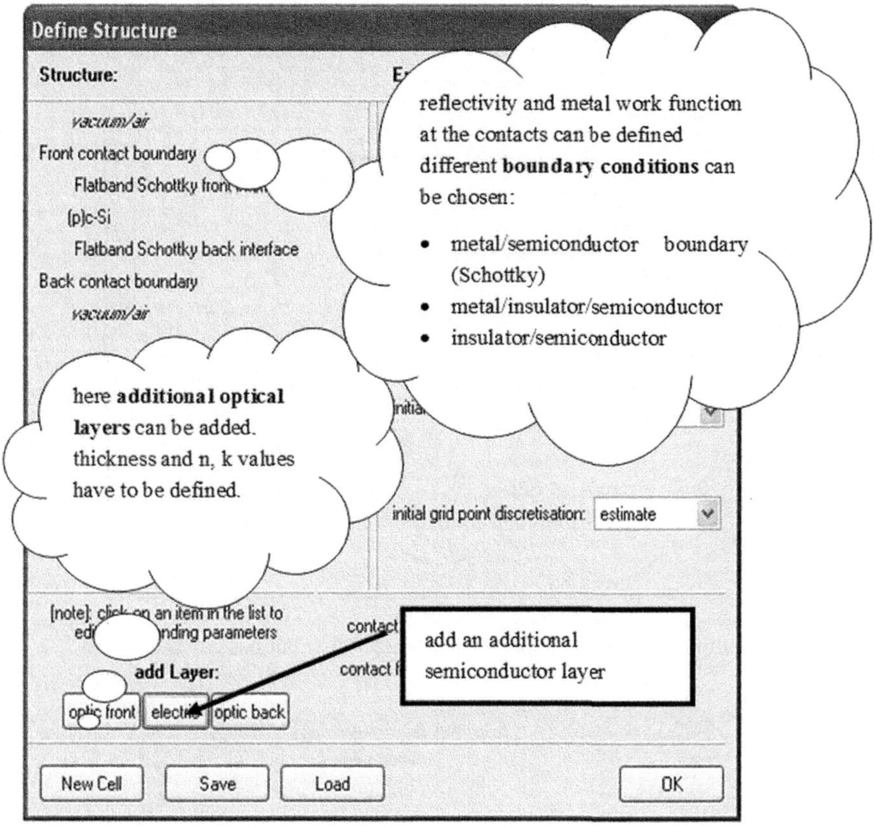

Fig. 14.7 Electrical and optical layers can be added and the boundary and interface conditions have to be defined.

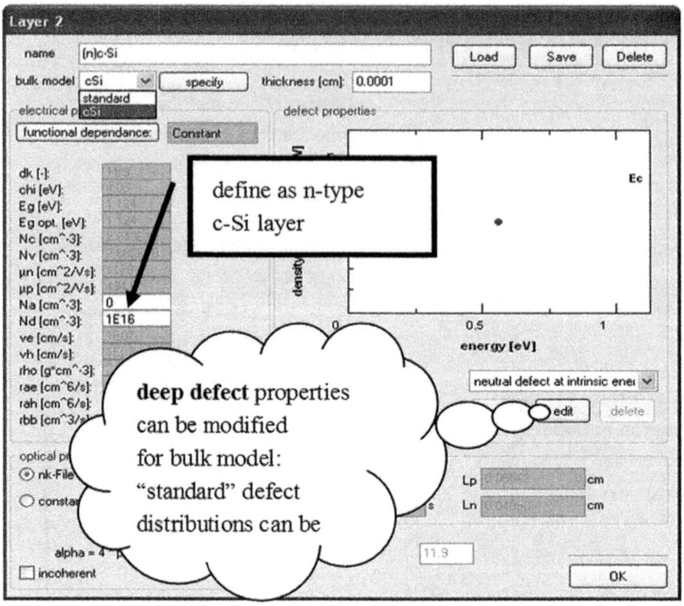

Fig. 14.8 In addition to the material parameters, defect distributions can be defined.

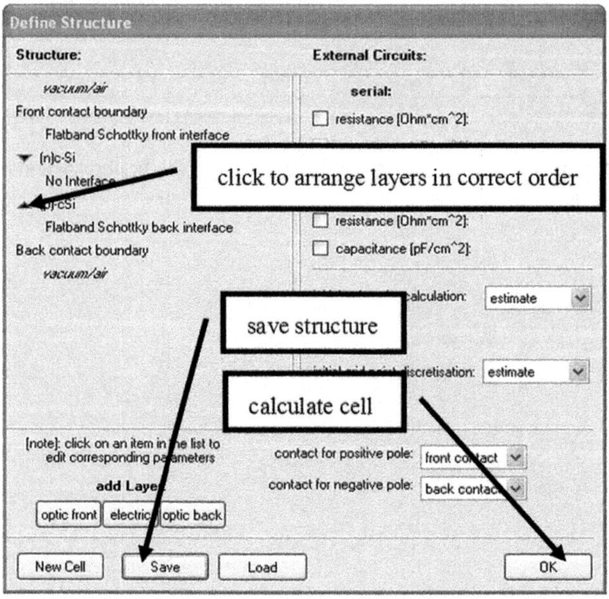

Fig. 14.9 The layer order can be rearranged, the structure saved and the simulation started.

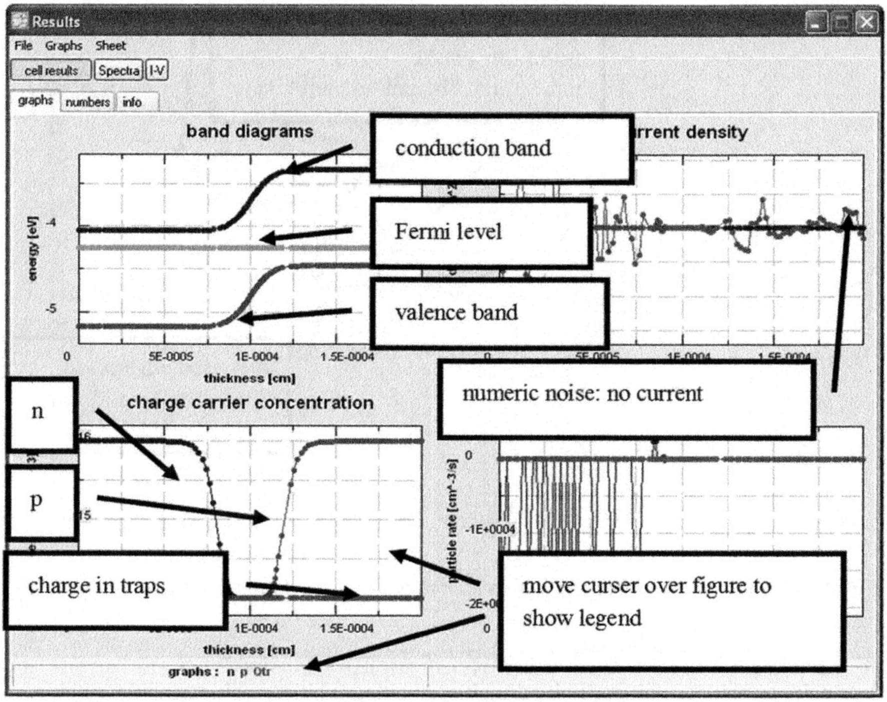

Fig. 14.10 The "Results" window shows band diagrams, carrier and currents densities as well as recombination and generation.

14.2.2 Analyzing Simulation Results

By clicking on the "OK" button in the “Define Structure” window the simulation for equilibrium conditions is started automatically and the "Results" window pops up (see Fig. 14.10). Typically, band diagrams, carrier densities, current densities and recombination/generation plots are displayed. The results for all calculated quantities can be saved as a table (see "File" menu). By clicking in one of the plots a new window opens (see Fig. 14.11) and the data can be analyzed in more detail. It is, for example, possible to measure distances and to carry out numerical integrations.

The data can be displayed as a figure or as a table and one can easily switch between the two visualizations as can be seen in Fig. 14.12. Furthermore, one can display the results of the optical calculations if calculations under illumination have been carried out.

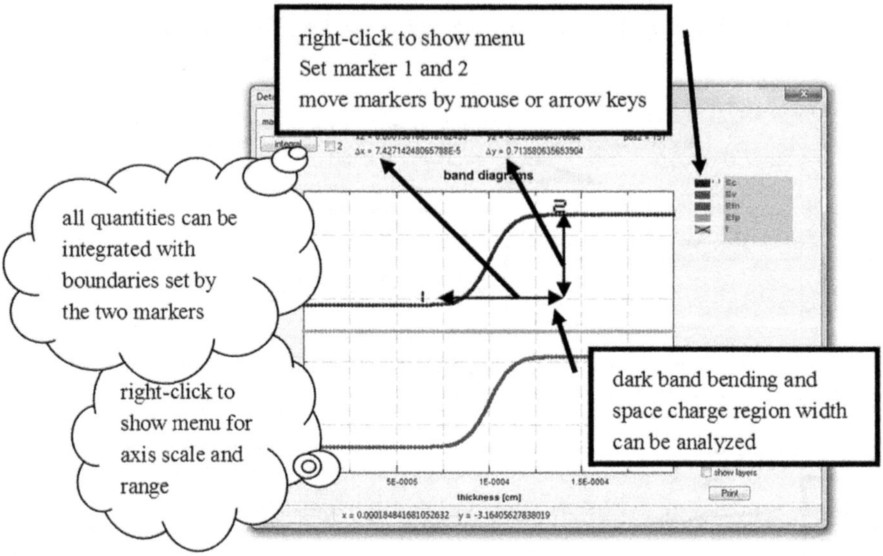

Fig. 14.11 There are several tools with which to analyze the data in detail.

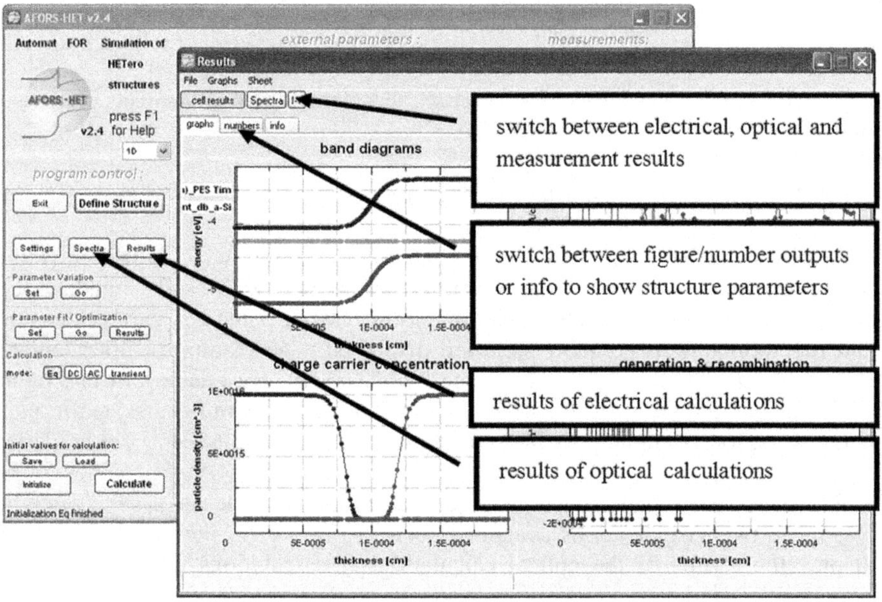

Fig. 14.12 In the "Results" window it is possible to switch between the results of electrical and optical calculation and to show the results of a simulated measurement. Furthermore, the data can be displayed as a figure or as a table.

14.2.3 A p/n-Junction under Illumination

To calculate the p/n-junction under illumination, the DC calculation mode has to be switched on. It is then possible to carry out calculations under non-equilibrium conditions and to select constant external illumination and constant current or voltage (see Fig. 14.13).

There are different possibilities for simulating a cell under illumination. You can either choose "monochromatic" illumination and define a photon flux and a wavelength or you can choose "spectral" illumination and load a "spectral file" or a "generation file" (sample files can be found in the "Spectra" folder of your AFORS-HET installation).

Simulation results for the p/n-junction illuminated with an AM1.5 spectrum calculated under open circuit and short circuit conditions can be seen in Fig. 14.14.

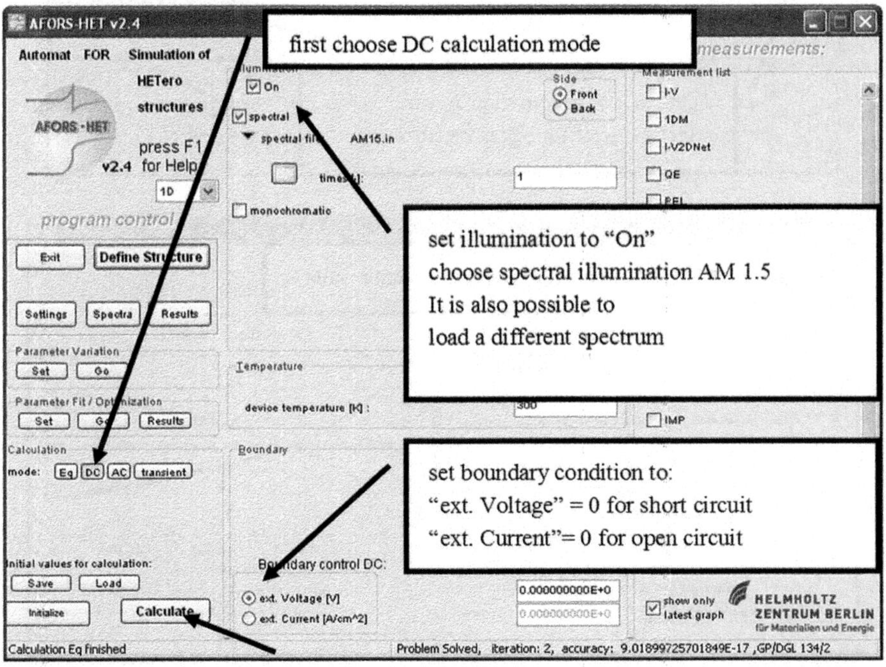

Fig. 14.13 The DC calculation mode allows for the selection of constant illumination and constant current or voltage.

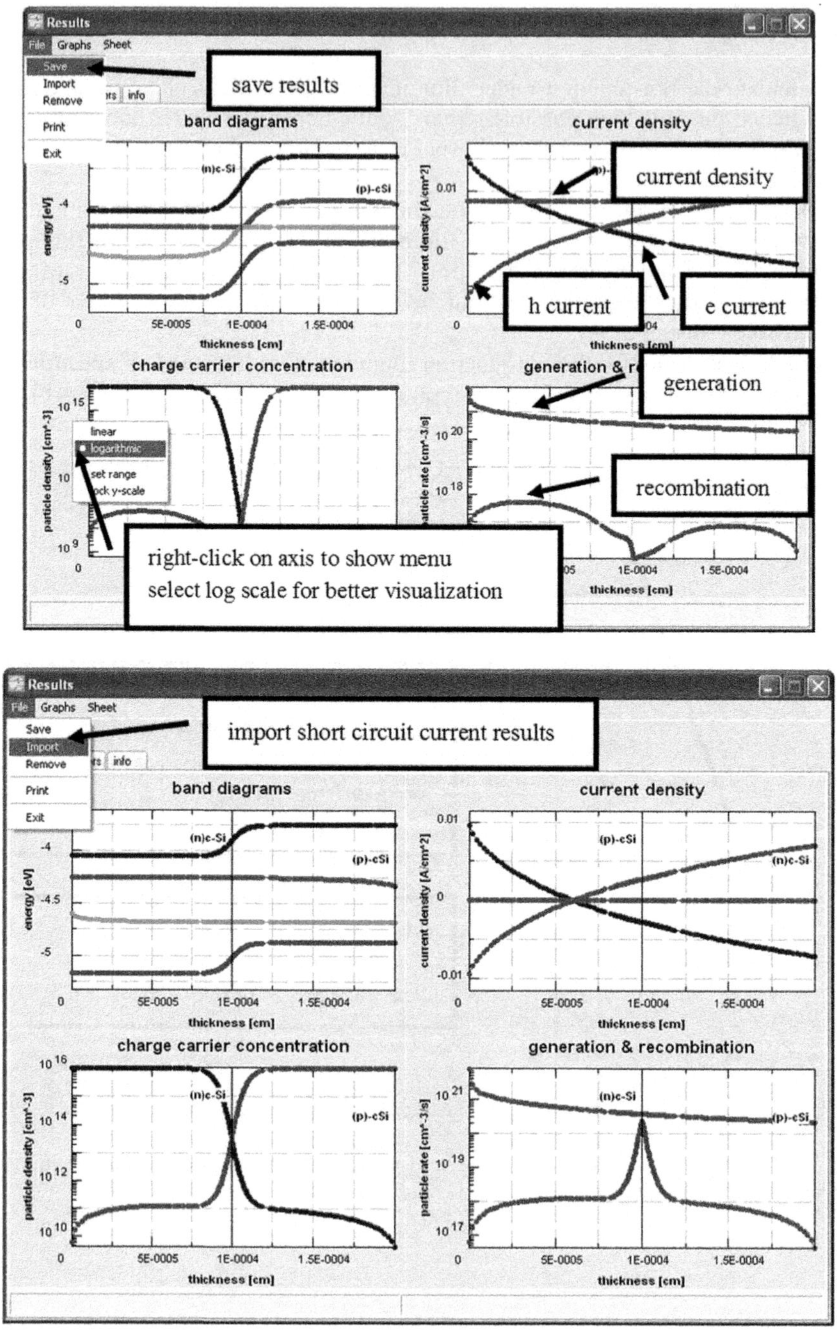

Fig. 14.14 Simulation results for short circuit (above) and open circuit (below) conditions. Results can be stored and imported.

14.2.4 Simulating an I-V Curve

AFORS-HET is capable of simulating many different types of measurements, as can be seen on the right-hand side of the main window. To set up measurements for example of an I-V curve, one has to follow the steps depicted in Fig. 14.15. Clicking on the "calc I-V" button will start the simulation and the results will be shown. These can be analyzed and saved as shown in Fig. 14.11 and 14.14. In Fig. 14.16 it is explained how to simulate an I-V curve under illumination and the results are shown, comparing them to the simulation of a dark I-V curve. As can be seen, the structure that has been defined works as a solar cell even though its efficiency is low. In the next chapter it will be explained how the solar cell can be optimized.

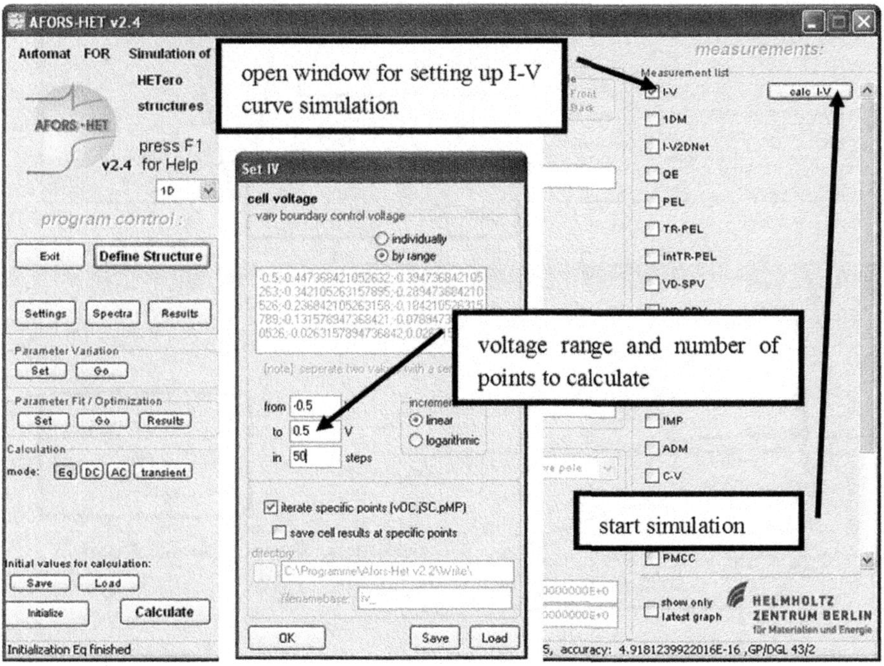

Fig. 14.15 Steps to set up the simulation of an I-V curve.

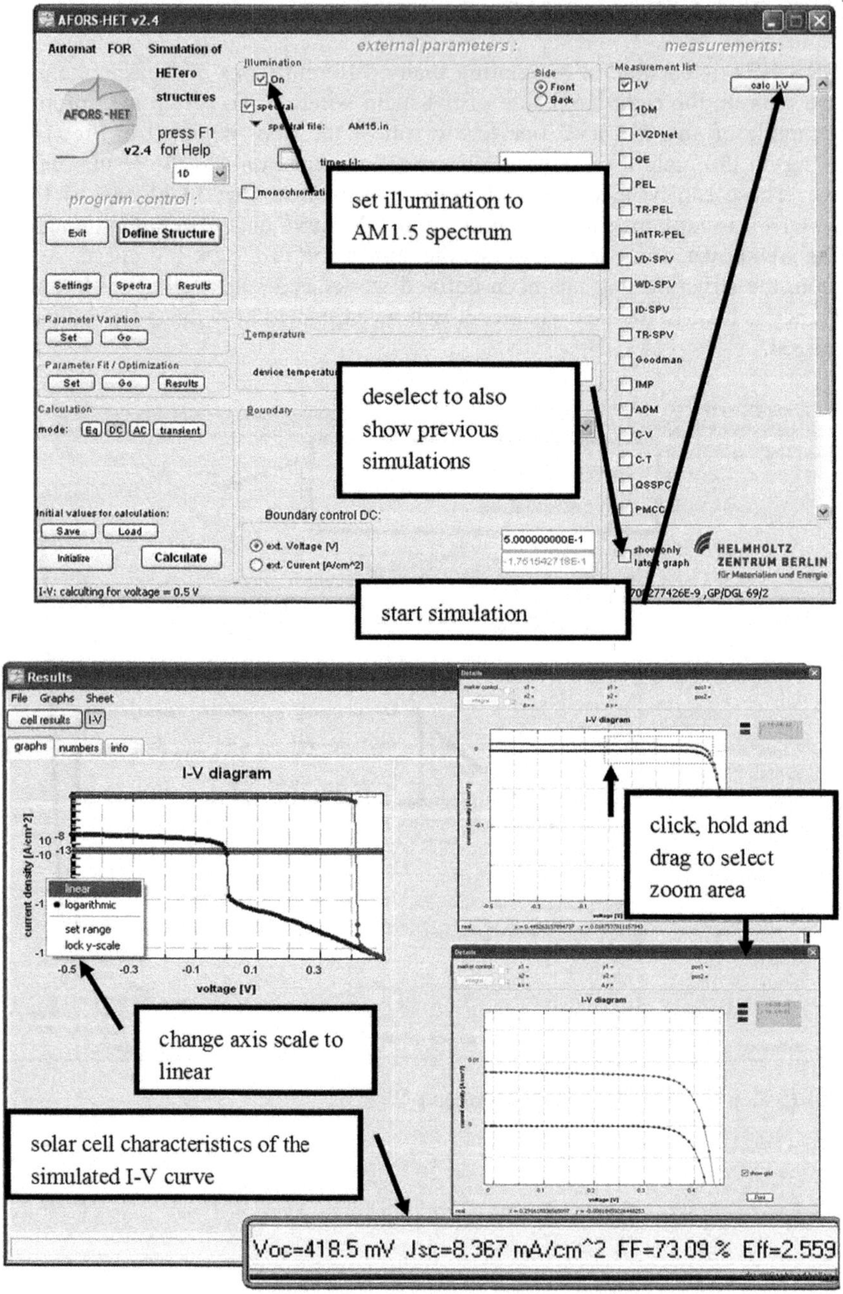

Fig. 14.16 Setting up the simulation of an I-V curve under illumination (above) and comparing it to a dark I-V curve (below).

14.2.4 *Optimizing Solar Cells - from a c-Si to an a-Si:H/c-Si Cell*

The efficiency of the simple cell can be enhanced in several optimization steps. The starting point is the cell defined above with a 1µm n-type emitter layer with a doping level of N_d=10^{16}cm^{-3} and a 1µm p-type absorber layer with N_a=10^{16}cm^{-3}. Firstly the absorber thickness should be enhanced to 100µm, secondly the emitter doping level should be enhanced to N_d=10^{19}cm^{-3} and finally an additional 1µm thick back surface field (BSF) layer with a doping level of N_a=10^{19}cm^{-3} should be added at the rear contact (see Fig. 14.7 on how to add an additional electric layer). Furthermore to simulate a realistic cell, an external circuit with parallel and serial resistors has to be added to model poor quality contacts and shunts in the device (see Fig. 14.5 on how to add resistors). In Fig. 14.17 the I-V curves under illumination calculated with AFORS-HET are shown and the influences of the different modifications become clear.

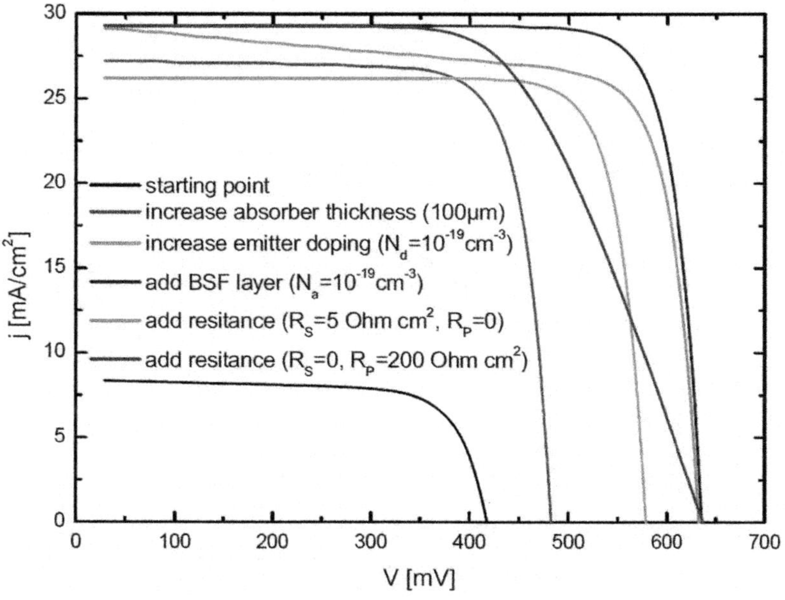

Fig. 14.17 I-V curves under illumination for different device optimization steps.

As can be seen in Fig. 14.17 an enhancement in short circuit current can be observed when the absorber thickness is increased. This is due to the fact that more photons can be absorbed in a thicker layer. When the doping level is increased, the dark band bending in the structure is increased and thus a higher open circuit voltage can be achieved. When adding a BSF layer the recombination at the rear contact is reduced and we gain open circuit voltage as well as short circuit current. Parallel and serial resistors lead to a decrease in fill factor.

To further enhance efficiency the silicon homojunction cell defined above could be transformed into a silicon heterojunction solar cell. To achieve this, the emitter and BSF layer need to be defined as *a*-Si:H instead of *c*-Si layers. This means changing all material parameters such as band gap and mobilities and to select defect distributions of band tails and dangling bonds typical for *a*-Si:H. As discussed in section 14.1 the exact *a*-Si:H parameters depend strongly on the deposition conditions of those layers. As a starting point we suggest parameters as listed in Table 14.1 for device grade *a*-Si:H layers and a device structure as shown in Fig. 14.2. In AFORS-HET all the defect distributions necessary to describe *a*-Si:H layers i.e. exponentially decaying distributions to model the band tails and Gauss distributions to model the dangling bond defects can be defined (Fig. 14.18).

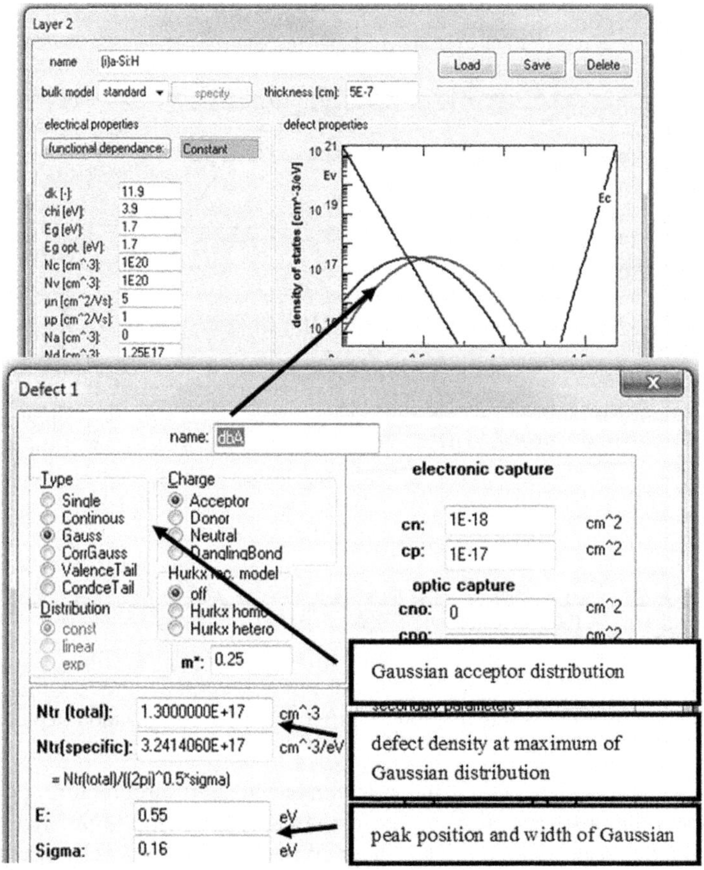

Fig. 14.18 Definition of defects and defect distributions in AFORS-HET. In the "layer" window (above) all defined distributions can be seen, whereas in the "defect" window (below) the details for the definition of the acceptor part of the Gaussian distributed dangling bond defects are shown.

A very helpful tool when optimizing solar cells is the “parameter variation” function which can be found at the left-hand side of the main window. With this tool it is for example possible to define a range of emitter doping levels for which to calculate solar cell characteristics and thus data as shown in Fig. 14.3 can be generated automatically. Basically all material parameters as well as all external parameters can be varied and all calculated cell parameters such as charge carrier densities and electric potential but also cell currents and recombination rates as well as quantities derived from the simulation of complete measurements can be selected as output.

14.2.5 Settings for the Numerical Solver

By clicking on the "settings" button in the AFORS-HET main window the settings for the numerical solver can be selected (see Fig. 14.19). By means of the break conditions the ratio of accuracy and computation speed can be adjusted. A crucial role for the accuracy of the calculations is played by the discretization in space (grid) and in energy (delta E). The grid, which is generated automatically when a structure is defined, can be modified by the user to achieve, for example, a finer resolution at the interfaces (see Fig. 14.20). A very fine grid is necessary if advanced interface models such as thermionic emission are used.

With the "optical" tab different optical models of varying precision can be chosen as they are discussed in Chapter 13. With the "measurement numeric" tab the accuracy of the extraction of solar cell characteristics from simulated I-V curves can be defined.

While calculating, AFORS-HET reports on the numerical solver in the main window (see Fig. 14.21), thus giving warnings in case of convergence problems. When calculating complex structures, such convergence problems may be encountered. This is a common difficulty when solving semiconductor equations numerically. It is therefore in general essential to make a good initial guess at the values. When a new structure is defined or the “initiate” button is pressed, AFORS-HET automatically chooses initial values for the calculation. Otherwise the previously calculated solution is taken as an initial value for the next calculation. It is also possible to load and save initial values (see Fig. 14.21). If convergence problems are encountered, it is often helpful to start with a more simple structure or less extreme external parameters and then gradually change the structure or external parameters to more extreme cases, always using the previous solution as the initial value for the next more difficult calculation.

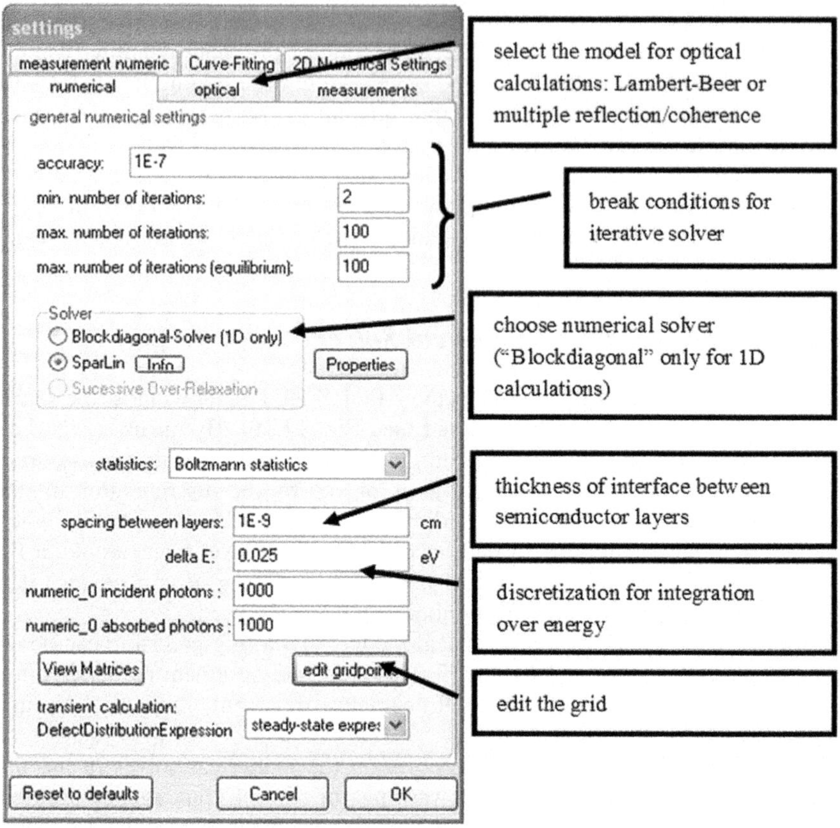

Fig. 14.19 Settings for the numerical solver.

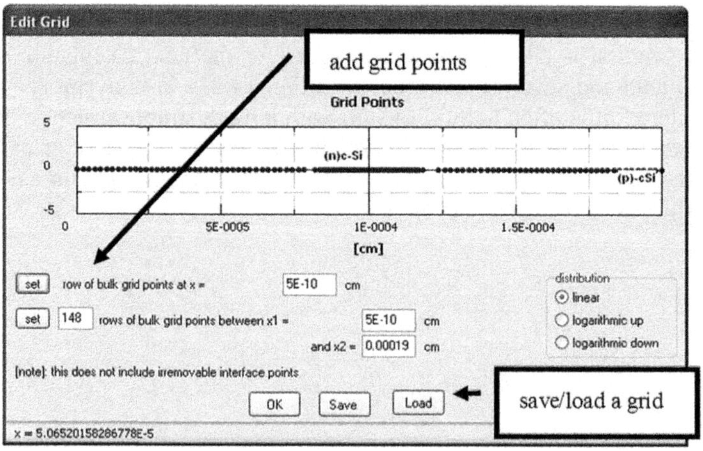

Fig. 14.20 Setting up a user-defined grid.

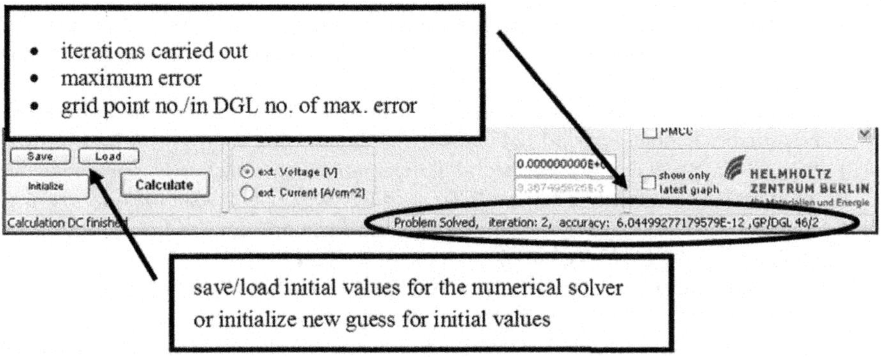

Fig. 14.21 Report on the numerical solver in the AFORS-HET main window.

References

[1] Kanevce, A., Metzger, W.K.: The role of amorphous silicon and tunneling in heterojunction with intrinsic thin layer (HIT) solar cells. J. Appl. Phys. 105, 094507–094507 (2009)

[2] Rahmouni, M., Datta, A., Chatterjee, P., Damon-Lacoste, J., Ballif, C., Roba i Cabarrocas, P.: Carrier transport and sensitivity issues in heterojunction with intrinsic thin layer solar cells on n-type crystalline silicon: A computer simulation study. J. Appl. Phys. 107, 054521 (2010)

[3] Froitzheim, A., Brendel, K., Elstner, L., Fuhs, W., Kliefoth, K., Schmidt, M.: Interface recombination in heterojunctions of amorphous and crystalline silicon. J. Non-Cryst. Sol. 299-302, 663–667 (2002)

[4] Stangl, R., Froitzheim, A., Schmidt, M., Fuhs, W.: Design criteria for amorphous/crystalline silicon heterojunction solar cells - a simulation study. In: Proceedings of the 3rd World Conference in Photovoltaic Energy Conversion, Osaka, Japan, pp. 1005–1008 (2004)

[5] Stangl, R., Froitzheim, A., Elstner, L., Fuhs, W.: Amorphous/crystalline silicon heterojunction solar cells, a simulation study. In: Proceedings of the 17th European Photovoltaic Solar Energy Conference, Munich, Germany, p.1383 (2001)

[6] Schmidt, M., Korte, L., Laades, A., Stangl, R., Schubert, C., Angermann, H., Conrad, E., Maydell, K.: Physical aspects of a-si:h/c-si heterojunction solar cells. Thin Solid Films 515, 7475–7480 (2007)

[7] Kleider, J.P., Gudovskikh, A.S., Roca i Cabarrocas, P.: Determination of the conduction band offset between hydrogenated amorphous silicon and crystalline silicon from surface inversion layer conductance measurements. Appl. Phys. Lett. 92, 162101 (2008)

[8] Tsunomura, Y., Yoshimine, Y., Taguchi, M., Baba, T., Kinoshita, T., Kanno, H., Sakata, H., Maruyama, E., Tanaka, M.: Twenty-two percent efficiency HIT solar cell. Sol. Energy Mater. Sol. Cells 93, 670–673 (2009)

[9] Clugston, D., Basore, P.: PC1D version 5: 32-bit solar cell modeling on personal computers. In: Proceedings of the 26th IEEE Photovoltaic Specialists Conference, Anaheim, USA (1997)

[10] Masetti, G., Severi, R., Solmi, S.: Modeling of Carrier Mobility Against Carrier Concentration in Arsenic-, Phosphorus-, and Boron-doped Silicon. IEEE Trans. Electron Devices 30, 764 (1983)
[11] Kerr, M.J., Cuevas, A.: General parameterization of Auger recombination in crystalline silicon. J. Appl. Phys. 91, 2473–2480 (2002)
[12] Sah, C.-T., Shockley, W.: Electron-hole recombination statistics in semiconductors through flaws with many charge conditions. Phys. Rev. 109, 1103 (1958)
[13] Halpern, V.: The statistics of recombination via dangling bonds in amorphous silicon. Philos. Mag. B 54, 473–482 (1986)
[14] Stutzmann, M., Biegelsen, D., Street, R.: Detailed investigation of doping in hydrogenated amorphous silicon and germanium. Phys. Rev. B 35, 5666–5701 (1987)
[15] Froitzheim, A., Stangl, R., Elstner, L., Schmidt, M., Fuhs, W.: Interface recombination in amorphous/crystalline silicon solar cells, a simulation study. In: Proceedings of the 29th IEEE Photovoltaic Specialists Conference, New Orleans, USA, pp. 1238–1241 (2003)
[16] Tanaka, M., Taguchi, M., Matsuyama, T., Sawada, T., Tsuda, S., Nakano, S., Hanafusa, H., Kuwano, Y.: Development of new a-Si/c-Si heterojunction solar cells: ACJ-HIT (artificially constructed junction-heterojunction with intrinsic thin-layer). Jpn. J. Appl. Phys. 31, 3518–3522 (1992)
[17] Fujiwara, H., Kondo, M.: Effects of a-Si:H layer thicknesses on the performance of a-Si:H/c-Si heterojunction solar cells. J. Appl. Phys. 101, 054516–054519 (2007)
[18] Taguchi, M., Terakawa, A., Maruyama, E., Tanaka, M.: Obtaining a higher Voc in hit cells. Prog. Photovoltaics Res. Appl. 13, 481–488 (2005)
[19] De Wolf, S., Beaucarne, G.: Surface passivation properties of boron-doped plasma-enhanced chemical vapor deposited hydrogenated amorphous silicon films on p-type crystalline si substrates. Appl. Phys. Lett. 88 (2006)
[20] Korte, L., Schmidt, M.: Investigation of gap states in phosphorous-doped ultra-thin a-si:h by near-uv photoelectron spectroscopy. J. Non-Cryst. Sol. 354, 2138–2143 (2008)
[21] De Wolf, S., Kondo, M.: Nature of doped a-Si:H/c-Si interface recombination. J. Appl. Phys. 105, 103707 (2009)

Chapter 15
Two-Dimensional Simulations of Interdigitated Back Contact Silicon Heterojunctions Solar Cells

Djicknoum Diouf, Jean-Paul Kleider, and Christophe Longeaud

Laboratoire de Génie Electrique de Paris,
CNRS UMR8507, CNRS UMR8507; SUPELEC;
Univ Paris-Sud; UPMC Univ Paris 06; 11 rue Joliot-Curie, Plateau de Moulon,
F-91192 Gif-sur-Yvette Cedex, France

Abstract. Interdigitated back contact silicon heterojunction (IBC-SiHJ) solar cells that combine the amorphous silicon/crystalline silicon (a-Si:H/c-Si) heterojunction- and interdigitated back contact (IBC) concepts are very promising in order to reach the highest one-junction efficiencies. In this chapter, a comparative two-dimensional simulation study has been done on the IBC-SiHJ structure based on n-type and p-type crystalline silicon by varying the values of the following parameters: minority carrier lifetime in c-Si, c-Si thickness, c-Si doping concentration, surface recombination velocity, density of defect states at the a-Si:H/c-Si hetero-interface and rear side geometry. The influence of these parameters has been tested by generating the current-voltage characteristics under illumination. Results indicate that the key parameters to achieve high efficiency are high c-Si substrate quality, low surface recombination velocity especially at the front surface, and a low recombining a-Si:H/c-Si interface. The width of the gap region (spacing between the back-surface field (BSF) and the emitter) must be kept as small as possible to avoid recombination of minority carriers in the bulk c-Si. For IBC-SiHJ based on n-type c-Si, the optimum geometry corresponds to a minimum size BSF region and a maximum size emitter region while for IBC-SiHJ based on p-type c-Si a BSF width equivalent to around 30% of the pitch is an optimum.

15.1 Introduction

The amorphous silicon/crystalline silicon heterojunction (SiHJ) solar cell using thin layers of hydrogenated amorphous silicon (a-Si:H) deposited at low temperature on a crystalline silicon (c-Si) substrate is one of the most interesting technological solutions for the photovoltaic market, basically due to the excellent performance and the simple low–temperature process [1,2]. In spite of achieving high efficiency (23%), the standard double heterojunction solar cells are limited

W.G.J.H.M. van Sark et al. (Eds.): Physics & Tech. of Amorphous-Crystalline, EM, pp.483–519.
springerlink.com

by optical absorption and reflection at the front surface. To overcome those limitations, the concept of interdigitated back contact (IBC) is very promising [3,4] and has reached 24,2 % at lab scale for conventional doped silicon solar cells without using heterojunctions [5]. Having all contacts at the rear side of the structure gives additional freedom for optimizing cell performance compared to the conventional front contacted structures. The junctions are on the rear side and can be optimized for electrical performance while the front side is designed for optimum optical performance. The combination of the interdigitated back contact concept with the silicon heterojunctions technique attracted much interest only recently, with a first report in 2007 [6]. So far the best efficiency obtained on this new solar cell structure is 15.7 % [7], wich is considerably lower than the 24.2% efficiency obtained with the diffused homojunction IBC cells. So far, all research groups are still struggling with the problem of processing and structuring the a-Si:H/c-Si rear side of the cell while maintaining excellent passivation of a-Si:H, and the achieved cell fill factor is still relatively low [8-10]. Thus, simulations will be helpful to identify key factors for achieving high efficiency with this type of cells.

The modelling of the IBC-SHJ solar cell requires device simulation software operating in two dimensions and incorporating amorphous silicon. AFORS-HET is the reference software for modelling heterojunctions a-Si:H/c-Si but up to now only a one dimension model is available [11, see also Chapters 13 and 14 in this book]. To study this innovative structure, we use ATLAS two-dimensional (2-D) device simulation software that provides accurate bulk and interface defects modelling and allows one to also model amorphous silicon [12].

We first present the geometrical structure of the reference IBC-SiHJ solar cell. Then, we present the general framework of the simulation by specifying the different physical models used in this work. Finally, the simulation results of the IBC-SiHJ solar cell will be presented and discussed to determine the important parameters whose adjustment is expected to achieve high efficiency.

15.2 Device Structure

15.2.1 Geometry of the Structure

The back contact cells differ from conventional structures in that all contacts are on the back side (not illuminated side) of the cell. The structure of IBC–SiHJ solar cell is shown in Fig. 15.1. On the front of the c-Si substrate an optical layer is placed thay plays the role of anti-reflective coating layer like silicon nitride (SiN_x). This front surface is subject to the illumination. On the rear side, we have an interdigitated structure of hydrogenated amorphous silicon layers alternately n-type and p-type to play the role of emitter or back surface field (BSF) according to the c-Si substrate doping. The emitter and the BSF are covered by metal contacts. They are separated by an insulator layer whose role will be studied later.

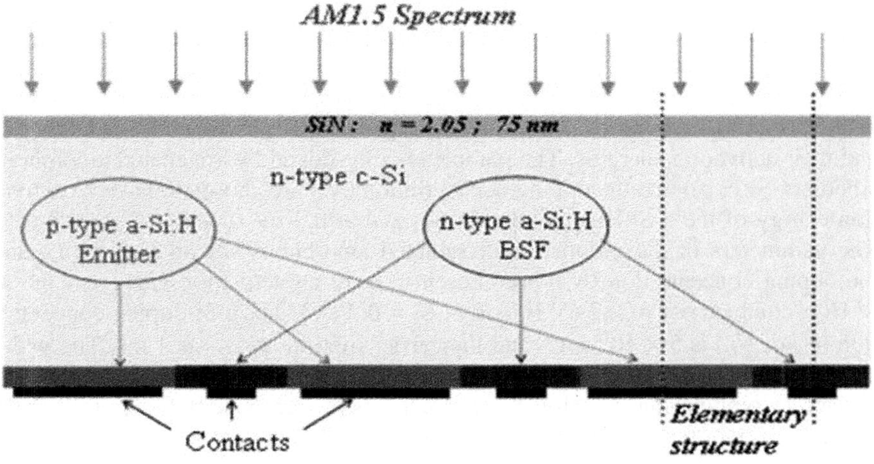

Fig. 15.1 Sketch of the complete structure of an IBC-SiHJ cell.

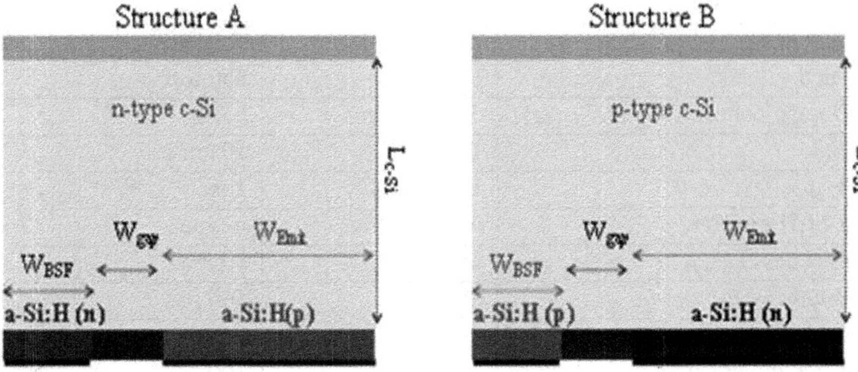

Fig. 15.2 Geometry of two basic structures of the reference IBC-SiHJ solar cell.

The IBC-SiHJ solar cell presents a periodic structure as we can see in Fig. 15.1. The periodicity of the structure allows us to use an elementary structure as shown in Fig. 15.2 that will serve as a basis for optimizing the performance of this type of cell. The geometrical and material parameters of the simulated elementary structure device were chosen in agreement with our own fabrication process [13]. The width of this elementary structure (pitch) is equal to half the distance between two electrodes with the same polarity. We used two reference structures shown in Fig. 15.2: one based on n-type c-Si and the second based on p-type c-Si. The study and optimization of IBC-SiHJ solar cell are presented simultaneously for this two reference structures. This will allow us to identify the differences due to the doping type of c-Si substrate.

The parameters of the doped hydrogenated amorphous silicon (a-Si:H) have been chosen to be in agreement with the experimentally measured conductivities and their activation energies. The parameters of p-doped hydrogenated amorphous silicon (a-Si:H(p)) and the doping concentration (N_A) are chosen to have an activation energy of the a-Si:H(p) conductivity fixed at 0.3 eV (i.e. E_F - E_V = 0.3 eV). The parameters of the n-doped hydrogenated amorphous silicon (a-Si:H(n)) and the doping concentration (N_D) are chosen to have an activation energy of the a-Si:H(n) conductivity of 0.2 eV (i.e. $E_C - E_F$ = 0.2 eV). The c-Si doping concentration (N_A or N_D) is 5×10^{15} cm^{-3} and the carrier lifetime $\tau_{0,\,SRH}$ is 1 ms. The width of the spacing between the emitter and the BSF called gap region is 200 μm. The surface recombinaison velocity, S_{FAV}, of carriers at front side is fixed at 10 cm/s. Table 15.1 gives the parameters used for our two reference cells.

Table 15.1 Parameters of IBC-SiHJ reference solar cells.

c-Si Substrate	n-type	p-type
thickness $L_{c\text{-}Si}$	250 μm	250 μm
pitch	850 μm	850 μm
Doping concentration (N_D, N_A)	5×10^{15} cm^{-3}	5×10^{15} cm^{-3}
S_{FAV}	10 cm/s	10 cm/s
$\tau_{0,\,SRH}$	1 ms	1 ms
a-Si:H emitter	p-type	n-type
thickness $L_{a\text{-}Si:H}$	10 nm	10 nm
Width W_{emit}	500 μm	500 μm
E_C-E_F or E_F-E_V	0.3 eV	0.2 eV
Metal coverage	100 %	100 %
Type of contact	Ohmic flat band	Ohmic flat band
a-Si:H BSF	n-type	p-type
thickness $L_{a\text{-}Si:H}$	10 nm	10 nm
Width W_{BSF}	150 μm	150 μm
E_C-E_F or E_F-E_V	0.2 eV	0.3 eV
Metal coverage	100 %	100 %
Type of contact	Ohmic flat band	Ohmic flat band

15.2.2 Meshing of the Structure

To simulate properly the behaviour of the structure, it is essential to apply an adapted mesh. A mesh as thin as possible applied to the whole structure ensures good accuracy of calculations but requires greater computation time to simulate the behaviour of this structure. It is therefore necessary to find a compromise between computational time and accuracy of the calculation. To reach this compromise, we applied a fine mesh only in areas where changes in physical quantities are important and a coarse mesh in areas where these quantities vary little. Thus, the mesh is refined in the critical areas that are the front surface (strong absorption), the c-Si/a-Si:H hetero-interfaces and the areas around frontiers of the various layers in which the variations of physical quantities are important. In the middle of c-Si substrate, a coarse mesh is used as the physical quantities vary very little. Figure 15.3 represents the mesh used to simulate IBC-SiHJ solar cell.

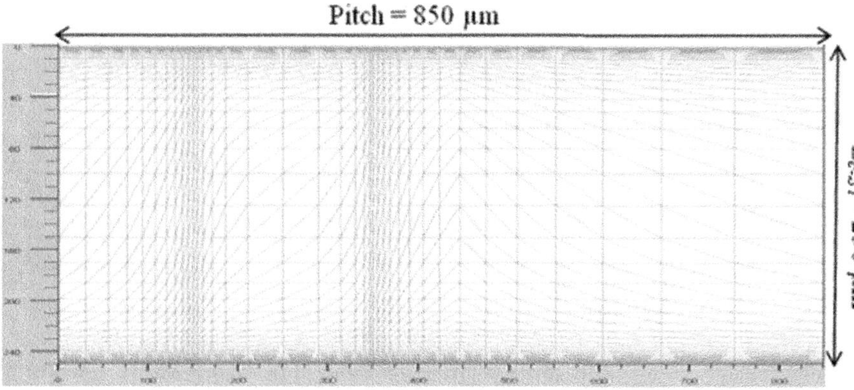

Fig. 15.3 Meshing of IBC-SiHJ reference solar cells. The mesh is fine at interfaces and coarse in the middle of the c-Si substrate.

15.2.3 Band Diagrams at Equilibrium

Figures 15.4 and 15.5 show the band diagrams at thermodynamic equilibrium at both a-Si:H /c-Si hetero-interfaces for the reference cell on p-type c-Si and on n-type c-Si, respectively. The conduction band offset (ΔE_C) is 0.15 eV and the valence band offset (ΔE_V) is 0.43 eV.

These band diagrams are extracted from the ATLAS simulation software that uses Anderson's model [14, 15].

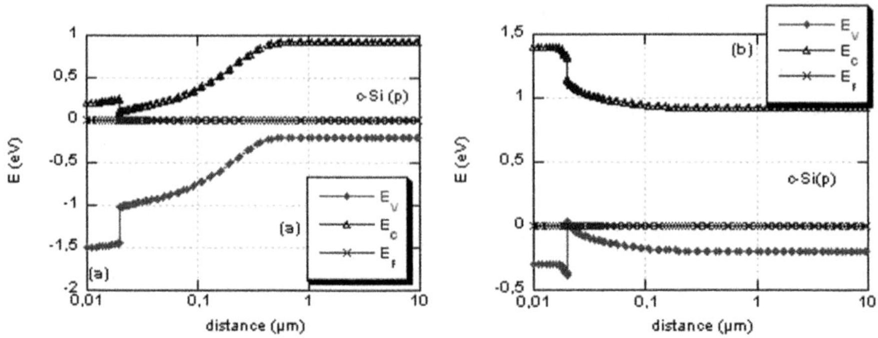

Fig. 15.4 Band diagram at thermodynamic equilibrium (a) of the a-Si: H (n) / c-Si (p) emitter heterojunction and (b) of the a-Si:H(p) / c-Si(p) BSF heterojunction for the p-type reference cell.

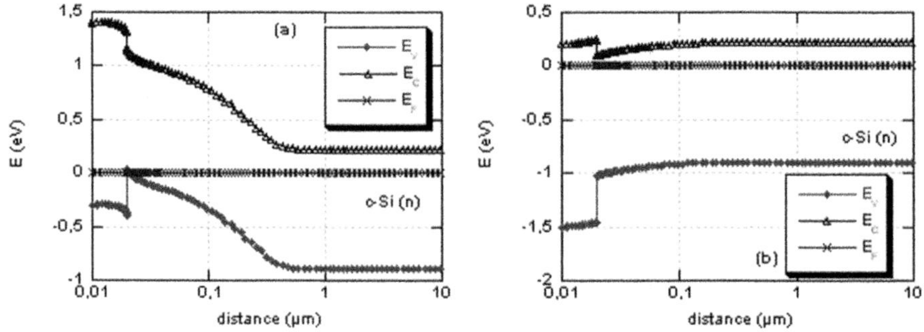

Fig. 15.5 Band diagrams at thermodynamic equilibrium (a) of the a-Si:H(p) / c-Si(n) emitter heterojunction and (b) of the a-Si:H(n) / c-Si(n) BSF heterojunction for the n-type reference cell.

15.3 Physical Models

The simulation is based on the solution of the three governing semiconductor equations: Poisson's equation, and electron and hole continuity equations. The Boltzmann statistic is used for carriers with the drift-diffusion model in ATLAS.

A good simulation of a given structure requires the introduction of adapted physical models. This choice determines the operation of the simulated structure and therefore its performance. It is therefore important to introduce appropriate settings in the physical models considered in order to take into account all the factors that can influence the behaviour of the structure.

Boltzmann's statistics is sufficient for non-degenerate semiconductors. At the a-Si:H/c-Si interface, a strong inversion layer can exist [15]. In this strong inversion layer the equilibrium Fermi level is very close to the band edge (see for example

Fig. 15.5, left). We checked that using Fermi-Dirac's statistics instead of Boltzmann's one did not lead to significant changes in the solar cell parameters. We also assume complete dopant ionization conditions, which is obviously the case since we consider only simulations at $T = 300$ K.

A brief review of the optical and recombination models considered to simulate the functioning of IBC-SiHJ solar cells is made in the following.

15.3.1 Optical Model

The optical generation rate is calculated at each grid point using

$$G_L(y) = \eta_0 \frac{P^*_{t,\,y=0}}{h\nu} \alpha\, e^{-\alpha\, y}, \tag{15.1}$$

where:

- P^* is the ray intensity factor, which contains the cumulative effects of reflections, transmissions, and loss due to absorption over the ray path
- η_o is the internal quantum efficiency, which represents the number of carrier pairs generated per absorbed photon
- y is the distance from the surface being taken as the origin
- h is Planck's constant
- λ is the wavelength
- c is the speed of light
- α is the absorption coefficient given by

$$\alpha = \frac{4\pi}{\lambda} k, \tag{15.2}$$

where k is the imaginary part of the optical index of refraction.

The incident irradiance is the AM1.5 spectrum. It is introduced in ATLAS through a file (.txt) giving the incident power for each wavelength of the spectrum. For each material, we must also introduce a file (.txt) giving the real part n and imaginary part k of the refractive index for each individual wavelength: n is used to calculate the reflection associated to this material and k is used to calculate the absorption in this material.

15.3.2 Recombination Models Applied to c-Si

15.3.2.1 Shockley-Read-Hall Recombination Model

The Shockley-Read-Hall recombination is universally used to describe recombination processes within the forbidden band gap of semi-conductor materials. This is essentially a two-step process; the theory was first derived by Shockley and Read [16] and by Hall [17]. The Shockley-Read-Hall recombination is modelled in ATLAS as follows:

$$R_{SRH} = \frac{(pn - n_i^2)}{\tau_p\left[n + n_i \exp\left(\frac{E_t - E_i}{k_B T}\right)\right] + \tau_n\left[p + n_i \exp\left(\frac{E_t - E_i}{k_B T}\right)\right]}, \tag{15.3}$$

where E_t is the trap energy level and E_i the intrinsic Fermi level, T is the lattice temperature, n_i the intrinsic carrier concentration, n and p are the electron and hole concentrations, respectively, and τ_n and τ_p are the electron and hole lifetimes, respectively, that depend on the c-Si doping concentration, $N_{c\text{-}Si}$, as proposed by Kendall [18]:

$$\tau_{SRH\ n(p)} = \frac{\tau_{0,SRH\ n(p)}}{1 + \frac{N_{c-Si}}{10^{16}}}, \tag{15.4}$$

where $\tau_{0,SRH\ n(p)}$ is a user defined value and corresponds to the lifetime at low doping density.

15.3.2.2 Auger Recombination Model

Auger recombination occurs through a three particle transition whereby a mobile carrier is either captured or emitted. The underlying physics for such processes is unclear and normally a more qualitative understanding is sufficient [19]. Auger recombination is commonly modelled using the expression [20]:

$$R_{Auger} = C_n\left(pn^2 - n_0 n_i^2\right) + C_p\left(np^2 - p_0 n_i^2\right), \tag{15.5}$$

where C_n is the Auger coefficient for electrons and C_p the Auger coefficient for holes, n (p) the concentration of electrons (holes), n_0(p_0) the corresponding values at equilibrium.

The Auger coefficients C_n and C_p are calculated according to the Auger recombination model proposed by Kerr and Cuevas [21]. Assuming $n_0 n_{ie}^2$ and $p_0 n_{ie}^2$ negligible and under low-injection conditions, the Auger lifetime corresponding to this process is defined by:

$$\tau_{Aug,low} = \frac{1}{1.8 \times 10^{-24} N_D^{1.65}} \tag{15.6}$$

for n-type c-Si, and

$$\tau_{Aug,low} = \frac{1}{6 \times 10^{-25} N_A^{1.65}} \tag{15.7}$$

for p-type c-Si.

15.3.2.3 Surface Recombination

The charge carriers (electrons or holes) can be generated or recombine at the surface of c-Si in addition to bulk generation-recombination. The surface recombination can be even more important than the bulk recombination. The calculation of the rate of surface recombination is an extension of the SRH theory by introducing the surface recombination velocities (S_{n0} or S_{P0}). The recombination rate is calculated as follows:

$$R_{surf} = \frac{pn - n_i^2}{\tau_n^{eff}\left[n + n_i \exp\left(\frac{E_t - E_i}{k_B T}\right)\right] + \tau_p^{eff}\left[p + n_i \exp\left(\frac{E_t - E_i}{k_B T}\right)\right]} \quad (15.8)$$

where τ_n^{eff} is the effective lifetime for electrons and τ_p^{eff} is the effective lifetime for holes that are expressed by:

$$\frac{1}{\tau_n^{eff}} = \frac{1}{\tau_n^i} + \frac{d_i}{A_i} S_{n0}, \quad \frac{1}{\tau_p^{eff}} = \frac{1}{\tau_p^i} + \frac{d_i}{A_i} S_{p0}, \quad (15.9)$$

where τ_n^i is the bulk lifetime for electrons calculated at an interface node i, τ_p^i is the bulk lifetime for holes calculated at an interface node i, d_i and A_i represent the thickness and area, respectively, at a surface node *i*.

15.3.2.4 Mobility Model

In ATLAS, the carrier mobility was taken dependent on the doping concentration. Masetti *et al.* [22] modelled the dependence of mobility on carrier concentration over a range of 8 orders from approximately 10^{13} cm^{-3} to 10^{21} cm^{-3} and this model gives results very close to that of Muller and Kamins [23]. They found that their model required different parameter sets for the electron mobility in arsenic and phosphorous n-doped silicon. The functional form of the electron mobility and hole mobility is expressed as:

$$\mu_p = \mu_{p,\min 1} \exp\left(-\frac{P_{c,p}}{N}\right) + \frac{\mu_{p,\max 1} - \mu_{p,\min 2}}{1 + \left(\frac{N}{C_{R,p}}\right)^{\alpha p}} - \frac{\mu_{p,1}}{1 + \left(\frac{C_{S,p}}{N}\right)^{\beta p}} \quad (15.10)$$

$$\mu_n = \mu_{n,\min 1} \exp\left(-\frac{P_{c,n}}{N}\right) + \frac{\mu_{n,\max 1} - \mu_{n,\min 2}}{1 + \left(\frac{N}{C_{R,n}}\right)^{\alpha n}} - \frac{\mu_{n,1}}{1 + \left(\frac{C_{S,n}}{N}\right)^{\beta n}} \quad (15.11)$$

where N is the total doping concentration, P_C, C_R and C_S are reference doping concentrations.The parameters for Masetti's model are set in Table 15.2.

Table 15.2 Parameters used in Masetti's model.

Parameters	Values	Parameters	Values	Unit
$\mu_{p,min1}$	44.9	$\mu_{n,min1}$	52.2	$cm^2. V^{-1}.s^{-1}$
$\mu_{p,min2}$	0.0	$\mu_{n,min2}$	52.2	$cm^2. V^{-1}.s^{-1}$
$\mu_{p,max}$	470.5	$\mu_{n,max}$	1417	$cm^2. V^{-1}.s^{-1}$
$\mu_{p,1}$	29.0	$\mu_{n,1}$	43.4	$cm^2. V^{-1}.s^{-1}$
$P_{C,p}$	9.23×10^{16}	$P_{C,n}$	0.0	cm^{-3}
$C_{R,p}$	2.23×10^{17}	$C_{R,n}$	9.68×10^{16}	cm^{-3}
$C_{S,p}$	6.1×10^{20}	$C_{S,n}$	3.34×10^{20}	cm^{-3}
α_p	0.719	α_n	0.68	-
β_p	2.0	β_n	2.0	-

15.3.3 Recombination in Amorphous Silicon

Amorphous silicon is a disordered material that contains a large number of defect states in the band gap. To describe this material, it is assumed that the density of states (DOS) is composed of four bands which are energetically distributed within the bandgap:

- two exponentially decaying band tail states (a donor-like valence band tail and an acceptor-like conduction band tail) that model the disorder in the amorphous silicon structure, and
- two Gaussian distributions of mid-gap states (one acceptor-like and the other donor-like) separated by a correlation energy U=0.2 eV that model the silicon dangling bonds, which is a typical approximation used to describe the amphoteric nature of dangling bonds defects [24, 25].

The parameters of hydrogenated amorphous silicon depend on deposition conditions and on the type of doping. The energy position and the maximum of the two Gaussian distributions of deep defects and their influence on the final properties of the amorphous silicon layer (conductivity, activation energy) depend on the equilibrium Fermi level position, as stated in the defect-pool model [26-28]. We use this general approach to properly fix the distributions of deep states, see Fig. 15.6.

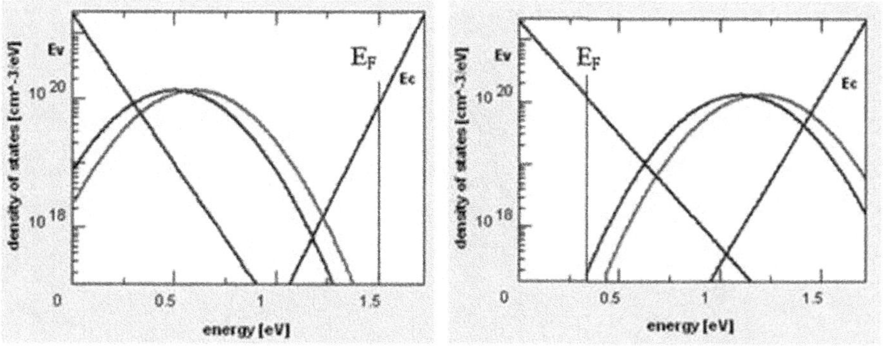

Fig 15.6 DOS of hydrogenated amorphous silicon: n-doped (left) and p-doped (right).

As for crystalline silicon, we define the electronic parameters associated to the amorphous silicon layer. We chose a band gap energy of 1.7 eV, an electron affinity of 3.9 eV, electron and hole mobilities of 20 $cm^2\ V^{-1}\ s^{-1}$ and 5 $cm^2\ V^{-1}\ s^{-1}$, respectively, effective densities of states for the conduction and valence bands of $2 \times 10^{20}\ cm^{-3}$, and a relative dielectric permittivity of 11.9.

To validate the TFT module of ATLAS which allows us to define the distribution of states in the band gap, we used our reference software AFORS-HET. We introduce the same parameters of a-Si:H (p-type or n-type) and obtained the same Fermi level position in the band gap with the two softwares. From the optical point of view, the same text file and the same generation model are used. Thus, using the TFT module of ATLAS we could produce a-Si:H layers having exactly the same properties as defined in AFORS-HET. Note that the recombination rate given by eq. (15.3) is valid for a discrete state located at E_t where N_t is the concentration of states at energy E_t. With continuous distributions of states like in amorphous silicon, N_t has to be replaced by the DOS at energy E_t, $N(E)$ and the expression has to be integrated over energies E_t in the band gap to get the total recombination rate.

15.3.4 Interface Defects States

For more realistic modelling of the a-Si:H / c-Si heterojunction, it is essential to introduce defect states at the hetero-interface and to take account of their influence. For that purpose, we introduce a thin (d_{int} =1 nm) defective layer of crystalline silicon at the interface [29]. The distribution of states of these defect states is taken as a Gaussian placed at mid-gap, as shown in Fig. 15.7. The character (acceptor or donor) is taken the same as for the c-Si doping substrate.

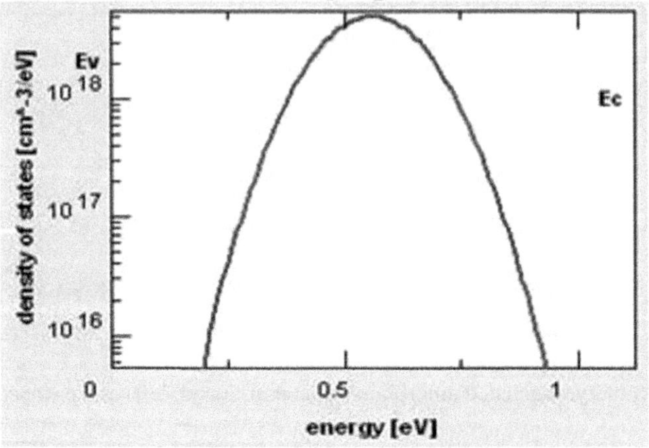

Fig. 15.7 DOS shape of the 1 nm thick interfacial defective layer.

The surface defect density, N_{SS} (cm^{-2}), is equal to $d_{int} \times N_{int}$, where N_{int} (cm^{-3}) is the integral over the entire band gap of the distribution of states in the thin (d_{int}=1 nm) defective interface layer:

$$N_{SS} = d_{\text{int}} \times N_{\text{int}} = 1nm \times \int_{E_V}^{E_C} N_{\text{int}}(E)dE\,. \tag{15.12}$$

The recombination rate due to these interface defects depends not only on their density N_{SS} but also on the capture cross-section σ_n of electrons or σ_p of holes. To discuss the effect of interface defects, the N_{SS} value is not sufficient. It is more complete to give the couple (N_{SS}; σ_n (or σ_p)) to properly describe the influence of interface defects.

15.4 Simulation Results

The behaviour of IBC-SiHJ solar cells depends on several parameters. These parameters, related to the materials and to the geometry of the cell structure, are interdependent, which makes it essential to study the influence of all them. This study will thus determine the critical and important parameters in the functioning of IBC-SiHJ solar cells and in the optimization of their performance. To study the influence of a given parameter, we vary its value while fixing all the others. For each variation of this given parameter, we generate the current-voltage curve, I(V), under AM1.5 illumination, and so extract the output characteristics: the open circuit voltage, V_{OC}, the short-circuit current density, J_{SC}, the fill factor FF and conversion efficiency η of the cell. The simulation results obtained with both reference cells are given in Table 15.3 and provide a reference for later comparison.

Table 15.3 Output performance parameters of the two reference cells. Note that the same minority carrier lifetime (1 ms) has been used for both n-type and p-type c-Si, which explains the better efficiency found for p-type c-Si, contrary to literature results, as will be explained later in the text.

c-Si	V_{OC} (mV)	J_{SC} (mA/cm^2)	η (%)	FF (%)
n-type	718	33,34	19,85	83,0
p-type	721	35,24	20,72	81,5

We will first study the influence of the c-Si substrate by varying the lifetime of minority carriers (substrate quality), the doping concentration (N_A or N_D) and the thickness of the substrate. The influence of front surface passivation will be studied later. Then we will focus on the impact of the presence of interface defect states at the c-Si / a-Si:H hetero-interfaces. To complete this overview, the optimization of the geometry of the rear side will be performed by studying the influence of the width of emitter, as well as BSF and gap regions.

15.4.1 Impact of c-Si Substrate

15.4.1.1 Effect of Minority Carrier Lifetimes, $\tau_{0,\,SRH}$

IBC-SiHJ solar cells are characterized by the position of contacts on the rear side (non-illuminated side). To collect the carriers, it is necessary that they reach the rear side of the cell without recombining. Thus, the importance of the c-Si substrate quality can be understood intuitively. The lifetime of minority carriers is the parameter that determines the c-Si substrate quality and is universally used. In the following, specifying the carrier lifetime in the c-Si substrate will refer to the lifetime of minority carriers, which determines the magnitudes of recombination rate. Indeed, with a higher lifetime, a free carrier travels a higher distance in the c-Si substrate before it recombines. The average distance that a minority carrier can travel without recombining is the diffusion length $L_{D,\,min}$ related to lifetime by the following terms:

$$L_{D,n} = \sqrt{D_n \tau_n} = \sqrt{\frac{k_B T}{q} \mu_n \tau_n}\ , \tag{15.13}$$

$$L_{D,p} = \sqrt{D_p \tau_p} = \sqrt{\frac{k_B T}{q} \mu_p \tau_p}\ , \tag{15.14}$$

for electrons and holes, respectively.

Figure 15.8 shows the impact of minority carrier lifetime, $\tau_{0,\,SRH}$, on the output characteristics of our two reference cells. Whatever the type doping of c-Si substrate, we observe an improvement in all output parameters (V_{OC}, J_{SC}, FF, and thus η) with higher minority carrier lifetime. This result can be understood naturally because with higher minority carrier lifetime, the diffusion length is higher and thus minority carriers will more likely reach the back surface without be recombining. So, the better the quality of c-Si substrate, the better is the performance of IBC-SiHJ solar cells.

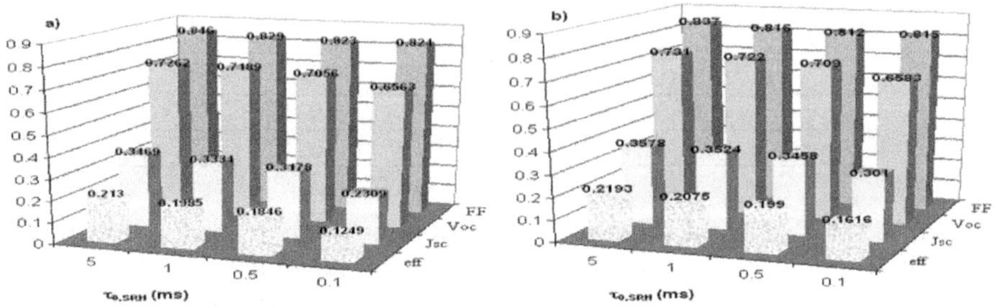

Fig. 15.8 Evolution of V_{OC} (V), J_{SC} (mA/mm^2), FF and η. (a) of the n-type reference cell and (b) of the p-type reference cell for different lifetime $\tau_{0,\ SRH}$ carriers. No defects in hetero-interfaces.

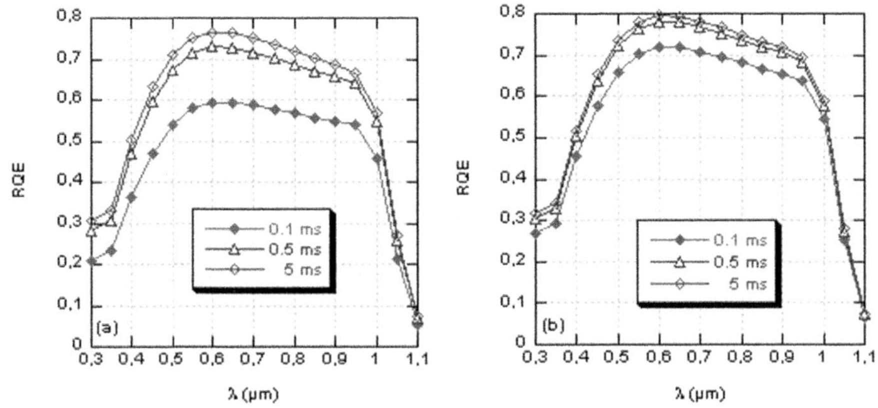

Fig. 15.9 External quantum efficiency (EQE) (a) of the n-type reference structure and (b) of the p-type reference structure for three different values of $\tau_{0,\ SRH}$ shown in the figure.

The evolution of the external quantum efficiencies of the two reference cells is shown in Fig. 15.9. The external quantum efficiency is improved for higher lifetimes. This result confirms the performance improvement with higher substrate quality.

The ratio ($L_{D\ min}/L_{c\text{-}Si}$) between the diffusion length of minority carriers ($L_{D\ min}$) and the substrate thickness ($L_{c\text{-}Si}$) is a good indicator of possible performance of rear-contact cells. In literature, a ratio ($L_{D\ min}/L_{c\text{-}Si}$) between 3 and 4 is given as sufficient to ensure satisfactory collection of carriers generated at the front surface and to obtain a maximum efficiency. It must be noted that this is valid only when using alignment technologies such as photolithography where the pitch can be

30-40 times smaller than a low cost alignment technology like screen printing or metallic masks. In our case, the dimensions of the reference structure correspond to the use of low cost alignment technologies. In Fig. 15.10, the evolution of the performance of the two reference structures with the diffusion length of minority carriers is shown.

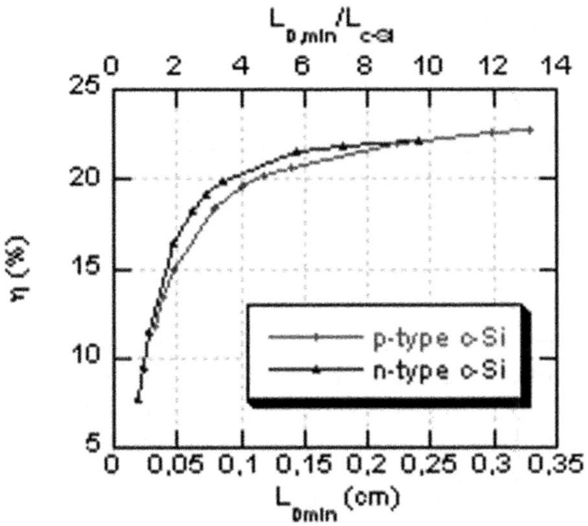

Fig. 15.10 Evolution of the performance of the two reference structures based on the diffusion length of minority carriers in c-Si.

For an n-type c-Si substrate, the efficiency begins to saturate when $L_{D\ min}$ is greater than 1.5 mm, which gives us a ratio $L_{D\ min}/L_{c\text{-}Si}$ of about 6. For a p-type substrate, the efficiency begins to saturate when $L_{D\ min}$ is greater than 2.5 mm, which gives us a ratio $L_{D\ min}/L_{c\text{-}Si}$ of about 10. The purpose of having a good substrate is therefore justified even with large pitch structures when using a low cost technology alignment.

In literature, the debate on the best doping type (n or p) of c-Si substrate has turned to the advantage of using n-type wafers. Indeed, the best experimental results on conventional silicon heterojunction solar cells (one dimension, 1D) were obtained with n-type c-Si by SANYO's group [1, 2]. The main reason given for these results is the fact that the band diagram of c-Si(n) / a-Si:H(p) heterojunction is more suitable (large ΔE_V) and allows for better V_{OC} [30, 31]. In the rear contacted cells, the critical parameter is the diffusion length of minority carriers $L_{D\ min}$. Indeed, as we can see in Fig. 15.10, this parameter is crucial for the performance of the IBC-SiHJ solar cell. The diffusion length of a carrier is given by the eqs. (15.13) and (15.14). It is a function of the lifetime of minority carrier and its mobility. For the same lifetime for electrons and holes, the diffusion length of electrons exceeds that of holes for the simple reason that the electron mobility is three times higher than that of holes. Comparing the two types of doping (n or p) will depend on the parameter chosen to do this. If we compare in terms of:

- **The minority carrier lifetime τ_{SRH}:** The results presented in Fig. 15.9 show that the best conversion efficiencies are obtained with the p-type reference cell. The V_{OC} obtained with the p-type reference cell is slightly higher (about 4 mV) than for n-type reference cell. The J_{SC} obtained with the p-type reference cell is well above (by 2 mA/cm^2). Indeed, the electrons are minority carriers in the p-type c-Si and have a three times higher mobility than holes. Therefore $L_{D\ min}$ is greater (for the same τ_{min}) with p-type c-Si, which explains the difference observed on the J_{SC} value. The fill factor is however slightly better with the reference structure of n-type. This reflects a more favourable band diagram obtained with n-type c-Si.
- **The diffusion length of minority carriers $L_{D,\ min}$:** The observation of Fig. 15.10 shows that the efficiency is higher with n-type c-Si for a given value of $L_{D,\ min}$. This is mainly due to a higher V_{OC} related to a more suitable band diagram for the c-Si(n) / a-Si:H(p) heterojunction with the chosen DOS in a-Si:H, doping in c-Si (5×10^{15} cm^{-3} was taken for both n- and p-type) and band offsets (larger ΔE_V compared to ΔE_C).

The use of n-type c-Si substrate for conventional (1D) heterojunction cells is suggested from better experimental results. Unlike IBC-SiHJ solar cells, conventional heterojunction cells are less sensitive to the lifetime of minority carriers. Indeed, the p-n (n-p) heterojunction in the conventional structures is located at the front and as the majority of carriers are generated in the first micrometers of c-Si substrate, it is understandable that the influence of minority carrier lifetime is less critical as is the case of IBC-SiHJ cells. The electron mobility is typically three times higher than that of holes, which is an inherent advantage of p-type c-Si substrates. Using a p-type c-Si substrate may thus prove more advantageous in the case of IBC-SiHJ cells. However, in practice it appears more difficult to obtain high p-type quality substrates ($\tau_{0,\ SRH}$ > 1 ms) due to a more peculiar production process.

15.4.1.2 Effect of c-Si Substrate Thickness, $L_{c\text{-}Si}$

One way to lower the manufacturing costs of silicon heterojunction cells is to reduce the amount of material used by reducing the thickness of the c-Si substrate. However, this solution is only viable if the reduction in the substrate thickness (amount of matter) does not result in a consequent decrease of cell performance. The thickness of c-Si generally used is about 250 μm. In this part, we shall study the impact of reducing c-Si substrate thickness on the output characteristics of the IBC-SiHJ solar cells.

<u>Effect on V_{OC}</u>

In Fig. 15.11, we show that V_{OC} increases when the c-Si substrate is thinner. Thus, a gain of 20 mV is obtained from 250 μm c-Si thickness to 100 μm for high substrate quality ($\tau_{0,\ SRH}$ = 5 ms) and the gain is 40 mV for low-quality substrate ($\tau_{0,\ SRH}$ = 0.1 ms). The V_{OC} enhancement with using thinner c-Si wafer is due to lower bulk recombination, which explains why this enhancement is more important with low quality substrate.

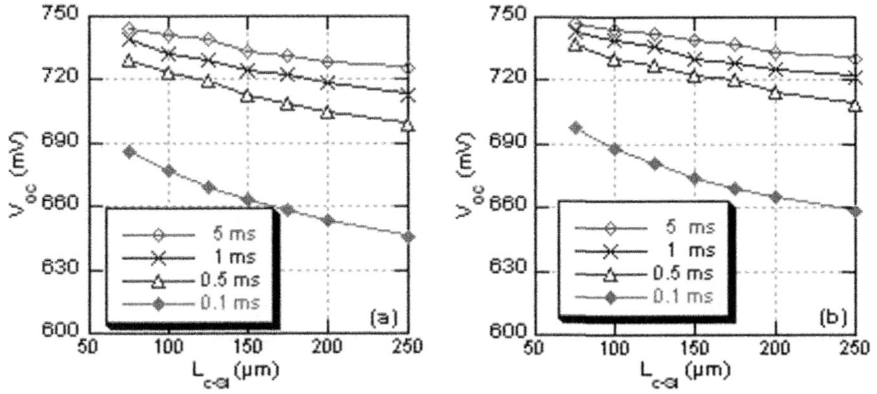

Fig. 15.11 Evolution of V_{OC} according to the substrate thickness $L_{c\text{-}Si}$ (a) of the n-type reference cell and (b) of the p-type reference cell for different values of $\tau_{0,\,SRH}$ shown in the figure. No defects at hetero-interfaces.

<u>Effect on J_{SC}</u>

In Fig. 15.12, the evolution of short circuit current density with the substrate thickness of crystalline silicon is shown. By reducing the c-Si thickness, the absorption of longer wavelength photons decreases, which induces less photogenerated electron-hole pairs in the substrate.

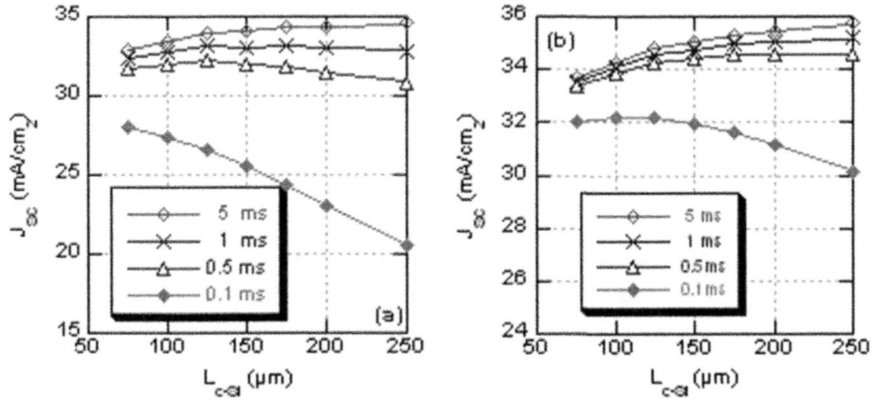

Fig. 15.12 Evolution of J_{SC} according to the substrate thickness $L_{c\text{-}Si}$ (a) of the n-type reference cell and (b) of the p-type reference cell for different values of $\tau_{0,\,HRS}$ shown in the figure. No defects at hetero-interfaces.

For good quality substrates ($\tau_{0,\ SRH}$> 1 ms), the thickness reduction results in a decrease in J_{SC} due to a decreased absorption of low energy photons. For poor quality substrate ($\tau_{0,\ SRH}$ = 0.1 ms), we can see in Fig. 15.12 an improvement of J_{SC} with lower $L_{c\text{-}Si}$. This improvement of J_{SC} is explained by a decrease in bulk recombination. Indeed, with lower lifetime and thicker c-Si substrate, bulk recombination predominates. Reducing $L_{c\text{-}Si}$ appears to be interesting for low substrate quality. For high quality substrates, a thickness reduction must be accomplished with an improved light trapping in order to avoid the loss in J_{SC}.

Effect on efficiency η

In Fig. 15.13, the evolution of the efficiency with $L_{c\text{-}Si}$ for various $\tau_{0,\ SRH}$ is shown. This evolution depends on minority carrier lifetime determined by $\tau_{0,\ SRH}$. If we reduce substrate thickness $L_{c\text{-}Si}$ from 250 μm to 75 μm:

- A drop in efficiency is observed for IBC-SiHJ solar cells with high c-Si substrate quality (large $\tau_{0,\ SRH} \geq 5$ ms). A maximal decrease in efficiency of 6% is calculated for the two reference cells for $\tau_{0,\ SRH} = 5$ ms.
- For c-Si substrates of medium quality, efficiency improves, reaches his maximum and decreases. $L_{c\text{-}Si}$ thickness corresponding to maximum efficiency is increasingly reduced with lower $\tau_{0,\ SRH}$.
- A clear improvement in performance is observed for poor c-Si substrate quality ($\tau_{0,\ SRH} \leq 0.1$ ms).

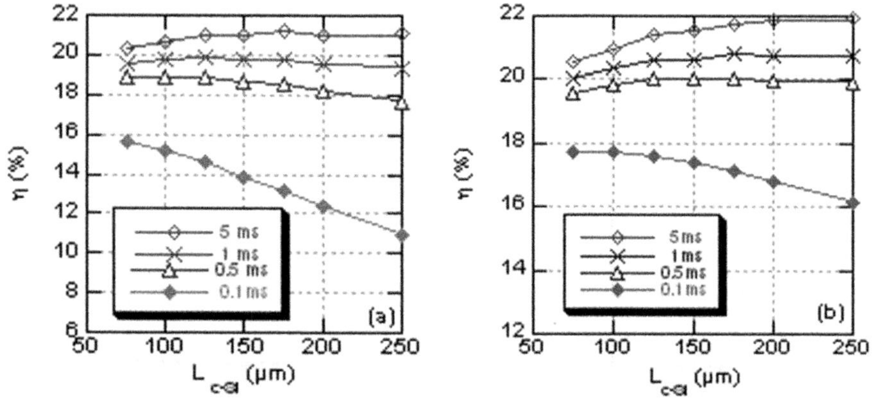

Fig. 15.13 Evolution of efficiency, η, according to the substrate thickness $L_{c\text{-}Si}$ (a) of the reference n-type cell and (b) of the reference p-type cell for different values of $\tau_{0,\ SRH}$. No defects at hetero-interfaces.

These results are very interesting in view of reducing manufacturing costs. On the one hand, the reduction of thickness for high substrate quality is justified because the ratio (material cost / efficiency loss) is largely in favour of reducing the c-Si substrate thickness. On the other hand, using a "bad" quality substrate (i.e.

multi-crystalline silicon) is found to be a rather interesting solution because then the ratio (manufacturing cost / performance) is very favourable.

These findings supporting the use of less thick c-Si substrate ($L_{c\text{-}Si}$ < 250 μm) are further strengthened with the assumption of total reflection on the back side. Indeed, by using metals with good reflective properties (metal like Ag), the low-energy photons reflected can further contribute to generate electron-hole pairs and result in an improvement of J_{SC} for a c-Si substrate with lower thickness.

15.4.1.3 Effect of c-Si Doping Concentration

Despite several studies on the link between resistivity and lifetime in c-Si [32-34], the influence of c-Si resistivity on the performance of heterojunction solar cells is still not very clear. The equations governing the dependence of bulk resistivity upon recombination are often simplified. In the following, we try to find out the influence of the different doping-related parameters in solar cell performances. Assuming good passivation at the front and rear surfaces and low recombining a-Si:H/c-Si interfaces, the bulk lifetime is determined by SRH and Auger recombination process and these processes depend on the doping concentrations (N_D or N_A). Assuming complete ionization, the c-Si resistivity which is very commonly used in the literature is related to the doping concentration by:

$$\rho = \frac{1}{q\mu_n N_D}, \tag{15.15}$$

$$\rho = \frac{1}{q\mu_p N_A}, \tag{15.16}$$

depending on the doping type. The carrier mobility was taken dependent on the doping concentration according to [21]. The bulk lifetime is defined as:

$$\frac{1}{\tau_{bulk}} = \frac{1}{\tau_{SRH}} + \frac{1}{\tau_{Aug}}, \tag{15.17}$$

where τ_{SRH} is the lifetime associated with the SRH recombination and τ_{Aug} is the lifetime associated with Auger recombination; it is depicted in Fig. 15.14.

τ_{bulk} is calculated taking account of both SRH and Auger contributions (surface recombination is neglected) as function of doping concentration. For doping concentrations below 10^{16} cm^{-3} (which corresponds to N_{SRH}), the SRH recombination mechanism is the predominant one. The lifetime is mainly determined by the SRH process, whatever the type of doping as seen in Fig. 15.14.

For an n-type substrate, the Auger recombination becomes significant when concentrations are above 10^{16} cm^{-3} and even becomes the dominant recombination process for $N_D > 5 \times 10^{16}$ cm^{-3}. The Auger recombination process is less intense for a p-type substrate since the Auger coefficient C_p is three times smaller than C_n. Auger recombination in p-type substrate is important for $N_A > 3 \times 10^{16}$ cm^{-3}.

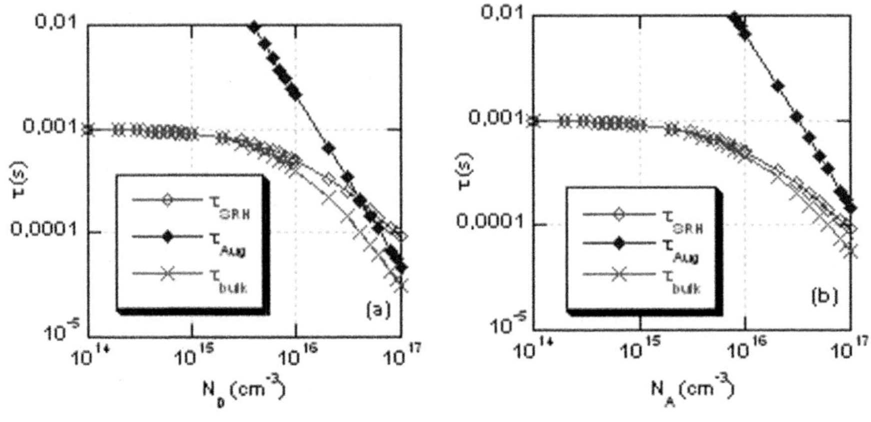

Fig. 15.14 Impact of SRH and Auger recombination mechanisms on the bulk lifetime τ_{bulk} (a) using n-type c-Si and (b) using p-type c-Si, depending on the doping concentrations N_D and N_A, respectively. $N_{SRH} = 10^{16}$ cm^{-3} and $\tau_{0,\,SRH} = 1$ ms.

Effect on V_{OC}

The open-circuit voltage normally increases with doping concentration as a consequence of the increase of the built-in voltage. However, dominance of Auger recombination above a given doping concentration leads to a decrease of V_{OC}, so that there is an optimum value, see Fig. 15.15.

In an n-type c-Si substrate, the optimum is at $N_D = 10^{16}$ cm^{-3}, whereas in a p-type substrate it is at $N_A = 3 \times 10^{16}$ cm^{-3} because Auger recombination is less intense. The variation of V_{OC} with the doping concentration is less important when using lower quality substrates.

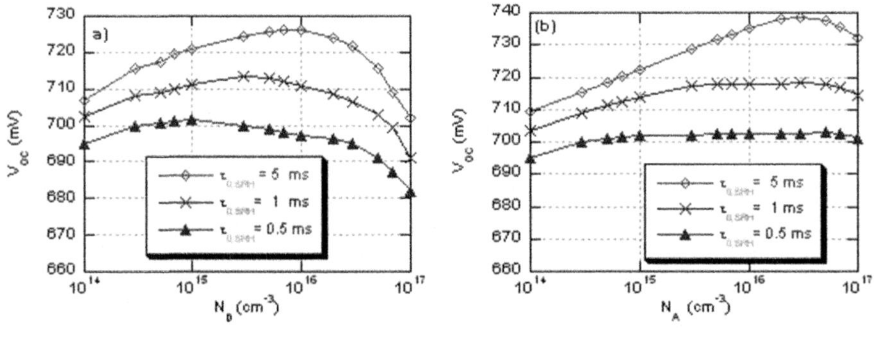

Fig. 15.15 Variations of the open-circuit voltage, V_{OC}, as function of the c-Si doping concentration, (a) for the n-type reference cell and (b) for the p-type reference cell for different substrate lifetimes. No defects at c-Si/a-Si:H hetero-interfaces.

Effect on J_{SC}

The short-circuit current density decreases when the doping concentration increases as we observe in Fig. 15.16. Whatever the doping type of c-Si, the lifetime of carriers and their mobility decrease with higher doping concentration. These dependencies explain the decrease in J_{SC} with higher doping concentration. We note that this decrease is more pronounced with a lower quality substrate.

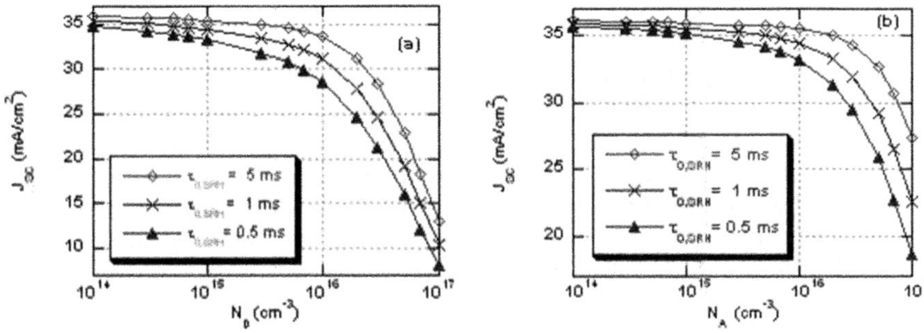

Fig. 15.16 Variations of the short-circuit current density as function of the c-Si doping concentration (a) for the n-type reference cell and (b) for the p-type reference cell for different lifetimes. No defects at c-Si/a-Si:H hetero-interfaces.

Effect on *FF*

By increasing the c-Si doping concentration (N_A or N_D), the fill factor is improved, see Fig. 15.17. This improvement is related to the resistivity of the c-Si substrate. Indeed, when increasing the doping concentration, the resistivity of the substrate decreases, which enhances conduction properties in the c-Si substrate and thus the fill factor.

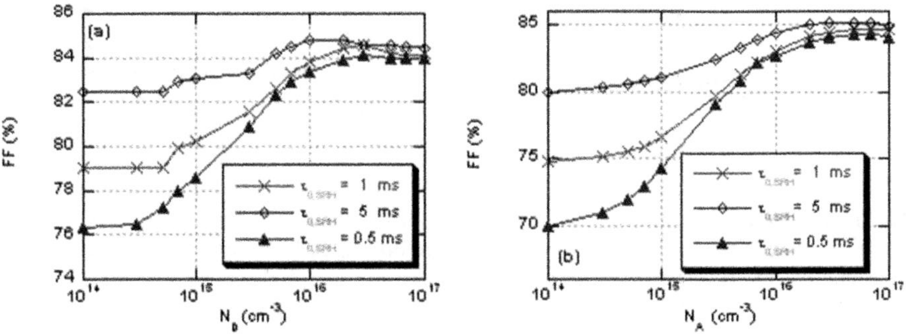

Fig. 15.17 Variations of the fill factor, *FF*, as function of the c-Si doping concentration (a) for the n-type reference cell and (b) for the p-type reference cell for different lifetimes. No defects at c-Si/a-Si:H hetero-interfaces.

The simulations were carried out without considering series resistance at the contacts which explains the very good values obtained.

Effect on efficiency, η

The parameters V_{OC}, J_{SC} and FF do not evolve in the same direction when the doping concentration varies. These various changes result in the existence of an optimum c-Si doping concentration corresponding to the maximum efficiency of IBC-SiHJ solar cells, as shown in Fig. 15.18. This optimum depends on the type of c-Si doping:

- **n-type c-Si substrate:** the optimal doping concentration is around $N_D = 2 \times 10^{15}$ cm^{-3} ($\rho = 2.5$ Ω.cm) for $\tau_{0,\,SRH} = 1$ ms and tends to decrease with lower substrate quality. However, we note that the efficiency varies very little on a wide range around this optimum. If we define the optimal range as the range in doping concentration wherein the cell efficiency is at least 95% of its maximum value, we find that the optimal range corresponds to levels below 10^{16} cm^{-3} which corresponds to resistivities above 0.58 Ω.cm. With an n-type c-Si doping, it is necessary to use c-Si substrates resistive for better performance.

- **p-type c-Si substrate:** the optimal doping concentration is $N_A = 10^{16}$ cm^{-3} ($\rho = 1.36$ Ω.cm). The optimum doping concentration range is 10^{15}–3×10^{16} cm^{-3}, which corresponds to resistivities in the range 0.51–12.7 Ω.cm.

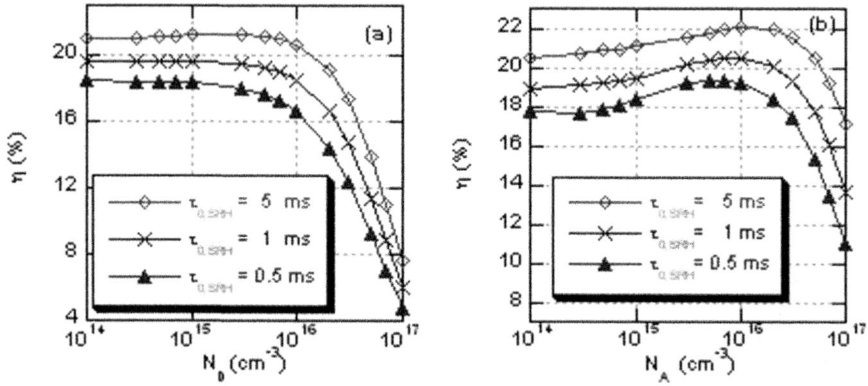

Fig. 15.18 Variation of the efficiency, η, as function of the c-Si doping concentration (a) for the n-type reference cell and (b) for the p-type reference cell for different lifetimes. No defects at c-Si/a-Si:H hetero-interfaces.

The optimum ranges of c-Si resistivity (or doping concentration) have been determined for both IBC-SiHJ reference cells. These optimum ranges depend on the geometry of the rear side which is based on the alignment technology used. The dimensions of our reference structure correspond to characteristic sizes obtained with low cost alignment technology such as screen printing.

15.4.2 Front Surface Passivation

The front surface passivation of the c-Si substrate is a key factor to achieve high efficiency IBC-SiHJ cells. It is estimated by the surface recombination velocity, S_{FAV}, of carriers at this front side. In Fig. 15.19, we show the impact of S_{FAV} on the current-voltage curves of our two reference cells. Increased S_{FAV} correspond to increased recombination rates of carriers at the front surface. Indeed, the majority of electron-hole pairs are photo-generated in the first micrometers of c-Si substrate while contacts are located on the rear side, so it is crucial that these photogenerated carriers do not recombine so that they can diffuse to the rear of the cell. As we can see in Fig. 15.19, an increase of S_{FAV} results in a degradation of J_{SC} and therefore in a degradation of the cell performance. So when S_{FAV} varies from 10 cm/s to 50 cm/s, the efficiency of IBC-SiHJ cells based on p-type c-Si drops from 20.4% to 19.4%, and from 19.3% to 17.2% for the IBC-SiHJ cell based on n-type c-Si. A good passivation of the c-Si front surface (S_{FAV} low as possible) is a key factor to obtain good performance with this type of cell.

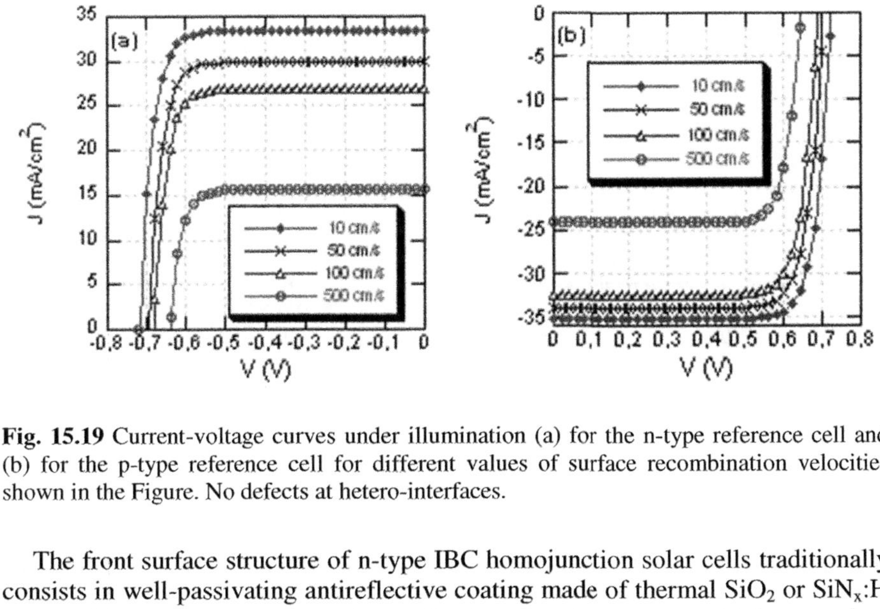

Fig. 15.19 Current-voltage curves under illumination (a) for the n-type reference cell and (b) for the p-type reference cell for different values of surface recombination velocities shown in the Figure. No defects at hetero-interfaces.

The front surface structure of n-type IBC homojunction solar cells traditionally consists in well-passivating antireflective coating made of thermal SiO_2 or SiN_x:H layers [35, 36]. The insertion of a lightly doped n+ layer at the c-Si front surface can be used on this cell structure [37] to create a so-called Front Surface Field (FSF) which shields the minority carriers from surface recombination [38]. The FSF also enhances lateral current transport in IBC cells by decreasing the base resistivity [39]. This effect is particularly interesting for industrial solar cells using large cell pitches and thinner substrates. For IBC cells which use the low temperature a-Si:H/c-Si heterojunctions (IBC-SiHJ cells) the front surface structure mostly consists in an a-Si:H/AR stack. Here the thin a-Si:H layer passivates the surface whereas an SiN_x:H [8,13] or TCO (Transparent Conductive Oxide) [40] layer is used for light trapping. Besides the thermal budget aspect, the main advantage of such low temperature front surface scheme is the possibility to fabricate it at any stage of the cell processing. However IBC-SiHJ cells face the same challenges regarding transport enhancement in the base and creation of an FSF as indicated above for homojunction cells.

As seen in Fig. 15.20, there is a strong dependence of cell performance with surface recombination velocity of carriers in classic IBC-SiHJ solar cell (without FSF layer). Since most of the carriers are generated near the front side, while the p-n heterojunction is at the back side, poor passivation means high S_{FAV}, and this will cause important carrier recombination before they can reach the back side. The figure shows the positive effect of an FSF layer for the passivation of the front side. Indeed, there is no degradation in cell performance with increasing minority carrier surface recombination velocity up to 5000 cm/s. This result is very important in industrial conditions where it is difficult to obtain good passivation on large areas. When using an a-Si:H FSF, recombination can take place at the a-SiH/c-Si interface. Presence of a high defect states density will annihilate the positive effect of the FSF layer on front surface passivation [41].

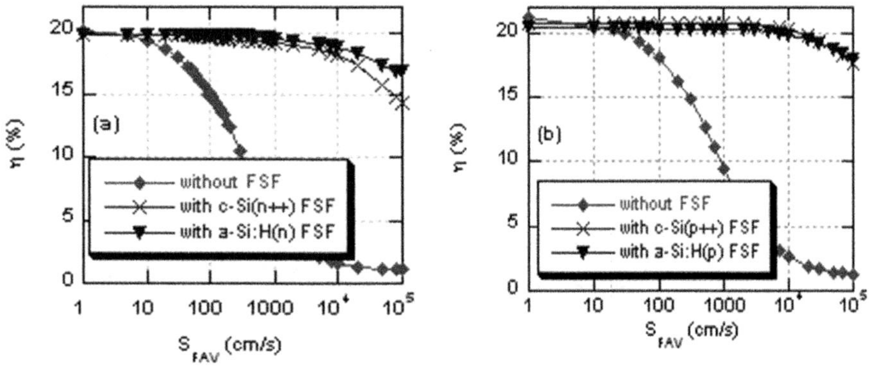

Fig. 15.20 Influence of surface recombination velocities on the front side without and with FSF layers for (a) n-type reference cell and (b) p-type reference cell. No interface states were taken into account for the a-Si:H FSF.

15.4.3 Impact of Defect States at c-Si / a-Si:H Hetero-Interfaces

The study of the influence of the c-Si substrate was performed by considering perfect c-Si/a-Si:H hetero-interfaces at rear side (without interface states). This ideal case of a-Si:H / c-Si hetero-interface is not realistic because there are always surface defect states that may be recombination centers for free carriers. It is therefore important to determine the influence of defect states on each hetero-interface. The definition used for these interface states has been described in section 3.4.

Based on eq. (15.3), the rate of SRH recombination through interface defect states is given by:

$$R_{\mathrm{int},SRH} = \int_{E_V}^{E_C} \frac{\sigma_n \sigma_p v_{th} \left(np - n_i^2\right)}{\sigma_n \left(n + n_i \exp\left(\frac{E - E_i}{k_B T}\right)\right) + \sigma_p \left(p + n_i \exp\left(\frac{E_i - E}{k_B T}\right)\right)} N_{\mathrm{int}}(E)\, dE, \tag{15.18}$$

where $N_{int}(E)$ is the density of interface states located at E, and σ_p and σ_n are the capture cross sections of electrons and holes.

This expression of $R_{int,\, SRH}$ can be simplified depending on the considered c-Si / a-Si:H hetero-interface [42]. If we consider the c-Si(n) / a-Si:H (n) hetero-interface, we have an accumulation of electrons at the c-Si surface (if the density of interface defects is not so high that it completely modifies the band diagram compared to the ideal defect-free case). Then p can be negligible compared to n and thus the recombination rate can be simplified into:

$$R_{\mathrm{int},SRH} \approx \sigma_p\, p\, v_{th} \int N_{\mathrm{int}}(E)\, dE = \sigma_p\, p\, v_{th} N_{SS}\,. \tag{15.19}$$

Thus only the very few holes at c-Si interface determine the recombination.

If we consider the c-Si(n) / a-Si:H (p) hetero-interface, we have a strong electron inversion at the c-Si surface. Then n can be negligible compared to p and thus the recombination rate can be simplified into:

$$R_{\mathrm{int},SRH} \approx \sigma_n\, n\, v_{th} \int N_{\mathrm{int}}(E)\, dE = \sigma_n\, n\, v_{th} N_{SS}\,. \tag{15.20}$$

Thus only the very few electrons at c-Si interface determine the recombination.

In this section, we will study the impact of interface defect states on the functioning of the IBC-SiHJ solar cell. Firstly, we consider the c-Si/BSF hetero-interface as perfect (without defects) and introduce defects at the c-Si/emitter hetero-interface. Secondly, we consider the c-Si/emitter interface as perfect (without defects) and introduce defects at the c-Si/BSF interface. This procedure will determine separately the influence of the density of interface states N_{SS} on each a-Si: H/c-Si interface. We will also look at the influence of capture cross sections.

15.4.3.1 At the c-Si/emitter Hetero-Interface

We have introduced interface defect states only at the c-Si/emitter hetero-interface, i.e. c-Si (n)/a-Si:H (p) or c-Si (p)/a-Si:H (n) hetero-interfaces. The c-Si/BSF (n-n or p-p) hetero-interface is considered as perfect (defect free). The capture cross sections of electrons and holes are set at $\sigma = \sigma_n = \sigma_p = 10^{-15}$ cm^2.

The introduction of interface states can have a direct impact on the band diagram of the c-Si/a-Si:H heterojunction.

In Fig. 15.21, the influence on the band diagram of the p-n heterojunction at thermodynamic equilibrium is shown. In both c-Si(p) / a-Si:H(n) and c-Si(n) / a-Si:H(p) cases, changes in the band diagram occur mainly when $N_{SS} > 10^{12}$ cm^{-2}.

In Fig. 15.22 we show the variation of the open circuit voltage and fill factor with N_{SS}. For the n-type c-Si substrate, we note a degradation of V_{OC} for N_{SS} values above 10^{12} cm^{-2}. For the p-type substrate, V_{OC} begins to drop already for lower defect densities, when $N_{SS} > 10^{11}$ cm^{-2}. This degradation of V_{OC} for high N_{SS} is explained by recombination through the interface states that becomes increasingly dominant. It is worth noting that the c-Si(n) / a-Si:H(p) structure is more robust against interface states density than the c-Si(p) / a-Si:H(n) structure, for which values slightly above 10^{11} cm^{-2} already induce a significant drop of V_{OC}, while the effect on the equilibrium band bending is negligible.

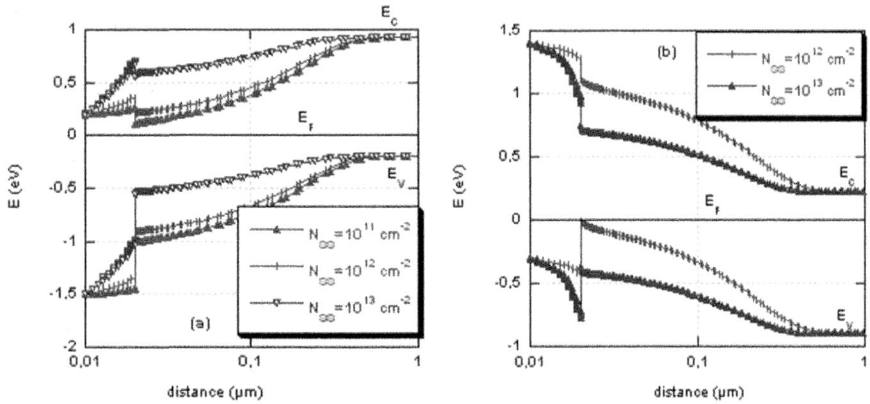

Fig. 15.21 Evolution of band diagram at thermodynamic equilibrium (a) of the c-Si (p) / a-Si: H (n) emitter heterojunction of the p-type reference structure and (b) of the c-Si (n) / a-Si:H (p) emitter heterojunction of the n-type reference structure for different values of N_{SS}. The zero energy is taken at the Fermi level.

As can also be seen in Fig. 15.22, the fill factor follows the same global trend, although it is a little bit less sensitive than V_{OC}.

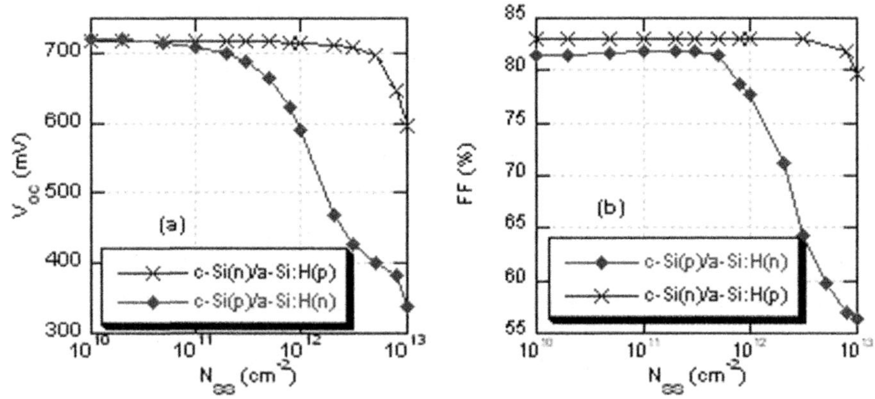

Fig. 15.22 Evolution (a) V_{OC} and (b) fill factor as a function of the interface state density, N_{SS}, placed at the emitter heterojunction only (the BSF heterojunction being free of interface defects) for the two reference cells.

Finally, the evolution of J_{SC} is not shown here, because it remains constant in the two IBC-SiHJ reference cells. The presence of defects at the c-Si/emitter hetero-interface does not affect J_{SC}.

The passivation of c-Si/emitter hetero-interfaces is very important to obtain a high V_{OC} (>700 mV). A large defect density at these hetero-interfaces results in a higher recombination through these states, which limits the performance of the cell.

15.4.3.2 At the c-Si/BSF Hetero-Interface

In this section we have introduced interface defect states only at the c-Si/BSF (n-n or p-p) hetero-interface for each reference cell, i.e. the c-Si(p) /a-Si:H(p) or c-Si(n) / a-Si:H(n) hetero-interfaces. The c-Si/emitter (p-n or n-p) hetero-interfaces are considered as perfect (defect free).

A large defect density at this interface results in a degradation of V_{OC}, as can be seen in Fig. 15.23. This degradation is however much weaker than in the case where defects are introduced at the emitter interface (see above).

We also note in Fig. 15.23 that interface defects induce degradation of J_{SC} (and also on FF, not shown in the figure because the effect on FF is weak).

When comparing the two types of structures, we note that defects at the BSF hetero-interface have a greater impact on c-Si(n) than on c-Si(p). This is likely to be related to the larger valence band offset with respect to the conduction band one.

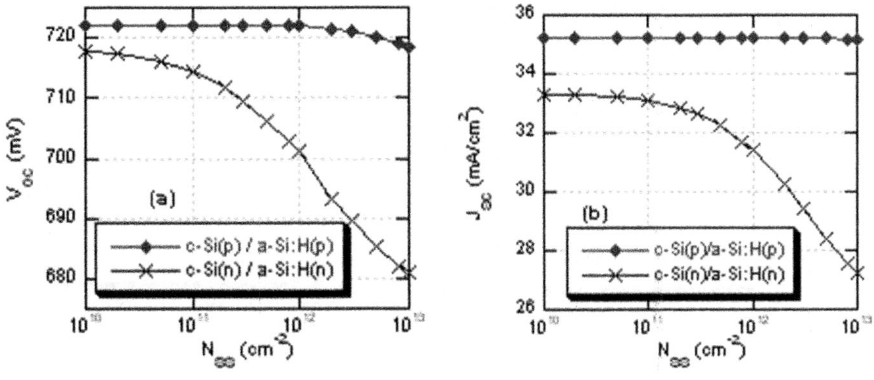

Fig. 15.23 Evolution of (a) V_{OC} and (b) J_{SC} as function of N_{SS} at the c-Si(p)/a-Si:H(p) hetero-interface for the p-type reference structure and at c-Si(n)/a-Si:H(n) hetero-interface for the n-type reference structure.

To summarize the effect of interface defects at the a-Si:H/c-Si hetero-interfaces:

- J_{SC} is degraded only by the presence of a high density of defects at the c-Si/BSF hetero-interface
- V_{OC} and FF are degraded by the presence of a high density of defects at any c-Si/a-Si:H hetero-interface (emitter or BSF), but the degradation is much more important for defects at the emitter interface.

Good passivation ($N_{SS} < 10^{11}$ cm^{-2}) of any c-Si/a-Si:H hetero-interface is therefore a critical factor to obtain high efficient IBC-SiHJ solar cells.

As usually a larger emitter width as compared to the BSF width is recommended (as described later in sections 15.4.4.1 and 15.4.4.3) the use of n-type c-Si wafers is more favourable as the structure on (n) c-Si is less sensitive to interface defects than the (p) c-Si one.

15.4.3.3 Influence of the Capture Cross Section of Electrons, σ_n, and of Holes, σ_p

Recombination induced by defect states depends as on interface defect density but also depends on the capture cross sections of electrons (σ_n) or holes (σ_p), as can be seen from eq. (15.18).

The influence of capture cross sections of carriers is considered by fixing N_{SS} at 5×10^{12} cm^{-2} at one hetero-interface (the other being defect-free) and varying either σ_n or σ_p from 10^{-16} cm^2 to 10^{-14} cm^2, the capture cross section for the other type of carriers (σ_n or σ_p, respectively) being set at 10^{-15} cm^2. This interval corresponds to most of the possible values of capture cross sections found in the literature. This will allow us to determine the influence of the capture cross section of each carrier on recombination at each c-Si / a-Si:H hetero-interface. Figure 15.24 shows the variation of V_{OC} following this approach.

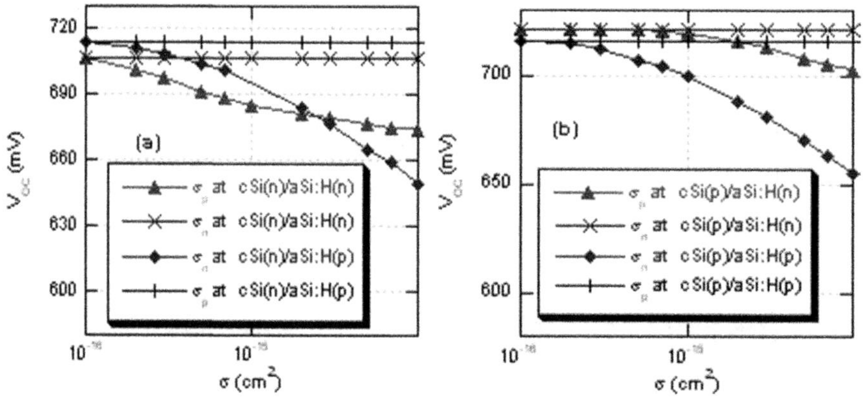

Fig. 15.24 Evolution of V_{OC} as a function of either the capture cross section of electrons, σ_n, or the capture cross section of holes, σ_p, (the other capture cross section being set at 10^{-15} cm^2) on each hetero-interface (a) for the n-type reference cell and (b) for the p-type reference cell. In all cases, $N_{SS} = 5 \times 10^{12}$ cm^{-2} at the indicated hetero-interface, the other hetero-interface being free of defects.

For the n-type reference cell, we observe that:

- At the c-Si(n) / a-Si:H(n) hetero-interface, σ_p has an effect on V_{OC} while σ_n has no influence. We have an accumulation of electrons at the hetero-interface.

With larger σ_p, recombination at the interface states increases as suggested by eq. (15.19) the observed decrease in V_{OC}.

- At the c-Si(n) / a-Si:H(p) hetero-interface, σ_n has an effect on V_{OC} while σ_p has no influence. We have a strong inversion layer at the c-Si surface of this hetero-interface. Thus, larger σ_n, causes an increase of recombination as suggested by eq. (15.20) explaining the observed decrease in V_{OC}.

The same reasoning is used for the p-type reference structure. With the accumulation of holes at the c-Si surface of the c-Si(p) / a-Si:H(p) hetero-interface, recombination is influenced by σ_n. With the strong inversion at the c-Si surface of the c-Si(p) / a-Si:H(n) hetero-interface, recombination is influenced by σ_p.

It is worth discussing the advantage of using n-type or p-type c-Si with respect to capture cross sections of defects. Indeed, it is well known that in bulk c-Si, capture cross sections of electrons are larger than that of holes for most metal defect impurities [44, 45], e.g. σ_n=5×10^{-14} cm^2 and σ_p=7×10^{-17} cm^2 for interstitial Fe. This plays in favour of using n-type c-Si wafers, where bulk minority carriers are holes. Regarding interface defects, things are less straightforward. Indeed, as shown above, at the BSF hetero-interface, due to an accumulation layer, the solar cell performance is limited by the capture of bulk-minority carriers. On the other hand, at the emitter hetero-interface, due to the presence of a strong inversion layer, the solar cell performance is limited by the capture of bulk-majority carriers. Since the width of the emitter should be larger than that of the BSF (as described later in sections 15.4.4.1 and 15.4.4.3), p-type wafers would be preferable if interface defects followed the same trend as bulk metal impurities, i.e. if the capture cross section of electrons would be larger than that of holes. However, interface defects are silicon dangling bonds rather than metal impurities. The values of capture cross sections of silicon dangling bonds are not precisely known, and a large spreading is found in the literature. For Si/SiO_2 interfaces, the σ_n/σ_p ratio of states close to midgap was found larger than 1, between 3 and 1000, depending on the interface defect density [45]. There are several reports in a-Si:H with quite different values [46-50]. Owing to the amphoteric nature of silicon dangling bonds, one should compare the electron capture cross section of positively charged dangling bonds, σ_n^+, that should determine the recombination rate at the c-Si(n) / a-Si:H(p) interface, to the hole capture cross section of negatively charged dangling bonds, σ_p^-, that should determine the recombination rate at the c-Si(p) / a-Si:H(n) interface. From the work of Street *et al.* [48, 49], and provided the free carrier mobilities be the same in doped and undoped a-Si:H, σ_n^+ and σ_p^- can be estimated at 2×10^{-14} cm^2 and 4-8×10^{-15} cm^2, respectively, so that the σ_n^+/σ_p^- is between 2.5 and 5. On the contrary, Olibet *et al.* obtained values of 2.5×10^{-15} cm^{2}and 5×10^{-14} cm^2 for σ_n^+ and σ_p^-, respectively, giving a ratio σ_n^+/σ_p^- of 0.05 [50]. However, these values were obtained from fits to quasi-steady state photoconductivity measurements that need a lot of parameters and assumptions. To summarize this discussion, the true values of electron and hole capture cross sections for defects at the a-Si:H/c-Si interface are still unknown, and it is not really possible to conclude on the preference of using n-type or p-type wafers from this point of view.

15.4.4 Rear Side Geometry Optimization

The geometry of the rear side is also one of the research areas to optimize the performance of IBC-SiHJ solar cells. The rear geometry consists of alternating layers of n-doped a-Si:H and p-doped a-Si:H separated by a gap insulating region. The widths of the emitter, W_{emit}, of the BSF, W_{BSF}, and their spacing (gap region), W_{gap}, may have an impact on IBC-SiHJ solar cells.

15.4.4.1 Influence of BSF Width

To study the influence of the BSF width, we varied W_{BSF} while holding constant the gap region width and the emitter width. By increasing the BSF width, the

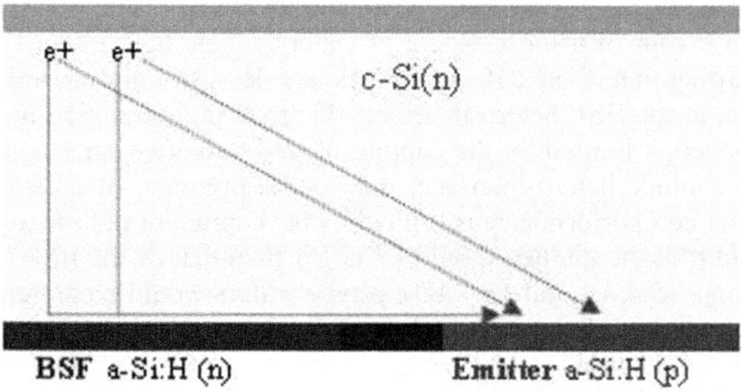

Fig. 15.25 Illustration of the increased distance for minority carriers (holes denoted e +) to reach the emitter junction increasing their chances to recombine.

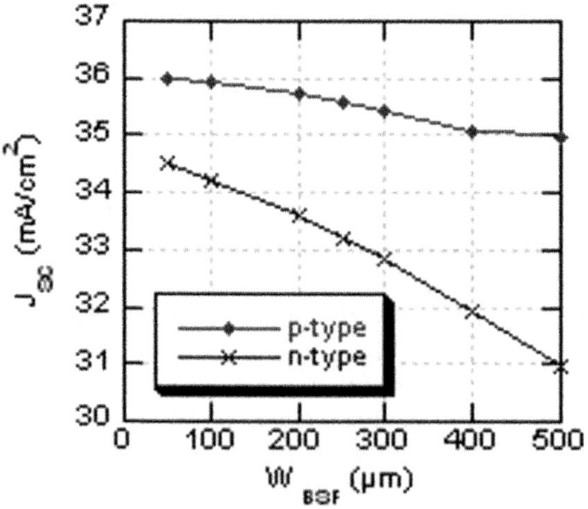

Fig. 15.26 Influence of BSF width on J_{SC} for the two reference cells.

minority carriers photogenerated in c-Si above the middle of the BSF travel a greater distance before reaching the p-n emitter heterojunction (see Fig. 15.25), which increases their chances to recombine.

Thus, increasing the BSF width results in a decrease of J_{SC} as we can see in Fig. 15.26. The c-Si/BSF hetero-interface can also be a source of recombination in the presence of high interface defect states density ($N_{SS} > 10^{11}$ cm^{-2}) as we have seen in section 15.3.2.2, which will accentuate the decrease of J_{SC} observed in Fig. 15.24. The BSF should be taken as narrow as possible to reach best J_{SC} values.

15.4.4.2 Influence of Gap Width, W_{gap}

We continue to optimize the geometry of the rear side studying the influence of the width of the spacing between the emitter and the BSF. As for the BSF region, increased gap width corresponds to an additional lateral distance to travel for

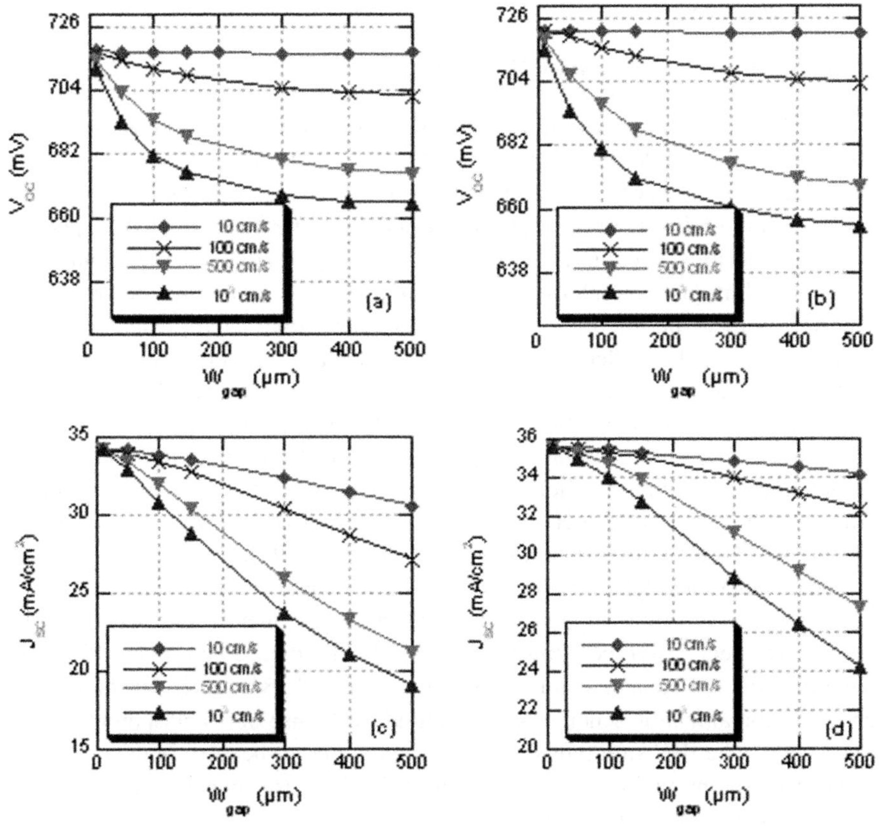

Fig. 15.27 Influence of gap width between BSF and emitter regions, W_{gap}, and of surface recombination velocity at the rear surface, S_{FAR}, on V_{OC} (a) for the n-type reference structure and (b) for the p-type reference structure; Influence on J_{SC} (c) for the n-type reference structure and (d) for the p-type reference structure.

minority carriers photogenerated in c-Si above the BSF region. This additional distance increases their chances to recombine before reaching the emitter. In addition, the interface with the c-Si substrate and the gap region can also be recombining. We have therefore extended the study by considering different surface passivation through the surface recombination velocity (S_{FAR}) of carriers for each value of W_{gap}.

In Fig. 15.27, the impact of the gap region width (W_{gap}) and of the passivation (S_{FAR}) on V_{OC} and on J_{SC} is shown. If we have an excellent passivation of the c-Si surface within the gap region, increasing W_{gap} results in a decrease of J_{SC} only. The V_{OC} is affected by increasing W_{gap} with lower surface passivation (higher S_{FAR}). Indeed, poor passivation means more recombination of minority carriers. The gap width will be taken as short as possible with good surface passivation to avoid further damaging the cell performance. To remove this gap region would be an attractive solution but this would lead to internal cell short circuit, thus making this solution unsuited.

15.4.4.3 Influence of Emitter Width, W_{emit}

The study of the emitter width influence is also important to optimize the geometry of the rear side. We therefore varied W_{emit} by fixing W_{gap} and W_{BSF}. Unlike the decrease of J_{SC} observed with higher W_{BSF} and W_{gap}, increasing W_{emit} results in an improvement of J_{SC}, see Fig. 15.28(a). Indeed, with the increase of W_{emit} photogenerated minority carriers in the c-Si are more likely located above the emitter and have no additional lateral distance to travel to reach the p-n heterojunction. On the contrary, the majority carriers have an additional lateral distance to travel before they can be collected at the BSF contact. It therefore appears that an additional series resistance will depend on the c-Si resistivity. This explains the decrease of the fill factor observed in Fig. 15.28(b) with larger W_{emit}.

With the different evolutions of J_{SC} and of the fill factor, the optimum W_{emit} corresponding to maximum efficiency is 400 microns for the p-type reference cell while for the n-type IBC-SiHJ reference substrate, a wider emitter, e.g. 1000 microns thick, gives better efficiency as seen in Fig. 15.28(c).

The contact metallization scheme can have an impact on IBC-SiHJ cell (results not shown here). With high contact resistance, low metal coverage, fill factor losses have been observed. But for IBC-SiHJ cells there is no trade-off between metallization and contact shading (except for bifacial IBC cells). Furthermore the choice of metallization schemes is not limited by the use of a Transparent Conductive Oxide. The IBC structure should thus allow one to better minimize the series resistance by optimizing contact grid metallization, as compared to the conventional vertical cell structure.

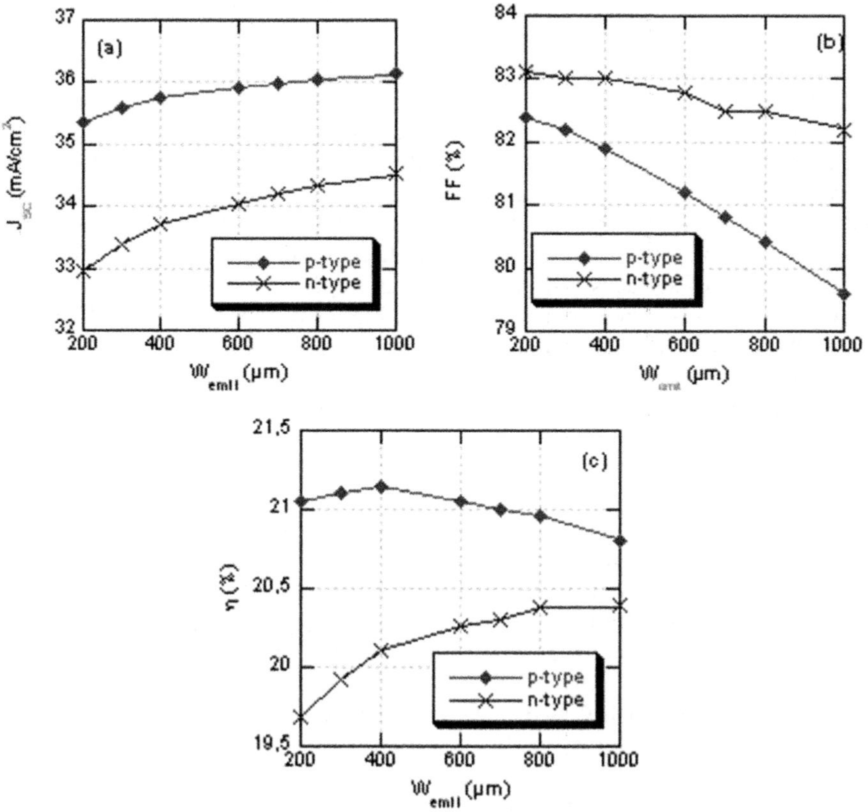

Fig. 15.28 Influence of emitter width on (a) J_{SC}, (b) the fill factor and (c) the cell efficiency for both IBC-SiHJ reference cells.

15.4 Conclusions

2D numerical simulations were performed in order to study the IBC-SiHJ solar cells. The impact of several electronic and geometrical parameters on the IBC-SiHJ solar cell output characteristics has been determined. This study allows us to find out the optimal properties of c-Si substrate and to optimize the rear side geometry of this type of cell that leads to achieve high efficiency.

The performances of the IBC-SiHJ are very sensitive to the c-Si substrate quality, i.e. the minority carrier lifetime or the minority carrier diffusion length. The latter has to be much larger than the wafer thickness for the minority carriers mainly photogenerated on the front side to reach the rear side and to be collected. For a c-Si wafer thickness of 250 µm, minority carrier lifetimes above 5 ms are required to achieve the best cell efficiencies.

We determined the optimal range of c-Si doping concentrations (N_D or N_A), corresponding to 95% of maximum efficiency, to be $N_D < 10^{16}$ cm^{-3} (or $\rho_{c\text{-}Si} >$ 0.6 Ω.cm) for n-type c-Si and 10^{15} cm^{-3} $< N_D < 3 \times 10^{16}$ cm^{-3} (0.51 Ω.cm $< \rho_{c\text{-}Si} <$ 12.7 Ω.cm) for p-type c-Si. From this point of view, the IBC-SiHJ are quite flexible. Our simulations also show this type of cell to be suitable for the use of thinner c-Si wafers ($L_{c\text{-}Si} < 250$ μm) with no significant loss of performance, which is important for reducing the manufacturing costs. By decreasing the c-Si wafer thickness, one can also put up with degraded c-Si quality, but light trapping then becomes the main issue in order to keep high short circuit current.

Passivation of surfaces and interfaces is an important issue. The c-Si front surface passivation is very important. Using a-Si:H as a front surface field layer can be a solution if the c-Si/a-Si:H hetero-interface is free of defects ($N_{SS} < 10^{11}$ cm^{-2}).

Passivation of emitter and BSF c-Si/a-Si:H hetero-interfaces is also a key factor. We showed that high interface defect densities ($N_{SS} > 10^{11}$ cm^{-2}) lead to a degradation of cell performance. We showed that the influence on V_{OC} and J_{SC} also depends on whether defects lie at the emitter or BSF interface of the IBC-SiHJ solar cell.

Optimization of the rear side geometry has to be done. The width of the gap region (spacing between the BSF and the emitter) must be kept as small as possible to avoid recombination of minority carriers in the bulk c-Si. For IBC-SiHJ based on n-type c-Si, the optimum geometry corresponds to a minimizing BSF region and a maximizing emitter region while for IBC-SiHJ based on p-type c-Si a BSF width equivalent to around 30% of the pitch is an optimum.

The performance of IBC-SiHJ solar cells can be impacted by a large number of parameters which requires careful control of all stages of the manufacturing process. However, if these parameters are well controlled, and with appropriate light trapping (in order to increase the short circuit current above 40 mA/cm^2) the IBC-SiHJ cell can lead to efficiencies larger than 25%.

References

[1] Taira, S., Yoshimine, Y., Baba, T., Taguchi, M., Kanno, H., Kinoshita, T., Sakata, H., Maruyama, E., Tanaka, M.: Our approaches for achieving HIT solar cells with more than 23 % efficiency. In: Proceedings of the 22nd European Photovoltaic Solar Energy Conference, Milano, Italy, pp. 932–935 (2007)
[2] Mishima, T., Taguchi, M., Sakata, H., Maruyama, E.: Development status of high-efficiency HIT solar cells. Sol. Energy Mater. Sol. Cells (2010), doi:10.1016/j.solmat.2010.04.030
[3] van Kerschaver, E., Beaucarne, G.: Back-contact Solar Cells: A Review. Prog. Photovolt. Res. Appl. 14, 107–123 (2006)
[4] Nichiporuk, O.: Simulation, fabrication et analyse de cellules photovoltaïques à contacts arrière interdigités. Thèse INSA Lyon (2005)
[5] Sunpower press release (2010), http://investors.sunpowercorp.com/releasedetail.cfm?ReleaseID=482133
[6] Lu, M., Bowden, S., Das, U., Birkmire, R.: a-Si/c-Si heterojunction for interdigitated back conctact solar cell. In: Proceedings of the 22nd European Photovoltaic Solar Energy Conference, Milano, Italy, pp. 924–927 (2007)

[7] Desrues, T., Souche, F., Vandeneynde, A., Muñoz, D., Ozanne, A.-S., Ribeyron, P.-J.: Emitter optimization for interdigitated back contact (IBC) silicon heterojunction (Si-HJ) solar cells. In: Proceedings of the 22nd European Photovoltaic Solar Energy Conference, Valencia, Spain (2010)

[8] Tucci, M., Serenelli, L., Salza, E., de Iullis, S., Geerlings, L.J., Caputo, D., Ceccarelli, M.: Novel scheme of amorphous/crystalline silicon heterojunction solar cell. In: Proceedings of the 23rd European Photovoltaic Solar Energy Conference, Valencia, Spain, pp. 1749–1752 (2008)

[9] Stangl, R., Bivour, M., Conrad, E., Didschuns, I., Korte, L., Lips, K., Schmidt, M.: A novel high efficiency buried grid rear contact amorphous/crystalline silicon heterojunction solar cell concept. In: Proceedings of the 22nd European Photovoltaic Solar Energy Conference, Milano, Italy, pp. 870–874 (2007)

[10] Desrues, T., Ribeyron, P.-J., Vandeneynde, A., Ozanne, A.-S., Munoz, D., Souche, F., Denis, C., Heslinga, D., Diouf, D., Kleider, J.-P.: Progress in contacting a-Si:H/c-Si heterojunction solar cells and its application to interdigitated back contact structure. In: Proceedings of the 24th European Photovoltaic Solar Energy Conference, Hamburg, Germany (2009)

[11] Stangl, R., Kriegel, M., Schmidt, M.: AFORS-HET, Version 2.2, a numerical computer simulation program for simulation of heterojunction solar cells and measurements. In: Proceedings of the 4th World Conference on Photovoltaic Energy Conversion, Hawaii, USA, pp. 1350–1353 (2006)

[12] User's manual for ATLAS from Silvaco International, version 5. 12.1

[13] Desrues, T., Ribeyron, P.-J., Vandeneynde, A., Ozanne, A.-S., Souche, F., Veschetti, Y., Bettinelli, A., Roca i Cabarrocas, P., Labrune, M., Diouf, D., Kleider, J.-P., Lemiti, M.: New process integration for interdigitated back contact (IBC) a-Si:H/c-Si heterojunction solar cells. In: Proceedings of the 23rd European Photovoltaic Solar Energy Conference, Valence, Spain, pp. 1673–1676 (2008)

[14] Anderson, R.L.: Experiments on Ge-GaAs heterojunctions. Solid-State Electron. 5, 341–351 (1962)

[15] Kleider, J.P.: Band lineup theories and the determination of band offsets from electrical measurements. chapter 12 in this book

[16] Shockley, W., Read, W.T.: Statistics of the Recombination of Holes and Electrons. Phys. Rev. 87, 835–842 (1952)

[17] Hall, R.N.: Electron Hole Recombination in Germanium. Phys. Rev. 87, 387 (1952)

[18] Kendall, D.L., de Vries, D.B.: Diffusion in silicon. In: Haberecht, R.R. (ed.) Semiconductor Silicon, p. 358. The Electrochemical Society, New York (1969)

[19] Law, M.E., Salley, E., Long, M., Burk, D.E.: Self-Consistent Model of Minority-Carrier Lifetime, Diffusion Length, and Mobility. IEEE Elect. Dev. Lett. 12, 401–403 (1991)

[20] Fossum, J.G., Lee, D.S.: A Physical Model for the Dependence of Carrier Lifetime on Doping Density in Nondegenerate Silicon. Solid State Electronics 25, 741–747 (1982)

[21] Kerr, M.J., Cuevas, A.: General parameterization of Auger recombination in crystalline silicon. J. Appl. Phys. 91, 2473–2480 (2002)

[22] Masetti, G., Severi, M., Solmi, S.: Modeling of Carrier Mobility Against Carrier Concentration in Arsenic, Phosphorous and Boron doped Silicon. IEEE Trans. Electron. Dev. 30, 764–769 (1983)

[23] Muller, R.S., Kamens, T.I.: Device Electronics for Integrated Circuits. Wiley, New York (1986)

[24] Thorpe, M.F., Weaire, D.: Electronic Density of States of Amorphous Si and Ge. Phys. Rev. 26, 1581–1584 (1971)
[25] Anderson, P.W.: Absence of Diffusion in Certain Random Lattices. Phys. Rev. 109, 1492–1505 (1958)
[26] Winer, K.: Defect formation in a-Si:H. Phys. Rev. B 41, 12150–12161 (1990)
[27] Powell, M.J., Deane, S.C.: Improved defect-pool model for charged defects in amorphous silicon. Phys. Rev. B 48, 10815–10827 (1993)
[28] Powell, M.J., Deane, S.C.: Defect-pool model and the hydrogen density of states in hydrogenated amorphous silicon. Phys. Rev. B 53, 10121–10132 (1996)
[29] Gudovskikh, A.S., Ibrahim, S., Kleider, J.P., Damon-Lacoste, J., Roca i Cabarrocas, P., Veschetti, Y., Ribeyron, P.-J.: Determination of band offsets in a-Si:H/c-Si heterojunctions from capacitance-voltage measurements: capabilities and limits. Thin Solid Films 515, 7481–7485 (2007)
[30] Cotter, J.E., Guo, J.H., Cousins, P.J., Abbott, M.D., Chen, F.W., Fisher, K.C.: P-Type Versus n-Type Silicon Wafers: Prospects for High-Efficiency Commercial Silicon Solar Cells. IEEE Trans. Elec. Dev. 53, 1893–1901 (2006)
[31] Yamaguchi, M., Ohshita, Y., Arafune, K., Sai, H., Tachibana, M.: Present status and future of crystalline silicon solar cells in Japan. Solar Energy 80, 104–110 (2006)
[32] Jacoboni, C., Canali, C., Ottaviani, G., Quaranta, A.A.: A review of some charge transport properties of silicon. Solid State Electron. 20, 77–89 (1977)
[33] Del Alamo, J.A., Swanson, R.M.: Measurement of Steady-State Minority Carrier Transport Parameters in Heavily Doped n-Type Silicon. Solid State Electron. 30, 1127 (1987)
[34] Kerr, M.J.: Surface, emitter and bulk recombination in silicon and development of silcon nitride passivated solar cells. PhD Thesis (June 2002)
[35] Swanson, R.M., Beckwith, S.K., Crane, R.A., Eaides, W.D., Wark, Y.O., Sinton, R.A., Swiiwiun, S.E.: Point contact silicon solar cells. IEEE Trans. Elect. Dev. 31, 661–664 (1984)
[36] Engelhart, P., Harder, N.-P., Merkle, A., Grischke, R., Meyer, R., Brendel, R.: RISE: 21.5% Efficient Back Junction Silicon Solar Cell with Laser Technology as a Key Processing Tool. In: Proceedings of the 4th World Conference on Photovoltaic Energy Conversion (WCPEC), Waikoloa, Hawaii, USA, pp. 900–904. IEEE, Los Alamitos (2006)
[37] Mulligan, W.P., Rose, D.H., Cudzinovic, M.J., De Ceuster, D.M., McIntosh, K.R., Smith, D.D., Swanson, R.M.: Manufacture of solar cells with 21% efficiency. In: Proceedings of the 19th European Photovoltaic Solar Energy Conference, Paris, France, pp. 387–390 (2004)
[38] Gruenbaum, P.E., King, R.R., Swanson, R.M.: Photoinjected hot-electron damage in silicon point-contact solar cells. J. Appl. Phys. 66, 6110–6114 (1989)
[39] De Ceuster, D.M., Cousins, P., Rose, D., Vicente, D., Tipones, P., Milligan, W.: Low cost, high volume production of >22% efficiency silicon solar cells. In: Proceedings of the 22nd European Photovoltaic Solar Energy Conference, Rome, Italy, pp. 816–819 (2007)
[40] Lu, M., Bowden, S., Das, U., Birkmire, R.: Interdigitated back contact silicon heterojunction solar cell and the effect of front surface passivation. Appl. Phys. Lett. 91, 063507 (2007)
[41] Diouf, D., Kleider, J.P., Desrues, T., Ribeyron, P.J.: 2D simulations of interdigitated back contact heterojunctions solar cells based on n-type crystalline silicon. Energy Procedia 2, 59–62 (2010)

[42] Kleider, J.P., Gudovskikh, A.S.: Characterization of amorphous/crystalline silicon interfaces from electrical measurements. In: Mater. Res. Symp. Proc., vol. 1066, pp. 75–86 (2008)
[43] Macdonald, D., Geerligs, L.J.: Recombination activity of interstitial iron and other transition metal point defects in p- and n-type crystalline silicon. Appl. Phys. Lett. 85, 4061–4063 (2004)
[44] Martinuzzi, S., Palais, O., Pasquinelli, M., Ferrazza, F.: N-type multicrystalline silicon wafers and rear junction solar cells. Eur. Phys. J. Appl. Phys. 32, 187–192 (2005)
[45] Cooper Jr, J.A., Schwartz, R.J.: Electrical characteristics of the $Si0_2$-Si interface near midgap and in weak inversion. Solid-State Electron. 17, 641–654 (1974)
[46] Beck, N., Wyrsch, N., Hof, C., Shah, A.: Mobility lifetime product—A tool for correlating a-Si:H film properties and solar cell performances. J. Appl. Phys. 79, 9361–9368 (1996)
[47] Meaudre, M., Meaudre, R.: Method for the determination of the capture cross sections of electrons from space-charge-limited conduction in the dark and under illumination in amorphous semiconductors. Appl. Phys. Lett. 85, 245–247 (2004)
[48] Street, R.A.: Trapping parameters of dangling bonds in hydrogenated amorphous silicon. Appl. Phys. Lett. 41, 1060–1062 (1982)
[49] Street, R.A., Zesch, J., Thompson, M.J.: Effects of doping on transport and deep trapping in hydrogenated amorphous silicon. Appl. Phys. Lett. 43, 672–674 (1983)
[50] Olibet, S., Vallat-Sauvain, E., Ballif, C.: Model for a-Si:H/c-Si interface recombination based on the amphoteric nature of silicon dangling bonds. Phys. Rev. B 76, 035326 (2007)

Chapter 16
Technology and Design of Classical and Heterojunction Back Contacted Silicon Solar Cells

Niels E. Posthuma, Barry J. O'Sullivan, and Ivan Gordon

imec, Belgium

Abstract. Ever since the first proposal of Interdigitated Back Contact (IBC) silicon solar cells in 1975, this type of cell has been under development as a means to reach high energy conversion efficiencies. Since no metal contacts are present on the front of the cell, IBC cells in general have a high generated current density (J_{sc}). Apart from this obvious advantage, IBC cells also have advantages related to the integration in modules. The series interconnection between various cells can be done at module level, without the need for connecting the front of one cell to the rear of the next one, as is the case in two-side contacted cells. IBC solar cell efficiencies of 21 to 24 percent have been shown on large area industrially produced cells. Another successful high efficiency concept is the heterojunction emitter solar cell, where the junctions are realized by application of intrinsic and doped amorphous silicon (a-Si) layers on high quality mono-crystalline silicon bulk material. The cells realized with a-Si heterojunctions have high open-circuit voltage values thanks to the excellent passivating properties of the a-Si layers. Combining the IBC concept with heterojunction junctions using thin high quality substrates has the potential of reaching solar cell efficiencies over 25 percent.

Classical IBC cells have been studied for many years. Some of the more important aspects include substrate quality, front and rear surface passivation, rear-junction design and design and structure of the metallization. In literature several types of IBC cells have been reported, where a large variety of processes are described utilizing technologies ranging from lab-scale to industrially applicable methods. In recent years IBC research has focused on development of low cost and industrial technologies suited for IBC cell production on large scale and exploring the way towards the use of thinner and thinner silicon substrates.

Although results have been reported on heterojunction emitter structures for almost twenty years, it is only in the last five years that the implementation of the heterojunction emitter at the rear of the wafer has received much research interest. For this reason, many of the papers that have been published on this subject display cell structures and processing that are not optimized, and are typically fabricated on small area cells. Efficiencies currently are in the order of 12 to 16

W.G.J.H.M. van Sark et al. (Eds.): Physics & Tech. of Amorphous-Crystalline, EM, pp. 521–537.
springerlink.com

percent. These proof-of-concept cells are just the start of a new development and a rapid evolution in the efficiencies is expected in the months and years ahead.

16.1 Introduction

High efficiency silicon solar cells have been a topic of research for many years [1, 2]. Initially this research consisted of exploring the limits of the applied material and investigation of the cell design and process steps. Subsequently, in a second phase transferring the obtained knowledge to produce solar cells at a competitive cost and at increased scale became the focus.

Several silicon solar cell concepts have been introduced over the last decades reaching energy conversion efficiencies well above 20 percent under AM1.5 illumination conditions. The concepts that have reached these high efficiencies are the Passivated Emitter and Rear Locally diffused (PERL) [3, 4], the Interdigitated Back Contact (IBC) [5] and Heterojunction with an Intrinsic Thin layer (HIT) [6] solar cells. The highest efficiency that has been obtained at lab scale is 25% using the PERL solar cell concept [4]. Large area commercially available n-type IBC and HIT cells are produced with efficiencies of around 22 percent [7, 8].

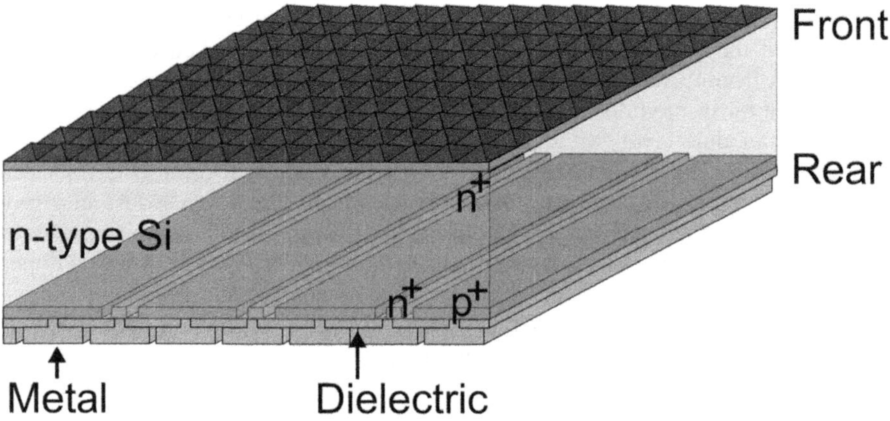

Fig. 16.1 Schematic illustration of an IBC solar cell.

Back contacted solar cells in general have specific advantages over cell structures where contacts are located on the front of the cell. A schematic illustration of an IBC cell is shown in Fig. 16.1. First of all, since no metal contacts are present at the front of the cell, IBC cells in general have a high generated current density (J_{sc}). Apart from this obvious advantage, IBC cells also have advantages related to the integration in modules. The series interconnection between various cells can be done at module level, without the need for connecting the front of one cell to the rear of the next one, as is the case in two-side contacted cells. This is especially important for solar cells with reduced substrate thickness as these interconnections create mechanical stress, causing cell breakage. Placing IBC cells on a module plate with a pre-defined contact grid reduces the complexity of the module fabrication and the cells can be stacked closely together, since no space is needed for these interconnections. This results in a uniform attractive appearance of the realized modules.

Apart from IBC cells, two other concepts for making back contacted solar cells are the Emitter Wrap Through (EWT) and Metal Wrap Through (MWT) cells [5]. In these concepts the emitter is still at the front of the cell, but the contacts to this emitter are brought to the rear of the cell by via holes. These concepts are especially attractive when applying silicon substrates with a limited carrier diffusion length.

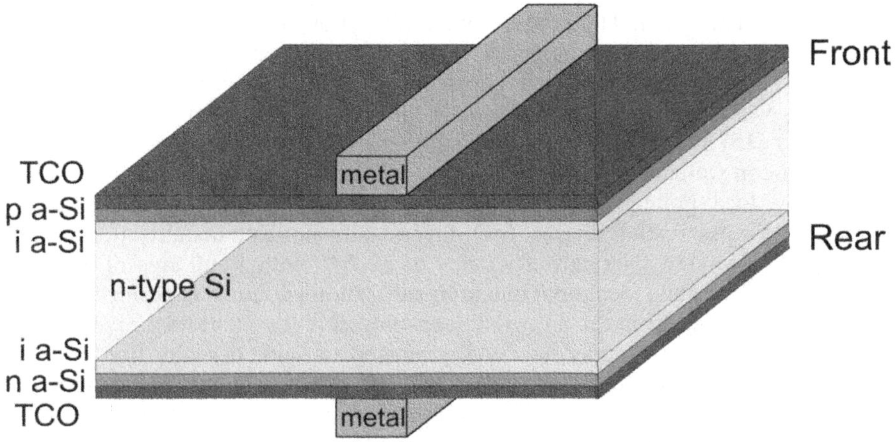

Fig. 16.2 Schematic illustration of a heterojunction emitter solar cell.

An illustration of a heterojunction emitter solar cell is shown in Fig. 16.2. The heterojunction cells are realized by applying a layer of doped amorphous silicon. In most cases an intrinsic amorphous silicon layer is deposited between the crystalline silicon and the doped a-Si layer, to enhance the surface passivation. A transparent conductive oxide is present on the amorphous silicon emitter layer to solve issues with the low lateral conductivity in amorphous silicon. The heterojunction emitter solar cells have the advantage of having a very high open-circuit voltage (V_{oc}), with values up to 747 mV [9, 10]. This high level of V_{oc} is related to the excellent passivating properties of the amorphous silicon applied to create the junctions. Besides the passivating properties, a second advantage of using amorphous silicon is that the temperature budget needed to deposit the layers is relatively low. The use of amorphous silicon as the emitter layer, with its energy gap of 1.7-1.8 eV [11], also facilitates separation and collection of the photogenerated carriers.

With these two main assets in mind, a high J_{sc} for IBC cells and a high V_{oc} for a-Si heterojunction cells, the combination of these two cell concepts is an attractive proposition. An additional benefit of a back-contact heterojunction cell is the absence of a transparent conductive oxide from the front side of the wafer, and its absorption losses. Combining the high current density and open-circuit voltage should in principle lead to very high efficient back contacted heterojunction silicon solar cells.

In this chapter the technology and design of classical and heterojunction back contact cells will be reviewed. In section 16.2 classical IBC cells will be discussed after which in section 16.3 the heterojunction IBC cells realized so far will be described. In the final section a brief outlook for heterojunction IBC cell integration will be given.

16.2 Design and Technology of Classical IBC Cells

The idea of a back contacted solar cell was initially proposed by Schwartz and Lammert in 1975 [12]. The contacts were placed at the rear of the cell to overcome issues with the conduction of high currents when applying concentrated sunlight [13, 14]. The use of back contacted solar cells for non-concentrating applications was extensively investigated at lab scale in the 1980's at the Stanford University [15] and at the Université Catholique de Louvain [16]. The use of high quality silicon wafers in combination with dedicated rules and equipment for clean processing, local point contacts with optimized local phosphorous and boron doping, excellent passivating thermal SiO_2 layers and a suitable metallization scheme led to an impressive solar cell efficiency of 22.7 % with a cell area of 35.7 cm^2 [17]. Patterning of the local doping and metallization pattern was performed by lithography. Starting from the knowledge developed at the Stanford University, the company SunPower, founded in 1985, managed to start, upscale, fine tune and significantly decrease the process cost of the IBC production process using n-type Si substrates. Over the years the solar cell performance continued to improve. Currently SunPower is producing large area n-type IBC cells with average efficiencies around 22 to 23 percent, with a record cell efficiency of 24.2 percent on large area 125 mm n-type CZ mono-crystalline silicon wafers [18].

More and more research groups and companies have started to study both n- and p-type IBC solar cell concepts in detail [19,20,21,22,23]. In the last decade the focus of research has been more on developing cost effective process technologies for IBC cells, experimenting with slightly lower quality silicon base material and reducing the wafer thickness. New concepts based on self-aligning metallization processes and application of laser technologies have been demonstrated. Research teams of ISFH, Fraunhofer ISE and Q-cells have shown efficiencies in the order of 21 to 22 percent on lab scale. UNSW applied their laser grooving technology for IBC cells, reaching almost 20 percent. Sharp has demonstrated IBC cell pilot line production on CZ n-type wafers with a top efficiency of 20.1 percent. A short overview of various IBC cell concepts presented in literature is given in Table 16.1.

Table 16.1 Overview of IBC solar cell efficiencies reported in literature.

Organization	Ref.	Year	Substrate	Area (cm^2)	Efficiency (%)
Stanford University	[17]	1990	FZ, n-type	35.7	22.7
ISFH	[19]	2006	FZ, p-type	4	22.0
Q-cells, ISFH, ISE	[20]	2006	CZ, n-type	4	21.0
UNSW	[23]	2007	FZ, n-type	8	19.9
Sharp	[22]	2008	CZ, n-type	157	20.1
SunPower	[18]	2010	CZ, n-type	155	24.2

In this section various properties of IBC cells will be described, both on design and on technology level.

16.2.1 Substrate Quality

The silicon substrate quality has a rather large impact on the performance of IBC cells. Since the emitter is placed at the rear in this type of cell, the generated carriers, the majority of which are created close to the front surface, have to move to the rear junctions. Furthermore the carriers also have to move laterally, to reach either the emitter or the back surface field. In case the bulk material quality is not sufficient, the generated carriers will recombine before they reach these rear junctions. A general rule of thumb is that the diffusion length of the minority charge carriers should be three times the thickness of the cell. For example for a cell thickness of 200 μm a diffusion length of 600 μm is needed. Such a diffusion length corresponds roughly to a minority carrier lifetime of 140 μs. 2-Dimensional simulation results, presented in various publications [24, 20, 25,26], show that for an optimal cell performance lifetimes well above 500 μs will be essential (see Fig. 16.3). Of course reducing the substrate thickness will allow for decreased material quality.

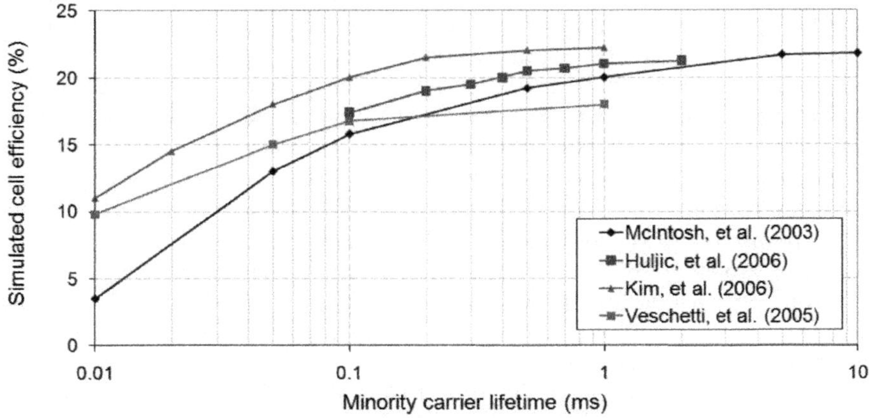

Fig. 16.3 Simulated dependence of the efficiency of an IBC cell on the minority carrier lifetime [24,21,25,26].

For IBC cells n-type silicon material will be preferential over p-type. An advantage of n-type material is that most impurities have a larger capture cross section for electron capture than for hole capture, which makes n-type material less sensitive to common impurities, such as iron. In case CZ silicon is used, no light induced degradation will take place in n-type material, while boron oxygen complexes will degrade the performance of p-type CZ solar cells after illumination. A lower effective surface recombination velocity can be reached on n-type surfaces due to the lower capture cross section of holes compared to electrons using thermal oxide or silicon nitride surface passivation. In IBC cells this is beneficial for passivating the front surface field of an n-type cell, but disadvantageous for the emitter surface at the rear of the cell. A disadvantage of n-type material is that the minority carriers (holes) have a lower mobility compared to electrons in p-type.

In terms of substrate resistivity two counteracting effects should be considered. A high doping level of the substrate will result in higher conductance, important for the movement of the generated carriers. A low doping level on the other hand will result in less recombination losses compared to higher doping. In case the injection level is higher than the doping level, the base enters in high injection regime and further resistive losses are avoided.

The highest IBC cell efficiencies have been achieved on n-type FZ, PV-FZ or CZ material (see Table 16.1), although good results have also been obtained on p-type wafers. So far, mostly cell thicknesses of 160 to 250 μm are used. In initial designs mostly low substrate doping levels were used (100 Ω·cm), such that the cells operated in high injection regime. This was also related to the initial use of IBC cells in concentrator applications. Most recent IBC cell data on large wafer sizes is obtained using a substrate resistivity in the order of 1 to 10 Ω·cm.

16.2.2 Surface Passivation

Front surface passivation relies largely on the application of a front surface field (FSF) [27], see Fig. 16.4. In case no FSF is used the quality of the front side surface passivation should be extremely good, with a surface recombination velocity below 10 cm/s, to obtain the optimum cell efficiency. Adding a front surface field will provide electrical shielding of the front, such that the recombination velocity required is less sensitive. Recombination velocities in the order of $1 \cdot 10^3$ to $1 \cdot 10^4$ cm/s can still result in the good performance, although for the best performance a saturation current density J_o of less than 50 fA/cm^2 is needed. The doping profile of an optimized FSF is approximately 1 μm deep and has a surface dopant concentration of around $1 \cdot 10^{18}$ cm^{-3} with a sheet resistance above 200 Ω/sq [17]. As shown in Fig. 16.4, at low surface recombination velocity values a clear degradation can be seen in cell efficiency with high doping levels in the FSF, related to absorption losses in this area.

For rear surface passivation the type of layer should be carefully chosen, since it has to passivate both the emitter and the back surface field (BSF). Preferably this passivating layer should have a very low density of interface states on both n- and p-type surfaces. This should lead to a rear saturation current density J_0 of less than 100 fA/cm^2. Thermal silicon oxide is a suitable candidate although in case boron is used as an emitter dopant, care should be taken not to disturb the doping profile when growing the thermal oxide. Possibly a retrograde boron profile is obtained, since at high temperature boron easily diffuses into the oxide layer. Passivation layers with a certain amount of fixed charge, such as silicon nitride or aluminum oxide can also be an interesting option, as a single layer or combined in passivating stacks. Having either fixed negative or positive charge will be mostly beneficial for one type of dopant, and care should be taken not to shunt the cells by electrically connecting the emitter and BSF due to the fixed and induced charge present in the dielectric material and at the silicon/dielectric interface.

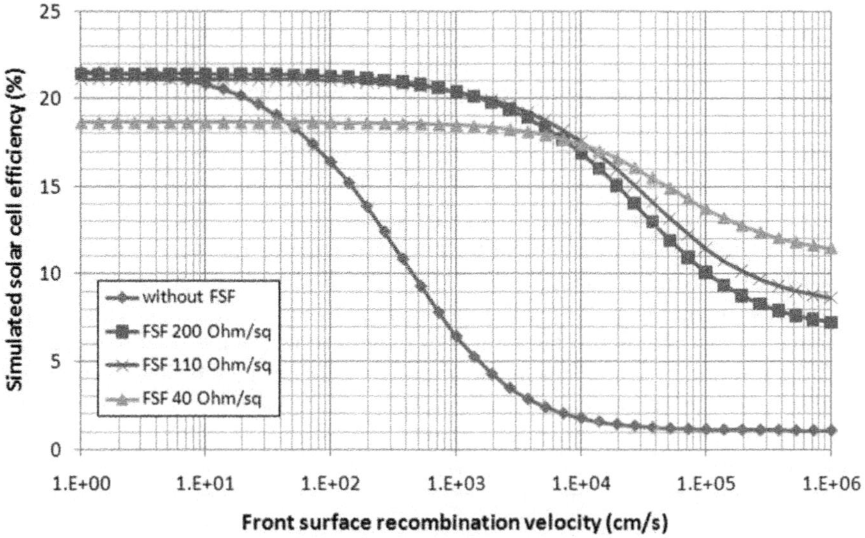

Fig. 16.4 Simulated cell efficiency as a function of the front surface recombination velocity of a rear junction cell with and without front surface field (FSF) simulated using PC-1D.

In most IBC cell designs thermal silicon oxide is used, both to passivate the rear as well as the front of the cell. In some cases the front side is passivated by silicon nitride. In the Oblique-Evaporation-of-Contact (OECE) cell [28], a metal-insulator-semiconductor (MIS) contacting scheme is applied using aluminum on a very thin oxide layer to shield the metal contact from the solar cell surface.

Recently the concept of a buried emitter has been introduced in IBC cells [29,30]. By placing the p-type emitter within the surface and covering the top layer by phosphorous, most of the rear side surface becomes n-type, which is easier to passivate efficiently. IBC solar cells with the buried emitter concept have been successfully realized, reaching an efficiency of 21.8% on lab-scale (3.97 cm^2) on n-type CZ Si wafers [29].

16.2.3 Structuring of the Rear Junctions

The layout of the rear side pattern, both the dimension of the junctions and of the metallization, is also an important aspect for IBC cells. The emitter should cover a large part of the rear side, between 70 to 90 percent [31]. By using lower emitter coverage (50-60 percent), the generated photocurrent density will reduce, unless the dimensions of emitter and base are small, which can also result in high values of J_{sc} [31]. However, in the last case the surface passivation of the non-doped base-material should be extremely, preferably less than 5 cm/s.

The pitch of the rear junction pattern, defined as the distance from one end of the emitter to the next emitter area, should be chosen such that the generated carriers can reach the emitter or BSF before recombining in the bulk. The largest distance a carrier has to diffuse occurs when it is generated above the middle of the emitter area. For a diffusion length of 600 μm, assuming a decent material quality, a pitch of around 1.5 mm can be tolerated. In case of high quality substrates having a bulk minority carrier lifetime of more than 1 ms, the pitch can be as large as 3.5 mm. The maximum pitch size is also related to the resistivity of the substrate. For highly resistive material the fill factor (FF) of the solar cell will be limited in case of a large pitch size due to the series resistance. According to simulation [31], the FF of an IBC cell with a bulk resistivity of 10 Ω·cm will reduce rather fast to values below 75 percent for pitch sizes above 1.5 mm, while for a base resistivity of 1 Ω·cm the theoretical FF stays above 80 percent. According to these simulations, including a FSF will assist in improving the lateral conductance of generated carries and reduce the impact of the pitch size on the FF. Of course how much the FSF helps to improve the conductivity depends on the sheet resistance of the FSF and the resistivity of the substrate. In most cases the pitch size is also largely determined by the applied process technology. In case of a lithography process the dimensions can be as small as a few micrometer, while screen-printing limits the minimal pitch to a few 100 micrometer. Using laser technology a wide range of dimensions is possible as low as a few 10's of micrometers.

The gap between the emitter and the BSF should be very well passivated and be preferably as small as possible. Especially for cells with low emitter coverage the effect of a gap with surface recombination velocities above 100 cm/s is significant. Alternatively also a small overlap of both junction types can be allowed, resulting in locally compensated regions, leaving no un-doped regions on the rear surface.

In most designs reported in literature, the pitch is between 1.2 and 2 mm and emitter coverage of around 70 to 80% [32, 20, 21]. On the exact layout often not much detailed information is given, since these parameters are critical to achieve the best cell efficiencies. For high efficiency lab-scale IBC cells the structuring of the junctions is mostly done by using lithography in combination with wet-etching or using dielectric masks. For industrial type of cells printing of etch-pastes or etch resists [32] or laser ablation of a dielectric mask [19, 21] are applied. In case of laser ablation it is critical not to damage the silicon surface. In some cases a combination of laser ablation and wet chemical etching is used, in order to remove the local damage created by the laser treatment. Different types of lasers are under investigation to assess the impact of the treatment and to minimize the laser damage.

16.2.4 Metallization

In order to keep a low level of surface recombination at the rear of the IBC cell, the metal semiconductor contact area should be limited as much as possible. The opening in the dielectric passivation layer can be defined by lithography and wet chemical etching, by laser ablation, by printing etch pastes or other means. As described in the previous section, it is important not to damage the silicon while locally removing the passivation layer. The location of the contact holes with respect to the edges of the junctions should be chosen such to minimize the distance the carriers have to travel to these contact openings and thereby reduce recombination and resistive losses.

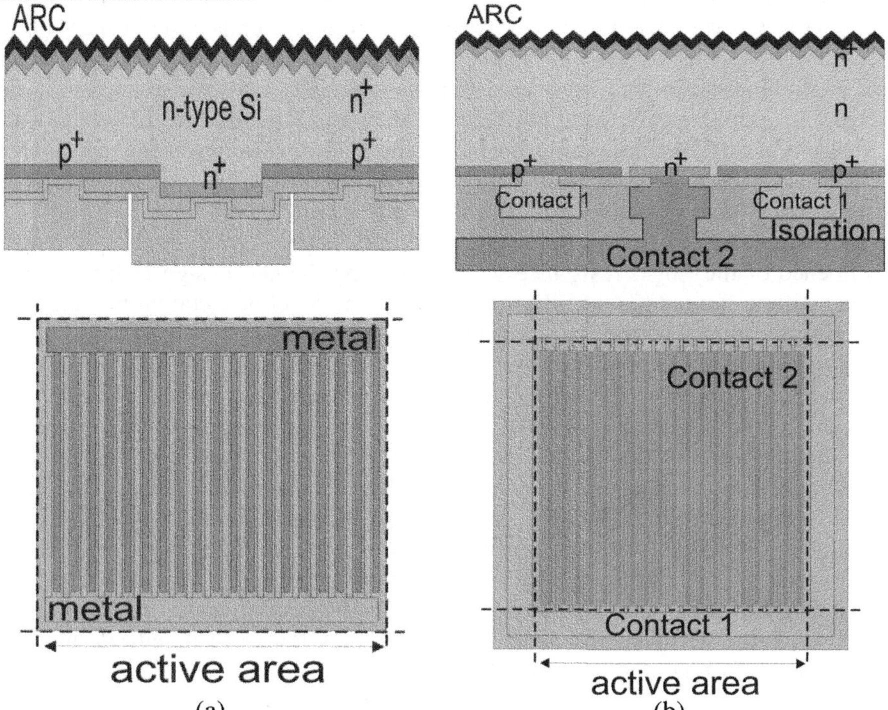

Fig 16.5. Schematic illustration of (a) a 1-level metallization and (b) a 2-level metallization scheme for IBC cells.

The metallization of IBC cells can be done in one or in multiple levels. A one metal level configuration (see Fig. 16.5a) is the easiest in terms of the amount process steps, although in this case the dimensions of the metal pattern are very important. As the resistance of a metal line (finger) strongly increases with cell size, it is important to make the width of the finger as broad as possible. Having broad fingers has the advantage of a low line resistance without the need for thick metal stacks, but also of obtaining good reflection properties at the rear surface. Ideally the metal fingers connecting the emitter and the BSF should be equally broad. However, since the emitter coverage is around 70 to 90 percent, this means that the contacts of the BSF are covering the emitter area, only separated by the passivation dielectric stack. To avoid shunting between emitter and BSF, this dielectric should also be electrically insulating and pinhole free.

Various options can be chosen in terms of the use of busbars. A first option is not to use any busbars at all. In this case the interconnection will be done at module level [22]. The fingers of the cell are aligned with the pattern on the module plate that has a predefined metal scheme, such to interconnect the fingers themselves and also the various cells. It is critical that the alignment of the cells on the module plate is done accurately such that the emitter and BSF contact lines are not shorted. An advantage of this concept is that the metal pattern on cell level is less critical in terms of line resistance. A second option is to place the busbars of the cell outside the active cell area. This can be done as a means to limit the influence of recombination in the busbar area. This is especially attractive for research devices, since in this case there will be non active area limiting the efficiency of the final modules. The third option is placing the busbars within the active cell area. Careful design and an insulating and passivating dielectric layer separating the metal from the semiconductor are essential in this case.

An advantage of the two level metallization design (see Fig. 16.5b) is that the influence of the finger resistance while upscaling the cell design to larger wafer sizes will be limited [33,34,]. By placing a second level of metal on the interdigitated finger grid separated by an insulating layer, the fingers will be connected at multiple places over the wafer area, such that the length of the current path through the fingers will be limited. These two metallization layers can be connected to large metal pads or rings, such that these contacts can be interconnected on module level. Adding this second level of metallization will increase the amount of process steps and the complexity, leading to a more expensive cell [35].

As described in the IBC solar cell patent of SunPower, patterning of the interdigitated metallization pattern is done on industrial level by using screen-printed plating resist [32]. First a seed layer is applied over the whole rear surface. Subsequently the plating resist is screen-printed, which is defining the area's where no electro-plating will take place. After plating a thick layer of Cu and Sn, the resist is removed and the seed layer is etched, using the thick metal layer as the patterning layer. In this industrial design no overlap between the base contact metal and the emitter region is allowed, to avoid the risk of creating shunts at these locations.

An alternative to using screen-printed resists is making use of topography for defining the separation between the emitter and BSF contacts [36, 19]. The step-height created to define the IBC cell fingers is also used to isolate the emitter and BSF contacts. On the vertical edges of the etched fingers less metal is deposited, such that performing a metal etch results in separation of the contacts.

Laser processing in combination with metal plating is used in the Interdigitated Backside Buried Contact (IBBC) cell concept [37, 23]. In this process the emitter is diffused over the whole rear surface and covered by a dielectric. Laser grooves are realized for the emitter and BSF contacts and additional heavy diffusions are done in these grooves for additional field effect shielding of the contacts. Within the grooves, nickel is selectively electrolessly plated followed by silicidation and copper plating.

In the IBC cell concept called Oblique-Evaporation-of-Contact (OECE) [28], the process contains local mechanical grooving of the rear side to obtain ridges with a height of more than 100 μm. These grooves are used for self-aligned metallization on only one side of the groove by using the shadowing effect of the grooves themselves, placing the sample under the right angle in the evaporation system. Furthermore a metal-on-insulator-semiconductor (MIS) contact consisting of Al/SiO_x/silicon is used. For this concept a best cell efficiency of 21.5 percent is reported for a 4 cm^2 area cell on p-type FZ silicon.

16.3 Design and Technology of Heterojunction IBC Cells

Although results have been reported on heterojunction emitter structures for almost twenty years, it's only in the last five years that the implementation of the heterojunction emitter at the rear of the wafer has received much research interest, with the first report of such a structure in 2007 [38]. For this reason, many of the papers that have been published on this subject display cell structures and processing that are not optimized, and are typically fabricated on small area cells, indeed the largest area that has been reported is 25 cm^2 [39, 40]. Most of the results have been obtained on float zone silicon, of both p- and n-type. Furthermore, patterning of the rear side of the cell is mostly achieved with photolithography, to demonstrate the proof of concept of various cell structures. However, the efficiencies reported so far are up to a maximum of around 15 to 16 percent (see Table 16.2), which is significantly lower than those for diffused emitter IBC cells (>22%) as described in the previous section. Currently, the main challenge to achieve efficiency improvements is to make the interdigitated pattern of p- and n-type amorphous silicon and combine this with a good surface preparation right before the deposition. In this section, a review of the technologies reported to date on heterojunction IBC cells will be presented.

Table 16.2 Overview of reported heterojunction IBC cell efficiencies.

Organization	Ref.	Year	Substrate	Area (cm^2)	Efficiency (%)
Univ. of Delaware	[38]	2007	FZ, n-type	1.32	11.8
HZB	[42,43]	2008	FZ, p-type	1	13.9
INES	[39,40]	2010	FZ, n-type	25	15.7
ENEA	[41]	2008	CZ, p-type	6.25	15
University of Toronto	[41]	2009	FZ, n-type	1	8.1
imec	[45]	2010	CZ, n-type	1	15.2

A generalized schematic of an heterojunction IBC cell is shown in Fig. 16.6. On the textured front side of the cell an anti-reflective coating (for example silicon nitride) is deposited, often combined with a passivating layer of a-Si. Losses of incident light, due to absorption in the amorphous silicon layer places limitations on this layer thickness. As an alternative, a front surface field could be applied, as is standard for classical IBC cells. In most cases the application of a FSF is omitted to limit the processing temperature.

At the rear of the cell the interdigitated amorphous silicon layers are applied. In the schematic shown in Fig. 16.6 an intrinsic a-Si layer is applied to passivate the rear surface, like the classical 2-side contacted heterojunction solar cells. As will be described in more detail below, currently a large variety of stacks are used, either with or without an intrinsic layer. Furthermore, sometimes a passivation layer is applied in-between the BSF and the emitter regions. Also, a conductive barrier layer can be inserted between the amorphous silicon layer and metal contact to shield the amorphous silicon from degrading effects of the metal, enhance conductivity and improve the reflection properties at the rear. Patterning can be done by lithography, shadow masking or other more industrial techniques. Key issues in the development of the rear side of the heterojunction IBC cells are the junction design, the quality of the c-Si/a-Si interface, the use of intrinsic amorphous silicon, isolation between n- and p-type a-Si layers and the metal contacts to the a-Si layers.

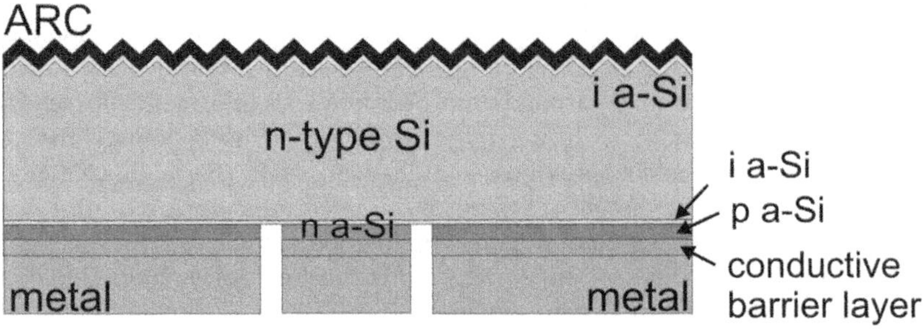

Fig 16.6. Schematic illustration of a heterojunction IBC cell.

The first results on a heterojunction IBC cell was reported by Lu et al in 1997, [38], where p^+ a-Si and n^+ a-Si emitter and BSF regions, respectively, were patterned with photolithography to achieve 11.8% efficient cells. It is noted that the front side of the cell was not textured, and no passivation layer was present between the BSF and emitter regions.

In the 'Back Enhanced Heterostructure with InterDigitated Contact' (BEHIND) cell structure published in 2008 by Tucci *et al.* of ENEA [41], CZ p-type Si was used. A CrSi layer was formed on the emitter and BSF a-Si regions to enhance conductivity. The processing scheme did not involve lithography, but instead used metal shadow masks to pattern the doped a-Si regions. The V_{oc} on these cells was 695 mV, the highest reported thus far on such heterojunciton IBC cells, and efficiency as high as 15%.

In the work of Desrues *et al.* of INES [39], non-textured n-CZ wafers are used, and intrinsic amorphous silicon is located between the a-Si emitter and BSF regions to avoid shunting. A transparent conductive oxide is present between the BSF and the aluminum base contact. This processing also avoids lithography by using metal masks for patterning. It is shown that there are significant series resistance effects in contacting the interdigitated devices, which are strongly dependent on the percentage of the emitter surface area which is covered by metal, but nonetheless efficiencies of almost 13% are achieved. By further optimization of the buffer layer recently 15.7% has been reached with a V_{oc} of 678 mV [40].

Two innovative cell structures have been presented by Stangl *et al.* at HZB [42,43]. In the first of their structures, the 'Rear Emitter Crystalline/Amorphous Silicon Heterojunction' (RECASH) cell, the Al base contact is deposited first, and insulated with an oxide layer, prior to emitter fabrication, and contacting through a TCO layer. Such cells have demonstrated efficiencies of 13.9%, but it is noted that there are issues with interface states at the Si/a-Si interface, which due to the presence of the insulating oxide precludes the use of HF etching. Another structure proposed is the 'Point Rear Emitter Crystalline/Amorphous Silicon Heterojunction' (PRECASH) solar cell. The emitter and emitter contact layers are deposited, and lithographically patterned to open holes, where, firstly, SiO_2 insulation is used to avoid shunting between the emitter and subsequently deposited BSF region. This oxide is patterned lithographically, and the BSF and metal contact are subsequently deposited. Such cells have demonstrated efficiencies of 7.15 % (albeit using non-optimised front surface processes), but present an alternative route to maximize the active area by using point contact instead of line contacts as in all other works.

The 'Back Amorphous-Crystalline silicon Heterojunction' (BACH) solar cell, proposed by Hertanto *et al.* from the University of Toronto was reported first in 2009 [44], and presents another approach, whereby thermal SiO_2 is used as a passivation layer, and also as an anti-reflection layer. Openings in this are created by lithography for deposition of the p^+ a-Si emitter first, before use of a SiN as a protective mask, into which further openings are patterned for the back surface field. Such cells result in efficiencies of between 7.3 and 8.1 %, with variations in the wafer resistivity, and front surface processing. The fill factor values vary between 75 and 78 %, and are among the higher reported to date on heterojunction IBC structures, indicating limited series resistance in the metallization processing.

In the work of imec [45], a non-optimum rear surface patterning is presented, where passivating intrinsic, and p^+ doped amorphous silicon layers, and a TCO, are deposited across the wafer, and selectively (lithography-aided) removed to enable deposition of the base contact. This structure does not use a BSF, but instead uses an Al layer, deposited directly on 1-5 Ω.cm, in what is effectively a Schottky contact. Despite this non-ideal structure, best cell efficiencies as high as 15.2% have been measured, aided by a short circuit current density J_{sc} of around 38 mA/cm^2, both of which are, to the authors' knowledge, the highest values reported presently for such cell structures.

To summarize the results that have been reported in this field, it is clear that there is room for improvement for the processing schemes and indeed cell structures. Much research and development work is required before such structures can be fabricated industrially. However, given the results that have been reported up to this point, on what are effectively proof-of-concept cells, and the continued interest in this field, a rapid evolution of the efficiencies in the months and years ahead can be expected.

16.4 Outlook for Heterojunction IBC Cell Integration

In the first sections of this chapter an overview is presented of a variety of technology and design aspects of IBC silicon solar cells reported in literature. A significant amount of the knowledge available for classical IBC cells with doped junctions within the silicon substrate is also applicable to heterojunction IBC solar cells. In this last section, an outlook is given of possible integration routes for heterojunction IBC solar cells.

First of all, the processing techniques for reaching high efficiencies, controlling the cleanliness of the silicon wafers and the metal contamination level of each process step, is similarly important for both cell concepts.

The front side of a heterojunction IBC cell can be textured similarly to the processes developed for industrial silicon solar cells. For the front surface passivation, two main options can be identified. The first option is to use the classical approach, where a front surface field or floating junction is applied by doping the silicon wafer at elevated temperatures. This FSF is combined with a passivating and anti-reflection layer such as for example silicon nitride. Alternative passivating layers are aluminum oxide and silicon dioxide, possibly also combined with silicon nitride. The advantage of the application of a FSF is the reduced impact of the surface recombination velocity on the solar cell efficiency (see Fig. 16.4) that is proven to be stable over the lifetime of the solar cell. As mentioned before, an additional reported advantage of the FSF is that it improves the lateral conductivity of the cell structure, depending on the substrate resistivity.

A second option targeting low temperature processing could omit the use of a front surface field or floating junction. A single thin layer of intrinsic amorphous silicon can be deposited directly on the textured silicon front surface. In this case a stable surface recombination velocity below 10 cm/s is required to avoid degradation of the solar cell efficiency. Alternative low temperature passivating layers such as for example silicon dioxide, silicon nitride and aluminum oxide that provide a sufficiently low and stable recombination velocity can also be applied. A heterojunction FSF consisting of intrinsic and doped a-Si is also an alternative option. The total a-Si layer thickness should be kept as thin as possible (< 15nm) not to absorb too much light. This thin amorphous silicon layer can be combined with an ARC such as low temperature deposited silicon nitride.

For the rear of the IBC cell, the technology and design of heterojunction cells is similar to classical IBC cells in many aspects. The design rules for junction dimension, emitter coverage will hold also for heterojunction IBC cells in general. Existing technologies for patterning and metallization can also be applied for heterojucntion cell integration.

Very specific for heterojunction IBC cells is of course first of all the application of the amorphous silicon itself. The quality of the a-Si/c-Si interface will largely determine the final solar cell efficiency. A key aspect here is the control over this interface in combination with patterning of the interdigitated junctions. Specific issues to be investigated in more detail are the application of an intrinsic thin layer between the c-Si and doped a-Si layer, the use or need for suitable conductive (barrier) layers in-between the a-Si layers and the metal contact and furthermore the type of metal that is used. This layer between the metal and the amorphous silicon will provide additional conductivity and at the same time serve as a barrier to avoid degradation of the a-Si layer induced by the metal contact. In many of the reported heterojunction IBC cells the FF of the cells is limiting the cell efficiency. Selecting the right type of interlayer and metal contact will therefore be crucial in the further development of this cell type. For the metalisation schemes and processing used, it is necessary to ensure that the temperatures do not exceed those of the a-Si deposition process (around 200 °C). This may necessitate specific pastes or precursors for screen printing, jetting or other metallization techniques, and subsequent curing steps, for instance.

Another aspect to keep in mind is the integration of the cells into a module. The metal busbars will be connected to the module plate having a predefined metallization pattern interconnecting the various cells. In order not to degrade the amorphous silicon quality the module fabrication process should be done at sufficiently low temperatures.

References

[1] Green, M.A., Zhao, J., Wang, A., Wenham, S.R.: Very High Efficiency Silicon Solar Cells—Science and Technology. IEEE TED 46, 1940–1947 (1999)

[2] Glunz, S.W.: High-Efficiency Crystalline Silicon Solar Cells. Advances in OptoElectronics (2007), doi:10.1155/2007/97370

[3] Zhao, J., Wang, A., Green, M.A.: 24.5% efficiency silicon PERT cells on MCZ substrates and 24.7% efficiency PERL cells on FZ substrates. Progress in Photovoltaics 7, 471–474 (1999)

[4] Green, M.A.: The Path to 25% Silicon Solar Cell Efficiency: History of Sili-con Cell Evolution. Progress in Photovoltaics: Research and Applications 17, 183–189 (2009)

[5] Van Kerschaver, E., Beaucarne, G.: Back-contact Solar Cells: A Review. Progress in Photovoltaics: Research and Applications 14, 107–123 (2006)

[6] Taguchi, M., Tsunomura, Y., Inoue, H., et al.: High efficiency HIT solar cells on thin (< 100 mm) silicon wafer. In: Proceedings of the 24th European PVSEC, Hamburg, pp. 1690–1693 (2009)

[7] De Ceuster, D., Cousins, P., Rose, D., et al.: Low Cost, high volume produc-tion of >22% efficiency silicon solar cells. In: Proceedings of the 22nd European PVSEC, Milan, pp. 816–819 (2007)

[8] Tsunomura, Y., Yoshimine, Y., Taguchi, M., et al.: Twenty-two percent efficiency HIT solar cell. SolMAT 93, 670–673 (2009)

[9] Kanno, H., Ide, D., Tsunomura, Y., et al.: Over 22% efficient HIT solar cell. In: Proceedings of the 23rd EU-PVSEC, Valencia, pp. 1136–1139 (2008)

[10] Sakata, H., Tsunomura, Y., Inoue, H.: R&D progress of next-generation very thin HIT solar cells. In: Proceedings of the 25th EU-PVSEC, Valencia, pp. 1102–1105 (2010)

[11] Ray, S., Hazra, S.: Preparation of high and low bandgap amorphous silicon by PECVD and their application in solar cell. In: Proceedings of the 25th IEEE PVSC, Crystal City, pp. 1077–1080 (1996)

[12] Schwartz, R.J., Lammert, M.D.: Silicon solar cells for high concentration applications. IEEE Electron Devices Meeting 21, 350–352 (1975)

[13] Garner, C.M., Nasby, R.D., Sexton, F.W.: An interdigitated back contact solar cell with high current collection. IEEE EDL 1, 256–258 (1980)

[14] Sinton, R.A., Kwark, Y., Swirhun, S., Swanson, R.M.: Silicon Point Contact Concentrator Solar Cells. IEEE EDL 6, 405–407 (1985)

[15] King, R.R., Sinton, R.A., Swanson, R.M.: Doped surfaces in one sun, point-contact solar cells. Appl. Phys. Let. 54, 1460–1462 (1989)

[16] Verlinden, P., van de Wiele, F., Stehelin, G., Floret, F., David, J.P.: High efficiency interdigitated back contact solar cells. In: Proceedings of the IEEE PVSC, Orlando, pp. 405–410 (1987)

[17] King, R.R.: Studies of oxide-passivated emitters in silicon and applications to solar cells. PhD thesis Stanford University (1990)

[18] Cousins, P.J., Smith, D.D., Luan, H.-C., et al.: Generation 3: improved performance at lower cost. In: 35th IEEE PVSEC, Hawaii (2010)

[19] Engelhart, P., Harder, N.P., Grischke, R., et al.: Laser Structuring for Back Junction Silicon Solar Cells. Progress in Photovoltaics: Research and Applications 15, 237–243 (2007)

[20] Granek, F., Hermle, M., Huljíc, D., et al.: Enhanced lateral current transport via the front n+ diffused layer of n-type high-efficiency back-junction back-contac silicon solar cells. Progress in Photovoltaics: Research and Applications 17, 47–56 (2009)

[21] Huljíc, D.M., Zerres, M., Mohr, A., et al.: Development of a 21% back-contact monocrystalline solar cell for large-scale production. In: Proceedings of the 21st EU-PVSEC, Dresden, pp. 765–768 (2006)

[22] Nakamura, K., Kohira, M., Abiko, Y., et al.: Development of back contact Si solar cell and module in pilot production line. In: Proceedings of the 23rd EU-PVSEC, Valencia, pp. 1006–1009 (2008)

[23] Guo, J., Cotter, J.E., McIntosh, K.R., et al.: Edge passivation for small-area, high efficiency solar cells. In: Proceedings of the 22nd EU-PVSEC, Milan, pp. 1348–1351 (2007)

[24] McIntosh, K.R., Cudzinovic, J., Smith, D.D., et al.: The choice of silicon wafer for the production of low-cost rear-contact solar cells. In: Proceedings of the 3rd World PVSC Conference, Osaka, pp. 971–974 (2003)

[25] Kim, D.S., Meemongkolkiat, V., Ebong, A., et al.: 2D-modeling and development of interdigitated back contact solar cells on low-cost substrates. In: Proceedings of the 4th World PVSC, Hawaii, pp. 1417–1420 (2006)

[26] Veschetti, Y., Muller, J.C., Quang, N.L., et al.: Investigation of an industrial process for development of rear contact solar cell. In: Proceedings of the 20th EU-PVSEC, Barcelona, pp. 841–844 (2005)

[27] Granek, F., Reichel, C., Hermle, M., et al.: Front surface passivation of n-type high efficiency back-junction silicon solar cells using front surface field. In: Proceedings of the 23rd EU-PVSEC, Milan, pp. 1262–1265 (2007)

[28] Müller, J.W., Merkle, A., Hezel, R.: The back-OECO solar cell: a rear con-tacted, bifacially sensitive and industrially feasible solar cells with efficiencies of 21.5%. In: Proceedings of the 20th EU-PVSEC, Barcelona, pp. 1020–1023 (2005)

[29] Mertens, V., Bordihn, S., Larionova, Y., et al.: The buried emitter solar cell concept: interdigitated back-junction structure with virtually 100% emitter coverage of the cell area. In: Proceedings of the 24th EU-PVSEC, Hamburg, pp. 934–936 (2009)

[30] Harder, N.P., Mertens, V., Brendel, R.: Numerical simulations of buried emitter back-junction solar cells. Progress in Photovoltaics: Research and Applications 17, 253–263 (2009)

[31] Hermle, M.: Analyse neuartiger Silizium- und III-VSolarzellenmittels Simulation und Experiment. PhD thesis Universität Konstanz/Fraunhofer ISE (2008)

[32] US Patent: Solar cell and method of manufacture, Sunpower Corporation, US, 7, 339,110 B1 (2008)

[33] Verlinden, P., Swanson, R.M., Sinton, R.A., Kane, D.E.: Multilevel metallization for large area point-contact solar cells. In: Proceedings of the 20th IEEE PVSC, Las Vegas, pp. 532–537 (1988)

[34] Sinton, R.A., Verlinden, P.J., Crane, R.A., et al.: Large-area 21% efficient Si solar cells. In: Proceedings of the 23rd IEEE PVSC

[35] Cudzinovic, M.J., McIntosh, K.R.: Process simplifications to the pegasus solar cell – sunpower's high-efficiency bificial silicon solar cell. In: Proceedings of the 29th IEEE PVSC, New Orleans, pp. 70–73 (2002)

[36] Sinton, R.A., Swanson, R.M.: Simplified backside-contact solar cells. IEEE TED 37, 348–352 (1990)

[37] Guo, J., Cotter, J.E.: Laser-grooved backside contact solar cells with 680-mV open-circuit voltage. IEEE TED 51, 2186–2192 (2004)

[38] Lu, M., Bowden, S., Das, U., Birkmore, R.: Interdigitated back contact silicon heterojunction solar cell and the effect of front surface passivation. Applied Physics Letters 91, 063507 (2007)

[39] Desrues, T., Ribeyron, P.-J., Vandeneynde, A., Ozanne, A.-S., et al.: Progress in contacting a-Si:H/c-Si heterojunction solar cells and its application to interdigitated back contact structure. In: Proceedings of the 24th EU-PVSEC, Hamburg, pp. 2202–2205 (2009)

[40] Desrues, T., Souche, F., Vandeneynde, A., et al.: Emitter optimization for interdigitated back contact (IBC). In: Proceedings of the 25th EU-PVSEC, Valencia, pp. 2374–2377 (2010)

[41] Tucci, M., Serenelli L., Salza E., et al.: Behind (Back Enhanced Heterostructure With Interdigitated Contact) Solar Cell. In: Proceedings of the 23rd EU-PVSEC, Valencia, pp. 1749–1752 (2008)

[42] Stangl R., Haschke J., Bivour M., Korte L., Schmidt M., Lips K., Rech B.: Planar rear emitter back contact silicon heterojunction solar cells. Solar Energy Materials & Solar Cells 93, 1900–1903 (2009)

[43] Stangl, R., Haschke, J., Bivour, M., Schmidt, M., Lips, K., Rech, B.: Planar rear emitter back contact Amorphous/crystalline silicon heterojunction solar cells (RECASH / PRECASH). In: Proceedings of the 33rd IEEE PVSC, San Diego, pp. 1–6 (2008)

[44] Hertanto A., Liu H., Yeghikyan D., Gangadhar Rayaprol B., Kherani N.P., Zukotynski S.: Back Amorphous-Crystalline Silicon Heterojunction (BACH) Photovoltaic Device. In: Proceedings of the 34rd IEEE PVSC, Philadelphia, pp. 1767–1770 (2009)

[45] O'Sullivan B.J., Bearda T., Qiu Y., Robbelein J., Gong C., Posthuma N.E., Gordon I., Poortmans J.: Interdigitated rear contact solar cells with amorphous silicon heterojunction emitter. In: Proceedings of the 35rd IEEE PVSC, Honolulu, pp. 3549–3552 (2010)

Chapter 17
a-Si:H/c-Si Heterojunction Solar Cells: A Smart Choice for High Efficiency Solar Cells

Delfina Muñoz, Thibaut Desrues, and Pierre-Jean Ribeyron

CEA-INES, Savoie Technolac - 50 avenue du lac Léman - BP258 F-73375 LE BOURGET DU LAC – CEDEX

Abstract. In this chapter, we start by a short presentation of the state-of-the art of the energy market to understand the evolution of the energetic demand and the role of photovoltaic technology in the near future. Moreover, we present all the actual industrial high efficiency solar cells among which is located the heterojunction technology. Then, we talk about the key points that define the technology, the main bottlenecks and the main solutions found at INES research group on heterojunction devices. Also, we show our best results obtained recently and some guidelines to improve still more the efficiency of the devices. Finally, we finish by a summary of the main advantages of this technology taking into account all the parameters described above.

17.1 PV Context 2010

In 2009, solar energy sources represented only 0.05 % of the total energy produced in the world as shown in Fig. 17.1. Even if renewable energy sources production are growing rapidly, it has to be noted that the energy production worldwide is still mostly (83 %) dependant on fossil fuel energy sources. The continuous growing demand of energy worldwide and the necessity to reduce the greenhouse gases emissions impose to increase dramatically decarbonised energy sources into the energy mix.

In this context, Photovoltaic energy will expand drastically in the coming years and will represent a growing part into the energetic distribution worldwide. The most important parameter that should be considered for PV expansion is energy cost in terms of €/W and thus €/kWh. As soon as the cost of PV will rich grid parity with respect to other energy sources, PV should then even more rapidly expand and provoke a massive paradigm shift. Thus, the solar PV market has been booming over the last two decades and it is forecasted to continue this trend in the coming years (Fig. 17.2). By the end of 2008, the global cumulative capacity installed was approaching 15 GW. Today, Europe is leading with more than 9 GW

W.G.J.H.M. van Sark et al. (Eds.): Physics & Tech. of Amorphous-Crystalline, EM, pp. 539–572.
springerlink.com

representing over 65% of the Global cumulative PV installed capacity. Japan (2.1 GW) and the US (1.2 GW) are following behind, representing 15% and 8%, respectively, of the Global cumulative PV power installed.

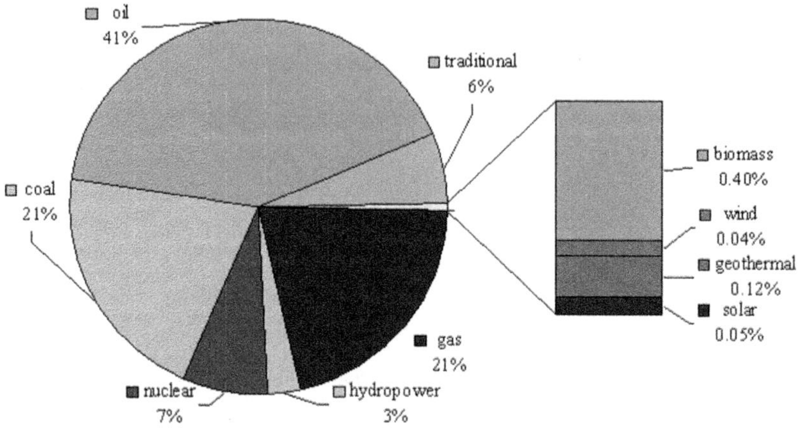

Fig. 17.1 Energy sources in the worldwide energy mix.

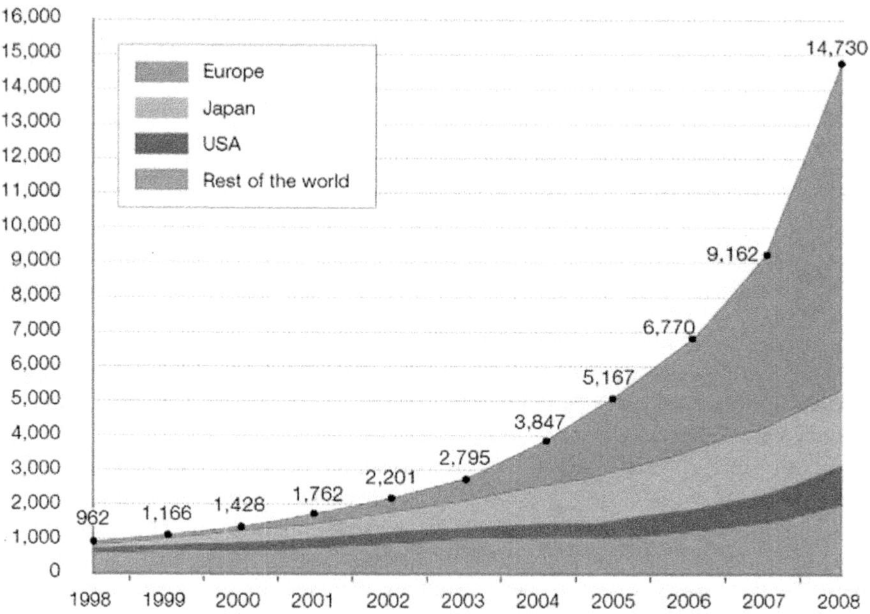

Fig. 17.2 PV installation in MW for Europe, Japan, USA and rest of the world from 1998 to 2008 [1].

The estimations for the future show that these aggressive trends will go on in the world with a large worldwide expansion of PV (Fig. 17.3). Particularly, Europe will represent an important market share.

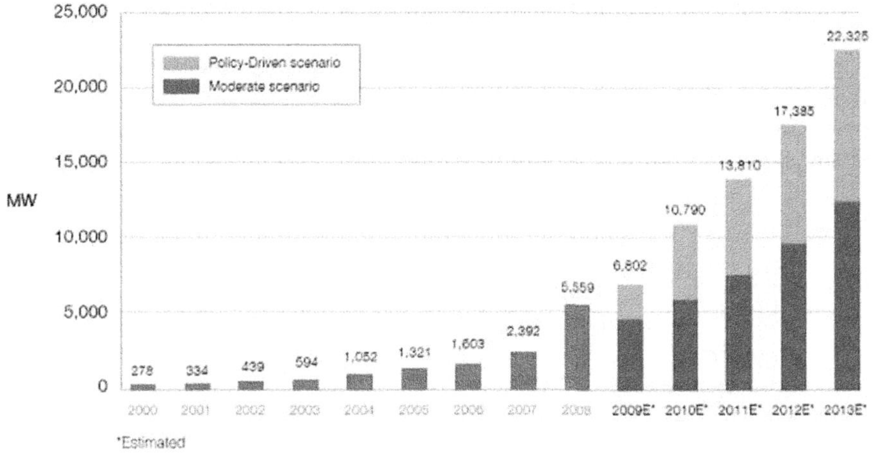

Fig. 17.3 Estimated PV installation from 2009 to 2013 with two different scenarios (Policy driven/ Moderate [1]).

Despite the rapid extension of the thin film technology and production, crystalline silicon will remain the dominant technology, due to its high production volume, maturity, and large availability of silicon feedstock and capability of reducing costs (Fig. 17.4).

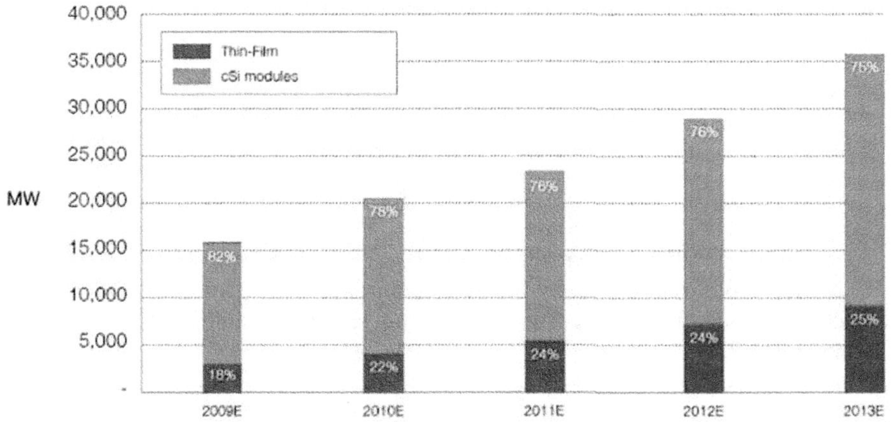

Fig. 17.4 Market share estimation of the different technologies [1].

In Fig. 17.5 the breakdown of the estimated global cost is shown of a given PV silicon wafer technology. Silicon feedstock and module processing represent an important part of the overall costs. If one wants to reduce the cost (€/W), the different costs have to be decreased, such as material costs, processes and systems. Nevertheless, the only parameter that can have an immediate impact on these different costs is the amount of Watt produced. Thus, the efficiency is a major parameter in order to reduce the overall costs of a given PV silicon technology. Indeed, the efficiency has a global impact, not only at the module level but on the whole system. This means that every effort to improve the efficiency at the cell and module level has a direct impact on the whole system cost.

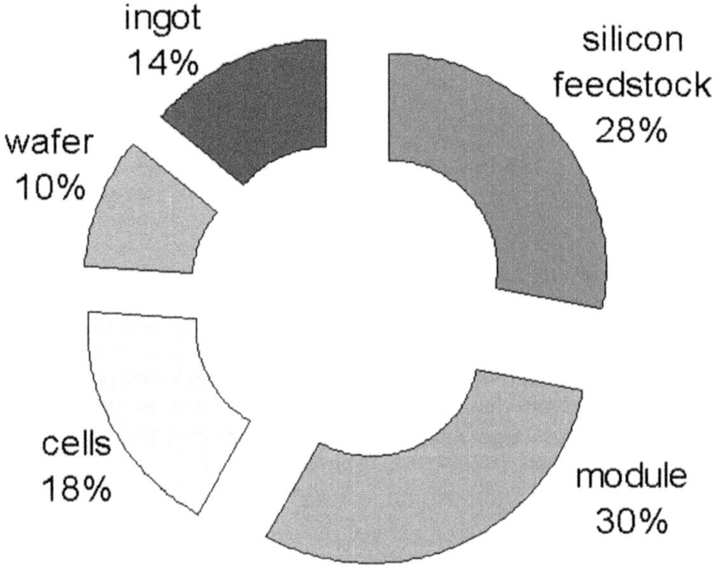

Fig. 17.5 Overall costs of a given PV Silicon technology.

In the same way, higher efficiencies allow the corresponding technology to improve its energetic intensity (W/m²). Thus, for applications were the space available is limited, this is a major advantage.

Thus, it is clear that since the discovery of the photovoltaic effect, the race for higher efficiency has been a major point for every PV technology, from organic PV, thin films, crystalline PV to concentrated PV since 1980 and even earlier (Fig. 17.6).

In this context, it is clear that amongst the different c-Si technologies, the high performance ones will strongly increase their Market share (Fig. 17.7). Then, worldwide, the research and development on PV crystalline silicon is focusing on over 20 % silicon solar cell and module efficiency target.

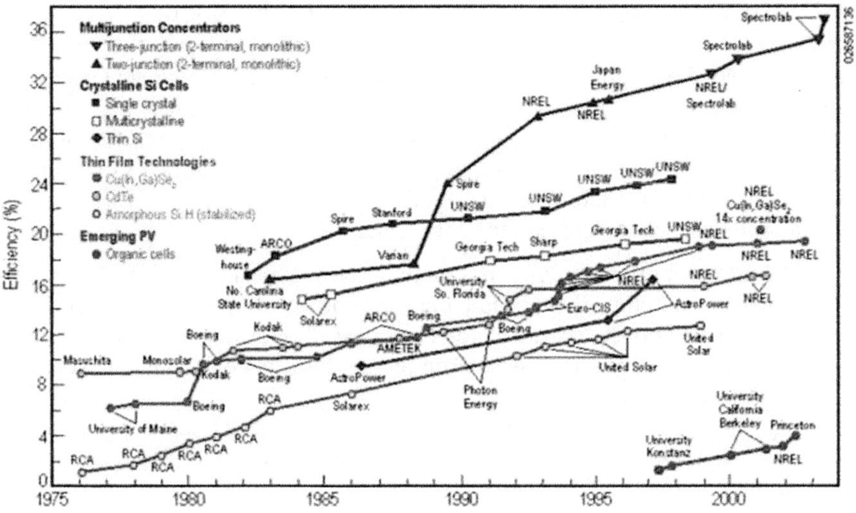

Fig. 17.6 Efficiency Race for various PV technologies [2].

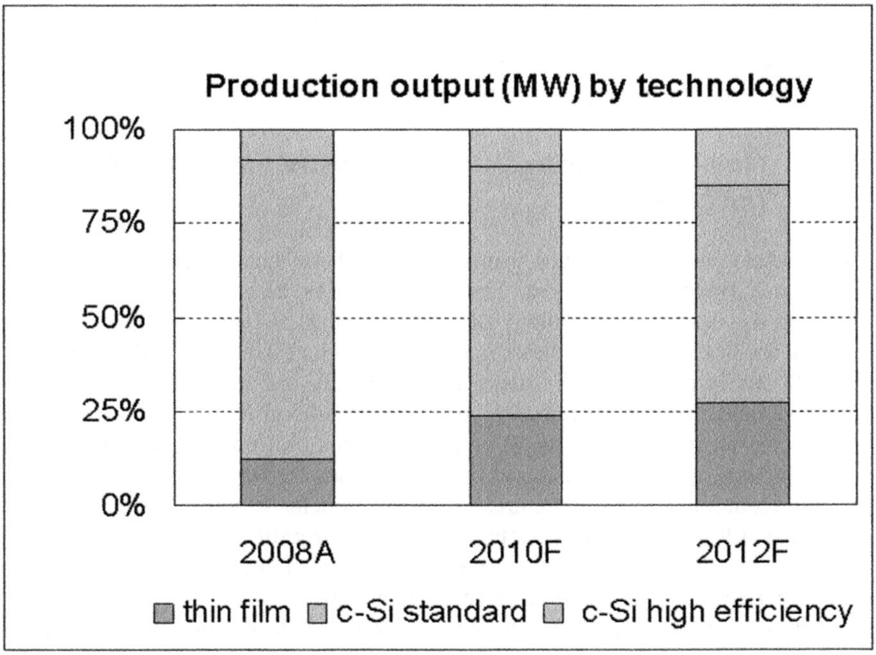

Fig. 17.7 Actual Market (A) share and forecast (F) of high efficiency silicon solar cells.

17.2 Overview on High Efficiency c-Si Technologies

Since the early development of c-Si solar cells, researchers aim to enhance the device efficiency. The theoretical upper limit for c-Si cells efficiency was calculated by Kerr et al. [3] to be about 29%, whereas a "practical" limit of 27% was proposed by Swanson [4]. Those values are given for an illumination of one sun with AM1.5 spectrum. The main interests in fabricating high efficiency devices – apart from the scientific challenge- are the following:

1. using more efficient cells in a module allow a cost reduction of the whole power generating system. For the same power output, module fabrication and installation costs are indeed reduced. For example, del Canizo et al. calculated savings around 10% on the module cost by enhancing solar cells efficiency by 10% relative [5].
2. the more efficient solar module are, the less area is needed to produce a certain amount of power. This is of particular interest in BIPV applications where useable surfaces can be limited.

The highest efficiencies can be obtained for what is called "laboratory cells", i.e. devices with small area and very complex structures like the PERL (passivated emitter and rear locally diffused cell) [6]. The challenge is to succeed in fabricating large area devices showing high efficiencies with low cost processes. Up to now, only the HIT (heterojunction with intrinsic thin-layer solar cell) cells from Sanyo and the IBC (interdigitated back-contact) structures from SunPower Corp show $\geq 20\%$ efficiencies while being industrially viable.

17.2.1 On p-Type Substrates: The Standard Structure Pushed to the Limit

The standard cell design results from a trade-off between a reasonable device efficiency and low processing cost. The main part of c-Si cells production is today based on this standard technology developed in the 1970s on p-type c-Si wafers. The standard cell design has three main technological features. At the front side a low reflectivity is obtained by texturing the surface and covering it with an antireflective coating (ARC) such as Hydrogenated Silicon Nitride SiN_x:H. The rear side electrode is obtained by an Al layer which creates a Back Surface Field (BSF) during the contact firing. Front and back electrodes are deposited with the screen-printing technique. These structures allow efficiencies of about 15% while being fabricated at low cost and this explains their success until now.

Since the early 1980s, many research teams try to get closer to the 27% limit by making the device structure more efficient and thus more complex. Most of these improvements aimed to enhance the optical confinement and lower recombination phenomena in the c-Si bulk and at the surface. The UNSW (University of New South Wales) made a lot of research in this field and created different structures: PESC (Passivated Emitter Solar Cell [7]), PERT (Passivated Emitter Rear Totally diffused) [6, 8, 9] and PERL (Passivated Emitter Rear Locally diffused) (see Fig. 17.8). The PERL structure reached 25% efficiency which is the highest value ever obtained for c-Si solar cells. These 4 cm^2 devices, shown on Fig 17.8, were fabricated with microelectronics processes such as photolithography and thermal oxidation on p-type FZ wafers (1Ωcm and 450 μm thick). Many technological improvements were necessary to reach these results:

- "Inverted pyramids" texturing
- Double ARC
- Low area front contact grid
- Selective Emitter
- Thermal oxide for surface passivation
- Local BSF

Within those « high efficiency technologies », the most widespread are the Selective Emitter (SE) and Local BSF (LBSF). In contrary to standard cells where the doping of the emitter and BSF regions is uniform, these advanced structures use different doping levels under the contact and in the non-contacted area. A heavy doping is used under the contact to reduce series resistance whereas a lower doping (or no doping at all for the BSF) is made between the contacts. A doped zone showing a higher sheet resistance means indeed less carrier recombination than a heavily doped one. These technological improvements enhance particularly the V_{OC} and J_{SC} values by reducing recombination in the doped zones.

Reducing the front contact grid width is a key point to reduce shadowing and enhance J_{SC} values. An interesting way to fabricate narrow grid lines is the "buried contacts" technology [9]. The Electro-Plating and Light-Induced Plating [10] methods are also promising but not industrialized up to now.

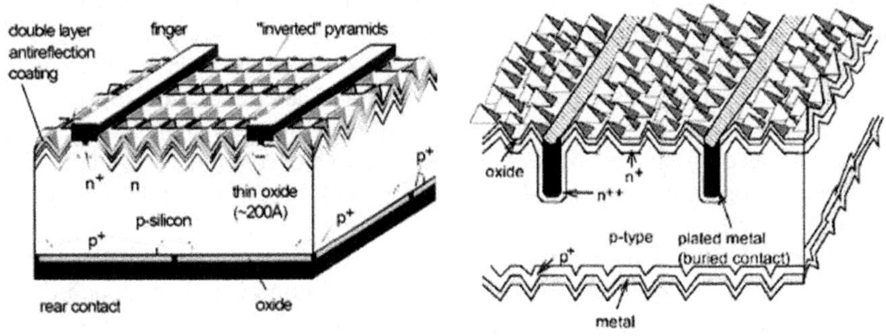

Fig. 17.8 Sketches of the PERL [6] and Buried Contact [11] structures.

Because of its complicated fabrication process, the PERL cell cannot be industrialized without being simplified. The challenge is to avoid patterning steps as much as possible, and replace photolithography steps by low cost techniques. Screen-printing, laser and plating technologies seem to be the most promising to achieve high efficiency on p-type c-Si industrial solar cells. Suntech – a Chinese cell producer - recently showed about 19% cell efficiency on large area solar cells fabricated in their pilot line [12]. This result is the best obtained on industrial p-type cells and show that producing ≥ 20% efficiency solar cells based on these substrates seems now to be feasible. However, modelling results show that with p-type CZ (less expensive) substrates the cell efficiency cannot significantly surpass this 20% threshold. To achieve higher cell efficiencies at an industrial level, n-type material seems to be more promising [13].

17.2.2 On n-Type Substrates: Back to the Future?

The first c-Si solar cells fabricated in the 1950s used n-type substrates (Phosphorus-doped). At the time these devices were dedicated to spatial applications more than terrestrial ones. P-type substrates showed higher radiation tolerance in space and were therefore preferred to n-type ones although lower efficiency was obtained. Since a few years, n-type materials are regaining interest because of their better electronic quality compared to p-type c-Si. N-type c-Si is indeed less sensitive to chemical and crystallographic defects [14]. Moreover, CZ and multicrystalline c-Si suffer less from LID (Light induced degradation) on n-type than on p-type material [15]. The LID is due to B-O pairs that appear under illumination and act as recombination centres. It means that high efficiency solar cells can be fabricated on CZ n-type c-Si whereas only FZ – i.e. more expensive - p-type c-Si can be used.

In 1999 the UNSW fabricated high efficiency solar cells on n-type FZ c-Si. It was shown that Rear Emitter (RE) structures could achieve higher efficiency than Front Emitter (FE) ones [6]. An efficiency of 22.7% was achieved on RE cells whereas a slightly lower 21.9% was reached for the FE structure (22 cm^2). The p^+ emitter caused indeed more recombination when it was located at the front side. In 2010, Glunz et al. demonstrated 23.9% efficiency on small area FE cells [13]. This could be achieved as a result of Al_2O_3 passivation of the front emitter, whereas thermal oxidation was used previously at UNSW. Naber et al. developed an industrial process on large area n-type wafers for this cell design [16]. The front Boron emitter is here passivated by a wet-chemical oxide covered by Silicon Nitride. This technology allowing 18.5% efficiency will be commercialized by Yingli Solar as “PANDA” cells.

The industrialization of n-type cells seems to be easier for the RE structure. They can be fabricated with exactly the same equipment than standard p-type cells, but here the Al layer forms an emitter at the rear side and the n^+ diffusion a Front Surface Field (FSF) – see Fig. 17.9. Efficiencies above 18% have been achieved on large area (148.6 cm^2) CZ c-Si by UNSW [17] and Fraunhofer ISE [13] with an Al-alloyed rear emitter. However, RE devices have to be fabricated on high quality substrates with very low front surface recombination velocities [18].

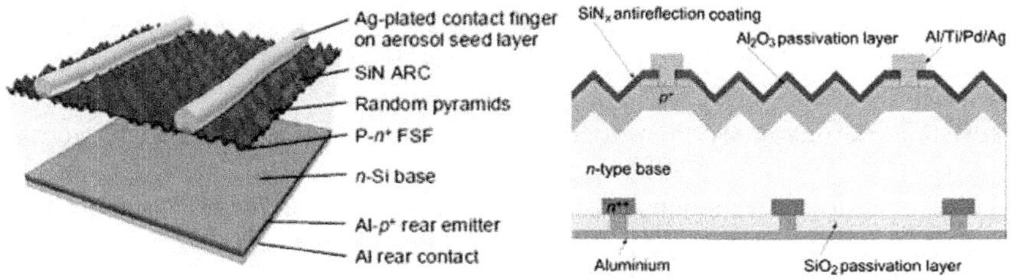

Fig. 17.9 N-type structures. Left: Rear Emitter; Right: Front Emitter [13].

17.2.3 Interdigitated Back Contact Cells

Devices having both contacts at the rear side are of main interest to reach very high efficiencies. The IBC structure was presented in 1975 by Schwartz et al to be used in concentrating PV [19]. It has no front grid shadowing and can be optimized to avoid any resistive loss. Moreover it can be easier to fabricate a module with IBC cells due to their coplanar interconnection.

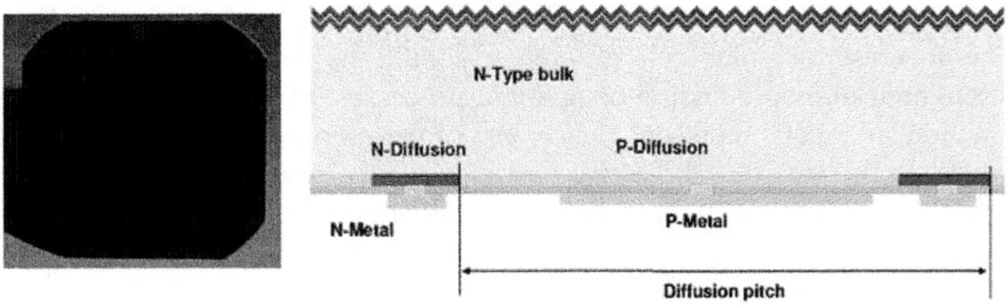

Fig. 17.10 IBC structure. Left: Front side view (no grid); Right: Sketch of the device [20].

SunPower Corp already begun in 1993 to industrialize IBC cells with a fabrication process based on photolithography steps. Five different masks were used at the time [21]. This process has been since then been adapted to lower cost and allow for large area wafers achieving an average efficiency of 22.4% on the last production line [20] - see Fig. 17.10. To reach such results, many parameters have to be carefully controlled for the process optimization:

- The front surface passivation level has to be extremely high. The Front Surface Recombination Velocity (FSRV) must stay below 10 $cm.s^{-1}$ to avoid J_{SC} losses. Thermal oxidation associated with Silicon Nitride is typically adapted for this use, and a light n^+ diffusion (Front Surface Field – FSF) also shows some advantages (less sensitivity to the FSRV [21], enhancement of lateral current transport [23]).
- The carrier diffusion length has to be high enough to allow carriers to be collected at the rear side. Thus to achieve more than 25% efficiency the c-Si substrate has to have lifetime values larger than 10 ms [24]

- The substrate resistivity and the rear geometry (pitch, see Fig. 17.9) have to be simultaneously optimised to achieve high FF values. Industrial wafers are usually thin (150 à 200 μm) and the pitch value quite high [20]. Lateral series resistance effects may then appear and cause FF drop if no FSF is applied [23].
- Low saturation current density (J_0) has to be achieved in the doped zones (surface and bulk) and at the contacts. The "point contact" technology limits the contact area where J_0 value are usually high (1000 $fA.cm^{-2}$) [25].

In 2010 an efficiency of 24.2% has been achieved by SunPower with a new cell design including passivating contacts [24]. An excellent V_{OC} value of 721 mV is obtained, showing that homojunction contacts can now almost reach the same passivation level as heterojunction ones.

Other research institutes also work on IBC structures, until now only at the lab level (4 cm^2 cells). Dicker et al. [26] and Engelhart et al. [27] obtained about 22% efficiency on p-type substrates, whereas a better result of 22.7% was obtained on n-type c-Si [13].

17.2.4 Heterojunction Solar Cells

The increase of solar cell efficiency should be realized using simple, high-throughput mass-production compatible processes. In this context, silicon heterojunction solar cells represent clearly one of the most promising options in the near future. In this devices, the active part is basically produced by a low temperature growth (180-220°C) of ultra-thin layers of amorphous silicon, (constituting the "emitter", the base-contact, and at the same time surface passivation layers), onto both sides of a thin crystalline silicon wafer-base.

At this stage, some issues become essential, such as low temperature cell processing, excellent surface passivation and cost effective processes with high throughput. Thus, heterojunction solar cell technology concentrates the following advantages:

- High efficiency potential, up to 23% (standard design, large area cell); potential up to 25% with advanced structures (IBC-HJ)
- Because of the low temperature process, thin wafers can be used (down to 100 microns) and lower breakage can be achieved than in conventional high-temperature processing. Thinner wafers also mean reduced energy payback time
- Due to its high surface passivation level, the use of thinner wafers is well adapted to the technology, by reducing costs (lower Si material consumption with reduced efficiency loss)
- The newly developed mass production equipment sold for TFT-LCD (thin film transistor-liquid crystal display) industry (amorphous and microcrystalline Si) could be readily adapted for the commercial production of high efficiency silicon heterojunction cells at very low processing costs

Sanyo Electric Co., Ltd. (Japan) was the first to report 20 % efficiency on a 1 cm^2 cell with "Heterojunction Intrinsic Thin film" silicon in 1994 [28], the so-called HIT concept, first published in 1992 . This new kind of technology has allowed the Japanese company to reach a significant market share of PV sales, being the only industrial company commercially using silicon heterojunctions. They produce top quality modules reaching very high efficiencies at industrial level (18.2 to 20 % efficiency at the cell level, 16 to 17.5% at the module level). Currently, the record efficiency for large cells (100 cm^2) is 23 % with very high open circuit voltages (739 mV obtained by using simplified, low cost, low temperature process) [29].

Then, elsewhere in the world a great interest on heterojunction solar cells technology has been started. Recent results for other companies in Asia, show efficiencies higher than 19 % for industrial size cells [29b]. Also, at the laboratory size, heterojunction devices are simulated and studied principally in AIST and Tokyo Inst Technology in Japan [29c, 29d]; the Sungkyunkwan University of Korea [29e] and the Institute of Electrical Engineering at the Chinese Academy of Sciences in China [29f].

In USA, the best results are obtained mainly by National Renewable Energy Laboratory NREL (Branz & co-workers, USA). NREL have demonstrated very high Voc (710 mV on n-type wafers) and quite high efficiencies on both n type and p-type (18.7%) and uses amorphous intrinsic thin layers at the interface between the doped layer and the bulk silicon to ensure excellent surface passivation [30].

In Europe, many organisations are working on heterojunction solar cells with different strategies. Their individual results are excellent with the best cell efficiencies after SANYO on large area surfaces and reveal an excellent scientific level.

Notably, INES has reached the highest efficiency in Europe, second to Sanyo so far on large area industrial wafers on N type c-Si and 125 pseudo square solar cells with up to 19.6 % (20 % on 100 cm^2) and Voc up to 718 mV [31]. This performance has been possible thanks to the equipment loaned recently by JUSUNG Engineering to INES.

HZB has reached very high efficiencies in Europe, with 17.4% on p-type c-Si and 19.8 % on n-type c-Si wafers with pyramidal surface texturization [32], on 1 cm^2 wafers. EPFL Neuchâtel has recently reached 20 % on textured substrates [33] and high-open circuit voltage of 720 mV on 4 cm^2 solar cells [34]. On the Industrial side, Roth and Rau develop a process on large area wafers with efficiencies up to 19 % on 125PS wafers and Voc up to 726 mV and 21% on 4cm^2 cells [34b].

Finally, it is worth to remark that a recent interest has risen for Si-HJ cells with IBC structure. This device could avoid the absorption in the front TCO and a-Si:H emitter layer. Such a new device has already been introduced in 2007 [35] but its efficiency is up to now very low (about 15% [36]). However this new device could be theoretically more efficient than IBC cells having homojunction contacts while being fabricated with less patterning steps.

17.3 Key Points of a-Si:H/c-Si Heterojunction Solar Cells

In this part we will focus on the analysis of the key parameters which will allow to reach the high efficiency of a-Si:H/c-Si heterojunction solar cells. Following the Pareto principle, we will focus only on the most important technology aspects impacting each individual parameter on the efficiency (η) equation:

$$\eta = V_{oc} \times J_{sc} \times FF \tag{17.1}$$

where V_{oc} is the open circuit of the solar cell, J_{sc} is the short-circuit current density and FF is the fill factor of the device. In Fig. 17.11 we show a simplified structure of a standard heterojunction solar cell taking into account all the layers and interfaces on the device.

Fig. 17.11 Structure of a standard heterojunction solar cell.

Even if the quality of the deposited layers is important to guarantee the high efficiency of the devices, one of the most critical aspects is the role of the interfaces on the performance of the solar cell. Indeed, the thickness of the layers (especially the amorphous silicon ones) is in the nm range. Then, any defects in the first nm or any damaging in the last nm of the layers is susceptible of playing an important role on the solar cell parameters, especially the FF. Despite the good results obtained on the finished devices, the physical understanding of interfaces in the whole solar cell is still limited. Nevertheless, some efforts have been done to understand the heart of the structure, the interface a-Si:H/c-Si by means of advanced characterisation or simulation studies [32, 33, 37, 38, 39].

Moreover, because of the low temperature constraint (<200ºC), the cumulative fabrication process can easily affects the layers and interfaces deposited previously. Table 17.1 shows qualitatively the impact of the different layers and interfaces plotted in Fig. 17.11 on the different parameters used.

Table 17.1 Qualitative impact of the different layers and interfaces forming a standard HJ solar cell on the characteristic parameters used to evaluate the efficiency of the device (- none, x-low, xx-medium, xxx-strong).

layer	V_{oc}	J_{sc}	FF
front-metal grid	-	xxx	xxx
interface 1 metal/TCO	-	x	xx
front - TCO	x	xxx	xxx
interface 2 TCO/a-Si:H	x	x	xx
front a-Si:H	xx	xxx	xx
interface 3 a-Si:H /c-Si	xxx	-	xx
c-Si	xx	xxx	x
interface 4 c-Si/a-Si:H	xxx	-	x
back a-Si:H	xx	-	x
interface 5 a-Si:H/TCO	x	x	x
back TCO	x	x	x
interface 6 TCO/metal	-	x	xx
back metal	-	x	xx

As it is clearly observed, the V_{oc} of the device is very sensitive to the interface a-Si:H/c-Si and the amorphous layers. By contrast, the J_{sc} is mainly affected by the front-side of the cell (TCO-a-Si:H-metallisation) and the optical confinement achieved especially by the texturization of the c-Si substrate. On the contrary, it is worthy to remark that all the layers and interfaces play an important role on the FF, especially because of the direct impact of the series resistance (R_s). It is, without any doubt, the most difficult parameter to analyze.

17.3.1 Open – Circuit Voltage

As previously mentioned, the V_{oc} is the outstanding parameter that differentiates the HJ technology from all the c-Si ones. The high Voc values are mainly reached by the higher built-in voltage achievable due to the presence of the abrupt heterojunction between the a-Si:H film and the c-Si wafer, allowing values over 740 mV, which are not possible with an homojunction. Moreover, considering technological aspects, high V_{oc} are mainly obtained when controlling the c-Si surface quality, the first nanometres of a-Si:H and the electric field by a-Si:H layers and TCOs.

17.3.1.1 The c-Si Surface Quality

On HJ solar cells, where the interface between crystalline silicon and the amorphous layer is used to form the p/n junction, the substrate surface directly becomes part of the electronic interface [40]. Moreover, in the case of structured substrates (pyramids from alkaline texturization), interface defects become increasingly critical to the recombination losses of charge carriers on silicon interfaces.

The interface recombination velocity is mainly affected by the density and the character of interface states (D_{it}). These states are localized in an interlayer extended over only a few Angstroms and result from silicon dangling bond defects with different back-bond configurations. The relation between interface state densities and structural imperfections at silicon surfaces has been intensively investigated, because the reduction of their densities is a main concern in microelectronic device technology [41].

Thus, a-Si:H emitter and BSF deposition requires c-Si substrates with undamaged (roughness, cracks), contamination-free and chemically stable silicon surfaces. So the density of rechargeable surface states has to be reduced by preparing atomically flat silicon surfaces and well-ordered silicon surfaces through the application of special cleaning procedures. It was shown that the ideally H-terminated surface is characterized by a very low density of surface states ($<10^{10}$ $cm^{-2}eV^{-1}$) [42].

Many techniques are used to analyze D_{it}. Among them, it is worth to remark the non-destructive and highly surface-sensitive spectroscopic techniques as the surface photovoltage [42] and the modelling on electrical measurements of the capacity of the junction [37]. The analysis by Quasi-Steady-State Photoconductance (QSSPC) measurements [43] of the passivation quality of a symmetrical (i) a-Si:H layer deposited on the c-Si substrate is sufficient to compare the surface quality of the samples.

Standard cleanings used in microelectronic technology (for instance RCA [44]) have been applied successfully for HJ solar cells [45, 46]. Anyhow, as shown in Table 17.1, they are not enough to guarantee high values of effective lifetime and implicit –Voc remains under 720 mV on the devices especially from a robustness point of view. A multistep RCA cleaning gives better results (see Table 17.) as suggested also by Angermann [45] and is especially interesting because of its robustness. Other alternative cleanings based on HF/HNO_3 have been successfully studied by [45] with very good results. Table 17.2 summarizes different essays done to compare all the treatments cited before.

It is important to remark that all these cleaning procedures require high grade and high cost equipment and facilities to avoid any recontamination, so they are expensive and difficult to integrate into a production line. Thus, efforts have to be done to simplify the cleanings without loosing efficiency on the device and process robustness.

Moreover, the densities of interface states on the initially H-terminated silicon surfaces were found to increase drastically during the first monolayer of oxide growth. Depending on the cleaning efficiency, the duration of the initial phase of oxidation in air ranges from a couple of minutes up to some hours on

Table 17.2 QSSPC results of a symmetrical 50nm (i) a-Si:H layer deposited on (n) c-Si substrates after different cleaning procedures on polished and two different textured substrates performed at INES. The results show the high quality passivation level on textured wafer achieved by means of a simplified cleaning not based on the RCA procedure.

wafer	cleaning	implied-V_{oc} (mV)	ΔV_{oc} (±) (mV)
Polished FZ (111) (n) c-Si 300µm 1-5 Ωcm	no	< 580	-
	HF last	725	4
	RCA+HF last	728	6
Textured CZ (100) (n) c-Si 200µm 1-5 Ωcm	no	<580	-
	HF last	650	20
	RCA	< 580	-
	RCA + HF	720	17
	RCA+ rinsing recontamination + HF	670	20
	multistep RCA + HF	732	6
	alternative-industrial +HF	739	4
Textured FZ (100) (n) c-Si 200µm 1-5 Ωcm	multistep RCA + HF	739	6
	alternative-industrial +HF	742	4

H-terminated surfaces of polished or texturized, (100) and (111) substrates. Consequently, the time between ex-situ H-passivation by wet processes and the a-Si:H deposition has to be as short as possible [47].

17.3.1.2 The First a-Si:H nm Deposition

As mentioned in the previous section, in order to produce a high efficiency heterojunction silicon solar cell, a very low defect density at the interface is needed and it is normally achieved by wet chemical treatments. But a second physical demand needs to be satisfied. A very low defect density in the first atom layers of the amorphous silicon is assumed to be important. Two main characteristics will define the properties of this first nm of the layer:

- The deposition technique to guarantee a soft plasma a-Si:H layer deposition.
- A layer with very low density of gap defect states.

The PECVD-deposition technique is predominantly used for deposition of amorphous hydrogenated silicon, although others techniques like Hot-Wire CVD have also interesting approaches. For instance, the absence of ion bombardment in

Hot-wire CVD allows depositing very high passivation quality films with a controlled and soft interface [48, 49].

Then, the control of the deposition conditions is critical especially in this first nm of the layers to maintain the best passivation level and thus, the best V_{oc} on the device.

To obtain the lowest density of defect states on the interface, the insertion of an intrinsic (i)a-Si:H buffer layer between the substrate and the a-Si:H emitter and/or the a-Si:H back surface field has been reported as an essential step to achieve high open circuit voltages [50]. The reason for this finding is an improved effective passivation of the c-Si surface, because intrinsic a-Si:H has a lower density of gap defect states. Taking into account that doped a-Si layers have mid-gap state densities above 10^{18} $cm^{-2}eV^{-1}$ [51], they provoke dark tunnelling leakage currents. By inserting the intrinsic a-Si:H spacer, the midgap defects level is reduced to only 10^{15} to 10^{16} $cm^{-2}eV^{-1}$ and avoids any tunnelling leakage current.

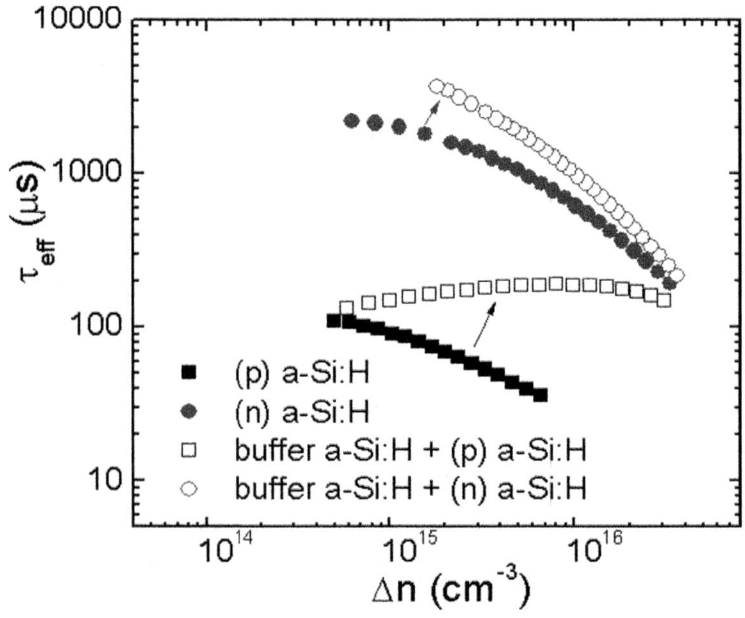

Fig. 17.12 QSSPC measurements for the best layers and stacks used in this work in a symmetrical configuration (same layer in both sides of the wafer). It is important to remark the big improvement observed when adding the buffer layers between the doped a-Si:H and the c-Si substrate [31].

High V_{oc} well above 720mV and efficiencies above 22% were only reported with this thin intrinsic layer [52]. Thus, the quality of this layer is assumed to be the key-feature to achieve exceptionally good performance. In Fig. 17.12, QSSPC measurements of 4 different symmetrical configurations are shown: (p) a-Si:H, buffer+(p) a-Si:H, (n) a-Si:H, buffer + (n) a-Si:H. It is important to remark that for the (p)-type layer, the passivation results are very low (implicit-V_{oc} < 610mV), and to obtain higher implicit-V_{oc} values (V_{oc} > 650mV) it is mandatory to introduce the thin (i) a-Si:H buffer layer. Even with this layer, the passivation level is lower compared to a simple (n) a-Si:H layer. This result is expected considering that (n) a-Si:H are less defective layers (less interface states) and allow a better interface than boron doped ones. Anyhow, as pointed out in Fig. 17.12, an increase of effective lifetime (τ_{eff}) is observed (implicit-V_{oc} =705 mV) but, as expected, is less pronounced considering that simple (n) a-Si:H layers already surpass 700mV of implicit Voc.

It is worth to mention other approaches to substitute successfully the (i) a-Si:H layer by other passivation layers. For instance, a-SiC:H or a-SiO:H which allow better optical properties, maintaining high passivation levels as we will discuss later [53, 54].

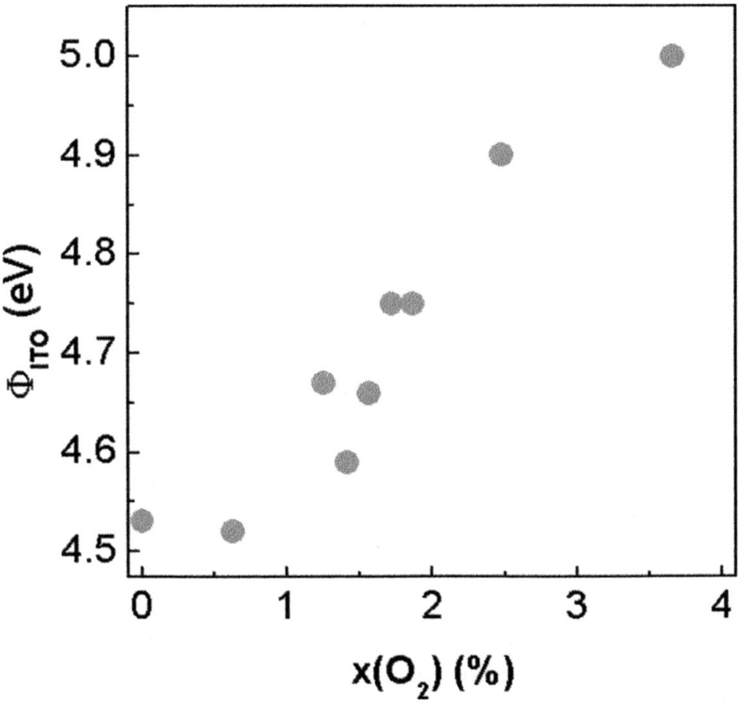

Fig. 17.13 Variation of the work function of ITO layers with the oxygen content measured by Capacitance-Voltage.

17.3.1.3 The Electric Field

To obtain the highest V_{oc}, the Fermi level in the thin amorphous doped layers (n or p-doped) should be as close to the conduction or valence band as possible and the work function of the transparent conducting oxides (TCO) should be adapted to avoid undesirable band bending as suggested by [55, 56, 57]. Due to the extremely thin emitter layer used in this kind of device, the built-in potential available cannot be merely defined by the difference between the work function of the emitter and the base, but it depends on TCO work function too. As a consequence, because of the correlation between built-in potential and open circuit voltage, the TCO work function strongly affects the solar cell performance. In Fig. 17.13, it is shown an example of how the ITO can vary its work function as function of its oxygen content measured by Capacitance-Voltage as proposed by [58].

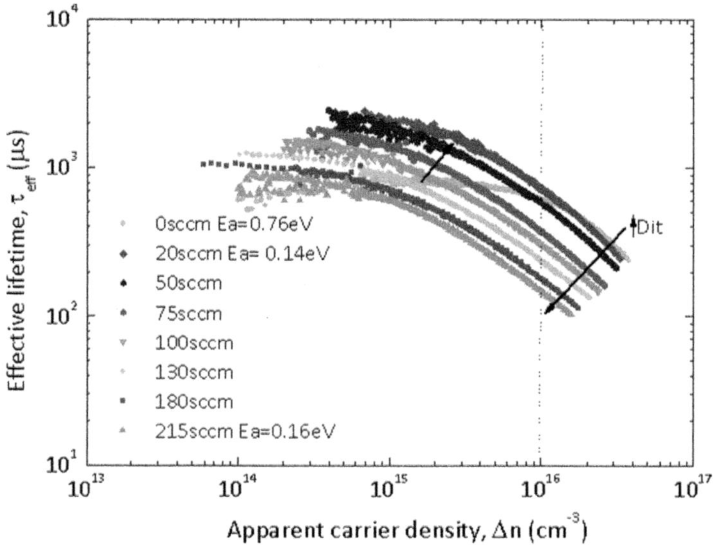

Fig. 17.14 QSSPC measurements of (n) a-Si:H layers with different doping. For higher levels of carrier density ($\Delta n > 10^{16}$ cm^{-3}) there is a correlation between the interface state ensity and the decrease of the curve with the addition of the doping gas (increase of defects).

On the other hand, the optimal gas phase doping concentration for an a-Si:H/c-Si solar cell is obtained when the Fermi energy in the a-Si:H layer is sufficiently close to the conduction or valence band and the defect density is still considerably low. It has to be taken into account that while on the one hand the band bending increases with increasing doping concentration, on the other hand the number of gap states in the a-Si:H layer and interface states increases as shown in Fig. 17.14. The activation energy measured for the layers (E_a) show that in our case, 20 sccm of PH_3 is enough to guarantee a good compromise on the doping of the layer. Furthermore, when we add more doping gas, we do not improve the electric field but the D_{it} increases.

17.3.2 Short-Circuit Current

There are two main well-established methods for minimizing reflection loss and so enhance the short-circuit current of a solar cell: the texturization of 100 oriented crystalline silicon wafers and an appropriate antireflection coating (ARC). Moreover, in the case of HJ solar cells, it has to be taken into account the non negligible absorption looses of the amorphous layers forming the emitter. To all of this, it is important to add the need to minimize the shadowing of the metallization pattern to allow the maximum of light entering the cell.

17.3.2.1 Texturization

Surface texturization of (100)-oriented crystalline silicon wafers is a frequently used technique in modern solar cell processing to reduce optical reflections. Many successful approaches based on texturing silicon surface by alkaline solutions (NaOH, KOH) [59, 60] have been reported. Alkaline etchants at low concentration

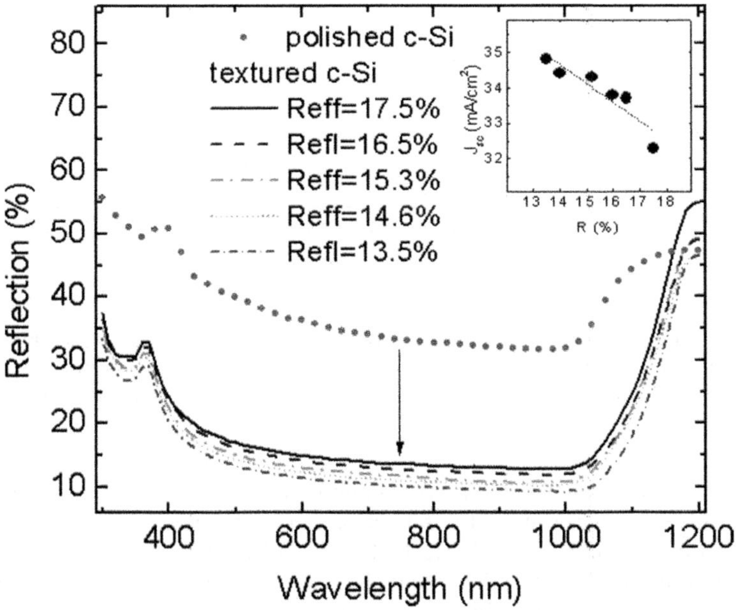

Fig. 17.15 Reflectivity measurements for different texturization processes on 200μm c-Si wafers. The 300μm polished c-Si sample (•) is plotted for comparison. In the inset, a correlation between the reflectivity and the short-circuit current measured on the HJ cell.

in water expose Si (111) faces resulting in square-based pyramids randomly distributed over the cell surface (see Fig. 17.8). This texture will enhance light trapping capability by increasing both coupling of lights into the cell and reflectivity [61] of lights trying to escape from the cell [62].

Figure 17.15 shows the reflection spectra of polished (100) c-Si (300μm thick) and different textured ones (200μm). Solar effective reflectivity (R_{efl}) values calculated between 300 and 1200nm are also reported. Best R_{efl} of about 13.5% are achieved with an optimized texturization process.

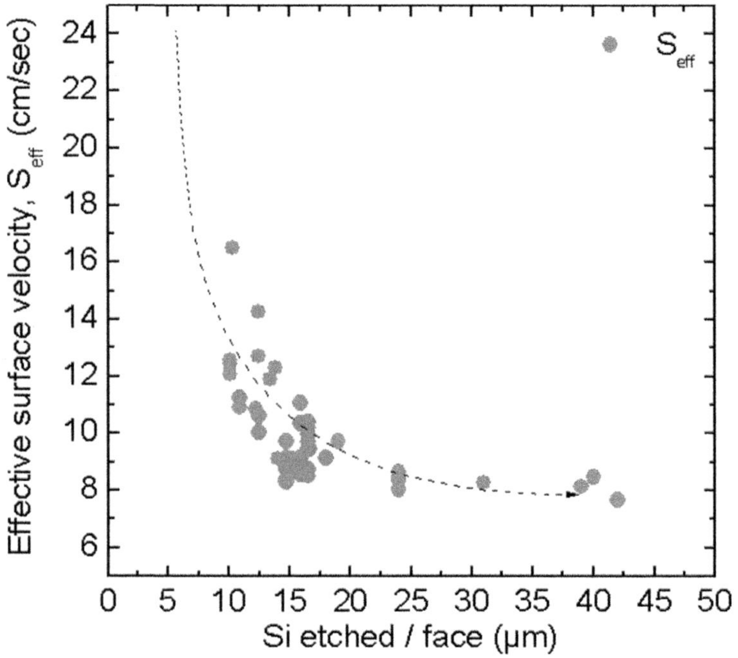

Fig. 17.16 Effective surface recombination velocity against the Silicon etched by face during alkaline texturization. When starting with as cut wafers, it is necessary to etch at least 15 microns of Silicon by face to surpass the damaged zone and guarantee a good surface state.

Since heterojunction solar cell processing involves deposited layers, the quality of the surface becomes critical for the performance of the device. On the one hand, there is an impact on V_{oc} (see Table 17.1) caused by surface effects such as residual metal impurities or by the incomplete removal of the damaged zone (see Fig. 17.16). On the other hand, there are conformity and uniformity constraints; impacting Voc and Jsc. Thus, flat surfaces or a large distribution of pyramid sizes affects strongly the reflectivity generating optical losses. Furthermore, on the flat areas (see for instance Fig. 17.17a) the layers are thicker (normal deposition) and though absorption increases. Current density losses between a bad (Fig. 17.17a) and a good texture (Fig. 17.17b) can be higher than 3mA/cm^2 for the same cell configuration.

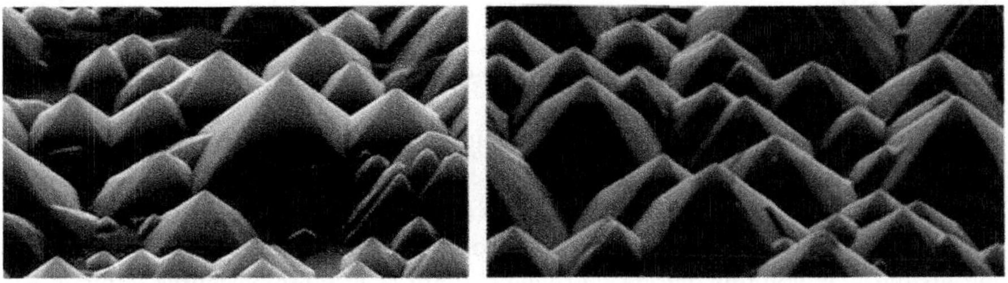

Fig. 17.17 SEM measurements of two different randomly textured surfaces, (a) with flat surfaces and (b) with a 100% coverage by pyramids.

17.3.2.2 Absorption in a-Si:H layers

It is known that photon absorption in the emitter of a heterojunction solar cell leads to a considerable current loss due to the high recombination in this layer [63, 64]. Therefore, it is necessary to minimize light absorption in the window layer or to reduce the recombination in order to improve the efficiency of the solar cell. One direct attempt is to widen the optical band gap and suppresses absorption by adding carbon or oxygen in the emitter [65, 66]. However, the inclusion of the impurities on the a-Si:H layers induces a deterioration of the electrical properties because of the increase of defects states [66].

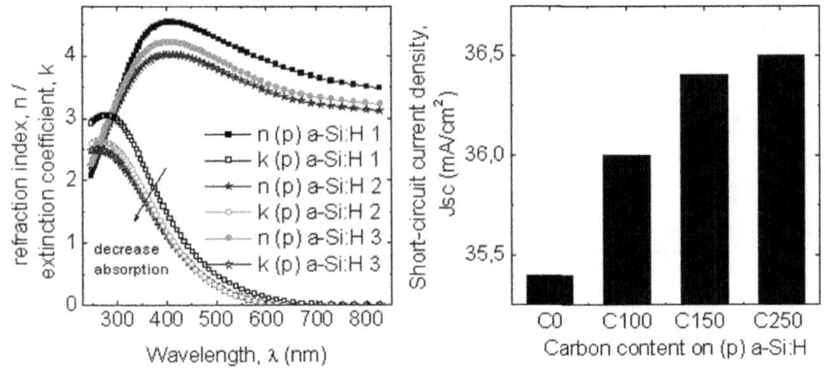

Fig. 17.18 Refractive index and extinction coefficient obtained by spectroscopic ellipsometry using the Tauc-Lorentz model for three different carbon doped (p) a-Si:H layers (a). (b) shows the short-circuit current obtained on the finished solar cells on (n) c-Si with different carbon contents on the doped emitter layer.

In Fig. 17.18a we show the spectroscopic ellipsometry measurements (n,k) of different (p) a-Si:H layers with different optical properties achieved, in this case, adding CH_4 to the gas mixture with different total dilution (C0, C100 and C250 sccm). The absorption in the layer decreases when increasing the carbon content in the layer, impacting on the J_{sc} with a gain of more than 1mA/cm^2 (Fig. 17.18b).

17.3.2.3 Transparent Conductive Oxides

In a-Si:H/c-Si HJ solar cells, it is mandatory to find material deposited at low temperature which combines good antireflection and collection of charges properties. Transparent conducting oxides (TCO) as In_2O_3:Sn (ITO) films have been used widely on HJ solar cells since low resistive films with high transparency in the visible wavelengths can be fabricated easily at low temperatures. Optical and electrical properties optimization is a difficult task. Best compromise must also consider the metallization pattern and his shadowing.

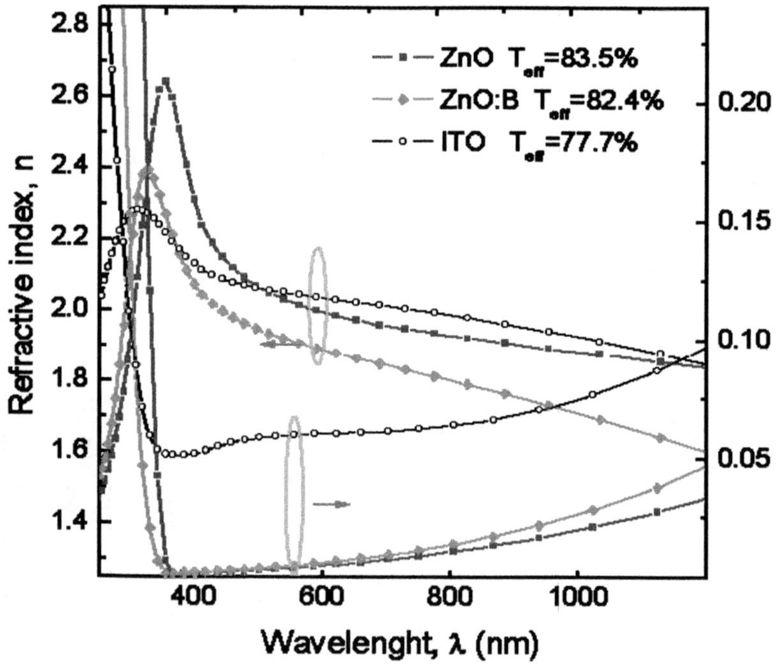

Fig. 17.19 Refractive index and extinction coefficient measured by ellipsometry for an ITO, a ZnO and a ZnO:B.

The key parameter for the reduction of optical losses is the development of a high-quality TCO with low carrier concentration (*N*) and high carrier mobility (μ) to maintain high conductivities ($<5\cdot10^{-3}\Omega$cm). For instance, standard ITO layers with $\mu\approx35$cm^2/Vs and $N\approx2\cdot10^{20}$cm^{-3} show a large absorption induced by free carriers in the near-infrared (NIR) wavelengths [9], which results in enhanced light absorption within the TCO in the HJ solar cell (see Fig. 17.19). The free carrier absorption also leads to a decrease in refractive index and an increase on the extinction coefficient of the TCO in the visible and NIR wavelengths [9], which in turn enhances the light reflection at the TCO/a-Si:H interface.

One alternative option to ITO is to use ZnO:Al (AZO) or ZnO:B layers. As shown by ellipsometry measurements on Fig. 17.19, the transmission of the layer are higher and hence, absorption is generally lower. By contrast, is this case, ZnO:B layers reach only resistivities around 10^{-3} Ω.cm. This layers can be easily used at the back side of the device, were lateral conduction is not necessary (except on bifacial configuration), enhancing the Jsc more than 1mA/cm^2 compared to ITO ones. On the contrary, they are not suitable on the front side because of their electrical limitations.

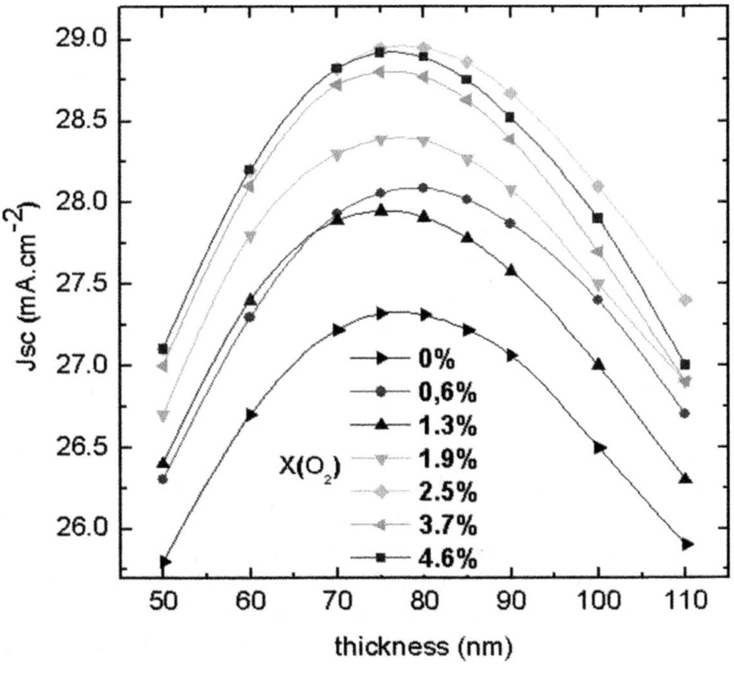

Fig. 17.20 Optilayer simulations of a c-Si/a-Si/ITO stack using a semi-infinite model. We have varied the oxygen content on ITO layers to obtain different optical properties.

Furthermore, the TCO optimization has to be done also in terms of thickness to maximize the antireflection effect and the photon collection. In Fig. 17.20, simulations by Optilayer® software [67] show that on polished c-Si there is a maximum of J_{sc} obtained of about 75-85nm of ITO. This maximum slightly evolves with the oxygen content, and though, with the optical properties of the layer [68] in the whole spectral range. A non-optimized TCO can drop the J_{sc} more than 4mA/cm^2.

17.3.3 Fill Factor

The fill factor (FF) is the most difficult parameter to analyze in a solar cell because it depends on all layers and interfaces. This is especially critical in the case of HJ solar cells because of the high amount of layers and interfaces of the device (see table 1). All of them have an impact on the main parameter affecting the FF, the series resistance (R_s). Though, to minimize the R_s, both, layers optimization and integration analysis, have to be done. Moreover, as we will see later on, in the majority of the cases a gain in FF generates a loose on J_{sc} or V_{oc}. Thus, a compromise between parameters and layers has to be found in terms of global efficiency.

Nevertheless, there are some critical points that are mainly affecting the R_s values: the less conductive layers (buffers) and the interface a-Si:H/c-Si, the lateral collection (TCO-metallization) and the low temperature metal contacts.

17.3.3.1 The Buffer Layer and the Interfaces

Actually, if the a-Si:H buffer layer is very thin, it can play an important role in the whole Rs of the device due to his low conductivity. As we see in Fig. 17.21, the sensibility of all parameters with only some nanometers variation is huge. When varying the thickness of the buffer layer from 5 to 10 nm, even if V_{oc} increases due to the less defective layer [69] there is a maximum of efficiency around 7 nm in this particular case mainly guided by a drastic fall in the FF.

Another key factor identified as critical to avoid the S-shape curve which drastically lowers the FF is the presence of Schottky barriers at the interfaces surrounding the (p) a- Si:H emitter. On the one hand, a non-optimized interface between the p-doped a-Si:H and ITO generate a barrier if there is a poor effective doping density or/and a too thin doped layer or/and a reduced work function of the TCO [70]. On the other hand, it is demonstrated by [38, 71] that the quality of the interface (D_{it}) and the valence band offset between a-Si:H/c-Si provoked by a non optimized a-Si:H can strongly affect the carrier collection. In both cases, the final efficiency of the device is reduced even if very high V_{oc} and J_{sc} values are possible.

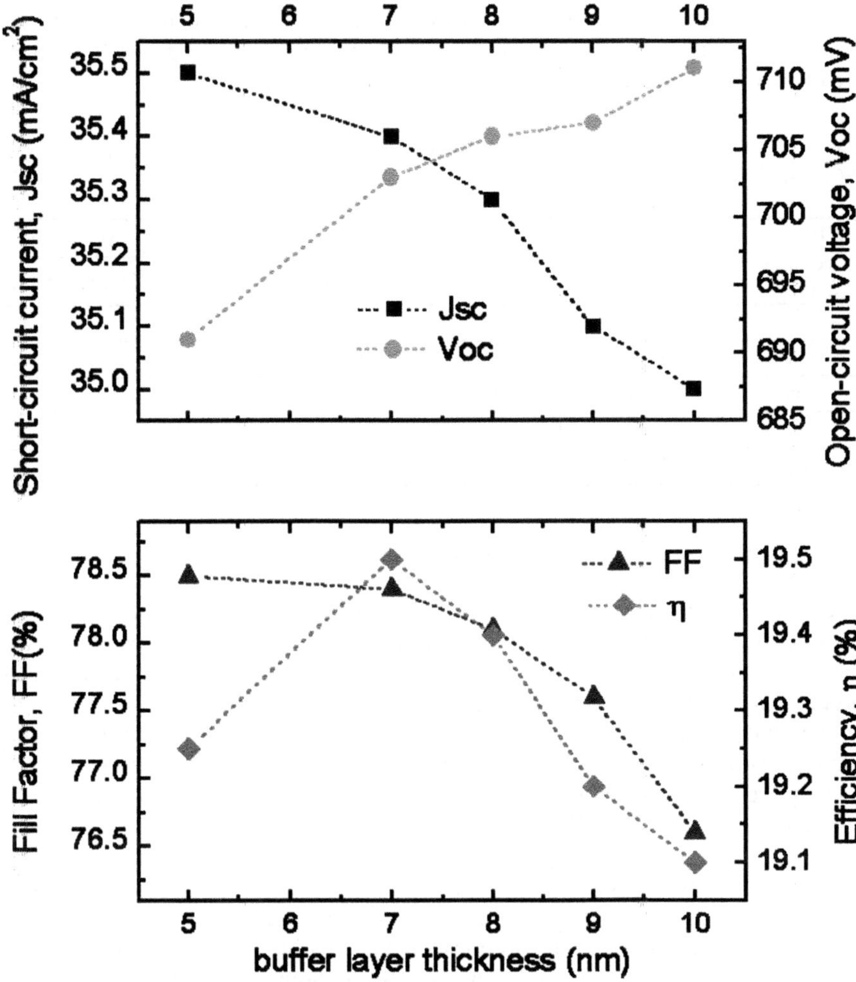

Fig. 17.21 Impact of the front buffer layer thickness on textured HJ solar cell parameters (V_{oc}, J_{sc}, FF, η)

17.3.3.2 The Lateral Collection (TCO-Metallization Pattern)

The series resistance of the solar cell can be, as a simplified scheme, decomposed on a vertical R_{sv} and a lateral R_{sl} component. R_{sl} is only affected by the lateral conduction of the charges in the TCOs to reach the metallization fingers on the front side (and eventually on the back side on bifacial devices). Though, the closer the fingers are, the lower the its value. As seen before, the shadowing losses are enormous with a dense grid, so the best solution is a compromise between the shadowing and the conduction/transparency of the TCO.

In Table 17.3, an example is shown of five sets of HJ solar cells with different ITO layers varying the oxygen ratio to obtain a different electrical/optical compromise. The impact on the couple Jsc/FF and the direct effect of the R_s on the FF are clearly observed. The R_s values are calculated following [72].

Table 17.3 Comparison of five types of ITO layers with different oxygen contents giving a different electrical/optical compromise and solar cells parameters obtained.

Oxygen content [%]	ρ	n	k	J_{sc}	V_{oc}	FF	η	R_s
	[Ω cm]	633		[mA/cm^2]	[mV]	[%]	[%]	[Ωcm^2]
0	6.60E-04	1.86	0.166	31.2	695	76	17	1
1.9	4.76E-04	1.89	0.041	35.2	698	77	19	0.86
2.5	7.90E-04	1.95	0.042	35.6	699	76	19	1.06
3.7	1.24E-03	2.01	0.057	35.8	700	76	19	1.22
4.5	2.12E-03	2.03	0.056	35.8	700	74	19	1.5

17.3.3.3 The Low Temperature Contacts

The low temperature constraint (<200ºC) on the HJ fabrication process is especially critical for the metallisation performed always at the last steps of the whole process. Thus, it is mandatory to change the standard industrial screen printed process, where the silver pastes are annealed at more than 800ºC to reach good conductivity and contact resistance values, by low temperature screen-printed pastes. Even though a lot of work is done in terms of pastes development, up to now, the resistivity of the low temperature metallization is still lower than the high temperature ones (in the 10^{-6} Ω·cm range). So, this is translated into a FF loss of about 1% absolute or even more. Nevertheless, one solution to decrease this vertical series resistance is to increase the fingers thickness (aspect ratio). But in screen-printed technology, it is extremely complicated to increase the thickness without increasing the width, and consequently, without a repercussion on the J_{sc} values. Moreover, for industrial concerns (throughput, cost), there is a limitation on the quantity of paste used and number of prints. Thus, a compromise has to be found between electrical gain and industrialisation.

As an example, in Fig. 17.22, we demonstrate the feasibility of up to 4 successive prints, with very limited loss in final line width. An example of a SEM picture of a grid line after two successive prints is also presented in the insert. In this example, an aspect ratio of 0.36 is obtained.

As expected, we confirm a clear improvement of line resistance with the number of prints. It seems that final line resistance gain starts to saturate between 3 to 4 successive prints. This electrical gain has been tested on HJ solar cells. Batches with 2 prints are compared with batches with 3 prints on front side grid. Results are presented in Fig. 17.23.

Moreover, other low-temperature techniques are suitable for HJ solar cell metallisation as electro-deposition, PVD or evaporation. Anyhow, all of them have constraints in term of up-scaling and/or cost. At INES, new printing technologies are studied to enable only one single print for industrial applications.

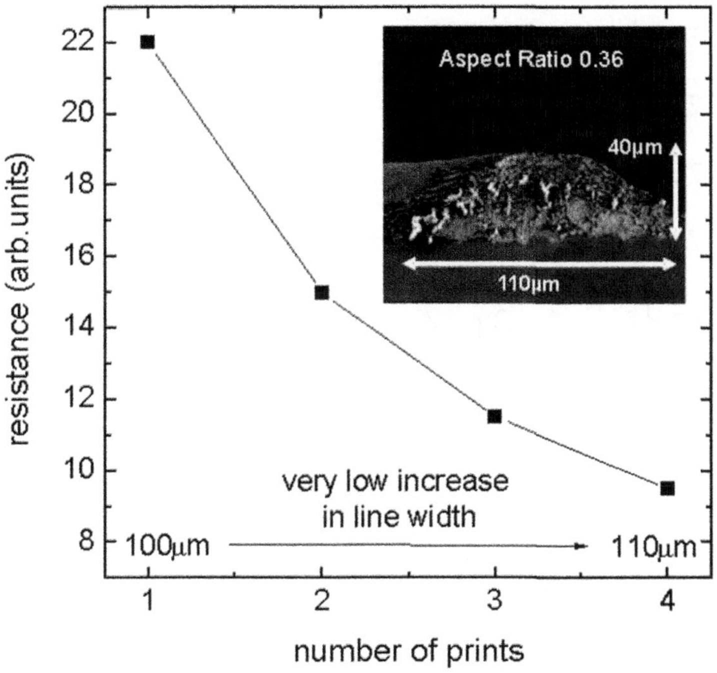

Fig. 17.22 Morphological and electrical demonstration of 4 successive screen prints without excessive loss in line width. A SEM picture of a finished grid line is also presented in the inset showing an aspect ratio of 0.36 for a two successive print process.

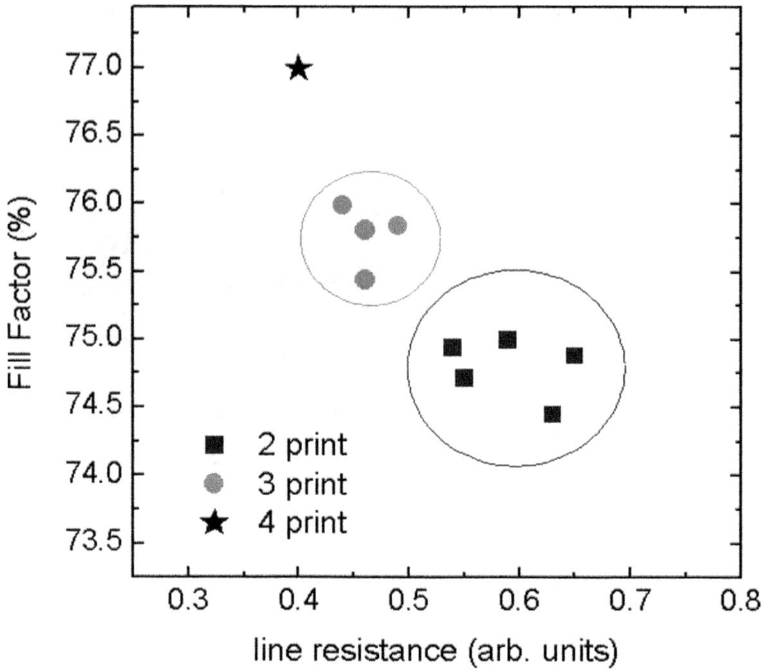

Fig. 17.23 Impact of front grid number of prints. The higher the number of prints, the higher fill factors are obtained, which is linked to a gain in line resistances.

17.4 Is a-Si:H/c-Si Heterojunction Technology a Smart Choice for High Efficiency Solar Cells?

After the technological key parameters described in the previous chapter, we can ask again if the a-Si:H/c-Si heterojunction solar cells technology is a smart choice for high efficiency solar cells. And, if we solve all the constraints mentioned before, we can say that it indeed is a smart choice for the following reasons:

- In terms of **efficiency**, because 23% is already demonstrated with V_{oc} up to 740 mV.
- In terms of **industrial process** because of simple process steps with equipment already available (LCD industry).
- In terms of **energetic productivity** (> 10%) because it is higher than homojunction cells (RCC and PERL).
- In terms of **module integration** because of possible bifacial integration due to the symmetric structure.
- In terms of **process temperature** because of the lower temperature (<200ºC) with improvement of low temperature screen printing pastes.
- In terms of **potential for the future** (thin wafers) because it is a unique technology for which V_{oc} is increasing with thinner wafers with room for improvements (I_{sc}) if infrared light management is improved.

- In terms of **material use,** even if there are still some constraints for ITO (Indium); it can be successfully replaced by other TCOs.
- In terms of **cost** because high efficiencies are combined with thinner wafers so, there are double cost reduction possibilities.

References

[1] EPIA (European Photovoltaic Industry Association): Global market outlook for photovoltaics until 2013 (2009), http://www.epia.org/publications/epiapublications (accessed October 15, 2010)
[2] NREL, http://www.nrel.gov
[3] Kerr, M.J., Campbell, P., Cuevas, A.: Lifetime and efficiency limits of crystalline silicon solar cells. In: Conf. Rec. 29th IEEE Photovolt. Spec. Conf., New Orleans, USA, pp. 438–441 (2002)
[4] Swanson, R.M.: Approaching the 29 % limit efficiency of silicon solar cells. In: Proc. 31st IEEE Photovolt. Spec. Conf., Lake Buena Vista, USA, pp. 889–894 (2005)
[5] del Canizo, C., del Coso, G., Sinke, W.C.: Crystalline Silicon Solar Module Technology: Towards the 1 euro Per Watt-Peak Goal. Prog. Photovolt.: Res. & Appl. 17, 199–209 (2009)
[6] Zhao, J., Wang, A., Green, M.A.: 24.5% Efficiency silicon PERT cells on CZ substrates and 24.7% efficiency PERL cells on FZ substrates. Prog. Photovolt.: Res. & Appl. 7, 471–474 (1999)
[7] Blakers, A.W., Green, M.A.: 20% Efficiency Silicon Solar Cells. Appl. Phys. Lett. 48, 215–217 (1986)
[8] Zhao, J., Wang, A.H., Green, M.A.: 24.5% efficiency PERT silicon solar cells on SEH MCZ substrates and cell performance on other SEH CZ and FZ substrates. Sol. Energy Mat. Sol. Cells 66, 27–36 (2001)
[9] Wenham, S.: Buried-contact silicon solar cells. Prog. Photovolt.: Res. & Appl. 1, 3–10 (1993)
[10] Mette, A., Schetter, C., Wissen, D., Lust, S., Glunz, S.W., Willeke, G.: Increasing the Efficiency of Screen-Printed Silicon Solar Cells by Light-Induced Silver Plating. In: Conf. Rec. 4th IEEE World Conf. Photovolt. Energy Convers., Hawai, USA, vol. 1, pp. 1056–1059 (2006)
[11] Wenham, S.R., Zhao, J., Dai, X., Wang, A., Green, M.A.: Surface passivation in high efficiency silicon solar cells. Sol. Energy Mat. Sol. Cells 65, 377–384 (2001)
[12] Shi, Z., Wenham, S., Ji, J.: Mass production of the innovative PLUTO solar cell technology. In: Conf. Rec. 34th IEEE Photovolt. Spec. Conf., Philadelphia, USA, pp. 1922–1926 (2009)
[13] Glunz, S.W., Benick, J., Biro, D., Bivour, M., Hermle, M., Pysch, D., Rauer, M., Reichel, C., Richter, A., Rudiger, M., Schmiga, C., Suwito, D., Wolf, A., Preu, R.: n-type silicon - enabling efficiencies > 20% in industrial production. In: Conf. Rec. 35th IEEE Photovolt. Spec. Conf., Hawai, USA, pp. 50–56 (2010)
[14] Cotter, J.E., Guo, J.H., Cousins, P.J., Abbott, M.D., Chen, F.W., Fisher, K.C.: P-Type Versus n-Type Silicon Wafers: Prospects for High-Efficiency Commercial Silicon Solar Cells. IEEE Trans. Electron Devices 53(8), 1893–1901 (2006)
[15] Cuevas, A., Kerr, M.J., Samundsett, C., Ferrazza, F., Coletti, G.: Millisecond minority carrier lifetimes in n-type multicrystalline silicon. Appl. Phys. Lett. 81(26), 4952–4954 (2002)

[16] Naber, R.C.G., Guillevin, N., Burgers, A.R., Geerligs, L.J., Weeber, A.W.: ECN n-Type Silicon Solar Cell Technology: An Industrial Process that Yields 18.5%. In: Proc. 24th Eur. Photovolt. Sol. Energy Conf., Hamburg, Germany, pp. 2177–2179 (2009)
[17] Mai, L., Hameiri, Z., Tjahjono, B.S., Wenham, S.R., Sugianto, A., Edwards, M.B.: Rear junction laser doped solar cells on CZ n-type silicon. In: Conf. Rec. 34th IEEE Photovolt. Spec. Conf., Philadelphia, USA, pp. 1811–1815 (2009)
[18] Mihailetchi, V.D., Geerligs, L.J., De Iuliis, S.: N-type silicon solar cell with Al back junction: results and modeling. In: Proc. 16th Cryst. Silicon Workshop, Colorado, USA, pp. 233–236 (2006)
[19] Schwartz, R.J., Lammert, M.D.: Silicon Solar Cell for high concentration application. In: Proc. IEEE Int. Electron Devices Meet., pp. 350–352 (1975)
[20] De Ceuster, D., Cousins, P., Rose, D., Vicente, D., Tipones, P., Mulligan, W.: Low Cost, high volume production of >22% efficiency silicon solar cells. In: Proc. 22nd Eur. Photovolt. Sol. Energy Conf., Milano, Italy, pp. 816–819 (2007)
[21] Sinton, R.A., Verlinden, P.J., Swanson, R.M., Crane, R.A., Wickham, K., Perkins, J.: Improvements in silicon backside contact solar cells for high-value one-sun applications. In: Proc. 13th Eur. Photovolt. Sol. Energy Conf., Nice, France, pp. 1586–1589 (1995)
[22] Granek, F., Reichel, C., Hermle, M., Huljić, D.M., Schultz, O., Glunz, S.W.: Front surface passivation of n-type high-efficiency back-junction silicon solar cells using front surface field. In: Proc. 22nd Eur. Photovolt. Sol. Energy Conf., Milano, Italy, pp. 1262–1265 (2007)
[23] Granek, F., Hermle, M., Huljic, D.M., Schultz-Wittmann, O., Glunz, S.W.: Enhanced Lateral Current Transport Via the Front N+ Diffused Layer of N-type High-efficiency Back-junction Back-contact Silicon Solar Cells. Prog. Photovolt.: Res. & Appl. 17, 47–56 (2008)
[24] Cousins, P.J., Smith, D.D., Hsin-Chiao, L., Manning, J., Dennis, T.D., Waldhauer, A., Wilson, K.E., Harley, G., Mulligan, W.P.: Generation 3: Improved performance at lower cost. In: Conf. Rec. 35th IEEE Photovolt. Spec. Conf., Hawai, USA, pp. 275–278 (2010)
[25] Swanson, R.M.: Point contact solar cells: theory and modelling. In: Proc. 18th IEEE Photovolt. Spec. Conf., Las Vegas, USA, pp. 604–610 (1985)
[26] Dicker, J., Schumacher, O., Warta, W., Glunz, S.W.: Analysis of one-sun monocrystalline rear-contacted silicon solar cells with efficiencies of 22.1%. J. Appl. Phys. 91, 4335–4343 (2002)
[27] Engelhart, P., Harder, N.P., Neubert, T., Plagwitz, H., Fischer, B., Meyer, R., Brendel, R.: Laser processing of 22 % efficient back-contacted silicon solar cells. In: Proc. 21st Eur. Photovolt. Sol. Energy Conf., Dresden, Germany, pp. 773–777 (2006)
[28] Sawada, T., Terada, N., Tsuge, S., Baba, T., Takahama, T., Wakisaka, K., Tsuda, S., Nakano, S.: High-efficiency a-Si/c-Si heterojunction solar cell. In: Conf. Rec. 24th IEEE Photovolt. Spec. Conf., Waikoloa, USA, vol. 2, pp. 1219–1226 (1994)
[29] Maruyama, E., Terakawa, A., Taguchi, M., Yoshimine, Y., Ide, D., Baba, T., Shima, M., Sakata, H., Tanaka, M.: Sanyo's Challenges to the Development of High-efficiency HIT Solar Cells and the Expansion of HIT Business. In: Conf. Rec. 4th IEEE World Conf. Photovolt. Energy Convers., Hawai, USA, pp. 1455–1460 (2006)
[29b] Nakamura, K., Isaka, T., Funakoshi, Y., Tonomura, Y., Machida, T., Okamoto, K.: Development of 20% efficiency mass production Si solar cells. In: Proc. 20th Eur. Photovolt. Sol. Energy Conf., Barcelona, Spain, pp. 789–792 (2005)
[29c] Fujiwara, H., Kaneko, T., Kondo, M.: Application of hydrogenated amorphous silicon oxide layers to c-Si heterojunction solar cells. Appl. Phys. Lett. 91, 133508-3 (2007)

[29d] Sritharathikhun, J., Jiang, F., Miyajima, S., Yamada, A., Konagai, M.: Optimization of p-Type Hydrogenated Microcrystalline Silicon Oxide Window Layer for High-Efficiency Crystalline Silicon Heterojunction Solar Cells. Jpn. J. Appl. Phys. 48, 101603-5 (2009)

[29e] Dao, V.A., Heo, J., Choi, H., Kim, Y., Park, S., Jung, S., Lakshminarayan, N., Yi, J.: Simulation and study of the influence of the buffer intrinsic layer, back-surface field, densities of interface defects, resistivity of p-type silicon substrate and transparent conductive oxide on heterojunction with intrinsic thin-layer (HIT) solar cell. Sol. Energy 84, 777–783 (2010)

[29f] Zhao, L., Zhou, C., Li, H., Diao, H., Wang, W.: Characterization on the Passivation Stability of HF Aqueous Solution Treated Silicon Surfaces for HIT Solar Cell Application by the Effective Minority Carrier Lifetime Measurement. Chin. J. Phys. 48(3), 392–399 (2010)

[30] Wang, Q., Page, M.R., Iwaniczko, E., Xu, Y., Roybal, L., Bauer, R., To, R., Yuan, H.C., Duda, A., Hasoon, F., Yan, Y.F., Levi, D., Meier, D., Branz, H.M., Wang, T.H.: Efficient heterojunction solar cells on p-type crystal silicon wafers. Appl. Phys. Lett. 96, 013507-3 (2010)

[31] Muñoz, D., Ozanne, A.S., Harrison, S., Danel, A., Souche, F., Denis, C., Favier, A., Desrues, T., Martin de Nicolás, S., Nguyen, N., Hickel, P.E., Mur, P., Salvetat, T., Moriceau, H., Le-Tiec, Y., Kang, M.S., Kim, K.M., Janin, R., Pesenti, C., Blin, D., Nolan, T., Kashkoush, I., Ribeyron, P.J.: Towards high efficiency on full wafer a-Si:H/c-Si heterojunction solar cells: 19.6% on 148 cm2. In: Conf. Rec. 35th IEEE Photovolt. Spec. Conf., Hawai, USA, pp. 39–43 (2010)

[32] Conrad, E., Korte, L., von. Maydell, L.K., Angermann, H., Schubert, C., Stangl, R., Schmidt, M.: Development and optimization of a-Si:H/c-Si heterojunction solar cells completely processed at low temperatures. In: Proc. 21st Eur. Photovolt. Sol. Energy Conf., Dresden, Germany, pp. 784–787 (2006)

[33] Fesquet, L., Olibet, S., Vallat-Sauvain, E., Shah, A., Ballif, C.: High quality surface passivation and heterojunction fabrication by VHF-PECVD deposition of amorphous silicon on crystalline Si: theory and experiments. In: Proc. 22nd Eur. Photovolt. Sol. Energy Conf., Milano, Italy, pp. 1678–1681 (2007)

[34] Olibet, S., Monachon, C., Hessler-Wyser, A., Vallat-Sauvain, E., De Wolf, S., Fesquet, L., Damon-Lacoste, J., Ballif, C.: Textured Silicon Heterojunction Solar Cells With Over 700 mV Open-Circuit Voltage Studied by Transmission Electron Microscopy. In: Proc. 23rd Eur. Photovolt. Sol. Energy Conf., Valencia, Spain, pp. 1140–1144 (2008)

[34b] Lachenal, D., Andrault, Y., Bätzner, D., Guerin, C., Kobas, M., Mendes, B., Strahm, B., Tesfai, M., Wahli, G., Buechel, A., Descoeudres, A., Choong, G., Bartlome, R., Barraud, L., Zicarelli, F., Bôle, P., Fesquet, L., Damon-Lacoste, J., De Wolf, S., Ballif, C.: High Efficiency Silicon Heterojunction Solar-Cell Activities in Neuchâtel, Switzerland. In: Proc. 23rd Eur. Photovolt. Sol. Energy Conf., Valencia, Spain, pp. 1272–1275 (2010)

[35] Lu, M., Bowden, S., Das, U., Birkmire, R.: Interdigitated back contact silicon heterojunction solar cell and the effect of front surface passivation. Appl. Phys. Lett. 91, 063507-3 (2007)

[36] Tucci, M., Serenelli, L., Salza, E., Pirozzi, L., de Cesare, G., Caputo, D., Ceccarelli, M., Martufi, P., De Iuliis, S., Geerligs, L.J.: BEHIND (Back Enhanced Heterostructure with INterdigitateD contact) solar cell. In: Proc. 23rd Eur. Photovolt. Sol. Energy Conf., Valencia, Spain, pp. 1749–1752 (2008)

[36b] Desrues, T., Souche, F., Vandeneynde, A., Muñoz, D., Ozanne, A.S., Ribeyron, P.J.: Emitter Optimization for Interdigitated Back Contact (IBC) Silicon Heterojunction (Si-HJ) Solar Cells. In: Proc. 25th Eur. Photovolt. Sol. Energy Conf., Valencia, Spain, pp. 2374–2377 (2010)
[37] Kleider, J.P., Chouffot, R., Gudovskikh, A.S., Roca i Cabarrocas, P., Labrune, M., Ribeyron, P.J., Bruggemann, R.: Electronic and structural properties of the amorphous/crystalline silicon interface. Thin Solid Films 517(23), 6386–6391 (2009)
[38] Das, U., Hegedus, S., Zhan, L., Appel, J., Rand, J., Birkmine, R.: Investigation of hetero-interface and junction properties in silicon heterojunction solar cells. In: Conf. Rec. 35th IEEE Photovolt. Spec. Conf., Hawai, USA, pp. 1358–1362 (2010)
[39] Schmidt, M., Korte, L., Laades, A., Stangl, R., Schubert, C., Angermann, H., Conrad, E., von Maydell, K.: Physical aspects of a-Si:H/c-Si hetero-junction solar cells. Thin Solid Films 515, 7475–7480 (2007)
[40] Leibovitch, M., Kronik, L., Fefer, E., Korobov, V., Shapiraa, Y.: Constructing band diagrams of semiconductor heterojunctions. Appl. Phys. Lett. 66(4), 457–459 (1995)
[41] Hattori, T.: Ultraclean Surface Processing of Silicon Wafers, p. 437. Springer, Heidelberg (1998)
[42] Angermann, H., Rappich, J., Korte, L., Sieber, I., Conrad, E., Schmidt, M., Hubener, K., Polte, J., Hauschild, J.: Wet-chemical passivation of atomically flat and structured silicon substrates for solar cell. Appl. Surf. Sci. 254(12), 3615–3625 (2008)
[43] Sinton, R.A., Cuevas, A.: Contactless determination of current–voltage characteristics and minority carrier lifetimes in semiconductors from quasi-steady-state photoconductance data. Appl. Phys. Lett. 69, 2510–2512 (1996)
[44] Kern, W.: The Evolution of Silicon Wafer Cleaning Technology. J. Electrochem. Soc. 137, 1887–1891 (1990)
[45] Angermann, H., Henrion, W., Rebien, M., Röseler, A.: Wet-chemical preparation and spectroscopic characterization of Si interfaces. Appl. Surf. Sci. 235, 322–339 (2004)
[46] Wang, Q., Page, M.R., Iwaniczko, E., Xu, Y.Q., Roybal, L., Bauer, R., To, B., Yuan, H.C., Duda, A., Yan, Y.F.: Crystal silicon heterojunction solar cells by hot-wire CVD. In: Conf. Rec. 33rd IEEE Photovolt. Spec. Conf., San Diego, USA (2008)
[47] Tardif, F., Joly, J.P., Courteaux, A., Straube, U., Danel, A., Kamarinos, G.: Detection of noble metals in HF based chemistries by microwave photoconductivity decay (μ-PCD). The Electrochemical Society Proceedings 97-22, 385–393 (1997)
[48] Wang, Q., Page, M.R., Iwaniczko, E., Xu, Y., Roybal, L., Bauer, R., Levi, D., Yan, Y., Wang, T., Branz, H.M.: High Open-Circuit Voltage in Silicon Heterojunction Solar Cells. In: Materials Research Society Symposium Proceedings, vol. 989, pp. 41–46. Materials Research Society, Warrendale (2007)
[49] Martin, I., Munoz, D., Voz, C., Vetter, M., Alcubilla, R., Damon-Lacoste, J., Cabarrocas, P.R., Villar, F., Bertomeu, J., Andreu, J.: Comparison of (n+) a-Si:H /(p) c-Si Heterojunction Emitters using a-Si:H Films Deposited by PECVD or HWCVD. In: Conf. Rec. IEEE 4th World Conf. Photovolt. Energy Conv., pp. 1091–1094 (2006)
[50] Taguchi, M., Terakawa, A., Maruyama, E., Tanaka, M.: Obtaining a higher V_{OC} in HIT cells. Prog. Photovolt: Res. Appl. 13, 481–488 (2005)
[51] Pierz, K., Fuhs, W., Mell, H.: Correlation between defect density and Fermi-level position in a-Si:H. J. Non-Cryst. Solids 114, 651–653 (1989)
[52] Taguchi, M., Tsunomura, Y., Inoue, H., Taira, S., Nakashima, T., Baba, T., Sakata, H., Maruyama, E.: High-Efficiency HIT Solar Cell on Thin Silicon Wafer (< 100 <Emphasis Type="Italic">μm) In: Proc. 24th Eur. Photovolt. Sol. Energy Conf., Hamburg, Germany, pp. 1690–1693 (2009)

[53] Schropp, R.E.I., Daey Ouwens, J., Von Der Linden, M.B., Van Der Werf, C.H.M., Van Der Weg, W.F., Alkemede: Defect and band gap engineering of amorphous silicon solar cells. In: Proc. on Amorphous Silicon Technology, PFA, p. 797 (1993)

[54] Sritharathikhun, J., Yamamoto, H., Miyajima, S., Yamada, A., Konagai, M.: Optimization of Amorphous Silicon Oxide Buffer Layer for High-Efficiency p-Type Hydrogenated Microcrystalline Silicon Oxide/n-Type Crystalline Silicon Heterojunction Solar Cells. Jpn. J. Appl. Phys. 47(11), 8452–8455 (2008)

[55] Centurioni, E., Iencinella, D.: Role of front contact work function on amorphous silicon/crystalline silicon heterojunction solar cell performance. IEEE Electron Device Lett. 24, 177–179 (2003)

[56] Zhao, L., Zhou, C.L., Li, H.L., Diao, H.W., Wang, W.J.: Role of the work function of transparent conductive oxide on the performance of amorphous/crystalline silicon heterojunction solar cells studied by computer simulation. Phys. Stat. Sol (a) 205(5), 1215–1221 (2008)

[57] Stangl, R., Froitzheim, A., Schmidt, M., Fuhs, W.: Design criteria for amorphous/crystalline silicon heterojunction solar cells - a simulation study. In: Proc. 3rd World Conf. Photovolt. Energy Conv., Osaka, Japan, pp. 1005–1008 (2003)

[58] Muñoz, D.: Silicon heterojunction solar cells obtained by Hot-Wire CVD. PhD thesis, Universitat Politècnica de Catalunya (2008)

[59] Yerokhov, Y., Hezel, R., Lipinski, M., Ciach, R., Nagel, H., Mylyanych, A., Panek, P.: Cost-effective methods of texturing for silicon solar cells. Sol. Energy Mater. Sol. Cells 72, 291–298 (2002)

[60] Zhao, J., Wang, A., Green, M.A., Ferraza, F.: 19.8% efficient "honeycomb" textured multicrystalline and 24.4% monocrystalline silicon solar cells. Appl. Phys. Lett. 73, 1991–1993 (1998)

[61] Yagi, T., Uraoka., Y., Fuyuki, T.: Ray-trace simulation of light trapping in silicon solar cell with texture structures. Sol. Energy Mater. Sol. Cells 90, 2647–2656 (2006)

[62] Campbell, P., Green, M.A.: High performance light trapping textures for monocrystalline silicon solar cells. Sol. Energy Mater. Sol. Cells 65, 369–375 (2001)

[63] Street, R.A.: Hydrogenated Amorphous Silicon. Cambridge University Press, Cambridge (1991)

[64] Street, R.A.: Technology and Applications of Amorphous Silicon. Springer, New York (2000)

[65] Racine, B., Ferrari, A.C., Morrison, N.A., Hutchings, I., Milne, W.I., Robertson, J.: Properties of amorphous carbon–silicon alloys deposited by a high plasma density source. J. Appl. Phys. 90, 5002–5012 (2001)

[66] Mueller, T., Duengen, W., Ma, Y., Job, R., Scherff, M., Fahrner, W.R.: Investigation of the emitter band gap widening of heterojunction solar cells by use of hydrogenated amorphous carbon silicon alloys. J. Appl. Phys. 102, 074505-7 (2007)

[67] Tikhonravov, A.V., Trubetskov, M.K.: OptiLayer Thin Film Software, `http://www.optilayer.com`

[68] Bender, M., Seelig, W., Daube, C., Frankenberger, H., Ocker, B., Stollenwerk, J.: Dependence of oxygen flow on optical and electrical properties of DC-magnetron sputtered ITO films. Thin Solid Films 326, 72–77 (1998)

[69] Mueller, T., Schwertheim, S., Scherff, M., Zastrow, U., Huang, Y., Fahrner, W.R.: Crystalline silicon surface passivation by PECVD-deposited hydrogenated amorphous silicon oxide layers [a-SiOx:H]. In: Proc. 22nd Eur. Photovolt. Sol. Energy Conf., Milano, Italy, pp. 1553–1555 (2007)

[70] Kanevce, A., Metzger, W.K.: The role of amorphous silicon and tunneling in heterojunction with intrinsic thin layer (HIT) solar cells. J. Appl. Phys. 105, 094507-7 (2009)
[71] Datta, A., Rahmouni, M., Nath, M., Boubekri, R., Roca i Cabarrocas, P., Chatterjee, P.: Insights gained from computer modeling of heterojunction with instrinsic thin layer "HIT" solar cells. Sol. Energy Mat. Sol. Cells 94(9), 1457–1462 (2010)
[72] Pysch, D., Mette, A., Glunz, S.W.: A review and comparison of different methods to determine the series resistance of solar cells. Sol. Energy Mat. Sol. Cells 91, 1698–1706 (2007)

Author Index

Index

GPSR Compliance

The European Union's (EU) General Product Safety Regulation (GPSR) is a set of rules that requires consumer products to be safe and our obligations to ensure this.

If you have any concerns about our products, you can contact us on ProductSafety@springernature.com

In case Publisher is established outside the EU, the EU authorized representative is:

Springer Nature Customer Service Center GmbH
Europaplatz 3
69115 Heidelberg, Germany

Batch number: 10755053

Printed by Printforce, the Netherlands